DIGITAL SIGNAL PROCESSING

DIGITAL SIGNAL PROCESSING

Alan V. Oppenheim

Department of Electrical Engineering
Massachusetts Institute of Technology

Ronald W. Schafer

Bell Telephone Laboratories
Murray Hill, New Jersey

PRENTICE-HALL, INC., Englewood Cliffs, New Jersey.

Library of Congress Cataloging in Publication Data

OPPENHEIM, ALAN V.
Digital Signal Processing.

Includes bibliographical references.
1. Signal theory (Telecommunication) 2. Digital electronics. I. SCHAFER, RONALD W. joint author. II. Title.
TK5102.5.0245 621.3819'58'2 74-17280
ISBN 0-13-214635-5

10 9 8 7 6 5 4 3 2

Printed in the United States of America

PRENTICE-HALL INTERNATIONAL, INC., *London*
PRENTICE-HALL OF AUSTRALIA, PTY. LTD., *Sydney*
PRENTICE-HALL OF CANADA, LTD., *Toronto*
PRENTICE-HALL OF INDIA PRIVATE LIMITED, *New Delhi*
PRENTICE-HALL OF JAPAN, INC., *Tokyo*

To Phyllis and Dorothy

Contents

Preface

This book has grown out of our teaching and research activities in the field of digital-signal processing. It is designed, primarily, as a text for a senior or first-year graduate level course. The notes on which this book is based have been used for a one-semester introductory course in the M.I.T. Department of Electrical Engineering as well as for a continuing education course at Bell Laboratories. The notes were also used for three years in a condensed two-week summer course at M.I.T. and, in latter stages of development, at a number of universities around the country for one-semester courses.

A typical one-semester course would cover Chapters 1, 2, and 3, in depth, and selected fundamental topics from Chapters 4, 5, 6, and 7. The remainder of the text, in conjunction with supplemental reading, forms the basis for a second-semester course on advanced topics and applications.

Essential to learning a subject of this nature is the actual practice in working out new results and the application of the results to the solutions of real problems. Therefore, an important part of the text is a collection of approximately 250 homework problems. These problems are designed to extend results developed in the text, to develop some results that were referred to in the text, and to illustrate applications to practical problems. Solutions to the problems are available to instructors through the publisher. A self-study course consisting of a set of video-tape lectures and a study guide have been developed to accompany this text. Further information about the lecture tapes and study guide may be obtained through the M.I.T. Center for Advanced Engineering Study.

It is assumed that the reader has a background of advanced calculus, including an introduction to complex variable theory and an exposure to linear system theory for continuous-time signals, including Laplace and Fourier transforms, as taught in most electrical and mechanical engineering curricula. With this background the book is self-contained. In particular, no prior experience with discrete-time signals, z-transforms, discrete Fourier transforms and the like is assumed.

In Chapter 1, we begin with the definition of discrete-time signals and the class of linear shift-invariant systems. The representation of this class of systems in both the time domain, using the convolution sum, and the frequency domain, using the Fourier transform, is presented. In Chapter 2, we generalize the treatment of the Fourier transform to a discussion of the representation of discrete-time systems and signals in terms of the z-transform. Most of Chapter 2 deals with the definition and properties of the z-transform and the system function representation of linear shift-invariant systems. In both Chapters 1 and 2, we conclude with a brief generalization of the one-dimensional ideas and results to two dimensions. In Chapter 3, we introduce the discrete Fourier transform. This transformation forms the basis for many digital signal processing techniques and is inherently a discrete-time concept. In addition to developing the discrete Fourier transform and its properties, we introduce the notion of implementing discrete convolution explicitly by means of the discrete Fourier transform. Thus, the first three chapters present the fundamental ideas in discrete-time signal representation.

The discussion in Chapter 1 introduces the class of linear shift-invariant systems which are representable by linear constant coefficient difference equations. In Chapter 4, we discuss the implementation of this class of systems as digital networks composed of adders, delay elements, and coefficient multipliers. Much of the chapter is concerned with developing a variety of the important digital filter structures. The flow graph and matrix representations introduced in this chapter lead to a theory of digital networks. Included is a discussion of Tellegen's theorem for digital networks and some of the resulting network sensitivity relations.

In Chapter 5 we present the basic issues involved in digital filter design and also present the more common and useful digital filter design methods. These include a number of analytical methods as well as algorithmic or computer-aided design techniques.

In continuous-time linear system theory, the Fourier transform is primarily an analytical tool for representing signals and systems. In contrast, in the discrete-time domain, many signal processing systems and algorithms involve the explicit computation of the Fourier transform. Chapter 6 is directed toward a discussion of the computation of the discrete Fourier transform. Most of the discussion centers on fast Fourier transform algorithms for both one-dimensional and multi-dimensional sequences.

In Chapter 7, we introduce the discrete Hilbert transform. This transform arises in a variety of practical applications, including inverse filtering, complex representations for real bandpass signals, and many others. It also has particular significance for a class of signal processing techniques referred to as homomorphic signal processing that is discussed in Chapter 10.

In the discussion in the first seven chapters, it is assumed that the discrete-time signals with which we are dealing are deterministic signals. Another important class of discrete-time signals is the class of random signals characterized primarily by the properties of averages—such as correlation functions and their Fourier transforms, spectral density functions. Thus, in Chapter 8 we introduce some of the basic concepts concerning discrete random signals. While random signals arise in a variety of situations, we make particular use of the results of Chapter 8 in Chapter 9. For most of the discussion up to Chapter 9, signals are assumed to be discrete in time, but continuous in amplitude. In representing signals in a digital computer or with digital hardware, it is necessary to quantize the amplitudes. The effect of this amplitude quantization depends strongly on the choice of arithmetic and word-length as well as the signal processing application. Many of the finite register length effects can be characterized using the results of Chapter 8. Thus, in Chapter 9, we consider the representation of quantization effects in terms of additive noise and analyze these effects for both digital filtering and fast Fourier transform algorithms.

In Chapter 10, we introduce a class of signal processing techniques referred to as homomorphic signal processing. This class of techniques, although nonlinear, is based upon a generalization of the linear techniques that was the focus of the early chapters of this book, and can be understood in terms of the ideas presented in the earlier chapters. Thus, in addition to introducing a new class of signal processing techniques, this chapter serves as a vehicle for utilizing—in a sophisticated way—many of the previous results developed in this book. In addition, we present in this chapter a number of applications of homomorphic signal processing that we hope will give the reader a strong indication of the breadth of applications of digital signal processing in general.

The final chapter introduces another important class of applications of digital signal processing techniques; the estimation of the spectrum of a random signal. Many of the techniques discussed throughout this book are useful in this class of problems. In Chapter 11, we do not attempt to treat this topic exhaustively, but rather offer it as an elementary introduction to a very complicated subject.

In the selection and preparation of the material presented, we have attempted to focus on fundamental principles that will be widely applicable to the diverse fields where signal processing is required and have avoided details that we felt to be inappropriate for this textbook.

Many of these details, together with additional applications, can be found in the companion book by Rabiner and Gold.†

Throughout the preparation of this book, we have enjoyed a close collaboration with a number of people who we feel fortunate to count as both colleagues and friends. In particular, much of this book has been greatly influenced by a close association with Dr. B. Gold and Mr. C. M. Rader of the M.I.T. Lincoln Laboratory; Dr. J. L. Flanagan, Dr. J. F. Kaiser, and Dr. L. R. Rabiner, of Bell Laboratories; and Prof. T. G. Stockham, Jr. of the University of Utah. Drs. Gold and Flanagan, in particular, have not only had a direct influence on this book but have also had a profound effect upon the personal and professional growth of A. V. O. and R. W. S. respectively. Our debt to them is gratefully acknowledged. Also, one of us (R. W. S.), has enjoyed a long and fruitful collaboration in research with Dr. L. R. Rabiner that has influenced the book in many ways.

Our students, too, have had a significant impact on this book. Specifically, Dr. R. Mersereau has helped in teaching the course at M.I.T. and has been our best critic. His thoughtful comments have improved the book significantly. Dr. R. E. Crochiere has also aided in teaching the course and has made numerous helpful suggestions. In addition, his doctoral research has significantly influenced Chapter 4. D. Johnson, M. R. Portnoff, and V. Zue have served as teaching assistants and have offered many criticisms and corrections that have improved the book. H. Hersey, in addition to serving as a teaching assistant, has prepared solutions to accompany the text.

Many of our own students, as well as students and professors at other institutions, have read and criticized versions of the manuscript. Their comments have been most helpful and are greatly appreciated. We also wish to thank S. Bates, J. Dubnowski, R. Kuc, G. Kopec, J. McClellan, E. Singer, B. Stuck, and a number of the people mentioned above for assisting in reading final page proofs.

We wish to thank M.I.T. and Bell Laboratories for providing support and a stimulating environment in which the ideas of this book could be developed. Part of the book was written while one of the authors (A. V. O.) was a Guggenheim Fellow at the Speech Communication Laboratory in Grenoble, France. The support of the Guggenheim Foundation and the hospitality of the Speech Communication Laboratory during this period is gratefully acknowledged.

Finally, we express thanks to Miss Judy Johnson for help in preparing the manuscript and to Miss Monica Edelman, Mrs. Anita Caswell, and Mrs. Monica Pettyjohn for their contributions to typing of various drafts of the book.

ALAN V. OPPENHEIM/RONALD W. SCHAFER

† L. R. Rabiner and B. Gold, *Theory and Application of Digital Signal Processing*, Prentice-Hall, Inc., Englewood Cliffs, N.J., 1975.

DIGITAL SIGNAL PROCESSING

Introduction

Digital signal processing, a field which has its roots in 17th and 18th century mathematics, has become an important modern tool in a multitude of diverse fields of science and technology. The techniques and applications of this field are as old as Newton and Gauss and as new as digital computers and integrated circuits.

Digital signal processing is concerned with the representation of signals by sequences of numbers or symbols and the processing of these sequences. The purpose of such processing may be to estimate characteristic parameters of a signal or to transform a signal into a form which is in some sense more desirable. The classical numerical analysis formulas, such as those designed for interpolation, integration, and differentiation, are certainly digital signal processing algorithms. On the other hand, the availability of high speed digital computers has fostered the development of increasingly complex and sophisticated signal processing algorithms, and recent advances in integrated circuit technology promise economical implementations of very complex digital signal processing systems.

Signal processing, in general, has a rich history, and its importance is evident in such diverse fields as biomedical engineering, acoustics, sonar, radar, seismology, speech communication, data communication, nuclear science, and many others. In many applications, as, for example, in EEG and EKG analysis or in systems for speech transmission and speech recognition, we may wish to extract some characteristic parameters. Alternatively, we may wish to remove interference, such as noise, from the signal or to modify the signal to present it in a form which is more easily interpreted by an expert. As another example, a signal transmitted over a communications channel is generally perturbed in a variety of ways, including channel distortion, fading, and the insertion of background noise. One of the objectives at the receiver is to compensate for these disturbances. In each case, processing of the signal is required.

Signal processing problems are not confined, of course, to one-dimensional signals. Many picture processing applications require the use of two-dimensional signal processing techniques. This is the case, in x-ray enhancement, the enhancement and analysis of aerial photographs for detection of forest fires or crop damage, the analysis of satellite weather photos, and the enhancement of television transmissions from lunar and deep-space probes. Seismic data analysis as required in oil exploration, earthquake measurements and nuclear test monitoring also utilizes multidimensional signal processing techniques.

Until recently, signal processing has typically been carried out using analog equipment. Some exceptions to this were evident in the 1950s, particularly in areas where sophisticated signal processing was required. This was the case, for example, in the analysis of some geophysical data which could be recorded on magnetic tape for later processing on a large digital computer. This class of problems was one of the first examples of signal processing using digital computers. This type of signal processing could not generally be done in real-time; for example, minutes or even hours of computer time were often required to process only seconds of data. Even so, the flexibility of the digital computer made this alternative extremely inviting.

During this same period the use of digital computers in signal processing also arose in a different way. Because of the flexibility of digital computers, it was often useful to simulate a signal processing system on a digital computer before implementing it in analog hardware. In this way, a new signal processing algorithm, or system, could be studied in a flexible experimental environment before committing economic and engineering resources to constructing it. Typical examples of such simulations were the vocoder simulations carried out at Lincoln Laboratory and at Bell Laboratories. In the implementation of an analog channel vocoder, the filter characteristics often affect the quality of the resulting speech signal in unpredictable ways. Through computer simulations, these filter characteristics were adjusted and the quality of a system evaluated prior to construction of the analog equipment.

In all of the above examples of signal processing using digital computers, the computer offered tremendous advantages in flexibility. However, the processing could not always be done in real-time. Consequently, a prevalent attitude at that time was that the digital computer was being used to *approximate*, or *simulate*, an analog signal processing system. In keeping with that style, early work on digital filtering was very much concerned with ways in which a filter could be programmed on a digital computer so that with analog-to-digital conversion of the signal, followed by the digital filtering, followed by digital-to-analog conversion, the overall system approximated a good analog filter. The notion that digital systems might, in fact, be practical for the actual implementation of signal processing in speech communication or radar processing or any of the variety of other applications seemed at the most optimistic times to be highly speculative. Speed, cost, and size were, of course, three of the important factors in favor of the use of analog components.

As signals were being processed on digital computers, there was a natural tendency to experiment with increasingly sophisticated signal processing algorithms. Some of these algorithms grew out of the flexibility of the digital computer and had no apparent practical implementation in analog equipment. Thus, many of these algorithms were treated as interesting, but somewhat impractical, ideas. An example of a class of algorithms of this type was the set of techniques referred to as cepstrum analysis and homomorphic filtering. It had been clearly demonstrated on digital computers that these techniques could be applied to advantage in speech bandwidth compression systems, deconvolution, and echo removal. Implementation of these techniques requires the explicit evaluation of the inverse Fourier transform of the logarithm of the Fourier transform of the input. The required accuracy and resolution of the Fourier transform were such that analog spectrum analyzers were not practical. The development of such signal processing algorithms made the notion of all-digital implementation of signal processing systems even more tempting. Active work began on the investigation of digital vocoders, digital spectrum analyzers, and other all-digital systems, with the hope that eventually such systems would become practical.

The evolution of a new point of view toward digital signal processing was further accelerated by the disclosure in 1965 of an efficient algorithm for computation of Fourier transforms. This class of algorithms has come to be known as the fast Fourier transform or FFT. The implications of the FFT were significant from a number of points of view. Many signal processing algorithms which had been developed on digital computers required processing times several orders of magnitude greater than real-time. Often this was tied to the facts that spectrum analysis was an important component of the signal processing and that no efficient means had been known for implementing it. The fast Fourier transform algorithm reduced the computation time of the Fourier transform by orders of magnitude. This permitted the implementation of increasingly sophisticated signal processing algorithms with processing times that allowed interaction with the system. Furthermore, with the realization that the fast Fourier transform algorithm might, in fact, be implementable in special purpose digital hardware, many signal processing algorithms which previously had appeared to be impractical began to appear to have practical implementations with special purpose digital hardware.

Another important implication of the fast Fourier transform algorithm was tied to the fact that it was an inherently discrete-time concept. It was directed toward the computation of the Fourier transform of a discrete-time signal or sequence and involved a set of properties and mathematics that were exact in the discrete-time domain—it was not simply an approximation to a continuous-time Fourier transform. The importance of this was that it had the effect of stimulating a reformulation of many signal processing concepts and algorithms in terms of discrete-time mathematics and these techniques then formed an exact set of relationships in the discrete-time domain. This represented a shift away from the notion that signal processing on a digital computer was merely an approximation to analog signal processing techniques. With this shift in point of view there emerged a strong interest in the new or reborn field of digital signal processing.

The techniques and applications of digital signal processing are expanding at a tremendous rate. With the advent of large scale integration and the resulting reduction in cost and size of digital components, together with increasing speed, the class of applications of digital signal processing techniques is growing. Special purpose digital filters can now be implemented at sampling rates in the megahertz range. Special purpose processors for implementing the fast Fourier transform at high data rates are commercially available. Simple digital filters have been integrated on circuit chips. Almost all current discussions of speech bandwidth compression systems are directed toward all digital implementation because these are now the most practical. Digital processors also form an integral part of many modern radar and sonar systems. In addition to the development of special purpose digital signal processing hardware, there are available special programmable digital signal processing computers whose architecture is matched to signal processing problems. Such computers are finding application in real-time signal processing as well as for real-time simulations directed toward the development of special purpose digital hardware.

The importance of digital signal processing appears to be increasing with no visible sign of saturation. Indeed, the future development of the field is likely to be even more dramatic than the course of development that we have just described. The impact of digital signal processing techniques will undoubtedly promote revolutionary advances in some fields of application. A notable example is in the area of telephony, where digital techniques promise dramatically increased economy and flexibility in implementing switching and transmission systems.

Because of the direction of the evolution of this field, we are convinced that digital signal processing techniques should be treated in their own right rather than as an approximation to analog signal processing. While we assume in this book that the reader is familiar with continuous-time linear system theory and signal representation, we begin in the first chapter by defining the set of signals and systems with which we are concerned and from there develop our techniques. Most of our emphasis is on the class of linear shift-invariant systems, the counterpart of linear time-invariant systems in the continuous-time case. In Chapter 10, however, we will expand this set of ideas to a generalization of this class of systems.

A conscious effort has been made in the presentation of the material to avoid forcing results from analog signal processing into the framework of digital signal processing. As will become evident from the first chapters, many of the ideas and results in digital signal processing have direct counterparts in analog signal processing. For example, convolution is an important tool in representing linear shift-invariant systems. Similarly, the frequency domain and Fourier analysis play an essential role. While there are strong similarities between the analog and digital domains, there are also strong and important differences. Thus, although intuition developed in previous exposure to analog signal processing theory and practice may at times be useful, we caution the reader against letting this intuition interfere with his understanding of digital signal processing.

While digital signal processing is a dynamic, rapidly growing field, its fundamentals are well formulated. Many of these fundamentals stem from the classical numerical analysis techniques developed in the 1600s. Important refinements of the techniques which provide the foundation for digital signal processing are evident in the development and treatment of sampled-data control systems in the 1940s and 1950s, and again in the development of the field of digital signal processing in its present form. We have chosen the set of topics presented in this book to enable the reader to develop a firm base in the fundamentals and to begin to develop an appreciation for the wide scope of applications and future directions for the field.

1

Discrete-Time Signals and Systems

1.0 Introduction

A *signal* can be defined as a function that conveys information, generally about the state or behavior of a physical system. Although signals can be represented in many ways, in all cases the information is contained in a pattern of variations of some form. For example, the signal may take the form of a pattern of time variations or a spatially varying pattern. Signals are represented *mathematically* as functions of one or more independent variables. For example, a speech signal would be represented mathematically as a function of time and a picture would be represented as a brightness function of two spatial variables. It is a common convention, and one that will be followed in this book, to refer to the independent variable of the mathematical representation of a signal as time, although it may in fact not represent time.

The independent variable of the mathematical representation of a signal may be either continuous or discrete. *Continuous-time* signals are signals that are defined at a continuum of times and thus are represented by continuous variable functions. *Discrete-time signals* are defined at discrete times and thus the independent variable takes on only discrete values; i.e., discrete-time signals are represented as sequences of numbers. As we will see, signals such as speech or pictures may have either a continuous or a discrete variable representation, and if certain conditions hold, these representations are entirely equivalent.

In addition to the fact that the independent variables can be either continuous or discrete, the signal amplitude may be either continuous or discrete. *Digital signals* are those for which both time and amplitude are discrete. Continuous-time, continuous-amplitude signals are sometimes called *analog signals.*

In almost every area of science and technology, signals must be processed to facilitate the extraction of information. Thus, the development of signal processing techniques and systems is of great importance. These techniques usually take the form of a transformation of a signal into another signal that is in some sense more desirable than the original. For example, we may wish to design transformations for separating two or more signals that have been combined in some way; we may wish to enhance some component or parameter of a signal; or we may wish to estimate one or more parameters of a signal. Signal processing systems may be classified along the same lines as signals. That is, *continuous-time systems* are systems for which both the input and output are continuous-time signals and *discrete-time systems* are those for which the input and output are discrete-time signals. Similarly *analog systems* are systems for which the input and output are analog signals and *digital systems* are those for which the input and output are digital signals. *Digital signal processing*, then, deals with transformations of signals that are discrete in both amplitude and time. This chapter, and in fact the major part of this book, deals with discrete-time rather than digital signals and systems. The effects of discrete amplitude are considered in detail in Chapter 9.

Discrete-time signals may arise by sampling a continuous-time signal or they may be generated directly by some discrete-time process. Whatever the origin of the discrete-time signals, digital signal processing systems have many attractive features. They can be realized with great flexibility using general-purpose digital computers, or they can be realized with digital hardware. They can, if necessary, be used to simulate analog systems or, more importantly, to realize signal transformations impossible to realize with analog hardware. Thus, digital representations of signals are often desirable when sophisticated signal processing is required.

In this chapter we consider the fundamental concepts of discrete-time signals and signal processing systems first for one-dimensional signals and then for two-dimensional signals. We shall place the most emphasis on the class of linear shift-invariant discrete-time systems. It will be true in this chapter and succeeding ones that many of the properties and results that we derive will be similar to properties and results for linear time-invariant continuous-time systems as presented in a variety of excellent texts [1–3]. In fact, it is possible to approach the discussion of discrete-time systems by treating sequences as analog signals that are impulse trains. This approach, if implemented carefully, can lead to correct results and in fact forms the basis for much of the classical discussion of sampled data systems (see, for example, [4–6]). In many present digital signal processing applications, however, not all sequences arise from sampling a continuous-time signal. Furthermore, many discrete-time systems are not simply approximations to corresponding analog systems. Therefore, rather than attempt to force results from analog system theory into a discrete framework, we shall derive similar results starting within a framework and with notation suitable to discrete-time systems. Discrete-time signals will be related to analog signals only when necessary.

1.1 Discrete-Time Signals—Sequences

In discrete-time system theory, we are concerned with processing signals that are represented by sequences. A sequence of numbers x, in which the nth number in the sequence is denoted $x(n)$, is formally written as

$$x = \{x(n)\}, \qquad -\infty < n < \infty \tag{1.1}$$

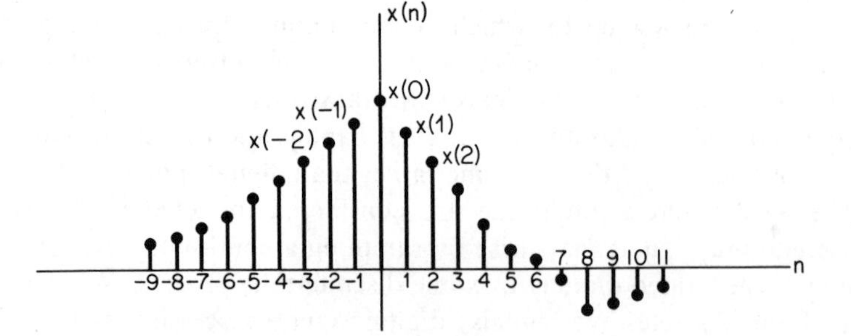

Fig. 1.1 Graphical representation of a discrete-time signal.

Although sequences do not always arise from sampling analog waveforms, for convenience we shall refer to $x(n)$ as the "nth sample" of the sequence. Also, although strictly speaking $x(n)$ denotes the nth number in the sequence, the notation of Eq. (1.1) is often unnecessarily cumbersome, and it is convenient and unambiguous to refer to "the sequence $x(n)$." Discrete-time signals (i.e., sequences) are often depicted graphically as shown in Fig. 1.1. Although the abscissa is drawn as a continuous line, it is important to recognize that $x(n)$ is only defined for integer values of n. It is *not* correct to think of $x(n)$ as being zero for n not an integer; $x(n)$ is simply undefined for non-integer values of n.

Some examples of sequences are shown in Fig. 1.2. The *unit-sample sequence*, $\delta(n)$, is defined as the sequence with values

$$\delta(n) = \begin{cases} 0, & n \neq 0 \\ 1, & n = 0 \end{cases}$$

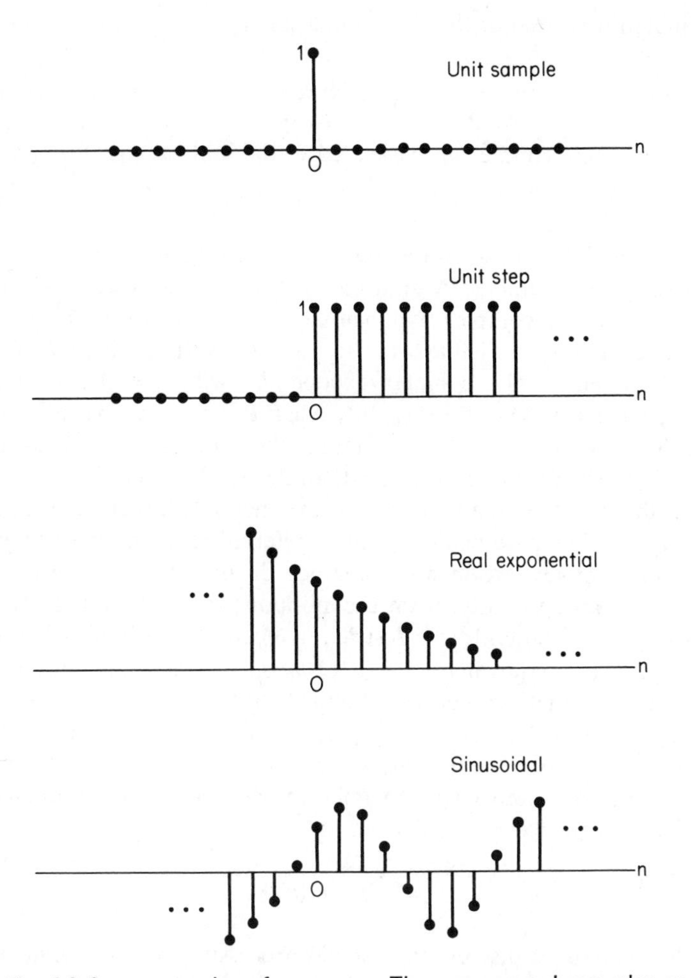

Fig. 1.2 Some examples of sequences. The sequences shown play an important role in the analysis and representation of discrete-time signals and systems.

As we will see shortly, the unit-sample sequence plays the same role for discrete-time signals and systems that the unit impulse function does for continuous-time signals and systems. For convenience the unit-sample sequence is often referred to as a *discrete-time impulse*, or simply as an *impulse*. It is important to note that a discrete-time impulse does not suffer from the mathematical complications that a continuous-time impulse suffers from. Its definition is simple and precise. The unit-step sequence, $u(n)$, has values

$$u(n) = \begin{cases} 1, & n \geq 0 \\ 0, & n < 0 \end{cases}$$

The unit step is related to the unit sample as

$$u(n) = \sum_{k=-\infty}^{n} \delta(k) \tag{1.2}$$

Similarly, the unit sample can be related to the unit step as

$$\delta(n) = u(n) - u(n-1) \tag{1.3}$$

A real exponential sequence is any sequence whose values are of the form a^n, where a is a real number. A sinusoidal sequence has values of the form $A \cos(\omega_0 n + \phi)$. A complex exponential sequence is of the form $e^{(\sigma+j\omega_0)n}$.

A sequence $x(n)$ is defined to be *periodic* with period N if $x(n) = x(n+N)$ for all n. The complex exponential with $\sigma = 0$ and sinusoidal sequences have a period of $2\pi/\omega_0$ only when this real number is an integer. If $2\pi/\omega_0$ is not an integer but a rational number, the sinusoidal sequence will be periodic but with a period longer than $2\pi/\omega_0$. If $2\pi/\omega_0$ is not a rational number, the sinusoidal and complex exponential sequences will not be periodic at all. The parameter ω_0 will be referred to as the frequency of the sinusoid or complex exponential whether or not they are periodic. The frequency ω_0 can be chosen from a continuous range of values. However, there is no loss of generality in restricting ω_0 to be continuous in the range $0 \le \omega_0 \le 2\pi$ (or equivalently $-\pi \le \omega_0 \le \pi$) since the sinusoidal or complex exponential sequences obtained by varying ω_0 in the range $2\pi k \le \omega_0 \le 2\pi(k+1)$ are exactly identical for any k to those obtained by varying ω_0 in the range $0 \le \omega_0 \le 2\pi$.

It is sometimes convenient to refer to the *energy* in a sequence. The energy $\mathcal{E}$ in a sequence $x(n)$ is defined as

$$\mathcal{E} = \sum_{n=-\infty}^{\infty} |x(n)|^2$$

In the analysis of discrete-time signal processing systems, sequences are manipulated in several basic ways. The product and sum of two sequences x and y are defined as the sample by sample product and sum, respectively;

$$x \cdot y = \{x(n)y(n)\} \qquad \text{(product)}$$

$$x + y = \{x(n) + y(n)\} \qquad \text{(sum)}$$

Multiplication of a sequence x by a number α is defined as

$$\alpha \cdot x = \{\alpha x(n)\}$$

A sequence y is said to be a delayed or shifted version of a sequence x if y has values

$$y(n) = x(n - n_0)$$

where n_0 is an integer.

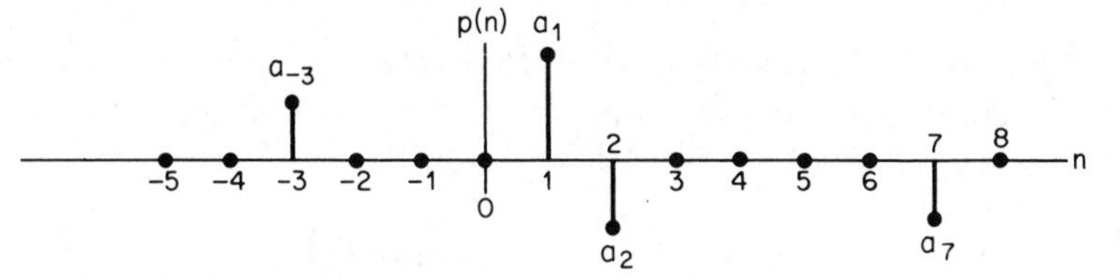

Fig. 1.3 Example of a sequence to be expressed as a sum of scaled, delayed unit samples.

An arbitrary sequence can be expressed as a sum of scaled, delayed unit samples. For example, the sequence $p(n)$ of Fig. 1.3 can be expressed as

$$p(n) = a_{-3}\,\delta(n+3) + a_1\,\delta(n-1) + a_2\,\delta(n-2) + a_7\,\delta(n-7)$$

More generally, an arbitrary sequence can be expressed as

$$x(n) = \sum_{k=-\infty}^{\infty} x(k)\,\delta(n-k) \tag{1.4}$$

1.2 Linear Shift-Invariant Systems

A *system* is defined mathematically as a unique transformation or operator that maps an input sequence $x(n)$ into an output sequence $y(n)$. This is denoted as

$$y(n) = T[x(n)]$$

and is often depicted as in Fig. 1.4.

Fig. 1.4 Representation of a transformation that maps an input sequence $x(n)$ into an output sequence $y(n)$.

Classes of discrete-time systems are defined by placing constraints on the transformation $T[\;]$. Because they are relatively easy to characterize mathematically and because they can be designed to perform useful signal processing functions, the class of linear shift-invariant systems will be studied extensively. In Chapter 10 we shall discuss a more general class of systems, of which linear systems are a special case.

The class of *linear systems* is defined by the principle of superposition. If $y_1(n)$ and $y_2(n)$ are the responses when $x_1(n)$ and $x_2(n)$, respectively, are the inputs, then a system is linear if and only if

$$T[ax_1(n) + bx_2(n)] = aT[x_1(n)] + bT[x_2(n)] = ay_1(n) + by_2(n) \tag{1.5}$$

for arbitrary constants a and b. We have seen that an arbitrary sequence $x(n)$ can be represented as a sum of delayed and scaled unit-sample sequences as in Eq. (1.4). This representation together with Eq. (1.5) suggests that a linear system can be completely characterized by its unit-sample response. Specifically, let $h_k(n)$ be the response of the system to $\delta(n-k)$, a unit sample occurring at $n = k$. Then from Eq. (1.4)

$$y(n) = T\left[\sum_{k=-\infty}^{\infty} x(k)\,\delta(n-k)\right]$$

Using Eq. (1.5) we can write

$$y(n) = \sum_{k=-\infty}^{\infty} x(k)T[\delta(n-k)] = \sum_{k=-\infty}^{\infty} x(k)h_k(n) \tag{1.6}$$

Thus, according to Eq. (1.6) the system response can be expressed in terms of the response of the system to $\delta(n-k)$. If only linearity is imposed, $h_k(n)$ will depend on both n and k, in which case the computational usefulness of Eq. (1.6) is limited. A more useful result is obtained if we impose the additional constraint of shift-invariance.

The class of *shift-invariant systems* is characterized by the property that if $y(n)$ is the response to $x(n)$, then $y(n-k)$ is the response to $x(n-k)$, where k is a positive or negative integer. When the index n is associated with time, shift-invariance corresponds to *time-invariance*. The property of shift-invariance implies that if $h(n)$ is the response to $\delta(n)$, then the response to $\delta(n-k)$ is simply $h(n-k)$. Thus, Eq. (1.6) becomes

$$y(n) = \sum_{k=-\infty}^{\infty} x(k)h(n-k) \tag{1.7}$$

Any linear shift-invariant system, then, is completely characterized by its unit-sample response $h(n)$.

Equation (1.7) is commonly called the *convolution sum*. If $y(n)$ is a sequence whose values are related to the values of two sequences $h(n)$ and $x(n)$ as in Eq. (1.7), we say that $y(n)$ is the *convolution of* $x(n)$ *with* $h(n)$ and denote this by the notation

$$y(n) = x(n) * h(n)$$

By a substitution of variables in Eq. (1.7) we obtain the alternative expression,

$$y(n) = \sum_{k=-\infty}^{\infty} h(k)x(n-k) = h(n) * x(n) \tag{1.8}$$

Thus the order in which two sequences are convolved is unimportant, and hence the system output is the same if the roles of the input and unit-sample response are reversed. In other words, a linear shift-invariant system with input $x(n)$ and unit-sample response $h(n)$ will have the same output as a linear shift-invariant system with input $h(n)$ and unit-sample response $x(n)$.

Two linear shift-invariant systems in cascade form a linear shift-invariant

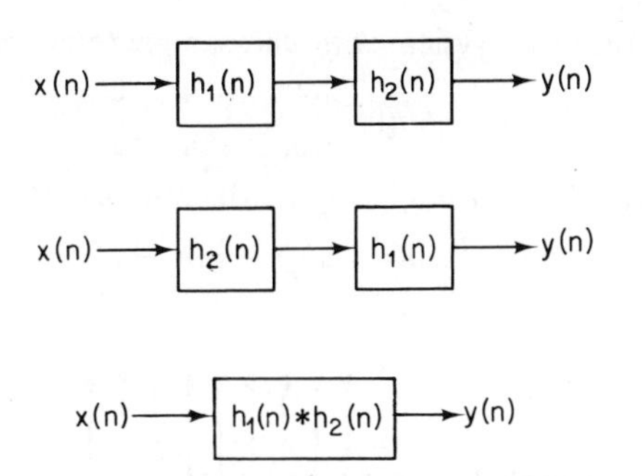

Fig. 1.5 Three linear shift-invariant systems with identical unit-sample responses.

system with a unit-sample response which is the convolution of the two unit-sample responses. Since the order in which two sequences are convolved is unimportant, it follows that the unit-sample response of a cascade combination of linear shift-invariant systems is independent of the order in which they are cascaded. This property is summarized in Fig. 1.5, where the three systems shown all have the same unit-sample response. It also follows from Eq. (1.7) or (1.8) that two linear shift-invariant systems in parallel are equivalent to a single system whose unit-sample response is the sum of the individual unit-sample responses. This is depicted in Fig. 1.6.

Although the convolution-sum expression is analogous to the convolution integral of continuous-time linear system theory, it must be emphasized that the convolution sum should not be thought of as an approximation to the convolution integral. In contrast to the convolution integral, which plays mainly a theoretical role in analog linear system theory, we shall see that in addition to its theoretical importance, the convolution sum may serve as a realization of a discrete-time linear system. Thus it is important to gain some insight into the properties of the convolution sum and a degree of manipulative skill in using the convolution sum in actual calculations.

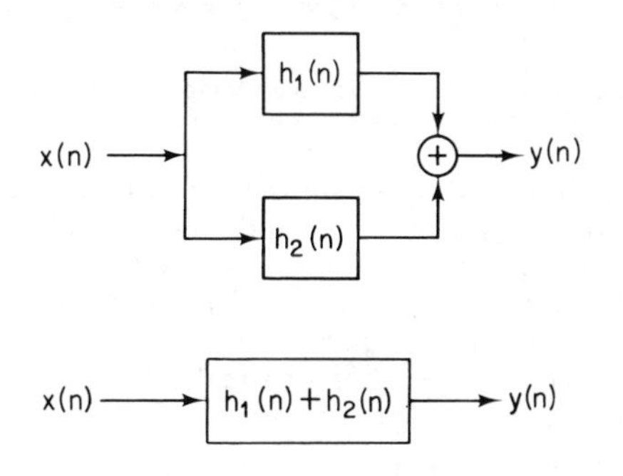

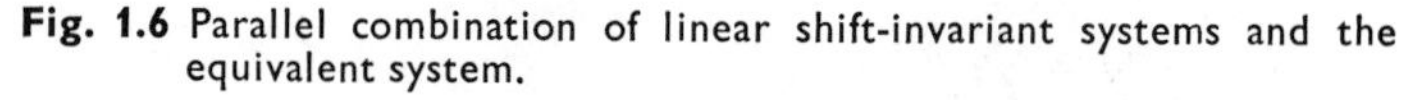

Fig. 1.6 Parallel combination of linear shift-invariant systems and the equivalent system.

EXAMPLE. Consider a system with unit-sample response

$$h(n) = \begin{cases} a^n, & n \geq 0 \\ 0, & n < 0 \end{cases}$$

or, equivalently, $h(n) = a^n u(n)$. To find the response to an input,

$$x(n) = u(n) - u(n - N)$$

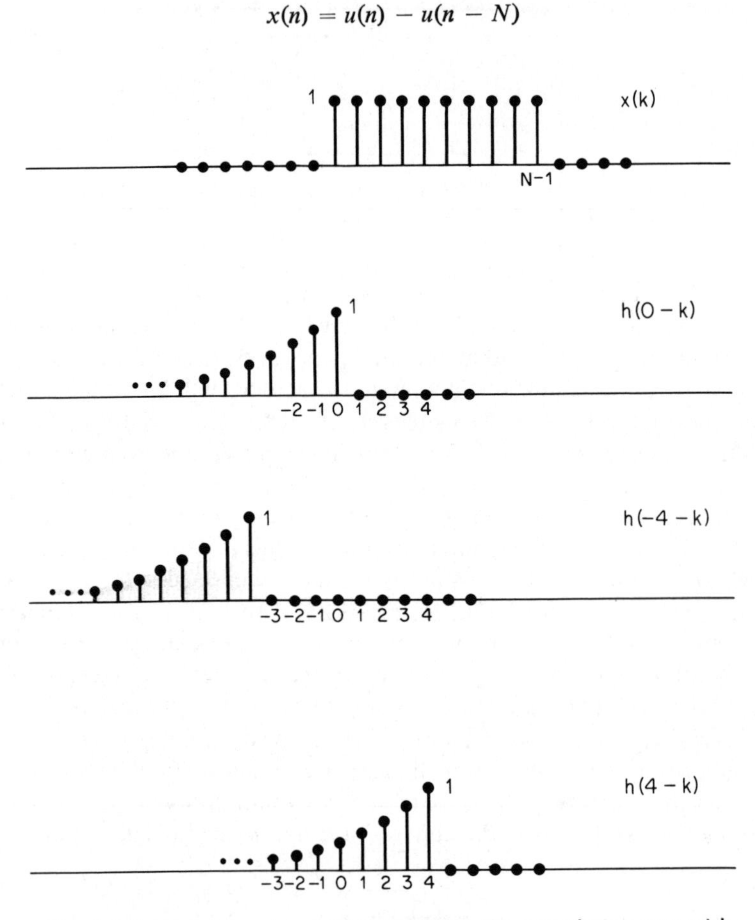

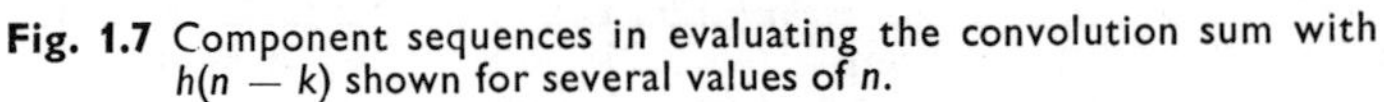

Fig. 1.7 Component sequences in evaluating the convolution sum with $h(n - k)$ shown for several values of n.

we note from Eq. (1.7) that to obtain the nth value of the output sequence we must form the product $x(k)h(n - k)$ and sum the values of the resulting sequence. The two component sequences are shown in Fig. 1.7 as a function of k, with $h(n - k)$ shown for several values of n. As we see in Fig. 1.7 for $n < 0$, $h(n - k)$ and $x(k)$ have no nonzero samples that overlap, and consequently $y(n) = 0$, $n < 0$. For n greater than or equal to zero but less than N, $h(n - k)$ and $x(k)$ have nonzero samples that overlap from $k = 0$ to $k = n$; thus for $0 \leq n < N$,

$$y(n) = \sum_{k=0}^{n} a^{n-k} = a^n \frac{1 - a^{-(n+1)}}{1 - a^{-1}}, \qquad 0 \leq n < N$$

For $N-1 \leq n$ the nonzero samples that overlap extend from $k = 0$ to $k = N-1$, and thus

$$y(n) = \sum_{k=0}^{N-1} a^{n-k} = a^n \frac{1-a^{-N}}{1-a^{-1}}, \qquad N \leq n$$

The response $y(n)$ is sketched in Fig. 1.8.

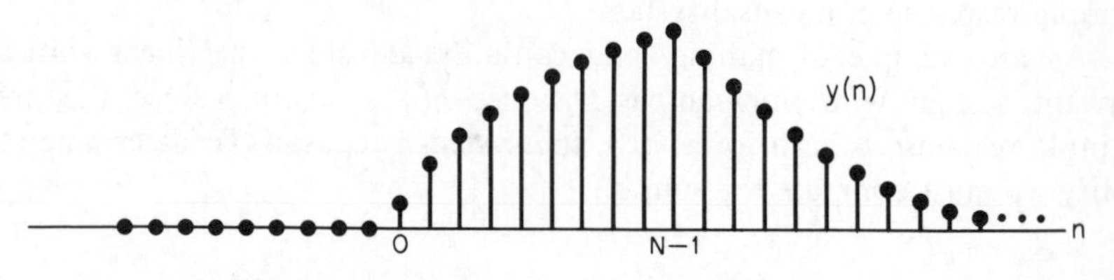

Fig. 1.8 Response of the system with unit-sample response $h(n) = a^n u(n)$ to the input $u(n) - u(n-N)$.

1.3 Stability and Causality

We have seen that the constraints of linearity and time-invariance define a class of systems that are represented by the convolution sum. The additional constraints of stability and causality define a more restricted class of linear time-invariant systems of practical importance.

We define a *stable system* as one for which every bounded input produces a bounded output. Linear shift-invariant systems are stable if and only if

$$S \triangleq \sum_{k=-\infty}^{\infty} |h(k)| < \infty \tag{1.9}$$

This can be shown as follows. If Eq. (1.9) is true and x is bounded, i.e. $|x(n)| < M$, for all n, then from Eq. (1.8)

$$|y(n)| = \left| \sum_{k=-\infty}^{\infty} h(k)x(n-k) \right| \leq M \sum_{k=-\infty}^{\infty} |h(k)| < \infty$$

Thus y is bounded. The converse is proved by showing that if $S = \infty$, then a bounded input can be found that will cause an unbounded output. Such an input is the sequence with values

$$x(n) = \begin{cases} \dfrac{h^*(-n)}{|h(-n)|} & h(n) \neq 0 \\ 0, & h(n) = 0 \end{cases}$$

where $h^*(n)$ is the complex conjugate of $h(n)$. $x(n)$ is clearly bounded. The value of the output at $n = 0$ is

$$y(0) = \sum_{k=-\infty}^{\infty} x(-k)h(k) = \sum_{k=-\infty}^{\infty} \frac{|h(k)|^2}{|h(k)|} = S$$

Thus if $S = \infty$, the output sequence is unbounded.

A *causal system* is one for which changes in the output do not precede changes in the input. That is, for a causal system, if $x_1(n) = x_2(n)$, $n < n_0$, then $y_1(n) = y_2(n)$, $n < n_0$. A linear shift-invariant system is causal if and only if the unit-sample response is zero for $n < 0$ (see Problem 11 of this chapter). For this reason it is sometimes convenient to refer to a sequence that is zero for $n < 0$ as a *causal sequence*, meaning that it could be the unit-sample response of a causal system.

As an example of stability and causality, consider the linear shift-invariant system with unit-sample response $h(n) = a^n u(n)$; since the unit-sample response is zero for $n < 0$, the system is causal. To determine stability we must compute the sum

$$S = \sum_{k=-\infty}^{\infty} |h(k)| = \sum_{k=0}^{\infty} |a|^k$$

If $|a| < 1$, the geometric series sums to $S = 1/(1 - |a|)$. However, if $|a| \geq 1$, the series diverges and hence the system is stable only for $|a| < 1$.

1.4 Linear Constant-Coefficient Difference Equations

In many applications a subclass of linear shift-invariant systems plays an important role. This subclass consists of those systems for which the input $x(n)$ and the output $y(n)$ satisfy an Nth-order linear constant-coefficient difference equation of the form

$$\sum_{k=0}^{N} a_k y(n-k) = \sum_{r=0}^{M} b_r x(n-r) \tag{1.10}$$

In general, a system of this class need not be causal. For example, consider the first-order difference equation

$$y(n) - ay(n-1) = x(n) \tag{1.11}$$

It is easily verified by direct substitution that if $x(n) = \delta(n)$, Eq. (1.11) is satisfied by either $y(n) = a^n u(n)$ or $y(n) = -a^n u(-n-1)$. The first solution corresponds to a causal filter that is stable if $|a| < 1$. The second solution is noncausal and is stable only if $|a| > 1$. It is common to interpret a difference equation such as Eq. (1.10) as characterizing a causal system, and we shall generally make this assumption unless we state otherwise.

Without additional information, a difference equation of the form of Eq. (1.10) does not uniquely specify the input–output relation of a system. This is a consequence of the fact that, as with differential equations, a family of solutions exists. For the difference equation (1.11), for example, if $y_1(n)$ satisfies the equation for $x(n) = x_1(n)$, then so does any solution of the form $y(n) = y_1(n) + ka^n$, where k is an arbitrary constant. More generally, any solution of Eq. (1.10) can have added to it a component that satisfies the

homogeneous difference equation (i.e., the difference equation with the right-hand side equal to zero) and the sum will still satisfy Eq. (1.10).

A system that satisfies a linear constant-coefficient difference equation will correspond to a linear shift-invariant system only if we choose the homogeneous component accordingly. If the system is causal, for example, we must specify initial rest conditions so that if $x(n) = 0$, $n < n_0$, then $y(n) = 0$, $n < n_0$. Throughout our discussions we shall generally assume, unless stated otherwise, that if a system satisfies a linear constant-coefficient difference equation, it will also satisfy the necessary conditions for it to be a linear shift-invariant system.

If we assume that the system is causal, a linear difference equation provides an explicit relationship between the input and output. This can be seen by rewriting Eq. (1.10) as

$$y(n) = -\sum_{k=1}^{N} \frac{a_k}{a_0} y(n-k) + \sum_{r=0}^{M} \frac{b_r}{a_0} x(n-r) \tag{1.12}$$

Thus the nth value of the output can be computed from the nth input value and the N and M past values of the output and input, respectively. As in the case of the convolution sum, the difference equation not only represents the system for theoretical purposes, but it may also serve as a computational realization of the system.

EXAMPLE. The first-order difference equation of Eq. (1.11) yields the recursion formula

$$y(n) = ay(n-1) + x(n)$$

To obtain the unit-sample response, let $x(n) = \delta(n)$ and assume initial rest conditions. Then

$$\begin{aligned} h(n) &= 0, \qquad n < 0 \\ h(0) &= ah(-1) + 1 = 1 \\ h(1) &= ah(0) = a \\ &\ \ \vdots \\ h(n) &= ah(n-1) = a^n \end{aligned}$$

Thus

$$h(n) = a^n u(n)$$

To obtain a different solution, let $x(n) = \delta(n)$, but assume now that $y(n) = 0$ for $n > 0$. From the difference equation of Eq. (1.11) we can write the recursion relation

$$y(n-1) = \frac{1}{a}[y(n) - x(n)]$$

or

$$y(n) = \frac{1}{a}[y(n+1) - x(n+1)]$$

Then

$$h(n) = 0, \qquad n > 0$$

$$h(0) = \frac{1}{a}[h(1) - x(1)] = 0$$

$$h(-1) = \frac{1}{a}[h(0) - x(0)] = -a^{-1}$$

$$h(-2) = \frac{1}{a}[h(-1) - x(-1)] = -a^{-2}$$

$$\vdots$$

$$h(n) = \frac{1}{a}\,h(n+1) = -a^n$$

Thus

$$h(n) = -a^n u(-n-1)$$

In general, a linear shift-invariant system may have a unit-sample response that is of finite duration or it may be of infinite duration. Because of the properties of some digital processing techniques, it is useful to distinguish between these two classes. If the unit-sample response is of finite duration, it will be referred to as a *finite impulse response* (*FIR*) *system*, and if the unit sample response is of infinite duration, it will be referred to as an *infinite impulse response* (*IIR*) *system*. If $N = 0$ in Eq. (1.10) so that

$$y(n) = \frac{1}{a_0}\left[\sum_{r=0}^{M} b_r x(n-r)\right]$$

then it corresponds to an FIR system. In fact, comparison with Eq. (1.8) shows that the above difference equation is identical to the convolution sum, and hence it follows directly that

$$h(n) = \begin{cases} \dfrac{b_n}{a_0}, & n = 0, 1, \ldots, M \\ 0, & \text{otherwise} \end{cases}$$

An FIR system can always be described by a difference equation of the form of Eq. (1.10) with $N = 0$. In contrast, for an IIR system, N must be greater than zero.

1.5 Frequency-Domain Representation of Discrete-Time Systems and Signals

In the previous sections we have introduced some of the fundamental concepts of the theory of discrete-time signals and systems. For linear shift-invariant systems we saw that a representation of the input sequence as a weighted sum of delayed unit-sample sequences leads to a representation of the output as a weighted sum of delayed unit-sample responses. Just as continuous-time signals may be represented in a number of different ways, so it is with discrete-time signals, and, just as for analog signals and systems, sinusoidal or complex exponential sequences play a particularly important role. This is because a fundamental property of linear shift-invariant systems is that the steady-state response to a sinusoidal input is sinusoidal of the same frequency as the input, with amplitude and phase determined by the system. It is this property of linear shift-invariant systems that makes representations of signals in terms of sinusoids or complex exponentials (i.e., Fourier representations) so useful in linear system theory.

To see this for discrete-time systems, suppose that the input sequence is $x(n) = e^{j\omega n}$ for $-\infty < n < \infty$, i.e., a complex exponential of radian frequency ω. Then using Eq. (1.8) the output is

$$\begin{aligned} y(n) &= \sum_{k=-\infty}^{\infty} h(k)e^{j\omega(n-k)} \\ &= e^{j\omega n} \sum_{k=-\infty}^{\infty} h(k)e^{-j\omega k} \end{aligned}$$

If we define

$$H(e^{j\omega}) = \sum_{k=-\infty}^{\infty} h(k)e^{-j\omega k} \tag{1.13}$$

we can write

$$y(n) = H(e^{j\omega})e^{j\omega n} \tag{1.14}$$

From Eq. (1.14) we see that $H(e^{j\omega})$ describes the change in complex amplitude of a complex exponential as a function of the frequency ω. The quantity $H(e^{j\omega})$ is called the *frequency response* of the system whose unit-sample response is $h(n)$. In general, $H(e^{j\omega})$ is complex and can be expressed in terms of its real and imaginary parts as

$$H(e^{j\omega}) = H_R(e^{j\omega}) + jH_I(e^{j\omega})$$

or in terms of magnitude and phase as

$$H(e^{j\omega}) = |H(e^{j\omega})|e^{j\arg[H(e^{j\omega})]}$$

Sometimes it will be convenient to refer to the *group delay* rather than the phase. The group delay is defined as the negative of the first derivative, with respect to ω, of the phase.

Since a sinusoid can be expressed as a linear combination of complex exponentials, the frequency response also expresses the response to a sinusoidal input. Specifically, consider

$$x(n) = A\cos(\omega_0 n + \phi) = \frac{A}{2}e^{j\phi}e^{j\omega_0 n} + \frac{A}{2}e^{-j\phi}e^{-j\omega_0 n}$$

From Eq. (1.14) the response to $(A/2)e^{j\phi}e^{j\omega_0 n}$ is

$$y_1(n) = H(e^{j\omega_0})\frac{A}{2}e^{j\phi}e^{j\omega_0 n}$$

If $h(n)$ is real, then from Eq. (1.7) the response to $(A/2)e^{-j\phi}e^{-j\omega_0 n}$ is the complex conjugate of the response to $(A/2)e^{j\phi}e^{j\omega_0 n}$. Thus the net response is

$$y(n) = \frac{A}{2}[H(e^{j\omega_0})e^{j\phi}e^{j\omega_0 n} + H(e^{-j\omega_0})e^{-j\phi}e^{-j\omega_0 n}]$$

or

$$y(n) = A\,|H(e^{j\omega_0})|\cos(\omega_0 n + \phi + \theta)$$

where $\theta = \arg[H(e^{j\omega_0})]$ is the phase response of the system at frequency ω_0.

$H(e^{j\omega})$ is a continuous function of ω. Furthermore, it is a periodic function of ω with period 2π. This property follows directly from Eq. (1.13) since $e^{j(\omega+2\pi)k} = e^{j\omega k}$. The fact that the frequency response has the same value at ω and $\omega + 2\pi$ simply means that the system responds identically to complex exponentials of these two frequencies. Since these two exponential sequences are indistinguishable, this behavior is eminently reasonable.

EXAMPLE. As an example of the evaluation of the frequency response, consider a system with unit-sample response

$$h(n) = \begin{cases} 1, & 0 \le n \le N-1 \\ 0, & \text{elsewhere} \end{cases} \tag{1.15}$$

as depicted in Fig. 1.9. The system function is

$$\begin{aligned} H(e^{j\omega}) &= \sum_{n=0}^{N-1} e^{-j\omega n} = \frac{1-e^{-j\omega N}}{1-e^{-j\omega}} \\ &= \frac{\sin(\omega N/2)}{\sin(\omega/2)}e^{-j(N-1)\omega/2} \end{aligned} \tag{1.16}$$

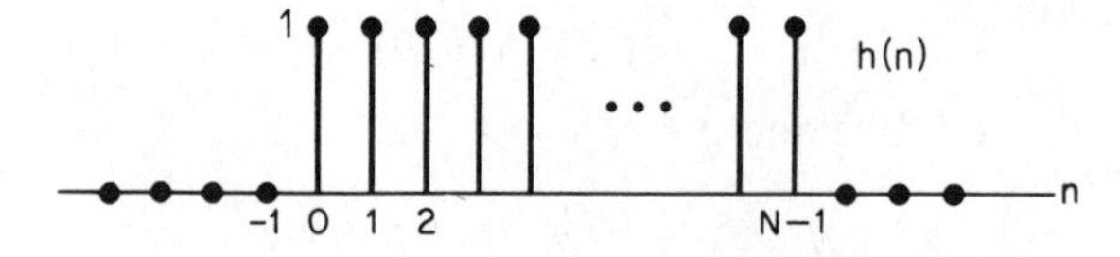

Fig. 1.9 Unit-sample response of the system for which the frequency response is to be computed.

The magnitude and phase of $H(e^{j\omega})$ are plotted in Fig. 1.10 for the case $N = 5$.

Since $H(e^{j\omega})$ is a periodic function of ω, it can be represented by a Fourier series. Equation (1.13) in fact, expresses $H(e^{j\omega})$ in the form of a Fourier series where the Fourier coefficients correspond to the unit-sample response $h(n)$. With that observation it follows that $h(n)$ can be evaluated

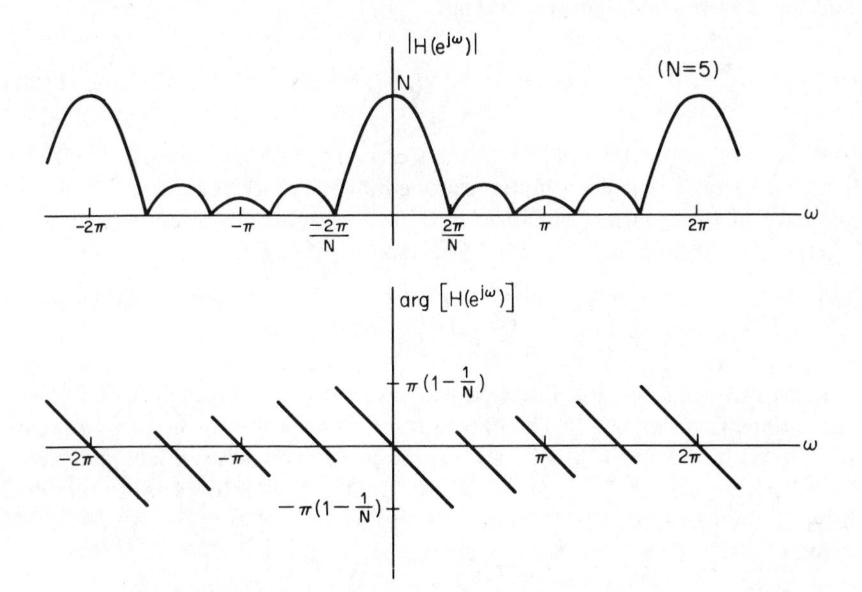

Fig. 1.10 Magnitude and phase of the frequency response of the system with unit-sample response of Fig. 1.9.

from $H(e^{j\omega})$ by means of the relation used to obtain the Fourier coefficients of a periodic function, [1–3] i.e.,

$$h(n) = \frac{1}{2\pi} \int_{-\pi}^{\pi} H(e^{j\omega}) e^{j\omega n} \, d\omega \tag{1.17}$$

where

$$H(e^{j\omega}) = \sum_{n=-\infty}^{\infty} h(n) e^{-j\omega n} \tag{1.18}$$

These equations have an alternative interpretation as a representation of the sequence $h(n)$. Specifically it is useful to consider Eq. (1.17) as representing the sequence $h(n)$ as a superposition (integral) of exponential signals whose complex amplitudes are determined by Eq. (1.18). Thus Eqs. (1.17) and (1.18) are a Fourier transform pair for the sequence $h(n)$, with Eq. (1.18) playing the role of a direct (analysis) transform of the sequence $h(n)$ and Eq. (1.17) being the inverse (synthesis) Fourier transform. Such a representation exists if the series in Eq. (1.18) converges.

The representation of a sequence by the transform of Eq. (1.18) is not restricted to the unit-sample response of a system but can be applied to any sequence provided that the series of Eq. (1.18) converges. Thus for a general sequence $x(n)$ we define the Fourier transform as

$$X(e^{j\omega}) = \sum_{n=-\infty}^{\infty} x(n) e^{-j\omega n} \tag{1.19}$$

and the inverse Fourier transform as

$$x(n) = \frac{1}{2\pi}\int_{-\pi}^{\pi} X(e^{j\omega})e^{j\omega n}\,d\omega \tag{1.20}$$

The series of Eq. (1.19) does not always converge, as, for example, if $x(n)$ is a unit step or a real or complex exponential sequence for all n. The convergence of the Fourier transform is subject to a variety of definitions and interpretations. If $x(n)$ is absolutely summable, i.e., if

$$\sum_{n=-\infty}^{\infty} |x(n)| < \infty$$

the series is said to be absolutely convergent and it converges uniformly to a continuous function of ω. Thus the frequency response for a stable system will always converge. If a sequence is absolutely summable, it will also have finite energy; i.e., $\sum_{n=-\infty}^{\infty} |x(n)|^2$ will be finite. This follows in a straightforward manner from the fact that

$$\sum |x(n)|^2 \leq [\sum |x(n)|]^2$$

It is not true, however, that a sequence with finite energy is absolutely summable. An example of a sequence that has finite energy but is not absolutely summable is the sequence

$$x(n) = \frac{\sin \omega_0 n}{\pi n}$$

If a sequence is not absolutely summable but has finite energy, one may employ a type of convergence in which the series converges so that the mean-square error is zero. The attendant Gibbs oscillations at a discontinuity [1–3] are of practical significance in filter design and will be discussed in a later chapter.

The fact that a sequence can be represented as a superposition of complex exponentials is of great importance in the analysis of linear shift-invariant systems. This is because of the principle of superposition and the fact that the response of such a system to a complex exponential is completely determined by the frequency response, $H(e^{j\omega})$. If we view Eq. (1.20) as a superposition of complex exponentials of incremental amplitude, then the response of a linear shift-invariant system to an input $x(n)$ is a corresponding superposition of responses due to each complex exponential that goes into representing the input. Since the response to each complex exponential is obtained by multiplying by $H(e^{j\omega})$,

$$y(n) = \frac{1}{2\pi}\int_{-\pi}^{\pi} H(e^{j\omega})X(e^{j\omega})e^{j\omega n}\,d\omega$$

Thus the Fourier transform of the output is

$$Y(e^{j\omega}) = H(e^{j\omega})X(e^{j\omega}) \tag{1.21}$$

This result has its counterpart in continuous-time linear system theory and can of course be derived in a more rigorous manner by simply evaluating the Fourier transform of the convolution sum

$$y(n) = \sum_{k=-\infty}^{\infty} h(n - k)x(k)$$

Although this more formal approach provides rigorous justification of Eq. (1.21) (see Problem 17 of this chapter), the previous discussion is meant to emphasize that Eq. (1.21) is a direct result of the special properties of linear shift-invariant systems.

To illustrate the results that we have been discussing, consider the following example.

EXAMPLE. *Ideal Lowpass Filter.* The ideal discrete-time lowpass filter has frequency response $H(e^{j\omega})$, as shown in Fig. 1.11. That is, for $-\pi < \omega < \pi$,

$$H(e^{j\omega}) = \begin{cases} 1, & |\omega| \leq \omega_{co} \\ 0, & \omega_{co} < |\omega| \leq \pi \end{cases}$$

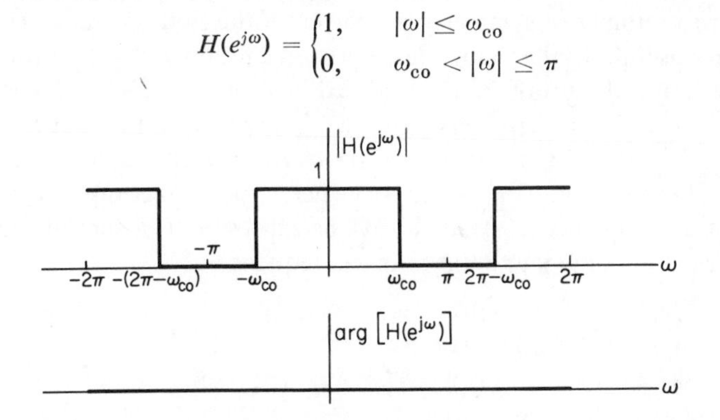

Fig. 1.11 Frequency response of an ideal discrete-time lowpass filter.

Since $H(e^{j\omega})$ is periodic, this specifies the frequency response for all ω. Such a system clearly removes all components of the input of frequency in the range $\omega_{co} < |\omega| \leq \pi$. The impulse response $h(n)$ is found from Eq. (1.17) as

$$h(n) = \frac{1}{2\pi}\int_{-\omega_{co}}^{\omega_{co}} e^{j\omega n}\, d\omega = \frac{\sin \omega_{co} n}{\pi n}$$

and is shown in Fig. 1.12 for $\omega_{co} = \pi/2$.

The ideal lowpass filter is an example of a system that is very effectively described in the frequency domain. It is easy to see that the system completely removes all frequencies in the input above the cutoff frequency ω_{co}. Clearly, the ideal lowpass filter is not causal; furthermore, it can be shown to be unstable in the strict sense of the definition of Sec. 1.3. Nevertheless, it

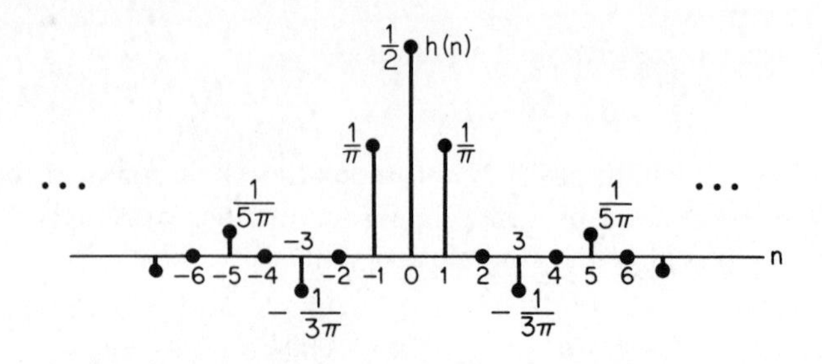

Fig. 1.12 Unit-sample response of ideal lowpass filter with cutoff frequency $\omega_{co} = \pi/2$.

is extremely important conceptually, and considerable effort will be expended in Chapter 5 in finding ways to design systems that approximate the behavior of the ideal lowpass filter.

1.6 Some Symmetry Properties of the Fourier Transform

There are a number of symmetry properties of the Fourier transform that are often very useful. In the following discussion a number of these properties are presented, and the proofs are considered in Problems 25–27 of this chapter.

A *conjugate symmetric sequence* $x_e(n)$ is defined as a sequence for which $x_e(n) = x_e^*(-n)$ and a *conjugate antisymmetric sequence* $x_o(n)$ is defined as a sequence for which $x_o(n) = -x_o^*(-n)$ where * denotes complex conjugation. An arbitrary sequence $x(n)$ can always be expressed as a sum of a conjugate symmetric and conjugate antisymmetric sequence as

$$x(n) = x_e(n) + x_o(n) \tag{1.22a}$$

where

$$x_e(n) = \tfrac{1}{2}[x(n) + x^*(-n)] \tag{1.22b}$$

and

$$x_o(n) = \tfrac{1}{2}[x(n) - x^*(-n)] \tag{1.22c}$$

A real sequence which is conjugate symmetric so that $x_e(n) = x_e(-n)$ is generally referred to as an *even sequence*, and a real sequence which is conjugate antisymmetric so that $x_o(n) = -x_o(-n)$ is generally referred to as an *odd sequence*.

A Fourier transform $X(e^{j\omega})$ can be decomposed into a sum of conjugate symmetric and conjugate antisymmetric functions as

$$X(e^{j\omega}) = X_e(e^{j\omega}) + X_o(e^{j\omega}) \tag{1.23a}$$

where

$$X_e(e^{j\omega}) = \tfrac{1}{2}[X(e^{j\omega}) + X^*(e^{-j\omega})] \tag{1.23b}$$

and

$$X_o(e^{j\omega}) = \tfrac{1}{2}[X(e^{j\omega}) - X^*(e^{-j\omega})] \tag{1.23c}$$

where $X_e(e^{j\omega})$ is conjugate symmetric and $X_o(e^{j\omega})$ is conjugate antisymmetric; i.e.,

$$X_e(e^{j\omega}) = X_e^*(e^{-j\omega})$$

and

$$X_o(e^{j\omega}) = -X_o^*(e^{-j\omega})$$

As with sequences, if a real function is conjugate symmetric, it is generally referred to as an *even function*, and if conjugate antisymmetric it is generally referred to as an *odd function.*

Let us first consider a general, complex sequence $x(n)$ with Fourier transform $X(e^{j\omega})$. It can be shown (see Problem 25) that the Fourier transform of $x^*(n)$ is $X^*(e^{-j\omega})$ and the Fourier transform of $x^*(-n)$ is $X^*(e^{j\omega})$. As a result of this, and using the fact that the Fourier transform of the sum of two sequences is the sum of the Fourier transforms, it follows that the Fourier transform of $\frac{1}{2}[x(n) + x^*(n)]$ or, equivalently, Re $[x(n)]$ is $\frac{1}{2}[X(e^{j\omega}) + X^*(e^{-j\omega})]$ or, equivalently, the conjugate symmetric part of $X(e^{j\omega})$. In a similar manner, $\frac{1}{2}[x(n) - x^*(n)]$ or, equivalently, j Im $[x(n)]$ has a Fourier transform that is the conjugate antisymmetric component $X_o(e^{j\omega})$. By considering the Fourier transform of $x_e(n)$ and $x_o(n)$, the conjugate symmetric and conjugate antisymmetric components of $x(n)$, it follows that the Fourier transform of $x_e(n)$ is Re $[X(e^{j\omega})]$ and the Fourier transform of $x_o(n)$ is j Im $[X(e^{j\omega})]$.

If $x(n)$ is a real sequence, these symmetry properties become particularly straightforward and useful. Specifically, for a real sequence, the Fourier transform is conjugate symmetric; i.e., $X(e^{j\omega}) = X^*(e^{-j\omega})$. Expressing $X(e^{j\omega})$ in terms of its real and imaginary parts as

$$X(e^{j\omega}) = \text{Re}\,[X(e^{j\omega})] + j\,\text{Im}\,[X(e^{j\omega})]$$

it then follows that

$$\text{Re}\,[X(e^{j\omega})] = \text{Re}\,[X(e^{-j\omega})]$$

and

$$\text{Im}\,[X(e^{j\omega})] = -\text{Im}\,[X(e^{-j\omega})]$$

i.e., the real part of the Fourier transform is an even function and the imaginary part is an odd function. In a similar manner, by expressing $X(e^{j\omega})$ in polar coordinates as

$$X(e^{j\omega}) = |X(e^{j\omega})|\, e^{j\arg[X(e^{j\omega})]}$$

it follows that for a real sequence $x(n)$, the magnitude of the Fourier transform $|X(e^{j\omega})|$ is an even function of ω and the phase given by arg $[X(e^{j\omega})|$ is an odd function of ω. Also for a real sequence, the even part of $x(n)$ transforms to Re $[X(e^{j\omega})]$ and the odd part of $x(n)$ transforms to j Im $[X(e^{j\omega})]$. All the above symmetry properties are summarized in Table 1.1.

Table 1.1

Sequence	*Fourier Transform*
$x(n)$	$X(e^{j\omega})$
$x^*(n)$	$X^*(e^{-j\omega})$
$x^*(-n)$	$X^*(e^{j\omega})$
Re $[x(n)]$	$X_e(e^{j\omega})$ [conjugate symmetric part of $X(e^{j\omega})$]
j Im $[x(n)]$	$X_o(e^{j\omega})$ [conjugate antisymmetric part of $X(e^{j\omega})$]
$x_e(n)$ [conjugate symmetric part of $x(n)$]	Re $[X(e^{j\omega})]$
$x_o(n)$ [conjugate antisymmetric part of $x(n)$]	j Im $[X(e^{j\omega})]$
The following properties apply only when $x(n)$ is real:	
any real $x(n)$	$X(e^{j\omega}) = X^*(e^{-j\omega})$ (Fourier transform is conjugate symmetric)
	Re $[X(e^{j\omega})] =$ Re $[X(e^{-j\omega})]$ (real part is even)
	Im $[X(e^{j\omega})] = -$Im $[X(e^{-j\omega})]$ (imaginary part is odd)
	$\lvert X(e^{j\omega})\rvert = \lvert X(e^{-j\omega})\rvert$ (magnitude is even)
	arg $[X(e^{j\omega})] = -$arg $[X(e^{-j\omega})]$ (phase is odd)
$x_e(n)$ [even part of $x(n)$]	Re $[X(e^{j\omega})]$
$x_o(n)$ [odd part of $x(n)$]	j Im $[X(e^{j\omega})]$

1.7 Sampling of Continuous-Time Signals

In previous sections of this chapter we have refrained from relating discrete-time and continuous-time signals and systems except for pointing out some of the similarities between important theoretical concepts. Often, however, discrete-time signals are derived from continuous-time signals by periodic sampling; consequently it is important to understand how the sequence so derived is related to the original signal.

Consider an analog signal $x_a(t)$ that has the Fourier representation [1-3]

$$x_a(t) = \frac{1}{2\pi}\int_{-\infty}^{+\infty} X_a(j\Omega)e^{j\Omega t}\,d\Omega \tag{1.24a}$$

$$X_a(j\Omega) = \int_{-\infty}^{\infty} x_a(t)e^{-j\Omega t}\,dt \tag{1.24b}$$

The sequence $x(n)$ with values $x(n) = x_a(nT)$ is said to be derived from $x_a(t)$ by periodic sampling, and T is called the *sampling period.* The reciprocal of T is called the *sampling frequency* or the *sampling rate.* In order to determine the sense in which $x(n)$ represents the original signal $x_a(t)$, it is convenient to relate $X_a(j\Omega)$, the continuous-time Fourier transform of $x_a(t)$, to $X(e^{j\omega})$, the discrete-time Fourier transform of the sequence $x(n)$. From Eq. (1.24a) we note that

$$x(n) = x_a(nT) = \frac{1}{2\pi}\int_{-\infty}^{\infty} X_a(j\Omega)e^{j\Omega nT}\, d\Omega \tag{1.25}$$

From the discrete-time Fourier transform we also obtain the representation

$$x(n) = \frac{1}{2\pi}\int_{-\pi}^{\pi} X(e^{j\omega})e^{j\omega n}\, d\omega \tag{1.26}$$

To relate Eqs. (1.25) and (1.26) it is convenient to express Eq. (1.25) as a sum of integrals over intervals of length $2\pi/T$, as in

$$x(n) = \frac{1}{2\pi}\sum_{r=-\infty}^{\infty}\int_{(2r-1)\pi/T}^{(2r+1)\pi/T} X_a(j\Omega)e^{j\Omega nT}\, d\Omega$$

Each term in the sum can be reduced to an integral over the range $-\pi/T$ to $+\pi/T$ by a change of variables to obtain

$$x(n) = \frac{1}{2\pi}\sum_{r=-\infty}^{\infty}\int_{-\pi/T}^{\pi/T} X_a\left(j\Omega + j\frac{2\pi r}{T}\right)e^{j\Omega nT}e^{j2\pi rn}\, d\Omega$$

If we interchange the order of integration and summation and note that $e^{j2\pi rn} = 1$ for all integer values of r and n, then we obtain

$$x(n) = \frac{1}{2\pi}\int_{-\pi/T}^{\pi/T}\left[\sum_{r=-\infty}^{\infty} X_a\left(j\Omega + j\frac{2\pi r}{T}\right)\right]e^{j\Omega nT}\, d\Omega \tag{1.27}$$

With the substitution $\Omega = \omega/T$, Eq. (1.27) becomes

$$x(n) = \frac{1}{2\pi}\int_{-\pi}^{\pi}\left[\frac{1}{T}\sum_{r=-\infty}^{\infty} X_a\left(\frac{j\omega}{T} + j\frac{2\pi r}{T}\right)\right]e^{j\omega n}\, d\omega$$

which is identical in form to Eq. (1.26). Therefore, we can make the identification

$$X(e^{j\omega}) = \frac{1}{T}\sum_{r=-\infty}^{\infty} X_a\left(\frac{j\omega}{T} + j\frac{2\pi r}{T}\right) \tag{1.28}$$

Alternatively, we can express Eq. (1.28) in terms of the analog frequency variable Ω as

$$X(e^{j\Omega T}) = \frac{1}{T}\sum_{r=-\infty}^{\infty} X_a\left(j\Omega + j\frac{2\pi r}{T}\right) \tag{1.29}$$

Equations (1.28) and (1.29) make very clear the relationship between the continuous-time Fourier transform and the Fourier transform of a sequence derived by sampling. For example, if $X_a(j\Omega)$ is as depicted in Fig. 1.13(a), then $X(e^{j\omega})$ will be as shown in Fig. 1.13(b) if $\Omega_o/2 > \pi/T$. In Fig. 1.13(b)

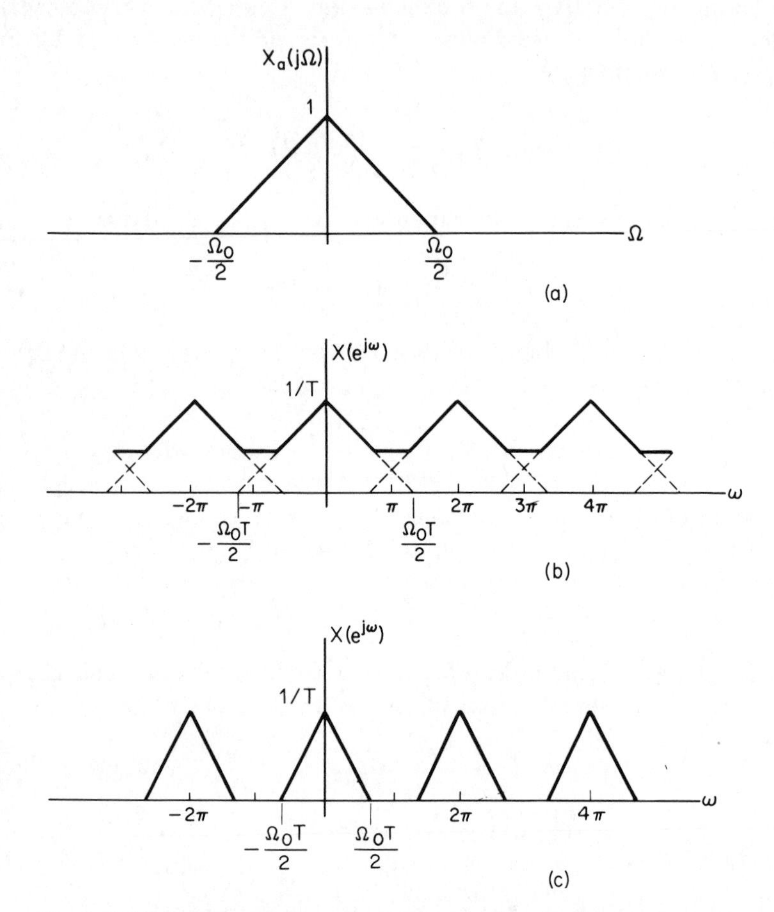

Fig. 1.13 (a) Fourier transform of a continuous-time signal. (b) Fourier transform of the discrete-time signal obtained by periodic sampling. The sampling period is large, so the periodic repetitions of the continuous-time transform overlap. (c) Same as (b), but with the sampling period small enough so that the periodic repetitions of the continuous-time transform do not overlap.

we note that if the sampling period is too large, the shifted versions of $X_a(j\omega/T)$ overlap. In this case the upper frequencies in $X_a(j\Omega)$ get reflected into the lower frequencies in $X(e^{j\omega})$. This phenomenon, where in effect a high-frequency component in $X_a(j\Omega)$ takes on the identity of a lower frequency, is called *aliasing*. From Fig. 1.13(c) it is clear that if $\Omega_o/2 < \pi/T$, i.e., we sample at a rate at least twice the highest frequency of $X_a(j\Omega)$, then

$X(e^{j\omega})$ is identical to $X_a(\omega/T)$ in the interval $-\pi \leq \omega \leq \pi$. In this case it is reasonable to expect that $x_a(t)$ can be recovered from the samples $x_a(nT)$ by an appropriate interpolation formula. This sampling rate is generally referred to as the *Nyquist rate*.

To derive the interpolation formula let us assume that $\Omega_o/2 < \pi/T$ as in Fig. 1.13(c).

Then

$$X(e^{j\Omega T}) = \frac{1}{T} X_a(j\Omega), \qquad -\pi/T \leq \Omega \leq \pi/T \tag{1.30}$$

From the continuous-time Fourier transform

$$x_a(t) = \frac{1}{2\pi} \int_{-\pi/T}^{\pi/T} X_a(j\Omega) e^{j\Omega t}\, d\Omega \tag{1.31}$$

Combining Eqs. (1.30) and (1.31) we can write

$$x_a(t) = \frac{1}{2\pi} \int_{-\pi/T}^{\pi/T} T X(e^{j\Omega T}) e^{j\Omega t}\, d\Omega$$

Since

$$X(e^{j\Omega T}) = \sum_{k=-\infty}^{\infty} x_a(kT) e^{-j\Omega Tk}$$

it then follows that

$$x_a(t) = \frac{T}{2\pi} \int_{-\pi/T}^{\pi/T} \left[\sum_{k=-\infty}^{\infty} x_a(kT) e^{-j\Omega Tk} \right] e^{j\Omega t}\, d\Omega$$

or, interchanging the order of summation and integration,

$$x_a(t) = \sum_{k=-\infty}^{\infty} x_a(kT) \left[\frac{T}{2\pi} \int_{-\pi/T}^{\pi/T} e^{j\Omega(t-kT)}\, d\Omega \right]$$

Evaluating the integral, we obtain

$$x_a(t) = \sum_{k=-\infty}^{\infty} x_a(kT) \frac{\sin (\pi/T)(t - kT)}{(\pi/T)(t - kT)} \tag{1.32}$$

Equation (1.32) provides an interpolation formula for recovering the continuous-time signal $x_a(t)$ from its samples. The representation of a continuous-time signal in the form of Eq. (1.32) is only valid for bandlimited functions and with T chosen sufficiently small so that no aliasing occurs.

Equation (1.32) can be thought of as an expansion of the continuous-time signal in the form

$$x_a(t) = \sum_{k=-\infty}^{\infty} c_k \phi_k(t) \tag{1.33}$$

where the coefficients c_k and the functions $\phi_k(t)$ are given by

$$c_k = x_a(kT) \tag{1.34a}$$

and

$$\phi_k(t) = \frac{\sin (\pi/T)(t - kT)}{(\pi/T)(t - kT)} \tag{1.34b}$$

In general there are many classes of functions $\phi_k(t)$ that can be used to express a continuous-time function in the form of Eq. (1.33), including sinusoidal functions, Laguerre functions, and Legendre polynomials. In any representation of the form of Eq. (1.33) the sequence of coefficients c_k can be regarded as a discrete-time signal that represents the continuous-time signal $x_a(t)$ [7]. Not all such representations are equally useful. A strong advantage with the choice of the functions $\phi_k(t)$ of Eq. (1.34b) is that the coefficients are easily obtained simply by sampling the continuous-time signal. A second advantage is that it preserves convolution in the sense that if $y_a(t)$ is the continuous-time convolution of $x_a(t)$ and $h_a(t)$, then $y_a(nT)$ will be the discrete convolution of $x_a(nT)$ and $h_a(nT)$ provided that the sampling period T is chosen small enough to avoid aliasing, i.e., so that the representation of Eq. (1.32) is valid. The advantage of a representation that preserves convolution is that it permits the simulation or implementation of a continuous-time linear time-invariant system by a discrete-time linear shift-invariant system. The disadvantage of the representation using the functions of Eq. (1.34b) is that it is restricted to bandlimited functions.

There are many other choices for the functions $\phi_k(t)$ of Eq. (1.33) that will provide a discrete representation of continuous-time signals which preserves convolution [8]. A necessary and sufficient condition for this to be true is considered in Problem 32 of this chapter.

1.8 Two-Dimensional Sequences and Systems

In the previous sections we focused the discussion on the representation of one-dimensional signals and systems. In this section we shall concentrate on the extension of some of those results to two-dimensional signals and systems. Many important signal processing problems involve the processing of multi-dimensional signals. All the properties of signals and systems that we have derived thus far in this chapter are easily extended to the multidimensional case. In the chapters that follow, however, it will generally not be the case that results for one-dimensional sequences and systems extend to the multi-dimensional case.

A two-dimensional sequence is a function of two integer variables and is often depicted graphically as shown in Fig. 1.14. As in the one-dimensional case, it is useful to define the unit-sample, unit-step, exponential, and sinusoidal sequences. The two-dimensional *unit-sample sequence* $\delta(m, n)$ is defined as the sequence that is zero except at the origin; i.e.,

$$\delta(m, n) = \begin{cases} 0, & m, n \neq 0 \\ 1, & m = n = 0 \end{cases}$$

The two-dimensional *unit-step sequence* $u(m, n)$ is unity in the first quadrant

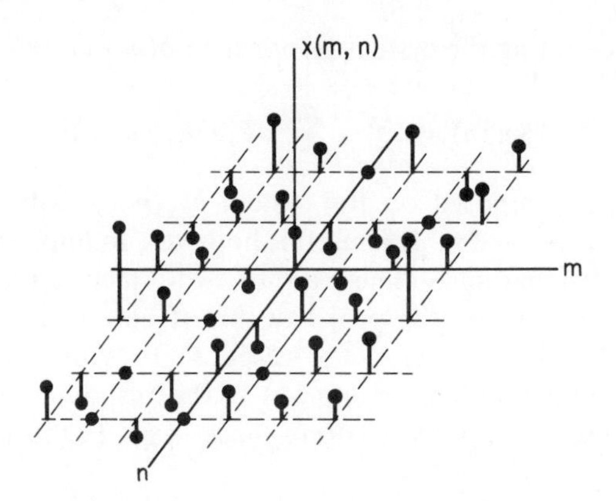

Fig. 1.14 Graphical representation of a two-dimensional sequence.

of the (m, n)-plane and zero elsewhere; i.e.,

$$u(m, n) = \begin{cases} 1, & m \geq 0, n \geq 0 \\ 0, & \text{otherwise} \end{cases}$$

A two-dimensional exponential sequence is of the form $a^m b^n$, and a two-dimensional sinusoidal sequence is of the form

$$A \cos(\omega_0 m + \phi) \cos(\omega_1 n + \theta).$$

A *separable sequence* is one that can be expressed as a product of one-dimensional sequences; i.e., $x(m, n)$ is separable if it can be expressed in the form

$$x(m, n) = x_1(m)x_2(n)$$

The unit-sample, unit-step, exponential, and sinusoidal sequences are all separable. An example of a sequence that is not separable is the sequence

$$x(m, n) = \cos(\omega_0 mn)$$

As with the one-dimensional case, an arbitrary two-dimensional sequence can be expressed as a linear combination of shifted unit samples as

$$x(m, n) = \sum_{k=-\infty}^{\infty} \sum_{r=-\infty}^{\infty} x(k, r)\, \delta(m - k, n - r)$$

On the basis of this equation a two-dimensional linear system can be described in terms of its responses to shifted unit samples. Specifically with $y(m, n) = T[x(m, n)]$, where $T[\ ]$ is the transformation for a linear system,

$$y(m, n) = T\left[\sum_{k=-\infty}^{\infty} \sum_{r=-\infty}^{\infty} x(k, r)\, \delta(m - k, n - r)\right]$$

$$= \sum_{k=-\infty}^{\infty} \sum_{r=-\infty}^{\infty} x(k, r) T[\delta(m - k, n - r)]$$

With $h_{k,r}(m, n)$ denoting the system response to $\delta(m - k, n - r)$,

$$y(m, n) = \sum_{k=-\infty}^{\infty} \sum_{r=-\infty}^{\infty} x(k, r)h_{k,r}(m, n) \tag{1.35}$$

With only linearity imposed on the system $h_{k,r}(m, n)$ will depend on the four variables k, r, m, and n. It is useful, however, to impose the additional constraint of shift-invariance. The class of two-dimensional shift-invariant systems is characterized by the property that if $y(m, n)$ is the response to $x(m, n)$, then $y(m - k, n - r)$ is the response to $x(m - k, n - r)$. With this constraint on the system, if $h(m, n)$ is the response to $\delta(m, n)$, then $h(m - k, n - r)$ is the response to $\delta(m - k, n - r)$. In this case, Eq. (1.35) becomes

$$y(m, n) = \sum_{k=-\infty}^{\infty} \sum_{r=-\infty}^{\infty} x(k, r)h(m - k, n - r) \tag{1.36}$$

Equation (1.36) is the *convolution sum* for a two-dimensional linear shift-invariant system. By a substitution of variables in Eq. (1.36) we obtain the alternative expression

$$y(m, n) = \sum_{k=-\infty}^{\infty} \sum_{r=-\infty}^{\infty} h(k, r)x(m - k, n - r) \tag{1.37}$$

Thus, as in the one-dimensional case, the operation of convolving two sequences is commutative; i.e., the order in which they are convolved is unimportant. This implies, among other things, that the unit-sample response of a cascade combination of linear shift-invariant systems is independent of the order in which they are cascaded.

A stable system is one for which every bounded input produces a bounded output. Paralleling the argument in Sec. 1.3 (see Problem 39 of this chapter), it can be shown that a two-dimensional linear shift-invariant system is stable if and only if

$$S \triangleq \sum_{k=-\infty}^{\infty} \sum_{r=-\infty}^{\infty} |h(k, r)| < \infty \tag{1.38}$$

A two-dimensional system is said to be causal if when two inputs $x_1(m, n)$ and $x_2(m, n)$ are equal for $(m < m_1, n < n_1)$, then the corresponding outputs $y_1(m, n)$ and $y_2(m, n)$ are equal for $(m < m_1, n < n_1)$. For a linear, shift-invariant system, causality implies that the unit-sample response is zero for $(m < 0, n < 0)$. Conversely, if the unit-sample response is zero for $(m < 0, n < 0)$, then the system is causal.

An important subclass of two-dimensional linear shift-invariant systems are those for which the input and output satisfy a linear constant-coefficient difference equation of the form

$$\sum_{k=0}^{M_1} \sum_{r=0}^{N_1} a_{kr} y(m - k, n - r) = \sum_{k=0}^{M_2} \sum_{r=0}^{N_2} b_{kr} x(m - k, n - r) \tag{1.39}$$

where, for linearity, we impose the additional constraint that if $x(m, n) = 0$ for all m and n then $y(m, n) = 0$ for all m and n. As with one-dimensional difference equations, the two-dimensional difference equation (1.39) can correspond to a causal or a non-causal system. If we assume that the system is causal, Eq. (1.39) can be expressed as a recursion relation; i.e.,

$$y(m, n) = \frac{1}{a_{00}}\left\{\sum_{k=0}^{M_2}\sum_{r=0}^{N_2} b_{kr}x(m-k, n-r) - \sum_{\substack{k=0 \\ k,r \neq 0 \text{ simultaneously}}}^{M_1}\sum_{r=0}^{N_1} a_{kr}y(m-k, n-r)\right\} \tag{1.40}$$

For example, if $x(m, n)$ is zero for $m < 0, n < 0$, then, as a result of causality, $y(m, n)$ is zero for $m < 0$, $n < 0$. This then provides a set of initial conditions to use in iterating Eq. (1.40).

For two-dimensional linear shift-invariant systems the response to a complex exponential of the form $e^{j\omega_1 m}e^{j\omega_2 n}$ is a complex exponential of the same complex frequencies. Specifically, with

$$x(m, n) = e^{j\omega_1 m}e^{j\omega_2 n}$$

then from the convolution sum,

$$\begin{aligned} y(m, n) &= \sum_{k=-\infty}^{\infty}\sum_{r=-\infty}^{\infty} h(k, r)e^{j\omega_1 m}e^{j\omega_2 n}e^{-j\omega_1 k}e^{-j\omega_2 r} \\ &= H(e^{j\omega_1}, e^{j\omega_2})e^{j\omega_1 m}e^{j\omega_2 n} \end{aligned}$$

where

$$H(e^{j\omega_1}, e^{j\omega_2}) = \sum_{k=-\infty}^{\infty}\sum_{r=-\infty}^{\infty} h(k, r)e^{-j\omega_1 k}e^{-j\omega_2 r} \tag{1.41}$$

$H(e^{j\omega_1}, e^{j\omega_2})$ is the frequency response of the two-dimensional system. It is a continuous function of ω_1 and ω_2 and is a periodic function of each of these variables with period 2π. It can be shown (see Problem 38 of this chapter) that if $h(m, n)$ is separable, then $H(e^{j\omega_1}, e^{j\omega_2})$ is separable; i.e., it can be expressed in the form

$$H(e^{j\omega_1}, e^{j\omega_2}) = H_1(e^{j\omega_1})H_2(e^{j\omega_2})$$

The sequence $h(m, n)$ can be recovered from the frequency response by the relation

$$h(m, n) = \frac{1}{4\pi^2}\int_{-\pi}^{\pi}\int_{-\pi}^{\pi} H(e^{j\omega_1}, e^{j\omega_2})e^{j\omega_1 m}e^{j\omega_2 n}\, d\omega_1\, d\omega_2 \tag{1.42}$$

More generally we define the two-dimensional Fourier transform of a sequence $x(m, n)$ as

$$X(e^{j\omega_1}, e^{j\omega_2}) = \sum_{k=-\infty}^{\infty}\sum_{r=-\infty}^{\infty} x(m, n)e^{-j\omega_1 m}e^{-j\omega_2 n} \tag{1.43}$$

with the inverse relation given by

$$x(m, n) = \frac{1}{4\pi^2} \int_{-\pi}^{\pi} \int_{-\pi}^{\pi} X(e^{j\omega_1}, e^{j\omega_2}) e^{j\omega_1 m} e^{j\omega_2 n} \, d\omega_1 \, d\omega_2 \tag{1.44}$$

By applying the transform of Eq. (1.43) to the convolution sum, it follows that the Fourier transforms of the input and output of a two-dimensional linear shift-invariant system are related by

$$Y(e^{j\omega_1}, e^{j\omega_2}) = H(e^{j\omega_1}, e^{j\omega_2}) X(e^{j\omega_1}, e^{j\omega_2}) \tag{1.45}$$

Summary

In this chapter we have considered a number of basic definitions relating to discrete-time signals and systems. In particular, we considered the definition of a set of basic sequences, the definition and representation of linear shift-invariant systems in terms of the convolution sum, and some implications of stability and causality. An important subclass of linear shift-invariant systems is that for which the input and output satisfy a linear constant-coefficient difference equation. The iterative solution of such equations was discussed and the classes of FIR and IIR systems defined.

An important means for the analysis and representation of linear shift-invariant systems lies in their frequency-domain representation. The response of a system to a complex exponential input was considered, leading to the definition of the frequency response. The relation between impulse response and frequency response was then interpreted as the Fourier transform pair. Some properties of the Fourier transform pair were then developed.

Although the material in this chapter was presented without direct reference to continuous-time signals, an important class of digital signal processing problems arise from sampling such signals. Consequently, in Sec. 1.7 we considered the relationship between continuous-time signals and sequences obtained by periodic sampling.

The chapter closed with a brief introduction to two-dimensional sequences and systems.

REFERENCES

1. E. A. Guillemin, *Theory of Linear Physical Systems*, John Wiley & Sons, Inc., New York, 1963.
2. A. Papoulis, *The Fourier Integral and Its Applications*, McGraw-Hill Book Company, New York, 1962.
3. S. Mason and H. J. Zimmermann, *Electronic Circuits, Signals and Systems*, John Wiley & Sons, Inc., New York, 1960.
4. J. R. Ragazzini and G. F. Franklin, *Sampled Data Control Systems*, McGraw-Hill Book Company, New York, 1958.
5. H. Freeman, *Discrete-Time Systems*, John Wiley & Sons, Inc., New York, 1965.

6. B. C. Kuo, *Discrete-Data Control Systems*, Prentice-Hall, Inc., Englewood Cliffs, N.J., 1970.
7. K. Steiglitz, "The Equivalence of Analog and Digital Signal Processing," *Inform. Control*, Vol. 8, No. 5, Oct. 1965, pp. 455–467.
8. A. V. Oppenheim and D. H. Johnson, "Discrete Representation of Signals," *Proc. IEEE*, Vol. 60, No. 6, June 1972, pp. 681–691.

PROBLEMS

1. Consider an arbitrary linear system with input $x(n)$ and output $y(n)$. Show that if $x(n) = 0$ for all n, then $y(n)$ must also be zero for all n.
2. For each of the sequences in Fig. P1.2, use discrete convolution to find the response to the input $x(n)$ of the linear shift-invariant system with unit-sample response $h(n)$.

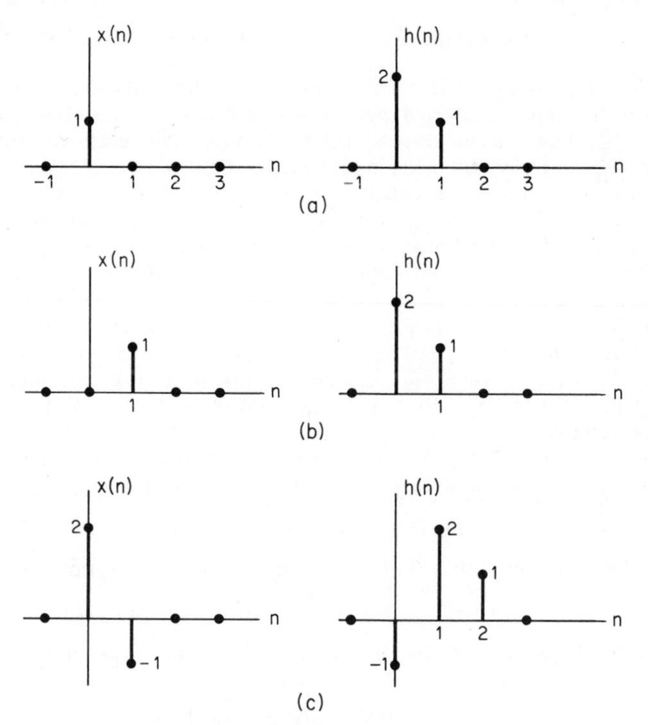

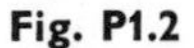
Fig. P1.2

3. By explicitly evaluating the convolution sum, evaluate the convolution $y(n) = x(n) * h(n)$ of the sequences

$$h(n) = \begin{cases} \alpha^n, & 0 \leq n < N \\ 0, & \text{elsewhere} \end{cases}$$

$$x(n) = \begin{cases} \beta^{n-n_0}, & n_0 \leq n \\ 0, & n < n_0 \end{cases}$$

A closed-form solution can be obtained.

4. Let $e(n)$ be an exponential sequence, i.e.,

$$e(n) = \alpha^n, \qquad \text{for all } n$$

and let $x(n)$ and $y(n)$ denote two arbitrary sequences. Show that

$$[e(n)x(n)] * [e(n)y(n)] = e(n)[x(n) * y(n)]$$

5. Let $x(n)$, $y(n)$, and $w(n)$ denote three arbitrary sequences. Show that
(a) discrete convolution is commutative; i.e.,

$$x(n) * y(n) = y(n) * x(n)$$

(b) discrete convolution is associative; i.e.,

$$x(n) * [y(n) * w(n)] = [x(n) * y(n)] * w(n)$$

(c) discrete convolution is distributive with respect to addition; i.e.,

$$x(n) * [y(n) + w(n)] = x(n) * y(n) + x(n) * w(n)$$

6. Consider a discrete-time linear shift-invariant system with unit-sample response $h(n)$. If the input $x(n)$ is a periodic sequence with period N, i.e., $x(n) = x(n + N)$, show that the output $y(n)$ is also a periodic sequence with period N.

7. If the output of a system is the input function multiplied by a complex constant, then that input function is called an *eigenfunction* of the system.
(a) Show that the function $x(n) = z^n$, where z is a complex constant, is an eigenfunction of a linear shift-invariant system.
(b) By constructing a counterexample, show that $z^n u(n)$ is *not* an eigenfunction of a linear shift-invariant system.

8. The unit-sample response of a linear shift-invariant system is known to be zero except in the interval $N_0 \le n \le N_1$. The input $x(n)$ is known to be zero except in the interval $N_2 \le n \le N_3$. As a result, the output is constrained to be zero except in some interval $N_4 \le n \le N_5$. Determine N_4 and N_5 in terms of N_0, N_1, N_2, and N_3.

9. By direct evaluation of the convolution sum, determine the step response of a linear shift-invariant system whose unit-sample response $h(n)$ is given by

$$h(n) = a^{-n}u(-n), \qquad 0 < a < 1$$

10. Consider a system with a finite-duration unit-sample response $h(n)$ such that

$$h(n) = 0, \qquad n < 0, N \le n, \qquad \text{where } N > 0$$

Show that if $|x(n)| \le B$, then a bound on the output is given by

$$|y(n)| \le B \sum_{k=0}^{N-1} |h(k)|$$

Show also that this bound can be attained; i.e., determine a sequence $x(n)$ with $|x(n)| \le B$ for which

$$y(n) = B \sum_{k=0}^{N-1} |h(k)|.$$

11. Causality of a system was defined in Sec. 1.3. From this definition show that for a linear shift-invariant system, causality implies that the unit-sample response $h(n)$ is zero for $n < 0$. Show also that if the unit-sample response is zero for $n < 0$, then the system will necessarily be causal.

12. For each of the following systems, determine whether or not the system is (1) stable, (2) causal, (3) linear, (4) shift-invariant.
(a) $T[x(n)] = g(n)x(n)$.
(b) $T[x(n)] = \sum_{k=n_0}^{n} x(k)$.
(c) $T[x(n)] = \sum_{k=n-n_0}^{n+n_0} x(k)$.
(d) $T[x(n)] = x(n - n_0)$.
(e) $T(x(n)] = e^{x(n)}$.
(f) $T[x(n)] = ax(n) + b$.
Justify your answer.

13. Consider a system with input $x(n)$ and output $y(n)$. The input–output relation for the system is defined by the following two properties: (1) $y(n) - ay(n - 1) = x(n)$; (2) $y(0) = 1$.
(a) Determine whether the system is shift-invariant.
(b) Determine whether the system is linear.
(c) Assume that the difference equation (property 1) remains the same but the value $y(0)$ is specified to be zero. Does this change your answer to either part (a) or (b)?

14. Consider the linear discrete-time shift-invariant system with unit-sample response

$$h(n) = \left(\frac{j}{2}\right)^n u(n), \qquad \text{where } j = \sqrt{-1}$$

Determine the steady-state response i.e., the response for large n to the excitation

$$x(n) = [\cos \pi n]u(n)$$

15. A discrete-time system is shown Fig. P1.15. The system transformation $y(n) = T[x(n)]$ is arbitrary and may be nonlinear and time varying. The only known property of the system is that it is well defined; i.e., the output for any given input is unique. Suppose that the input $x(n)$ is chosen as $x(n) = Ae^{j\omega n}$ and that some parameter P of the output is measured (e.g., the maximum amplitude). P will, in general, be a function of ω.

Let us consider P for different excitation frequencies. Show that P is *periodic in* ω and determine the period. Will a similar result be true in the continuous-time case?

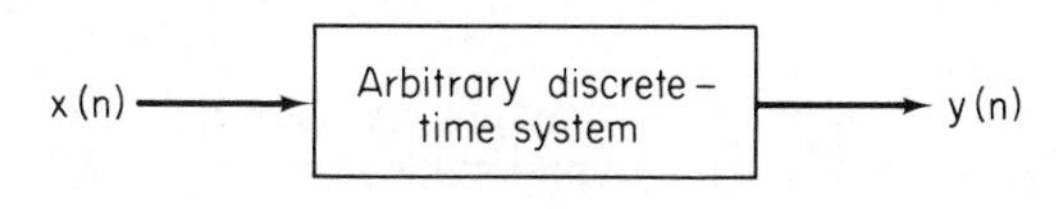

Fig. P1.15

16. In Sec. 1.5 the Fourier transform pair for a sequence was given as

$$H(e^{j\omega}) = \sum_{n=-\infty}^{\infty} h(n)e^{-j\omega n} \qquad \text{(Fourier transform)} \qquad \text{(P1.16-1)}$$

$$h(n) = \frac{1}{2\pi}\int_{-\pi}^{\pi} H(e^{j\omega})e^{j\omega n}\, d\omega \qquad \text{(inverse Fourier transform)} \qquad \text{(P1.16-2)}$$

(a) By substituting Eq. (P1.16-1) into Eq. (P1.16-2) and evaluating the integral, verify that the two equations are inverses of each other.
(b) Repeat part (a) by substituting Eq. (P1.16-2) into Eq. (P1.16-1).

17. In Sec. 1.5 it was argued intuitively that

$$Y(e^{j\omega}) = H(e^{j\omega})X(e^{j\omega}) \tag{P1.17-1}$$

where $Y(e^{j\omega})$, $H(e^{j\omega})$, and $X(e^{j\omega})$ are the Fourier transforms of the output $y(n)$, unit-sample response $h(n)$, and input $x(n)$ of a linear time-invariant system; i.e.,

$$y(n) = \sum_{n=-\infty}^{\infty} h(n-k)x(k) \tag{P1.17-2}$$

Verify Eq. (P1.17-1) by applying the Fourier transform to the convolution sum given in Eq. (P1.17-2).

18. (a) Consider a linear shift-invariant system with unit-sample response $h(n) = \alpha^n u(n)$, where α is real and $0 < \alpha < 1$. If the input is $x(n) = \beta^n u(n)$, $0 < |\beta| < 1$, determine the output $y(n)$ in the form $y(n) = (k_1\alpha^n + k_2\beta^n)u(n)$ by explicitly evaluating the convolution sum.
(b) By explicitly evaluating the transforms $X(e^{j\omega})$, $H(e^{j\omega})$, and $Y(e^{j\omega})$ corresponding to $x(n)$, $h(n)$, and $y(n)$ specified in part (a), show that

$$Y(e^{j\omega}) = H(e^{j\omega})X(e^{j\omega})$$

19. Let $x(n)$ and $X(e^{j\omega})$ denote a sequence and its Fourier transform. Show that

$$\sum_{n=-\infty}^{\infty} x(n)x^*(n) = \frac{1}{2\pi}\int_{-\pi}^{\pi} X(e^{j\omega})X^*(e^{j\omega})\,d\omega$$

This is one form of Parseval's theorem.

20. A causal linear shift-invariant system is described by the difference equation

$$y(n) - ay(n-1) = x(n) - bx(n-1)$$

Determine the value of b ($b \neq a$) such that this system is an allpass system; i.e., the magnitude of its frequency response is constant, independent of frequency.

21. Show that the sequence $[\sin(\pi n/2)]/\pi n$ is square summable but not absolutely summable.

22. $f(n)$ and $g(n)$ are real, causal, and stable sequences with Fourier transforms $F(e^{j\omega})$ and $G(e^{j\omega})$, respectively. Show that

$$\frac{1}{2\pi}\int_{-\pi}^{\pi} F(e^{j\omega})G(e^{j\omega})\,d\omega = \left\{\frac{1}{2\pi}\int_{-\pi}^{\pi} F(e^{j\omega})\,d\omega\right\}\left\{\frac{1}{2\pi}\int_{-\pi}^{\pi} G(e^{j\omega})\,d\omega\right\}$$

23. In the design of either analog or digital filters, we often approximate a specified magnitude characteristic, without particular regard to the phase. For example, standard design techniques for lowpass and bandpass filters are derived from consideration of the magnitude characteristics only.

In many filtering problems, one would ideally like the phase characteristics to be zero or linear. For causal filters it is impossible to have zero phase. However, for many digital filtering applications, it is not necessary that the unit-sample response of the filter be zero for $n < 0$ if the processing is not to be carried out in real time.

One technique commonly used in digital filtering when the data to be filtered are of finite duration and stored, for example, on a disc or magnetic tape, is to process the data forward and then backward through the same filter.

Let $h(n)$ be the unit-sample response of a causal filter with an arbitrary phase characteristic. Assume that $h(n)$ is real and denote its Fourier transform by $H(e^{j\omega})$. Let $x(n)$ be the data that we want to filter. The filtering operation is performed as follows:

(a) *Method A:*

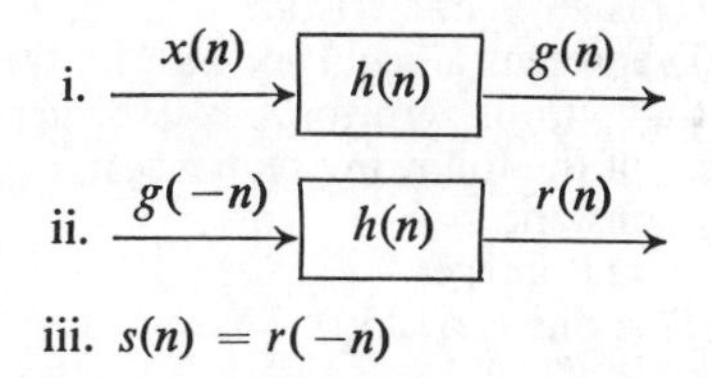

iii. $s(n) = r(-n)$

(1) Determine the overall unit-sample response $h_1(n)$ that relates $x(n)$ and $s(n)$, and show that it has a zero-phase characteristic.
(2) Determine $|H_1(e^{j\omega})|$ and express it in terms of $|H(e^{j\omega})|$ and $\arg[H(e^{j\omega})]$.

(b) *Method B:* Process $x(n)$ through the filter $h(n)$ to get $g(n)$. Also process $x(n)$ backward through $h(n)$ to get $r(n)$. The output $y(n)$ is then taken as the sum $g(n)$ plus $r(-n)$

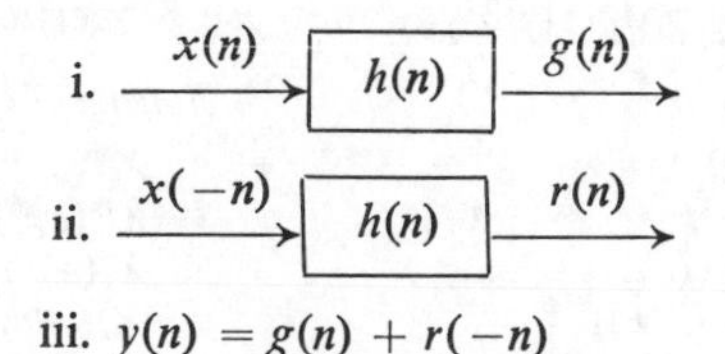

iii. $y(n) = g(n) + r(-n)$

This composite set of operations can be represented by a filter, with input $x(n)$, output $y(n)$, and unit-sample response $h_2(n)$.
(1) Show that the composite filter $h_2(n)$ has a zero-phase characteristic.
(2) Determine $|H_2(e^{j\omega})|$ and express it in terms of $|H(e^{j\omega})|$ and $\arg[H(e^{j\omega})]$.

(c) Suppose that we are given a sequence of finite duration, on which we would like to perform a bandpass, zero-phase filtering operation. Furthermore, assume that we are given the bandpass filter $h(n)$, with frequency response as specified in Fig. P1.23, which has the magnitude characteristic that we

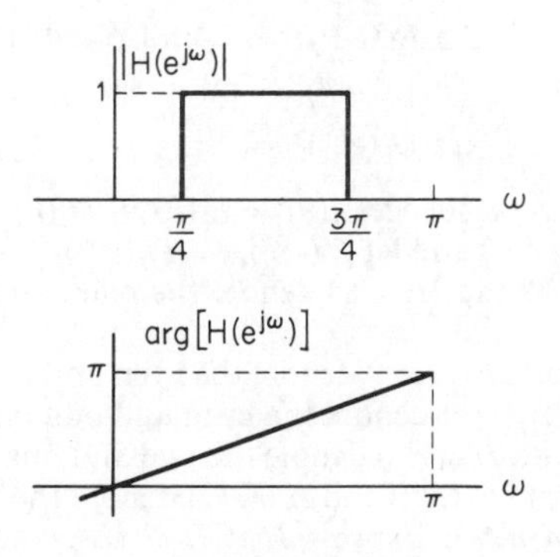

Fig. P1.23

desire but linear phase. To achieve zero phase, we could use either method (A) or (B). Determine and sketch $|H_1(e^{j\omega})|$ and $|H_2(e^{j\omega})|$. From these results which method would you use to achieve the desired bandpass filtering operation? Explain why. More generally, if $h(n)$ has the desired magnitude but a nonlinear phase characteristic which method is preferable to achieve a zero phase characteristic?

24. Let $x(n)$ and $X(e^{j\omega})$ represent a sequence and its transform. Do not assume that $x(n)$ is real or that $x(n)$ is zero for $n < 0$. Determine, in terms of $X(e^{j\omega})$, the transform of each of the following sequences:
(a) $kx(n)$, $k =$ any constant.
(b) $x(n - n_0)$, $n_0 =$ a real integer.
(c) $g(n) = x(2n)$. (This one is tricky!)
(d) $g(n) = \begin{cases} x(n/2), & n \text{ even}, \\ 0, & n \text{ odd}. \end{cases}$
(e) $x^2(n)$.

25. In Sec. 1.6 we stated a number of symmetry properties of the Fourier transform. All these properties follow in a relatively straightforward way from the transform pair. Below is a list of some of the properties stated. Prove that each is true. In carrying out the proof you may use the definition of the transform pair as given in Eqs. (1.19) and (1.20) and any previous property in the list. For example, in proving property 3 you may use properties 1 and 2.

Sequence	*Fourier Transform*
1. $x^*(n)$	$X^*(e^{-j\omega})$
2. $x^*(-n)$	$X^*(e^{j\omega})$
3. $\text{Re}\,[x(n)]$	$X_e(e^{j\omega})$
4. $j\,\text{Im}\,[x(n)]$	$X_o(e^{j\omega})$
5. $x_e(n)$	$\text{Re}\,[X(e^{j\omega})]$
6. $x_o(n)$	$j\,\text{Im}\,[X(e^{j\omega})]$

26. $x(n)$ and $X(e^{j\omega})$ denote a sequence and its Fourier transform. Determine, in terms of $x(n)$, the sequence corresponding to
(a) $X(e^{j(\omega-\omega_0)})$.
(b) $\text{Re}\,[X(e^{j\omega})]$.
(c) $\text{Im}\,[X(e^{j\omega})]$.

27. From the properties proved in Problem 25 above, show that for a real sequence $x(n)$, the following symmetry properties of the Fourier transform $X(e^{j\omega})$ hold:

$$\begin{aligned} \text{Re}\,[X(e^{j\omega})] &= \text{Re}\,[X(e^{-j\omega})] \\ \text{Im}\,[X(e^{j\omega})] &= -\text{Im}\,[X(e^{-j\omega})] \\ |X(e^{j\omega})| &= |X(e^{-j\omega})| \\ \arg\,[X(e^{j\omega})] &= -\arg\,[X(e^{-j\omega})] \end{aligned}$$

28. Consider a complex sequence $h(n) = h_r(n) + jh_i(n)$, where $h_r(n)$ and $h_i(n)$ are both real sequences, and let $H(e^{j\omega}) = H_R(e^{j\omega}) + jH_I(e^{j\omega})$ denote its transform, where $H_R(e^{j\omega})$ and $H_I(e^{j\omega})$ denote the real and imaginary parts, respectively, of $H(e^{j\omega})$.

Let $H_{ER}(e^{j\omega})$ and $H_{OR}(e^{j\omega})$ denote the even and odd parts of $H_R(e^{j\omega})$, and let $H_{EI}(e^{j\omega})$ and $H_{OI}(e^{j\omega})$ denote the even and odd parts of $H_I(e^{j\omega})$. Furthermore, let $H_A(e^{j\omega})$ and $H_B(e^{j\omega})$ denote the real and imaginary parts of the transform of $h_r(n)$, and let $H_C(e^{j\omega})$ and $H_D(e^{j\omega})$ denote the real and imaginary parts of the transform of $h_i(n)$. Express H_A, H_B, H_C, and H_D in terms of H_{ER}, H_{OR}, H_{EI}, and H_{OI}.

29. Two operations that are often encountered in signal processing systems are depicted in Fig. P1.29-1. The sampler retains the even-numbered points and

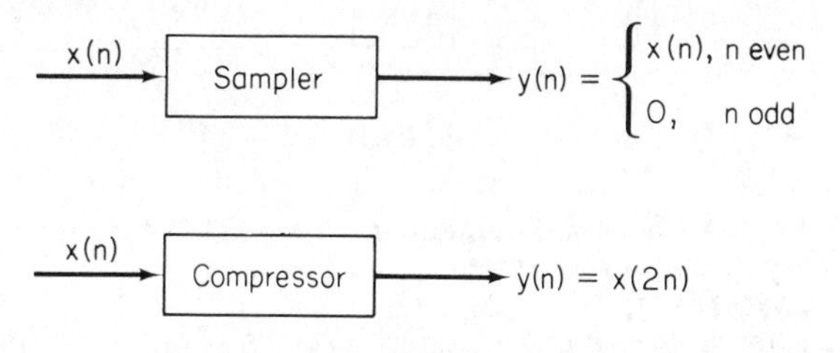

Fig. P1.29-1

sets the odd-numbered points in the sequence to zero. The compressor generates a sequence that consists only of the even-numbered points from the input.

In Fig. P1.29-2 is shown a system consisting of the cascade of a sampler, a

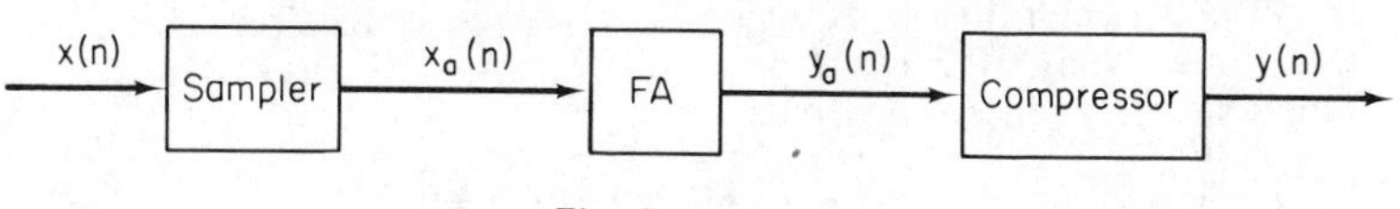

Fig. P1.29-2

digital filter (FA), and a compressor. In Fig. P1.29-3 is shown a system consisting of the cascade of a compressor and a digital filter (FB).

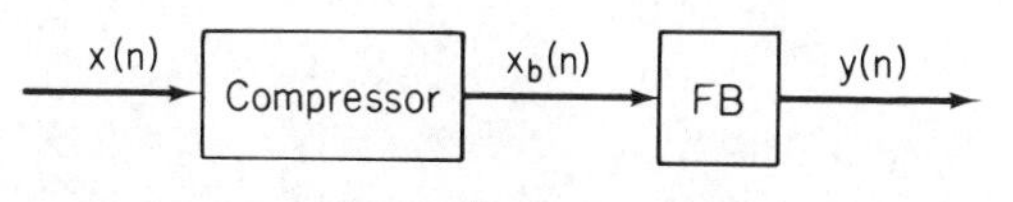

Fig. P1.29-3

The digital filters FA and FB are linear, causal and shift-invariant. The frequency response for FA is $H_A(e^{j\omega}) = 1/[1 - ae^{-j\omega}]$, where $0 < a < 1$ [i.e., $h(n) = a^n u(n)$]. Determine the frequency response for FB so that the systems of Figs. P1.29-2 and P1.29-3 are equivalent.

30. Let $h_a(t)$ denote the impulse response of a linear time-invariant continuous-time filter and $h_d(n)$ the unit-sample response of a linear shift-invariant discrete-time filter.

(a) If

$$h_a(t) = \begin{cases} e^{-at}, & t \geq 0 \\ 0, & t < 0 \end{cases}$$

determine the analog filter frequency response and sketch its magnitude.

(b) If $h_d(n) = h_a(nT)$ with $h_a(t)$ as in part (a), determine the digital filter frequency response and sketch its magnitude.

(c) For a given value of a, determine as a function of T the minimum magnitude of the digital filter frequency response.

31. One context in which digital filters are frequently used is in filtering band-limited analog data as shown in Fig. P1.31, where T represents the time between samples. (Assume that T is sufficiently small to prevent aliasing.) The overall system relating $x(t)$ to $y(t)$ is equivalent to an analog filter.

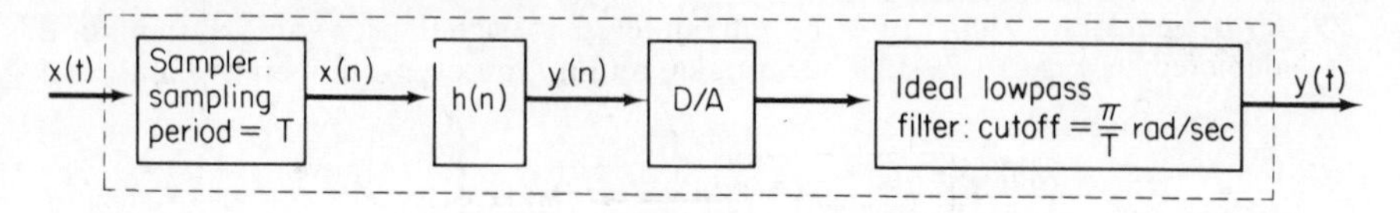

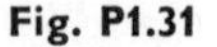
Fig. P1.31

(a) If $h(n)$ has a cutoff of $\pi/8$ rad, and if $1/T = 10$ kHz, what will be the cutoff frequency of the overall system?
(b) Repeat part (a) for $1/T = 20$ kHz.

32. As we indicated in Sec. 1.7, a continuous-time function $x_a(t)$ can generally be expressed in terms of a set of basis functions as

$$x_a(t) = \sum_{k=-\infty}^{\infty} c_k \phi_k(t) \tag{P1.32-1}$$

where the functions $\{\phi_k(t)\}$ are a set of linearly independent functions. The coefficients c_k can then be thought of as a sequence that represents the continuous-time signal. If $x_a(t)$ is bandlimited, then, as we saw in Sec. 1.7, the functions $\phi_k(t)$ can be chosen as

$$\phi_k(t) = \frac{\sin (\pi/T)(t - kT)}{(\pi/T)(t - kT)} \tag{P1.32-2}$$

One desirable property of the functions $\phi_k(t)$ is that they be chosen in such a way that the discrete representation of the continuous-time signals preserves convolution, i.e., if

$$x_{a1}(t) = \sum_{k=-\infty}^{\infty} c_{1k} \phi_k(t) \tag{P1.32-3a}$$

$$x_{a2}(t) = \sum_{k=-\infty}^{\infty} c_{2k} \phi_k(t) \tag{P1.32-3b}$$

$$x_{a3}(t) = \sum_{k=-\infty}^{\infty} c_{3k} \phi_k(t) \tag{P1.32-3c}$$

with

$$x_{a3}(t) = \int_{-\infty}^{\infty} x_{a1}(\tau) x_{a2}(t - \tau)\, d\tau \tag{P1.32-3d}$$

then

$$c_{3k} = \sum_{n=-\infty}^{\infty} c_{1n} c_{2(k-n)} \tag{P1.32-3e}$$

This then means that a linear time-invariant continuous-time system can be represented by a linear shift-invariant discrete-time system.

(a) Show that if equations (P1.32-3) hold, then $\Phi_k(j\Omega)$, the Fourier transform of $\phi_k(t)$, must satisfy the relation

$$\Phi_k(j\Omega)\Phi_n(j\Omega) = \Phi_{k+n}(j\Omega) \tag{P1.32-4}$$

(b) Show that the result in part (a) implies that $\Phi_k(j\Omega)$ is of the form

$$\Phi_k(j\Omega) = [G(j\Omega)]^{-k}$$

(c) Show that $1/T$ times the Fourier transform of the functions in Eq. (P1.32–2) satisfy Eq. (P1.32–4).

(d) Can you think of any other sets of linearly independent functions whose transforms satisfy Eq. (P1.32-4)? Give a few examples.

33. In many communications systems, information is transmitted by means of an alphabet of distinct, continuous signals, say $f_1(t), f_2(t), \ldots, f_m(t)$. If all the signals in the alphabet have the same energy, then a receiver channel tuned to receive, say $f_r(t)$ would compute the quantity $M_r = \int_{-\infty}^{+\infty} s(t)f_r(t)\,dt$, where $s(t)$, the received signal, is one of the signals in the alphabet.

(a) Show, using the Schwarz inequality, that M_r is maximum when $s(t) = f_r(t)$. (Note that we are talking here about continuous-time functions.) This means that to determine which signal was sent, we would compute M_r for each r. The largest value obtained would indicate which signal was transmitted.

(b) Suppose that, in a particular system, all the signals in the alphabet are bandlimited, so that the highest frequency is f_0 hertz. We wish to implement the receiver digitally; i.e., the received signals $s(t)$ and $f_r(t)$ are sampled with a sampling period of T to obtain the sequences $s(n)$ and $f_r(n)$ and the channel output is computed as

$$M_r = \sum_{n=-\infty}^{\infty} s(n)f_r(n)$$

What is the smallest sampling period T required in relation to f_0 to make this equivalent to the continuous system? The obvious guesses are that $1/T$ is either $f_0/4, f_0/2, f_0, 2f_0$, or $4f_0$. Fully justify your answer.

(c) Suppose that we computed M as specified in part (b) with T being *twice* as long as required. Are there conditions when M_r would still be a maximum when $s(n) = f_r(n)$? Think carefully about this one.

34. In Sec. 1.7 we derived the interpolation formula, Eq. (1.32), which expresses a bandlimited continuous-time signal, in terms of its samples. In particular, with $x_a(t)$ denoting a bandlimited continuous-time signal, it was shown that

$$x_a(t) = \sum_{k=-\infty}^{\infty} x_a(kT)\frac{\sin (\pi/T)(t - kT)}{(\pi/T)(t - kT)} \qquad \text{(P1.34-1)}$$

where T is less than 2π divided by the bandwidth of $x_a(t)$.

Often, in converting a sequence to a continuous-time signal, the sequence is first converted to a staircase function and then lowpass-filtered as depicted in Fig. P1.34. Determine the frequency response of the lowpass filter to recover $x_a(t)$.

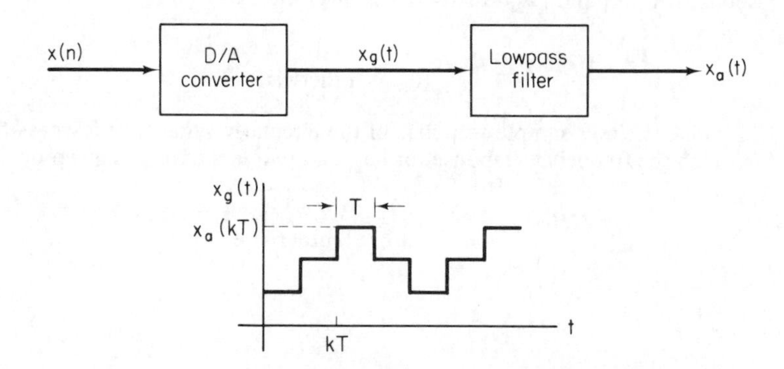

Fig. P1.34

35. Consider an analog signal

$$x_a(t) = s_a(t) + \alpha s_a(t - T)$$

Assume that the Fourier transform of $x_a(t)$ is bandlimited such that $X_a(j\Omega) = 0$ for $|\Omega| > \pi/T$ and that $x_a(t)$ is sampled at the Nyquist rate to obtain the sequence

$$x(n) = x_a(nT)$$

Find the unit sample response of a discrete-time system such that

$$x(n) = \sum_{k=-\infty}^{\infty} s(k)h(n - k)$$

where $s(n) = s_a(nT)$.

36. $u(m, n)$ and $\delta(m, n)$ denote the two-dimensional unit-step and unit-sample sequences, respectively.
(a) Express $u(m, n)$ in terms of $\delta(m, n)$.
(b) Express $\delta(m, n)$ in terms of $u(m, n)$.

37. Consider a two-dimensional linear shift-invariant system with input $x(m, n)$, output $y(m, n)$, and unit-sample response $h(m, n)$. Show that if $x(m, n)$ and $h(m, n)$ are both separable, then $y(m, n)$ will also be separable.

38. Let $x(m, n)$ denote a two-dimensional sequence and $X(e^{j\omega_1}, e^{j\omega_2})$ its Fourier transform. Show that if $x(m, n)$ is separable, then $X(e^{j\omega_1}, e^{j\omega_2})$ is separable.

39. By paralleling the argument carried out in Sec. 1.3 for the one-dimensional case, show that a necessary and sufficient condition for stability of a two-dimensional linear shift-invariant system with unit-sample response $h(k, r)$ is that

$$\sum_{k=-\infty}^{\infty} \sum_{r=-\infty}^{\infty} |h(k, r)| < \infty$$

40. In Problem 19, page 38, Parseval's theorem for one-dimensional sequences was derived. Derive the corresponding relation for two-dimensional sequences.

41. Determine the frequency response of the two-dimensional filter for which the unit-sample response is

$$h(m, n) = \begin{cases} 1, & |m| < M \text{ and } |n| < N \\ 0, & \text{otherwise} \end{cases}$$

42. Determine the unit-sample response of the lowpass filter for which the frequency response for $|\omega_1|$ and $|\omega_2|$ less than π is given by

$$H(e^{j\omega_1}, e^{j\omega_2}) = \begin{cases} 1, & |\omega_1| < a \text{ and } |\omega_2| < b \\ 0, & \text{otherwise} \end{cases}$$

43. Determine the unit-sample response of the circularly symmetric lowpass filter for which the frequency response for $|\omega_1|$ and $|\omega_2|$ less than π is given by

$$H(e^{j\omega_1}, e^{j\omega_2}) = \begin{cases} 1, & \sqrt{\omega_1^2 + \omega_2^2} < A \\ 0, & \text{otherwise} \end{cases}$$

2

The z-Transform

2.0 Introduction

In continuous-time system theory the Laplace transform can be considered as a generalization of the Fourier transform. In a similar manner it is possible to generalize the Fourier transform for discrete-time signals and systems, resulting in what is commonly referred to as the z-transform. The z-transform plays an important role in the analysis and representation of discrete-time linear shift-invariant systems. In this chapter we shall define the z-transform representation of a sequence and study in detail how the properties of a sequence are related to the properties of its z-transform.

In discussing the z-transform we shall call upon a number of results from the theory of complex variables. In applying these results we shall take pains to be precise but will not attempt to maintain a high degree of mathematical rigor.

2.1 z-Transform

The z-transform $X(z)$ of a sequence $x(n)$ is defined as

$$X(z) = \sum_{n=-\infty}^{\infty} x(n)z^{-n} \tag{2.1}$$

where z is a complex variable. It will sometimes be convenient to denote the z-transform of a sequence $x(n)$ as $\mathfrak{Z}[x(n)]$. In some contexts it is useful to refer to the z-transform as given by Eq. (2.1) as the *two-sided z-transform* and consider also the *one-sided z-transform* defined as

$$X_I(z) = \sum_{n=0}^{\infty} x(n)z^{-n}$$

Clearly, if $x(n) = 0$ for $n < 0$, the one-sided and two-sided z-transforms are equivalent, but not otherwise. In this book we shall not have occasion to utilize the one-sided transform but consider only the two-sided z-transform as defined by Eq. (2.1). A discussion of the one-sided z-transform can be found in a variety of texts (see, for example, [1,2]).

By expressing the complex variable z in polar form as $z = re^{j\omega}$, Eq. (2.1) has an interpretation in terms of the Fourier transform as defined in Chapter 1. Specifically, with z expressed in this form, Eq. (2.1) becomes

$$X(re^{j\omega}) = \sum_{n=-\infty}^{\infty} x(n)(re^{j\omega})^{-n}$$

or

$$X(re^{j\omega}) = \sum_{n=-\infty}^{\infty} x(n)r^{-n}e^{-j\omega n} \tag{2.2}$$

Thus, according to Eq. (2.2), the z-transform of $x(n)$ can be interpreted as the Fourier transform of $x(n)$ multiplied by an exponential sequence. For $r = 1$, i.e., for $|z| = 1$, the z-transform is equal to the Fourier transform of the sequence.

As we discussed in Chapter 1, the power series representing the Fourier transform does not converge for all sequences. Similarly, the z-transform does not converge for all sequences or for all values of z. For any given sequence the set of values of z for which the z-transform converges is called the *region of convergence*. As we stated in Sec. 1.5, uniform convergence of the Fourier transform requires that the sequence be absolutely summable. Applying this to Eq. (2.2) we require that

$$\sum_{n=-\infty}^{\infty} |x(n)r^{-n}| < \infty \tag{2.3}$$

It should be clear from Eq. (2.3) that because of the multiplication of the sequence by the real exponential r^{-n}, it is possible for the z-transform to converge even if the Fourier transform does not. For example, the sequence $x(n) = u(n)$ is not absolutely summable, and consequently the Fourier transform does not converge. However, $r^{-n}u(n)$ is absolutely summable if $|r| > 1$, and consequently the z-transform for the unit step exists with a region of convergence $1 < |z| < \infty$.

In general the power series of Eq. (2.1) will converge in an annular region of the z-plane,

$$R_{x-} < |z| < R_{x+} \tag{2.4}$$

where in general R_{x-} can be as small as zero and R_{x+} can be as large as infinity. For example, the region of convergence of the z-transform of the sequence $x(n) = u(n)$ is defined by $R_{x-} = 1$, $R_{x+} = \infty$.

A power series of the form of Eq. (2.1) is a *Laurent series.* As such, a number of elegant and powerful theorems from the theory of complex functions (as presented, for example, in [3]) can be employed in the study of the z-transform. A Laurent series, and therefore the z-transform, represents an analytic function at every point inside the region of convergence, and hence the z-transform and all its derivatives must be continuous functions of z inside the region of convergence.

In Sec. 1.5 we discussed the fact that some sequences exist which are not absolutely summable but have finite energy, and that in those cases it is possible to consider the Fourier transform to exist if one accepts convergence in the sense that the mean-square error approaches zero. An example of such a sequence is the impulse response of an ideal lowpass filter. In such a case the z-transform does not exist, as is suggested by the fact that for the ideal lowpass filter the Fourier transform is not a continuous function and consequently is not analytic. In cases where the Fourier transform exists but the z-transform does not, we may still refer to the Fourier transform as the z-transform evaluated for $|z| = 1$, although this is, strictly speaking, not correct.

An important class of z-transforms are those for which $X(z)$ is a rational function, i.e., a ratio of polynomials in z. The roots of the numerator polynomial are those values of z for which $X(z) = 0$ and are referred to as the *zeros of* $X(z)$. Values of z for which $X(z)$ is infinite are referred to as the *poles of* $X(z)$. The poles of $X(z)$ for finite values of z are the roots of the denominator polynomial. In addition, poles may occur at $z = 0$ or $z = \infty$. For rational z-transforms there are a number of important relationships between the location of poles of $X(z)$ and the region of convergence of the z-transform. Clearly no poles of $X(z)$ can occur within the region of convergence since the z-transform does not converge at a pole. Furthermore, as we will justify in a later discussion, the region of convergence is bounded by poles.

It is often convenient to display the z-transform graphically by means of a pole–zero plot in the z-plane. For example, consider the sequence

$$x(n) = a^n u(n)$$

The z-transform is given by

$$X(z) = \sum_{n=-\infty}^{\infty} a^n u(n) z^{-n}$$

$$= \sum_{n=0}^{\infty} (az^{-1})^n$$

which converges to

$$X(z) = \frac{1}{1 - az^{-1}}, \qquad \text{for } |z| > |a|$$

By rewriting $X(z)$ as a ratio of polynomials in z, we see that $X(z)$ has a zero at $z = 0$ and a pole at $z = a$. This is displayed in the z-plane in Fig. 2.1,

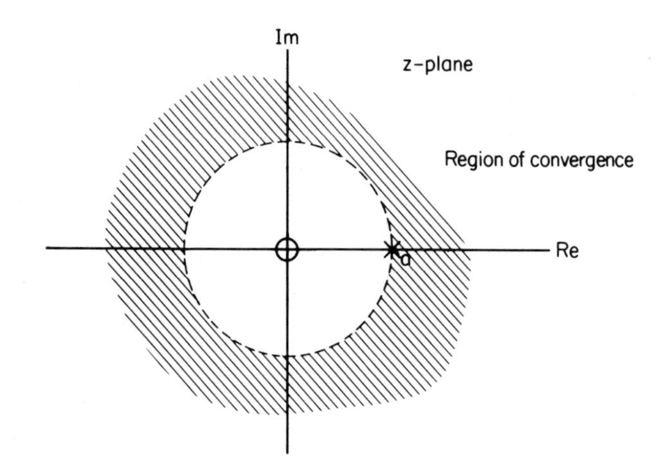

Fig. 2.1 Pole–zero plot and region of convergence in the z-plane for the z-transform of the sequence $a^n u(n)$.

where the zero is denoted by 0 and the pole is denoted by x. The region of convergence is indicated by the shaded region and includes the entire z-plane for $|z| > a$.

The properties of the sequence $x(n)$ determine the region of convergence of $X(z)$. To see how the region of convergence relates to the sequence, it is convenient to consider a number of special cases.

1. Finite-Length Sequences. Suppose that only a finite number of sequence values are nonzero, so that

$$X(z) = \sum_{n=n_1}^{n_2} x(n)z^{-n} \tag{2.5}$$

where n_1 and n_2 are finite integers. Convergence of this expression requires simply that $|x(n)| < \infty$ for $n_1 \leq n \leq n_2$. Then z may take on all values except $z = \infty$ if $n_1 < 0$ and $z = 0$ if $n_2 > 0$. Thus finite-length sequences have a region of convergence that is at least $0 < |z| < \infty$, and it may include either $z = 0$ or $z = \infty$.

2. Right-Sided Sequences. A right-sided sequence is one for which

$x(n) = 0$ for $n < n_1$. The z-transform of such a sequence is

$$X(z) = \sum_{n=n_1}^{\infty} x(n)z^{-n} \tag{2.6}$$

The region of convergence of the above series is the exterior of a circle. To see that this is true, suppose that the series is absolutely convergent for $z = z_1$, so that

$$\sum_{n=n_1}^{\infty} |x(n)z_1^{-n}| < \infty \tag{2.7}$$

If we consider the series $\sum_{n=n_1}^{\infty} |x(n)z^{-n}|$, we note that if $n_1 \geq 0$, then if $|z| > |z_1|$, each term is smaller than in the series of Eq. (2.7), and thus

$$\sum_{n=n_1}^{\infty} |x(n)z^{-n}| < \infty \qquad \text{for } |z| > |z_1|$$

If $n_1 < 0$, then we express the series as

$$\sum_{n=n_1}^{\infty} |x(n)z^{-n}| = \sum_{n=n_1}^{-1} |x(n)z^{-n}| + \sum_{n=0}^{\infty} |x(n)z^{-n}| \tag{2.8}$$

The first series on the right-hand side of Eq. (2.8) is finite for any finite value of z. The second series by the previous argument converges for $|z| > |z_1|$. Thus if R_{x-} is the smallest value of $|z|$ for which the series of Eq. (2.6) converges, then the series converges for

$$R_{x-} < |z|$$

with the exception of $z = \infty$ if $n_1 < 0$.† Thus right-sided sequences have a region of convergence that is the exterior of a circle with radius R_{x-}. We observe that if $n_1 \geq 0$, so that the sequence is causal, the z-transform will converge at $z = \infty$. Conversely, if $n_1 < 0$, it will not converge at $z = \infty$. Thus if the region of convergence of the z-transform is the exterior of a circle, it is a right-sided sequence. Furthermore, if it includes $z = \infty$, it is a causal sequence.

From Eq. (2.7) we note also that since the series converges, each term is bounded and thus there exists a finite constant A such that

$$|x(n)z_1^{-n}| < A, \qquad n \geq n_1 \tag{2.9}$$

Expressing $|z_1|$ as $|z_1| = r$, a positive number greater than R_{x-}, then

$$|x(n)| < Ar^n, \qquad n \geq n_1$$

and hence for convergence the sequence can grow no faster than an exponential as $n \to \infty$. If the region of convergence of $x(n)$ extends inside, the unit

† We note that R_{x-} is the radius of convergence of the series of negative powers of z in the z-transform of the sequence $x(n)$. This accounts for our use of the subscript $x-$.

circle so that r can be chosen less than unity, then $|x(n)|$ must approach zero at least as fast as an exponential as $n \to \infty$.

EXAMPLE. An example of a right-sided sequence is the sequence $x(n) = a^n u(n)$, which, as we saw previously, has the z-transform

$$X(z) = \frac{1}{1 - az^{-1}}, \qquad |z| > |a| \tag{2.10}$$

3. Left-Sided Sequences. A left-sided sequence is one for which $x(n) = 0$, $n > n_2$. The z-transform is

$$X(z) = \sum_{n=-\infty}^{n_2} x(n)z^{-n} \tag{2.11}$$

By changing the index of summation through the substitution $n = -m$, we obtain

$$X(z) = \sum_{m=-n_2}^{\infty} x(-m)z^{m}$$

Therefore, the results for right-sided sequences apply to this case with n replaced by $-n$ and with z replaced by z^{-1}. The region of convergence can be shown to be the interior of a circle, $|z| < R_{x+}$ except $z = 0$ if $n_2 > 0$.† If the z-transform of a left-sided sequence converges at $z = 0$, then the sequence is zero for $n \geq 0$. It also follows that if $X(z)$ converges for $|z| = r$, then

$$|x(n)| < Ar^n, \qquad n \leq n_2$$

where A is a finite constant. For convergence, then, the sequence can grow no faster than an exponential as $n \to -\infty$. If the region of convergence includes the unit circle, $x(n)$ must approach 0 as $n \to -\infty$.

EXAMPLE. As an example of a left-sided sequence, first consider $x(n) = -b^n u(-n - 1)$. The z-transform is

$$\begin{aligned} X(z) &= \sum_{n=-\infty}^{-1} - b^n z^{-n} \\ &= \sum_{n=1}^{\infty} - b^{-n} z^{n} \\ &= 1 - \sum_{n=0}^{\infty} b^{-n} z^{n} \end{aligned}$$

This series converges if $|b^{-1}z| < 1$ or $|z| < |b|$, in which case

$$X(z) = 1 - \frac{1}{1 - b^{-1}z} = \frac{z}{z - b}, \qquad \text{for } |z| < |b| \tag{2.12}$$

† In this case we use the subscript $x+$ to denote the radius of convergence of the series of positive powers of z in the z-transform of the sequence $x(n)$.

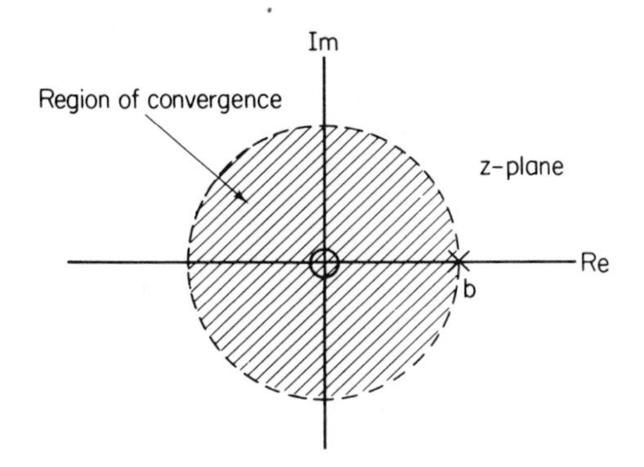

Fig. 2.2 Pole–zero plot and region of convergence in the z-plane for the z-transform of the sequence $-b^n u(-n-1)$.

The pole and zero and the region of convergence of $X(z)$ are indicated in Fig. 2.2.

Note that if $b = a$, the function $X(z)$ in Eq. (2.12) is identical to that of Eq. (2.10) in the previous example. This illustrates the extremely important fact that specifying the z-transform of a sequence requires not only the function $X(z)$ but also the region of convergence.

4. *Two-sided Sequences.* A two-sided sequence is one that extends from $n = -\infty$ to $n = +\infty$. In general we can write

$$X(z) = \sum_{n=-\infty}^{\infty} x(n)z^{-n}$$

$$= \sum_{n=0}^{\infty} x(n)z^{-n} + \sum_{n=-\infty}^{-1} x(n)z^{-n} \tag{2.13}$$

The first series is right-sided and converges for $R_{x-} < |z|$; the second series is left-sided and converges for $|z| < R_{x+}$. If $R_{x-} < R_{x+}$, there is a common region of convergence of the form

$$R_{x-} < |z| < R_{x+} \tag{2.14}$$

If $R_{x-} > R_{x+}$, there is no common region of convergence, and thus the series (2.13) does not converge. For the region of convergence as given in (2.14), the sequence $x(n)$ can grow no faster than an exponential in both directions and if

$$R_{x-} < 1 < R_{x+}$$

the sequence approaches zero exponentially in both directions. Such sequences have both a Fourier transform and a z-transform.

EXAMPLE. Consider the sequence

$$x(n) = \begin{cases} a^n, & n \geq 0 \\ -b^n, & n \leq -1 \end{cases} \tag{2.15}$$

where $|a| < |b|$. Using the results of the previous two examples,

$$X(z) = \frac{z}{z-a} + \frac{z}{z-b} = \frac{z(2z-a-b)}{(z-a)(z-b)} \tag{2.16}$$

where the region of convergence is

$$|a| < |z| < |b| \tag{2.17}$$

The poles and zeros and the region of convergence are shown in Fig. 2.3. The region of convergence is the overlap of the shaded regions.

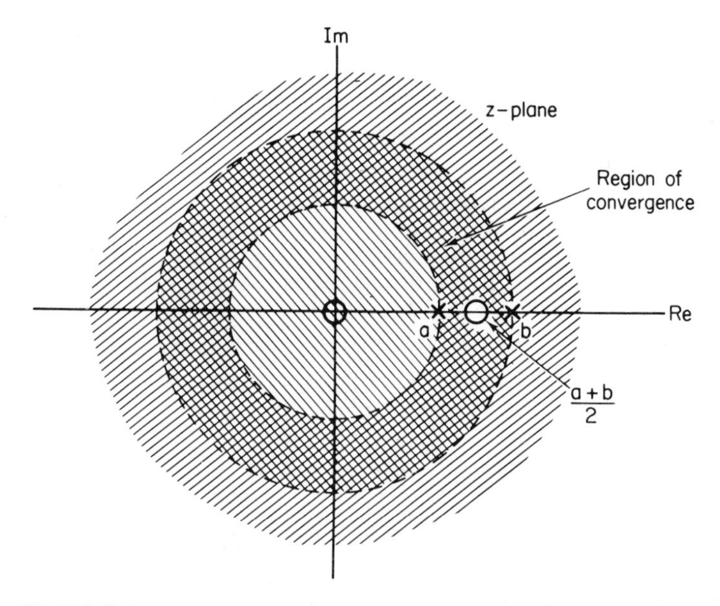

Fig. 2.3 Pole–zero plot and region of convergence for the sequence $x(n) = a^n u(n) - b^n u(-n-1)$.

2.2 Inverse z-Transform

The inverse z-transform relation can be derived by utilizing the *Cauchy integral theorem.* This theorem states that

$$\frac{1}{2\pi j}\oint_C z^{k-1}\,dz = \begin{cases} 1, & k = 0 \\ 0, & k \neq 0 \end{cases} \tag{2.18}$$

where C is a counterclockwise contour that encircles the origin.

The z-transform relation, as defined in Sec. 2.1, is given by

$$X(z) = \sum_{n=-\infty}^{\infty} x(n) z^{-n} \tag{2.19}$$

Multiplying both sides of Eq. (2.19) by z^{k-1} and integrating with a contour integral for which the contour of integration encloses the origin and lies entirely in the region of convergence of $X(z)$, we obtain

$$\frac{1}{2\pi j}\oint_C X(z)z^{k-1}\,dz = \frac{1}{2\pi j}\oint_C \sum_{n=-\infty}^{\infty} x(n)z^{-n+k-1}\,dz \tag{2.20}$$

Interchanging the order of integration and summation on the right-hand side of Eq. (2.20) (valid if the series is convergent) we obtain

$$\frac{1}{2\pi j}\oint_C X(z)z^{k-1}\,dz = \sum_{n=-\infty}^{\infty} x(n)\frac{1}{2\pi j}\oint_C z^{-n+k-1}\,dz \tag{2.21}$$

which from Eq. (2.18) becomes

$$\frac{1}{2\pi j}\oint_C X(z)z^{k-1}\,dz = x(k)$$

Therefore, the inverse z-transform relation is given by the contour integral

$$x(n) = \frac{1}{2\pi j}\oint_C X(z)z^{n-1}\,dz \tag{2.22}$$

where C is a counterclockwise closed contour in the region of convergence of $X(z)$ and encircling the origin of the z-plane. It should be stressed that in deriving Eq. (2.22) no assumption was made about whether k in Eq. (2.20) or n in Eq. (2.22) was positive or negative, and consequently Eq. (2.22) is valid for both positive and negative values of n.

For rational z-transforms contour integrals of the form of Eq. (2.22) are often conveniently evaluated using the residue theorem, i.e.,

$$\begin{aligned} x(n) &= \frac{1}{2\pi j}\oint X(z)z^{n-1}\,dz \\ &= \sum [\text{residues of } X(z)z^{n-1} \text{ at the poles inside } C] \end{aligned} \tag{2.23}$$

In general, if $X(z)z^{n-1}$ is a rational function of z, it may be expressed as

$$X(z)z^{n-1} = \frac{\psi(z)}{(z-z_0)^s} \tag{2.24}$$

where $X(z)z^{n-1}$ has s poles at $z = z_0$ and $\psi(z)$ has no poles at $z = z_0$. The residue of $X(z)z^{n-1}$ at $z = z_0$ is given by

$$\text{Res}\,[X(z)z^{n-1} \text{ at } z = z_0] = \frac{1}{(s-1)!}\left[\frac{d^{s-1}\psi(z)}{dz^{s-1}}\right]_{z=z_0} \tag{2.25}$$

In particular, if there is only a first-order pole at $z = z_0$, i.e., if $s = 1$, then

$$\text{Res}\,[X(z)z^{n-1} \text{ at } z = z_0] = \psi(z_0) \tag{2.26}$$

As an example of the use of the inverse transform relation, let us consider the inverse transform of

$$X(z) = \frac{1}{1 - az^{-1}} \qquad |z| > |a|$$

as derived in a previous example. Using Eq. (2.23) we obtain

$$x(n) = \frac{1}{2\pi j} \oint_C \frac{z^{n-1}}{1 - az^{-1}}\, dz = \frac{1}{2\pi j} \oint_C \frac{z^n\, dz}{z - a}$$

where the contour of integration, C, is a circle of radius greater than a. For $n \geq 0$, then, the contour of integration encloses only one pole at $z = a$. Consequently, for $n \geq 0$, $x(n)$ is given by

$$x(n) = a^n, \qquad n \geq 0$$

For $n < 0$, there is a multiple-order pole at $z = 0$ whose order depends on n. For $n = -1$, the pole is first order with a residue of $-a^{-1}$. The residue of the pole at $z = a$ is a^{-1}. Consequently, the sum of the residues is zero, and hence $x(-1) = 0$. For $n = -2$

$$\text{Res}\left[\frac{1}{z^2(z - a)} \text{ at } z = a\right] = a^{-2}$$

and

$$\text{Res}\left[\frac{1}{z^2(z - a)} \text{ at } z = 0\right] = -a^{-2}$$

and thus $x(-2) = 0$. Continuing this procedure it can be verified that for this example $x(n) = 0$, $n < 0$. As n becomes more negative the evaluation of the residue of the multiple-order pole at $z = 0$ becomes increasingly tedious. While Eq. (2.22) is valid for all n, its use for $n < 0$ is often cumbersome because of the multiple-order poles at $z = 0$.

This can be avoided by modifying Eq. (2.22) by means of a substitution of variables making it easier to apply for $n < 0$. Specifically, consider the substitution of variables $z = p^{-1}$, so that Eq. (2.22) becomes

$$x(n) = \frac{-1}{2\pi j} \oint_{C'} X\left(\frac{1}{p}\right) p^{-n+1} p^{-2}\, dp \tag{2.27}$$

Observe that since the contour in Eq. (2.22) is a counterclockwise contour, the contour in the above expression is a clockwise contour. Multiplying by -1 to reverse the direction of the contour, the above substitution of variables then leads to the expression

$$x(n) = \frac{1}{2\pi j} \oint_{C'} X\left(\frac{1}{p}\right) p^{-n-1}\, dp \tag{2.28}$$

If the contour C in Eq. (2.22) is a circle of radius r in the z-plane, then the contour C' in Eq. (2.28) is a circle of radius $1/r$ in the p-plane. The poles of $X(z)$ that were outside the contour C correspond now to poles of $X(1/p)$ that are inside the contour C', and vice versa. Additional poles and/or zeros may or may not appear at the origin and/or at infinity, but that is not crucial to the argument. For the specific example that we were considering previously, $x(n)$ in terms of this substitution of variables now becomes

$$x(n) = \frac{1}{2\pi j} \oint_{C'} \frac{p^{-n-1}}{1 - ap}\, dp$$

The contour of integration C' is now a circle of radius less than $1/a$. For $n < 0$ there are no singularities inside the contour of integration, and consequently it easily follows that for $n < 0$, $x(n) = 0$. Just as Eq. (2.22) was cumbersome (although certainly valid) for evaluating $x(n)$ for $n < 0$, this expression is likewise cumbersome (but still valid) for evaluating $x(n)$ for $n \geq 0$, because of the multiple-order poles that appear at the origin.

In many instances evaluation of Eq. (2.22) or (2.28) is unnecessarily difficult and involved. A number of special techniques are often easier to apply. In the remainder of this section we consider several such techniques.

Power Series. If the z-transform is available in power-series form, we can simply observe that the sequence value $x(n)$ is the coefficient of the term involving z^{-n} in the power series

$$X(z) = \sum_{n=-\infty}^{\infty} x(n) z^{-n}$$

If $X(z)$ is given as a closed-form expression, we can often derive an appropriate power series or employ a previously derived power-series expansion.

EXAMPLE. Consider the z-transform

$$X(z) = \log(1 + az^{-1}), \qquad |a| < |z| \tag{2.29}$$

Using the power-series expansion for $\log(1 + x)$, we obtain

$$X(z) = \sum_{n=1}^{\infty} \frac{(-1)^{n+1} a^n z^{-n}}{n}$$

Therefore, $x(n)$ is seen to be

$$x(n) = \begin{cases} (-1)^{n+1} \dfrac{a^n}{n}, & n \geq 1 \\ 0, & n \leq 0 \end{cases} \tag{2.30}$$

For rational z-transforms a power-series expansion can be obtained by long division.

EXAMPLE. Consider the z-transform

$$X(z) = \frac{1}{1 - az^{-1}}, \qquad |z| > |a| \tag{2.31}$$

Since the region of convergence is the exterior of a circle, the sequence is a right-sided sequence. Furthermore, since $X(z)$ approaches a finite constant as z approaches infinity, it is a causal sequence. Thus we divide so as to obtain a series in powers of z^{-1}. Carrying out the long division we obtain

$$\begin{array}{r|l} & 1 + az^{-1} + a^2z^{-2} + \cdots \\ \hline 1 - az^{-1} & 1 \\ & \underline{1 - az^{-1}} \\ & \quad az^{-1} \\ & \quad \underline{az^{-1} - a^2z^{-2}} \\ & \qquad\quad a^2z^{-2} \\ & \qquad\quad \cdots \end{array}$$

or

$$\frac{1}{1 - az^{-1}} = 1 + az^{-1} + a^2z^{-2} + \cdots$$

so that

$$x(n) = a^n u(n)$$

EXAMPLE. As another example we can consider the same ratio of polynomials as in Eq. (2.31) but with a different region of convergence; i.e.,

$$X(z) = \frac{1}{1 - az^{-1}}, \qquad |z| < |a| \tag{2.32}$$

Because of the region of convergence, the sequence is a left-sided sequence and since $X(z)$ at $z = 0$ is finite, the sequence is zero for $n > 0$. Thus we divide so as to obtain a series in powers of z as follows:

$$\begin{array}{r|l} & -a^{-1}z - a^{-2}z^2 - \cdots \\ \hline -a + z & z \\ & \underline{z - a^{-1}z^2} \\ & \quad a^{-1}z^2 \\ & \quad \cdots \end{array}$$

Therefore, $x(n) = -a^n u(-n - 1)$.

Partial-Fraction Expansion. Another technique that is often useful for rational z-transforms is to carry out a partial-fraction expansion and identify the inverse z-transform of the simpler terms. If $F(x)$ is a ratio of polynomials in the variable x with the order of the numerator less than the order of the denominator and with only first-order poles, it can be expressed in a partial-fraction expansion of the form

$$F(x) = \frac{P(x)}{Q(x)} = \sum_{k=1}^{N} \frac{A_k}{x - x_k} \tag{2.33}$$

where the x_k's are the poles of $F(x)$ and the A_k's are the residues at the poles; i.e.,

$$A_k = (x - x_k)F(x)\big|_{x=x_k} \tag{2.34}$$

If the order of the numerator is greater than the order of the denominator, then a polynomial is added to the right-hand side of Eq. (2.33) whose order is equal to the order of the numerator minus the order of the denominator. Thus, if the order of $P(x)$ is M and the order of $Q(x)$ is N with $M \geq N$, then Eq. (2.33) is replaced by

$$F(x) = B_{M-N}x^{M-N} + B_{M-N-1}x^{M-N-1} + \ldots + B_1 x + B_0 + \sum_{k=1}^{N} \frac{A_k}{x - x_k} \tag{2.35}$$

The B_i's can be obtained simply by long division and the A_k's are again obtained by means of Eq. (2.34). If $F(x)$ has multiple-order poles, Eq. (2.35) must be additionally modified. In particular, if $F(x)$ has a pole of order s at $x = x_i$, Eq. (2.35) becomes

$$F(x) = B_{M-N}x^{M-N} + B_{M-N-1}x^{M-N-1} + \ldots + B_1 x + B_0 + \sum_{k=1}^{N} \frac{A_k}{x - x_k} + \sum_{k=1}^{s} \frac{C_k}{(x - x_i)^k}$$

The coefficients A_k and B_i are obtained as before. The coefficients c_k are obtained by the relation

$$C_k = \frac{1}{(s-k)!}\left\{\frac{d^{s-k}}{dx^{s-k}}[x - x_i]^s F(x)\right\}_{x=x_i}$$

To apply the partial-fraction expansion to the z-transform we may consider the z-transform to be a ratio of polynomials in either z or in z^{-1}.

EXAMPLE. Consider a right-sided sequence with z-transform

$$X(z) = \frac{1}{(1 - az^{-1})(1 - bz^{-1})} = \frac{z^2}{(z-a)(z-b)} = \frac{a^{-1}b^{-1}}{(z^{-1} - (1/a))(z^{-1} - (1/b))}$$

Carrying out a partial-fraction expansion with $X(z)$ considered as a ratio of polynomials in z^{-1}, we obtain

$$X(z) = \frac{a^{-1}b^{-1}}{(z^{-1} - (1/a))(z^{-1} - (1/b))} = \frac{1}{(b-a)}\frac{1}{(z^{-1} - a^{-1})} + \frac{1}{(a-b)}\frac{1}{(z^{-1} - b^{-1})}$$

$$= \left(\frac{a}{a-b}\right)\left(\frac{1}{1 - az^{-1}}\right) + \left(\frac{b}{b-a}\right)\left(\frac{1}{1 - bz^{-1}}\right) \tag{2.36}$$

Since we assumed that the sequence is right sided, each of the terms in Eq. (2.36) corresponds to a right-sided sequence. These are first-order z-transforms which we can recognize from previous examples, and thus we can write by inspection of Eq. (2.36) that

$$x(n) = \frac{a}{a-b}a^n u(n) + \frac{b}{b-a}b^n u(n)$$

The same example is considered in Problem 4 of this chapter, with $X(z)$ considered as a ratio of polynomials in z rather than z^{-1}.

We should note that for left-sided or two-sided sequences the partial-fraction technique works equally well, but one must be careful to determine which poles correspond to right-sided sequences and which correspond to left-sided sequences.

2.3 z-Transform Theorems and Properties

In solving signal processing problems, it is important to have an understanding of and a facility in the use of the properties of the z-transform. In this section we discuss some of the most important results. Additional theorems and properties can be found in references [1] and [2].

2.3.1 *Region of Convergence of Rational z-Transforms*

As mentioned previously, for a sequence with a rational z-transform the region of convergence cannot contain any poles and is bounded by poles or by zero or infinity. The fact that it cannot contain any poles follows simply from the fact that by definition the z-transform does not converge at a pole. To argue that it is bounded by poles, let us consider first the case of a right-sided sequence and assume that the poles occur at $a_0, a_1, \ldots, a_N$, with a_N having the largest magnitude. We will assume, to simplify the argument, that all the poles are simple poles, although the argument is easily generalized. Then for n greater than some value n_0, the sequence consists of a sum of exponentials of the form

$$x(n) = \sum_{k=0}^{N} A_k(a_k)^n, \qquad n > n_0 \tag{2.37}$$

The region of convergence is determined by the set of values of z for which the sequence $x(n)z^{-n}$ is absolutely summable. Since a right-sided sequence of the form $(a_k)^n z^{-n}$ is absolutely summable for $|z| > |a_k|$ but not for $|z| \leq |a_k|$, it follows that the right-sided sequence of Eq. (2.37) has a region of convergence defined by $|z| > |a_N|$. Thus it is bounded on the inside by the pole with the largest magnitude and on the outside by infinity. By an identical argument it follows that for a left-sided sequence the region of convergence is bounded on the outside by the pole with the smallest magnitude and on the inside by zero. For two-sided sequences some of the poles contribute only for $n \geq 0$ and the rest only for $n \leq 0$. The region of convergence is bounded on the inside by the pole with the largest magnitude that contributes for $n \geq 0$ and on the outside by the pole with the smallest magnitude that contributes for $n \leq 0$. For example, in Fig. 2.4 is shown the same pole–zero plot with the four possible choices for the region of convergence. Figure 2.4(a) corresponds to a right-sided sequence, Fig. 2.4(b) to a left-sided sequence, and the remaining two correspond to two-sided sequences. In general the region of convergence is a connected region. We could not, for example, consider the region of convergence to be $|z| < |a|$ *and* $|z| > |c|$. If

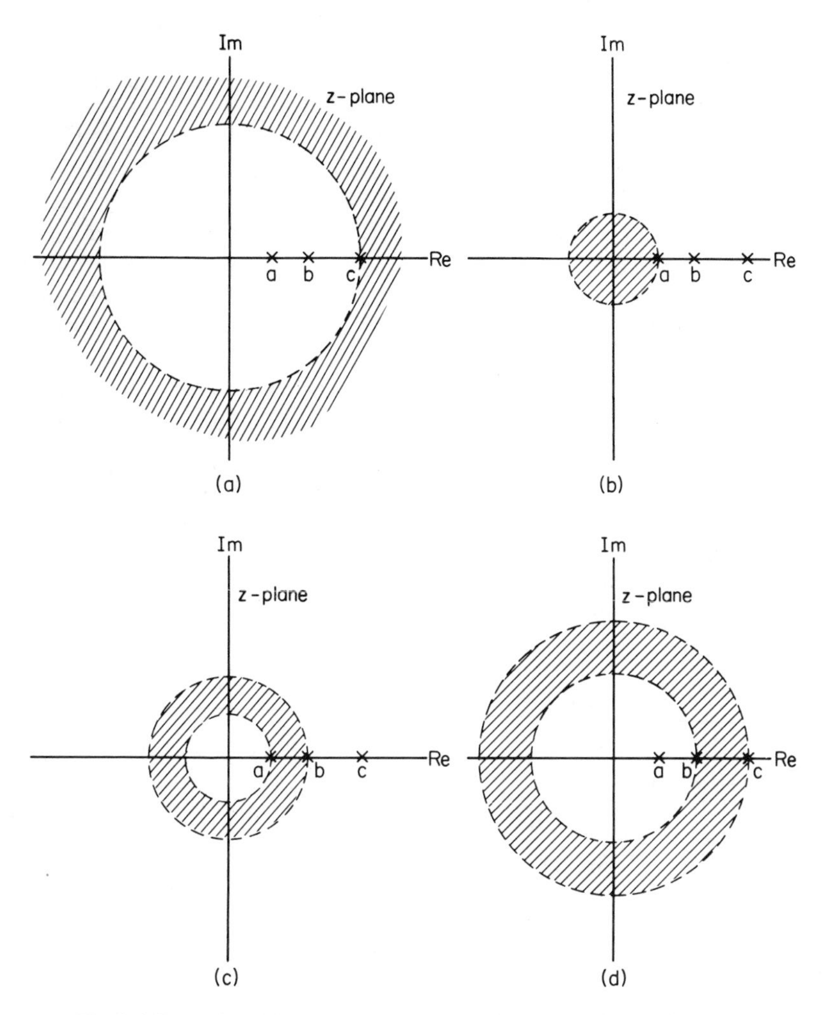

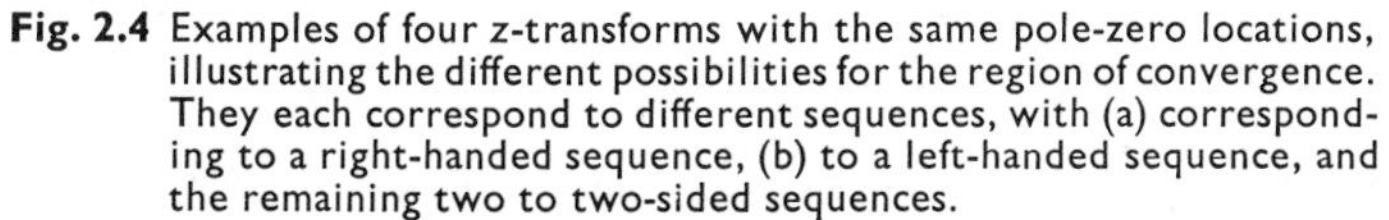
Fig. 2.4 Examples of four z-transforms with the same pole-zero locations, illustrating the different possibilities for the region of convergence. They each correspond to different sequences, with (a) corresponding to a right-handed sequence, (b) to a left-handed sequence, and the remaining two to two-sided sequences.

this were permissible, the inverse z-transform would yield different sequences, depending on where we chose to take the contour of integration.

2.3.2 Linearity

Consider two sequences $x(n)$ and $y(n)$ with z-transforms $X(z)$ and $Y(z)$, respectively; i.e.,

$$\mathcal{Z}[x(n)] = X(z), \qquad R_{x-} < |z| < R_{x+}$$

$$\mathcal{Z}[y(n)] = Y(z), \qquad R_{y-} < |z| < R_{y+}$$

Then

$$\mathcal{Z}[ax(n) + by(n)] = aX(z) + bY(z), \qquad R_- < |z| < R_+ \tag{2.38}$$

where the region of convergence is at least the overlap of the individual regions of convergence. For sequences with rational z-transforms, if the poles of $aX(z) + bY(z)$ are the union of the poles of $X(z)$ and $Y(z)$, then the region of convergence will be exactly equal to the overlap of the individual regions of convergence, and thus R_- will be the maximum of R_{x-} and R_{y-} and R_+ will be the minimum of R_{x+} and R_{y+}. If the linear combination is such that some zeros are introduced which cancel poles, then the region of convergence may be larger. A simple example of this occurs when $x(n)$ and $y(n)$ are of infinite duration but the linear combination is of finite duration. In this case the region of convergence of the linear combination is the entire z-plane, with the possible exception of zero and/or infinity. For example, the sequences $a^n u(n)$ and $a^n u(n-1)$ both have a region of convergence defined by $|z| > |a|$, but the sequence corresponding to the difference $[a^n u(n) - a^n u(n-1)]$ has a region of convergence that is the entire z-plane.

2.3.3 *Shift of a Sequence*

Consider a sequence $x(n)$ such that

$$\mathcal{Z}[x(n)] = X(z), \qquad R_{x-} < |z| < R_{x+}$$

Then for the sequence whose values are $x(n + n_0)$, we have (see Problem 10, of this chapter)

$$\mathcal{Z}[x(n + n_0)] = z^{n_0} X(z), \qquad R_{x-} < |z| < R_{x+} \tag{2.39}$$

The regions of convergence of $x(n)$ and $x(n + n_0)$ are identical, with the possible exception of $z = 0$ or $z = \infty$. For example, the sequence $\delta(n)$ has a z-transform that converges everywhere in the z-plane, but the z-transform of $\delta(n - 1)$ does not converge at $z = 0$ and the z-transform of $\delta(n + 1)$ does not converge at $z = \infty$. As we see from Eq. (2.39), for n_0 positive, zeros are introduced at $z = 0$ and poles at $z = \infty$; for n_0 negative, poles are introduced at the origin and zeros at infinity.

2.3.4 *Multiplication by an Exponential Sequence*

If a sequence $x(n)$ is multiplied by an exponential sequence a^n, where a may be complex, then (see Problem 10 of this chapter)

$$\mathcal{Z}[a^n x(n)] = X(a^{-1}z), \qquad |a| \cdot R_{x-} < |z| < |a| \cdot R_{x+} \tag{2.40}$$

If $X(z)$ has a pole at $z = z_1$, then $X(az^{-1})$ will have a pole at $z = az_1$. In general, all the pole–zero locations are scaled by a factor a. If a is a positive real number, this can be interpreted as a shrinking or expanding of the z-plane; i.e., the pole and zero locations change along radial lines in the z-plane. If a is complex with unity magnitude, the scaling corresponds to a rotation in the z-plane; i.e., the pole and zero locations change along circles centered at the origin.

2.3.5 Differentiation of $X(z)$

The derivative of the z-transform, multiplied by $-z$, is the z-transform of $x(n)$ linearly weighted; i.e.

$$\mathcal{Z}[nx(n)] = -z\frac{dX(z)}{dz}, \qquad R_{x-} < |z| < R_{x+} \tag{2.41}$$

2.3.6 Conjugation of a Complex Sequence

$$\mathcal{Z}[x^*(n)] = X^*(z^*), \qquad R_{x-} < |z| < R_{x+} \tag{2.42}$$

This follows in a straightforward way from the definition of the z-transform.

2.3.7 Initial Value Theorem

If $x(n)$ is zero for $n < 0$, then

$$x(0) = \lim_{z\to\infty} X(z) \tag{2.43}$$

This theorem is easily shown by considering the limit of each term in the series of Eq. (2.1), see Problem 16, of this chapter.

2.3.8 Convolution of Sequences

If $w(n)$ is the convolution of the two sequences $x(n)$ and $y(n)$, then the z-transform of $w(n)$ is the product of the z-transforms of $x(n)$ and $y(n)$; i.e., if

$$w(n) = \sum_{k=-\infty}^{\infty} x(k)y(n-k) \tag{2.44}$$

then $W(z) = X(z)Y(z)$. To show this we write

$$W(z) = \sum_{n=-\infty}^{\infty}\left[\sum_{k=-\infty}^{\infty} x(k)y(n-k)\right]z^{-n}$$

If we interchange the order of summation,

$$W(z) = \sum_{k=-\infty}^{\infty} x(k) \sum_{n=-\infty}^{\infty} y(n-k)z^{-n}$$

changing the index of summation in the second sum from n to $m = n - k$, we obtain

$$W(z) = \sum_{k=-\infty}^{\infty} x(k)\left[\sum_{m=-\infty}^{\infty} y(m)z^{-m}\right]z^{-k}$$

Thus for values of z inside the regions of convergence of both $Y(z)$ and $X(z)$ we can write

$$W(z) = X(z)Y(z), \qquad R_{y-} < |z| < R_{y+}, \quad R_{x-} < |z| < R_{x+} \tag{2.45}$$

where the region of convergence includes the intersection of the regions of convergence of $Y(z)$ and $X(z)$. If a pole that borders on the region of convergence of one of the z-transforms is canceled by a zero of the other, then the region of convergence of $W(z)$ will be larger.

EXAMPLE. Let $y(n) = a^n u(n)$ and $x(n) = u(n)$. The corresponding z-transforms are

$$Y(z) = \sum_{n=0}^{\infty} a^n z^{-n} = \frac{1}{1 - az^{-1}}, \qquad |z| > |a|$$

and

$$X(z) = \sum_{n=0}^{\infty} z^{-n} = \frac{1}{1 - z^{-1}}, \qquad |z| > 1$$

The transform of the convolution is then

$$W(z) = \frac{1}{(1 - az^{-1})(1 - z^{-1})}$$

$$= \frac{z^2}{(z - a)(z - 1)}, \qquad |z| > 1$$

The poles and zeros of $W(z)$ are plotted in Fig. 2.5 and the region of convergence is seen to be the overlap region.

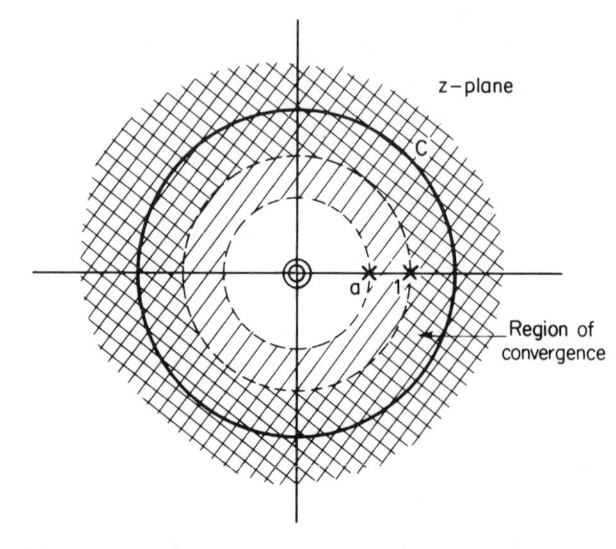

Fig. 2.5 Pole–zero plot for the z-transform of the convolution of the sequences $u(n)$ and $a^n u(n)$.

The sequence $w(n)$ can be obtained from the inverse z-transform

$$w(n) = \frac{1}{2\pi j} \oint_C \frac{z^{n+1}\, dz}{(z - a)(z - 1)}$$

where the contour C is chosen to be in the overlap region, as shown in Fig. 2.5. For $n \geq 0$ there are no poles at $z = 0$. Therefore, using the residue theorem,

$$\begin{aligned} w(n) &= \frac{(1)^{n+1}}{1 - a} + \frac{a^{n+1}}{a - 1} \\ &= \frac{1 - a^{n+1}}{1 - a}, \qquad n \geq 0 \end{aligned}$$

Although there are poles at $z = 0$, for $n < -1$, we need not evaluate the contour integral to see that $w(n)$ must be zero for $n < 0$. We simply note that since both $x(n)$ and $y(n)$ are zero for $n < 0$, $w(n)$ must be also. We note that, for this example, the region of convergence of $W(z)$ was the intersection of those for $x(n)$ and $y(n)$. If, however, we had chosen $y(n)$ such that

$$Y(z) = \frac{1 - z^{-1}}{1 - az^{-1}}, \qquad |z| > |a|$$

then the pole at $z = 1$ would be canceled and the region of convergence of $w(n)$ would extend inward to the pole at $z = a$.

2.3.9 Complex Convolution Theorem

We showed above that the z-transform of a convolution of sequences is the product of the z-transforms of the individual sequences. In the case of continuous-time signals and Fourier transforms, we are aware of the duality of the time and frequency domain. Specifically, in the continuous case, a convolution of time functions leads to a product of transforms, and similarly a convolution of transforms derives from a product of time functions. In the case of sequences and z-transforms, we would not expect an exact duality because of the fact that sequences are discrete while their transforms are continuous. However, a similar kind of relationship can be derived in which the z-transform of a product of sequences has a form similar to a convolution.

To derive the complex convolution theorem, let

$$w(n) = x(n)y(n)$$

so that

$$W(z) = \sum_{n=-\infty}^{\infty} x(n)y(n)z^{-n}$$

But

$$y(n) = \frac{1}{2\pi j} \oint_{C_1} Y(v)v^{n-1}\, dv$$

where C_1 is a counter-clockwise contour within the region of convergence of $Y(v)$. Then

$$W(z) = \frac{1}{2\pi j} \sum_{n=-\infty}^{\infty} x(n) \oint_{C_1} Y(v)\left(\frac{z}{v}\right)^{-n} v^{-1}\, dv$$

$$= \frac{1}{2\pi j} \oint_{C_1} \left[\sum_{n=-\infty}^{\infty} x(n)\left(\frac{z}{v}\right)^{-n} \right] v^{-1} Y(v)\, dv$$

or

$$W(z) = \frac{1}{2\pi j} \oint_{C_1} X\left(\frac{z}{v}\right) Y(v) v^{-1}\, dv \tag{2.46a}$$

where C_1 is a closed contour in the overlap of the regions of convergence of $X(z/v)$ and $Y(v)$. Alternatively, $W(z)$ can be expressed as

$$W(z) = \frac{1}{2\pi j} \oint_{C_2} X(v) Y\left(\frac{z}{v}\right) v^{-1}\, dv \tag{2.46b}$$

where C_2 is a closed contour in the overlap region between $X(v)$ and $Y(z/v)$. To determine the region of convergence to associate with $W(z)$, let us consider the regions of convergence for $X(z)$ and $Y(z)$, respectively, to be

$$X(z)\colon R_{x-} < |z| < R_{x+}$$

$$Y(z)\colon R_{y-} < |z| < R_{y+}$$

Then in Eq. (2.46a),

$$R_{y-} < |v| < R_{y+}$$

and

$$R_{x-} < \left|\frac{z}{v}\right| < R_{x+}$$

Combining these two expressions we require that

$$R_{x-}R_{y-} < |z| < R_{x+}R_{y+}$$

Again, in certain instances the region of convergence may be larger than this but will always include the region defined above and then extend inward or outward to the nearest poles.

To see that Eq. (2.46b) is indeed similar to a convolution, let the contour of integration be a circle with

$$v = \rho e^{j\theta} \qquad \text{and} \qquad z = re^{j\phi}$$

Equation (2.46b) becomes

$$W(re^{j\phi}) = \frac{1}{2\pi} \int_{-\pi}^{\pi} Y\left[\frac{r}{\rho}\, e^{j(\phi-\theta)}\right] X(\rho e^{j\theta})\, d\theta \tag{2.47}$$

which is similar in form to a convolution. In particular, with the exception of the limits on the integral, the above expression is identical to a convolution of $X(\rho e^{j\theta})$ and $Y(\rho e^{j\theta})$ considered as functions of θ. We note that these functions are periodic functions of θ and that the integration is carried out over only one period. A convolution of this type is often referred to as a *periodic convolution* and will play a particularly important role in Chapter 3.

In using the complex convolution theorem of Eq. (2.46a) or (2.46b), one of the primary sources of difficulty is in determining which poles of the integrand are inside the contour of integration and which are outside. A simple example in the use of the complex convolution theorem should serve to indicate the procedure.

EXAMPLE. Let $x(n) = a^n u(n)$ and $y(n) = b^n u(n)$. Then the z-transforms $X(z)$ and $Y(z)$ are, respectively,

$$X(z) = \frac{1}{1 - az^{-1}}, \qquad |z| > |a|$$

$$Y(z) = \frac{1}{1 - bz^{-1}}, \qquad |z| > |b|$$

Substituting into Eq. (2.46a), we obtain

$$W(z) = \frac{1}{2\pi j} \oint_{C_1} \frac{-(z/a)}{(v - z/a)} \frac{1}{v - b}\, dv$$

The integrand has two poles, one located at $v = b$ and the second at $v = z/a$. The contour of integration of this expression must be within the region of convergence of $Y(v)$, and consequently will enclose the pole at $v = b$. To determine whether it encloses the pole at $v = z/a$, we consider that the z-transform $X(z)$ is only valid for $|z| > |a|$. Therefore, the corresponding expression for $X(z/v)$ is only valid for $|z/v| > |a|$. Thus if

$$\left|\frac{z}{v}\right| > |a|$$

then

$$\left|\frac{z}{a}\right| > |v|$$

Consequently, the pole must always lie outside the contour of integration in v. The location of the poles and the contour of integration are indicated in Fig. 2.6, with a and b assumed to be real. Using the Cauchy residue theorem to evaluate $W(z)$, we obtain

$$W(z) = \frac{-z/a}{b - z/a}$$

$$= \frac{1}{1 - abz^{-1}}, \qquad |z| \geq ab$$

Observe that this expression results from only evaluating the residue at the pole inside the contour of integration. It is easily verified that if we had mistakenly considered the pole at z/a to be inside the contour of integration, the result of evaluating the integral would be identically zero.

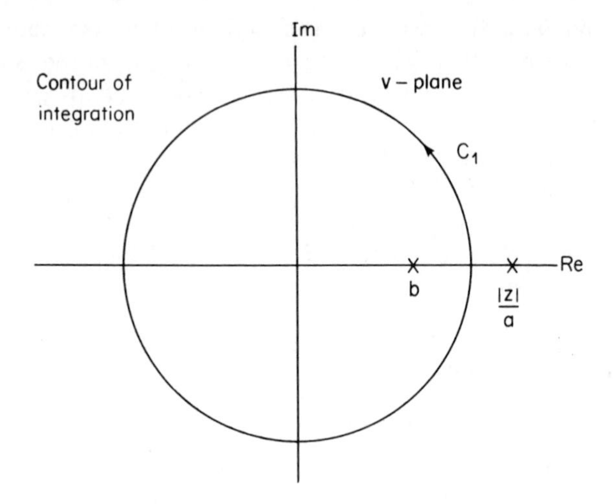

Fig. 2.6 Poles of the integrand and the contour of integration in an example of the use of the complex convolution theorem.

2.3.10 *Parseval's Relation*

In Problem 19 of Chapter 1, we considered Parseval's relation in terms of the Fourier transform. The generalization of this relation to the z-transform follows from the complex convolution theorem. In particular we consider two complex sequences, $x(n)$ and $y(n)$. Then Parseval's relation states that

$$\sum_{n=-\infty}^{\infty} x(n)y^*(n) = \frac{1}{2\pi j}\oint_C X(v)Y^*(1/v^*)v^{-1}\,dv \tag{2.48}$$

where the contour of integration is taken in the overlap of the regions of convergence of $X(v)$ and $Y^*(1/v^*)$. The above relation can be derived by defining a sequence $w(n)$ as

$$w(n) = x(n)y^*(n) \tag{2.49}$$

and noting that

$$\sum_{n=-\infty}^{\infty} w(n) = W(z)\big|_{z=1} \tag{2.50}$$

Then from Eq. (2.42) and the complex convolution theorem it follows that

$$W(z) = \frac{1}{2\pi j}\oint_C X(v)Y^*(z^*/v^*)v^{-1}\,dv$$

Thus applying Eqs. (2.49) and (2.50) we obtain Eq. (2.48). If $X(z)$ and $Y(z)$ converge on the unit circle, we can choose $v = e^{j\omega}$, and Eq. (2.48) becomes

$$\sum_{n=-\infty}^{\infty} x(n)y^*(n) = \frac{1}{2\pi}\int_{-\pi}^{\pi} X(e^{j\omega})Y^*(e^{j\omega})\,d\omega \tag{2.51}$$

2.3.11 Summary of Some z-Transform Theorems and Properties

In the above sections we presented and discussed a number of the theorems and properties of z-transforms. Many of these are useful in manipulating z-transforms, and consequently it is helpful to have them summarized for easy reference. The above properties and a number of other useful ones are presented in Table 2.1. The regions indicated are included in the region

Table 2.1

Sequence	z-Transform	
1. $x(n)$	$X(z)$	$R_{x-} < \|z\| < R_{x+}$
2. $y(n)$	$Y(z)$	$R_{y-} < \|z\| < R_{y+}$
3. $ax(n) + by(n)$	$aX(z) + bY(z)$	$\max[R_{x-}, R_{y-}] < \|z\| < \min[R_{x+}, R_{y+}]$
4. $x(n + n_0)$	$z^{n_0}X(z)$	$R_{x-} < \|z\| < R_{x+}$
5. $a^n x(n)$	$X(a^{-1}z)$	$\|a\|R_{x-} < \|z\| < \|a\|R_{x+}$
6. $nx(n)$	$-z\dfrac{dX(z)}{dz}$	$R_{x-} < \|z\| < R_{x-}$
7. $x^*(n)$	$X^*(z^*)$	$R_{x-} < \|z\| < R_{x+}$
8. $x(-n)$	$X(1/z)$	$1/R_{x+} < \|z\| < 1/R_{x-}$
9. Re $[x(n)]$	$\frac{1}{2}[X(z) + X^*(z^*)]$	$R_{x-} < \|z\| < R_{x+}$
10. Im $[x(n)]$	$\dfrac{1}{2j}[X(z) - X^*(z^*)]$	$R_{x-} < \|z\| < R_{x+}$
11. $x(n) * y(n)$	$X(z)Y(z)$	$\max[R_{x-}, R_{y-}] < \|z\| < \min[R_{x+}, R_{y+}]$
12. $x(n)y(n)$	$\dfrac{1}{2\pi j}\oint_C X(v)Y\left(\dfrac{z}{v}\right)v^{-1}\,dv$	$R_{x-}R_{y-} < \|z\| < R_{x+}R_{y+}$

of convergence, but in some cases the region of convergence may be larger than indicated.

2.4 System Function

In Chapter 1 we considered a description of a linear shift-invariant system in terms of the Fourier transform of the unit-sample response. As we saw, the Fourier transform of the unit-sample response corresponds to the frequency response of the system. Furthermore, in the frequency domain the input–output relation corresponds simply to a multiplication of the Fourier transform of the input and the Fourier transform of the unit-sample response.

More generally we can describe a linear shift-invariant system in terms of the z-transform of the unit-sample response. With $x(n)$, $y(n)$, and $h(n)$ denoting the input, output,, and unit-sample response, respectively, and $X(z)$, $Y(z)$, and $H(z)$ their z-transforms, it follows from the previous section that since

$$y(n) = x(n) * h(n)$$

then

$$Y(z) = X(z)H(z) \tag{2.52}$$

As with the Fourier transform, the input–output relation for a linear shift-invariant system corresponds to a multiplication of the z-transforms of the input and the unit-sample response.

The z-transform of the unit-sample response is often referred to as the *system function*. The system function evaluated on the unit circle (i.e., for $|z| = 1$) is the frequency response of the system.

It was shown in Chapter 1 that a necessary and sufficient condition for a system to be stable is that the unit-sample response $h(n)$ be absolutely summable. The region of convergence of the z-transform is defined by those values of z for which $h(n)z^{-n}$ is absolutely summable. Thus if the region of convergence of the system function includes the unit circle, the system is stable, and vice versa. Furthermore, we can state that for a stable and causal system, the region of convergence will include the unit circle and the entire z-plane outside the unit circle, including $z = \infty$.

When the system is describable by a linear constant-coefficient difference equation, the system function is a ratio of polynomials. To see that this is so, consider a system for which the input and output satisfy the general Nth-order difference equation

$$\sum_{k=0}^{N} a_k y(n-k) = \sum_{r=0}^{M} b_r x(n-r) \tag{2.53}$$

Applying the z-transform to each side of Eq. (2.53) we have

$$\mathcal{Z}\left[\sum_{k=0}^{N} a_k y(n-k)\right] = \mathcal{Z}\left[\sum_{r=0}^{M} b_r x(n-r)\right]$$

which from property 3 of Table 2.1 we can rewrite as

$$\sum_{k=0}^{N} a_k \mathcal{Z}[y(n-k)] = \sum_{r=0}^{M} b_r \mathcal{Z}[x(n-r)]$$

With $X(z)$ and $Y(z)$ denoting the z-transforms of $x(n)$ and $y(n)$, respectively, it follows from property 4 of Table 2.1 that

$$\mathcal{Z}[y(n-k)] = z^{-k}Y(z)$$

and

$$\mathcal{Z}[x(n-r)] = z^{-r}X(z)$$

Thus

$$\sum_{k=0}^{N} a_k z^{-k} Y(z) = \sum_{r=0}^{M} b_r z^{-r} X(z)$$

From Eq. (2.52), $H(z) = Y(z)/X(z)$, so

$$H(z) = \sum_{r=0}^{M} b_r z^{-r} \Big/ \sum_{k=0}^{N} a_k z^{-k} \tag{2.54}$$

Equation (2.54) expresses the functional form of the system function, and we note in particular that the coefficients in the numerator and denominator polynomials correspond, respectively, to the coefficients on the right- and left-hand sides of the difference equation (2.53).

Since Eq. (2.54) is a ratio of polynomials in z^{-1}, it can also be expressed in factored form as†

$$H(z) = \frac{A \prod_{r=1}^{M} (1 - c_r z^{-1})}{\prod_{k=1}^{N} (1 - d_k z^{-1})} \tag{2.55}$$

Each of the factors $(1 - c_r z^{-1})$ in the numerator of Eq. (2.55) contributes a zero at $z = c_r$ and a pole at $z = 0$. Similarly, each of the factors of the form $(1 - d_k z^{-1})$ in the denominator contributes a pole at $z = d_k$ and a zero at the origin. It is characteristic of the systems describable by linear constant-coefficient difference equations that their system functions are a ratio of polynomials, in z^{-1}. Consequently, to within the scale factor A in Eq. (2.55), the system function can be specified by a pole–zero pattern in the z-plane.

Equation (2.54) does not indicate the region of convergence of the system function. This is consistent with the fact that as we saw in Chapter 1, the difference equation does not uniquely specify the unit-sample response of a linear shift-invariant system. For the system function of Eq. (2.54) there are a number of choices for the region of convergence consistent with the requirements that they be annular regions bounded by, but not containing, poles. For a given ratio of polynomials, each possible choice for the region of convergence will lead to a different unit-sample response, but they will all correspond to the same difference equation. If we assume that the system is stable, then we would choose the annular region that includes the unit circle. If we assume that the system is causal, then we choose the region of convergence to be the exterior of a circle passing through the pole of $H(z)$ that is farthest from the origin. If the system is also stable, then, of course, all the

† Equation (2.55) assumes that b_0 and a_0 in Eq. (2.54) are not zero. In general, if $b_0, b_1, \ldots, b_{M_1}$ and $a_0, a_1, \ldots, a_{N_1}$ are all zero, then Eq. (2.55) would be expressed as

$$H(z) = \frac{Bz^{-M_1} \prod_{r=1}^{M-M_1} (1 - c_r z^{-1})}{z^{-N_1} \prod_{k=1}^{N-N_1} (1 - d_k z^{-1})}$$

poles lie inside the unit circle and the region of convergence will include the unit circle. For this reason it is often convenient, when describing the system function in terms of a pole-zero plot in the z-plane, to include the unit circle in the drawing to indicate whether the poles lie inside or outside the unit circle.

EXAMPLE. As a simple example consider a causal system characterized by the difference equation

$$y(n) = ay(n-1) + x(n)$$

The system function is

$$H(z) = \frac{1}{1 - az^{-1}} \tag{2.56}$$

and, by the causality assumption, the region of convergence is $|z| > |a|$, from which we note that the impulse response is

$$h(n) = a^n u(n)$$

In the special case that $N = 0$ in Eq. (2.54) or (2.53), the system has no poles except at $z = \infty$ and the system has a finite-duration unit-sample response. When N is greater than zero, the system has poles, each of which contributes an exponential sequence to the unit-sample response. Thus if the system function has poles, the unit-sample response is of infinite duration.

One of the advantages of a representation of the system function in terms of poles and zeros is that it leads to a useful geometric means for obtaining the system frequency response. We recall from a previous discussion that the response of the system to a sinusoidal excitation is describable in terms of the frequency response, i.e., the behavior of the system function on the unit circle. In particular, the response is sinusoidal with the same frequency as the input, and the amplitude of the output is equal to the amplitude of the input multiplied by the magnitude of the system function at the excitation frequency. The phase shift of the output is equal to the angle of the complex number that represents the system function at the excitation frequency. To determine the system function on the unit circle, corresponding to an excitation frequency ω_0, we would substitute $z = e^{j\omega_0}$ into Eq. (2.55). Let us consider, for example, a factor $(1 - c_r z^{-1})$, which contributes the zero and pole depicted in Fig. 2.7. At $z = e^{j\omega_0}$ the magnitude of the complex number represented by this factor is equal to the product of the length of the vector from the zero to the appropriate point on the unit circle, divided by the length of the vector from the pole to the same point on the unit circle. The angle of the complex number is equal to the angle of the zero vector minus the angle of the pole vector. The magnitude of the complex number represented by a product of such factors is the product of the magnitudes, and since the angle of the complex number is the sum of the angles, the total frequency response can be obtained by looking at the net effect of the zero and the pole vectors. For example, in Fig. 2.8 the pole–zero pattern for a

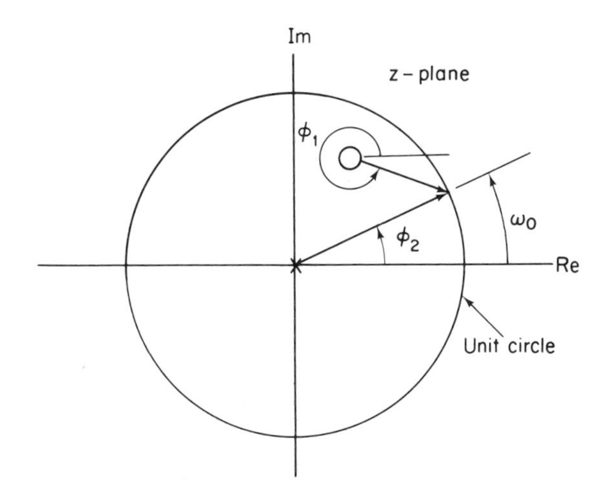

Fig. 2.7 Determination of the frequency response from the pole–zero pattern by geometric means.

second-order system is shown. It is clear from the behavior of the pole and zero vectors that the magnitude of the frequency-response peaks in the vicinity of the poles. From this geometric picture it should be clear also that poles or zeros at the origin offer no contribution to the magnitude of the frequency response and introduce only a linear component in the phase. These concepts are illustrated by Fig. 2.9, which shows the pole–zero diagram and the frequency response for a first-order difference equation corresponding

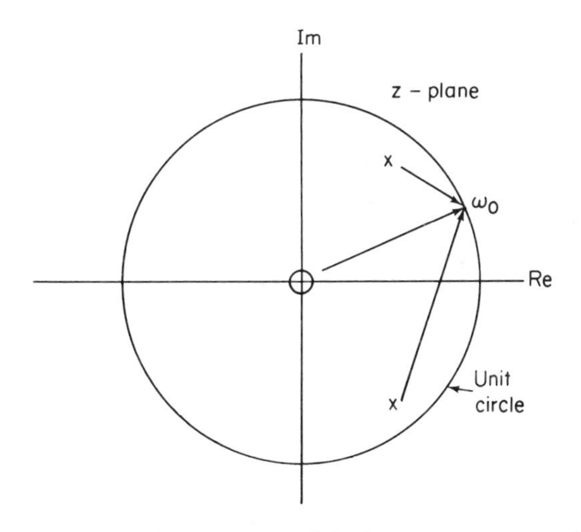

Fig. 2.8 Geometric determination of the frequency response of a second-order system.

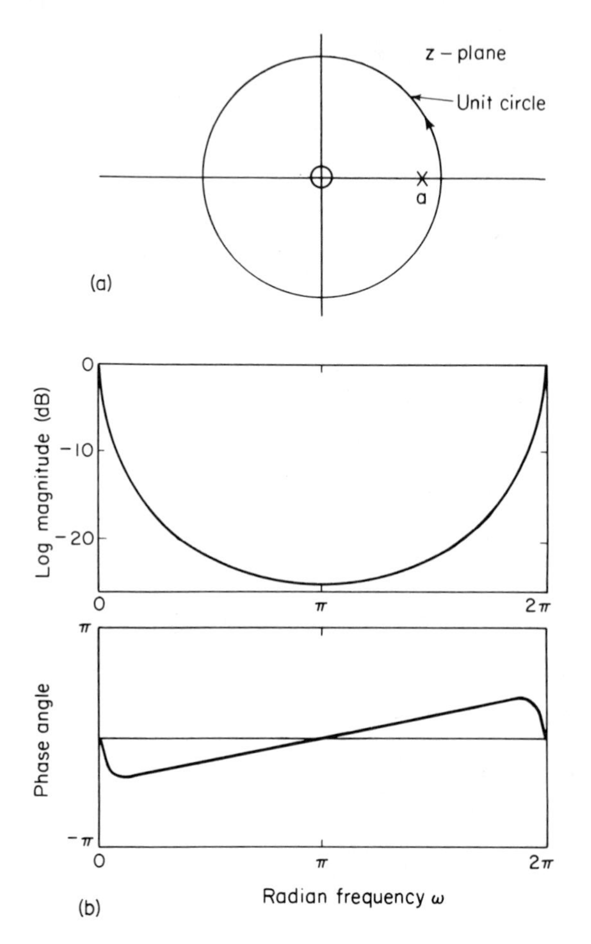

Fig. 2.9 Pole–zero pattern of a first-order filter and the corresponding frequency response.

to the system function $H(z) = 1/(1 - az^{-1})$ and impulse response $h(n) = a^n u(n)$.

As a second example, let the impulse response be a truncation of the impulse response of the previous example; i.e.,

$$h(n) = \begin{cases} a^n, & 0 \leq n \leq M - 1 \\ 0, & \text{elsewhere} \end{cases}$$

Then the system function is

$$H(z) = \sum_{n=0}^{M-1} a^n z^{-n} = \frac{1 - a^M z^{-M}}{1 - az^{-1}} \tag{2.57}$$

which can be written

$$H(z) = \frac{z^M - a^M}{z^{M-1}(z - a)}$$

Since the zeros of the numerator occur at

$$z_k = ae^{j(2\pi/M)k}, \qquad k = 0, 1, \ldots, M - 1$$

where a is assumed positive, the pole at $z = a$ is canceled by an identical zero. The pole–zero plot and corresponding frequency response for the case $M = 8$ are thus as shown in Fig. 2.10. Observe the peak at $\omega = 0$ ($z = 1$), where there are no zeros, and note the dips in the frequency response in the vicinity of each zero. These properties of the frequency response are easily deduced geometrically from the pole–zero plot.

2.5 Two-Dimensional z-Transform

In Chapter 1 we saw that many of the concepts and properties of discrete-time signals and systems in one dimension could be extended to multi-dimensional signals and systems. In previous sections of this chapter we have considered the z-transform as it applies in one dimension. In this section we consider the extension to two dimensions.

The two-dimensional z-transform $X(z_1, z_2)$ of a sequence $x(m, n)$ is defined as

$$X(z_1, z_2) = \sum_{m=-\infty}^{\infty} \sum_{n=-\infty}^{\infty} x(m, n) z_1^{-m} z_2^{-n} \tag{2.58}$$

where z_1 and z_2 are complex variables. With z_1 and z_2 expressed in polar form as

$$z_1 = r_1 e^{j\omega_1}$$

$$z_2 = r_2 e^{j\omega_2}$$

Equation (2.58) can be written as

$$X(r_1 e^{j\omega_1}, r_2 e^{j\omega_2}) = \sum_{m=-\infty}^{\infty} \sum_{n=-\infty}^{\infty} x(m, n) r_1^{-m} r_2^{-n} e^{-j\omega_1 m} e^{-j\omega_2 n} \tag{2.59}$$

Thus, as in the one-dimensional case, the two-dimensional z-transform can be interpreted as the two-dimensional Fourier transform of $x(m, n)$ multiplied by the two-dimensional exponential sequence $r_1^{-m} r_2^{-n}$. For $|z_1| = |z_2| = 1$, i.e., for $r_1 = r_2 = 1$, the z-transform is equal to the Fourier transform. For convergence of the two-dimensional z-transform we require that the sequence $x(m, n) z_1^{-m} z_2^{-n}$ be absolutely summable, i.e., that

$$\sum_{m=-\infty}^{\infty} \sum_{n=-\infty}^{\infty} |x(m, n) z_1^{-m} z_2^{-n}| < \infty \tag{2.60}$$

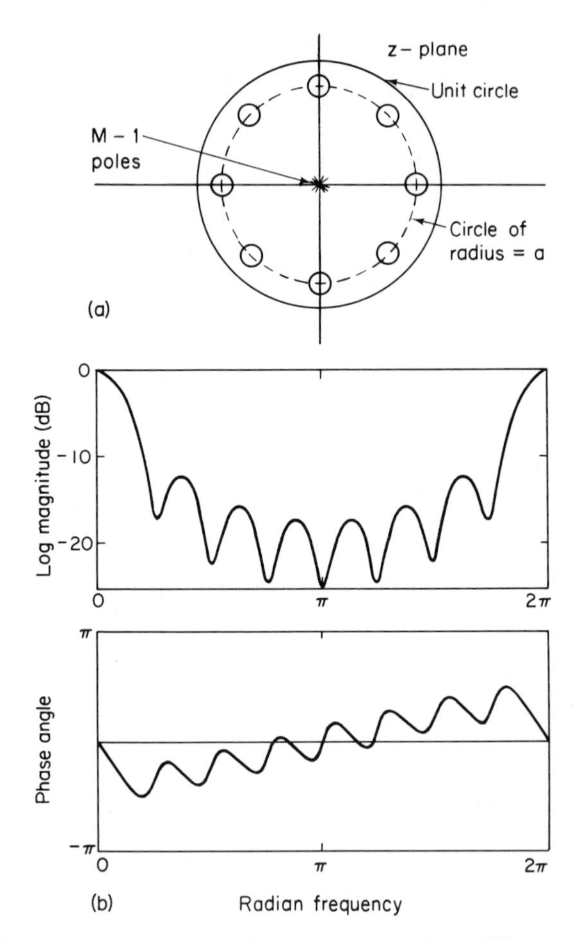

Fig. 2.10 Pole–zero pattern and frequency response for an FIR system with a unit-sample response that is a truncated version of the unit-sample response for the example illustrated in Fig. 2.9.

The set of values of z_1 and z_2 for which the two-dimensional z-transform converges defines the region of convergence.

The inverse two-dimensional z-transform can be obtained by applying the inverse Fourier transform relation of Eq. (1.44) to Eq. (2.59) as

$$x(m, n) = \left(\frac{1}{2\pi}\right)^2 \int_{-\pi}^{\pi} \int_{-\pi}^{\pi} X(r_1 e^{j\omega_1}, r_2 e^{j\omega_2}) r_1^m r_2^n e^{j\omega_1 m} e^{j\omega_2 n} \, d\omega_1 \, d\omega_2 \quad (2.61)$$

This can also be expressed as a contour integral in the form

$$x(m, n) = \left(\frac{1}{2\pi j}\right)^2 \oint_{C_1} \oint_{C_2} X(z_1, z_2) z_1^{m-1} z_2^{n-1} \, dz_1 \, dz_2 \quad (2.62)$$

where the contours C_1 and C_2 are closed contours encircling the origin and within the region of convergence. In contrast to the one-dimensional z-transform, it is generally very difficult to determine the region of convergence of $X(z_1, z_2)$ and consequently also the integration contours C_1 and C_2.

A two-dimensional z-transform $X(z_1, z_2)$ is said to be separable if it can be expressed in the form

$$X(z_1, z_2) = X_1(z_1)X_2(z_2)$$

$X(z_1, z_2)$ will only be separable if the sequence $x(m, n)$ from which it is derived is separable; i.e., $x(m, n) = x_1(m)x_2(n)$. In that case $X_1(z_1)$ and $X_2(z_2)$ are the one-dimensional z-transforms of $x_1(m)$ and $x_2(n)$, respectively.

All the properties of one dimensional z-transforms as summarized in Table 2.1 are easily extended to two dimensions as indicated in Table 2.2,

Table 2.2

Sequence	*z-Transform*
1. $x(m, n)$	$X(z_1, z_2)$
2. $y(m, n)$	$Y(z_1, z_2)$
3. $ax(m, n) + by(m, n)$	$aX(z_1, z_2) + bY(z_1, z_2)$
4. $x(m + m_0, n + n_0)$	$z_1^{m_0} z_2^{n_0} X(z_1, z_2)$
5. $a^m b^n x(m, n)$	$X(a^{-1}z_1, b^{-1}z_2)$
6. $mnx(m, n)$	$\dfrac{d^2X(z_1, z_2)}{dz_1\, dz_2}$
7. $x^*(m, n)$	$X^*(z_1^*, z_2^*)$
8. $x(-m, -n)$	$X(z_1^{-1}, z_2^{-1})$
9. Re $[x(m, n)]$	$\frac{1}{2}[X(z_1, z_2) + X^*(z_1^*, z_2^*)]$
10. Im $[x(m, n)]$	$\dfrac{1}{2j}[X(z_1, z_2) - X^*(z_1^*, z_2^*)]$
11. $x(m, n) * y(m, n)$	$X(z_1, z_2)Y(z_1, z_2)$
12. $x(m, n)y(m, n)$	$\left(\dfrac{1}{2\pi j}\right)^2 \oint_{C_1} \oint_{C_2} X\left(\dfrac{z_1}{v_1}, \dfrac{z_2}{v_2}\right) Y(v_1, v_2)v_1^{-1}v_2^{-1}\, dv_1\, dv_2$

and for those listed the proofs parallel those for the one-dimensional case.

The two-dimensional z-transform of a convolution of two two-dimensional sequences is the product of their z-transforms. Consequently, the input–output relation for a two-dimensional linear shift-invariant system, expressed in terms of the z-transform, corresponds to a multiplication of the z-transforms of the input and the unit-sample response. As in one dimension, the z-transform of the two-dimensional unit-sample response is referred to as the *system function*. The system function evaluated for $|z_1| = |z_2| = 1$ is the frequency response of the system.

When the system is describable by a linear constant-coefficient difference equation, the system function is a ratio of two-dimensional polynomials. In particular, with the input and output satisfying the difference equation

$$\sum_{k=0}^{M_1}\sum_{r=0}^{N_1} a_{kr}y(m-k, n-r) = \sum_{k=0}^{M_2}\sum_{r=0}^{N_2} b_{kr}x(m-k, n-r) \tag{2.63}$$

and applying the z-transform to both sides of Eq. (2.63) it follows, using the properties in Table 2.2, that

$$Y(z_1, z_2)\left[\sum_{k=0}^{M_1}\sum_{r=0}^{N_1} a_{kr}z_1^{-k}z_2^{-r}\right] = X(z_1, z_2)\left[\sum_{k=0}^{M_2}\sum_{r=0}^{N_2} b_{kr}z_1^{-k}z_2^{-r}\right]$$

so that the system function $H(z_1, z_2)$ is given by

$$H(z_1, z_2) = \frac{Y(z_1, z_2)}{X(z_1, z_2)} = \frac{\sum_{k=0}^{M_2}\sum_{r=0}^{N_2} b_{kr}z_1^{-k}z_2^{-r}}{\sum_{k=0}^{M_1}\sum_{r=0}^{N_1} a_{kr}z_1^{-k}z_2^{-r}} \tag{2.64}$$

In the one-dimensional case, when the system function consisted of a ratio of polynomials, it could be described in terms of its poles and zeros, i.e., the roots of the numerator and denominator polynomials. In contrast, a general two-dimensional polynomial cannot be factored. This represents a major difference between the one-dimensional and two-dimensional cases.

Comparing Eq. (1.38), which states the condition for stability of a two-dimensional linear shift-invariant system, and Eq. (2.60), which expresses the requirement on the convergence of the two-dimensional z-transform, we see that a system is stable if and only if the system function converges for $|z_1| = |z_2| = 1$. For one-dimensional causal systems stability was easily examined by regarding the pole locations. For that case a necessary and sufficient condition for stability is that all the poles of the system function lie inside the unit circle. In the special case that the unit-sample response or equivalently the system function is separable, the same condition can be applied to examine stability of a causal filter. This follows from the fact that for the separable case, with $h(m, n) = h_1(m)h_2(n)$, the unit-sample response will be absolutely summable if and only if $h_1(m)$ and $h_2(n)$ are absolutely summable. Thus a causal separable system will be stable if and only if the poles of $H_1(z_1)$ lie inside the unit circle in the z_1-plane and the poles of $H_2(z_2)$ lie inside the unit circle in the z_2-plane.

In the general case, a necessary and sufficient condition for a ratio of two-dimensional polynomials to correspond to a causal, stable system function is that the denominator polynomial cannot be zero if $|z_1|$ and $|z_2|$ are simultaneously greater than unity [4, 5]. We see that this is consistent with the condition discussed above for a separable system since, for example, a pole of $H(z_1)$ outside the unit circle, i.e., for $|z_1| > 1$, will result in the denominator polynomial of the two-dimensional system function being zero at that value of z_1 for all values of z_2 and therefore including values of z_2 for which $|z_2| > 1$.

One way of applying the above stability theorem is to factor the denominator polynomial as a polynomial in z_1, in which case the roots are a function of z_2. In this case the poles of $H(z_1, z_2)$ are a function of z_2 and we require that these poles can only lie outside the unit circle in the z_1-plane for values of z_2 for which $|z_2| < 1$.

EXAMPLE. Consider $H(z_1, z_2)$ of the form

$$H(z_1, z_2) = \frac{1}{1 - z_1^{-1} + 2z_2^{-1} - z_1^{-1}z_2^{-1}}$$

or

$$H(z_1, z_2) = \frac{1}{(1 + 2z_2^{-1}) - z_1^{-1}(1 + z_2^{-1})} \tag{2.65}$$

Thus the roots of the denominator polynomial are at

$$z_1 = \frac{1 + z_2^{-1}}{1 + 2z_2^{-1}} \tag{2.66}$$

For stability we require that in Eq. (2.66), for all values of z_2 for which $|z_2| \geq 1$, the magnitude of z_1 is less than unity, i.e., we require that for $|z_2| \geq 1$,

$$\left| \frac{1 + 2z_2^{-1}}{1 + z_2^{-1}} \right| < 1$$

or

$$|1 + z_2^{-1}| < |1 + 2z_2^{-1}|$$

Equivalently, this can be written as

$$|1 + z_2^{-1}|^2 < |1 + 2z_2^{-1}|^2$$

It is straightforward to verify that for $z_2^{-1} = -\frac{1}{3} + j\frac{1}{6}$ this inequality is not satisfied, and consequently the system function of Eq. (2.65) does not correspond to a stable system.

It is clear that while the stability condition discussed above is straightforward in theory, in practice it is difficult to apply. There are a number of other formulations of the stability conditions, which we will not present, that are theoretically less straightforward but are perhaps slightly simpler to apply.

Summary

In this chapter we have generalized much of the discussion in Chapter 1. In particular, we presented the definition of the z-transform and the region of convergence associated with right-sided, left-sided, and two-sided sequences. The inverse z-transform was then considered, including the evaluation of the inverse z-transform by complex integration, by partial-fraction expansion, and by the use of power series. Properties of the z-transform similar to those discussed in Chapter 1 for the Fourier transform were presented also.

The representation of linear shift-invariant systems in terms of the z-transform led to a discussion of the system function. For systems characterized by linear constant-coefficient difference equations, the system function is a ratio of polynomials, in which case it can be specified in terms of a pole-zero pattern in the z-plane. This representation leads, among other things, to a useful geometric means for obtaining the system frequency response.

The chapter closed with a brief introduction to the two-dimensional z-transform.

REFERENCES

1. J. R. Ragazzini and G. F. Franklin, *Sampled Data Control Systems*, McGraw-Hill Book Company, New York, 1958.
2. E. I. Jury, *Theory and Application of the z-Transform Method*, John Wiley & Sons, Inc., New York, 1964.
3. R. V. Churchill, *Complex Variables and Applications*, McGraw-Hill Book Company, New York, 1960.
4. J. L. Shanks, S. Treitel, and J. H. Justice, "Stability and Synthesis of Two-Dimensional Recursive Filters," *IEEE Trans. Audio Electroacoust.*, Vol. AU-20, No. 3, June 1972, pp. 115–128.
5. T. S. Huang, "Stability of Two-Dimensional Recursive Filters," *IEEE Trans. Audio Electroacoust.*, Vol. AU-20, No. 2, June 1972, pp. 158–163.

PROBLEMS

1. Determine the z-transform, including the region of convergence, for each of the following sequences:
   (a) $(\frac{1}{2})^n u(n)$.
   (b) $-(\frac{1}{2})^n u(-n-1)$.
   (c) $(\frac{1}{2})^n u(-n)$.
   (d) $\delta(n)$.
   (e) $\delta(n-1)$.
   (f) $\delta(n+1)$.
   (g) $(\frac{1}{2})^n [u(n) - u(n-10)]$.
2. Determine the z-transform of each of the following. Include with your answer the region of convergence in the z-plane and a sketch of the pole–zero pattern. Express all sums in closed form. α can be complex.
   (a) $x(n) = \alpha^{|n|}$, $0 < |\alpha| < 1$.
   (b) $x(n) = Ar^n \cos(\omega_0 n + \phi)u(n)$, $0 < r < 1$.

   (c) $$x(n) = \begin{cases} 1, & 0 \le n \le N-1, \\ 0, & N \le n, \\ 0, & n < 0. \end{cases}$$

   (d) $$x(n) = \begin{cases} n, & 0 \le n \le N, \\ 2N - n, & N+1 \le n \le 2N, \\ 0, & 2N \le n, \\ 0, & 0 > n. \end{cases}$$

   [*Hint* (easy way): First express $x(n)$ in terms of the $x(n)$ in part (c).]

3. Consider a sequence $y(n)$ given by $y(n) = x(n) * h(n)$, where $h(n) = (1 + j)^n u(n)$ and $|x(n)| \leq 1$. Is the sequence $|y(n)|$ necessarily bounded?

4. (a) Listed here are several z-transforms. For each one, determine the inverse z-transform using each of the three methods (contour integration, partial-fraction expansion, and long division) discussed in Sec. 2.2.

$$X(z) = \frac{1}{1 + \frac{1}{2}z^{-1}}, \qquad |z| > \tfrac{1}{2}$$

$$X(z) = \frac{1}{1 + \frac{1}{2}z^{-1}}, \qquad |z| < \tfrac{1}{2}$$

$$X(z) = \frac{1 - \frac{1}{2}z^{-1}}{1 + \frac{3}{4}z^{-1} + \frac{1}{8}z^{-2}}, \qquad |z| > \tfrac{1}{2}$$

$$X(z) = \frac{1 - \frac{1}{2}z^{-1}}{1 - \frac{1}{4}z^{-2}}, \qquad |z| > \tfrac{1}{2}$$

$$X(z) = \frac{1 - az^{-1}}{z^{-1} - a}, \qquad |z| > |1/a|$$

(b) Consider a right-sided sequence $x(n)$ with z-transform

$$X(z) = \frac{1}{(1 - az^{-1})(1 - bz^{-1})} = \frac{z^2}{(z - a)(z - b)}$$

In section 2.2 we considered the determination of $x(n)$ by carrying out a partial fraction expansion with $X(z)$ considered as a ratio of polynomials in z^{-1}.

Carry out a partial fraction expansion of $X(z)$ considered as a ratio of polynomials in z and from this expansion determine $x(n)$.

5. In Sec. 2.2 we discussed the fact that for $n < 0$ it is often more convenient to evaluate the inverse transform relation of Eq. (2.22) by using a substitution of variables $z = p^{-1}$ to obtain Eq. (2.28). If the contour of integration is the unit circle, for example, this has the effect of mapping the outside of the unit circle to the inside, and vice versa. It is worth convincing yourself, however, that without this substitution, we can obtain $x(n)$ for $n < 0$ by evaluating the residues at the multiple-order poles at $z = 0$.

Let

$$X(z) = \frac{1}{1 - \frac{1}{2}z}$$

where the region of convergence includes the unit circle.

(a) Determine $x(0)$, $x(-1)$, and $x(-2)$ by evaluating Eq. (2.22) explicitly, obtaining the residue for the poles at $z = 0$.

(b) Determine $x(n)$ for $n > 0$ by evaluating expression (2.22) and for $n \leq 0$ by evaluating Eq. (2.28).

6. Determine a sequence $x(n)$ whose z-transform is $X(z) = e^z + e^{1/z}$, $z \neq 0$.

7. Determine whether or not the function $F(z) = z^*$ can correspond to the z-transform of a sequence. Clearly explain your reasoning.

8. Suppose that $F(z)$ is a rational function, i.e.,

$$F(z) = \frac{N(z)}{D(z)}$$

where $N(z)$ and $D(z)$ are polynomials. Moreover, assume that $F(z)$ has no poles or zeros of multiplicity greater than unity. Let C be a simple closed contour, and let Z and P be, respectively, the number of zeros and poles of $F(z)$ enclosed by the contour. (Assume that there are no poles or zeros on C.)

(a) Show that

$$\frac{1}{2\pi j}\oint_C \frac{F'(z)}{F(z)}\,dz = Z - P$$

where $F'(z)$ is the derivative of $F(z)$.

(b) With $F(z)$ expressed in polar form as $F(z) = |F(z)|e^{j\arg F(z)}$, show that the change in $\arg[F(z)]$ as the contour C is traversed exactly once is $2\pi(Z-P)$.

(It can be shown that these results generalize in the case of multiple-order poles and zeros by counting poles and zeros according to their multiplicity; i.e., a second-order pole is counted twice.)

9. Let $X(z)$ denote a ratio of polynomials in z; i.e.,

$$X(z) = \frac{P(z)}{Q(z)}$$

Show that if $X(z)$ has a first-order pole at $z = z_0$, then

$$\text{Res}\ [X(z) \text{ at } z = z_0] = \frac{P(z_0)}{Q'(z_0)}$$

where $Q'(z_0)$ denotes the derivative of $Q(z)$ evaluated at $z = z_0$.

10. Show that if $X(z)$ is the z-transform of $x(n)$, then

(a) $z^{n_0}X(z)$ is the z-transform of $x(n + n_0)$.

(b) $X(a^{-1}z)$ is the z-transform of $a^n x(n)$.

(c) $-zX'(z)$ is the z-transform of $nx(n)$.

11. With $X(z)$ denoting the z-transform of $x(n)$, show that

$$\mathfrak{Z}[x^*(n)] = X^*(z^*).$$

$$\mathfrak{Z}[x(-n)] = X\left(\frac{1}{z}\right).$$

$$\mathfrak{Z}[\text{Re}\ x(n)] = \tfrac{1}{2}[X(z) + X^*(z^*)].$$

$$\mathfrak{Z}[\text{Im}\ x(n)] = \frac{1}{2j}\,[X(z) - X^*(z^*)].$$

12. Determine the z-transform of $n^2x(n)$ in terms of the z-transform of $x(n)$.

13. The autocorrelation sequence $c(n)$ of a sequence $x(n)$ is defined as

$$c(n) = \sum_{k=-\infty}^{\infty} x(k)x(n + k)$$

Determine the z-transform of $c(n)$ in terms of the z-transform of $x(n)$.

14. Let $x(n)$ denote a causal sequence; i.e., $x(n) = 0,\ n < 0$. Furthermore, assume that $x(0) \neq 0$.

(a) Show that there are no poles or zeros of $X(z)$ at $z = \infty$.

(b) Show that the number of poles in the finite z-plane equals the number of zeros in the finite z-plane. (The finite z-plane excludes $z = \infty$.)

15. Consider a finite impulse response filter with unit-sample response $h(n)$ of length $(2N + 1)$. If $h(n)$ is real and even, show that the zeros of the system function occur in mirror-image pairs about the unit circle; i.e., if $H(z) = 0$ for $z = \rho e^{j\theta}$, then $H(z) = 0$, also for $z = (1/\rho)e^{j\theta}$.

16. For a sequence $x(n)$, which is zero for $n < 0$, show that $\lim_{z \to \infty} X(z) = x(0)$. What is the corresponding theorem if the sequence is zero for $n > 0$?

17. Consider a sequence $x(n)$ for which the z-transform is

$$X(z) = \frac{\frac{1}{3}}{1 - \frac{1}{2}z^{-1}} + \frac{\frac{1}{4}}{1 - 2z^{-1}}$$

and for which the region of convergence includes the unit circle. Use the theorems derived in Problem 16 above to determine $x(0)$.

18. A *real* sequence $x(n)$ has all the poles and zeros of its z-transform inside the unit circle. Determine in terms of $x(n)$ a *real* sequence $x_1(n)$ not equal to $x(n)$ but for which $x_1(0) = x(0)$, $|x_1(n)| = |x(n)|$ and the z-transform of $x_1(n)$ has all its poles and zeros inside the unit circle.

19. $x(n)$ denotes a *finite-duration* sequence of length N with $x(n) = 0$, $n < 0$, and $x(n) = 0$, $n \geq N$. *Do not assume that $x(n)$ is real.* From the list choose the entry that corresponds to the number of distinct choices for $x(n)$ if the *magnitude* of its Fourier transform is specified.

(a) 1
(b) N
(c) $\log_2 N$
(d) 2^N
(e) ∞
(f) N^2
(g) $N(N - 1)$
(h) 2^{N-1}
(i) $N \cdot 2^N$
(j) $\log_2 (N - 1)$
(k) N^N
(l) $N!$
(m) $(N - 1)!$

20. Consider a set of N distinct sequences

$$S = \{x_\nu(n)\}, \qquad \nu = 0, 1, \ldots, N - 1$$

and the corresponding set of z-transforms

$$T = \{X_\nu(z)\}$$

where $X_\nu(z)$ is the z-transform of $x_\nu(n)$. The elements of S and T have the following properties:

(1) The x_ν are real.
(2) The x_ν are causal in the sense that $x_\nu(n) = 0$ for $n < 0$.
(3) The x_ν are stable; i.e., the region of convergence of $X_\nu(z)$ contains the unit circle.
(4) The X_ν are rational functions of z; i.e., they can be expressed as ratios of polynomials in z (or z^{-1}).
(5) The magnitudes of the X_ν are all equal on the unit circle; i.e., $|X_\mu(e^{j\omega})| = |X_\nu(e^{j\omega})|$ for all μ and ν and $-\pi < \omega < \pi$.
(6) S and T are complete; i.e., if $X_\nu \in T$ and $|X_\mu(e^{j\omega})| = |X_\nu(e^{j\omega})|$ for $-\pi < \omega < \pi$, then $X_\mu \in T$.

(a) Show that $X_\mu(z)$ can be expressed as

$$X_\mu(z) = R_{\mu\nu}(z)X_\nu(z)$$

where $R_{\mu\nu}(z)$ is a rational function of z (or z^{-1}) with unity magnitude on the unit circle; i.e.,

$$|R_{\mu\nu}(e^{j\omega})| = 1, \qquad -\pi < \omega \leq \pi$$

Is $R_{\mu\nu}(z)$ necessarily stable? Why?

(b) Denote the element of T having the fewest number of zeros outside the unit circle by $X_0(z)$. Using the initial value theorem, show that the element of S having the largest magnitude at the origin is $x_0(n)$; i.e.,

$$|x_0(0)| > |x_\nu(0)|, \qquad \text{for all } \nu \neq 0$$

(c) Suppose that the system function $H(z)$ for a stable causal filter is a rational function of z. Moreover, on the unit circle,

$$|H(e^{j\omega})| = |H(e^{-j\omega})|$$

and $\qquad -\pi < \omega \leq \pi$

$$\arg [H(e^{j\omega})] = -\arg [H(e^{-j\omega})]$$

However, all the zeros of $H(z)$ are not inside the unit circle. How would you construct from $H(z)$ a stable causal filter $G(z)$ with a real unit-sample response, which has all its zeros inside the unit circle and for which

$$|G(e^{j\omega})| = |H(e^{j\omega})|, \qquad -\pi < \omega \leq \pi?$$

[*Hint:* Use the result of part (a).]

21. Suppose that we are given a sequence $x(n)$ whose Fourier transform has the property that

$$X(e^{j\omega}) \begin{matrix} \neq 0, & |\omega| < \omega_c \\ = 0, & \omega_c < |\omega| < \pi \end{matrix}$$

(a) Show that if we define a new sequence $x_1(n)$ with values

$$x_1(n) = x(Mn), \qquad n = 0, \pm 1, \pm 2, \ldots$$

(i.e., we retain 1 of M samples), then

$$X_1(z) = \frac{1}{M} \sum_{l=0}^{M-1} X(z^{1/M} e^{-j(2\pi/M)l})$$

(b) For the case $\omega_c = \pi/M$, sketch the Fourier transform of $x_1(n)$. [Assume an arbitrary shape for $X(e^{j\omega})$.]

(c) Now suppose that we have a sequence $x_1(n)$ and that we define a new sequence $x_2(n)$ with values

$$x_2(n) = \begin{cases} x_1\left(\dfrac{n}{M}\right), & n = 0, \pm M, \pm 2M, \ldots \\ 0, & \text{elsewhere} \end{cases}$$

Then show that

$$X_2(z) = X_1(z^M)$$

(d) Sketch the Fourier transform of $x_2(n)$ for the $x_1(n)$ assumed in part (b).

(e) Using the previous results, show how the original sequence $x(n)$ can be recovered exactly from $x_2(n)$. What is the general relation between M and ω_c that ensures that $x(n)$ can be recovered?

22. Consider a linear shift-invariant system with impulse response $h(n)$ and input $x(n)$ given by

$$h(n) = \begin{cases} a^n, & n \geq 0 \\ 0, & n < 0 \end{cases}$$

$$x(n) = \begin{cases} 1, & 0 \leq n \leq (N-1) \\ 0, & \text{otherwise} \end{cases}$$

(a) Determine the output $y(n)$ by explicitly evaluating the discrete convolution of $x(n)$ and $h(n)$.

(b) Determine the output $y(n)$ by computing the inverse z-transform of the product of the z-transforms of the input and the unit-sample response.

23. Consider a linear shift-invariant discrete system for which the input $x(n)$ and output $y(n)$ are related by the first-order difference equation

$$y(n) + \tfrac{1}{2}y(n-1) = x(n)$$

From the following list choose *two* possible unit-sample responses for the system.

(a) $(-\frac{1}{2})^n u(n)$ (e) $(\frac{1}{2})^n u(n-1)$ (h) $(-\frac{1}{2})^n u(-n-1)$

(b) $(2^n)u(n)$ (f) $(-2)^n u(-n-1)$ (i) $\frac{1}{2}(-\frac{1}{2})^{n-1}u(-n-1)$

(c) $(n)^{1/2}u(n)$ (g) $(\frac{1}{2})^n u(n)$ (j) $2(-2)^{n-1}u(-n-1)$

(d) $(\frac{1}{2})^{-n}u(n)$

24. A causal linear shift-invariant system is described by the difference equation

$$y(n) = y(n-1) + y(n-2) + x(n-1)$$

(a) Find the system function $H(z) = Y(z)/X(z)$ for this system. Plot the poles and zeros of $H(z)$ and indicate the region of convergence.

(b) Find the unit-sample response of this system.

(c) You should have found this to be an unstable system. Find a stable (non-causal) unit-sample response that satisfies the difference equation.

25. Consider a linear shift-invariant discrete system with input $x(n)$ and output $y(n)$ for which

$$y(n-1) - \tfrac{5}{2}y(n) + y(n+1) = x(n)$$

The system may or may not be stable and/or causal.

By considering the pole–zero pattern associated with this difference equation, determine three possible choices for the unit-sample response of the system. Convince yourself that each choice satisfies the difference equation.

26. Consider a causal linear shift-invariant system with system function

$$H(z) = \frac{1 - a^{-1}z^{-1}}{1 - az^{-1}}$$

where a is real.

(a) For what range of values of a is the system stable?

(b) If $0 < a < 1$, plot the pole–zero diagram and shade the region of convergence.

(c) Show graphically in the z-plane that this system is an allpass system, i.e., that the magnitude of the frequency response is a constant.

(d) $H(z)$ is to be cascaded with a system $\hat{H}(z)$ so that the overall system function is unity. With $0 < a < 1$ and $\hat{H}(z)$ specified to be a stable system, determine its impulse response $\hat{h}(n)$.

27. Consider a linear discrete-time shift-invariant system with input $x(n)$ and output $y(n)$ for which

$$y(n-1) - \tfrac{10}{3}y(n) + y(n+1) = x(n)$$

The system is stable. Determine the unit-sample response.

28. Given here are four z-transforms. Determine which ones *could* be the transfer function of a discrete linear system which is *not necessarily stable* but for which the unit-sample response is zero for $n < 0$. Clearly state your reasons.

(a) $(1 - z^{-1})^2/(1 - \frac{1}{2}z^{-1})$.

(b) $(z - 1)^2/(z - \frac{1}{2})$.

(c) $(z - \frac{1}{4})^5/(z - \frac{1}{2})^6$.

(d) $(z - \frac{1}{4})^6/(z - \frac{1}{2})^5$.

29. Using z-transforms, determine the response of the discrete linear causal shift-invariant system with difference equation

$$y(n) - 2r\cos\theta y(n-1) + r^2 y(n-2) = x(n)$$

to an excitation

$$x(n) = \alpha^n u(n)$$

30. A sequence $x(n)$ is the output of a linear shift-invariant system whose input is $s(n)$. This system is described by the difference equation

$$x(n) = s(n) - e^{-8\alpha} s(n-8)$$

where $0 < \alpha$.

(a) Find the system function

$$H_1(z) = \frac{X(z)}{S(z)}$$

and plot its poles and zeros in the z-plane. Indicate the region of convergence.

(b) We wish to recover $s(n)$ from $x(n)$ with a linear shift-invariant system. Find the system function

$$H_2(z) = \frac{Y(z)}{X(z)}$$

such that $y(n) = s(n)$. Find all possible regions of convergence for $H_2(z)$ and, for each, tell whether or not the system is causal and stable.

(c) Find all possible choices for the unit-sample response $h_2(n)$ such that

$$y(n) = h_2(n) * x(n) = s(n)$$

31. In Problem 1.32 we discussed the representation of a continuous time function by a sequence corresponding to the coefficients in an expansion in terms of a set of basis functions. It is also useful to represent one sequence by another in a similar manner. Thus we can consider expanding a sequence $f(n)$ in terms of a set of sequences $\phi_k(n)$ as

$$f(n) = \sum_{k=0}^{\infty} g_k \phi_k(n) \qquad \text{(P2.31-1)}$$

The coefficients g_k in this expansion can then be considered as a new sequence which represents $f(n)$. In this problem we shall be concerned with sequences $f(n)$ which are zero for $n < 0$ and a particular choice for the functions $\phi_k(n)$. Specifically, we shall choose the sequences $\phi_k(n)$ such that their z-transforms $\Phi_k(z)$ are given by

$$\Phi_k(z) = \sum_{n=0}^{\infty} \phi_k(n) z^{-n} = \left(\frac{1 - az^{-1}}{z^{-1} - a}\right)^{-k}, \qquad k \geq 0 \qquad \text{(P2.31-2)}$$

with $|a| < 1$ and with the region of convergence for $\Phi_k(z)$ chosen such that $\phi_k(n)$ is zero for $n < 0$. Assume that the coefficient a is real.

(a) Let $F(z)$ denote the z-transform of $f(n)$ and let $G(w)$ denote the "w-transform" of g_k; i.e.,

$$F(z) = \sum_{n=0}^{\infty} f(n) z^{-n}$$

$$G(w) = \sum_{k=0}^{\infty} g_k w^{-k}$$

$G(w)$ and $F(z)$ are related by a substitution of variables; i.e.,

$$F(z) = G[R(z)]$$

Determine $R(z)$ and show that for $z = e^{j\omega}$, w is of the form $w = e^{j\theta}$, where ω and θ are real. It is not necessary to explicitly evaluate θ as a function of ω.

This shows, then, that the Fourier transforms of the original sequence $f(n)$ and the new sequence g_k are related by a mapping of the frequency axis. Since θ as a function of ω is nonlinear, transforming the sequence $f(n)$ to the sequence g_k corresponds to distorting the frequency axis nonlinearly. The remainder of this problem is concerned with how the sequence g_k can be obtained.

(b) It can be shown that the sequences $\phi_k(n)$ corresponding to Eq. (2.31-2) satisfy the relationship

$$\sum_{n=0}^{\infty} n\phi_k(n)\phi_r(n) = \begin{cases} 0, & k \neq r \\ k, & k = r \end{cases}$$

As a consequence of this relationship, g_k for $k > 0$ can be obtained from $f(n)$ according to the relation

$$g_k = \frac{1}{k}\sum_{n=0}^{\infty} n\phi_k(n)f(n) \tag{P2.31-3}$$

Let us assume that $f(n)$ is of finite duration; i.e., $f(n) = 0$, $n < 0$, and $n > (N - 1)$. According to Eq. (P2.31-3), g_k for $k > 0$ can be obtained by filtering $f(-n)$ with a linear shift-invariant digital filter. Determine the unit sample response of the filter in terms of $\phi_k(n)$ and determine the system function (i.e., the z-transform of the unit-sample response) in terms of $\Phi_k(z)$. Also, specify how to obtain g_k from the filter output. There is a different filter for each value of k.

The procedure for obtaining g_k developed in part (b) can only be used for $k > 0$ because of the factor $1/k$ in Eq. (P2.31-3). To determine g_0 we note from Eq. (P2.31-1) that

$$f(0) = \sum_{k=0}^{\infty} g_k\phi_k(0)$$

or

$$g_0 = \frac{1}{\phi_0(0)} f(0) - \sum_{k=1}^{\infty} g_k\phi_k(0) \tag{P2.31-4}$$

(c) (1) Determine $\phi_k(0)$.

(2) As in part (b), we assume that $f(n)$ is of finite duration. Show from Eq. (P2.31-4) that g_0 can be obtained by filtering $f(-n)$ with a linear shift-invariant digital filter, and specify how to obtain g_0 from the filter output. Express the *unit-sample response* of this filter as a summation involving $\phi_k(n)$.

(3) From the summation in part (2), determine *in closed form* the *system function* of the digital filter of part (2). *Hint:* The result is a simple first-order filter.

(*Note:* The results of this problem are discussed in reference [8] of Chapter 1.)

32. As discussed in Sec. 2.5, a necessary and sufficient condition for stability of a two-dimensional system function is that the denominator polynomial cannot be zero if $|z_1|$ and $|z_2|$ are simultaneously greater than or equal to unity.

Consider the class of first-order two-dimensional digital filters for which the transfer function $H(z_1, z_2)$ is expressible in the form

$$H(z_1, z_2) = \frac{1}{1 - az_1^{-1} - bz_2^{-1}}$$

Show that for this class of filters, a necessary and sufficient condition for stability is that

$$|a| + |b| < 1$$

3

The Discrete Fourier Transform

3.0 Introduction

In Chapters 1 and 2 we discussed the representation of sequences and linear shift-invariant systems in terms of the Fourier and z-transforms. For the special case in which the sequence to be represented is of finite duration, i.e., has only a finite number of nonzero values, it is possible to develop an alternative Fourier representation, referred to as the *discrete Fourier transform* (*DFT*). As we shall discuss in this chapter, the DFT is a Fourier representation of a finite-length sequence which is itself a sequence rather than a continuous function, and it corresponds to samples equally spaced in frequency of the Fourier transform of the signal. In addition to its importance theoretically as a Fourier representation of sequences, the DFT plays a central role in the implementation of a variety of digital signal processing algorithms, as a result of the existence of an efficient algorithm for the computation of the DFT [1, 2]. This algorithm will be discussed in detail in Chapter 6.

There are a number of points of view that can be taken toward the derivation and interpretation of the DFT representation of a finite-duration sequence. We have chosen in this chapter to base our presentation on the relationship between finite-length and periodic sequences, so we shall consider first the Fourier series representation of periodic sequences. While the Fourier representation of periodic sequences is important in its own right, we shall also apply those results to the representation of finite-length sequences.

This is done by constructing a periodic sequence for which each period is identical to the finite-length sequence. As we shall see, the Fourier series representation of this periodic sequence corresponds to the DFT of the finite-length sequence.

3.1 Representation of Periodic Sequences— The Discrete Fourier Series

Consider a sequence $\tilde{x}(n)$ that is periodic† with period N so that $\tilde{x}(n) = \tilde{x}(n + kN)$ for any integer value of k. Such a sequence cannot be represented by its z-transform, since there is no value of z for which the z-transform will converge. It is possible, however, to represent $\tilde{x}(n)$ in terms of a Fourier series, that is, by a sum of sine and cosine sequences or equivalently complex exponential sequences with frequencies that are integer multiples of the fundamental frequency $2\pi/N$ associated with the periodic sequence. In contrast to Fourier series for continuous periodic functions, there are only N distinct complex exponentials having a period that is an integer submultiple of the fundamental period N. This is a consequence of the fact that the complex exponential

$$e_k(n) = e^{j(2\pi/N)nk} \tag{3.1}$$

is periodic in k with a period of N. Thus $e_0(n) = e_N(n)$, $e_1(n) = e_{N+1}(n)$, etc., and consequently the set of N complex exponentials in Eq. (3.1) with $k = 0, 1, 2, \ldots, N - 1$ define all the distinct complex exponentials with frequencies that are an integer multiple of $2\pi/N$. Thus the Fourier series representation of a periodic sequence, $\tilde{x}(n)$, need only contain N of these complex exponentials and hence has the form

$$\tilde{x}(n) = \frac{1}{N}\sum_{k=0}^{N-1} \tilde{X}(k)e^{j(2\pi/N)nk} \tag{3.2}$$

The multiplying constant $1/N$ was included for convenience and, of course, has no important effect on the nature of the representation. To obtain the coefficients $\tilde{X}(k)$ from the periodic sequence $\tilde{x}(n)$, we use the fact that

$$\frac{1}{N}\sum_{n=0}^{N-1} e^{j(2\pi/N)nr} = \begin{cases} 1, & \text{for } r = mN,\ m \text{ an integer} \\ 0, & \text{otherwise} \end{cases} \tag{3.3}$$

Therefore, multiplying both sides of Eq. (3.2) by $e^{-j(2\pi/N)nr}$ and summing from $n = 0$ to $N - 1$, we obtain

$$\sum_{n=0}^{N-1} \tilde{x}(n)e^{-j(2\pi/N)nr} = \frac{1}{N}\sum_{n=0}^{N-1}\sum_{k=0}^{N-1} \tilde{X}(k)e^{j(2\pi/N)(k-r)n}$$

† Henceforth we shall use the tilde to denote periodic sequences whenever it is important to clearly distinguish periodic and aperiodic sequences.

Or, interchanging the order of summation on the right-hand side of the equation,

$$\sum_{n=0}^{N-1} \tilde{x}(n)e^{-j(2\pi/N)nr} = \sum_{k=0}^{N-1} \tilde{X}(k)\left[\frac{1}{N}\sum_{n=0}^{N-1} e^{j(2\pi/N)(k-r)n}\right]$$

So that, using Eq. (3.3),

$$\sum_{n=0}^{N-1} \tilde{x}(n)e^{-j(2\pi/N)nr} = \tilde{X}(r)$$

Thus the coefficients $\tilde{X}(k)$ in Eq. (3.2) are obtained by the relation

$$\tilde{X}(k) = \sum_{n=0}^{N-1} \tilde{x}(n)e^{-j(2\pi/N)nk} \tag{3.4}$$

We note that the sequence $\tilde{X}(k)$ as given by Eq. (3.4) is periodic with a period of N; i.e., $\tilde{X}(0) = \tilde{X}(N)$, $\tilde{X}(1) = \tilde{X}(N + 1)$, etc. This is, of course, consistent with the fact that the complex exponentials of Eq. (3.1) are only distinct for $k = 0, 1, \ldots, N - 1$, and thus there can only be N distinct coefficients in the Fourier series representation of a periodic sequence.

The Fourier series coefficients can be interpreted to be a sequence of finite length, given by Eq. (3.4) for $k = 0, \ldots, N - 1$ and zero otherwise, or as a periodic sequence defined for all k by Eq. (3.4). Clearly both of these interpretations are equivalent. It is generally more convenient to interpret the Fourier series coefficients $\tilde{X}(k)$ as a periodic sequence. In this way there is a duality between the time and frequency domains for the Fourier series representation of periodic sequences. Equations (3.2) and (3.4) together can be viewed as a transform pair and will be referred to as the discrete Fourier series (DFS) representation of a periodic sequence. For convenience in notation these equations will generally be written in terms of W_N defined as

$$W_N = e^{-j(2\pi/N)}$$

Thus the DFS analysis and synthesis pair are expressed as

$$\tilde{X}(k) = \sum_{n=0}^{N-1} \tilde{x}(n)W_N^{kn} \tag{3.5}$$

$$\tilde{x}(n) = \frac{1}{N}\sum_{k=0}^{N-1} \tilde{X}(k)W_N^{-kn} \tag{3.6}$$

where both $\tilde{X}(k)$ and $\tilde{x}(n)$ are periodic sequences.

The periodic sequence $\tilde{X}(k)$ has a convenient interpretation as samples on the unit circle, equally spaced in angle, of the z-transform of one period of $\tilde{x}(n)$. To obtain this relationship, let $x(n)$ represent one period of $\tilde{x}(n)$; i.e., $x(n) = \tilde{x}(n)$ for $0 \leq n \leq N - 1$ and $x(n) = 0$, otherwise as shown in Fig. 3.1. Then $X(z)$, the z-transform of $x(n)$, is given by

$$X(z) = \sum_{n=-\infty}^{\infty} x(n)z^{-n}$$

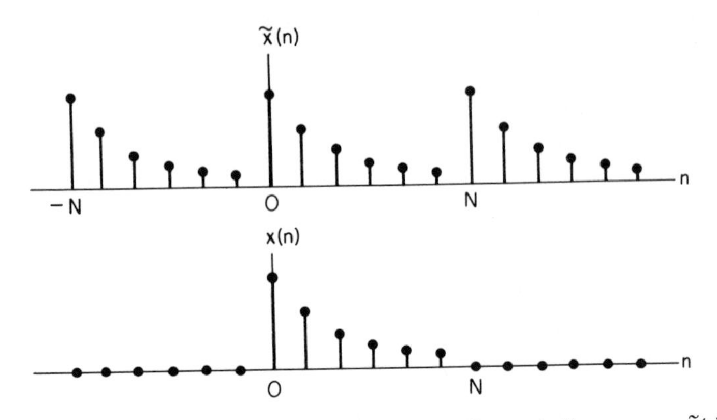

Fig. 3.1 Finite-length sequence $x(n)$ equal to the periodic sequence $\tilde{x}(n)$ over one period and zero otherwise.

or, since $x(n) = 0$ outside the range $0 \le n \le N-1$, $X(z)$ is

$$X(z) = \sum_{n=0}^{N-1} x(n)z^{-n} \tag{3.7}$$

Comparing Eqs. (3.5) and (3.7), we see then that $X(z)$ and $\tilde{X}(k)$ are related by

$$\tilde{X}(k) = X(z)\big|_{z=e^{j(2\pi/N)k}=W_N^{-k}} \tag{3.8}$$

This then corresponds to sampling the z-transform $X(z)$ at N points equally spaced in angle around the unit circle, with the first such sample occurring at $z = 1$. These points on the unit circle are indicated in Fig. 3.2.

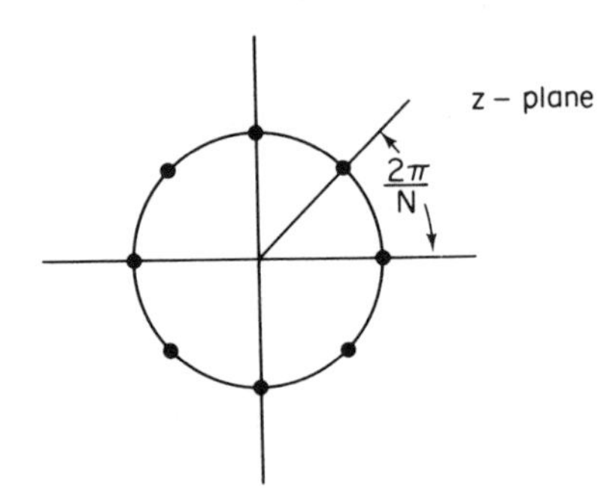

Fig. 3.2 Points in the z-plane at which the z-transform of one period of a periodic sequence equals the Fourier series coefficients.

EXAMPLE. As an illustration of Fourier series representation for a periodic sequence, let us consider the sequence $\tilde{x}(n)$ given in Fig. 3.3. From Eq. (3.5),

$$\tilde{X}(k) = \sum_{n=0}^{4} W_{10}^{nk} = \sum_{n=0}^{4} e^{-j(2\pi/10)nk}$$

$$= e^{-j(4\pi k/10)} \frac{\sin(\pi k/2)}{\sin(\pi k/10)} \tag{3.9}$$

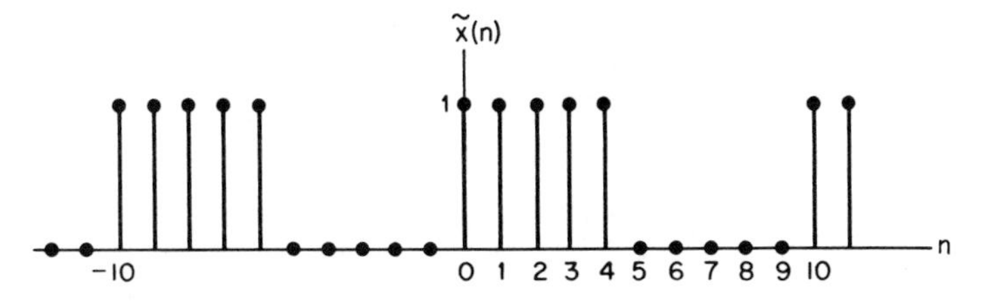

Fig. 3.3 Periodic sequence for which the Fourier series representation is to be computed.

The magnitude and phase of the periodic sequence $\tilde{X}(k)$ given by Eq. (3.9) are sketched in Fig. 3.4. The z-transform evaluated on the unit circle of one period of $\tilde{x}(n)$ is given by

$$X(e^{j\omega}) = e^{-j2\omega} \frac{\sin (5\omega/2)}{\sin (\omega/2)}$$

It is easily verified that Eq. (3.8) is satisfied in this case. The magnitude and phase of $X(e^{j\omega})$ are sketched in Fig. 3.5. It is important to note in particular the fact that the sequences in Fig. 3.4(a) and (b) correspond to samples of Fig. 3.5(a) and (b), respectively.

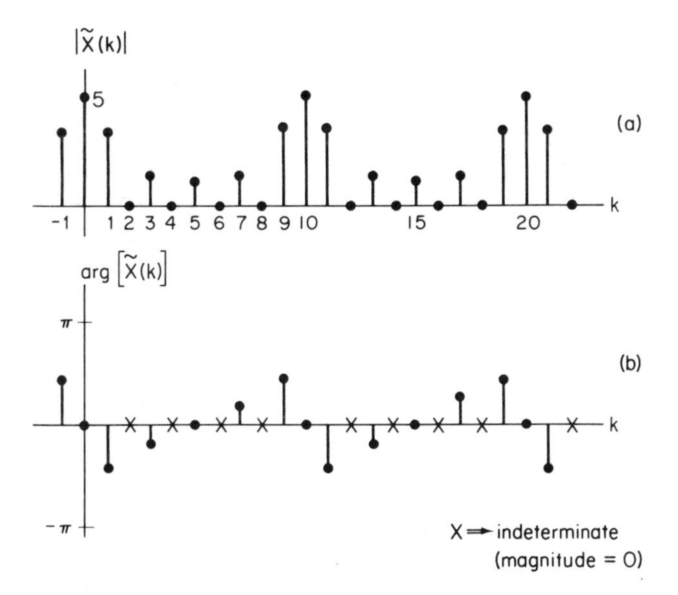

Fig. 3.4 Magnitude and phase of the Fourier series coefficients of the sequence of Fig. 3.3.

3.2 Properties of the Discrete Fourier Series

Just as with the Fourier and Laplace transforms in the case of continuous-time signals, and with the z-transform in the case of discrete-time aperiodic

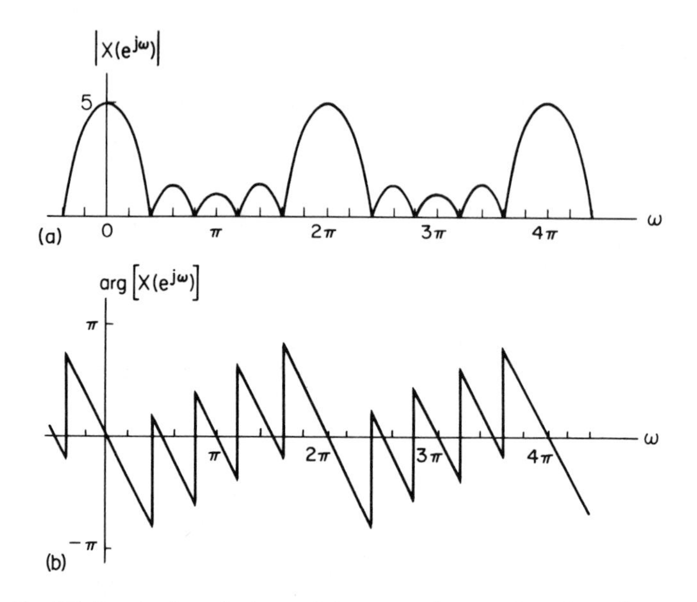

Fig. 3.5 Magnitude and phase of the z-transform evaluated on the unit circle of one period of the sequence in Fig. 3.3.

sequences, there are certain properties of the discrete Fourier series that are of fundamental importance to its successful use in signal processing problems. In this section we summarize these important properties, proofs of which are considered in Problem 2 of this chapter. It is not surprising that we will find that many of the basic properties are analogous to properties of the z-transform. However, as we shall be careful to point out, the periodicity of both $\tilde{x}(n)$ and $\tilde{X}(k)$ results in some important distinctions. Furthermore, there is an exact duality between the time and frequency domains with the DFS representation that does not exist with the z-transform representation of sequences.

3.2.1 *Linearity*

If two periodic sequences $\tilde{x}_1(n)$ and $\tilde{x}_2(n)$, both with periods equal to N, are combined according to

$$\tilde{x}_3(n) = a\tilde{x}_1(n) + b\tilde{x}_2(n)$$

then the coefficients in the DFS representation of $\tilde{x}_3(n)$ are given by

$$\tilde{X}_3(k) = a\tilde{X}_1(k) + b\tilde{X}_2(k) \tag{3.10}$$

where all sequences are periodic with period N.

3.2.2 *Shift of a Sequence*

If a periodic sequence $\tilde{x}(n)$ has Fourier coefficients $\tilde{X}(k)$, then the shifted sequence $\tilde{x}(n + m)$ can easily be shown to have the coefficients $W_N^{-km}\tilde{X}(k)$.

Clearly any shift that is greater than the period (i.e., $m \geq N$) cannot be distinguished in the time domain from a shorter shift, $m' = m$ modulo N.

Because of the fact that the Fourier series coefficients of a periodic sequence are a periodic sequence, a similar result applies to a shift in the Fourier coefficients. Specifically, the values of the periodic sequence $\tilde{X}(k + l)$ are the Fourier coefficients of the sequence $W_N^{nl}\tilde{x}(n)$, where l is an integer.

3.2.3 Symmetry Properties

As with the Fourier transform as discussed in Chapter 1, there are a number of symmetry properties of the DFS representation of a periodic sequence. The derivation of these properties is similar in style to those in Chapter 1 and is left as an exercise (see Problem 2 of this chapter). The resulting properties are summarized below.

For a general complex sequence $\tilde{x}(n)$ with Fourier coefficients $\tilde{X}(k)$, the Fourier coefficients for $\tilde{x}^*(n)$ are $\tilde{X}^*(-k)$ and for $\tilde{x}^*(-n)$ are $\tilde{X}^*(k)$. As a consequence, the DFS for Re $[\tilde{x}(n)]$ is $\tilde{X}_e(k)$, the conjugate symmetric part of $\tilde{X}(k)$ and the DFS for j Im $[\tilde{x}(n)]$ is $\tilde{X}_o(k)$, the conjugate antisymmetric part of $\tilde{X}(k)$. Furthermore, the DFS for $\tilde{x}_e(n)$ is Re $[\tilde{X}(k)]$ and the DFS for $\tilde{x}_o(n)$ is j Im $[\tilde{X}(k)]$. It follows from this that for $\tilde{x}(n)$ real, Re $[\tilde{X}(k)]$ is an even sequence and Im $[\tilde{X}(k)]$ is an odd sequence. Also the magnitude of $\tilde{X}(k)$ is even and the phase is odd. Furthermore, for a real sequence, Re $[\tilde{X}(k)]$ is the DFS for $\tilde{x}_e(n)$ and j Im $[\tilde{X}(k)]$ is the DFS for $\tilde{x}_o(n)$.

3.2.4 Periodic Convolution

Let $\tilde{x}_1(n)$ and $\tilde{x}_2(n)$ be two periodic sequences of period N with the discrete Fourier series denoted by $\tilde{X}_1(k)$ and $\tilde{X}_2(k)$, respectively. We wish to determine the sequence $\tilde{x}_3(n)$ for which the DFS is $\tilde{X}_1(k) \cdot \tilde{X}_2(k)$. To derive this relation, we note that

$$\tilde{X}_1(k) = \sum_{m=0}^{N-1} \tilde{x}_1(m) W_N^{mk}$$

$$\tilde{X}_2(k) = \sum_{r=0}^{N-1} \tilde{x}_2(r) W_N^{rk}$$

so that

$$\tilde{X}_1(k)\tilde{X}_2(k) = \sum_{m=0}^{N-1} \sum_{r=0}^{N-1} \tilde{x}_1(m)\tilde{x}_2(r) W_N^{k(m+r)}$$

Then

$$\tilde{x}_3(n) = \frac{1}{N} \sum_{k=0}^{N-1} W_N^{-nk} \tilde{X}_1(k)\tilde{X}_2(k)$$

$$= \sum_{m=0}^{N-1} \tilde{x}_1(m) \sum_{r=0}^{N-1} \tilde{x}_2(r) \left[\frac{1}{N} \sum_{k=0}^{N-1} W_N^{-k(n-m-r)}\right]$$

Let us consider $\tilde{x}_3(n)$ for $0 \leq n \leq N - 1$. Then we observe that

$$\frac{1}{N}\sum_{k=0}^{N-1} W_N^{-k(n-m-r)} = \begin{cases} 1, & \text{for } r = (n-m) + lN \\ 0, & \text{otherwise} \end{cases}$$

where l is any integer. This results in

$$\tilde{x}_3(n) = \sum_{m=0}^{N-1} \tilde{x}_1(m)\tilde{x}_2(n-m) \tag{3.11a}$$

Equation (3.11a) states that $\tilde{x}_3(n)$ is obtained by combining $\tilde{x}_1(n)$ and $\tilde{x}_2(n)$ in a manner reminiscent of a convolution. It is important to note, however, that in contrast to the convolution of aperiodic sequences, the sequences $\tilde{x}_1(m)$ and $\tilde{x}_2(n - m)$ in Eq. (3.11a) are periodic in m with period N and consequently so is their product. Also, the summation is carried out only over one period. This type of convolution is commonly referred to as a periodic convolution. By a simple change of summation index in Eq. (3.11a)

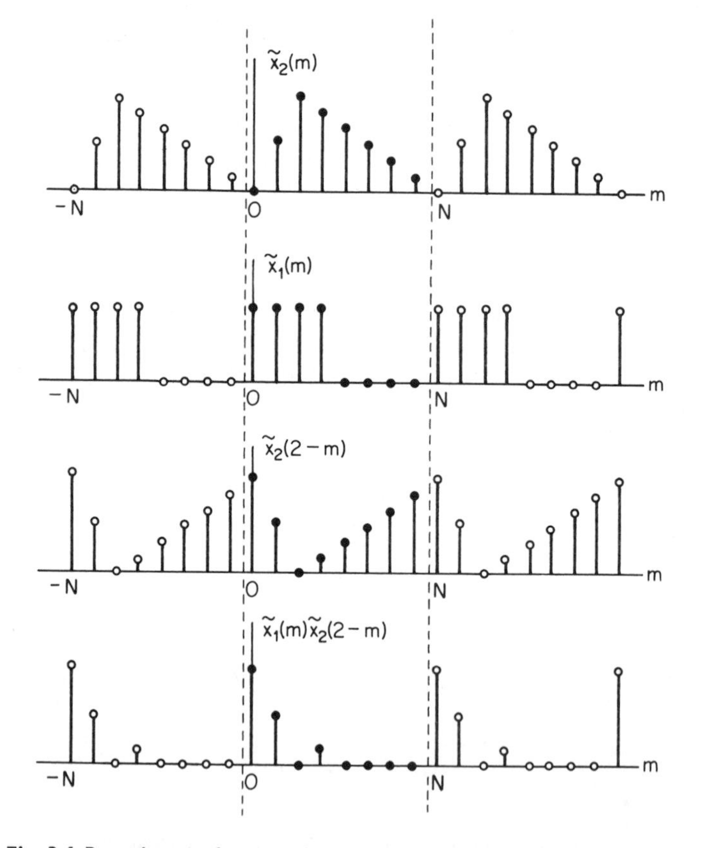

Fig. 3.6 Procedure in forming the periodic convolution of two periodic sequences.

we can show that

$$\tilde{x}_3(n) = \sum_{m=0}^{N-1} \tilde{x}_2(m)\tilde{x}_1(n-m) \tag{3.11b}$$

An illustration of the procedure in forming the periodic convolution of two periodic sequences corresponding to Eq. (3.11a) is given in Fig. 3.6. With this type of convolution, as one period slides out of the interval over which it is evaluated, the next period slides in.

If the roles of time and frequency are interchanged, we obtain a result almost identical to the previous result. That is, the periodic sequence

$$\tilde{x}_3(n) = \tilde{x}_1(n)\tilde{x}_2(n)$$

where $\tilde{x}_1(n)$ and $\tilde{x}_2(n)$ are periodic sequences each with period N, has the Fourier coefficients given by

$$\tilde{X}_3(k) = \frac{1}{N}\sum_{l=0}^{N-1} \tilde{X}_1(l)\tilde{X}_2(k-l) \tag{3.12}$$

corresponding to $1/N$ times the periodic convolution of $\tilde{X}_1(k)$ and $\tilde{X}_2(k)$.

3.3 Summary of Properties of the DFS Representation of Periodic Sequences

As with the Fourier transform and z-transform it is often useful to refer to the properties of the DFS discussed in Sec. 3.2. These properties are summarized in Table 3.1.

Table 3.1

Periodic Sequence (period N)	*DFS Coefficients*
1. $\tilde{x}(n)$	$\tilde{X}(k)$ periodic with period N
2. $\tilde{y}(n)$	$\tilde{Y}(k)$ periodic with period N
3. $a\tilde{x}(n) + b\tilde{y}(n)$	$a\tilde{X}(k) + b\tilde{Y}(k)$
4. $\tilde{x}(n+m)$	$W_N^{-km}\tilde{X}(k)$
5. $W_N^{ln}\tilde{x}(n)$	$\tilde{X}(k+l)$
6. $\sum_{m=0}^{N-1} \tilde{x}(m)\tilde{y}(n-m)$ (periodic convolution)	$\tilde{X}(k)\tilde{Y}(k)$
7. $\tilde{x}(n)\tilde{y}(n)$	$\frac{1}{N}\sum_{l=0}^{N-1} \tilde{X}(l)\tilde{Y}(k-l)$
8. $\tilde{x}^*(n)$	$\tilde{X}^*(-k)$
9. $\tilde{x}^*(-n)$	$\tilde{X}^*(k)$
10. Re $[\tilde{x}(n)]$	$\tilde{X}_e(k)$ [conjugate symmetric part of $\tilde{X}(k)$]
11. j Im $[\tilde{x}(n)]$	$\tilde{X}_o(k)$ [conjugate antisymmetric part of $\tilde{X}(k)$]
12. $\tilde{x}_e(n)$ [conjugate symmetric part of $\tilde{x}(n)$]	Re $[\tilde{X}(k)]$
13. $\tilde{x}_o(n)$ [conjugate antisymmetric part of $\tilde{x}(n)$]	j Im $[\tilde{X}(k)]$

The following properties apply only when $\tilde{x}(n)$ is real:

14. Any real $\tilde{x}(n)$ $\quad \begin{cases} \tilde{X}(k) = \tilde{X}^*(-k) \\ \text{Re}\,[\tilde{X}(k)] = \text{Re}\,[\tilde{X}(-k)] \\ \text{Im}\,[\tilde{X}(k)] = -\text{Im}\,[\tilde{X}(-k)] \\ |\tilde{X}(k)| = |\tilde{X}(-k)| \\ \arg\,[\tilde{X}(k)] = -\arg\,[\tilde{X}(-k)] \end{cases}$

15. $\tilde{x}_e(n)$ $\quad \text{Re}\,[\tilde{X}(k)]$

16. $\tilde{x}_o(n)$ $\quad j\,\text{Im}\,[\tilde{X}(k)]$

3.4 Sampling the z-Transform

We have seen in Sec. 3.1 that the values $\tilde{X}(k)$ in the DFS representation of a periodic sequence are identical to samples of the z-transform of a single period of $\tilde{x}(n)$ at N equally spaced points on the unit circle. In this section we consider more generally the relation between an aperiodic sequence with z-transform $X(z)$ and the periodic sequence for which the DFS coefficients correspond to samples of $X(z)$ equally spaced in angle around the unit circle. Toward this end, consider an aperiodic sequence $x(n)$ whose z-transform

$$X(z) = \sum_{n=-\infty}^{\infty} x(n)z^{-n} \tag{3.13}$$

has a region of convergence that includes the unit circle, a condition that is always met for finite-length sequences. If we evaluate the z-transform at N equally spaced points around the unit circle as shown in Fig. 3.7, we obtain the periodic sequence

$$\tilde{X}(k) = X(z)\big|_{z=W_N^{-k}} = \sum_{n=-\infty}^{\infty} x(n)W_N^{kn} \tag{3.14}$$

where $W_N = e^{-j(2\pi/N)}$.

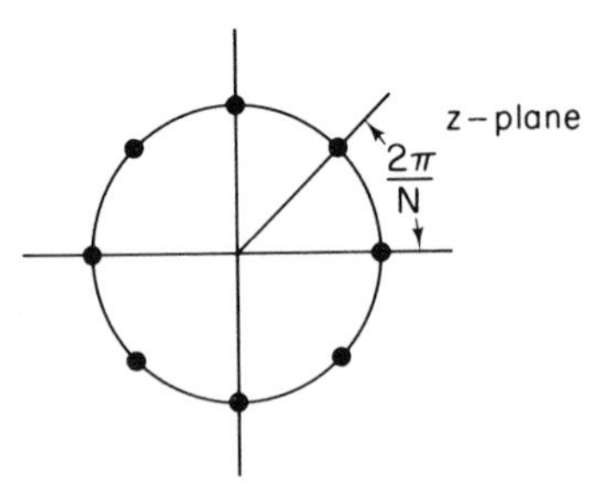

Fig. 3.7 Points on the unit circle at which $X(z)$ is sampled to obtain a periodic sequence $\tilde{X}(k)$.

We have just seen that there is a unique relationship between a periodic sequence $\tilde{X}(k)$ and a periodic sequence $\tilde{x}(n)$ given by

$$\tilde{x}(n) = \frac{1}{N}\sum_{k=0}^{N-1} \tilde{X}(k)W_N^{-kn} \tag{3.15}$$

To inquire as to the relationship between the periodic sequence $\tilde{x}(n)$ and the original sequence $x(n)$, let us substitute the values of $\tilde{X}(k)$ from Eq. (3.14) into Eq. (3.15), obtaining

$$\tilde{x}(n) = \frac{1}{N}\sum_{k=0}^{N-1}\sum_{m=-\infty}^{\infty} x(m)W_N^{km}W_N^{-kn}$$

Interchanging the order of summation yields

$$\tilde{x}(n) = \sum_{m=-\infty}^{\infty} x(m)\left[\frac{1}{N}\sum_{k=0}^{N-1} W_N^{-k(n-m)}\right]$$

Using Eq. (3.3), it follows that

$$\frac{1}{N}\sum_{k=0}^{N-1} W_N^{-k(n-m)} = 1 \qquad \text{for} \qquad m = n + rN$$

and zero otherwise, so that

$$\tilde{x}(n) = \sum_{r=-\infty}^{\infty} x(n + rN) \tag{3.16}$$

Thus, the resulting periodic sequence is formed from the aperiodic sequence overlapping successive repetitions of the aperiodic sequence. This is reminiscent of the relationship discussed in Sec. 1.7 between the Fourier transform of a continuous-time signal and the Fourier transform of the discrete-time signal obtained by periodic sampling as evidenced by the similarity between Eqs. (1.29) and (3.16). From Eq. (3.16) we see that if the aperiodic sequence $x(n)$ is of finite duration less than N, then each period of $\tilde{x}(n)$ will be a replica of $x(n)$; if it is of duration greater than N, there will be an overlap of nonzero samples, resulting in aliasing similar to that discussed in Sec. 1.7. As a result, it follows that if $x(n)$ is of duration less than N, it can be recovered exactly from $\tilde{x}(n)$ by simply extracting one period of $\tilde{x}(n)$. Equivalently, then, a finite-duration sequence of length N (or less) can be represented exactly by N samples of its z-transform on the unit circle. Since the original sequence $x(n)$ can be recovered from the N samples of $X(z)$ on the unit circle, it is clear that $X(z)$ can be recovered from these same N samples. If $x(n)$ is zero for $n \geq N$, then

$$X(z) = \sum_{n=0}^{N-1} x(n)z^{-n} \tag{3.17}$$

Since $x(n) = \tilde{x}(n)$ for $0 \leq n \leq N - 1$, we can substitute Eq. (3.15) into Eq. (3.17), obtaining

$$X(z) = \sum_{n=0}^{N-1} \frac{1}{N}\sum_{k=0}^{N-1} \tilde{X}(k)W_N^{-kn}z^{-n}$$

Interchanging the order of summation we obtain

$$X(z) = \frac{1}{N}\sum_{k=0}^{N-1} \tilde{X}(k)\left[\sum_{n=0}^{N-1} (W_N^{-k} z^{-1})^n\right]$$

which can be written as

$$\begin{aligned} X(z) &= \frac{1}{N}\sum_{k=0}^{N-1} \tilde{X}(k)\frac{1 - z^{-N}}{1 - W_N^{-k} z^{-1}} \\ &= \frac{1 - z^{-N}}{N}\sum_{k=0}^{N-1} \frac{\tilde{X}(k)}{1 - W_N^{-k} z^{-1}} \end{aligned} \tag{3.18}$$

This equation expresses $X(z)$, the z-transform of a finite-duration sequence of length N, in terms of N "frequency samples" of $X(z)$ on the unit circle. As will be shown in a later chapter, this expression is the basis for one possible realization of a system having a finite-duration unit-sample response. By substituting $z = e^{j\omega}$ we can show that Eq. (3.18) becomes

$$X(e^{j\omega}) = \sum_{k=0}^{N-1} \tilde{X}(k)\Phi\left(\omega - \frac{2\pi}{N}k\right) \tag{3.19}$$

where

$$\Phi(\omega) = \frac{\sin(\omega N/2)}{N\sin(\omega/2)} e^{-j\omega[(N-1)/2]} \tag{3.20}$$

The function $\sin(\omega N/2)/[N\sin(\omega/2)]$ is plotted in Fig. 3.8 for $N = 5$. Note that the function $\Phi(\omega)$ has the property

$$\Phi\left(\frac{2\pi}{N}k\right) = \begin{cases} 0, & k = 1, 2, \ldots, N-1 \\ 1, & k = 0 \end{cases}$$

so that

$$X(e^{j\omega})\big|_{\omega=(2\pi/N)k} = \tilde{X}(k), \qquad k = 0, 1, \ldots, N-1 \tag{3.21}$$

i.e., the interpolation is exact at the original sample points, as expected.

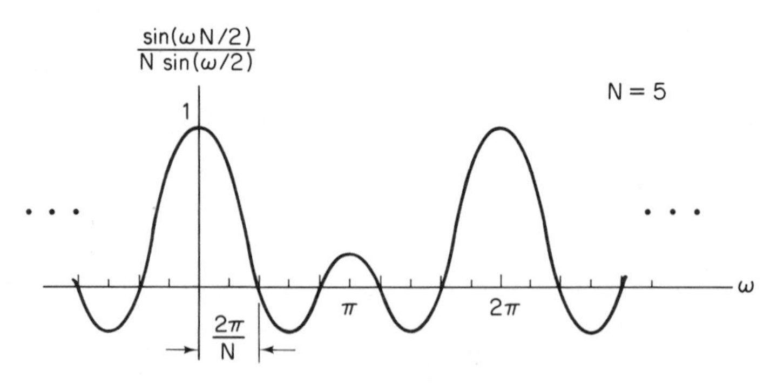

Fig. 3.8 Function $\sin(\omega N/2)/[N\sin(\omega/2)]$, as defined by Eq. (3.20) for $N = 5$.

3.5 Fourier Representation of Finite-Duration Sequences— The Discrete Fourier Transform

In the previous section we considered the representation of periodic sequences in terms of the discrete Fourier series. With the correct interpretation, the same representation can be applied to finite-duration sequences. The resulting Fourier representation for finite-duration sequences will be referred to as the discrete Fourier transform (DFT).

The results of the previous section suggest two points of view toward the Fourier representation of finite-duration sequences. Specifically, we can represent a finite-duration sequence of length N by a periodic sequence with period N, one period of which is identical to the finite-duration sequence.† In the sense that the periodic sequence has a unique DFS representation, so does the original finite-duration sequence, since we can compute a single period of the periodic sequence, and thus the finite-length sequence, from the DFS.

An alternative point of view is suggested by the previous section, where we showed that a finite-duration sequence can be exactly represented by samples of its z-transform. Specifically we showed that the periodic sequence obtained by sampling the z-transform at N equally spaced points on the unit circle is identical to the discrete Fourier series coefficients of the periodic sequence formed as above. The sequence corresponding to these samples of the z-transform was shown to be a periodically repeated version of the original sequence such that when N samples of the z-transform are used, no "aliasing" occurs. Thus both these viewpoints lead to the representation of a finite-duration sequence as one period of a periodic sequence.

Let us consider a finite-duration sequence $x(n)$ of length N so that $x(n) = 0$ except in the range $0 \leq n \leq (N - 1)$. Clearly a sequence of length M less than N can also be considered to be of length N with the last $(N - M)$ points in the interval having amplitude zero, and sometimes it will be convenient to do this. The corresponding periodic sequence of period N, for which $x(n)$ is one period, will be denoted by $\tilde{x}(n)$ and is given by

$$\tilde{x}(n) = \sum_{r=-\infty}^{\infty} x(n + rN) \tag{3.22a}$$

Since $x(n)$ is of finite length N there is no overlap between the terms $x(n + rN)$ for different values of r. Thus Eq. (3.22a) can alternatively be written as‡

$$\tilde{x}(n) = x(n \text{ modulo } N) \tag{3.22b}$$

† For simplicity we generally assume that the nonzero region is $0 \leq n \leq N - 1$; however, this is arbitrary and results can be derived that accommodate any interval of N samples.

‡ With n expressed as

$$n = n_1 + n_2 N$$

where $0 \leq n_1 \leq N - 1$, n modulo N is equal to n_1.

For convenience we shall use the notation $((n))_N$ to denote (n modulo N) and with this notation Eq. (3.22b) is expressed as

$$\tilde{x}(n) = x((n))_N \tag{3.23a}$$

The finite-duration sequence $x(n)$ is obtained from $\tilde{x}(n)$ by extracting one period; i.e.,

$$x(n) = \begin{cases} \tilde{x}(n), & 0 \le n \le N-1 \\ 0, & \text{otherwise} \end{cases}$$

Again for notational convenience it is useful to define the rectangular sequence $\mathcal{R}_N(n)$ given by

$$\mathcal{R}_N(n) = \begin{cases} 1, & 0 \le n \le N-1 \\ 0, & \text{otherwise} \end{cases}$$

With this notation the above equation can be expressed as

$$x(n) = \tilde{x}(n)\mathcal{R}_N(n) \tag{3.23b}$$

As defined in Sec. 3.1, the discrete Fourier series coefficients $\tilde{X}(k)$ of the periodic sequence $\tilde{x}(n)$ are themselves a periodic sequence with period N. To maintain a duality between the time and frequency domains, we shall choose the Fourier coefficients that we associate with a finite-duration sequence to be a finite-duration sequence corresponding to one period of $\tilde{X}(k)$. Thus with $X(k)$ denoting the Fourier coefficients that we associate with $x(n)$, $X(k)$ and $\tilde{X}(k)$ are related by

$$\tilde{X}(k) = X((k))_N \tag{3.24a}$$

$$X(k) = \tilde{X}(k)\mathcal{R}_N(k) \tag{3.24b}$$

From Sec. 3.1 $\tilde{X}(k)$ and $\tilde{x}(n)$ are related by

$$\tilde{X}(k) = \sum_{n=0}^{N-1} \tilde{x}(n)W_N^{kn} \tag{3.25a}$$

$$\tilde{x}(n) = \frac{1}{N}\sum_{k=0}^{N-1} \tilde{X}(k)W_N^{-kn} \tag{3.25b}$$

Since the summations in Eqs. (3.25a) and (3.25b) involve only the interval between 0 and $(N-1)$, it follows from Eqs. (3.23), (3.24), and (3.25) that

$$X(k) = \begin{cases} \displaystyle\sum_{n=0}^{N-1} x(n)W_N^{kn}, & 0 \le k \le N-1 \\ 0, & \text{otherwise} \end{cases} \tag{3.26a}$$

$$x(n) = \begin{cases} \displaystyle\frac{1}{N}\sum_{k=0}^{N-1} X(k)W_N^{-kn}, & 0 \le n \le N-1 \\ 0, & \text{otherwise} \end{cases} \tag{3.26b}$$

The transform pair given in equations (3.26) will be referred to as the discrete Fourier transform (DFT), with Eq. (3.26a) representing the analysis transform and (3.26b) the synthesis transform. We note that on the basis of the discussion in Sec. 3.4, the DFT of a finite-duration sequence corresponds to equally spaced samples on the unit circle of its z-transform. It must be emphasized that just as the distinction between a finite-duration sequence of length N and a periodic sequence of period N is minor in the sense that both are defined by N values, the distinction between Eqs. (3.25) and (3.26) is relatively minor. However, as we will see in the next section, it is always important to remember that where DFT relations are concerned, a finite length sequence is represented as one period of a periodic sequence.

3.6 Properties of the Discrete Fourier Transform

In this section we consider a number of properties of the DFT for finite-duration sequences. These properties are essentially similar to those presented in Sec. 3.2 for periodic sequences and result from the *implied* periodicity in the DFT representation of finite-duration sequences. Our purpose in this section is to reconsider these properties from the point of view of a finite length sequence defined only in the interval $0 \leq n \leq N-1$.

3.6.1 *Linearity*

If two finite-duration sequences $x_1(n)$ and $x_2(n)$ are linearly combined, as in

$$x_3(n) = ax_1(n) + bx_2(n)$$

then the DFT of $x_3(n)$ is

$$X_3(k) = aX_1(k) + bX_2(k)$$

Clearly if $x_1(n)$ has duration N_1 and $x_2(n)$ has duration N_2, then the maximum duration of $x_3(n)$ will be $N_3 = \max[N_1, N_2]$. Thus in general the DFTs must be computed with $N = N_3$. If, for example, $N_1 < N_2$, then $X_1(k)$ is the DFT of the sequence $x_1(n)$ augmented by $N_2 - N_1$ zeros. That is,

$$X_1(k) = \sum_{n=0}^{N_1-1} x_1(n)W_{N_2}^{kn}, \qquad 0 \leq k \leq N_2 - 1$$

$$X_2(k) = \sum_{n=0}^{N_2-1} x_2(n)W_{N_2}^{kn}, \qquad 0 \leq k \leq N_2 - 1$$

3.6.2 *Circular Shift of a Sequence*

Consider a sequence $x(n)$ as shown in Fig. 3.9(a), its periodic counterpart $\tilde{x}(n)$ as shown in Fig. 3.9(b), and $\tilde{x}(n+m)$, the result of shifting $\tilde{x}(n)$ by m samples as indicated in Fig. 3.9(c). The finite-duration sequence, which we shall denote by $x_1(n)$, obtained by extracting one period of $\tilde{x}(n+m)$ in the

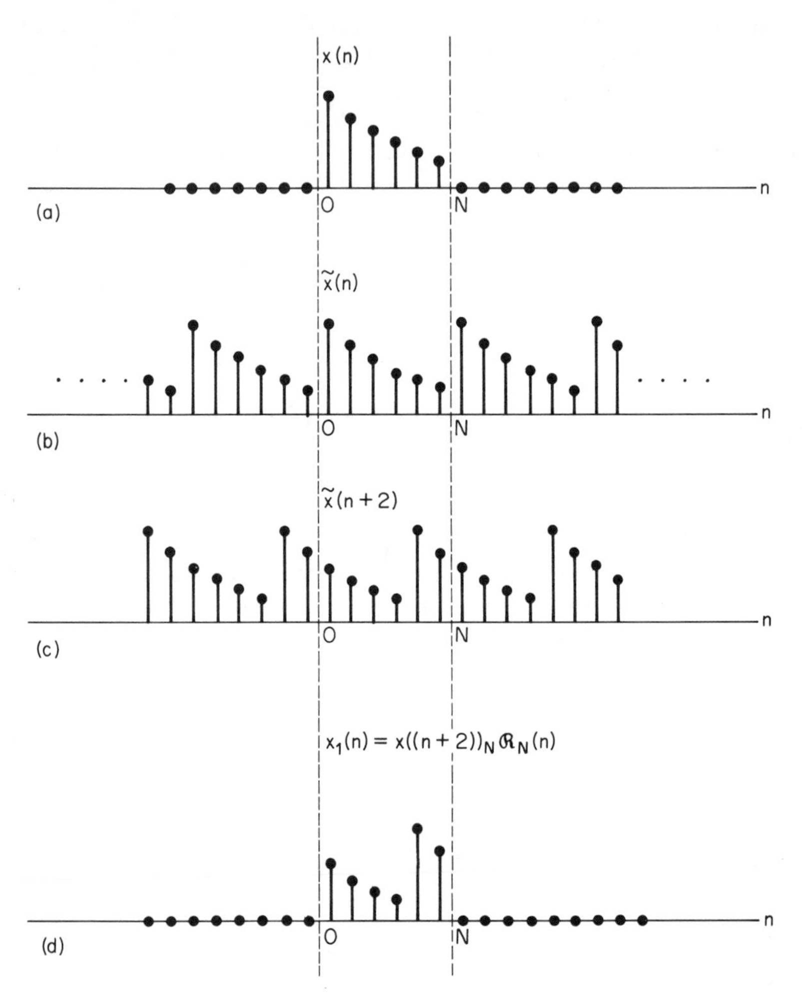

Fig. 3.9 Circular shift of a sequence.

range $0 \leq n \leq N - 1$, is shown in Fig. 3.9(d). Comparison of Fig. 3.9(a) and (d) indicates clearly that $x_1(n)$ does *not* correspond to a linear shift of $x(n)$, and in fact both sequences are confined to the interval between 0 and $(N - 1)$. By reference to Fig. 3.9(b) and (c) we see that in shifting the periodic sequence and examining the interval between 0 and $(N - 1)$, as a sample leaves the interval an identical sample enters the interval at the other end. Thus we can imagine forming $x_1(n)$ by shifting $x(n)$ in such a way that as a sample leaves the interval 0 to $(N - 1)$ at one end it enters at the other end.

A useful interpretation of such a shift is to imagine the finite-duration sequence $x(n)$ displayed around the circumference of a cylinder in such a way that the cylinder has a circumference of exactly N points. As we travel around the circumference of the cylinder a number of times, the sequence that we see will, in fact, be the periodic sequence $\tilde{x}(n)$. A linear shift of the periodic sequence $\tilde{x}(n)$ corresponds then to a *rotation* of the cylinder. Such a shift of a sequence is generally referred to as a circular shift. More formally we can use Eqs. (3.23a) and (3.23b) to express the circular shift. Specifically,

$$\tilde{x}_1(n) = \tilde{x}(n + m) = x((n + m))_N$$

Thus, from Eq. (3.23b),

$$x_1(n) = x((n + m))_N \mathcal{R}_N(n)$$

We wish now to relate the DFT of $x(n)$ and the DFT of $x_1(n)$. Referring to Sec. 3.2.2 we recall that with $\tilde{X}(k)$ and $\tilde{X}_1(k)$ denoting the DFS of the periodic sequences $\tilde{x}(n)$ and $\tilde{x}_1(n) = \tilde{x}(n + m)$,

$$\tilde{X}_1(k) = W_N^{-km} \tilde{X}(k) \tag{3.27}$$

Consequently, from Eq. (3.24b) it follows that

$$X_1(k) = W_N^{-km} X(k) \tag{3.28}$$

Because of the duality between the time and frequency domains, a similar result holds when a circular shift is applied to the DFT coefficients. Specifically, with $X(k)$ and $X_1(k)$ denoting the DFT of $x(n)$ and $x_1(n)$, respectively, if

$$X_1(k) = X((k + l))_N \mathcal{R}_N(k) \tag{3.29}$$

then

$$x_1(n) = W_N^{ln} x(n) \tag{3.30}$$

3.6.3 Symmetry Properties

In Chapter 1 we discussed the decomposition of an arbitrary sequence into the sum of a conjugate symmetric and conjugate antisymmetric component and presented a set of symmetry relations for the Fourier transform. In the consideration of the symmetry properties of the DFT for finite-duration sequences we cannot in general use the definitions of the conjugate symmetric and conjugate antisymmetric components as given in Sec. 1.6 since, for a sequence $x(n)$ of duration N, the conjugate symmetric component $x_e(n)$ and the conjugate antisymmetric component $x_o(n)$ are both of duration $(2N - 1)$. We note, however, that for a periodic sequence $\tilde{x}(n)$, with period N, the conjugate symmetric and antisymmetric components are still periodic with period N. This suggests the decomposition of $x(n)$ into the two finite-duration sequences of duration N corresponding to one period of the conjugate symmetric and conjugate antisymmetric components of $\tilde{x}(n)$. We shall denote these components of $x(n)$ as $x_{ep}(n)$ and $x_{op}(n)$. Thus with

$$\tilde{x}(n) = x((n))_N \tag{3.31}$$

and

$$\tilde{x}_e(n) = \tfrac{1}{2}[\tilde{x}(n) + \tilde{x}^*(-n)] \tag{3.32}$$

and

$$\tilde{x}_o(n) = \tfrac{1}{2}[\tilde{x}(n) - \tilde{x}^*(-n)] \tag{3.33}$$

we define $x_{ep}(n)$ and $x_{op}(n)$ as

$$x_{ep}(n) = \tilde{x}_e(n)\mathcal{R}_N(n) \tag{3.34a}$$

$$x_{op}(n) = \tilde{x}_o(n)\mathcal{R}_N(n) \tag{3.34b}$$

or, equivalently,

$$x_{ep}(n) = \tfrac{1}{2}[x((n))_N + x^*((-n))_N]\mathcal{R}_N(n) \tag{3.35a}$$

$$x_{op}(n) = \tfrac{1}{2}[x((n))_N - x^*((-n))_N]\mathcal{R}_N(n) \tag{3.35b}$$

Clearly $x_{ep}(n)$ and $x_{op}(n)$ are not equivalent to $x_e(n)$ and $x_o(n)$ as defined by Eq. (1.22). However, it can be shown that (see Problem 17 of this chapter)

$$x_{ep}(n) = [x_e(n) + x_e(n - N)]\mathcal{R}_N(n) \tag{3.36a}$$

and

$$x_{op}(n) = [x_o(n) + x_o(n - N)]\mathcal{R}_N(n) \tag{3.36b}$$

In other words, $x_{ep}(n)$ and $x_{op}(n)$ can be generated by aliasing $x_e(n)$ and $x_o(n)$ into the interval $0 \le n \le N - 1$. The sequences $x_{ep}(n)$ and $x_{op}(n)$ will be referred to as the *periodic conjugate symmetric* and *periodic conjugate anti-symmetric components* of $x(n)$. When $x_{ep}(n)$ and $x_{op}(n)$ are real, they will be referred to as the periodic even and periodic odd components, respectively. This choice of terminology is somewhat misleading since the sequences $x_{ep}(n)$ and $x_{op}(n)$ are *not* periodic sequences but represent one period of the periodic sequences $\tilde{x}_e(n)$ and $\tilde{x}_o(n)$.

Equations (3.35a) and (3.35b) define $x_{ep}(n)$ and $x_{op}(n)$ in terms of $x(n)$. The inverse relation, expressing $x(n)$ in terms of $x_{ep}(n)$ and $x_{op}(n)$, can be obtained by using Eqs. (3.32) and (3.33) to write that

$$\tilde{x}(n) = \tilde{x}_e(n) + \tilde{x}_o(n)$$

Thus

$$x(n) = \tilde{x}(n)\mathcal{R}_N(n) = [\tilde{x}_e(n) + \tilde{x}_o(n)]\mathcal{R}_N(n) = \tilde{x}_e(n)\mathcal{R}_N(n) + \tilde{x}_o(n)\mathcal{R}_N(n) \tag{3.37}$$

Combining Eqs. (3.34) and (3.37),

$$x(n) = x_{ep}(n) + x_{op}(n) \tag{3.38}$$

The symmetry properties of the DFT now follow in a straightforward way by applying the results discussed in Sec. 3.2.3. Thus we consider a finite-duration sequence $x(n)$ of length N for which the DFT is $X(k)$. Then the

DFT of $x^*(n)$ is $X^*((-k))_N\mathcal{R}_N(k)$ and the DFT of $x^*((-n))_N\mathcal{R}_N(n)$ is $X^*(k)$. The DFT for Re $[x(n)]$ is $X_{ep}(k)$ and the DFT for j Im $[x(n)]$ is $X_{op}(k)$. Similarly, the DFT for $x_{ep}(n)$ is Re $[X(k)]$ and the DFT for $x_{op}(n)$ is j Im $[X(k)]$. It follows, then, that for $x(n)$ real, Re $[X(k)]$ and $|X(k)|$ are periodic even sequences and Im $[X(k)]$ and arg $[X(k)]$ are periodic odd sequences. Also for a real sequence Re $[X(k)]$ is the DFT for $x_{ep}(n)$ and j Im $[X(k)]$ is the DFT for $x_{op}(n)$.

3.6.4 Circular Convolution

In Sec. 3.2.4 we found that multiplication of the DFS coefficients of two sequences corresponds to a periodic convolution of the sequences. Here we consider the finite-duration sequences $x_1(n)$ and $x_2(n)$, both of duration N, with DFTs $X_1(k)$ and $X_2(k)$, and we wish to determine the sequence $x_3(n)$ for which the DFT coefficients are $X_1(k)X_2(k)$. To determine $x_3(n)$ we can simply apply the results of Sec. 3.2.4. Specifically, $x_3(n)$ corresponds to one period of $\tilde{x}_3(n)$, which is given by Eq. (3.11). Thus

$$\begin{aligned} x_3(n) &= \left[\sum_{m=0}^{N-1} \tilde{x}_1(m)\tilde{x}_2(n-m)\right]\mathcal{R}_N(n) \\ &= \left[\sum_{m=0}^{N-1} x_1((m))_N x_2((n-m))_N\right]\mathcal{R}_N(n) \end{aligned} \tag{3.39}$$

Equation (3.39) differs from a linear convolution of $x_1(n)$ and $x_2(n)$ as defined by Eq. (1.7) in some important respects. For a *linear convolution*, the basic operation involves multiplying $x_1(n)$ by a reversed and linearly shifted version of $x_2(n)$, and then summing the values in the product. To obtain successive values of the sequence representing the convolution, the two sequences are successively shifted relative to each other. In contrast, for the convolution as given by Eq. (3.39) we can imagine displaying one of the sequences around the circumference of a cylinder with a circumference of exactly N points. The second sequence is time reversed and also displayed on the circumference of a cylinder with a circumference of N points. If we imagine placing one cylinder inside the other, then successive values in the convolution can be obtained by multiplying the values on one cylinder by the corresponding values on the second cylinder and forming the sum of the resulting N products. To generate successive values in the convolution, one cylinder is rotated relative to the other. A little thought should make it clear that this is exactly equivalent to first constructing two periodic sequences and then implementing the convolution as described by Eq. (3.11) and illustrated in Fig. 3.6. With this kind of interpretation of the convolution, such a convolution is often referred to as a circular convolution. The N-point circular convolution of two sequences $x_1(n)$ and $x_2(n)$ will be represented notationally by $x_1(n) \circledN x_2(n)$.

EXAMPLE. A simple example of circular convolution is provided by the result of Sec. 3.6.2. Let $x_2(n)$ be a finite-duration sequence of length N, and

$$x_1(n) = \delta(n - n_0)$$

where $n_0 < N$. Clearly $x_1(n)$ can be considered as the finite-duration sequence

$$x_1(n) = \begin{cases} 0, & 0 \leq n < n_0 \\ 1, & n = n_0 \\ 0, & n_0 < n \leq N - 1 \end{cases}$$

The DFT of $x_1(n)$ is

$$X_1(k) = W_N^{kn_0}$$

If we form the product

$$X_3(k) = W_N^{kn_0} X_2(k)$$

we see from Sec. 3.6.2 that the finite-duration sequence corresponding to $X_3(k)$ is the sequence $x_2(n)$ rotated to the right in the interval $0 \leq n \leq N - 1$ by n_0

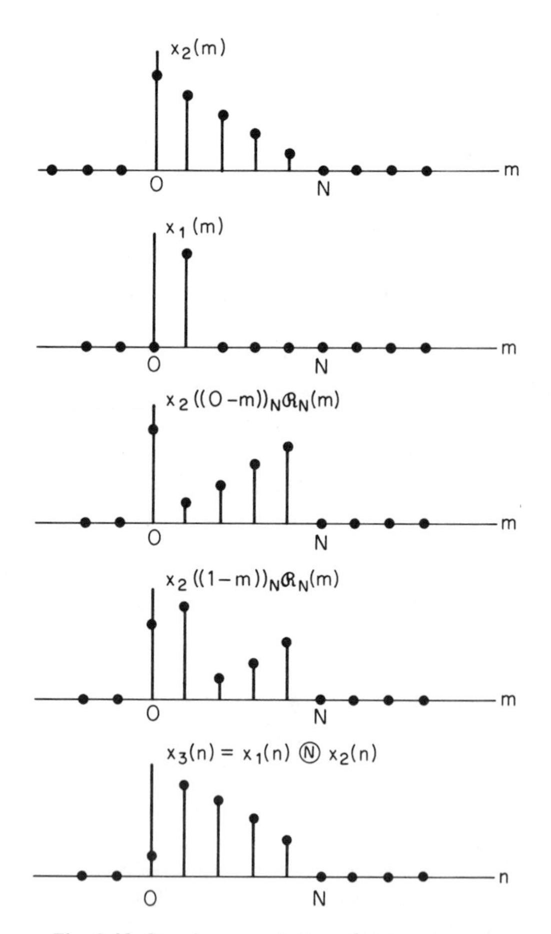

Fig. 3.10 Circular convolution of two sequences.

samples. That is, the circular convolution of a sequence $x_2(n)$ with a single delayed unit sample results in a rotation of the sequence $x_2(n)$ in the interval $0 \le n \le N - 1$. This example is illustrated in Fig. 3.10 for $N = 5$ and $n_0 = 1$. Here we have shown the sequences $x_2(m)$ and $x_1(m)$ and then $x_2((0 - m))_N$ and $x_2((1 - m))_N$. The last sequence shown is the result of the circular convolution of $x_1(n)$ and $x_2(n)$.

EXAMPLE. As another example of a circular convolution, let

$$x_1(n) = x_2(n) = \begin{cases} 1, & 0 \le n \le N - 1 \\ 0, & \text{otherwise} \end{cases}$$

Then

$$\begin{aligned} X_1(k) = X_2(k) &= \sum_{n=0}^{N-1} W_N^{kn} \\ &= N, \quad k = 0 \\ &= 0, \quad \text{otherwise} \end{aligned}$$

Thus

$$X_3(k) = X_1(k)X_2(k) = \begin{cases} N^2, & k = 0 \\ 0, & \text{otherwise} \end{cases}$$

and we see that

$$x_3(n) = N, \qquad 0 \le n \le N - 1$$

This is depicted in Fig. 3.11. Clearly, as the sequence $x_2(n)$ is rotated with respect to $x_1(n)$, the sum of products $x_1(m)x_2(n - m)$ will always be equal to N, as shown in Fig. 3.11. It is, of course, possible to consider $x_1(n)$ and $x_2(n)$ as $2N$-point sequences by augmenting them with N zeros. If we then perform a $2N$-point

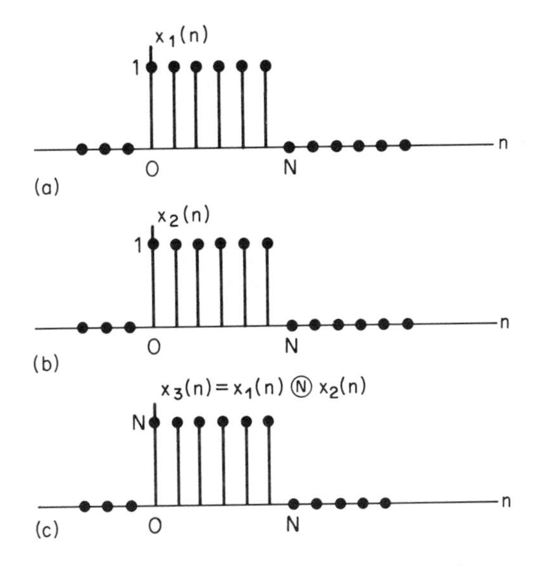

Fig. 3.11 N-point circular convolution of two rectangular sequences of duration N.

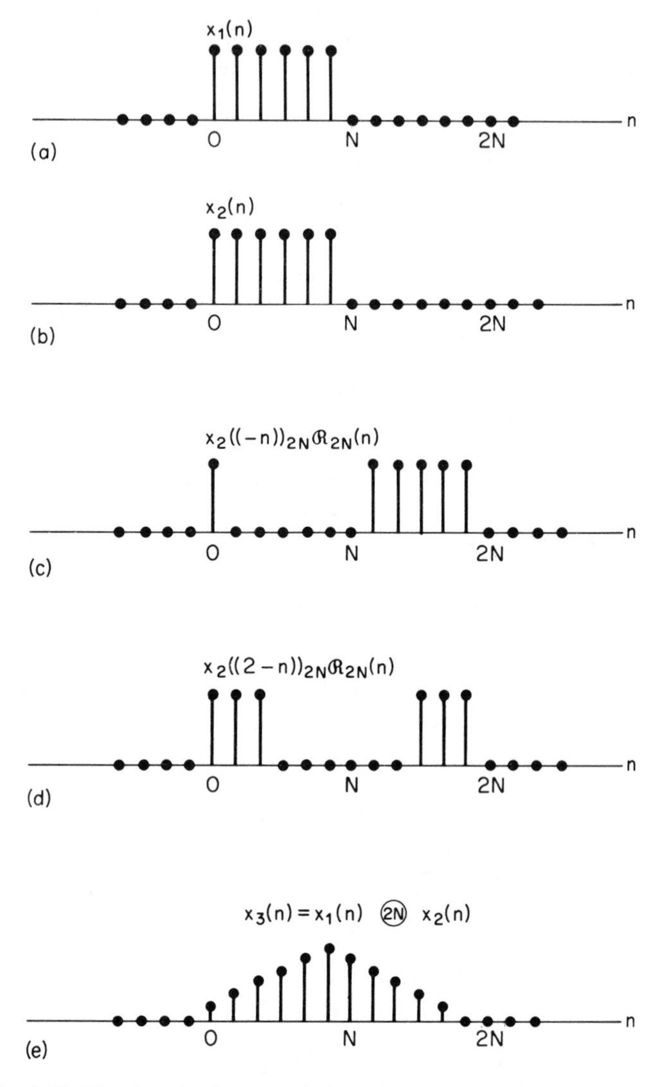

Fig. 3.12 $2N$-point circular convolution of two rectangular sequences of duration N.

circular convolution of the augmented sequences, we obtain the sequence in Fig. 3.12, which can be seen to be identical to the linear convolution of the finite-duration sequences $x_1(n)$ and $x_2(n)$.

This example points to a useful interpretation of circular convolution. Consider two finite-duration sequences $x_1(n)$ and $x_2(n)$ with Fourier transforms

$$X_1(e^{j\omega}) = \sum_{n=0}^{N-1} x_1(n)e^{-j\omega n}$$

$$X_2(e^{j\omega}) = \sum_{n=0}^{N-1} x_2(n)e^{-j\omega n}$$

The sequence $x_3(n)$ corresponding to the product

$$X_3(e^{j\omega}) = X_1(e^{j\omega})X_2(e^{j\omega})$$

is

$$x_3(n) = \sum_{m=0}^{N-1} x_1(m)x_2(n - m)$$

corresponding to the linear convolution of $x_1(n)$ and $x_2(n)$. The resulting sequence is of duration $2N - 1$ samples. Now the DFTs,

$$X_1(k) = \sum_{n=0}^{N-1} x_1(n)W_N^{nk}$$

$$X_2(k) = \sum_{n=0}^{N-1} x_2(n)W_N^{nk}$$

are samples of the Fourier transforms $X_1(e^{j\omega})$ and $X_2(e^{j\omega})$ at frequencies $\omega_k = 2\pi k/N$, a sampling rate adequate to represent $x_1(n)$ and $x_2(n)$ without aliasing in the time domain. The sequence $x_4(n)$ corresponding to

$$X_4(k) = X_1(k)X_2(k)$$

is

$$x_4(n) = \left[\sum_{r=-\infty}^{\infty} x_3(n + rN)\right]\mathcal{R}_N(n) \tag{3.40}$$

Since $x_3(n)$ has duration $2N - 1$, it is clear that $x_4(n)$ will be an aliased version of $x_3(n)$. This can be seen by comparing Figs. 3.11(c) and 3.12(e). Figure 3.11(c) corresponds to the N-point circular convolution and Fig. 3.12(e) corresponds to the $2N$-point circular convolution or, equivalently, the linear convolution of the two sequences. According to Eq. (3.40) if we alias Fig. 3.12(e) modulo N, we will obtain the sequence of Fig. 3.11(c). That this is so is seen by adding the second half of the triangular sequence of Fig. 3.12(e) to the first half and multiplying the result by $\mathcal{R}_N(n)$.

3.7 Summary of Properties of the Discrete Fourier Transform

The above properties of the discrete Fourier transform are summarized in Table 3.2.

Table 3.2

Finite-Length Sequence (length N)	*DFT*
1. $x(n)$	$X(k)$
2. $y(n)$	$Y(k)$
3. $ax(n) + by(n)$	$aX(k) + bY(k)$
4. $x((n+m))_N \mathcal{R}_N(n)$	$W_N^{-km} X(k)$
5. $W_N^{ln} x(n)$	$X((k+l))_N \mathcal{R}_N(k)$
6. $\left[\sum_{m=0}^{N-1} x((m))_N y((n-m))_N\right] \mathcal{R}_N(n)$	$X(k)Y(k)$
7. $x(n)y(n)$	$\frac{1}{N}\left[\sum_{l=0}^{N-1} X((l))_N Y((k-l))_N\right] \mathcal{R}_N(k)$
8. $x^*(n)$	$X^*((-k))_N \mathcal{R}_N(k)$
9. $x^*((-n))_N \mathcal{R}_N(n)$	$X^*(k)$
10. $\text{Re}\,[x(n)]$	$X_{ep}(k) = \frac{1}{2}[X((k))_N + X^*((-k))_N]\mathcal{R}_N(k)$
11. $j\,\text{Im}\,[x(n)]$	$X_{op}(k) = \frac{1}{2}[X((k))_N - X^*((-k))_N]\mathcal{R}_N(k)$
12. $x_{ep}(n)$	$\text{Re}\,[X(k)]$
13. $x_{op}(n)$	$j\,\text{Im}\,[X(k)]$
The following properties apply only when $x(n)$ is real:	
14. Any real $x(n)$	$\begin{cases} X(k) = X^*((-k))_N \mathcal{R}_N(k) \\ \text{Re}\,[X(k)] = \text{Re}\,[X((-k))_N]\mathcal{R}_N(k) \\ \text{Im}\,[X(k)] = -\text{Im}\,[X((-k))_N]\mathcal{R}_N(k) \\ \lvert X(k)\rvert = \lvert X((-k))_N\rvert \mathcal{R}_N(k) \\ \arg\,[X(k)] = -\arg\,[X((-k))_N]\mathcal{R}_N(k) \end{cases}$
15. $x_{ep}(n)$	$\text{Re}\,[X(k)]$
16. $x_{op}(n)$	$j\,\text{Im}\,[X(k)]$

3.8 Linear Convolution Using the Discrete Fourier Transform

As will be shown in a later chapter, highly efficient algorithms are available for computing the discrete Fourier transform of a finite-duration sequence. For this reason, it is computationally efficient to consider implementing a convolution of two sequences by computing their discrete Fourier transforms, multiplying, and computing the inverse discrete Fourier transform. In most applications, we are interested in implementing a linear convolution of two sequences. This is certainly true, for example, when we wish to filter a sequence such as a speech waveform or a radar signal. As we have seen in the preceding section, the multiplication of discrete Fourier transforms corresponds to a circular convolution of the sequences. If we are interested in obtaining a linear convolution, then we must ensure that circular convolution has the effect of a linear convolution. The key to the method for doing this is in evidence in the second example in Sec. 3.6.4.

Let us first consider two N-point sequences, $x_1(n)$ and $x_2(n)$, and let $x_3(n)$ denote their linear convolution, i.e.,

$$x_3(n) = \sum_{m=0}^{N-1} x_1(m)x_2(n-m)$$

It is straightforward to verify that $x_3(n)$ is of length $2N - 1$; i.e., it can have at most $2N - 1$ nonzero points. If it is obtained by multiplying the discrete Fourier transforms of $x_1(n)$ and $x_2(n)$, then each of these discrete Fourier transforms, $X_1(k)$ and $X_2(k)$, must also have been computed on the basis of $2N - 1$ points. Thus if we define

$$\begin{aligned} X_1(k) &= \sum_{n=0}^{2N-2} x_1(n)W_{2N-1}^{nk} \\ X_2(k) &= \sum_{n=0}^{2N-2} x_2(n)W_{2N-1}^{nk} \\ x_3(n) &= \frac{1}{2N-1}\left[\sum_{k=0}^{2N-2} [X_1(k)X_2(k)]W_{2N-1}^{-nk}\right]\mathcal{R}_{2N-1}(n) \end{aligned} \tag{3.41}$$

Then $x_3(n)$ will be the linear convolution of $x_1(n)$ and $x_2(n)$. Of course, a linear convolution would also be achieved if the discrete Fourier transform were computed on the basis of more than $2N - 1$ points but would not in general be achieved if the DFTs were computed on the basis of fewer points. As another way of viewing this procedure for achieving a linear convolution, we note that computation of the DFT on the basis of $2N - 1$ points corresponds to a Fourier series for a periodic sequence constructed from $x_1(n)$ and $x_2(n)$ in such a way that the last $N - 1$ points in each period are zero. These periodic sequences are illustrated in Fig. 3.13. This figure also illustrates the process of obtaining the periodic convolution and we observe that because of the additional zeros added in each period, the nonzero values in one period of $\tilde{x}_1(n)$ are only engaged with nonzero values in a single period of $\tilde{x}_2(n)$.

In general we may wish to convolve two sequences of unequal lengths. If $x_1(n)$ has duration N_1 and $x_2(n)$ has duration N_2, then their convolution will have length $N_1 + N_2 - 1$. Thus in this case we would multiply the discrete Fourier transforms computed on the basis of $N \geq N_1 + N_2 - 1$.

The above procedure permits the computation of the linear convolution of two finite-duration sequences using the discrete Fourier transform. In some applications we would like to convolve a finite-duration sequence with a sequence of indefinite duration, as, for example, in filtering a speech

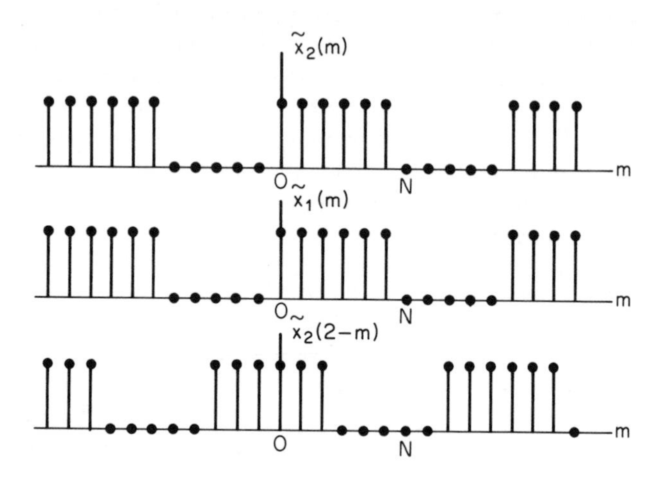

Fig. 3.13 Periodic sequences of period $(2N - 1)$ constructed from N-point finite-duration sequences. The last $(N - 1)$ points in each period are zero.

waveform. While theoretically we could store the entire waveform and then implement the procedure discussed above on the basis of a DFT for a large number of points, such a DFT is generally too large to compute. Another consideration is that for this method of filtering no filtered points can be computed until all the input points have been collected. Generally we would like to avoid such a large delay in the processing. To do this while still using the discrete Fourier transform, the signal to be filtered must be segmented into sections of length L [3,4]. Each section can then be convolved with the finite duration unit-sample response and the filtered sections fitted together in the appropriate way. Such a technique of block filtering can then be implemented using the discrete Fourier transform as before.

To illustrate the procedure, and to develop the procedure for fitting the filtered sections together, consider the unit-sample response $h(n)$ of length M and the signal $x(n)$ depicted in Fig. 3.14. Let us decompose $x(n)$ into a sum of sections, each section having only L nonzero points, with the kth section denoted by $x_k(n)$,

$$x_k(n) = \begin{cases} x(n), & kL \leq n \leq (k+1)L - 1 \\ 0, & \text{otherwise} \end{cases} \tag{3.42}$$

Then $x(n)$ is equal to the sum of the $x_k(n)$; i.e.,

$$x(n) = \sum_{k=0}^{\infty} x_k(n) \tag{3.43}$$

and $x(n)$ convolved with $h(n)$ is equal to the sum of the $x_k(n)$ convolved with $h(n)$; i.e.,

$$x(n) * h(n) = \sum_{k=0}^{\infty} x_k(n) * h(n) \tag{3.44}$$

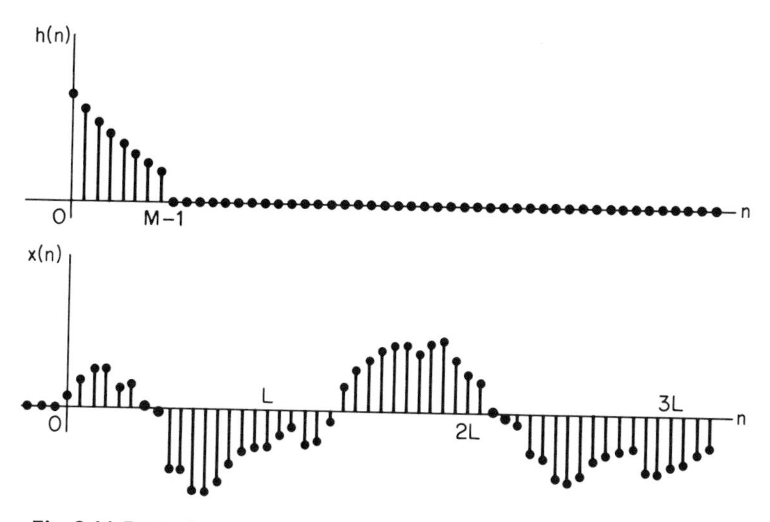

Fig. 3.14 Finite-duration unit-sample response $h(n)$ and signal $x(n)$ to be filtered.

Since the $x_k(n)$ have only L nonzero points and $h(n)$ is of length M, each of the terms $[x_k(n) * h(n)]$ in the sum is of length $L + M - 1$. Thus the linear convolution $x_k(n) * h(n)$ must be obtained using a $(L + M - 1)$-point DFT. Since the beginning of each input section is separated from its neighbors by L points and each filtered section is of length $(L + M - 1)$, the nonzero points in the filtered sections will overlap by $(M - 1)$ points in carrying out the sum required by Eq. (3.44). This is illustrated in Fig. 3.15. In Fig. 3.15(a), the input sections, $x_k(n)$, are depicted, and in Fig. 3.15(b) the filtered sections, $x_k(n) * h(n)$, are shown. The input waveform is reconstructed by adding the waveforms in Fig. 3.15(a), and the filtered result, $x(n) * h(n)$, is constructed by adding the filtered sections depicted in Fig. 3.15(b). This procedure for constructing the filtered output from filtered sections is often referred to as the *overlap-add method*, corresponding to the fact that the filtered sections are overlapped and added to construct the output. The overlapping results from the fact that the linear convolution of each section with the unit-sample response in general is longer than the section length.

An alternative procedure, commonly referred to as the *overlap-save method*, corresponds to implementing a circular convolution of $h(n)$ with $x_k(n)$, and identifying that part of the circular convolution that corresponds to a linear convolution. In particular, if we consider the circular convolution of the M-point unit-sample response with an N-point section, the first $M - 1$ points are incorrect while the remaining points are identical to those that would be obtained had we implemented a linear convolution. In this case, then, we would section $x(n)$ into sections of length N so that each input section overlapped the preceding section by $M - 1$ points. That

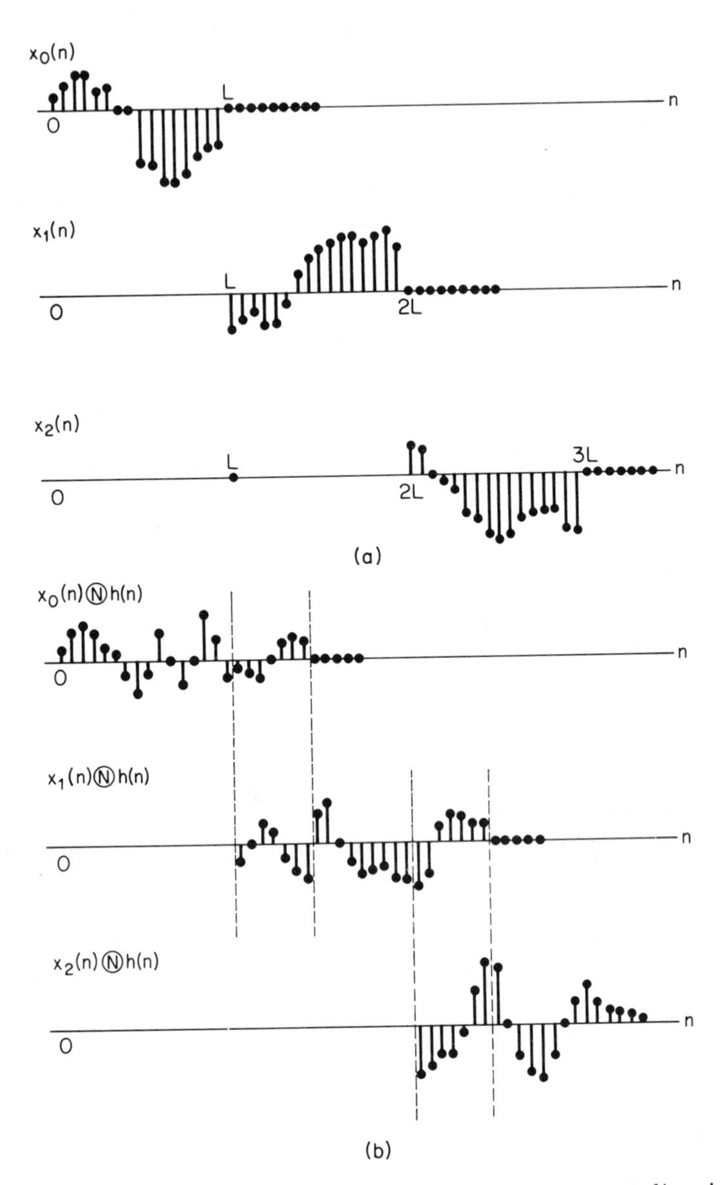

Fig. 3.15 (a) Decomposition of $x(n)$ into nonoverlapping sections of length L; (b) result of convolving each section with $h(n)$.

is, we define sections $x_k(n)$ as

$$x_k(n) = x(n + k(N - M + 1)), \quad 0 \leq n \leq N - 1$$

where in this case we have defined the time origin for each section to be at the beginning of that section rather than at the origin of $x(n)$. This method of sectioning is depicted in Fig. 3.16(a). The circular convolution of each section with $h(n)$ is denoted $y'_k(n)$. These sequences are depicted in Fig. 3.16(b). That portion of each output section in the region $0 \leq n \leq M - 2$ is the part that must be discarded. The remaining points from successive sections are then abutted to construct the final filtered output. That is,

$$y(n) = \sum_{k=0}^{\infty} y'_k(n - k(N + M - 1))$$

where

$$y_k(n) = \begin{cases} y'_k(n), & M - 1 \leq n \leq N - 1 \\ 0, & \text{otherwise} \end{cases}$$

This procedure, the overlap-save method, gets its name from the fact that each succeeding input section consists of $N - M + 1$ new points and $M - 1$ points saved from the previous section.

3.9 Two-Dimensional Discrete Fourier Transform

In Chapters 1 and 2 we saw that many of the transform properties discussed for one-dimensional signals could be extended to multidimensional signals. A similar generalization applies to the discrete Fourier series and discrete Fourier transform.

The discrete Fourier transform representation of two-dimensional sequences is of considerable computational importance in the digital processing of two-dimensional signals such as photographs and seismic array data. In this section we give a short discussion of the two-dimensional DFS and DFT that parallels the discussion of the previous sections of this chapter.

Let us begin by considering the definition of a two-dimensional periodic sequence. We say that a sequence is periodic in the row index with period M and in the column index with period N if

$$\tilde{x}(m, n) = \tilde{x}(m + qM, n + rN)$$

where q and r are arbitrary positive or negative integers. Such sequences have a Fourier series representation as a finite sum of complex exponentials in the form

$$\tilde{x}(m, n) = \frac{1}{MN} \sum_{k=0}^{M-1} \sum_{l=0}^{N-1} \tilde{X}(k, l) W_M^{-km} W_N^{-ln} \tag{3.45}$$

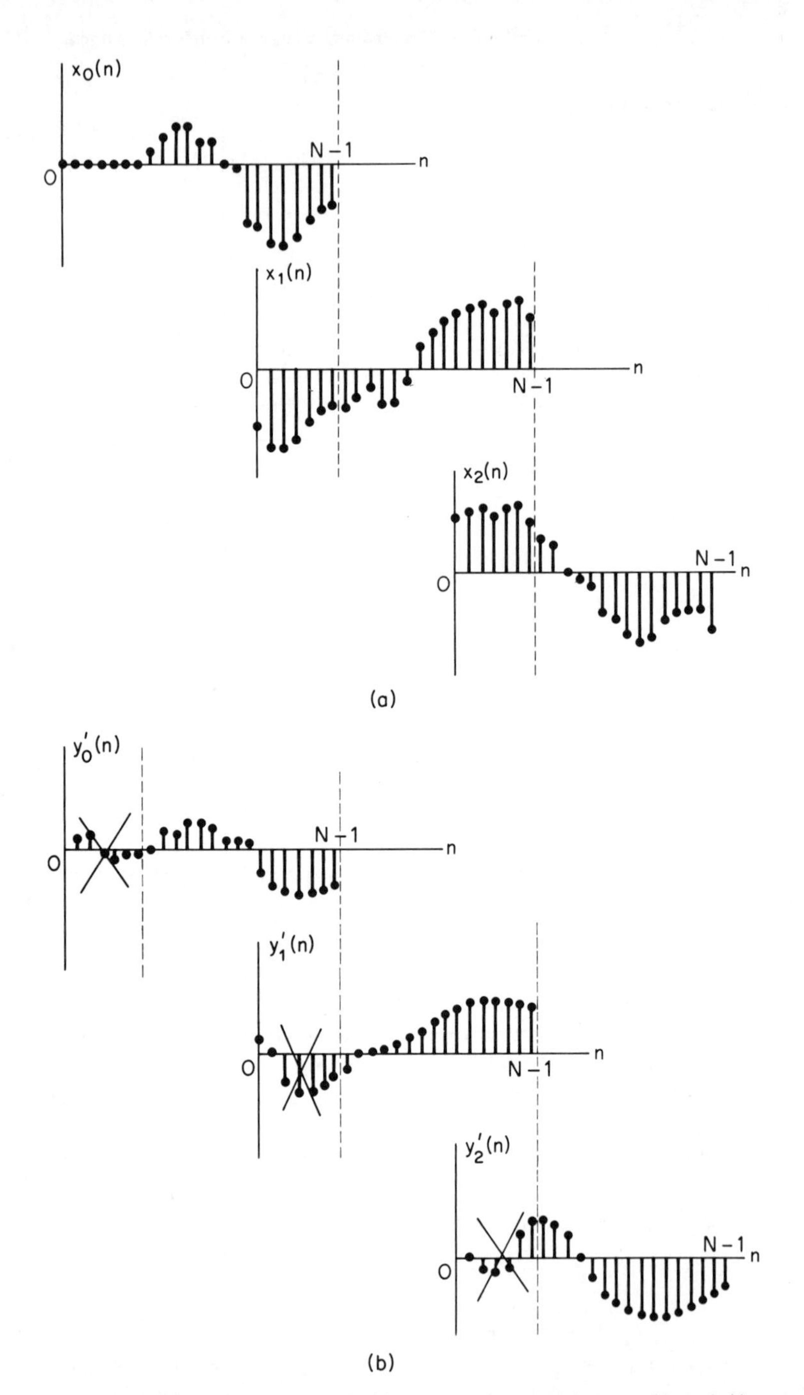

Fig. 3.16 (a) Decomposition of $x(n)$ into overlapping sections of length N; (b) result of circular convolution of each section with $h(n)$. The portions of each filtered section to be discarded in forming the linear convolution are indicated.

where $\tilde{X}(k, l)$ can be shown to be

$$\tilde{X}(k, l) = \sum_{m=0}^{M-1} \sum_{n=0}^{N-1} \tilde{x}(m, n) W_M^{km} W_N^{ln} \tag{3.46}$$

and

$$W_M = e^{-j(2\pi/M)}$$

$$W_N = e^{-j(2\pi/N)}$$

We can see from Eq. (3.46) that

$$\tilde{X}(k, l) = \tilde{X}(k + qM, l + rN)$$

for integer values of q and r, and thus $\tilde{X}(k, l)$ has the same periodicity as the sequence $\tilde{x}(m, n)$.

The one-dimensional discrete Fourier transform resulted from interpreting a finite-duration sequence as one period of a periodic sequence and applying the discrete Fourier series. In a similar manner we can apply the two-dimensional Fourier series to represent a two-dimensional sequence that is nonzero for only a finite area in the (m,n)-plane. Such a sequence will be referred to as a *finite-area sequence* and is the two-dimensional counterpart of a finite-duration sequence. The resulting Fourier representation is referred to as the two-dimensional discrete Fourier transform.

To develop the DFT for two-dimensional signals we consider a finite-area sequence $x(m, n)$ which is zero outside the interval $0 \leq m \leq M - 1$, $0 \leq n \leq N - 1$, i.e., it is of area (M, N), and construct the periodic sequence

$$\tilde{x}(m, n) = x[((m))_M, ((n))_N] \tag{3.47}$$

The original sequence $x(m, n)$ is recovered by extracting one period of $\tilde{x}(m, n)$, i.e.,

$$x(m, n) = \tilde{x}(m, n) \mathcal{R}_{M,N}(m, n) \tag{3.48}$$

where

$$\mathcal{R}_{M,N}(m, n) = \begin{cases} 1, & 0 \leq m \leq M - 1, 0 \leq n \leq N - 1 \\ 0, & \text{otherwise} \end{cases} \tag{3.49}$$

We then define the discrete Fourier transform of $x(m, n)$ to correspond to the Fourier series coefficients of $\tilde{x}(m, n)$. However, just as we did with one-dimensional sequences, we will maintain the duality between the time and frequency domains by interpreting the DFT coefficients to also be a finite-area sequence. Thus with $X(k, l)$ denoting the DFT of $x(m, n)$,

$$X(k, l) = \left[\sum_{m=0}^{M-1} \sum_{n=0}^{N-1} x(m, n) W_M^{km} W_N^{ln} \right] \mathcal{R}_{M,N}(k, l) \tag{3.50}$$

$$x(m, n) = \frac{1}{MN} \left[\sum_{k=0}^{M-1} \sum_{l=0}^{N-1} X(k, l) W_M^{-km} W_N^{-ln} \right] \mathcal{R}_{M,N}(m, n) \tag{3.51}$$

A useful interpretation of the two-dimensional DFT in terms of the one-dimensional DFT is placed in evidence by noting that the rectangular function $\mathcal{R}_{M,N}(k, l)$ is separable and hence can be written as

$$\mathcal{R}_{M,N}(k, l) = \mathcal{R}_M(k)\mathcal{R}_N(l) \tag{3.52}$$

Then Eq. (3.50) can be expressed as

$$X(k, l) = \left[\sum_{n=0}^{N-1} G(k, n)W_N^{ln}\right]\mathcal{R}_N(l) \tag{3.53a}$$

where

$$G(k, n) = \left[\sum_{m=0}^{M-1} x(m, n)W_M^{km}\right]\mathcal{R}_M(k) \tag{3.53b}$$

The function $G(k, n)$ corresponds to an M-point one-dimensional DFT for each value of n; i.e., it consists of N one-dimensional transforms, one for each *column* of $x(m, n)$. The two-dimensional DFT $X(k, l)$ is then obtained according to Eq. (3.53a) by implementing M one-dimensional transforms, one for each *row* of the sequence $G(k, n)$.

Equation (3.50) can alternatively be expressed as

$$X(k, l) = \left[\sum_{m=0}^{M-1} P(m, l)W_M^{km}\right]\mathcal{R}_M(k) \tag{3.54a}$$

where

$$P(m, l) = \left[\sum_{n=0}^{N-1} x(m, n)W_N^{ln}\right]\mathcal{R}_N(l) \tag{3.54b}$$

The function $P(m, l)$ then corresponds to a set of N-point transforms on the *rows* of $x(m, n)$. $X(k, l)$ is then obtained according to Eq. (3.54a) by transforming the *columns* of $P(m, l)$. In summary, then, the two-dimensional DFT can be implemented by using a one-dimensional transform first on the rows, then the columns, or vice versa. A similar interpretation can, of course, also be applied to the inverse DFT, as given by Eq. (3.51).

An interesting special case is a sequence that is separable, i.e., having the property that

$$x(m, n) = x_1(m)x_2(n) \tag{3.55}$$

In this case the function $G(k, n)$ in Eq. (3.53b) is $X_1(k)$, the one-dimensional DFT of $x_1(m)$, and is independent of n. The two-dimensional DFT is the product of $X_1(k)$ and $X_2(l)$, the DFT of $x_2(n)$, i.e.,

$$X(k, l) = X_1(k)X_2(l) \tag{3.56}$$

In this case evaluation of only one M-point DFT and one N-point DFT permits us to compute $X(k, l)$ for all k and l.

The two-dimensional discrete Fourier transform is clearly linear; i.e., if

$$x_3(m, n) = ax_1(m, n) + bx_2(m, n)$$

then

$$X_3(k, l) = aX_1(k, l) + bX_2(k, l)$$

where we have assumed $x_1(m, n)$ and $x_2(m, n)$ to have identical dimensions.

In the context of finite-duration one-dimensional sequences, we noted that a shift in the time domain should be interpreted as a rotation in the basic interval $0 \leq n \leq N-1$. In the case of the finite-area two-dimensional sequence $x(m+m_0, n+n_0)$, it can be shown that the corresponding DFT is $W_M^{-km_0} W_N^{-ln_0} X(k, l)$. In this case we interpret the shift in the spatial domain as a rotation of each column of the array by m_0 samples, followed by a rotation of each row of the new array by n_0 samples. This property, of course, has a symmetric counterpart where the spatial and frequency domains are interchanged.

As with the one-dimensional DFT, there is a set of symmetry properties for the two-dimensional DFT. Some of these are discussed in Problem 35 of this chapter and will not be considered further here.

An important application of the two-dimensional DFT is in the computation of convolutions for filtering. Consider two finite-area sequences $x_1(m, n)$ and $x_2(m, n)$, where $x_1(m, n)$ is of area (M_1, N_1) and $x_2(m, n)$ is of area (M_2, N_2). Let $X_1(k, l)$ and $X_2(k, l)$ denote the (M, N) DFT of $x_1(m, n)$ and $x_2(m, n)$, respectively, augmented by areas of zero samples if necessary. Then the product

$$X_3(k, l) = X_1(k, l)X_2(k, l) \tag{3.57}$$

corresponds to the sequence

$$x_3(m, n) = \sum_{q=0}^{M-1} \sum_{r=0}^{N-1} x_1[((q))_M, ((r))_N] x_2[((m-q))_M, ((n-r))_N] \; \mathcal{R}_{M,N}(m, n) \tag{3.58}$$

Equation (3.58) represents a periodic convolution of the periodic sequences $\tilde{x}_1(m, n)$ and $\tilde{x}_2(m, n)$ formed from $x_1(m, n)$ and $x_2(m, n)$ as in Eq. (3.47). In the context of finite-area sequences, it is a circular convolution in two dimensions. If we wish to obtain the linear two-dimensional convolution of $x_1(m, n)$ and $x_2(m, n)$, we must of course be certain that M and N are chosen so as to avoid aliasing as in the one-dimensional case. Convolution of a sequence spanning the area (M_1, N_1) with a sequence spanning the area (M_2, N_2) produces a sequence of area $[(M_1 + M_2 - 1), (N_1 + N_2 - 1)]$. Thus we must choose $M \geq M_1 + M_2 - 1$ and $N \geq N_1 + N_2 - 1$ in order to ensure that the circular convolution will be identical to the desired linear convolution.

This is depicted in Fig. 3.17, where we have shaded the nonzero regions of

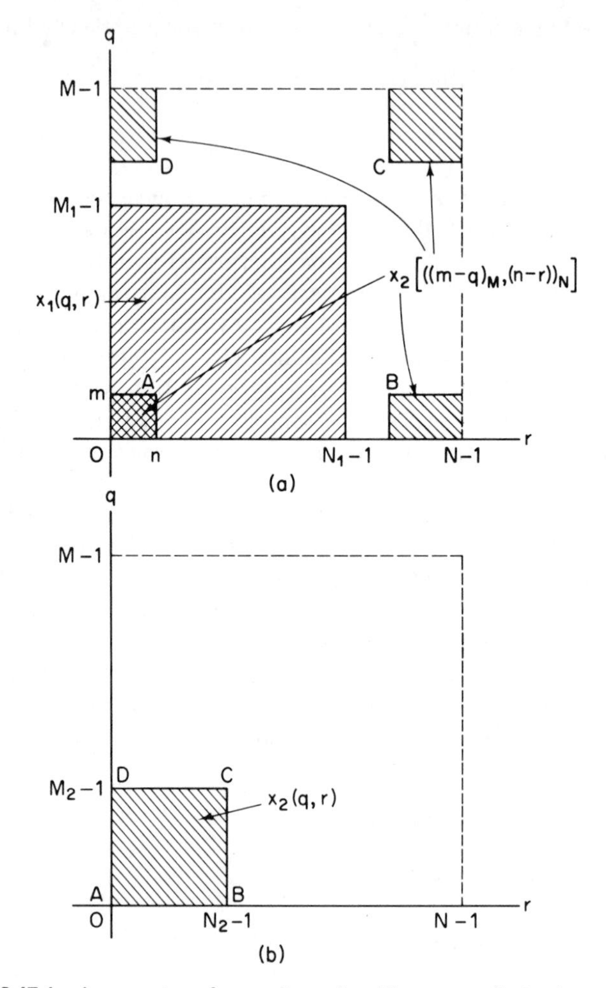

Fig. 3.17 Implementation of a two-dimensional linear convolution by means of a circular convolution: (a) $x_1(q,r)$ and $x_2[((m-q))_M, ((n-r))_N]$ (b) $x_2(q,r)$.

$x_1(m,n)$ and $x_2(m,n)$. In Fig. 3.17 we have superimposed $x_2(m-q, n-r)$ upon $x_1(q, r)$. Clearly if the above inequalities are satisfied, $x_2(m-q, n-r)$ will never "wrap around" and engage a nonzero portion of $x_1(q, r)$, and therefore the circular convolution will be identical to the desired linear convolution. If a small area is to be convolved with a much larger area, we can generalize the overlap-add or overlap-save methods of the previous section. Also, if one of the sequences to be convolved is separable, the two-dimensional convolution can be accomplished by repeated evaluation of one-dimensional convolutions. Separable finite-duration sequences are often used in filtering applications because of the computational simplifications discussed above.

Summary

In this chapter we have discussed a Fourier representation of finite-length sequences referred to as the discrete Fourier transform. This representation was based on the relationship between finite-length and periodic sequences. Specifically, with a periodic sequence constructed for which each period is identical to the finite-length sequence, the discrete Fourier transform of the finite-length sequence corresponds to the discrete Fourier series of the periodic sequence. Thus we first discussed the Fourier series representation of periodic sequences, including the properties of the Fourier series and the interpretation of the coefficients as samples on the unit circle, equally spaced in angle, of the z-transform of one period of the periodic sequences. The discrete Fourier series was then applied to the representation of finite-duration sequences. This included a discussion of the properties of the discrete Fourier transform and the consideration of the implementation of linear convolution using the discrete Fourier transform. The chapter concluded with a brief introduction to the two-dimensional discrete Fourier transform.

REFERENCES

1. J. W. Cooley and J. W. Tukey, "An Algorithm for the Machine Computation of Complex Fourier Series," *Math. Computation*, Vol. 19, Apr. 1965, pp. 297–301.
2. B. Gold and C. Rader, *Digital Processing of Signals*, McGraw-Hill Book Company, New York, 1969.
3. H. D. Helms, "Fast Fourier Transform Method of Computing Difference Equations and Simulating Filters," *IEEE Trans. Audio Electroacoust.*, Vol. 15, No. 2, 1967, pp. 85–90.
4. T. G. Stockham, "High Speed Convolution and Correlation," 1966 Spring Joint Computer Conference, *AFIPS Proc.*, Vol. 28, 1966, pp. 229–233.

PROBLEMS

1. Consider the first-order linear system defined by the difference equation $y(n) = ay(n-1) + x(n)$, where the coefficient a is between zero and unity. The input $x(n)$ is restricted to be a periodic sequence with period N; i.e., $\tilde{x}(n) = \tilde{x}(n + kN)$ for any integer value of k. The filter output is assumed to have reached steady state. Determine, in terms of the coefficient a, the unit-sample response for a *finite impulse response* filter which for this class of inputs yields an output $\tilde{y}(n)$ indistinguishable in steady state from the infinite impulse response filter defined by the above difference equation.
2. (a) In Sec. 3.3.3 we stated a number of symmetry properties of the discrete Fourier series for periodic sequences. We list here some of the properties stated. Prove that each of the properties listed is true. In carrying out the proof you may use the definition of the discrete Fourier series and any

previous property in the list. For example, in proving property 3 you may use properties 1 and 2.

Sequence	Discrete Fourier Series
1. $\tilde{x}(n + m)$	$W_N^{-km}\tilde{X}(k)$
2. $\tilde{x}^*(n)$	$\tilde{X}^*(-k)$
3. $\tilde{x}^*(-n)$	$\tilde{X}^*(k)$
4. Re $[\tilde{x}(n)]$	$\tilde{X}_e(k)$
5. j Im $[\tilde{x}(n)]$	$\tilde{X}_o(k)$

(b) From the properties proved in part (a), show that for a real periodic sequence $\tilde{x}(n)$, the following symmetry properties of the discrete Fourier series hold:

(1) Re $[\tilde{X}(k)]$ = Re $[\tilde{X}(-k)]$.
(2) Im $[\tilde{X}(k)]$ = $-$Im $[\tilde{X}(-k)]$.
(3) $|\tilde{X}(k)| = |\tilde{X}(-k)|$.
(4) arg $\tilde{X}(k)$ = $-$arg $\tilde{X}(-k)$.

3. In Fig. P3.3 are shown several periodic sequences $\tilde{x}(n)$. These sequences can be expressed in a Fourier series as

$$\tilde{x}(n) = \sum_{k=0}^{N-1} X(k)e^{j(2\pi/N)kn}$$

(a) For which sequences can the time origin be chosen such that all the $\tilde{X}(k)$ are real?
(b) For which sequences can the time origin be chosen such that all the $\tilde{X}(k)$ (except $X(0)$) are imaginary?
(c) For which sequences does $\tilde{X}(k) = 0$, $k = \pm 2, \pm 4, \pm 6$, etc.?

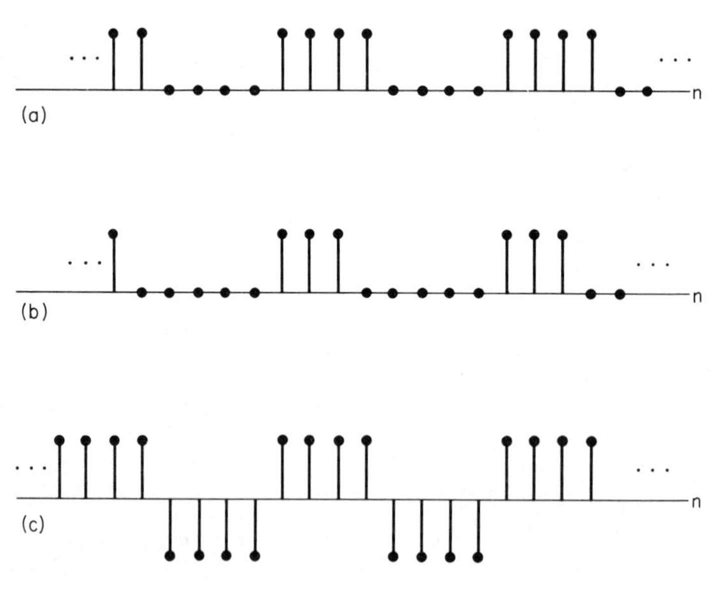

Fig. P3.3

4. If $\tilde{x}(n)$ is a periodic sequence with period N, it is also periodic with period $2N$. Let $\tilde{X}_1(k)$ denote the DFS coefficients of $\tilde{x}(n)$ considered as a periodic sequence with period N and $\tilde{X}_2(k)$ denote the DFS coefficients of $\tilde{x}(n)$ considered as a periodic sequence with period $2N$. $\tilde{X}_1(k)$ is, of course, periodic with period N and $\tilde{X}_2(k)$ is periodic with period $2N$. Determine $\tilde{X}_2(k)$ in terms of $\tilde{X}_1(k)$.

5. Consider two periodic sequences $\tilde{x}(n)$ and $\tilde{y}(n)$. $\tilde{x}(n)$ has period N and $y(n)$ has period M. The sequence $\tilde{w}(n)$ is defined as $\tilde{w}(n) = \tilde{x}(n) + \tilde{y}(n)$.
(a) Show that $\tilde{w}(n)$ is periodic with period MN.
(b) Since $\tilde{x}(n)$ has period N, its DFS coefficients $\tilde{X}(k)$ also have period N. Similarly, since $\tilde{y}(n)$ has period M, its DFS coefficients $\tilde{Y}(k)$ also have period M. The DFS coefficients of $\tilde{w}(n)$, $\tilde{W}(k)$, have period MN. Determine $\tilde{W}(k)$ in terms of $\tilde{X}(k)$ and $\tilde{Y}(k)$. You may find it helpful to refer to the results of Problem 4 above.

6. $\tilde{x}(n)$ denotes a periodic sequence with period N and $\tilde{X}(k)$ denotes its discrete Fourier series coefficients. The sequence $\tilde{X}(k)$ is also a periodic sequence with period N. Determine, in terms of $\tilde{x}(n)$, the discrete Fourier series coefficients of $\tilde{X}(k)$.

7. Compute the DFT of each of the following finite-length sequences considered to be of length N.
(a) $x(n) = \delta(n)$.
(b) $x(n) = \delta(n - n_0)$, where $0 < n_0 < N$.
(c) $x(n) = a^n$, $0 \leq n \leq N - 1$.

8. In Fig. P3.8 is shown a finite-length sequence $x(n)$. Sketch the sequence $x((-n))_4$.

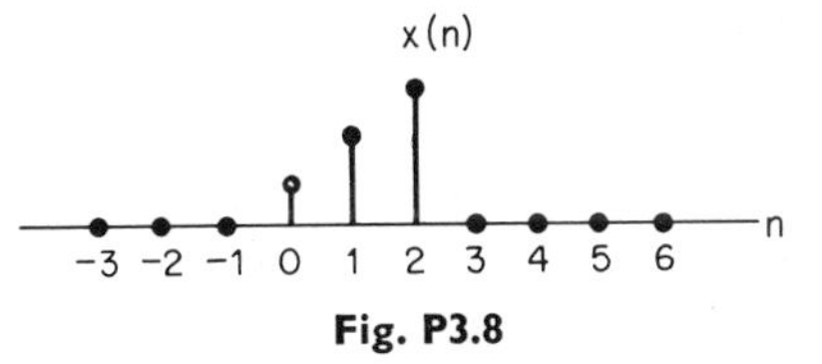

Fig. P3.8

9. $x(n)$ denotes a finite-length sequence of length N. Show that

$$x((-n))_N = x((N - n))_N$$

10. Analog data to be spectrum-analyzed are sampled at 10 kHz and the DFT of 1024 samples computed. Determine the frequencey spacing between spectral samples. Justify your answer.

11. The DFT of a finite-duration sequence corresponds to samples of its z-transform on the unit circle. For example, the DFT of a 10-point sequence $x(n)$

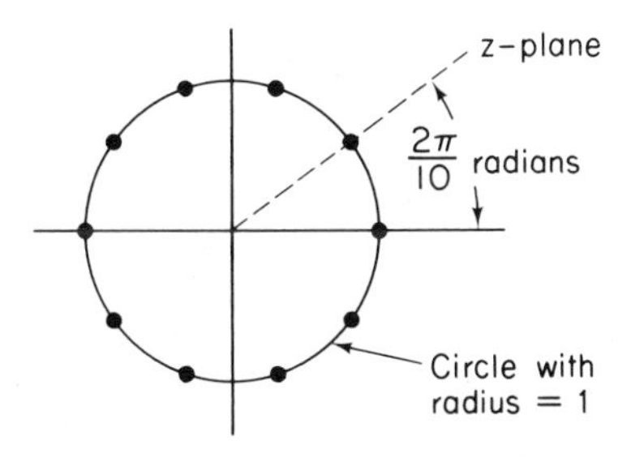

Fig. P3.11-1

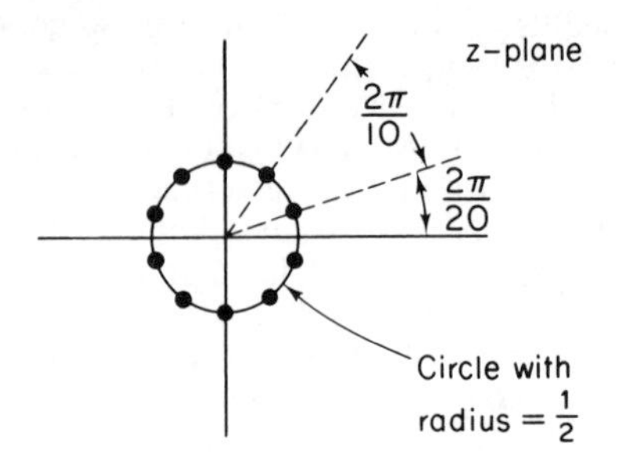

Fig. P3.11-2

corresponds to samples of $X(z)$ at the 10 equally spaced points indicated in Fig. P3.11-1. We wish to find the equally spaced samples of $X(z)$ on the contour shown in Fig. P3.11-2; i.e., $X(z)|_{z=0.5e^{j[(2\pi k/10)+(\pi/10)]}}$. Show how to modify $x(n)$ to obtain a sequence $x_1(n)$ such that the DFT of $x_1(n)$ corresponds to the desired samples of $X(z)$.

12. In Sec. 3.7.3 we stated a number of symmetry properties of the DFT, some of which are listed here. Prove that each of the properties listed is true. In carrying out the proof, you may use the definition of the DFT and any previous property in the list.

Sequence	*DFT*
1. $x((n+m))_N \mathcal{R}_N(n)$	$W_N^{-km} X(k)$
2. $x^*(n)$	$X^*((-k))_N \mathcal{R}_N(k)$
3. $x^*((-n))_N \mathcal{R}_N(n)$	$X^*(k)$
4. Re $[x(n)]$	$X_{\text{ep}}(k)$
5. j Im $[x(n)]$	$X_{\text{op}}(k)$

13. Using the properties in Problem 12 above, show that for a real sequence the following symmetry properties of the DFT hold:
(1) $\text{Re}\,[X(k)] = \text{Re}\,[X((-k))_N]\mathcal{R}_N(k)$.
(2) $\text{Im}\,[X(k)] = -\text{Im}\,[X((-k))_N]\mathcal{R}_N(k)$.
(3) $|X(k)| = |X((-k))_N|\mathcal{R}_N(k)$.
(4) $\arg\,[X(k)] = -\arg\,[X((-k))_N]\mathcal{R}_N(k)$.

14. Let $X(k)$ denote the N-point DFT of the N-point sequence $x(n)$.
(a) Show that if $x(n)$ satisfies the relation

$$x(n) = -x(N-1-n)$$

then

$$X(0) = 0$$

(b) Show that with N even and if

$$x(n) = x(N-1-n)$$

then

$$X\left(\frac{N}{2}\right) = 0$$

15. Let $X(k)$ denote the N-point DFT of an N-point sequence $x(n)$. $X(k)$ itself is an N-point sequence. If the DFT of $X(k)$ is computed to obtain a sequence $x_1(n)$, determine $x_1(n)$ in terms of $x(n)$.

16. Show from Eqs. (3.26) that with $x(n)$ as an N-point sequence and $X(k)$ as its N-point DFT,

$$\sum_{n=0}^{N-1} |x(n)|^2 = \frac{1}{N} \sum_{k=0}^{N-1} |X(k)|^2$$

This is commonly referred to as *Parseval's relation for the DFT.*

17. In Chapter 1 the conjugate symmetric and conjugate antisymmetric components of a sequence $x(n)$ were defined, respectively, as

$$x_e(n) = \tfrac{1}{2}[x(n) + x^*(-n)]$$
$$x_o(n) = \tfrac{1}{2}[x(n) - x^*(-n)]$$

In Sec. 3.63 we found it convenient to define the periodic conjugate symmetric and periodic conjugate antisymmetric components of a sequence of finite duration N as

$$x_{ep}(n) = \tfrac{1}{2}[x((n))_N + x^*((-n))_N]\mathcal{R}_N(n)$$
$$x_{op}(n) = \tfrac{1}{2}[x((n))_N - x^*((-n))_N]\mathcal{R}_N(n)$$

(a) Show that $x_{ep}(n)$ can be related to $x_e(n)$ and $x_{op}(n)$ can be related to $x_o(n)$ by the relations

$$x_{ep}(n) = [x_e(n) + x_e(n - N)]\mathcal{R}_N(n)$$
$$x_{op}(n) = [x_o(n) + x_o(n - N)]\mathcal{R}_N(n)$$

(b) $x(n)$ is considered to be a sequence of length N, and in general $x_e(n)$ cannot be recovered from $x_{ep}(n)$ and $x_o(n)$ cannot be recovered from $x_{op}(n)$. Show that with $x(n)$ considered as a sequence of length N, but with $x(n) = 0$ $n > N/2$ that $x_e(n)$ can be obtained from $x_{ep}(n)$ and $x_o(n)$ can be obtained from $x_{op}(n)$.

18. A finite-duration sequence $x(n)$ of length 8 has the eight-point DFT $X(k)$ shown in Fig. P3.18-1. A new sequence $y(n)$ of length 16 is defined by

$$y(n) = \begin{cases} x\left(\dfrac{n}{2}\right), & n \text{ even} \\ 0, & n \text{ odd} \end{cases}$$

From the list in Fig. P3.18-2, choose the sketch corresponding to the 16-point DFT of $y(n)$.

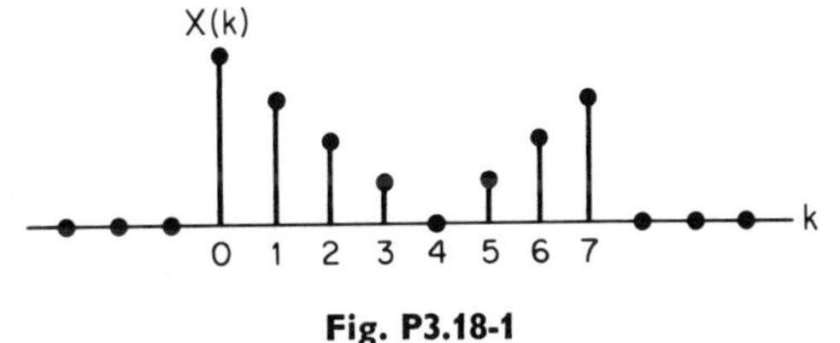

Fig. P3.18-1

19. One of the ways of implementing a discrete circular convolution of two sequences of finite duration is by multiplying their DFTs and computing the inverse DFT of the result. In particular, with $X(k)$, $Y(k)$, and $H(k)$ denoting

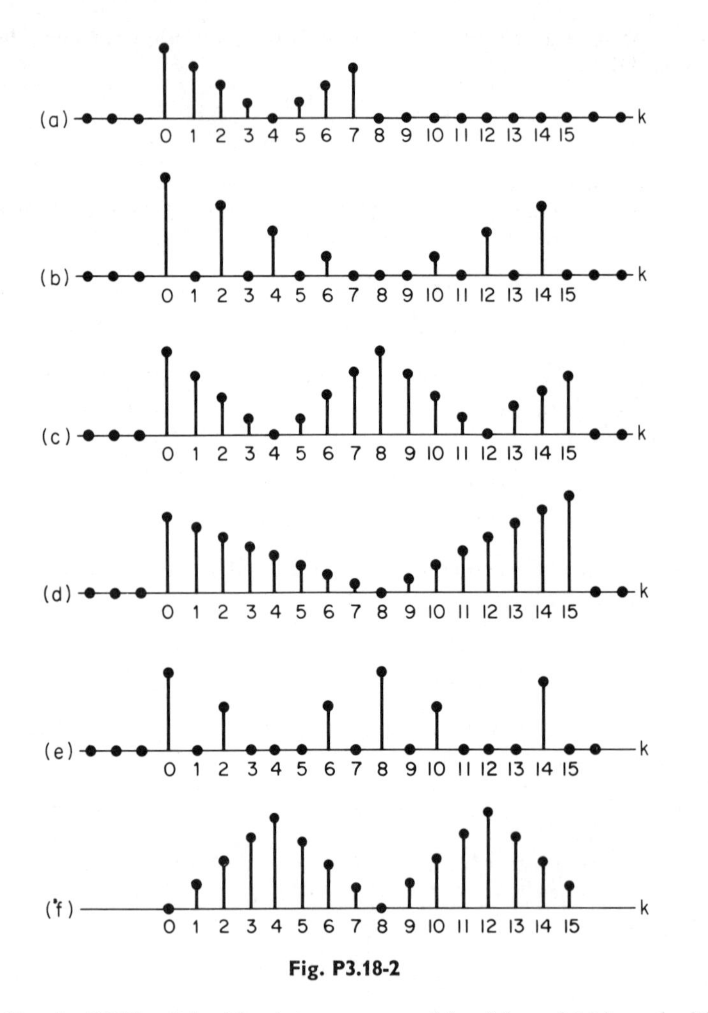

Fig. P3.18-2

the N-point DFTs of the N-point sequences $x(n)$, $y(n)$, and $h(n)$, and with

$$Y(k) = X(k)H(k) \tag{P3.19-1}$$

$$y(n) = \sum_{k=0}^{N-1} x(m)h[((n-m))_N], \qquad n = 0, 1, \ldots, N-1 \tag{P3.19-2}$$

In implementing a circular convolution in this way it is generally useful to have an upper bound on the output sequence. If one of the two sequences $x(n)$ or $h(n)$ is known, then $y(n)$ can be bounded by

$$|y(n)| \leq |x|_{\max} \sum_{m=0}^{N-1} |h(m)| \tag{P3.19-3}$$

or

$$|y(n)| \leq |h|_{\max} \sum_{m=0}^{N-1} |x(m)| \tag{P3.19-4}$$

In this problem we want to generate a bound on $y(n)$ without knowing either of the sequences $x(n)$ or $h(n)$ but with the constraints that $|x(n)| \leq 1$ and $|H(k)| \leq 1$. The results of this problem, and some additional discussion, are presented in a paper by A. V. Oppenheim and C. J. Weinstein, "A Bound on the Output of a Circular Convolution," *IEEE Trans. Audio Electroacoust.*, June 1969, pp. 344–348.

(a) Using Parseval's relation as derived in Problem 16, page 125, show that with $|x(n)| \leq 1$ and $|H(k)| \leq 1$, and with $x(n)$ and $y(n)$ related by Eqs. (P3.19-3) and (P3.19-4).

$$\sum_{n=0}^{N-1} |y(n)|^2 \leq \sum_{n=0}^{N-1} |x(n)|^2$$

(b) Combining the result in part (a) with an upper bound on $\sum_{n=0}^{N-1} |x(n)|^2$, show that

$$\sum_{n=0}^{N-1} |y(n)|^2 \leq N$$

and therefore that $|y(n)| \leq \sqrt{N}$.

It can be shown that the bound derived in part (b) is a least upper bound if $x(n)$ and $h(n)$ are complex sequences, i.e., a choice for each exists, consistent with the stated constraints, so that at least one point in the output sequence $y(n)$ has the value $\sqrt{N}$.

20. Consider a finite-duration sequence $x(n)$, which is zero for $n < 0$ and $n \geq N$, where N is even. Let the z-transform of $x(n)$ be denoted by $X(z)$. Listed here are two tables. In Table P3.20-1 are seven sequences obtained from $x(n)$. In Table P3.20-2 are nine sequences obtained from $X(z)$. For each sequence in Table P3.20-1, find its DFT in Table P3.20-2. The size of the transform considered must be greater than or equal to the length of the sequence $g_k(n)$. *For purposes of illustration only* assume that $x(n)$ can be represented by the envelope shown in Fig. P3.20.

21. Let $f(t)$ be a *real-valued, bandlimited, periodic* continuous-time function. The period of $f(t)$ is P, so $f(t) = f(t + sP)$ for any integer s. The only nonzero terms in the complex Fourier series representation of $f(t)$ correspond to frequencies between $2\pi M/P$; i.e.,

$$f(t) = \sum_{r=-M}^{M} a_r e^{j(2\pi rt/P)}$$

Furthermore, a_M is real.

A sequence $x_1(n)$ is generated by sampling $f(t)$ with a sampling period T_1, where

$$x_1(n) = f(nT_1)$$

and

$$T_1 = \frac{P}{2M}$$

Let $X_1(k)$ denote the discrete Fourier transform of one period of $x_1(n)$ starting at $n = 0$; i.e.,

$$X_1(k) = \sum_{n=0}^{2M-1} x_1(n) e^{-j(2\pi nk/2M)}$$

From $x_1(n)$ we would like to obtain a sequence $x_2(n)$ that corresponds to sampling $f(t)$ *twice* as fast. In other words,

$$x_2(n) = f(nT_2)$$

Table P3.20-1

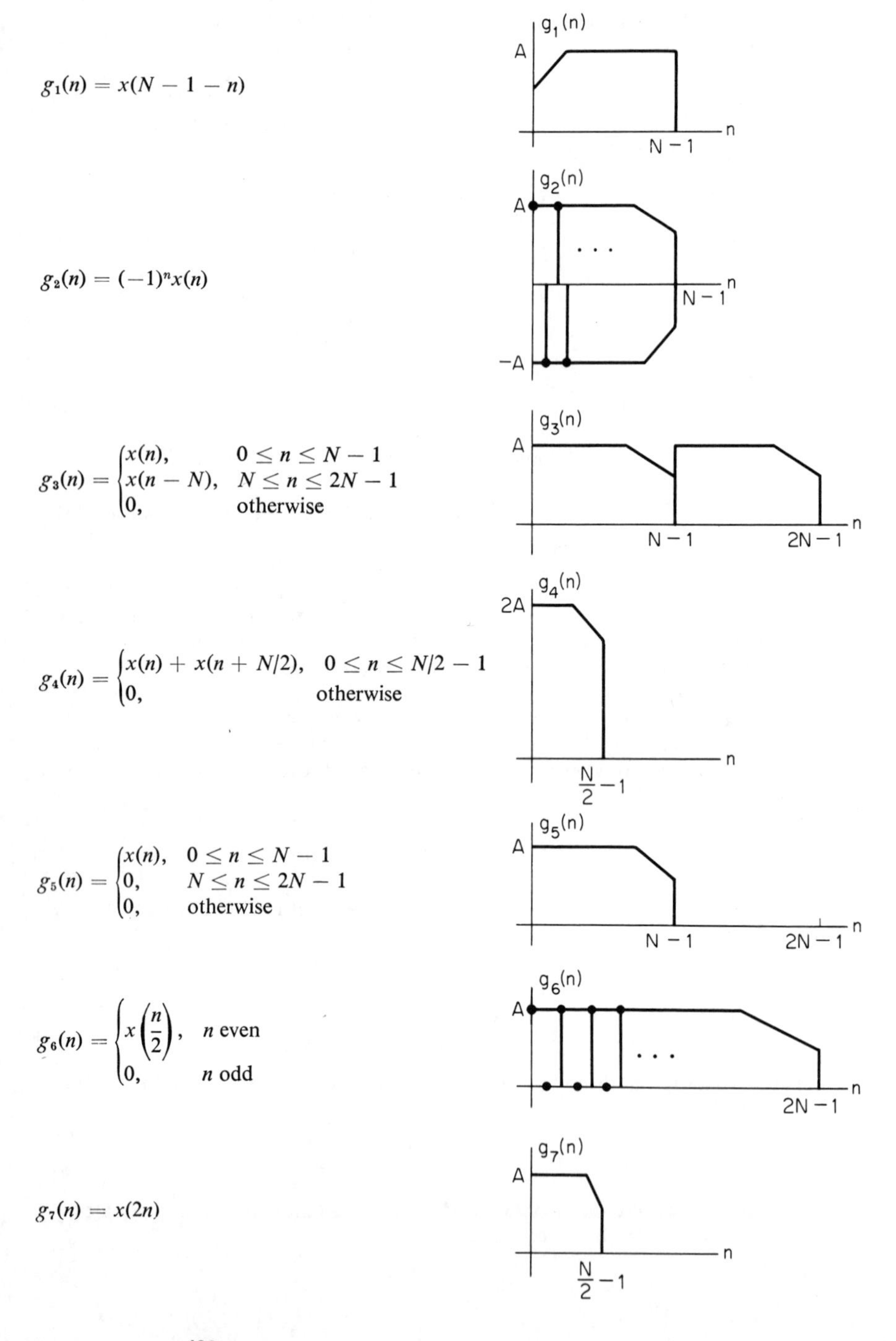

$$g_1(n) = x(N - 1 - n)$$

$$g_2(n) = (-1)^n x(n)$$

$$g_3(n) = \begin{cases} x(n), & 0 \leq n \leq N - 1 \\ x(n - N), & N \leq n \leq 2N - 1 \\ 0, & \text{otherwise} \end{cases}$$

$$g_4(n) = \begin{cases} x(n) + x(n + N/2), & 0 \leq n \leq N/2 - 1 \\ 0, & \text{otherwise} \end{cases}$$

$$g_5(n) = \begin{cases} x(n), & 0 \leq n \leq N - 1 \\ 0, & N \leq n \leq 2N - 1 \\ 0, & \text{otherwise} \end{cases}$$

$$g_6(n) = \begin{cases} x\left(\frac{n}{2}\right), & n \text{ even} \\ 0, & n \text{ odd} \end{cases}$$

$$g_7(n) = x(2n)$$

Table P3.20-2

$$H_1(k) = X(e^{j2\pi k/N})$$
$$H_2(k) = X(e^{j2\pi k/2N})$$
$$H_3(k) = \begin{cases} 2X(e^{j2\pi k/2N}), & k \text{ even} \\ 0, & k \text{ odd} \end{cases}$$
$$H_4(k) = X(e^{j2\pi k/(2N-1)})$$
$$H_5(k) = 0.5[X(e^{j2\pi k/N}) + X(e^{j2\pi(k+N/2)/N})]$$
$$H_6(k) = X(e^{j4\pi k/N})$$
$$H_7(k) = e^{j2\pi k/N}X(e^{-j2\pi k/N})$$
$$H_8(k) = X(e^{j(2\pi/N)(k+N/2)})$$
$$H_9(k) = X(e^{-j2\pi k/N})$$

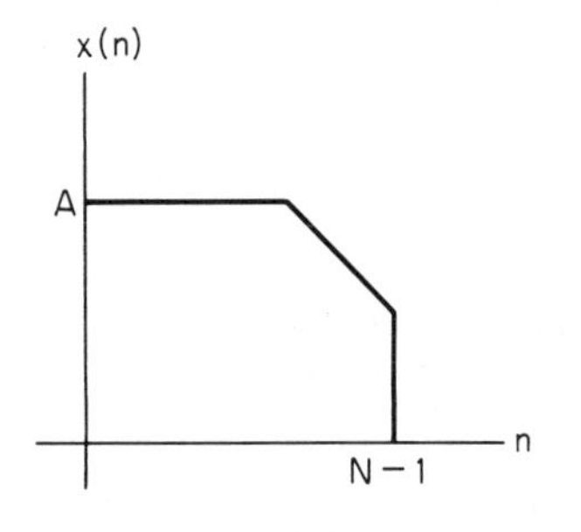

Fig. P3.20

where $T_2 = T_1/2 = P/4M$. Let $X_2(k)$ designate the discrete Fourier transform of one period, starting at $n = 0$, of the periodic sequence $x_2(n)$. Determine how one can obtain $X_2(k)$ directly from $X_1(k)$. Clearly justify your answer.

22. Let $x(n)$ denote an infinite-duration sequence with z-transform $X(z)$, and $x_1(n)$ a finite-duration sequence of length N whose N-point DFT is denoted by $X_1(k)$. Determine the relation between $x(n)$ and $x_1(n)$ if $X(z)$ and $X_1(k)$ are related by

$$X_1(k) = X(z)|_{z=W_N^{-k}}, \qquad k = 0, 1, 2, \ldots, N-1$$

where $W_N = e^{-j(2\pi/N)}$.

23. Let $X(e^{j\omega})$ denote the Fourier transform of the sequence $x(n) = (\frac{1}{2})^n u(n)$. Let $y(n)$ denote a finite-duration sequence of length 10; i.e., $y(n) = 0$, $n < 0$, and $y(n) = 0$, $n \geq 10$. The 10-point DFT of $y(n)$, denoted by $Y(k)$, corresponds to 10 equally spaced samples of $X(e^{j\omega})$; i.e., $Y(k) = X(e^{j2\pi k/10})$. Determine $y(n)$.

24. Let $x(n)$ and $y(n)$ denote the input and output of a stable, causal IIR system with difference equation of the form

$$y(n) = \sum_{k=1}^{p} a_k y(n-k) + x(n) \qquad \text{(P3.24-1)}$$

We wish to determine, using an N-point DFT, N samples of the frequency response of the system $H(e^{j\omega})$ equally spaced on the unit circle, i.e., for $\omega_k = (2\pi/N)k$, $k = 0, 1, \ldots, N-1$. One possible approach is to generate the impulse response and apply the results of Problem 23. For the class of systems characterized by Eq. (P3.24-1), there is a simpler possibility.

(a) Assuming that $p < N$, show how to compute the required N samples of $H(e^{j\omega})$ from the coefficients of the difference equation (P3.24-1) using a single N-point DFT and some simple arithmetic.

(b) Does this result generalize to the class of systems characterized by the class of difference equations of the form

$$y(n) = \sum_{k=1}^{p} a_k y(n-k) + \sum_{k=0}^{q} b_k x(n-k)?$$

25. Consider a finite-duration sequence $x(n)$ of length N so that $x(n) = 0$ for $n < 0$ and for $n > N - 1$. We want to compute samples of its z-transform $X(z)$ at M equally spaced points around the unit circle. One of the samples is to be at $z = 1$. The number of samples M is *less than* the duration of the sequence N; i.e., $M < N$. Determine and justify a procedure for obtaining the M samples of $X(z)$ by computing *only once* the M-point DFT of an M-point sequence obtained from $x(n)$.

26. Consider two finite-duration sequences $x(n)$ and $y(n)$ where both are zero for $n < 0$ and with

$$x(n) = 0, \qquad n \geq 8$$
$$y(n) = 0, \qquad n \geq 20$$

The 20-point DFTs of each of the sequences are multiplied and the inverse DFT computed. Let $r(n)$ denote the inverse DFT. Specify which points in $r(n)$ correspond to points that would be obtained in a *linear* convolution of $x(n)$ and $y(n)$.

27. We want to filter a very long string of data with an FIR filter whose unit-sample response is 50 samples long. We wish to implement this filter with an FFT using the *overlap-save technique*. To do this: (1) The input sections must be overlapped by V samples; and (2) from the output due to each section we must extract M samples such that when these samples from each section are butted together, the resulting sequence is the desired filtered output. Assume that the input segments are 100 samples long and that the size of the DFT is 128 ($= 2^7$) points. Further assume that the output sequence from the circular convolution is indexed from point 0 to point 127.

(a) Determine V.

(b) Determine M.

(c) Determine the index of the beginning and the end of the M points extracted; i.e., determine which of the 128 points from the circular convolution is extracted to be abutted with the result from the previous section.

28. A modified discrete Fourier transform (MDFT) has been proposed (J. L. Vernet, "Real Signals Fast Fourier Transform: Storage Capacity and Step Number Reduction by Means of an Odd Discrete Fourier Transform," *Proc. IEEE*, Oct. 1971, pp. 1531–1532) which computes samples of the z-transform on the unit circle offset from those computed by the DFT. In particular, with $X_M(k)$ denoting the MDFT of $x(n)$,

$$X_M(k) = X(z)\big|_{z=e^{j[2\pi k/N + \pi/N]}}, \qquad k = 0, 1, 2, \ldots, N-1$$

Assume that N is even.

(a) The N-point MDFT of a sequence $x(n)$ corresponds to the N-point DFT of a sequence $x_M(n)$ which is easily constructed from $x(n)$. Determine $x_M(n)$ in terms of $x(n)$.

(b) If $x(n)$ is real, all the points in the DFT are not independent, since the DFT is conjugate symmetric; i.e., $X(k) = X^*((-k))_N \mathscr{R}_N(k)$. Similarly, if $x(n)$ is real, all the points in the MDFT are not independent. Determine, for $x(n)$ real, the relationship between points in $X_M(k)$.

(c) (1) Let $R(k) = X_M(2k)$; i.e., $R(k)$ contains the even-numbered points in $X_M(k)$. From your answer in part (b), show that $X_M(k)$ can be recovered from $R(k)$.

(2) $R(k)$ can be considered to be the $N/2$-point MDFT of an $N/2$-point sequence $r(n)$. Determine a simple expression relating $r(n)$ directly to $x(n)$.

According to parts (b) and (c), the N-point MDFT of a real sequence $x(n)$ can be computed by forming $r(n)$ from $x(n)$ and then computing the $N/2$-point MDFT of $r(n)$. The next two parts are directed at showing the MDFT can be used to implement a linear convolution.

(d) Consider three sequences, $x_1(n)$, $x_2(n)$, and $x_3(n)$, all of length N. Let $X_{1M}(k)$, $X_{2M}(k)$, and $X_{3M}(k)$, respectively, denote the MDFTs of the three sequences. If

$$X_{3M}(k) = X_{1M}(k)X_{2M}(k)$$

express $x_3(n)$ in terms of $x_1(n)$ and $x_2(n)$. Your expression must be of the form of a single summation over a "combination" of $x_1(n)$ and $x_2(n)$ in the same style as (but not identical to) a circular convolution.

(e) It is convenient to refer to the result in part (d) as a modified circular convolution. If the sequences $x_1(n)$ and $x_2(n)$ are both zero for $n \geq N/2$, show that the modified circular convolution of $x_1(n)$ and $x_2(n)$ is identical to the linear convolution of $x_1(n)$ and $x_2(n)$.

29. We wish to implement a digital lowpass filter by sectioning the input, computing the DFT, multiplying by the DFT of the filter impulse response, computing the inverse DFT, and fitting the sections together. The number of nonzero points in the unit-sample response is M and the length of an input section is N.

There are two methods proposed for obtaining the $(N + M - 1)$-point DFT $H(k)$ that represents the filter. In both methods we begin with an M-point DFT $H_M(k)$ given by

$$H_M(k) = \begin{cases} 1, & 0 \leq k < M/4 \\ 1, & 3M/4 < k \leq M - 1 \\ 0, & \text{otherwise} \end{cases}$$

Assume that M is divisible by 4. The M-point inverse DFT of $H_M(k)$ is denoted by $h_M(n)$.

Method A: $H(k)$ is the $(N + M - 1)$-point DFT of $h_A(n)$ defined as

$$h_A(n) = \begin{cases} h_M(n), & 0 \leq n \leq M - 1 \\ 0, & M - 1 < n \leq N + M - 2 \end{cases}$$

Method B: $H(k)$ is the $(N + M - 1)$-point DFT of $h_B(n)$ defined as

$$h_B(n) = \begin{cases} h_M(n), & 0 \leq n \leq M/2 - 1 \quad \text{(assume } M \text{ even)} \\ h_M(n - N + 1), & N + M/2 - 1 \leq n \leq N + M - 2 \\ 0, & \text{otherwise} \end{cases}$$

(a) Sketch $h_A(n)$ and $h_B(n)$. The input is sectioned by overlapping $M - 1$ points as depicted in Fig. P3.29.

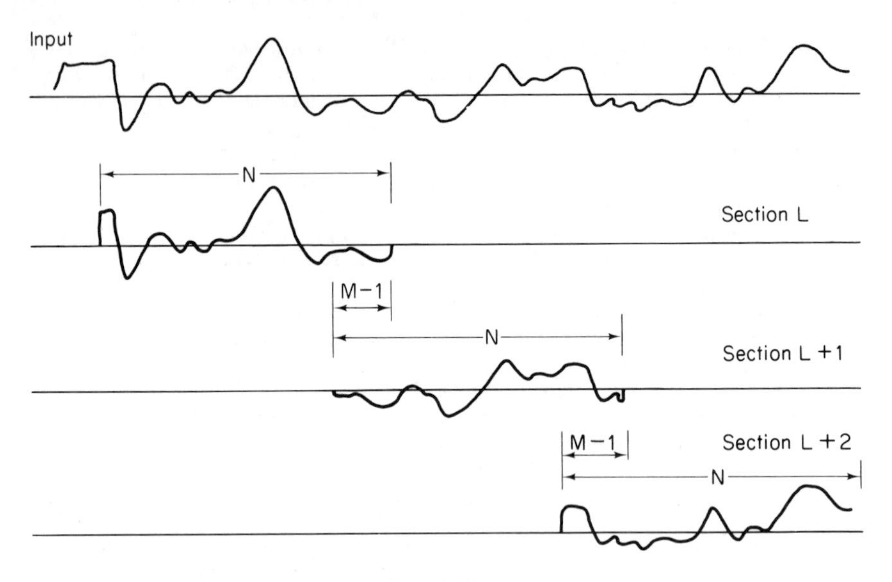

Fig. P3.29

Two methods are proposed for fitting the filtered sections back together.
Method 1: The filtered sections are put together by abutting only the last N points.
Method 2: The filtered sections are put together by discarding the last $M/2$ points and the first $(M/2) - 1$ points and abutting the remaining N points.

(b) With each of the two methods for putting sections together there are the two possibilities for obtaining $H(k)$. Call these four choices A-1, A-2, B-1, and B-2. Will all four result in a linear time-invariant filter? Justify your answer.

(c) For each of the four choices that results in a linear time-invariant filter, determine and sketch the unit-sample response. Which of these is the "best" lowpass filter?

30. The problem often arises in which a signal $x(n)$ has been filtered by a linear time-invariant system that results in a distorted signal $y(n)$ and we wish to recover the original signal. This can often be done by processing $y(n)$ with a linear time-invariant system whose impulse response is such that the overall impulse response of the two systems in cascade is a unit sample. This is generally referred to as *inverse filtering*.

We have discussed in Sec. 3.8 the procedure for implementing an FIR filter using the DFT. The procedure involves, in part, multiplying the DFT of the input, $X(k)$ (or input sections), by $H(k)$, the DFT of the system unit sample response to produce $Y(k)$, the DFT of the output. It is often suggested, incorrectly, that the sequence $h_1(n)$, whose DFT is $1/H(k)$, is the impulse response of the inverse filter. The purpose of this problem is to indicate why that suggestion is incorrect.

Consider a linear time-invariant system with impulse response $h(n)$ given by

$$h(n) = \delta(n) - \tfrac{1}{2}\delta(n - n_0)$$

as sketched in Fig. P3.30. This system is an idealized example of a system that would introduce reverberation. Assume that $N = 4n_0$.

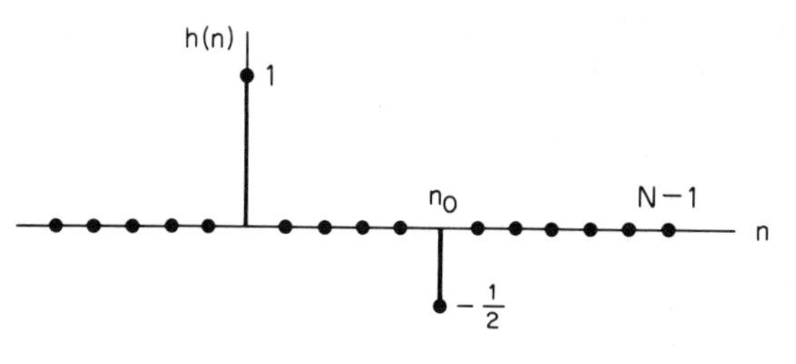

Fig. P3.30

(a) Determine the N-point DFT $H(k)$ of $h(n)$.
(b) Now consider the N-point DFT $H_1(k)$ of a sequence $h_1(n)$ specified by

$$H_1(k) = \frac{1}{H(k)} \qquad k = 0, 1, \ldots, N-1$$

Determine $h_1(n)$. [*Hint:* If you have trouble evaluating the IDFT summation directly, express $H_1(k)$ as a polynomial in $W_N^{n_0 k}$ and observe that the $h_1(n)$ are the coefficients of W_N^{nk}.]
(c) Sketch $h_1(n)$ as determined in part (b).
(d) By evaluating the linear convolution of $h(n)$ and $h_1(n)$, show that $h(n) * h_1(n)$ is *not* a unit sample $\delta(n)$ and, consequently, $h_1(n)$ is *not* the unit sample response for the inverse system.
(e) Compute and sketch the N-point *circular* convolution of $h(n)$ and $h_1(n)$.
(f) Determine the unit-sample response $h_i(n)$ of the inverse system for $h(n)$. This can be done in a number of ways. One is to note that with $H(z)$ and $H_i(z)$ denoting the z-transforms of $h(n)$ and $h_i(n)$, $H_i(z) = 1/H(z)$. The inverse z-transforms of $H_i(z)$ can then be evaluated by long division.
(g) Refer to Problem 23 and determine and verify numerically the relationship between $h_1(n)$ and $h_i(n)$.

31. We have shown that FIR filters can be implemented using the DFT. We have also shown that IIR filters can be implemented recursively. In this problem we shall consider implementing IIR filters using the DFT.

In particular, let us assume that $x(n)$ represents the input sequence to a linear, shift-invariant, stable, causal system which is characterized by the unit-sample response $h(n)$. Assume that $x(n) = 0$, $n < 0$. We shall denote the

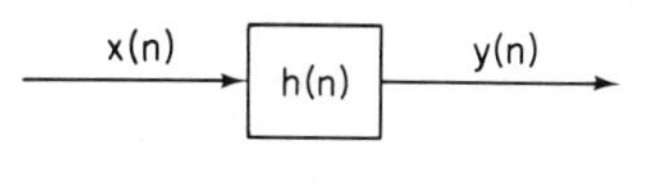

Fig. P3.31

output sequence by $y(n)$ (Fig. P3.31). We shall assume further that $H(z)$, the z-transform of $h(n)$, has only poles. Therefore, we can write

$$H(z) = \frac{1}{D(z)} = \frac{1}{1 + \sum_{i=1}^{Q} d_i z^{-i}} = \sum_{i=1}^{Q} \frac{A_i}{1 - z_i z^{-1}} \qquad \text{(P3.31-1)}$$

where $\{z_i, i = 1, 2, \ldots, Q\}$ represent the poles of the system (by assumption all simple poles) and $\{A_i, i = 1, 2, \ldots, Q\}$ represent the associated residues.

First we must partition $x(n)$, $h(n)$, and $y(n)$ into N-point sections, where $N \geq Q + 1$. Doing this we can define

$$x_m(n) = \begin{cases} x(n), & mN \leq n < (m+1)N \\ 0, & \text{otherwise} \end{cases}$$

$$h_m(n) = \begin{cases} h(n), & mN \leq n < (m+1)N \\ 0, & \text{otherwise} \end{cases}$$

$$y_m(n) = \begin{cases} y(n), & mN \leq n < (m+1)N \\ 0, & \text{otherwise} \end{cases}$$

where m is a nonnegative integer that specifies the section. In this problem we shall show that we can write

$$w_m(n) = x_m(n) * h_0(n) + g(n) * y_{m-1}(n - N)$$

$$y_m(n) = \begin{cases} w_m(n) + w_{m-1}(n), & mN \leq n < (m+1)N \\ 0, & \text{elsewhere} \end{cases} \qquad \text{(P3.31-2)}$$

where $g(n)$ is a sequence of length $\leq N$.

(a) Assuming such a $g(n)$ can be found, explain the operation that must be performed to compute $y_m(n)$ in Eq. (P3.31-2), using DFT techniques. A simple flow chart is sufficient.

To verify Eq. (P3.31-2), we shall assume that a $g(n)$ exists and then show that (1) it is of finite duration and (2) it can be evaluated from parameters that are known.

(b) Show that

$$H(z) = \frac{Y(z)}{X(z)} = \frac{H_0(z)}{1 - z^{-N}G(z)} = \frac{1}{D(z)}$$

where $H_0(z)$ is the z-transform of $h_0(n)$, the first section of the unit-sample response.

(c) By explicitly evaluating $H_0(z)$ and by using the relationship that you derived in part (b), find $G(z)$ in terms of $D(z)$, $\{z_i\}$, $\{A_i\}$, N, and Q. Verify that it is the z-transform of a Q-point sequence. [*Note:* It is not necessary to evaluate $g(n)$.]

(d) Using the results of this problem, show that in the general case $H(z) = C(z)/D(z)$, where $C(z)$ and $D(z)$ are both polynomials in z^{-1}, the network can be implemented by using the DFT.

Comment: The four parts of this problem can all be worked independently. The significance of the result that you derived in this problem is that an IIR filter can be realized by means of circular convolutions and the DFT. If we use as a basis of comparison the number of multiplications required to implement the discrete system, then $D(z)$ would have to be a polynomial of about 95th degree before this technique would be preferable to a direct implementation of the difference equation relating $y(n)$ to $x(n)$. Thus this result is of more theoretical than practical interest.

This problem is based on a paper by B. Gold and K. L. Jordan, "A Note on Digital Filter Synthesis," *Proc. IEEE* (*Letters*), Vol. 56, Oct. 1968, pp. 1717–1718.

32. Equations (3.53) and (3.54) express two alternative ways of computing a two-dimensional DFT using a one-dimensional DFT. Equations (3.53) correspond to transforming each column of $x(m, n)$ and then each row of the result. Equations (3.54) correspond to transforming rows first and then columns. To obtain some insight into these two alternative procedures it is instructive to consider a simple example.

(a) Compute the two-dimensional DFT of the sequence $\delta(m, n-1)$ using equations (3.53).

(b) Repeat part (a) using equations (3.54).

33. Show that the two-dimensional DFT of a finite-area sequence corresponds to samples of the two-dimensional z-transform. Specify in particular the locations of these samples in (z_1, z_2)-space.

34. In Problem 16, Parseval's relation was verified for the one-dimensional DFT. Determine and verify the corresponding relationship for the two-dimensional DFT.

35. A set of symmetry properties for the two-dimensional DFT exists which is similar to that for the one dimension as derived in Sec. 3.6.3. These properties are again based on the decomposition of a sequence $x(m, n)$ into its conjugate symmetric and conjugate antisymmetric components. Specifically, paralleling Eqs. (3.31)–(3.35), we define

$$x_{ep}(m, n) = \tfrac{1}{2}\{x[((m))_M, ((n))_N] + x^*[((-m))_M, ((-n))_N]\}\mathcal{R}_{M,N}(m, n)$$

$$x_{op}(m, n) = \tfrac{1}{2}\{x[((m))_M, ((n))_N] - x^*[((-m))_M, ((-n))_N]\}\mathcal{R}_{M,N}(m, n)$$

(a) Show that the following properties hold for $x(m, n)$ and its DFT $X(k, l)$:

	Sequence	*DFT*
1.	$x^*(m, n)$	$X^*[((-k))_M, ((-l))]_N\mathcal{R}_{M,N}(k, l)$
2.	$x^*[((-m))_M, ((-n))_N]\mathcal{R}_{M,N}(m, n)$	$X^*(k, l)$
3.	$x_{ep}(m, n)$	$\text{Re}\,[X(k, l)]$
4.	$x_{op}(m, n)$	$j\,\text{Im}\,[X(k, l)]$
5.	$\text{Re}[x(m, n)]$	$X_{ep}(k, l)$
6.	$j\,\text{Im}\,[x(m, n)]$	$X_{op}(k, l)$

(b) Show that if $x(m, n)$ is real, then

(1) $\text{Re}\,[X(k, l)] = \text{Re}\,\{X[((-k))_M, ((-l))_N]\}\mathcal{R}_{M,N}(k, l)$.

(2) $|X(k, l)| = |X[((-k))_M, ((-l))_N]|\mathcal{R}_{M,N}(k, l)$

(3) $\text{Im}\,[X(k, l)] = -\text{Im}\,\{X[((-k))_M, ((-l))_N]\}\mathcal{R}_{M,N}(k, l)$.

4

Flow Graph and Matrix Representation of Digital Filters

4.0 Introduction

In Chapters 1 and 2 we discussed the representation of linear shift-invariant digital systems in terms of the difference equation relating the input and output sequences and the system function relating their z-transforms. In those chapters we were only concerned with the input–output relation of the system. In implementing a digital filter on a digital computer or with special-purpose hardware, the input–output relation must be converted to a computational algorithm. The algorithm is essentially specified in terms of a set of basic computations or elements. For the implementation of discrete-time systems described by linear constant-coefficient difference equations it is convenient to choose as these elements the basic operations of addition, delay, and multiplication by a constant. The computational algorithm for implementing the filter is then defined by a structure or network consisting of an interconnection of these basic operations. As an illustration consider a system with a system function of the form

$$H(z) = \frac{\sum_{k=0}^{M} b_k z^{-k}}{1 - \sum_{k=1}^{N} a_k z^{-k}} = \frac{Y(z)}{X(z)} \tag{4.1}$$

The difference equation relating input and output is easily written down directly from the system function and is given by

$$y(n) = \sum_{k=1}^{N} a_k y(n - k) + \sum_{k=0}^{M} b_k x(n - k) \tag{4.2}$$

We can interpret Eq. (4.2) directly as a computational algorithm in which the delayed values of the input are multiplied by the coefficients b_k, the delayed values of the output are multiplied by the coefficients a_k, and all the resulting products are added. Alternatively, as we shall see in this chapter, there are an infinite variety of structures that will result in the same relationship between the input samples $x(n)$ and the output samples $y(n)$.

In the next sections we present the description of filter structures in terms of block-diagram, flow-graph, and matrix notation. In addition, we consider a number of basic structures. Although two structures may be equivalent (with regard to their input–output characteristics) for infinite-precision representations of the coefficients and variables, they may have greatly different characteristics when the precision is limited. In this chapter we discuss the effects of finite-precision representation of the coefficients in the filter. The effects of truncation or rounding of intermediate computations are discussed in Chapter 9.

The representations for digital networks that we will use represent, in essence, the flow of signals in the realization of a digital filter. Important network properties exist, and in general a well-established theory for digital networks is available in the theory of linear signal flow graphs. In Sec. 4.6 one such important and useful property, Tellegen's theorem for signal flow graphs, is presented and discussed.

4.1 Signal Flow Graph Representation of Digital Networks

The realization of a digital filter requires that past values of the output, input, and intermediate sequences be available. This implies the need for delay or storage of these past values. Furthermore, we must provide means for multiplication of the delayed samples by the coefficients, and means for adding together the resulting products. The digital filter may be realized either by use of the storage registers, arithmetic unit, and control unit of a general-purpose computer, or special digital hardware may be designed to perform the required computations. In the first case, the filter structure may be thought of as a specification of a computational algorithm, from which a computer program is derived. In the latter case it is often convenient to think of the filter structure as specifying a hardware configuration.

Corresponding to the basic operations required for implementation of a digital filter, the basic elements required to represent a difference equation pictorially are an adder, a delay, and a constant multiplier. Commonly used symbols are shown in Fig. 4.1. Physically, Fig. 4.1(a) represents a

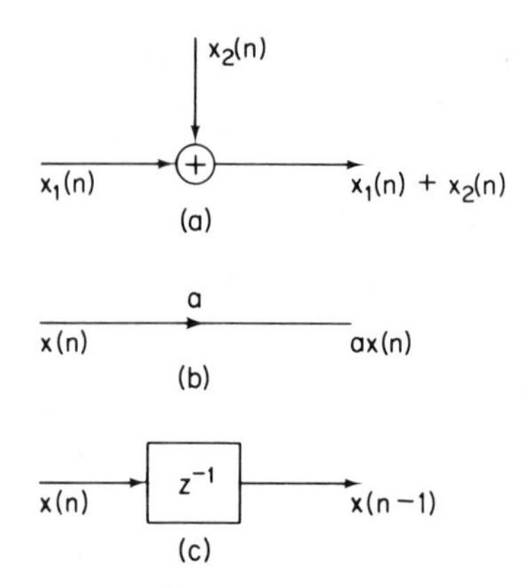

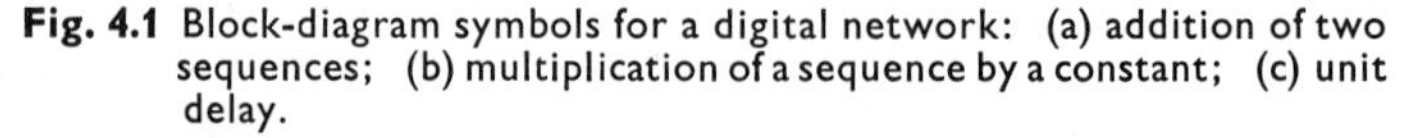
Fig. 4.1 Block-diagram symbols for a digital network: (a) addition of two sequences; (b) multiplication of a sequence by a constant; (c) unit delay.

means for adding together two sequences, Fig. 4.1(*b*) represents a means for multiplying a sequence by a constant, and Fig. 4.1(c) represents a means for storing the previous value of a sequence. The representation used for a single sample delay arises from the fact that the z-transform of $x(n - 1)$ is simply z^{-1} times the z-transform of $x(n)$.

As an example of the representation of a difference equation in terms of these elements, consider the second-order equation

$$y(n) = a_1 y(n - 1) + a_2 y(n - 2) + bx(n)$$

The digital network corresponding to this equation is shown in Fig. 4.2. In terms of a computer program, Fig. 4.2 shows explicitly that we must provide storage for the variables $y(n - 1)$ and $y(n - 2)$ and also the constants a_1, a_2, and b. Furthermore, we see that a given output sample is to be computed by forming the products $a_1 y(n - 1)$ and $a_2 y(n - 2)$, adding them together, and then adding the result to the product $bx(n)$. In terms of special digital hardware, Fig. 4.2 indicates that we must provide storage for the variables and constants, as well as means for multiplication and addition.

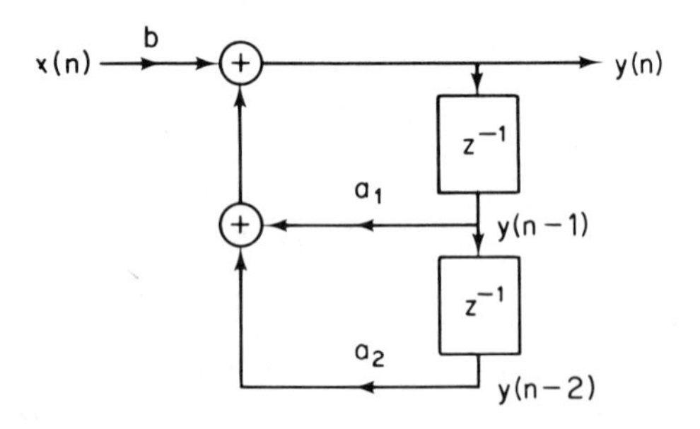

Fig. 4.2 Example of a digital network.

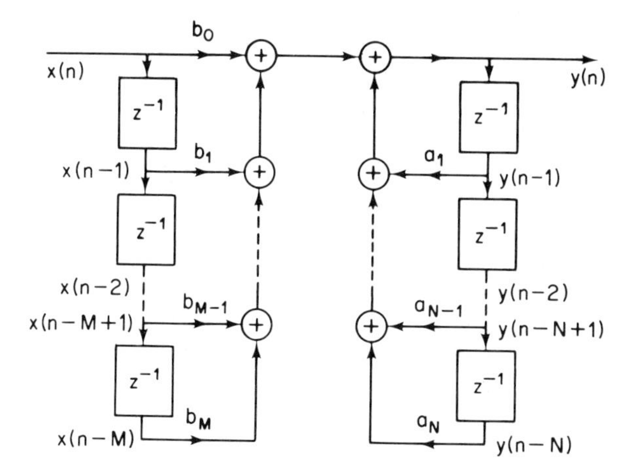

Fig. 4.3 Block-diagram representation for a general *N*th-order difference equation.

Thus diagrams such as Fig. 4.2 serve to depict the complexity of a digital filter algorithm and the amount of hardware required to realize the filter.

As a further example, a block-diagram representation for the general difference equation of Eq. (4.2) is shown in Fig. 4.3. The network of Fig. 4.3 is an explicit graphical representation of the difference equation (4.2). However, it can be rearranged or modified in a variety of ways without changing the overall transfer function. These different rearrangements correspond to different structures for implementing the filter.

In discussing different filter structures it is convenient to use the notation of linear signal flow graphs rather than block diagrams. A linear signal flow graph is essentially equivalent to a block-diagram representation with a few notational differences. In the remainder of this section we present the signal flow graph notation that we will use and in Sec. 4.2 we discuss the equivalence of signal flow graph and matrix representation of digital filters.

A signal flow graph is a network of directed branches that connect at nodes [1,2]. Associated with each node is a variable or node value. The value associated with node k is w_k. Branch (jk) denotes a branch originating at node j and terminating at node k, with the direction from j to k being indicated by an arrowhead on the branch. This is shown in Fig. 4.4. Each branch has an input signal and an output signal. The input signal from node j to branch (jk) is the node value w_j and the output signal from branch jk to node k is denoted v_{jk}. The dependence of a branch output upon the branch input is denoted

$$v_{jk} = f_{jk}[w_j] \tag{4.3}$$

That is, $f_{jk}[\ ]$ stands for the transformation of a branch input into a branch output.

To represent the injection of external inputs or sources into the graph we use *source nodes*. A source node has no entering branches. It is generally

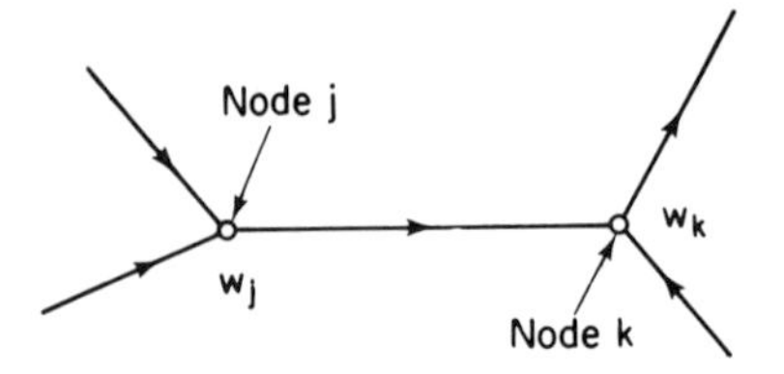

Fig. 4.4 Example of nodes and branches in a signal flow graph.

convenient to number the source nodes separately from the network nodes. The node value at source node j will be denoted as x_j and the output of a branch connecting source node j to network node k will be denoted as s_{jk}. A source node is depicted in Fig. 4.5.

Just as inputs to a graph can be obtained from a source node, we often find it convenient to extract outputs from a graph through *sink nodes*, i.e., nodes having only entering branches. A sink node is depicted in Fig. 4.6. The node value at sink node k will be denoted by y_k and the output of a branch connecting network node j to sink node k will be denoted as r_{jk}.

By definition, the node value at each node in a network is given by the sum of the outputs of all the branches entering the node. It is sometimes notationally convenient to assume that there is a branch in each direction between every pair of network nodes and that each source node is connected to each network node, although clearly some of the branch outputs may then be zero. With this notation, and assuming N network nodes in the graph numbered from 1 to N, M source nodes numbered from 1 to M, and P sink nodes numbered 1 to P the set of equations represented by the graph is

$$w_k = \sum_{\substack{j=1 \\ \text{(network} \\ \text{nodes)}}}^{N} v_{jk} + \sum_{\substack{j=1 \\ \text{(source} \\ \text{nodes)}}}^{M} s_{jk}, \qquad k = 1, 2, \ldots, N \tag{4.4a}$$

$$y_k = \sum_{\substack{j=1 \\ \text{(network} \\ \text{nodes)}}}^{N} r_{jk} \qquad k = 1, 2, \ldots, P \tag{4.4b}$$

As an example of how the above flow-graph concepts can be applied to the representation of a difference equation, consider the block diagram of the

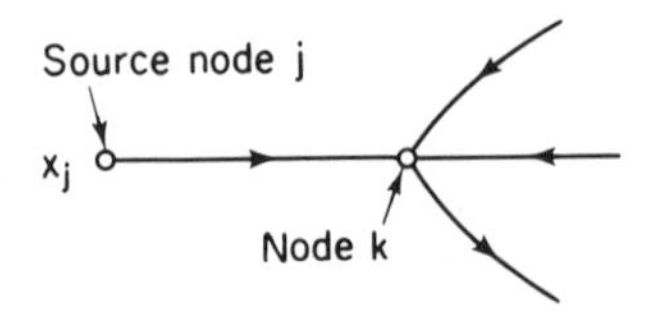

Fig. 4.5 Representation of a source node.

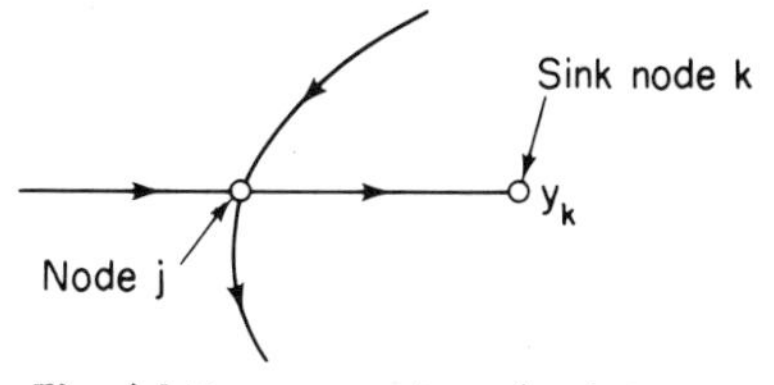

Fig. 4.6 Representation of a sink node.

first-order digital filter in Fig. 4.7(a). A signal flow graph corresponding to this network is shown in Fig. 7.4(b). In this case the branch variables are sequences. There is one source node connected to node 1 and one sink node connected to node 3. Writing equations (4.4) for this graph we obtain

$$w_1(n) = s_{11}(n) + v_{41}(n)$$

$$w_2(n) = v_{12}(n)$$

$$w_3(n) = v_{23}(n) + v_{43}(n)$$

$$w_4(n) = v_{24}(n)$$

$$y(n) = w_3(n)$$

From Fig. 4.7(b) we note that the branch outputs are

$$s_{11}(n) = x(n)$$

$$v_{12}(n) = f_{12}(w_1) = w_1(n)$$

$$v_{23}(n) = f_{23}(w_2) = w_2(n)$$

$$v_{43}(n) = f_{43}(w_4) = bw_4(n)$$

$$v_{41}(n) = f_{41}(w_4) = aw_4(n)$$

$$v_{24}(n) = f_{24}(w_2) = w_2(n-1) \qquad \text{(delay)}$$

$$y(n) = w_3(n)$$

These equations can be solved for $y(n)$ in terms of $x(n)$, yielding the single first-order difference equation,

$$y(n) = ay(n-1) + x(n) + bx(n-1)$$

We observe that all branches except one (branch 2,4) can be represented by a branch transmittance; i.e., the output signal is simply a constant times the branch input. Branch (2,4), however, is represented by a delay operator. Indeed, in general, $f_{jk}[\ ]$ denotes an operator that transforms an input sequence into a branch output sequence. In the case of linear time-invariant discrete-time systems characterized by difference equations, the signal flow

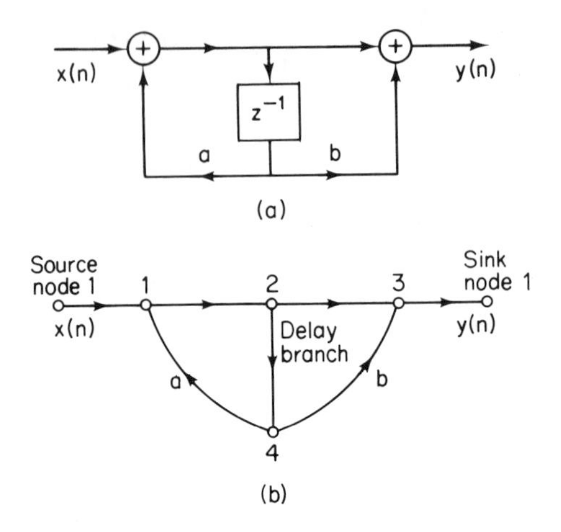

Fig. 4.7 (a) Block-diagram representation of a first-order digital filter; (b) structure of the signal flow graph corresponding to the block diagram in (a).

graph can also represent relationships among z-transforms. In this case, each branch can be characterized by its transfer function, i.e., a transmittance that is a function of z. Thus

$$V_{jk}(z) = F_{jk}(z)W_j(z) \tag{4.5}$$

For such graphs, the transmission of each branch will be placed next to the arrowhead that indicates the branch direction. For convenience, a branch with no transmittance explicitly indicated will be assumed to have unit transmittance. Also, it is convenient to sometimes indicate the node variables as sequences rather than z-transforms in which case it is understood that branch transmittances z^{-1} imply a unit delay of the input sequence. The graph of the previous example is shown again in Fig. 4.8 in the form to be used henceforth.

Comparison of Fig. 4.7(a) with Fig. 4.8 shows that there is direct correspondence between branches in the digital network and branches in the flow graph. In fact, the only important difference between the two is that nodes

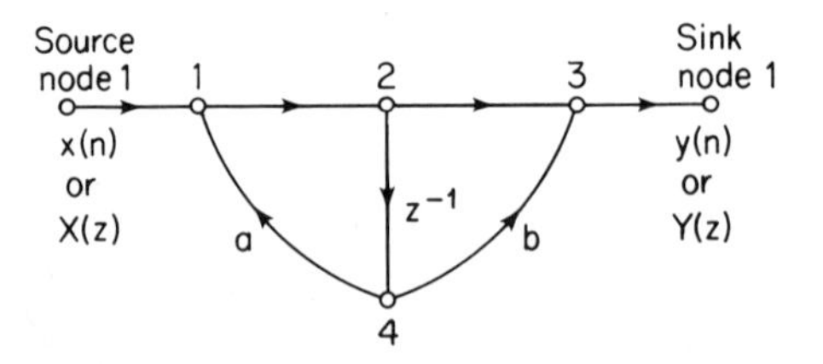

Fig. 4.8 Signal flow graph of Fig. 4.7(b) with the branch transmittances indicated.

in the flow graph represent both branch points and adders in the digital network. For example, nodes 1 and 3 correspond to adders and nodes 2 and 4 correspond to branch points in the original network. Signal flow graphs serve the purpose of pictorial representation of a discrete-time system and have the further advantage of lending themselves to graphical manipulations which provide insight into the behavior of the network.

When all the branches in a flow graph can be represented by transmittances, the set of equations represented by the graph is a linear set of equations, and manipulation of the graph is equivalent to manipulating this set of equations. Thus an alternative description of a linear flow graph is in terms of this set of equations, corresponding to a matrix representation, as considered in the next section.

4.2 Matrix Representation of Digital Networks

Equations (4.4) represented by a signal flow graph, expressed in terms of z-transforms, are

$$W_k(z) = \sum_{j=1}^{N} V_{jk}(z) + \sum_{j=1}^{M} S_{jk}(z), \qquad k = 1, 2, \ldots, N \tag{4.6a}$$

$$Y_k(z) = \sum_{j=1}^{N} R_{jk}(z) \qquad k = 1, 2, \ldots, P \tag{4.6b}$$

If the graph represents a linear shift-invariant system, then each branch can be represented by a transmittance. It will generally be convenient to assume that the branches from the source nodes to network nodes and from network nodes to sink nodes have a transmittance that is constant, independent of z. There is no loss of generality in this since if necessary a network node can be inserted connected directly to the source node, and this node in turn can have branches to other nodes with nonconstant transmittance. A similar procedure can be used, if necessary, for sink nodes. Then

$$V_{jk}(z) = F_{jk}(z)W_j(z) \tag{4.7}$$

and

$$S_{jk}(z) = b_{jk}X_j(z) \tag{4.8}$$

$$R_{jk}(z) = c_{jk}W_j(z) \tag{4.9}$$

Substituting Eqs. (4.7)–(4.9) into (4.6) we obtain the set of linear algebraic equations

$$W_k(z) = \sum_{j=1}^{N} F_{jk}(z)W_j(z) + \sum_{j=1}^{M} b_{jk}X_j(z) \tag{4.10a}$$

$$Y_k(z) = \sum_{j=1}^{N} c_{jk}W_j(z) \tag{4.10b}$$

These equations can be written compactly in matrix form as

$$\mathbf{W}(z) = \mathbf{F}^t(z)\mathbf{W}(z) + \mathbf{B}^t\mathbf{X}(z) \tag{4.11a}$$

$$\mathbf{Y}(z) = \mathbf{C}^t\,\mathbf{W}(z) \tag{4.11b}$$

where $\mathbf{W}(z)$ is a column vector of values $W_k(z)$, $k = 1, 2, \ldots, N$, $\mathbf{X}(z)$ is a column vector of values $X_j(z), j = 1, 2, \ldots, M$, and $\mathbf{Y}(z)$ is a column vector of values $Y_j(z)$ $j = 1, 2, \ldots, P$. The matrix $\mathbf{F}^t(z)$ is the transpose of the $N \times N$ matrix $\mathbf{F}(z)$ given by

$$\mathbf{F}(z) = \{F_{kj}(z)\} \tag{4.12}$$

For branches that are nonexistent in the flow graph, or equivalently have zero transmittance, the corresponding entry in the matrix $F_{kj}(z)$ is zero. $\mathbf{B}^t$ is the $N \times M$ transpose of the matrix

$$\mathbf{B} = \{b_{kj}\} \tag{4.13}$$

and $\mathbf{C}^t$ is the $P \times N$ transpose of the matrix $\mathbf{C} = \{c_{kj}\}$. The matrix transpose, denoted by t, is required in Eqs. (4.11) for consistency between the subscript conventions of flow graphs and matrices.

Equation (4.11a) can be solved for $\mathbf{W}(z)$ by matrix inversion as

$$\begin{aligned}\mathbf{W}(z) &= [\mathbf{I} - \mathbf{F}^t(z)]^{-1}\mathbf{B}^t\mathbf{X}(z)\\ &= \mathbf{T}^t(z)\mathbf{X}(z)\end{aligned} \tag{4.14a}$$

where

$$\mathbf{T}^t(z) = [\mathbf{I} - \mathbf{F}^t(z)]^{-1}\mathbf{B}^t = \{T_{jk}(z)\} \tag{4.14b}$$

$\mathbf{T}(z)$ is called the *transfer function matrix* of the system. As a consequence of Eq. (4.14), the signal at the kth node, $W_k(z)$, can be expressed as

$$W_k(z) = \sum_{j=1}^{M} T_{jk}(z)X_j(z) \tag{4.15}$$

i.e., each node variable is expressed as a linear combination of the sources. If only one source node (source node a) is nonzero, with value $X_a(z)$ and there is only one sink node with value $Y(z)$ so that $Y(z) = \mathbf{C}^t\mathbf{W}(z)$ then the output $Y(z)$ is given by

$$Y(z) = \mathbf{C}^t\mathbf{W}(z) = \mathbf{C}^t\mathbf{T}^t X_a(z) \tag{4.16}$$

so that the system is characterized by the system function

$$H(z) = \mathbf{C}^t\mathbf{T}^t \tag{4.17}$$

In the case for which each branch system function is at most first order, i.e., is either a constant multiplier or a constant multiplier together with a unit delay, the elements of the matrix $\mathbf{F}^t(z)$ in Eq. (4.11a) are either a constant or a constant times z^{-1}. It is convenient to separate the matrix elements that involve no delay from those that do. Thus we can express $\mathbf{F}^t(z)$ as

$$\mathbf{F}^t(z) = \mathbf{F}_c^t + z^{-1}\mathbf{F}_d^t \tag{4.18}$$

where $\mathbf{F}_c^t$ and $\mathbf{F}_d^t$ are $N \times N$ matrices. The matrix equation (4.11a) can therefore be expressed as

$$\mathbf{W}(z) = \mathbf{F}_c^t\mathbf{W}(z) + z^{-1}\mathbf{F}_d^t\mathbf{W}(z) + \mathbf{B}^t\mathbf{X}(z) \tag{4.19}$$

Likewise, the expression for $\mathbf{T}^t(z)$ becomes

$$\mathbf{T}^t(z) = [\mathbf{I} - \mathbf{F}_c^t - z^{-1}\mathbf{F}_d^t]^{-1}\mathbf{B}^t \tag{4.20}$$

where $\mathbf{I}$ is the identity matrix. Since $\mathbf{F}_c^t$ and $\mathbf{F}_d^t$ are constant, independent of z, the inverse z-transform of Eq. (4.19) is

$$\mathbf{w}(n) = \mathbf{F}_c^t\mathbf{w}(n) + \mathbf{F}_d^t\mathbf{w}(n-1) + \mathbf{B}^t\mathbf{x}(n) \tag{4.21a}$$

Also, Eq. (4.11b) implies that

$$\mathbf{y}(n) = \mathbf{C}^t\mathbf{w}(n) \tag{4.21b}$$

Equations (4.21) can, of course, be written directly from the flow graph, or, conversely, we can draw the flow graph directly from the set of equations.

EXAMPLE. As an example, consider the first-order system of Fig. 4.9. The set of equations implied by this graph is

$$\begin{bmatrix} w_1(n) \\ w_2(n) \\ w_3(n) \\ w_4(n) \end{bmatrix} = \begin{bmatrix} 0 & 0 & 0 & a_1 \\ 1 & 0 & 0 & 0 \\ 0 & b_0 & 0 & b_1 \\ 0 & 0 & 0 & 0 \end{bmatrix} \begin{bmatrix} w_1(n) \\ w_2(n) \\ w_3(n) \\ w_4(n) \end{bmatrix} + \begin{bmatrix} 0 & 0 & 0 & 0 \\ 0 & 0 & 0 & 0 \\ 0 & 0 & 0 & 0 \\ 0 & 1 & 0 & 0 \end{bmatrix} \begin{bmatrix} w_1(n-1) \\ w_2(n-1) \\ w_3(n-1) \\ w_4(n-1) \end{bmatrix} + \begin{bmatrix} 1 \\ 0 \\ 0 \\ 0 \end{bmatrix} [x(n)]$$

$$y(n) = [0 \quad 0 \quad 1 \quad 0] \begin{bmatrix} w_1(n) \\ w_2(n) \\ w_3(n) \\ w_4(n) \end{bmatrix} = w_3(n) \tag{4.22}$$

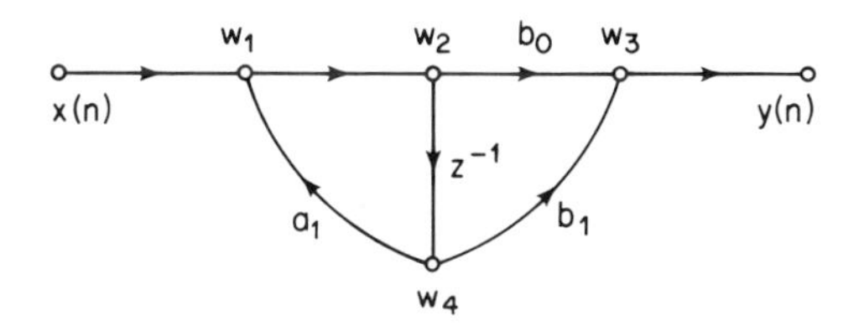

Fig. 4.9 Signal flow graph for a first-order system. The corresponding equations are those of equations (4.22).

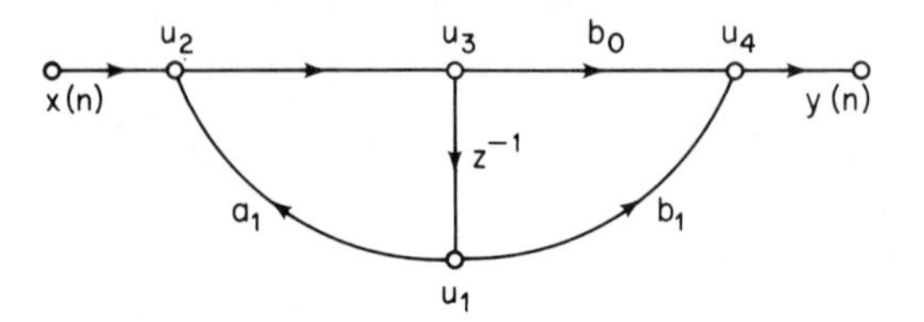

Fig. 4.10 Flow graph of Fig. 4.9 with the nodes renumbered.

It is clear that the form of the matrices $\mathbf{F}_c^t$ and $\mathbf{F}_d^t$ depends on the ordering of the equations or, equivalently, the numbering of the nodes. In Fig. 4.10 we show the graph of Fig. 4.9 with the node variables (denoted u_k) numbered in a different order. For this graph we obtain the set of equations

$$\begin{bmatrix} u_1(n) \\ u_2(n) \\ u_3(n) \\ u_4(n) \end{bmatrix} = \begin{bmatrix} 0 & 0 & 0 & 0 \\ a_1 & 0 & 0 & 0 \\ 0 & 1 & 0 & 0 \\ b_1 & 0 & b_0 & 0 \end{bmatrix} \begin{bmatrix} u_1(n) \\ u_2(n) \\ u_3(n) \\ u_4(n) \end{bmatrix} + \begin{bmatrix} 0 & 0 & 1 & 0 \\ 0 & 0 & 0 & 0 \\ 0 & 0 & 0 & 0 \\ 0 & 0 & 0 & 0 \end{bmatrix} \begin{bmatrix} u_1(n-1) \\ u_2(n-1) \\ u_3(n-1) \\ u_4(n-1) \end{bmatrix} + \begin{bmatrix} 0 \\ 1 \\ 0 \\ 0 \end{bmatrix} [x(n)]$$

$$y(n) = [0 \quad 0 \quad 0 \quad 1] \begin{bmatrix} u_1(n) \\ u_2(n) \\ u_3(n) \\ u_4(n) \end{bmatrix} = u_4(n) \tag{4.23}$$

By examining the flow graph of Fig. 4.9, or, equivalently, equations (4.22), it is clear that the node variables cannot be generated in sequence, i.e., first w_1, then w_2, etc., since, for example, w_4 is needed to compute w_1. On the other hand, the same flow graph with the nodes renumbered as in Fig. 4.10 can be computed in sequence.

In some cases there is no way of reordering the nodes in a flow graph to permit the node variables to be generated in sequence. A flow graph of this type is said to be *noncomputable*. A simple example of a noncomputable flow graph is shown in Fig. 4.11, where all the transmittances are constants. It must be emphasized that the fact that the flow graph is noncomputable does *not* mean that the set of equations representing the flow graph cannot be solved. It means that they cannot be solved directly for each node variable successively.

In the equations (4.23) corresponding to the flow graph of Fig. 4.10 we note that the matrix F_c^t has only zeros above the main diagonal and that furthermore the main diagonal elements are all zero. This is not true in equations (4.22), which correspond to the flow graph of Fig. 4.9. As shown in Problem 3 of this chapter, a necessary and sufficient condition for the computability of a flow graph is that the nodes can be numbered so that in the

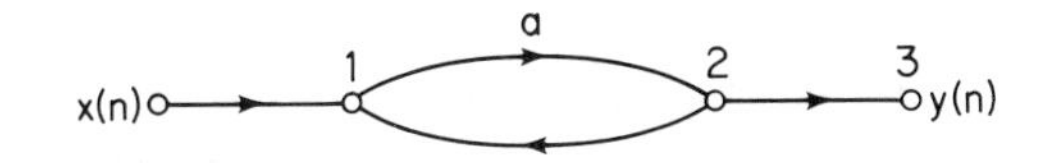

Fig. 4.11 Example of a noncomputable flow graph.

matrix F_c^t there are only zeros above the main diagonal and the main diagonal elements are zero. It can also be shown [3] that an equivalent necessary and sufficient condition for the computability of a flow graph is that there be no loops in the flow graph with no delay branches. In the flow graph of Fig. 4.11, for example, there is a loop with no delay, and consequently the graph is not computable.

The set of equations (4.23) can be obtained from (4.22) by permuting the node variables. In matrix notation this is accomplished by a linear transformation of the vector $\mathbf{w}(n)$ into the vector $\mathbf{u}(n)$; i.e.,

$$\mathbf{u}(n) = \mathbf{P}\mathbf{w}(n) \tag{4.24}$$

where $\mathbf{P}$ is a nonsingular $N \times N$ constant matrix. In the previous example,

$$\mathbf{P} = \begin{bmatrix} 0 & 0 & 0 & 1 \\ 1 & 0 & 0 & 0 \\ 0 & 1 & 0 & 0 \\ 0 & 0 & 1 & 0 \end{bmatrix} \tag{4.25}$$

Permuting the variables of a flow graph is a simple example of a more general principle that is placed in evidence by the matrix representation. More generally, if $\mathbf{P}$ is any nonsingular matrix, we can write

$$\mathbf{w}(n) = \mathbf{P}^{-1}\mathbf{u}(n) \tag{4.26}$$

and substitute this expression into equations (4.21) to obtain

$$\begin{aligned} \mathbf{u}(n) &= \mathbf{P}\mathbf{F}_c^t\mathbf{P}^{-1}\mathbf{u}(n) + \mathbf{P}\mathbf{F}_c^t\mathbf{P}^{-1}\mathbf{u}(n-1) + \mathbf{P}\mathbf{B}^t\mathbf{x}(n) \\ \mathbf{y}(n) &= \mathbf{C}^t\mathbf{P}^{-1}\mathbf{u}(n) \end{aligned} \tag{4.27}$$

Equations (4.27) are in the same form as equations (4.21) but correspond to a different flow graph or network representation. Thus there is a variety of network realizations of the same system function. This important fact serves as the basis for Secs. 4.3 and 4.4.

In equations (4.21) the present values of the node variables are expressed in terms of their present and past values; i.e., $w(n)$ is expressed in terms of $w(n)$ and $w(n-1)$. It is sometimes convenient to have network representations in which $w(n)$ is expressed explicitly in terms only of past values of the node variables and current values of input. This corresponds to the case in which the matrix $\mathbf{F}_c^t$ in Eq. (4.21) is equal to zero. With the network equations expressed in this form one can compute the value of the node vector $\mathbf{w}(n_1)$ at any time n_1 given the value of the node vector $\mathbf{w}(n_0)$ at some time n_0 and the input vector $\mathbf{s}(n)$ for $n_0 \leq n \leq n_1$ (see Problem 5 of this chapter). This is similar to what is commonly referred to as a state-variable representation, although in our formulation the number of states (nodes) is generally greater than the number of essential state variables necessary to represent the network. A state-variable representation of this type can be obtained from a representation in the form of Eqs (4.21). Specifically, Eq. (4.21a) can be written as

$$[\mathbf{I} - \mathbf{F}_c^t]\mathbf{w}(n) = \mathbf{F}_d^t\mathbf{w}(n-1) + \mathbf{B}^t\mathbf{x}(n)$$

Assuming that the matrix $[\mathbf{I} - \mathbf{F}_c^t]$ is nonsingular, we can solve for $\mathbf{w}(n)$ as

$$\mathbf{w}(n) = [\mathbf{I} - \mathbf{F}_c^t]^{-1}\mathbf{F}_d^t\mathbf{w}(n-1) + [\mathbf{I} - \mathbf{F}_c^t]^{-1}\mathbf{B}^t\mathbf{x}(n)$$

Defining

$$\mathbf{D} = [\mathbf{I} - \mathbf{F}_c^t]^{-1}\mathbf{B}^t \tag{4.28a}$$

and

$$\mathbf{A} = [\mathbf{I} - \mathbf{F}_c^t]^{-1}\mathbf{F}_d^t \tag{4.28b}$$

then we obtain the matrix representation

$$\mathbf{w}(n) = \mathbf{A}\mathbf{w}(n-1) + \mathbf{D}\mathbf{x}(n) \tag{4.29a}$$

$$\mathbf{y}(n) = \mathbf{C}^t\mathbf{w}(n) \tag{4.29b}$$

It can be shown (see Problem 4 of this chapter) that if the system is computable, the matrix $[\mathbf{I} - \mathbf{F}_c^t]$ is nonsingular. Thus we can find a matrix representation of the form of Eqs (4.29) for any computable system.

The transformation of node variables according to Eq. (4.26) corresponds to a transformation of the flow graph; i.e., it changes the structure of the digital network while leaving the input-output relationships the same. Since there are many transformations of the type of Eq. (4.26), there are many network realizations of the same system function. In practice there are a number of network configurations that are most commonly used. In the next section we consider some of the most common network forms. Although these networks are related to one another through linear transformations of the form of Eq. (4.26), we shall find it more convenient to arrive at these standard network forms by other means. To do this it is most convenient to consider IIR and FIR systems separately.

4.3 Basic Network Structures for IIR Systems

The previous two sections, which have dealt with the representation of linear shift-invarient discrete-time systems as networks, have made it clear that to each rational system function there corresponds a variety of different network configurations. One consideration in the choice between these different realizations is computational complexity. That is, networks with the fewest constant multipliers and the fewest delay branches are often most desirable, since multiplication is a time-consuming operation and each delay element corresponds to the use of a memory register. Consequently, a reduction in the number of constant multipliers means an increase in speed, and a reduction in the number of delays means a reduction in memory requirements. On the other hand, the effects of finite register length in actual hardware realizations of digital filters depend on the structure, and it is sometimes desirable to use a structure that does not have the minimum number of multipliers and delays but is less sensitive to finite-register-length effects. Thus it is important to discuss some of the more commonly used network forms. In this section we discuss IIR systems. In Sec. 4.5 we discuss FIR systems.

4.3.1 *Direct Form*

If we consider a rational system function of the form

$$H(z) = \frac{\sum_{k=0}^{M} b_k z^{-k}}{1 - \sum_{k=1}^{N} a_k z^{-k}} \tag{4.30}$$

then we recall that the input and output of such a system satisfy the difference equation

$$y(n) = \sum_{k=1}^{N} a_k y(n-k) + \sum_{k=0}^{M} b_k x(n-k) \tag{4.31}$$

Since this difference equation can be written down directly by inspection of the transfer function in the form of Eq. (4.30), the network corresponding to Eq. (4.31) is called the *direct form I* realization of the system characterized

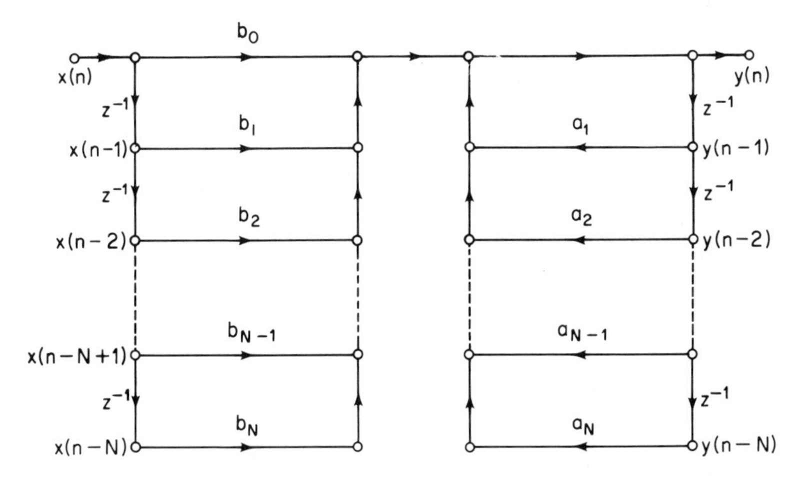

Fig. 4.12 Direct form I realization of an Nth-order difference equation.

by Eq. (4.30). This network is shown in Fig. 4.12. In drawing the network, we have, for convenience, assumed that $M = N$. Clearly if this is not the case, some of the branch transmittances in Fig. 4.12 will be zero. It should be observed that in Fig. 4.12 we have drawn the network graph so that each node has at most two inputs. Although this convention clearly results in more nodes than are necessary, it is consistent with the fact that in digital hardware or software realizations, the sum of more than two numbers is carried out by forming separate sums of two numbers; i.e., addition in digital hardware typically involves only two numbers at a time.

Since the set of coefficients b_k corresponds to the numerator polynomial, and the set of coefficients a_k corresponds to the denominator polynomial of $H(z)$, we can interpret Fig. 4.12 as consisting of a cascade of two networks, the first realizing the zeros and the second realizing the poles. In the case of linear shift-invariant systems, the overall input–output relation of a cascade is independent of the order in which subsystems are cascaded. This property suggests a second direct-form realization. Specifically, if we realize the poles of $H(z)$ first, corresponding to the right half of Fig. 4.12, and then realize the zeros, we obtain the network shown in Fig. 4.13.

We observe that the two strings of branches with transmission z^{-1} have the same input, and thus only one string is required. Thus the network in Fig. 4.13 can be redrawn as in Fig. 4.14. This network configuration is often referred to as the *direct form* II. We note that it has the minimum number of branches (M or N, whichever is greater) with transmission z^{-1}. That is, the minimum number of delay registers is required for this realization of the transfer function $H(z)$ in Eq. (4.30).

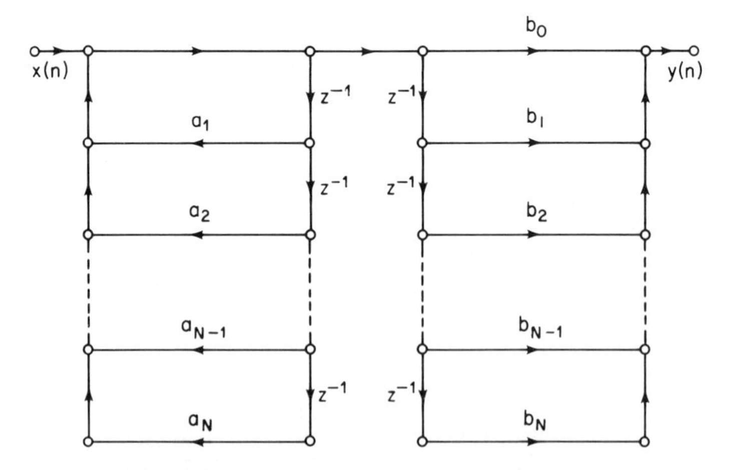

Fig. 4.13 Network of Fig. 4.12 with the order in which the poles and zeros are reverse cascaded.

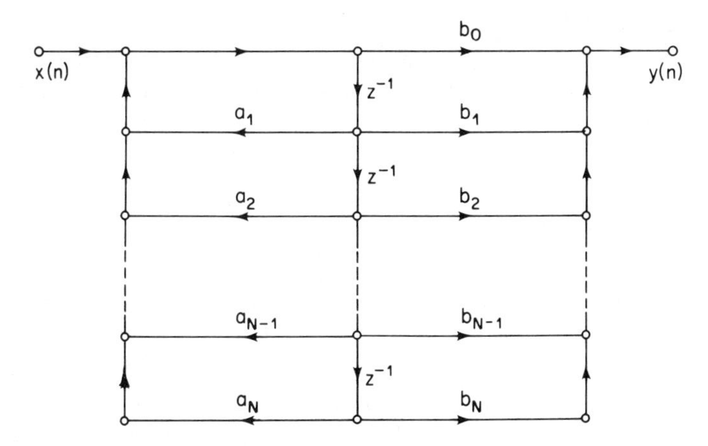

Fig. 4.14 Network of Fig. 4.13 with the two strings of delays combined into one. The resulting network form is referred to as the direct form II and has the minimum possible number of delays.

An example of another network structure having the minimum number of delays is obtained by writing the system function in the form

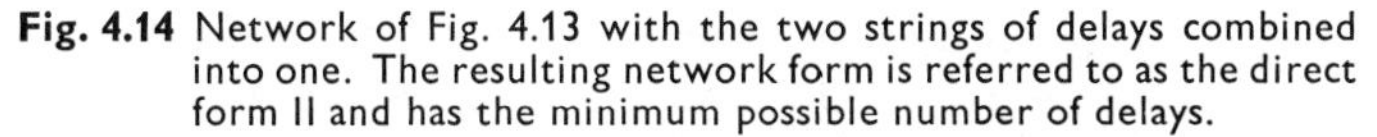

$$H(z) = b_0 + \sum_{r=1}^{M} \frac{(b_0 a_r + b_r) z^{-r}}{1 - \sum_{k=1}^{N} a_k z^{-k}} \tag{4.32}$$

The corresponding flow-graph is depicted in Fig. 4.15. There are many distinctly different forms that have this same minimum number of delays. Such network forms are often referred to as *canonic form* networks.

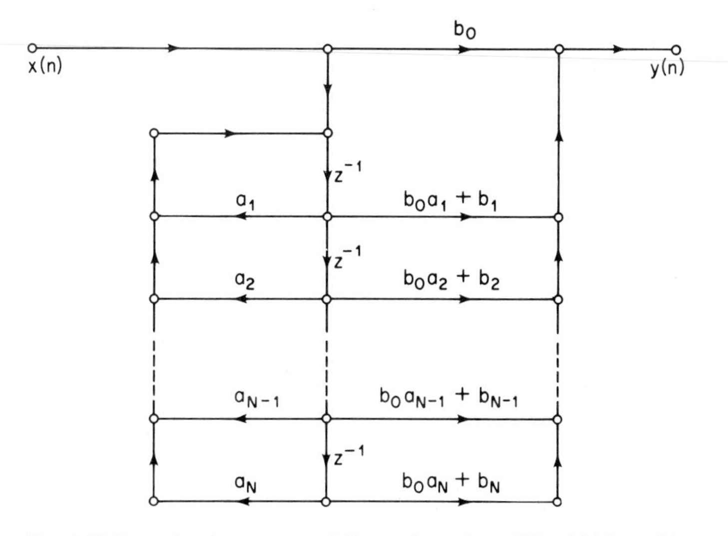

Fig. 4.15 Example of a structure different from that of Fig. 4.14 but which also has the minimum number of delays.

4.3.2 Cascade Form

The direct-form network structures were obtained directly from the system function $H(z)$ written in the form of Eq. (4.30). If we factor the numerator and denominator polynomials, we can write $H(z)$ as

$$H(z) = A \frac{\prod_{k=1}^{M_1} (1 - g_k z^{-1}) \prod_{k=1}^{M_2} (1 - h_k z^{-1})(1 - h_k^* z^{-1})}{\prod_{k=1}^{N_1} (1 - c_k z^{-1}) \prod_{k=1}^{N_2} (1 - d_k z^{-1})(1 - d_k^* z^{-1})} \tag{4.33}$$

where $M = M_1 + 2M_2$ and $N = N_1 + 2N_2$. In this expression the first-order factors represent real zeros at g_k and real poles at c_k, and the second-order factors represent complex-conjugate zeros at h_k and h_k^* and complex-conjugate poles at d_k and d_k^*. This represents the most general distribution of poles and zeros when all the coefficients a_k and b_k are real in Eq. (4.30). Equation (4.33) suggests a set of structures consisting of a cascade of first- and second-order subsystems. There is clearly considerable freedom in the choice of composition of the subsystems and in the order in which the subsystems are cascaded. In practice, however, it is important to implement the cascade realization using a minimum of storage, and hardware realizations are often implemented by time sharing or multiplexing a single second-order section. Thus it is convenient to think in general in terms of a cascade form based on the equation

$$H(z) = A \prod_{k=1}^{[(N+1)/2]} \frac{1 + \beta_{1k} z^{-1} + \beta_{2k} z^{-2}}{1 - \alpha_{1k} z^{-1} - \alpha_{2k} z^{-2}} \tag{4.34}$$

where $[(N + 1)/2]$ means the largest integer contained in $(N + 1)/2$. In this case we have assumed that $M \leq N$. In writing $H(z)$ in this form we have assumed that the real poles and real zeros have been combined in pairs. If there are an odd number of real zeros, one of the coefficients β_{2k} will be zero. Likewise if there are an odd number of real poles, one of the coefficients α_{2k} will be zero. The previous discussion of direct-form structures makes it clear that we can implement a cascade structure with minimum memory by using the direct form II realization of each second-order subsystem. A cascade realization of a sixth-order system using a direct form II realization of each second-order subsystem would appear as in Fig. 4.16.

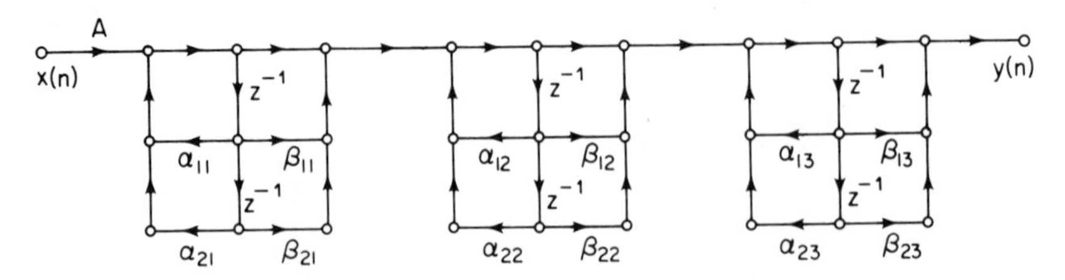

Fig. 4.16 Cascade structure with a direct form II realization of each second-order subsystem.

As we have pointed out, there is considerable flexibility in the manner in which the poles and zeros are paired together and in the order in which the resulting second-order subsystems are cascaded. It turns out that, although all such pairings and orderings are equivalent for infinite-precision arithmetic, they may differ considerably in practice, owing to finite-word-length effects.

4.3.3 Parallel Form

As an alternative to factoring the numerator and denominator polynomials of $H(z)$, we can express $H(z)$ as a partial-fraction expansion in the form

$$H(z) = \sum_{k=1}^{N_1} \frac{A_k}{1 - c_k z^{-1}} + \sum_{k=1}^{N_2} \frac{B_k(1 - e_k z^{-1})}{(1 - d_k z^{-1})(1 - d_k^* z^{-1})} + \sum_{k=0}^{M-N} C_k z^{-k} \quad (4.35)$$

If the coefficients a_k and b_k in Eq. (4.30) are real, then the quantities A_k, B_k, C_k, c_k, and e_k are all real. If $M < N$, then the term $\sum_{k=0}^{M-N} C_k z^{-k}$ is not included in Eq. (4.35).

The expression in Eq. (4.35) can be interpreted as a parallel combination of first- and second-order systems, or we may group the real poles in pairs so that we can write $H(z)$ as

$$H(z) = \sum_{k=0}^{M-N} C_k z^{-k} + \sum_{k=1}^{[(N+1)/2]} \frac{\gamma_{0k} + \gamma_{1k} z^{-1}}{1 - \alpha_{1k} z^{-1} - \alpha_{2k} z^{-2}} \quad (4.36)$$

A typical example of the parallel-form realization with $M = N$ is shown in Fig. 4.17.

The network structures discussed above represent the basic structures commonly encountered. There are, however, many others which correspond to the variety of ways in which the network equations or the numerator and denominator polynomials in the system function can be manipulated. Some examples can be found in references [4–8].

4.4 Transposed Forms

The theory of linear signal flow graphs provides a variety of manipulative procedures for transforming signal flow graphs into different forms while leaving the transmission between input and output unchanged. One of these procedures, called *flow-graph reversal* or *transposition*, leads to a set of transposed filter structures. Specifically, transposition of a flow graph is accomplished by reversing the directions of all branches in the network. For single input–single output systems, the resulting graph has the same transfer function as the original graph, with the input and output interchanged. This theorem is a direct consequence of Mason's gain formula for signal flow graphs [1], and it will be also shown to follow from a basic network theorem to be proved in a later section of this chapter.

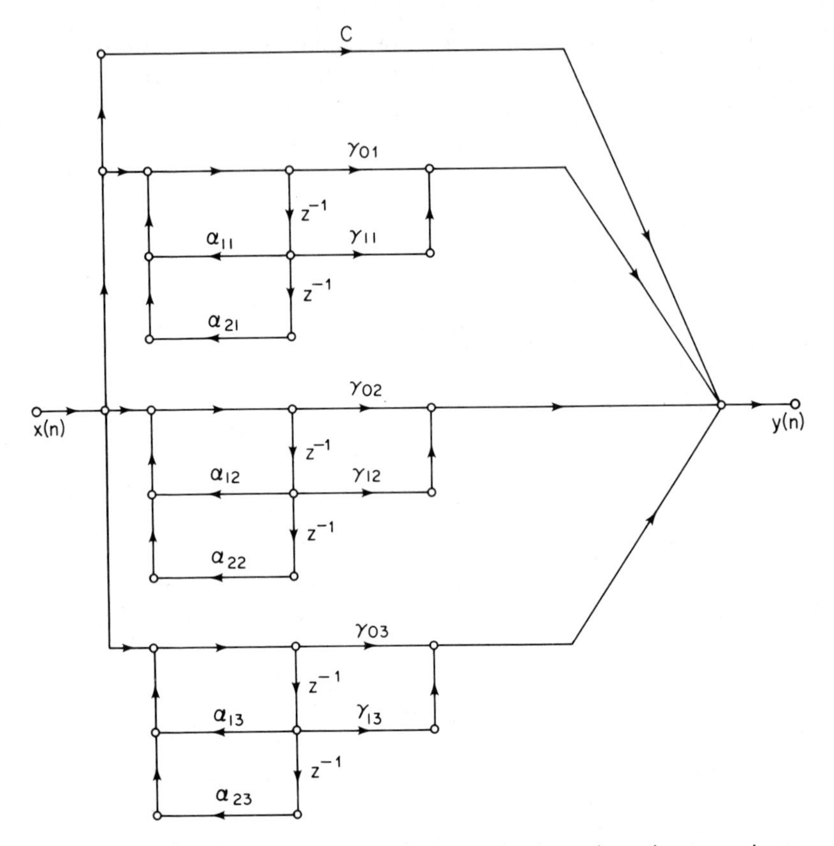

Fig. 4.17 Parallel-form realization with the real and complex poles grouped in pairs.

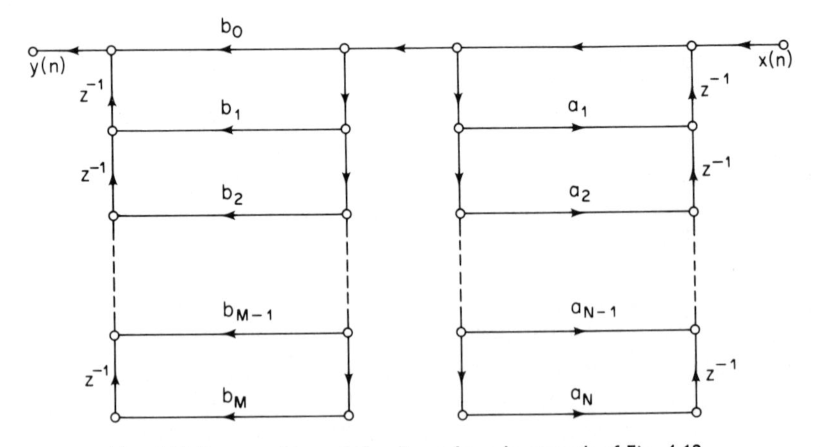

Fig. 4.18 Transposition of the direct form I network of Fig. 4.12.

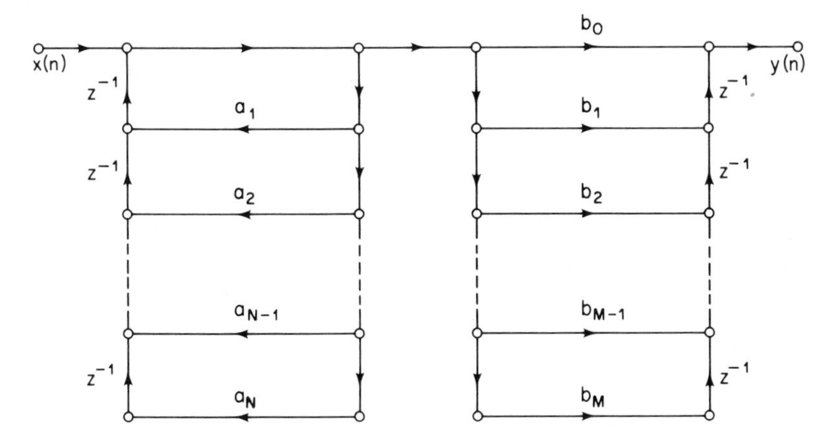

Fig. 4.19 Network of Fig. 4.16 redrawn to observe the convention that the input is on the left and the output on the right. The result is the transposed direct form I structure.

The result of applying this theorem to the direct form I network of Fig. 4.12 is shown in Fig. 4.18. If we redraw this structure with input on the left and output on the right, as is our convention, we obtain Fig. 4.19. Similarly, we obtain from Fig. 4.14 the transposed direct form II shown in Fig. 4.20.

Given the previously stated transposition theorem, it is clear that any network configuration can be transposed to yield another network form, with the same number of delay branches and the same number of coefficients.

4.5 Basic Network Structures for FIR Systems

The previous discussion has been directed toward realizations of systems having infinite-duration impulse responses. Of necessity, the realization of

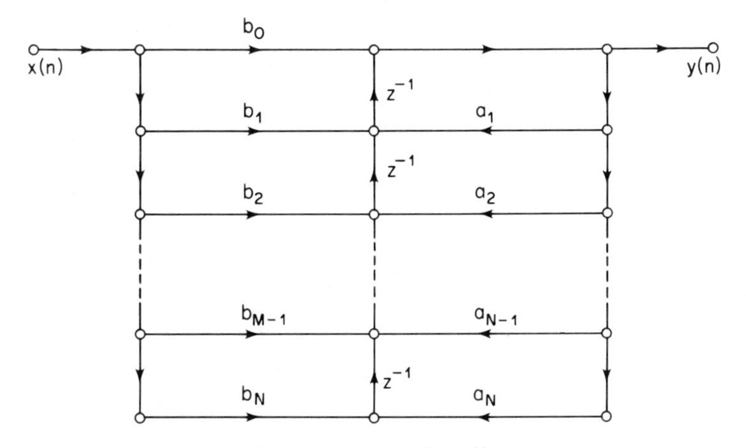

Fig. 4.20 Transposed direct form II structure.

such systems involves a recursive computational algorithm. In the case of causal systems with finite-duration impulse responses, realizations generally take the form of a nonrecursive computational algorithm. For such systems the system function is of the form

$$H(z) = \sum_{n=0}^{N-1} h(n)z^{-n} \tag{4.37}$$

That is, if the impulse response is N samples in duration, then $H(z)$ is a polynomial in z^{-1} of degree $N - 1$. Thus $H(z)$ has $N - 1$ poles at $z = 0$ and $N - 1$ zeros that can be anywhere in the finite z-plane. Just as for IIR systems, FIR systems have a variety of alternative forms of implementation. In this section we shall discuss the most important network forms for FIR systems.

4.5.1 Direct Form

The direct-form realization follows directly from the convolution sum relationship written in the form

$$y(n) = \sum_{k=0}^{N-1} h(k)x(n - k) \tag{4.38}$$

A signal flow graph for Eq. (4.38) is shown in Fig. 4.21. This structure can be

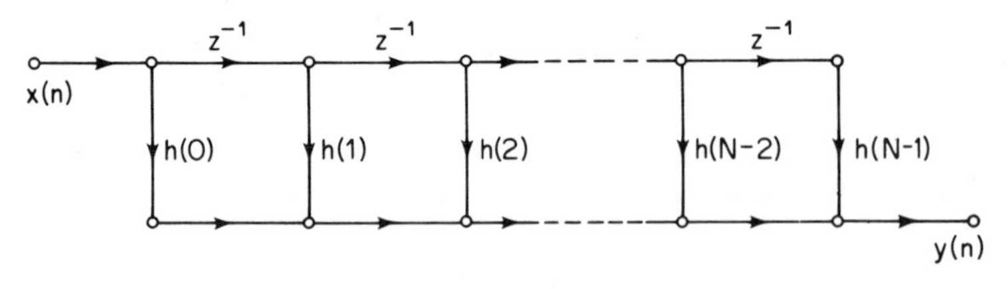

Fig. 4.21 Direct-form realization of an FIR system.

seen to be identical to that of Fig. 4.14 when all the coefficients a_k are zero. Thus the direct-form structure for FIR systems is a special case of the direct-form structure for IIR systems.

The network of Fig. 4.21 corresponds to the most straightforward ordering of the additions and multiplications implied by Eq. (4.38). Clearly there are many other ways that the computations can be organized, and thus

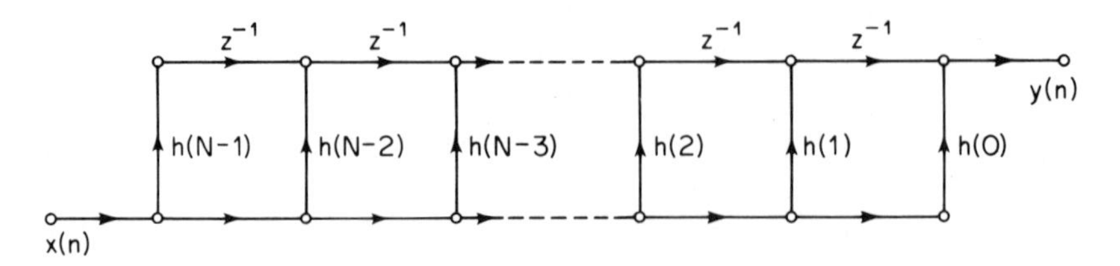

Fig. 4.22 Transposition of the network of Fig. 4.21.

there are many other theoretically equivalent network structures. For example, the transposition theorem introduced in the previous section can be applied to Fig. 4.21 to obtain the transposed direct-form structure of Fig. 4.22.

4.5.2 *Cascade Form*

An alternative to the direct form is suggested by writing $H(z)$ as a product of second-order factors as in

$$H(z) = \prod_{k=1}^{[N/2]} (\beta_{0k} + \beta_{1k}z^{-1} + \beta_{2k}z^{-2}) \tag{4.39}$$

where if N is even, one of the coefficients β_{2k} will be zero, corresponding to the fact that for N even, $H(z)$ has an odd number of real roots. The network

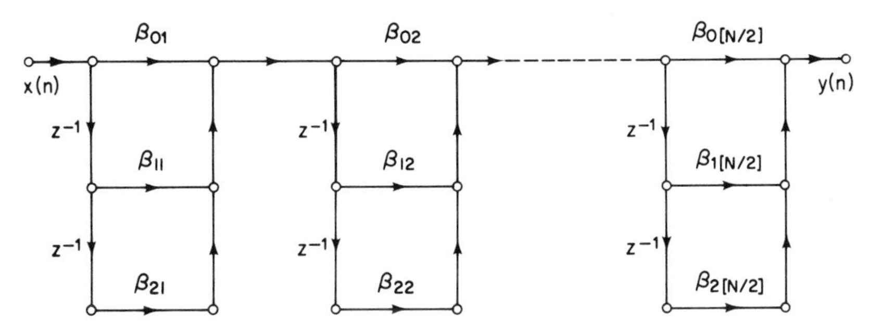

Fig. 4.23 Cascade-form realization of an FIR system.

corresponding to Eq. (4.39) is shown in Fig. 4.23, where we have realized each second-order factor in the direct form depicted in Fig. 4.21.

4.5.3 *Networks for Linear-Phase FIR Systems*

In many applications it is desirable to design filters to have linear phase. In this way, signals in the passband of the filter are reproduced exactly at the filter output except for a delay corresponding to the slope of the phase. One of the most important features of FIR systems is that they can be designed to have exactly linear phase. The unit-sample response for a causal FIR system with linear phase has the property that

$$h(n) = h(N - 1 - n) \tag{4.40}$$

To see that this condition implies linear phase, we write Eq. (4.37) as

$$\begin{aligned} H(z) &= \sum_{n=0}^{(N/2)-1} h(n)z^{-n} + \sum_{n=N/2}^{N-1} h(n)z^{-n} \\ &= \sum_{n=0}^{(N/2)-1} h(n)z^{-n} + \sum_{n=0}^{(N/2)-1} h(N-1-n)z^{-(N-1-n)} \end{aligned}$$

where N is assumed to be even. Using Eq. (4.40) we can write

$$H(z) = \sum_{n=0}^{(N/2)-1} h(n)[z^{-n} + z^{-(N-1-n)}] \tag{4.41}$$

If N is odd, it is easily shown that

$$H(z) = \sum_{n=0}^{[(N-1)/2]-1} h(n)[z^{-n} + z^{-(N-1-n)}] + h\left(\frac{N-1}{2}\right)z^{-[(N-1)/2]} \tag{4.42}$$

If we evaluate Eqs. (4.41) and (4.42) for $z = e^{j\omega}$, we obtain, for N even,

$$H(e^{j\omega}) = e^{-j\omega[(N-1)/2]}\left\{\sum_{n=0}^{(N/2)-1} 2h(n) \cos\left[\omega\left(n - \frac{N-1}{2}\right)\right]\right\}$$

and for N odd,

$$H(e^{j\omega}) = e^{-j\omega[(N-1)/2]}\left\{h\left(\frac{N-1}{2}\right) + \sum_{n=0}^{[(N-3)/2]-1} 2h(n) \cos\left[\omega\left(n - \frac{N-1}{2}\right)\right]\right\}$$

In both cases, the sums in brackets are real, implying a linear phase shift corresponding to delay of $(N-1)/2$ samples. [Note that for N even, $(N-1)/2$ is not an integer.]

Equations (4.41) and (4.42) imply direct-form network implementations which require $N/2$ (N even) or $(N+1)/2$ (N odd) multiplications rather than the N multiplications called for in the general case shown in Fig. 4.21. These networks are shown in Fig. 4.24 for N even and Fig. 4.25 for N odd. Transposed forms can, of course, be obtained from Figs. 4.24 and 4.25, as we have done before.

Imposing the symmetry condition of Eq. (4.40) on the coefficients of the polynomial $H(z)$ causes the zeros of $H(z)$ to occur in mirror-image pairs. That is, if z_0 is a zero of $H(z)$, then $1/z_0$ is also a zero of $H(z)$ (see Problem 15 of Chapter 2). Furthermore, if the coefficients $h(n)$ are real, then the zeros of $H(z)$ occur in complex-conjugate pairs. As a consequence, real zeros not on the unit circle occur in reciprocal pairs. Complex zeros not on the unit circle

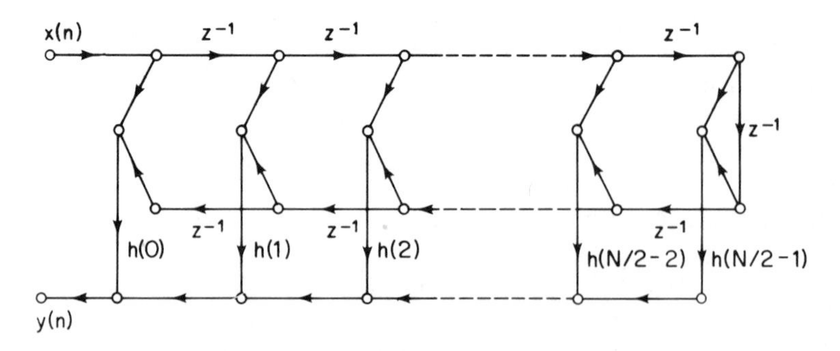

Fig. 4.24 Direct-form realization for an FIR system of even order with linear phase.

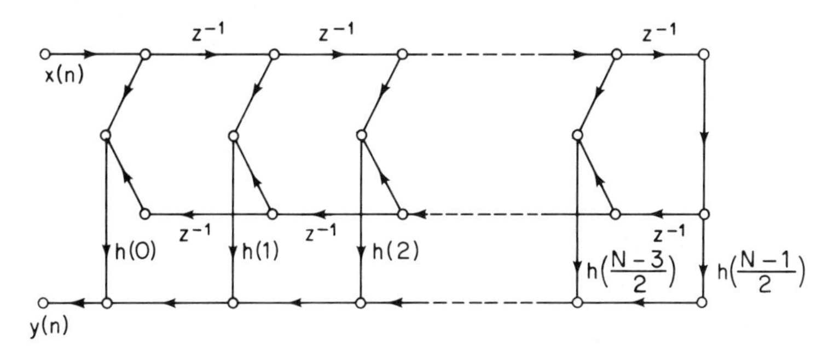

Fig. 4.25 Direct-form realization for an FIR system of odd order with linear phase.

occur in groups of four, corresponding to the complex conjugates and reciprocals. If a zero is on the unit circle, its reciprocal is also its conjugate. Consequently, complex zeros on the unit circle are conveniently grouped in pairs. Real zeros on the unit circle, i.e., zeros at $z = 1$ or $z = -1$, are their own reciprocal and complex conjugate and consequently are identified singly. The four cases are summarized in Fig. 4.26, where the zeros at z_1, z_1^*, $1/z_1$, and $1/z_1^*$ are considered as a group of four. The zeros at z_2 and $1/z_2$ are considered as a group of two, as are the zeros at z_3 and z_3^*. The zero at z_4 is considered singly. Corresponding to this group of zeros, $H(z)$ can be factored into a product of first-, second-, and fourth-order factors. Each of these factors is a polynomial whose coefficients have the same symmetry as does $H(z)$; i.e., each factor is a linear-phase polynomial. Therefore, we can obtain a realization in terms of a cascade of linear-phase systems of first, second, and fourth order. The first-order systems corresponding to a zero at either $z = \pm 1$ require no multiplications. The second-order factors will be of the form

$$1 + az^{-1} + z^{-2}$$

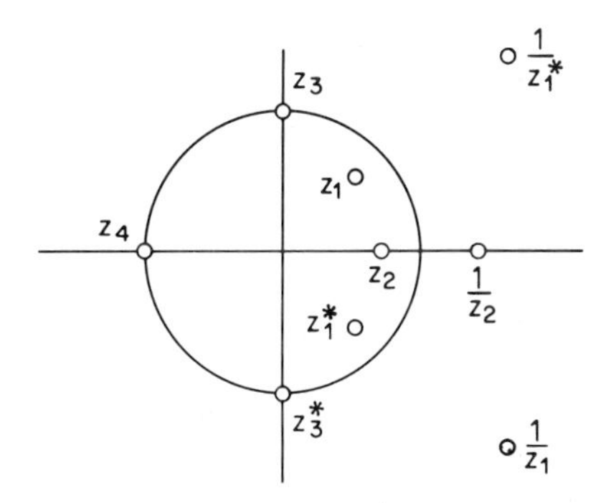

Fig. 4.26 Symmetry of zeros for a linear-phase FIR filter.

and thus will require only one multiplication. The fourth-order sections will have the form

$$a + bz^{-1} + cz^{-2} + bz^{-3} + az^{-4}$$

and will require three multiplications if realized as a linear-phase structure as in Fig. 4.25.

4.5.4 *Frequency-Sampling Structure*

In Chapter 3 we showed that the z-transform of a sequence of finite duration N can be represented in terms of N samples at equal spacing around the unit circle. For an FIR filter, Eq. (3.18) implies that the system function can be expressed as

$$H(z) = (1 - z^{-N})\frac{1}{N}\sum_{k=0}^{N-1}\frac{\tilde{H}(k)}{1 - W_N^{-k}z^{-1}} \tag{4.43}$$

where

$$W_N^{-k} = e^{j(2\pi k/N)}$$

and

$$\tilde{H}(k) = H(W_N^{-k}) = |\tilde{H}(k)|\, e^{j\theta(k)} \tag{4.44}$$

The quantities $\tilde{H}(k)$ are called *frequency samples* since they are simply samples of the frequency response of the system.†

Equation (4.43) suggests that an FIR system can be realized as a cascade

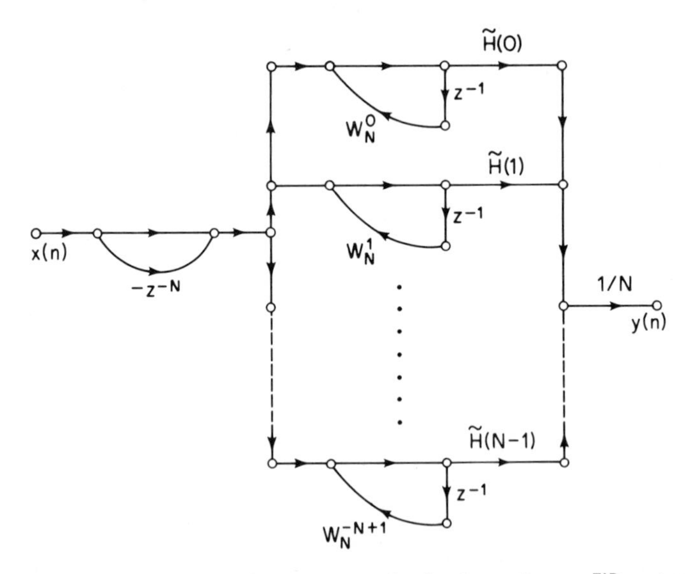

Fig. 4.27 Frequency-sampling structure for implementing an FIR system.

† The reader is reminded that, in keeping with the notation of Chapter 3, $\tilde{H}(k)$ is periodic with period N. $H(k)$, the DFT of $h(n)$, is one period of the periodic sequence $\tilde{H}(k)$.

of a simple FIR system with an IIR system as shown in Fig. 4.27. The system function of the FIR system is $1 - z^{-N}$, and the zeros of this system are at $z_k = \exp[j(2\pi/N)k]$. The IIR portion of Fig. 4.27 consists of a parallel combination of N complex first-order systems with poles at $z_k = \exp[j(2\pi/N)k]$. These first-order systems have poles exactly on the unit circle, the purpose of which is to cancel exactly one of the zeros of the FIR system. In practice, the stability difficulties attendant to poles on the unit circle are avoided by sampling the system function $H(z)$ on a circle of radius r, where r is very slightly less than unity [9]. In this case $H(z)$ has the representation

$$H(z) = (1 - r^N z^{-N}) \frac{1}{N} \sum_{k=0}^{N-1} \frac{\tilde{H}(k)}{1 - rW_N^{-k} z^{-1}} \tag{4.45}$$

where the exact representation of $H(z)$ by Eq. (4.45) requires that

$$\tilde{H}(k) = H(rW_N^{-k}) \tag{4.46}$$

In practice, however, r is chosen close to unity so that little error is incurred if Eq. (4.44) is used in place of Eq. (4.46).

In general, the frequency samples $\tilde{H}(k)$ are complex, as are the quantities W_N^{-k}. Therefore, realization of an FIR system as in Fig. 4.27 requires complex arithmetic operations. However, as we know from Chapter 3, if the impulse response samples $h(n)$ are real, then the frequency samples expressed in polar form satisfy the following symmetry conditions:

$$\begin{aligned} |\tilde{H}(k)| &= |\tilde{H}(N-k)| \\ \theta(k) &= -\theta(N-k), \qquad k = 0, 1, \ldots, N-1 \end{aligned} \tag{4.47}$$

where if $\tilde{H}(0) > 0$, then

$$\theta(0) = 0$$

Furthermore, since $(W_N^{-k})^* = W_N^{-(N-k)}$, the first-order networks in Fig. 4.27 occur in complex-conjugate pairs with the exception of the network with the pole at W_N^0 and, for N even, the network with the pole at $W_N^{-N/2}$. Consequently, the complex first-order networks can be grouped in complex-conjugate pairs and implemented as second-order networks with real coefficients. Specifically, assuming that N is even, we can express Eq. (4.45) as

$$H(z) = \frac{1 - r^N z^{-N}}{N} \left[\sum_{k=1}^{(N/2)-1} \frac{\tilde{H}(k)}{1 - rW_N^{-k} z^{-1}} + \sum_{k=N/2+1}^{N-1} \frac{\tilde{H}(k)}{1 - rW_N^{-k} z^{-1}} + \frac{\tilde{H}(0)}{1 - rz^{-1}} + \frac{\tilde{H}(N/2)}{1 + rz^{-1}} \right]$$

which, upon changing the index of summation in the second sum, becomes

$$H(z) = \frac{1 - r^N z^{-N}}{N} \left[\sum_{k=1}^{(N/2)-1} \left(\frac{\tilde{H}(k)}{1 - rW_N^{-k} z^{-1}} + \frac{\tilde{H}(N-k)}{1 - rW_N^{-N+k} z^{-1}} \right) + \frac{\tilde{H}(0)}{1 - rz^{-1}} + \frac{\tilde{H}(N/2)}{1 + rz^{-1}} \right] \tag{4.48}$$

Using Eq. (4.47) and the fact that $(W_N^{-k})^* = W_N^{-(N-k)}$, we can rewrite Eq. (4.48) as

$$H(z) = (1 - r^N z^{-N})\left[\sum_{k=1}^{(N/2)-1} \frac{2\,|\tilde{H}(k)|}{N} H_k(z) + \frac{\tilde{H}(0)/N}{1 - rz^{-1}} + \frac{\tilde{H}(N/2)/N}{1 + rz^{-1}}\right] \tag{4.49}$$

where

$$H_k(z) = \frac{\cos(\theta(k)) - rz^{-1}\cos[\theta(k) - 2\pi k/N]}{1 - 2rz^{-1}\cos(2\pi k/N) + r^2 z^{-2}} \tag{4.50}$$

This equation suggests the network structure shown in Fig. 4.28, where all the arithmetic operations now involve real numbers.

When N is odd, there will be no frequency sample for $k = N/2$. Thus the term involving $|\tilde{H}(N/2)|$ will be missing from Eq. (4.49) and from Fig. 4.28. If the system has linear phase, Eqs. (4.49) and (4.50) can be further

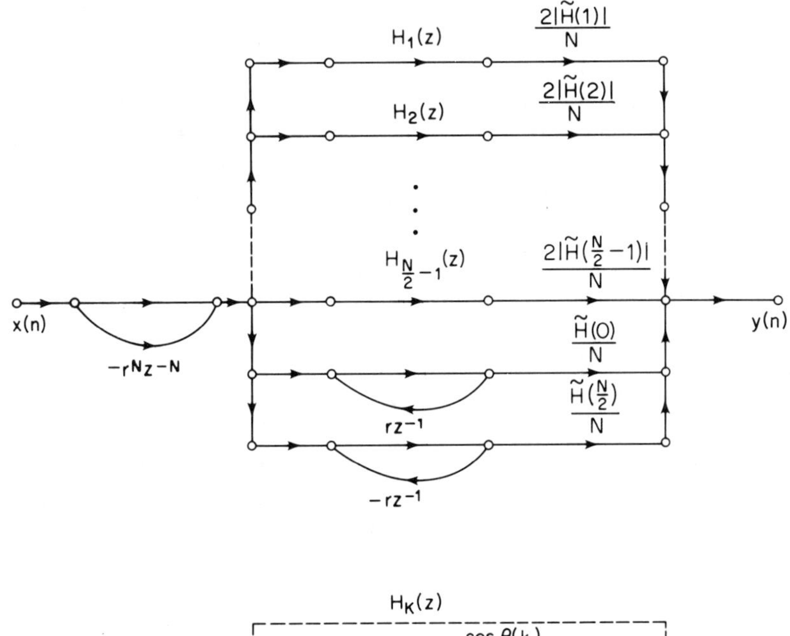

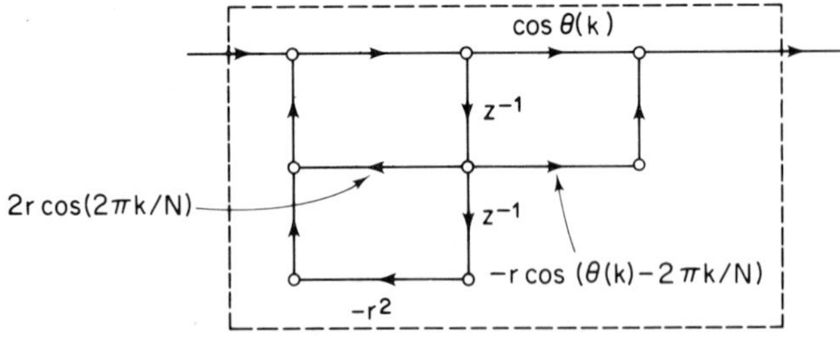

Fig. 4.28 Frequency-sampling structure with the sampling frequencies displaced from the unit circle and with the complex poles implemented in second-order sections.

simplified (see Problem 16). A similar structure can be derived by expressing $H(z)$ in terms of samples offset from the previous frequency samples by an angle of π/N. The derivation of this structure is outlined in Problem 17 of this chapter.

There are two principal advantages of frequency-sampling realizations. First, we note that the multipliers on the outputs of each second-order system in Fig. 4.28 are proportional to samples of the frequency response at equally spaced angles on the unit circle. These values can of course be obtained from the DFT of the impulse response. If the filter to be implemented is a frequency selective filter with one or more stopbands, it can be designed, as will be discussed in Chapter 5, so that the frequency samples in the stopbands are zero, thus reducing the number of second-order systems $H_k(z)$ that must be realized. If most of the frequency samples are zero, as in the case of a narrow-band lowpass or bandpass filter, then a frequency-sampling structure may require fewer multiplications than a direct-form realization. Of course, the frequency-sampling realization will always require more memory than a direct-form realization.

A second advantage stems from the observation that the poles and zeros of the filter structure depend only on the length of the impulse response. If an input is to be processed with a bank of FIR filters (i.e., several different impulse responses of length N), then a single realization of the factor $(1 - z^{-N})$, and of each second-order section, will serve for all the filters. Furthermore, the structure of Fig. 4.28 is highly modular, lending itself to time multiplexing of the second-order sections.

4.5.5 *Structures Based on Polynomial Interpolation Formulas*

As we have noted, the system function of an FIR system is a polynomial of degree $N - 1$ in the variable z^{-1}, where N is the length of the impulse response. It is well known that a polynomial of degree $N - 1$ is uniquely specified by its values at N distinct points. There are a variety of polynomial interpolation formulas, such as the Lagrange and Newton formulas, which specify a polynomial in terms of N values. Alternatively we can specify the value of the polynomial and its first N derivatives at some value of z^{-1} and construct the polynomial from its Taylor's series representation. Schuessler [10] has shown that such representations of a system function imply structures for implementation of FIR systems.

The frequency-sampling structure of the previous section is an example, where in this case the polynomial representing $H(z)$ is constructed by trigonometric interpolation between equally spaced points on the unit circle. A generalization of the frequency-sampling structure, called the *Lagrange structure*, is easily inferred from Eq. (4.43). First, we note that Eq. (4.43) can be expressed as

$$H(z) = P(z) \sum_{k=0}^{N-1} \frac{H(z_k)}{N(1 - z_k z^{-1})} \tag{4.51}$$

where

$$P(z) = \prod_{k=0}^{N-1} (1 - z_k z^{-1}) = 1 - z^{-N} \tag{4.52}$$

and

$$z_k = e^{j(2\pi/N)k}$$

It is easily shown that Eqs. (4.51) and (4.52) correspond to an $(N - 1)$st-degree polynomial in z^{-1} and that this polynomial gives the correct values at the sampling points (see Problem 14 of this chapter).

Equation (4.51) is a special case of the Lagrange interpolation formula when the sample points are equally spaced on the unit circle. In general, the sample points z_k may be chosen arbitrarily in the z-plane. In this case $H(z)$ has the representation

$$H(z) = P(z) \sum_{k=0}^{N-1} \frac{H(z_k)}{P_k(z_k)(1 - z_k z^{-1})} \tag{4.53}$$

where

$$P(z) = \prod_{k=0}^{N-1} (1 - z_k z^{-1}) \tag{4.54}$$

and

$$P_k(z) = \prod_{\substack{i=0 \\ i \neq k}}^{N-1} (1 - z_i z^{-1}) \tag{4.55}$$

It is easily verified that Eq. (4.53) represents a polynomial of degree $N - 1$ and that it gives the correct value at the sample points z_k. Thus it is an appropriate representation of the system function of an FIR system. The

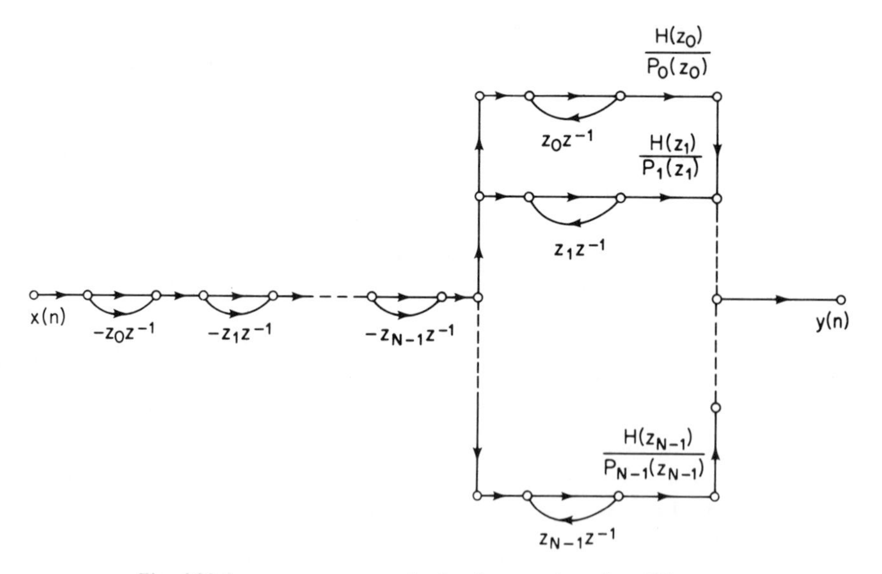

Fig. 4.29 Lagrange structure for implementation of an FIR system.

network structure suggested by Eq. (4.53) is depicted in Fig. 4.29. In particular we obtain a structure very similar to Fig. 4.27. If we choose the sampling points to occur in complex-conjugate pairs, and if the impulse response $h(n)$ is real, then we can combine complex-conjugate terms in the sum in Eq. (4.53) into second-order factors as in the frequency-sampling case and we would obtain a network similar to Fig. 4.28. Furthermore, we can multiply together factors in the polynomial $P(z)$ either in complex pairs or all together to obtain either a cascade-form or a direct-form realization of the nonrecursive part of the structure.

Schuessler [10] has discussed other FIR network structures based on the Newton and Hermite interpolation formulas and on Taylor's series expansions of $H(z)$. All these structures, including the frequency-sampling structure, in general require more multiplications and delays than either the direct or cascade forms. Thus, the utility of such structures lies in possible advantages with respect to sensitivity to quantization effects and in matching design procedures to system realization.

4.6 Parameter Quantization Effects

Linear shift-invariant systems are often used to perform a spectrum shaping or filtering operation. In the design of such systems, there are a number of important considerations: First, we must make a choice between an IIR system and an FIR system. Many factors may enter into this choice. For example, an IIR system may require the fewest delays and constant multipliers to meet a given set of filter specifications. On the other hand, the fact that FIR filters can have exactly linear phase is often a compelling argument in their favor.

Given the choice of system class, i.e., IIR or FIR, we must then determine a system function representation of the filter. The process of determining such system functions is the subject of Chapter 5. Finally, for realizing the system as a computer program or in hardware, a digital network or structure must be chosen. There are many considerations and tradeoffs involved in choosing a structure, such as the implied hardware or software complexity and difficulty in obtaining the parameters for the filter. Furthermore, almost all the design techniques developed up to the present time lead to a specification of the system function in terms of parameters that assumes unlimited precision. If we choose to realize the system by program in a general-purpose computer, the accuracy with which the parameters can be specified may be limited by the word length of the computer memory. In hardware realizations, it is, of course, desirable to minimize the length of the registers that must be provided to store the filter parameters.

Since the coefficients used in implementing a given filter will, in general, not be exact, then the poles and zeros of the system function will in general be different from the desired poles and zeros. This movement of the poles and zeros (zeros only in the FIR case) away from the desired locations results in a frequency response that differs from the desired frequency response; and, if the coefficient quantization errors are large, the system may fail to meet the desired design specifications. Furthermore, in the IIR case, one or more of the poles may move outside the unit circle, resulting in an unstable system that would be useless for the intended application. In general, the effect of coefficient quantization is highly dependent on the structure used to implement the system.

As we have seen, there is an infinite variety of network realizations that realize a given system function when the network parameters are represented with infinite precision. It is to be expected that some of these structures will be less sensitive than others to quantization of the parameters; i.e., the system function of the realization will be closer in some sense to the desired system function. Unfortunately, no systematic method has yet been developed for determining the best realization given constraints on the number of multipliers, word length, and the number of delays. In practice, the choice is generally limited to the network forms of Secs. 4.3, 4.4, and 4.5. In place of a detailed mathematical analysis of the parameter-sensitivity problem, a common practical approach is the use of simulations for determining acceptable quantization of the parameters of a given network. In the remainder of this section we shall discuss some simple results that provide insight into the problems of parameter sensitivity. It should be emphasized that the present understanding of the relationship between network structure and coefficient sensitivity is indeed meager. This remains an important and active research topic.

4.6.1 *Parameter Quantization Effects in IIR Systems*

Parameter quantization effects manifest themselves in deviations of the filter characteristics from the desired frequency response or equivalently in the movement of the poles and zeros away from their desired locations [11–13]. Thus one measure of the sensitivity of a given network realization to parameter quantization is the error in the location of the poles and zeros caused by an error in the constant multipliers in the network.

To indicate how parameter quantization affects the pole and zero locations, consider the system function expressed as

$$H(z) = \frac{\sum_{k=0}^{M} b_k z^{-k}}{1 - \sum_{k=1}^{N} a_k z^{-k}} \tag{4.56}$$

The coefficients a_k and b_k are the desired coefficients in a direct-form realization of the system. With quantized coefficients we, in fact, realize a system whose system function is

$$\hat{H}(z) = \frac{\sum_{k=0}^{M} \hat{b}_k z^{-k}}{1 - \sum_{k=1}^{N} \hat{a}_k z^{-k}} \tag{4.57}$$

where

$$\hat{a}_k = a_k + \Delta a_k$$

$$\hat{b}_k = b_k + \Delta b_k$$

Suppose that the poles of $H(z)$ are located at $z = z_i$, $i = 1, 2, \ldots, N$; i.e., expressed in factored form, the denominator polynomial in the system function is

$$P(z) = 1 - \sum_{k=1}^{N} a_k z^{-k} = \prod_{k=1}^{N} (1 - z_k z^{-1}) \tag{4.58}$$

Furthermore, let us define the poles of $\hat{H}(z)$ as $z_i + \Delta z_i$, where $i = 1, 2, \ldots,$ N. The error Δz_i can be expressed in terms of the errors in the coefficients as

$$\Delta z_i = \sum_{k=1}^{N} \frac{\partial z_i}{\partial a_k} \Delta a_k, \qquad i = 1, 2, \ldots, N \tag{4.59}$$

Using Eq. (4.58) and the fact that

$$\left(\frac{\partial P(z)}{\partial z_i}\right)_{z=z_i} \frac{\partial z_i}{\partial a_k} = \left(\frac{\partial P(z)}{\partial a_k}\right)_{z=z_i}$$

it follows that

$$\frac{\partial z_i}{\partial a_k} = \frac{z_i^{N-k}}{\prod_{\substack{l=1 \\ l \neq i}}^{N} (z_i - z_l)} \tag{4.60}$$

Equation (4.60) is a measure of the sensitivity of the ith pole to a change (error) in the kth coefficient in the denominator polynomial of $H(z)$. [This result is valid only for simple-order poles, as is evident from Eq. (4.58). The extension to multiple-order poles is straightforward.] Since for the direct form, the zeros depend only on the numerator coefficients b_k, an entirely analogous result can be obtained for the sensitivity of the zeros to errors in the b_k's.

A result of this form was first used by Kaiser [12,13] to show that if the poles (or zeros) are tightly clustered, it is possible that small errors in the coefficients can cause large shifts of the poles (or zeros). This can be seen by considering the denominator of Eq. (4.60). Each factor $(z_i - z_l)$ can be represented as a vector in the z-plane as shown in Fig. 4.30. The magnitude of the denominator of Eq. (4.60) is equal to the product of the lengths of the vectors from all the remaining poles to the pole z_i. Thus if poles (or zeros)

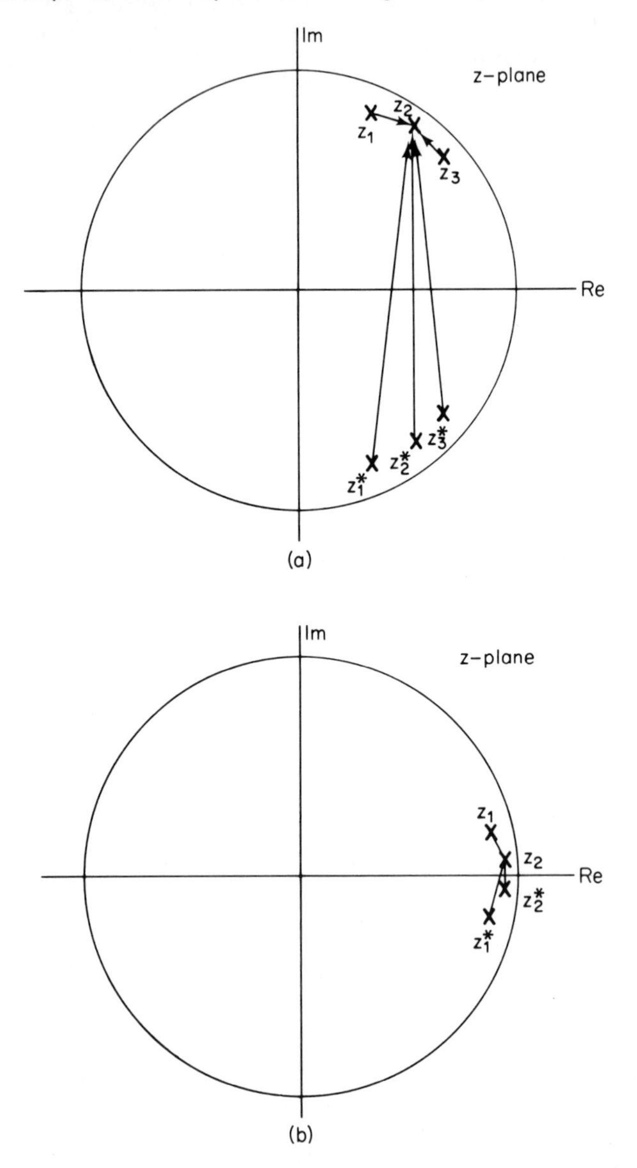

Fig. 4.30 Representation of the factors of Eq. (4.58) as vectors in the z-plane: (a) narrow bandpass filter; (b) narrow-bandwidth lowpass filter.

are tightly clustered as in Fig. 4.30(a), corresponding to a narrow bandpass filter or, as in Fig. 4.30(b), corresponding to a narrow-bandwidth lowpass filter, then we can expect the poles of the direct-form realization to be quite sensitive to quantization errors in the coefficients. Furthermore, it is evident that the larger the number of roots, the greater is the sensitivity.

The cascade and parallel forms, on the other hand, realize each pair of

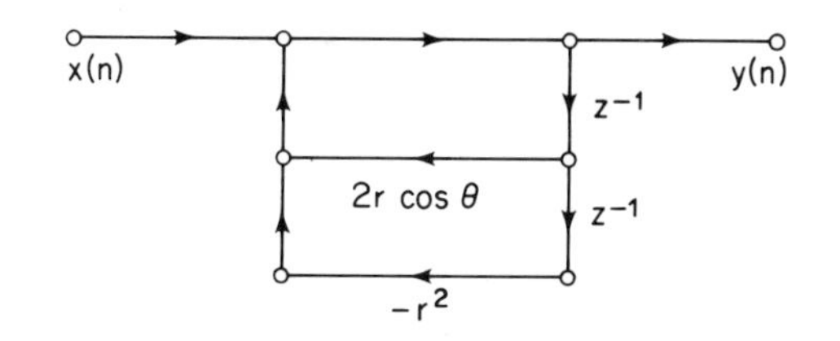

Fig. 4.31 Direct-form implementation of a complex-conjugate pole pair.

complex-conjugate poles separately. Thus the error in a given pole is independent of its distance from the other poles of the system. For this reason it can be stated that in general the cascade and parallel forms are to be preferred over the direct forms from the viewpoint of parameter quantization. This is particularly true in the case of narrow-band frequency selective filters with tightly clustered poles and zeros.

Even for the second-order systems used to implement the cascade and parallel forms, there remains some flexibility. Consider a complex-conjugate pole pair implemented using the direct form as in Fig. 4.31. With infinite-precision representation of the coefficients, this network has poles at $z = re^{j\theta}$ and $z = re^{-j\theta}$. However, if the multipliers $2r\cos\theta$ and $-r^2$ are quantized, clearly only a finite set of pole locations can be obtained. For a given quantization, the poles must lie on a grid in the z-plane defined by the intersection of concentric circles (corresponding to quantization of r^2), and vertical lines (corresponding to quantization of $2r\cos\theta$). Such a grid is illustrated in Fig. 4.32 for three-bit quantization; i.e., r^2 and $2r\cos\theta$ are

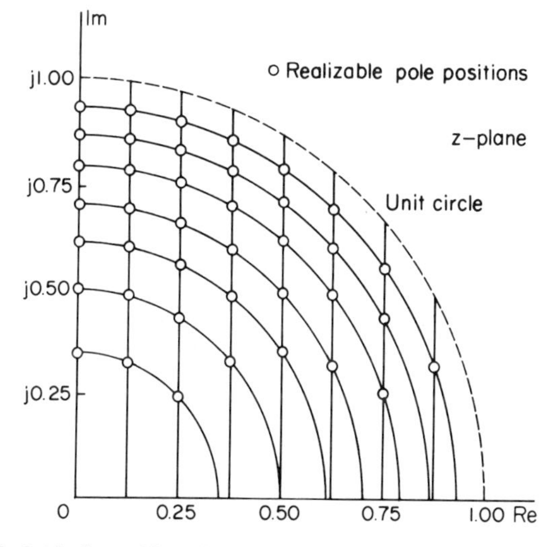

Fig. 4.32 Grid of possible pole locations for the network of Fig. 4.31 when r^2 and $2r\cos\theta$ are quantized to three bits.

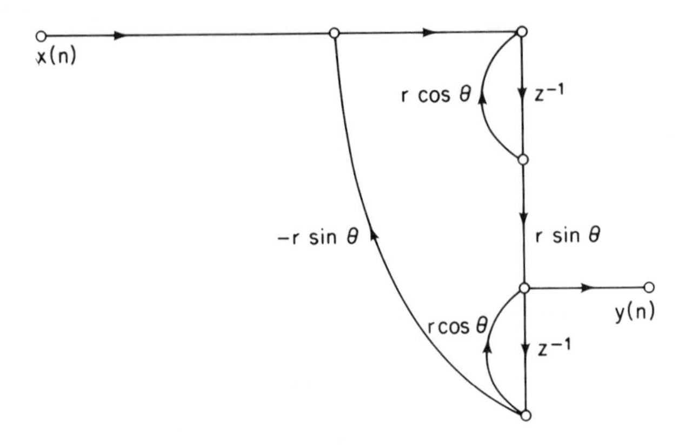

Fig. 4.33 Coupled-form implementation of a complex-conjugate pole pair.

restricted to eight different values. An alternative realization is the coupled form proposed by Gold and Rader [9], as shown in Fig. 4.33. It is easily shown (see Problem 1 of this chapter) that the system functions for the networks of Figs. 4.31 and 4.33 have the same poles for infinite-precision coefficients. For implementation of the network of Fig. 4.33 we must quantize $r \cos \theta$ and $r \sin \theta$; thus we can achieve a finite set of pole locations as shown in Fig. 4.34. Clearly we can obtain many more different structures, each having a different grid of possible pole locations. In practice, it is advantageous to choose a structure for which the corresponding grid is densest in the region of the z-plane where the poles are desired.

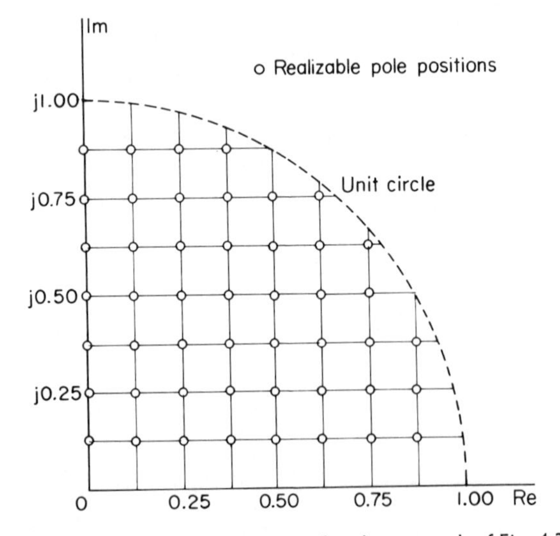

Fig. 4.34 Grid of possible pole locations for the network of Fig. 4.33 when the coefficients $r \cos \theta$ and $r \sin \theta$ are quantized to three bits.

The alternative structures discussed in Sec. 4.3 are of interest because of the possibility that they may offer lower sensitivity for realization of a given system function. At present, little is known about systematic ways of determining the best network realization for a given system function. An alternative to varying the network structure is to incorporate the network structure and parameter quantization into the approximation phase of the design. This is, in general, difficult to achieve, although the approach is certainly a promising area of research.

4.6.2 Parameter Quantization Effects in FIR Systems

We recall that for an FIR system we need only be concerned with the location of the zeros of the system function since, for a causal FIR system, all the poles lie at $z = 0$. Clearly, we could derive equations similar to Eq. (4.60) for the error in a zero due to errors in the coefficients (impulse response) for the direct-form realization of Sec. 4.4. The implication of this is the same; i.e., in general, we have more control over the zero locations in a cascade realization than in a direct-form realization.

Herrmann and Schuessler [14] have considered parameter quantization effects for the particular case of linear-phase systems. For the direct-form realizations of Figs. 4.24 and 4.25 it is clear that even if the coefficients are in error, the resulting system function will still be a mirror-image polynomial, and thus the system will still have linear phase. However, closely spaced zeros on the unit circle may move away from the unit circle, resulting in the system not meeting its design specifications. In the cascade form, if we use second-order sections of the form $(1 + az^{-1} + z^{-2})$ for each complex-conjugate pair of zeros on the unit circle, the zero can move only along the unit circle. Similarly, zeros at $z = \pm 1$ can be realized exactly, and real zeros either inside or outside the unit circle remain real. If the complex zeros off the unit circle were realized as second-order sections, we would find a grid of possible zero positions as in Fig. 4.32, where the grid would extend outside the unit circle. If we wish to maintain linear phase, we must ensure that for each zero inside the unit circle there is a conjugate reciprocal zero outside the unit circle. This can be done by expressing the fourth-order factor corresponding to zeros at $z = re^{\pm j\theta}$ and $z = (1/r)e^{\pm j\theta}$ as

$$\begin{aligned} 1 + d_1 z^{-1} + d_0 z^{-2} + d_1 z^{-3} + z^{-4} \\ = \frac{1}{r^2}(1 - 2r\cos\theta z^{-1} + r^2 z^{-2})(r^2 - 2r\cos\theta z^{-1} + z^{-2}) \end{aligned} \tag{4.61}$$

This corresponds to the subnetwork shown in Fig. 4.35. The grid of possible zero locations is shown again in Fig. 4.36, this time for five-bit quantization (32 different values of r^2 and $2r\cos\theta$).

This realization clearly requires more multipliers than are necessary for a fourth-order system. If the fourth-order sections are realized in the linear-phase direct form, we obtain the network of Fig. 4.37. The grid of possible

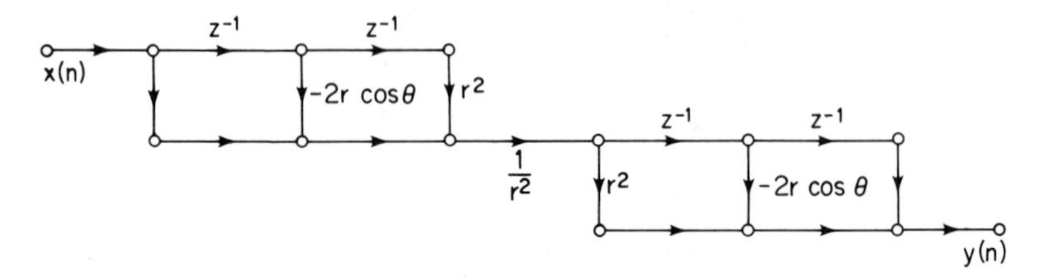

Fig. 4.35 Subnetwork to implement fourth-order factors in a linear-phase FIR system such that linear phase is maintained independent of parameter quantization.

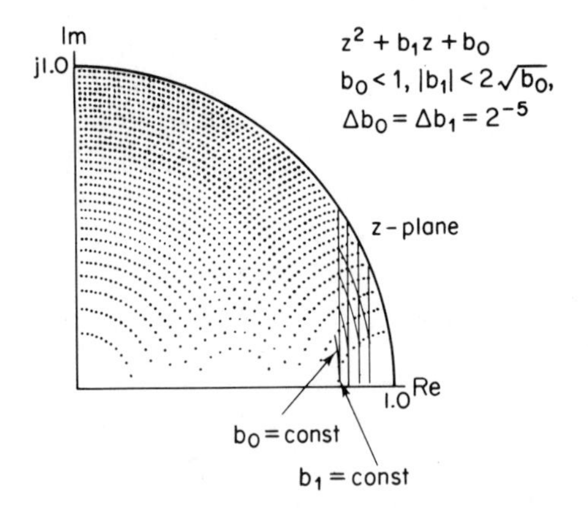

Fig. 4.36 Grid of possible zero locations for the subnetwork of Fig. 4.35 when the coefficients r^2 and $2r\cos\theta$ are quantized to five bits. (after Herrmann and Schuessler [14])

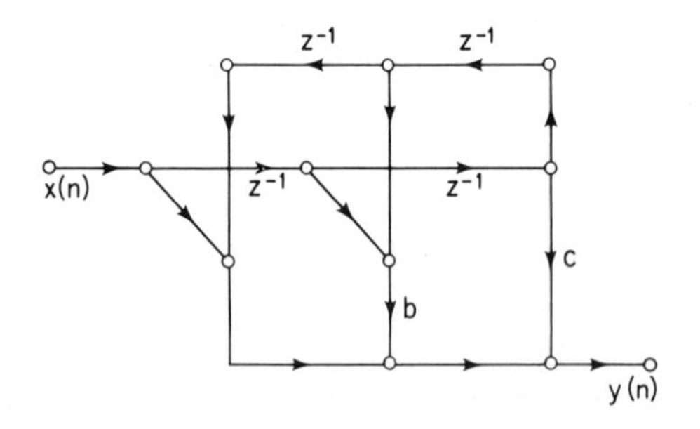

Fig. 4.37 Linear-phase direct-form realization for fourth-order factors in an FIR system.

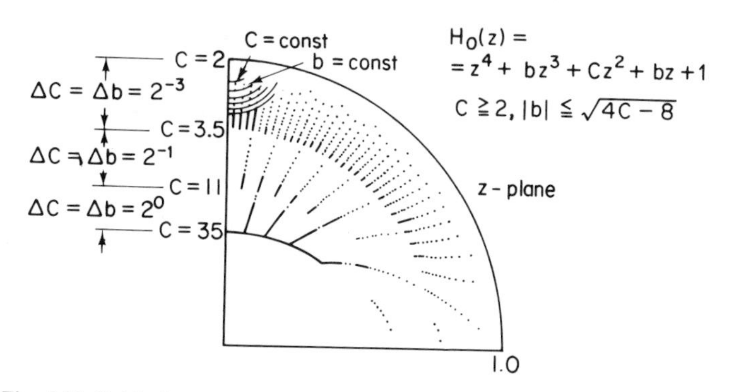

Fig. 4.38 Grid of possible zero locations for the subnetwork of Fig. 4.37 when the coefficients b and c are quantized to five bits. (after Herrmann and Schuessler [14])

zero locations for the direct form is shown in Fig. 4.38. While the use of fourth-order sections in linear-phase FIR filter implementation has the advantage that it maintains the linear-phase characteristics of the filter independent of coefficient quantization, it has been shown [14–16] that for a variety of filters the resulting filter characteristics are extremely sensitive to quantization. Consequently, it is often more desirable to implement linear-phase FIR filters as a cascade of second-order sections or in direct form.

The above discussion of sensitivity to parameter quantization has been based primarily on a consideration of the sensitivity of the pole and zero locations. It is also possible to analyze structures according to other definitions of sensitivity. Although this area remains an active research topic, some useful theorems relating to network sensitivity can be derived from the basic theory of linear signal flow graphs. We close this chapter with a discussion of an interesting and useful theorem for signal flow graphs, Tellegen's theorem. One consequence of this theorem is the transposition theorem for digital filters referred to in Sec. 4.4. Another consequence, as we shall see, are some sensitivity relations for digital filters.

4.7 Tellegen's Theorem for Digital Filters and Its Applications

Tellegen's theorem is an important basic theorem of conventional network theory [17]. The theorem is simple and general and part of its elegance derives from the fact that many network theorems, such as conservation of energy, reciprocity, etc., can be developed simply as consequences of Tellegen's theorem.

Since digital filter networks are not subject to Kirchhoff's laws, Tellegen's theorem in its most general form does not apply. However, a restricted form of Tellegen's theorem, referred to as the difference form, can be derived.

From this theorem a number of useful properties of digital networks can be developed.

A form of Tellegen's theorem for digital filters was first derived by Seviora and Sablatash [18]. An alternative form was proposed by Fettweis [19]. The treatment here and the results follow closely the approach taken by Fettweis. For the purposes of deriving Tellegen's theorem, it is convenient to use the notation of signal flow graphs as developed in Sec. 4.1.

In classical networks Tellegen's theorem is in the form of a relationship between the voltage distribution in one network and the current distribution in a second network, where the only relationship between the networks is that they have the same topology. In a similar manner, for signal flow graphs, we consider two flow graphs with the same topology but otherwise unrelated. By the same topology we mean that there is a one-to-one correspondence between the nodes in the two networks and between the branches (location and direction only). It is important to note, however, that the requirement that two networks be topologically equivalent is relatively minor. In particular, if we consider every flow graph to have a branch in each direction between every pair of nodes, with the transmission of some of the branches being zero, then any two flow graphs with the same number of nodes can be considered to be topologically equivalent.

In the following discussion it will be convenient to adopt the convention that each network is drawn in such a way that each network node has associated with it a source node connected to it by a branch with unity transmittance. Furthermore, we will assume that this source node is not connected to any other network nodes. The source nodes will be numbered to coincide with the associated network node so that source x_k is the input to network node k. If the source node and the corresponding branch are not drawn, this implies that the source node value is zero.

TELLEGEN'S THEOREM Consider two signal flow-graphs with the same topology. Let N denote the number of network nodes. The network node variables, branch outputs and source node values in the first network are denoted by w_k, v_{jk}, and x_j, respectively and in the second network by w'_k, v'_k, and x'_j. Then

$$\sum_{k=1}^{N}\sum_{j=1}^{N}(w'_k v_{jk} - w_k v'_{jk}) + \sum_{k=1}^{N}(w_k x'_k - w'_k x_k) = 0. \tag{4.62}$$

Proof. The proof of Eq. (4.62) follows almost directly from the definition of a signal flow graph. Recall that the branch outputs are related to the node variables and source inputs by

$$w_k = \sum_{j=1}^{N} v_{jk} + \sum_{j=1}^{M} s_{jk}$$

and thus, with our convention regarding source nodes

$$w_k = \sum_{j=1}^{N} v_{jk} + x_k \tag{4.63}$$

We begin by writing the identity

$$\sum_{k=1}^{N} (w_k w'_k - w'_k w_k) = 0 \tag{4.64}$$

Equation (4.62) follows in a straightforward manner by substituting Eq. (4.63) into Eq. (4.64).

Equation (4.62) will henceforth be referred to as Tellegen's theorem for signal flow graphs or, equivalently, for digital networks. It is important to recognize that its derivation depends only on Eq. (4.63); it does not require that the flow graph be linear. Furthermore, it is straightforward to show that if variables W_k, W'_k, V_{jk}, V'_{jk}, X_k, and X'_k are derived through a linear operation from w_k, w'_k, v_{jk}, v'_{jk}, x_k, and x'_k, respectively, then

$$\sum_{k=1}^{N} \sum_{j=1}^{N} (W_k V'_{jk} - W'_k V_{jk}) + \sum_{k=1}^{N} (W_k X'_k - W'_k X_k) = 0 \tag{4.65}$$

Thus Tellegen's theorem applies either to the sequence values [Eq. (4.62)] or to the z-transforms [Eq. (4.65)].

For the remainder of our discussion we will be concerned only with linear signal flow graphs that correspond to digital networks. To illuminate the notion of identical topologies and primed and unprimed variables (networks), consider the two networks shown in Fig. 4.39. To emphasize the

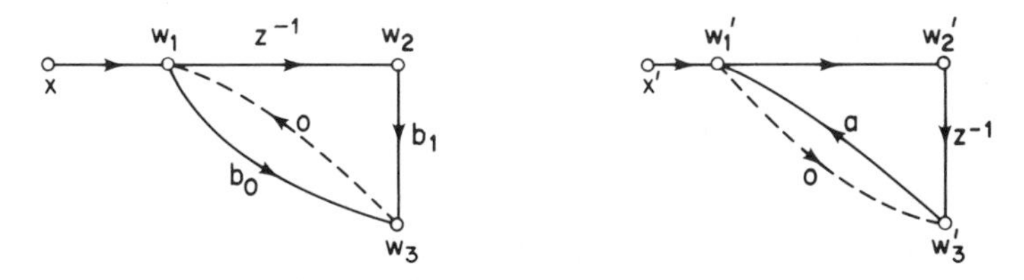

Fig. 4.39 Example of two networks that are topologically equivalent and thus satisfy Tellegen's theorem.

equivalence of the topology we have indicated by dashed lines a branch with zero transmittance. Normally, of course, such branches would not be drawn. It is a simple exercise to verify the validity of Tellegen's theorem for these two networks (see Problem 18 of this chapter).

4.7.1 *Reciprocal and Interreciprocal Digital Networks*

For passive analog networks consisting of interconnections of resistors, inductors, and capacitors, the notion of reciprocity plays an important role. For digital networks there exist corresponding notions of reciprocity and interreciprocity. For the notion of reciprocity we consider a given network excited by two different sets of sources. The z-transforms of the source node values for the two different sets will be denoted by X_k and X'_k. The value of the node variable of the kth node when the network is excited by the unprimed sources will be denoted by W_k. When the network is excited by the primed sources, this variable will be denoted by W'_k. Then the network is said to satisfy reciprocity if for any two signal distributions

$$\sum_{k=1}^{N} [W_k X'_k - W'_k X_k] = 0 \tag{4.66}$$

As one consequence of reciprocity consider any two nodes a and b in the network. Furthermore, let all unprimed source node values but X_a be zero and let all primed source node values but X'_b be zero. In addition, let $X_a = X'_b$. Then from Eq. (4.66)

$$W_b X'_b = W'_a X_a$$

or

$$W_b = W'_a \tag{4.67}$$

In other words, as a consequence of reciprocity, if we excite the graph at network node a and observe the output at node b, then for a reciprocal graph, the same excitation at node b will result in the same output at node a.

Most digital networks are not reciprocal. A related concept that is more useful with regard to digital networks is that of *interreciprocity*. In this case we consider two distinct signal flow graphs. Let X_k denote the source node values and W_k denote the node variables for one network and X'_k and W'_k the source node values and network node variables for the second network. Then the two networks are said to be interreciprocal if

$$\sum_{k=1}^{N} [W_k X'_k - W'_k X_k] = 0 \tag{4.68}$$

This equation is similar to Eq. (4.66), but it is important to keep in mind that for reciprocity the primed and unprimed networks differ only in the sources, whereas for interreciprocity both the sources and branch transmittances can differ in the primed and unprimed networks. A network that is reciprocal is also interreciprocal with itself. On the other hand, two networks may be interreciprocal without either network being reciprocal.

4.7.2 Proof of the Transposition Theorem

A useful property of digital networks is that they are interreciprocal with their transpose. To show this, recall from Section 4.4 that the transpose of a flow graph is generated by reversing the directions of all the branches but leaving their transmittances the same.

Let us consider a digital network where W_k denotes the node variable for the kth node. The transmission from node j to node k is denoted by F_{jk}; i.e.,

$$V_{jk} = F_{jk} W_j \tag{4.69}$$

In the transposed network, the node variable of the kth node is denoted by W'_k, and the branch transmittance between nodes j and k is denoted F'_{jk}, so that

$$V'_{jk} = F'_{jk}W'_j \tag{4.70}$$

By definition of the transposed network,

$$F'_{jk} = F_{kj}$$

To prove that a network and its transpose are interreciprocal, i.e., to show that Eq. (4.68) holds for the above conditions, we utilize the fact that a network and its transpose have the same topology so that Tellegen's theorem [Eq. (4.65)] holds. Thus

$$\sum_{j=1}^{N}\sum_{k=1}^{N}(W_kV'_{jk} - W'_kV_{jk}) + \sum_{k=1}^{N}(W_kX'_k - W'_kX_k) = 0 \tag{4.71}$$

Substituting Eqs. (4.69) and (4.70) into (4.71) we obtain

$$\sum_{j=1}^{N}\sum_{k=1}^{N}(W_kW'_jF'_{jk} - W'_kW_jF_{jk}) + \sum_{k=1}^{N}(W_kX'_k - W'_kX_k) = 0$$

or, equivalently,

$$\sum_{j=1}^{N}\sum_{k=1}^{N}W_kW'_jF'_{jk} - \sum_{j=1}^{N}\sum_{k=1}^{N}W'_kW_jF_{jk} + \sum_{k=1}^{N}(W_kX'_k - W'_kX_k) = 0 \tag{4.72}$$

Interchanging the indices of summation in the first double sum in Eq. (4.72) we obtain

$$\sum_{j=1}^{N}\sum_{k=1}^{N}(W'_kW_jF'_{kj} - W'_kW_jF_{jk}) + \sum_{k=1}^{N}(W_kX'_k - W'_kX_k) = 0 \tag{4.73}$$

But, since the primed and unprimed networks are transposes, $F'_{jk} = F_{jk}$, and therefore the double sum is zero and

$$\sum_{k=1}^{N}(W_kX'_k - W'_kX_k) = 0 \tag{4.74}$$

which proves that a network and its transpose are interreciprocal.

An interesting consequence of this fact is that for single input–single output networks, a network and its transpose have the same transfer function. This result can also be derived from Mason's gain formula [1] and was utilized in Secs. 4.3 and 4.4 as a means of deriving new network realizations.

To show this result from Eq. (4.74) consider any two nodes a and b. Furthermore, let all source nodes but X_a be zero in the original (unprimed) network and all source nodes but X'_b be zero in the transposed (primed) network. Then from Eq. (4.74)

$$W_bX'_b = W'_aX_a \tag{4.75}$$

From Eq. (4.75), if the same excitation is applied at node a in the original network and at node b in the transposed network, then the same response will be observed at node a in the transposed network as at node b in the original network.

4.7.3 Network-Sensitivity Formula

In Sec. 4.6 we discussed the problem of parameter sensitivity from the point of view of movements of the poles and zeros. Specifically we compared the direct-form networks to cascade and parallel realizations. For more complex networks, it is not as easy to obtain a general relationship between the poles and zeros and the network parameters. We can, with the aid of Tellegen's theorem, obtain a general expression for the sensitivity of the system functions of a given network to changes in the parameters of the network. Formulas of this type are useful, for example, in computer-aided analysis of digital networks.

For this discussion, it is convenient to define the system function between an arbitrary pair of nodes a and b in the network. Specifically, let us consider all source nodes but that connected to node a to be zero. Then the node variable $W_b(z)$ will be given by

$$W_b(z) = T_{ab}(z)X_a(z) \tag{4.76}$$

$T_{ab}(z)$ is then the system function from node a to node b. The sensitivity of this system function to changes in a branch transmittance $F_{nm}(z)$ is defined as

$$\frac{\partial T_{ab}(z)}{\partial F_{nm}(z)}$$

Fettweis [19] and Seviora and Sablatash [18] have shown that

$$\frac{\partial T_{ab}}{\partial F_{nm}} = T_{an}T_{mb} \tag{4.77}$$

where we have not explicitly indicated the functional dependence upon (z), and where T_{an} is the system function from node a to node n and T_{mb} is the system function from node m to node b.

To prove this result, we consider the three networks depicted in Fig. 4.40. The original network, shown in Fig. 4.40(a), is specified by the unprimed variables and for convenience is called the *unprimed network*. Figure 4.40(b) shows the transpose of the original network, called the *primed network*, and Fig. 4.40(c) shows the original system with a perturbation ΔF_{nm} in branch transmittance F_{nm}. This network is called the *double-primed network*. We assume that each network is excited by the same source X as depicted in Fig. 4.40.

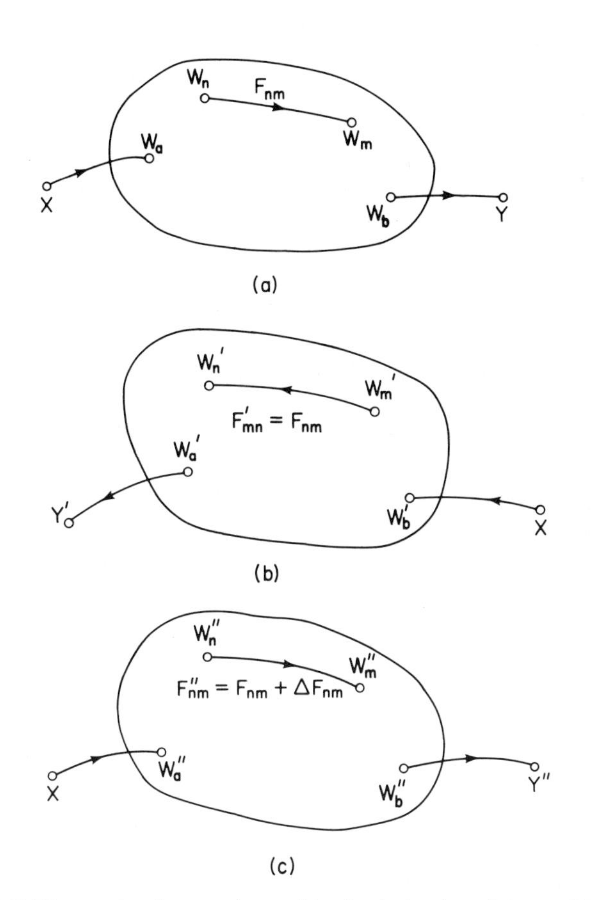

Fig. 4.40 Three related networks used in the derivation of the sensitivity relation Eq. (4.77): (a) original network; (b) transposed network (primed system); (c) original system with error in branch transmittance F_{nm} (double-primed system).

Using Eq. (4.65) for the primed and double-primed systems, we obtain

$$\sum_{k=1}^{N}\sum_{j=1}^{N}(W_k'V_{jk}'' - W_k''V_{jk}') + \sum_{k=1}^{N}(W_k'X_k'' - W_k''X_k') = 0 \tag{4.78}$$

By splitting the double sum in Eq. (4.78) into two double sums and interchanging the indices in the second resulting double sum, we obtain

$$\sum_{k=1}^{N}\sum_{j=1}^{N}(W_k'V_{jk}'' - W_j''V_{kj}') + \sum_{k=1}^{N}(W_k'X_k'' - W_k''X_k') = 0 \tag{4.79}$$

Using Eq. (4.79) and the fact that, by definition,

$$V_{jk}'' = F_{jk}''W_j''$$

and

$$V'_{jk} = F'_{jk}W'_j$$

we obtain

$$\sum_{k=1}^{N}\sum_{j=1}^{N} W''_j W'_k (F''_{jk} - F'_{kj}) + \sum_{k=1}^{N} (W'_k X''_k - W''_k X'_k) = 0 \tag{4.80}$$

From Fig. 4.40 we observe that the primed and double-primed networks are transposes except for the branch (nm). Thus, from the definition of the transpose,

$$F''_{jk} - F'_{kj} = 0, \qquad \text{for all } k \text{ and } j \text{ except for branch } (nm)$$

and

$$F''_{nm} - F'_{mn} = \Delta F_{nm}$$

Thus all the terms in the double sum of Eq. (4.80) are zero except one. Furthermore, since only one node has a nonzero source input in each network (node a for the double-primed network, node b for the primed network), we can write that

$$W''_n W'_m \Delta F_{nm} + X(W'_a - W''_b) = 0 \tag{4.81}$$

Now we express the node variables in Eq. (4.81) in terms of the source input X and appropriate transfer functions from the source input nodes as

$$W'_a = T'_{ba}X = T_{ab}X$$

$$W'_m = T'_{bm}X = T_{mb}X$$

$$W''_n = T''_{an}X$$

$$W''_b = T''_{ab}X = [T_{ab} + \Delta T_{ab}]X$$

where we have defined ΔT_{ab} as the change in the transfer function from input and output due to a change ΔF_{nm} in the branch transmittance F_{nm}. Substituting these equations into Eq. (4.81) we obtain

$$[T''_{an}T_{mb}\Delta F_{nm} + T_{ab} - T_{ab} - \Delta T_{ab}]X^2 = 0$$

Since this equation holds for all X, then

$$T''_{an}T_{mb}\Delta F_{nm} = \Delta T_{ab} \tag{4.82}$$

or

$$\frac{\Delta T_{ab}}{\Delta F_{nm}} = T''_{an}T_{mb} \tag{4.83}$$

Taking the limit of Eq. (4.83) as $\Delta F_{nm} \to 0$ we obtain

$$\lim_{\Delta F_{nm}\to 0}\left[\frac{\Delta T_{ab}}{\Delta F_{nm}}\right] = \lim_{\Delta F_{nm}\to 0}[T''_{an}T_{mb}] \tag{4.84}$$

As $\Delta F_{nm} \to 0$ the double-primed network approaches the unprimed network, so that

$$\frac{\partial T_{ab}}{\partial F_{nm}} = T_{an} T_{mb} \tag{4.85}$$

Thus we have obtained the desired sensitivity relation relating a change in the system function T_{ab} to a change in the branch transmittance F_{nm}. The desirable feature of this expression is that the sensitivity is expressed in terms of transfer functions of the network which can be computed using the matrix methods of Sec. 4.2. To determine a change ΔT_{ab} in the system function T_{ab} due to a large-scale change ΔF_{nm} in F_{nm}, we can use the Taylor's series expansion

$$\Delta T_{ab} = \frac{\partial T_{ab}}{\partial F_{nm}} \Delta F_{nm} + \frac{1}{2} \frac{\partial^2 T_{ab}}{\partial F_{nm}^2} (\Delta F_{nm})^2 + \ldots \tag{4.86}$$

The higher-order derivatives (sensitivities) of T_{ab} with respect to F_{nm} can be obtained from Eq. (4.85) by the chain rule of differentiation. Crochiere [20] has shown that this approach leads to an expression

$$\Delta T_{ab} = \frac{T_{an} T_{mb} \Delta F_{nm}}{1 - T_{mn} \Delta F_{nm}} \tag{4.87}$$

A derivation of this result is suggested in Problem 19 of this chapter.

Summary

This chapter was directed toward the representation of digital filters in terms of block diagrams, signal flow graphs, and matrix notation. After presenting the signal flow graph and matrix representations in general, we presented a number of basic structures for IIR and FIR filters.

One of the important considerations in the choice of a structure for implementation of a filter is the effect of finite-word-length coefficients. Thus we presented some of the parameter quantization considerations in the choice of filter structure. We then introduced Tellegen's theorem for digital filters. From this theorem several important properties of filter structures were developed, including the transposition theorem for signal flow graphs. The chapter concluded with the derivation of a network sensitivity formula using Tellegen's theorem. This formula provides a convenient means for computing sensitivity of filter structures and in addition leads to a large-scale sensitivity relation for filter structures.

REFERENCES

1. S. J. Mason and H. J. Zimmermann, *Electronic Circuits, Signals, and Systems*, John Wiley & Sons, Inc., New York, 1960.
2. Y. Chow and E. Cassignol, *Linear Signal-Flow Graphs and Applications*, John Wiley & Sons, Inc., New York, 1962.

3. R. E. Crochiere, "Digital Network Theory and Its Application to the Analysis and Design of Digital Filters," Ph.D. Thesis, Department of Electrical Engineering, MIT, 1974.
4. S. K. Mitra and R. J. Sherwood, "Canonic Realizations of Digital Filters Using the Continued Fraction Expansion," *IEEE Trans. Audio Electroacoust.*, Vol. AU-20, 1972, pp. 185–194.
5. A. Fettweis, "Digital Filter Structures Related to Classical Filter Networks," *Arch. Electronik Ubertragungstechnik*, Vol. 25, 1971, pp. 70–89.
6. A. Fettweis, "Some Principles of Designing Digital Filters Imitating Classical Filter Structures," *IEEE Trans. Circuit Theory*, Vol. CT-18, 1971, pp. 314–316.
7. R. E. Crochiere, "Digital Ladder Structures and Coefficient Sensitivity," *IEEE Trans. Audio Electroacoust.*, Vol. AU-20, 1972, pp. 240–246.
8. S. K. Mitra and R. J. Sherwood, "Digital Ladder Networks," *IEEE Trans. Audio Electroacoust.*, Vol. AU-21, 1973, pp. 30–36.
9. B. Gold and C. Rader, *Digital Processing of Signals*, McGraw-Hill Book Company, New York, 1969.
10. W. Schuessler, "On Structures for Nonrecursive Digital Filters," *Arch. Electronik Ubertragungstechnik*, Vol. 26, 1972, pp. 255–258.
11. J. B. Knowles and E. M. Olcayto, "Coefficient Accuracy and Digital Filter Response," *IEEE Trans. Circuit Theory*, Vol. CT-15, Mar. 1968, pp. 31–41.
12. J. F. Kaiser, "Digital Filters," Chapter 7 in *System Analysis by Digital Computer*, F. F. Kuo and J. F. Kaiser, John Wiley & Sons, Inc., New York, 1966.
13. J. F. Kaiser, "Some Practical Considerations in the Realization of Linear Digital Filters," *Proc. 3rd Allerton Conf. Circuit System Theory*, Oct. 20–22, 1965, pp. 621–633.
14. O. Herrmann and W. Schuessler, "On the Accuracy Problem in the Design of Nonrecursive Digital Filters," *Arch. Electronik. Ubertragungstechnik*, Vol. 24, 1970, pp. 525–526.
15. D. S. K. Chan, "Roundoff Noise in Cascade Realization of Finite Impulse Response Digital Filters," M.S. Thesis, Department of Electrical Engineering, MIT, June 1972.
16. D. S. K. Chan and L. Rabiner, "Analysis of Quantization Errors in the Direct Form for Finite Impulse Response Digital Filters," *IEEE Trans. Audio* Electroacoust., Vol. AU-21, Aug. 1973, pp. 354–366.
17. P. Penfield, Jr., R. Spence, and S. Duinker, *Tellegen's Theorem and Electric Networks*, MIT Press, Cambridge, Mass., 1970.
18. R. Seviora and M. Sablatash, "A Tellegen's Theorem for Digital Filters" *IEEE Trans. Circuit Theory* CT-18 1971, pp. 201–203.
19. A. Fettweis, "A General Theorem for Signal-Flow Networks, with Applications," *Arch. Electronik Ubertragungstechnik*, Vol. 25, 1971, pp. 557–561.
20. R. Crochiere, "Some Network Properties of Digital Filters," *Proc. 1973 Intern. Symp. Circuit Theory*, April 9–11, 1973, pp. 146–148.

PROBLEMS

1. Determine the system functions of the two networks in Fig. P4.1 and show that they have the same poles.
2. For the network shown in Fig. P4.2, determine all possible ordering of the nodes such that the node variables can be computed in sequence.
3. It was stated in Sec. 4.2 that a necessary and sufficient condition for the computability of a flow graph is that the nodes can be numbered so that in the matrix $\mathbf{F}_c^t$ there are only zeros above the main diagonal and the main-diagonal

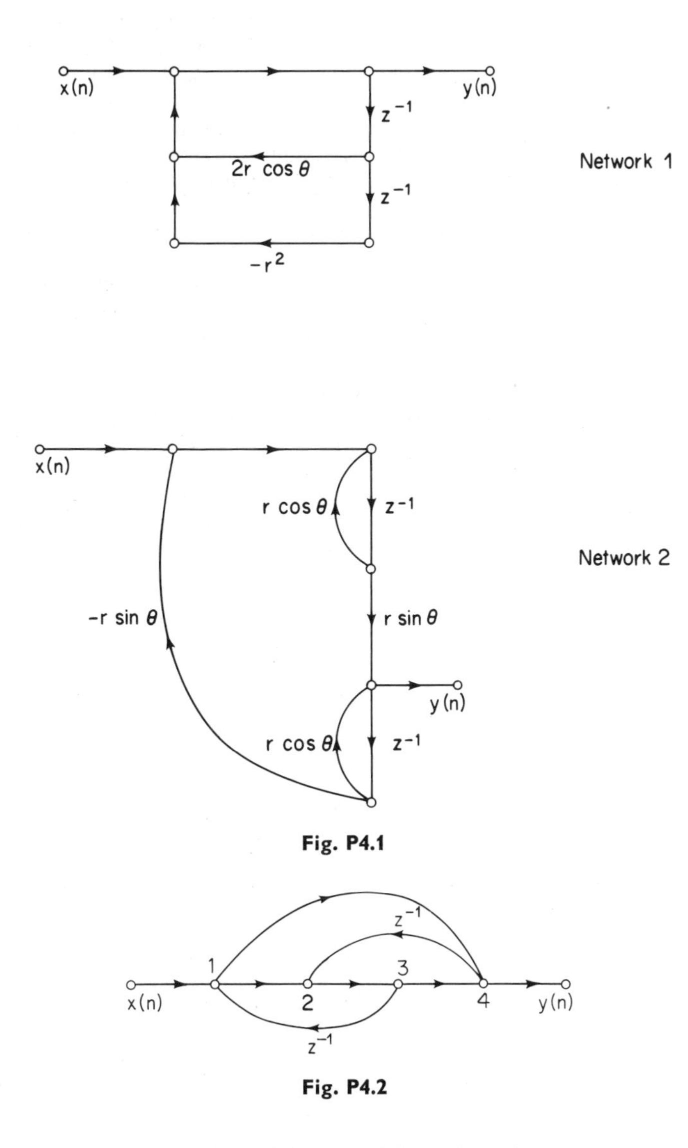

Fig. P4.1

Fig. P4.2

elements are zero. In this problem we wish to show that statement to be true.

Let c_{jk} denote the elements of the matrix $\mathbf{F}_c$.

(a) Assume that in the matrix $\mathbf{F}_c^t$ there are only zeros above the main diagonal and the main-diagonal elements are zero, so that $c_{jk} = 0, j \geq k$. Show that in this case the node variables $w_1(n)$, $w_2(n)$, . . . can be computed in sequence, i.e., first $w_1(n)$, then $w_2(n)$, etc. This requires showing that the computation of $w_i(n)$ does not utilize node variables $w_l(n)$ with $l \geq i$. It can, of course, utilize any elements of the vector $\mathbf{w}(n-1)$, since these correspond to the previous iteration of the flow graph.

Part (a) shows that the stated condition is a *sufficient* condition for computability. We now wish to show that it is also a necessary condition.

(b) Assume that in the matrix $\mathbf{F}_c^t$ there is at least one nonzero element above or on the main diagonal. Show that in this case the node variable *cannot* be computed in sequence.

Part (b) shows that the stated condition is a necessary condition since, if the nodes cannot be numbered so that $\mathbf{F}_c^t$ is zero above and on the main diagonal, then there is no ordering for which the node variables can be computed.

4. In deriving the matrix representation of equations (4.29) from (4.21) we assumed that the matrix $[\mathbf{I} - \mathbf{F}_c^t]$ is nonsingular. Using the results of problem 3 above, show that for a computable flow graph, this assumption will always be valid.

5. In equations (4.29) the matrix representation of a flow graph is in a form in which $\mathbf{w}(n)$ is expressed explicitly in terms only of past values of the node variables and current values of the input. In this problem we wish to modify that representation to obtain one in which the node vector $\mathbf{w}(n_1)$ at time n_1 can be computed from the node vector $\mathbf{w}(n_0)$ at time n_0 and the input vector $\mathbf{s}(n)$ for $n_0 \le n \le n_1$.

(a) Modify Eq. (4.29a) to express $\mathbf{w}(n)$ in terms of $\mathbf{w}(n-2)$ and the inputs $\mathbf{x}(n-1)$ and $\mathbf{x}(n)$.

(b) Let $n_1 = n_0 + M$, where M is constant. By generalizing the procedure and result in part (a), determine a matrix representation based on Eq. (4.29a) but in which $\mathbf{w}(n_1)$ is expressed in terms of $\mathbf{w}(n_0)$ and the inputs $\mathbf{x}(n_1 - M)$, $\mathbf{x}(n_1 - M + 1), \ldots, \mathbf{x}(n_1)$.

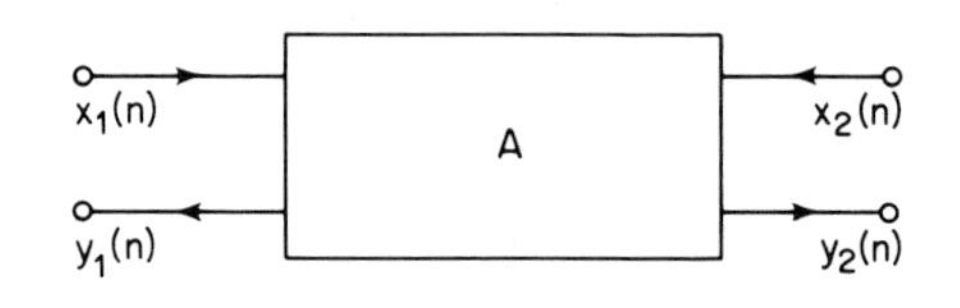

Fig. P4.6-1

6. In Fig. P4.6-1 is indicated a digital network A with two inputs, $x_1(n)$ and $x_2(n)$, and two outputs, $y_1(n)$ and $y_2(n)$. The network A can be described in terms of the two-port set of equations

$$Y_1(z) = H_1(z)X_1(z) + H_2(z)X_2(z)$$

$$Y_2(z) = H_3(z)X_1(z) + H_4(z)X_2(z)$$

where

$$H_1(z) = \frac{1}{1 - \frac{1}{2}z^{-1}}$$

$$H_2(z) = 1$$

$$H_3(z) = \frac{1 + 2z^{-1}}{1 + \frac{1}{2}z^{-1}}$$

$$H_4(z) = \frac{1}{1 + \frac{1}{2}z^{-1}}$$

(a) Draw a flow-graph implementation of the network. The transmittance of each branch must be a constant or a constant times z^{-1}. Higher-order functions of z^{-1} cannot be used as branch transmittances.

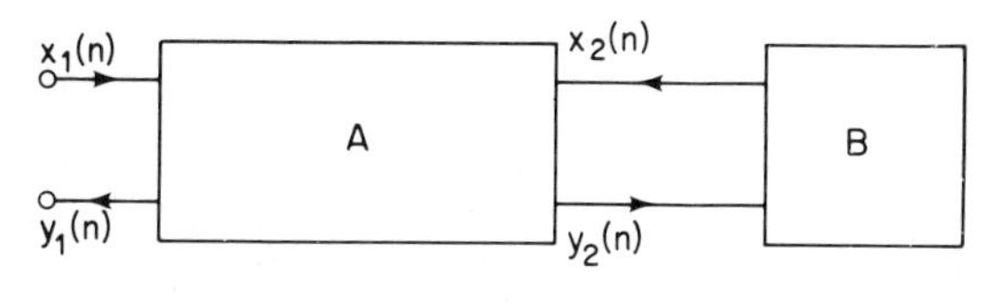

Fig. P4.6-2

(b) Write the set of equations corresponding to your network in part (a) in the form of equations (4.21).

(c) Determine an ordering for the nodes in the flow graph for which the flow graph is *not* computable and determine an ordering for the nodes so that the flow graph is computable.

(d) We wish now to connect a network B to the network A as indicated in Fig. P4.6-2. Network B has no internal source nodes and is characterized by the system function $X_2(z) = Y_2(z)H_B(z)$. Note that because of the notation used for network A, $y_2(n)$ denotes the *input* to network B and $x_2(n)$ is the output. By examining the flow graph in part (a) or the equations in part (b), determine a necessary and sufficient condition on $H_B(z)$ so that the overall system of Fig. P4.6-2 is computable. If you have difficulty, first try the two possibilities $H_B(z) = 1$ and $H_B(z) = z^{-1}$.

7. Consider the discrete-time linear causal system defined by the difference equation

$$y(n) - \tfrac{3}{4}y(n-1) + \tfrac{1}{8}y(n-2) = x(n) + \tfrac{1}{3}x(n-1)$$

Draw a signal flow graph to implement this system in each of the following forms:

(a) Direct form I.
(b) Direct form II.
(c) Cascade.
(d) Parallel.

For the cascade and parallel forms use only first-order sections.

8. For many applications it is useful to have a digital network that will generate a sinusoidal sequence. One possible way to do this is with a digital network whose unit-sample response is $e^{j\omega_0 n}u(n)$. The real and imaginary parts of this response are then $(\cos \omega_0 n)u(n)$ and $(\sin \omega_0 n)u(n)$, respectively.

In implementing a system with a complex unit-sample response, the real and imaginary parts are distinguished as separate outputs. By first writing the complex difference equation to produce the desired response, and equating real and imaginary parts, draw a digital network that will implement this system. The network that you draw can have only real coefficients in it. This implementation is often referred to as a *coupled-form oscillator*.

9. The system with transfer function $H(z) = (z^{-1} - a)/(1 - az^{-1})$ is an allpass system; i.e., the frequency response has unity magnitude.

(a) Draw a network realization of this system in direct II form.

(b) In implementing the network in part (a), the coefficients will, of course, be subjected to quantization. With quantized coefficients would the network in part (a) still correspond to an allpass network?

The difference equation relating input and output of the allpass system can be expressed as

$$y(n) - ay(n-1) = x(n-1) - ax(n)$$

or, equivalently,

$$y(n) = a[y(n-1) - x(n)] + x(n-1) \qquad \text{(P4.9-1)}$$

(c) Draw a network realization of Eq. (P4.9-1) that requires two delay branches but only one branch requiring a multiplication by other than $+1$ or -1.

(d) With quantized coefficients would the network in part (c) still correspond to an allpass network?

The primary disadvantage to the implementation in part (c) as compared with that in part (a) is that it requires two delays. In some applications, however, it is necessary to implement a cascade of allpass sections. For N allpass sections in cascade it is possible to utilize a realization of each in the form determined in part (c) but use only $(N + 1)$ delay branches. This is accomplished essentially by sharing a delay between sections.

(e) Consider the allpass system with transfer function

$$H(z) = \frac{z^{-1} - a}{1 - az^{-1}} \frac{z^{-1} - b}{1 - bz^{-1}} \qquad \text{(P4.9–2)}$$

Draw a network realization of this system by "cascading" two networks of the form obtained in part (c) in such a way that only three delay branches are required.

(f) With quantized coefficients, would the network in part (e) still correspond to an allpass network?

10. A class of digital filter structures based on continued fraction expansions has been proposed (S. K. Mitra and R. J. Sherwood, "Canonic Realizations of Digital Filters Using the Continued Fraction Expansion," *IEEE Trans. Audio Electroacoust.*, Vol. AU-20, 1972, pp. 185–194). Although there are a variety of forms of such structures, we wish in this problem to illustrate one particular form.

Consider a system function $H(z)$ in the form

$$H(z) = \frac{b_0 + b_1 z^{-1} + \ldots + b_M z^{-M}}{1 - a_1 z^{-1} - \ldots - a_N z^{-N}}$$

where we assume that $b_0 \neq 0$, $a_N \neq 0$, and $M \leq N$.

Multiplying numerator and denominator of $H(z)$ by z^N it can be expressed as

$$H(z) = \frac{b_0 z^N + b_1 z^{N-1} + \ldots + b_M z^{N-M}}{z^N - a_1 z^{N-1} - \ldots - a_N}$$

If we divide denominator into numerator, we obtain

$$H(z) = A_0 + G_0(z)$$

where $A_0 = b_0$ and $G_0(z)$ would in general have the form

$$G_0(z) = \frac{c_1 z^{N-1} + \ldots + c_M}{z^N - a_1 z^{N-1} - \ldots - a_N}$$

Now if $c_1 \neq 0$ and we divide numerator into denominator, we can express $G_0(z)$ as

$$G_0(z) = \frac{1}{A_1 + B_1 z + G_1(z)}$$

where $G_1(z)$ will have the form

$$G_1(z) = \frac{d_2 z^{N-2} + \ldots + d_N}{c_1 z^{N-1} + \ldots + c_M}$$

We can repeat the process of dividing numerator into denominator to obtain

$$G_1(z) = \frac{1}{A_2 + B_2 z + G_2(z)}$$

Thus, assuming that the set of rational functions $\{G_k(z)\}$, $k = 0, 1, \ldots, N$, obtained by the above process is such that the numerator is of degree $N - k - 1$ and the denominator is of degree $N - k$, then $H(z)$ can be expressed as

$$H(z) = A_0 + \cfrac{1}{A_1 + B_1 z + \cfrac{1}{A_2 + B_2 z + \cfrac{1}{\ddots + \cfrac{1}{A_N + B_N z}}}} \qquad \text{(P4.10-1)}$$

In order to implement a network realization based on Eq. (P4.10-1), we need only an implementation of the system function

$$G_k(z) = \frac{1}{A_{k+1} + B_{k+1} z + G_{k+1}(z)} \qquad \text{(P4.10-2)}$$

Multiplying numerator and denominator of Eq. (P4.10-2) by $(1/B_{k+1})z^{-1}$, we obtain

$$G_k(z) \frac{(1/B_{k+1})z^{-1}}{1 + (A_{k+1}/B_{k+1})z^{-1} + (1/B_{k+1})z^{-1} G_{k+1}(z)} \qquad \text{(P4.10-3)}$$

A network realization of Eq. (P4.10-3) is shown in Fig. P4.10.

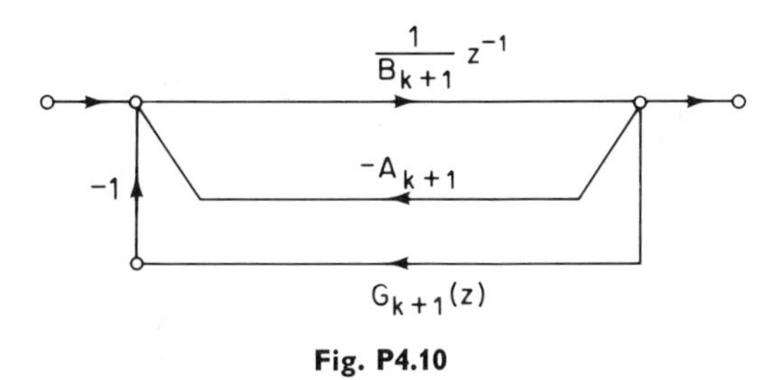

Fig. P4.10

(a) Since each successive $G_k(z)$ can be realized by a similar network, this suggests a complete structure for $H(z)$ expressed in the form of Eq. (P4.10-1). Assuming that N is odd, draw the network for such a structure. Each branch in this network must have a transmittance that is a constant or a constant times z^{-1}.

(b) For the second-order system with transfer function

$$H(z) = \frac{1}{1 - 2r \cos \theta z^{-1} + r^2 z^{-2}}$$

express $H(z)$ in the form of Eq. (P4.10-1).

(c) Draw the network realization of the system in part (b) in the form that you determined in part (a).

11. Speech production can be modeled as a linear system representing the vocal cavity, excited by puffs of air released through the vocal cords. In synthesizing speech on a digital computer, one approach is to represent the vocal cavity as a connection of cylindrical acoustic tubes with equal length but with different cross-sectional areas, as depicted in Fig. P4.11. Let us assume that we want

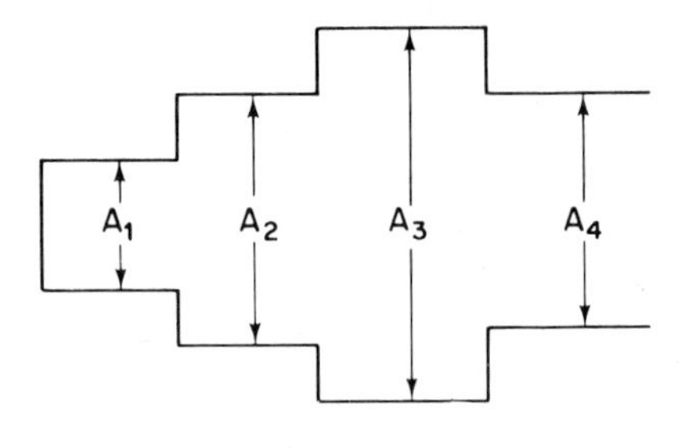

Fig. P4.11

to simulate this system in terms of the volume velocity representing air flow. The input is coupled into the vocal tract through a small constriction, the vocal cords. We will assume that the input is represented by a change in volume velocity at the left end, but that the boundary condition for traveling waves at the left end is that the net volume velocity must be zero. This is analogous to an electrical transmission line driven by a current source. The output is considered to be the volume velocity at the right end. We assume that each section is lossless.

At each interface between sections a forward-traveling wave is transmitted to the next section with one coefficient and reflected as a backward-traveling wave with a different coefficient. Similarly, a backward-traveling wave arriving at an interface is transmitted with one coefficient and reflected with a different coefficient. Specifically, if we consider a forward-traveling wave f_+ in a tube with cross-sectional area A_1 arriving at the interface with a tube of cross-sectional area A_2, then the forward-traveling wave transmitted is $(1 + \alpha)f_+$ and the reflected wave is αf_+, where

$$\alpha = \frac{A_2 - A_1}{A_1 + A_2}$$

Consider the length of each section to be 3.4 cm with the velocity of sound in air 34,000 cm/s. Draw a digital network that will implement the four-section tube in Fig. P4.11, with the output sampled at a 20-kHz rate.

In spite of the lengthy introduction, this is a reasonably straightforward problem. If you find it hard to think in terms of acoustic tubes, think in terms of transmission-line sections with different characteristic impedances. Just as with transmission lines, it is difficult to express the impulse response in closed form. Draw the network directly from physical considerations, in terms of forward- and backward-traveling pulses in each section.

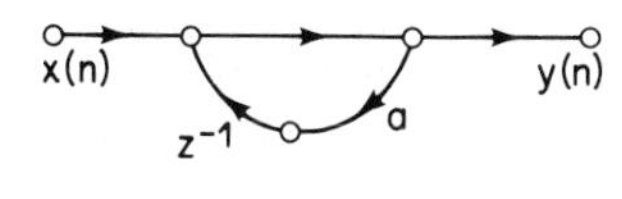

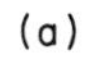

(a)

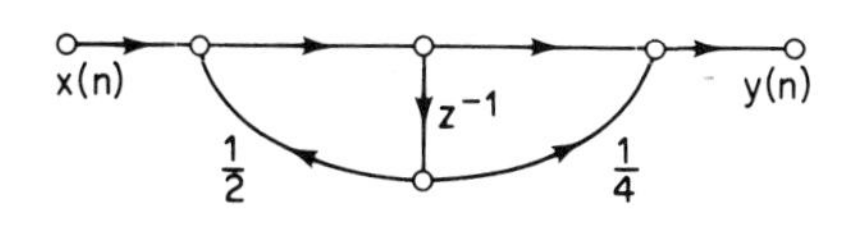

(b)

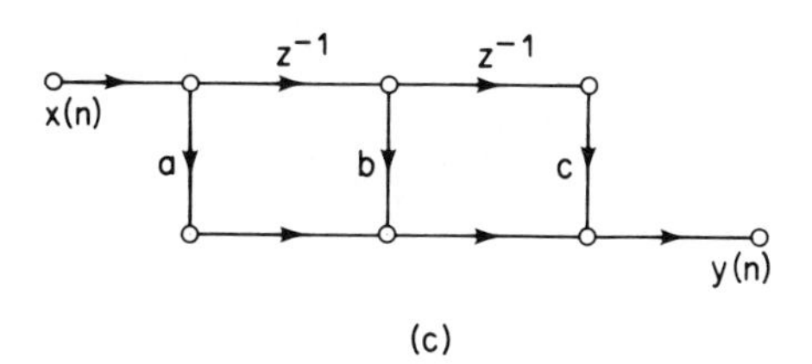

(c)

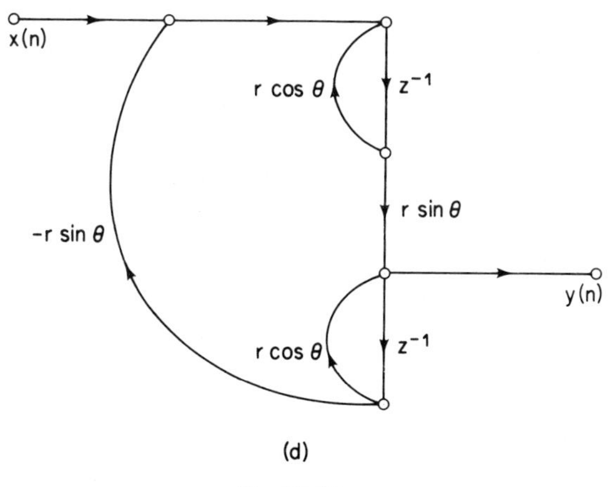

(d)

Fig. P4.12

12. In Fig. P4.12(a)–(d) several networks are shown. Determine the transpose of each and verify that in each case the original and transpose networks have the same transfer function.

13. In Fig. P4.13-1(a)–(f) six digital networks are shown. Determine which one of the last five [i.e., (b) through (f)] has the same transfer function as (a). You should be able to eliminate some of the possibilities by inspection.

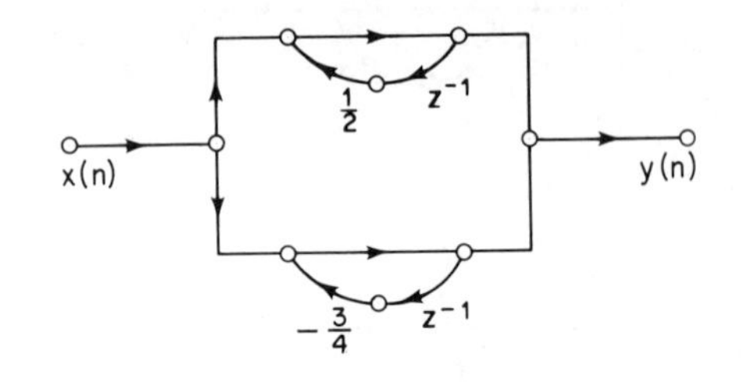

(a)

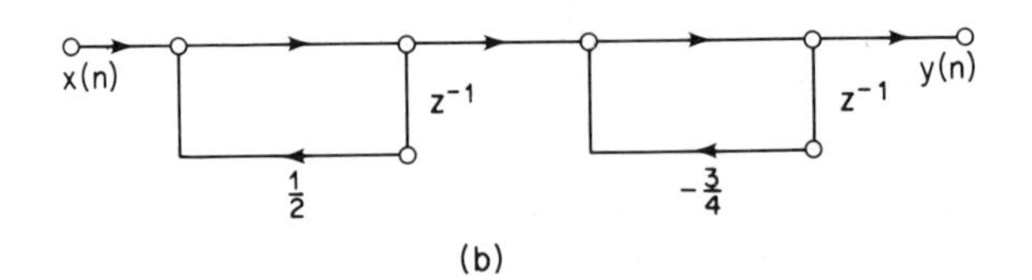

(b)

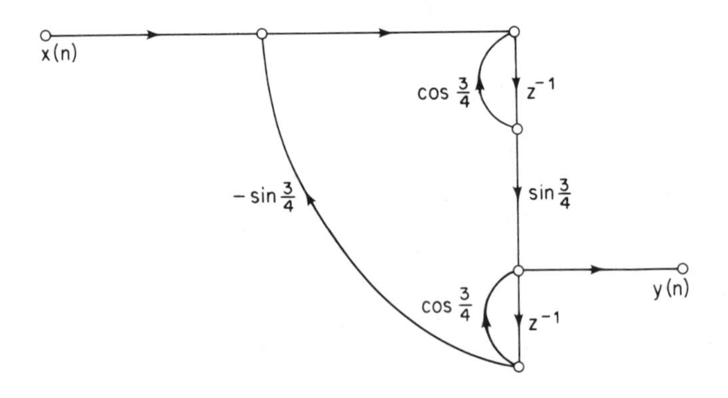

(c)

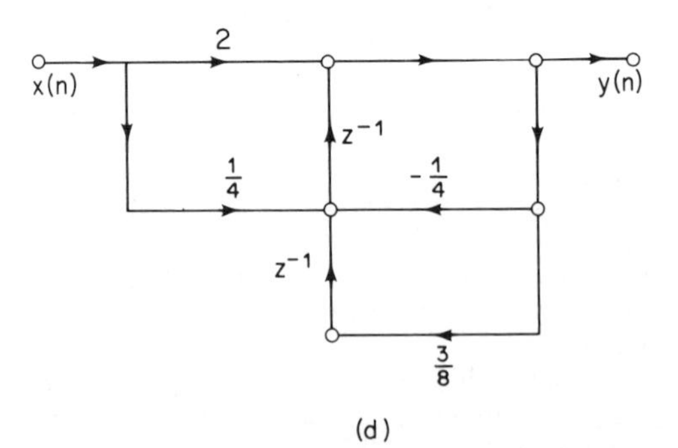

(d)

Fig. P4.13

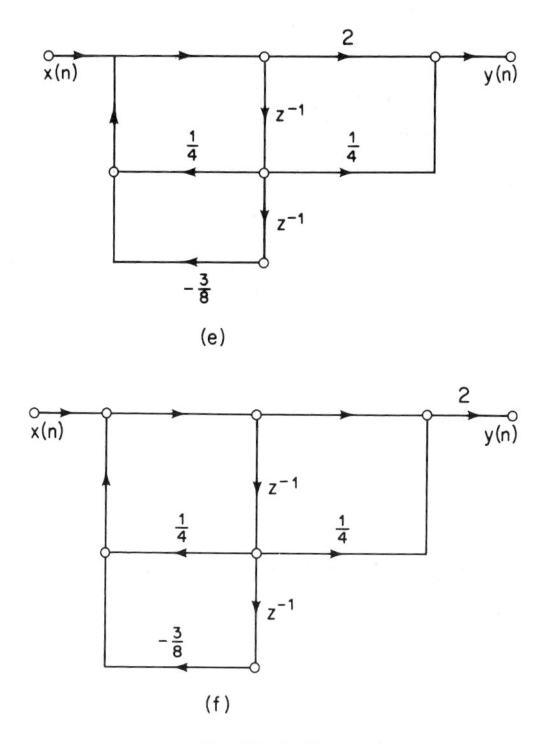

Fig. P4.13 (Contd.)

14. In Sec. 4.5.4 we derived the frequency-sampling structure for FIR filters based on an expansion of the system function $H(z)$ in terms of the DFT of the unit sample response or equivalently in terms of N samples at equal spacing around the unit circle. The expansion used was that given in Eq. (4.43), i.e.,

$$H(z) = (1 - z^{-N}) \frac{1}{N} \sum_{k=0}^{N-1} \frac{\tilde{H}(k)}{1 - W_N^{-k} z^{-1}} \qquad \text{(P4.14-1a)}$$

where

$$\tilde{H}(k) = H(z)|_{z=W_N{}^{-k}} \qquad \text{(P4.14-1b)}$$

To derive other FIR structures based on polynomial interpolation formulas, Eqs. (P4.14-1a) and (P4.14-1b) were then generalized in Sec. 4.5.5 to Eqs. (4.53), (4.54), and (4.55):

$$H(z) = P(z) \sum_{k=0}^{N-1} \frac{H(z_k)}{P_k(z_k)(1 - z_k z^{-1})} \qquad \text{(P4.14-2a)}$$

where

$$P(z) = \prod_{k=0}^{N-1} (1 - z_k z^{-1}) \qquad \text{(P4.14-2b)}$$

and

$$P_k(z) = \prod_{\substack{i=0 \\ i \neq k}}^{N-1} (1 - z_i z^{-1}) \qquad \text{(P4.14-2c)}$$

In this problem we wish to verify a number of statements made in Secs. 4.5.4 and 4.5.5 about Eqs. (P4.14-1) and (P4.14-2).

(a) Show that equations (P4.14-1) are a special case of equations (P4.14-2); i.e., show that if $z_k = W_N^{-k}$, then equations (P4.14-2) reduce to equations (P4.14-1). As a consequence we need only concentrate on the more general case of equations (P4.14-2).

(b) Show that $H(z)$ as expressed by equations (P4.14-2) is an $(N-1)$st degree polynomial in z^{-1}. To do this it is necessary to show that it has no poles other than at $z = 0$ and that it has no powers of z^{-1} higher than $(N-1)$.

(c) We wish finally to verify that $H(z)$ as expressed by equations (P4.14-2) gives the correct values at the sampling points in the z-plane; i.e., we wish to verify that for $z = z_i$, the right side of Eq. (P4.14-2a) reduces to $H(z_i)$. Show, using L'Hopital's rule, that

$$\lim_{z \to z_i} \left[P(z) \sum_{k=0}^{N-1} \frac{H(z_k)}{P_k(z_k)(1 - z_k z^{-1})} \right] = H(z_i)$$

15. Consider a finite impulse response filter for which the frequency response has the property that

$$H(e^{j\omega}) = |H(e^{j\omega})| e^{-j\omega n_0}$$

where n_0 is not necessarily an integer. Let N be the length of the unit-sample response. Recall that the unit-sample response is completely specified by N samples of $H(e^{j\omega})$ taken at $\omega = 2\pi k/N$, $k = 0, 1, \ldots, N-1$.

(a) Sketch $|H(e^{j\omega})|$ for the case $N = 15$, $n_0 = 0$, and

$$|H_k| = |H(e^{j(2\pi k/15)})| = \begin{cases} 1, & k = 0 \\ 1/2, & k = 1, 14 \\ 0, & \text{elsewhere} \end{cases}$$

(b) Write a general expression for $h(n)$ in terms of the H_k's. (Do not assume that $n_0 = 0$.)

(c) Sketch $h(n)$ for the cases (1) $n_0 = (N-1)/2 = 7$, and (2) $n_0 = N/2 = \frac{15}{2}$, where $|H_k|$ are as in part (a).

(d) Draw a complete flow-graph realization of this system when $N = 15$, $n_0 = \frac{15}{2}$, and the $|H_k|$ are as in part (a). This realization should be in the form of a recursive filter, i.e., a frequency-sampling filter.

16. A causal linear-phase FIR system has the property that $h(n) = h(N-1-n)$ for $n = 0, 1, \ldots, N-1$. This symmetry constraint was used in Sec. 4.5.3 to show that systems satisfying this constraint have linear-phase corresponding to a delay of $(N-1)/2$ samples. This constraint results in a significant simplification of the frequency-sampling realization of Eqs. (4.49) and (4.50).

(a) Using the above linear-phase constraint, show that, for N even, $\tilde{H}(N/2) = 0$.

(b) Determine an expression for $\theta(k)$ for $k = 0, 1, \ldots, N-1$ that is valid for N even. You may find it helpful to refer to the results in Sec. 4.5.3.

(c) Using the results of parts (a) and (b), show that for $h(n)$ linear phase and with N even, Eqs. (4.49) and (4.50) can be simplified to

$$H(z) = \frac{1 - z^{-N}}{N} \left[\sum_{k=1}^{(N/2)-1} \frac{(-1)^k |\tilde{H}(k)| \, 2 \cos(\pi k/N)(1 - z^{-1})}{1 - 2z^{-1} \cos(2\pi k/N) + z^{-2}} + \frac{\tilde{H}(0)}{1 - z^{-1}} \right]$$

(We have assumed for convenience that $r = 1$.)

(d) How is the above expression modified if N is odd?

(e) Draw a flow-graph representation of the system function derived in part (c)

(f) If enough of the coefficients $|\tilde{H}(k)|$ are zero, the frequency-sampling structure may require fewer arithmetic operations than the direct-form realization. How may coefficients can be nonzero and yet have fewer multiplications and additions than the direct form?

17. In Sec. 4.5.4 and Problem 16 above, we derived one class of frequency-sampling structures, based on sampling the frequency response of an FIR system at the frequencies $\omega_k = 2\pi k/N$, $k = 0, 1, \ldots, N-1$.

A second type of frequency sampling structure can be obtained by sampling the frequency response of an FIR system at the N frequencies

$$\omega_k = \frac{2\pi}{N}(k + \tfrac{1}{2}), \qquad k = 0, 1, \ldots, N-1$$

Let us define type 2 frequency samples as

$$\tilde{H}(k) = H(z)|_{z=e^{j\omega_k}}, \qquad k = 0, 1, \ldots, N-1$$

(a) Express the unit-sample response $h(n)$ of the system in terms of the type 2 frequency samples.

(b) Following the derivation of Sec. 3.5, show that $H(z)$ can be expressed as

$$H(z) = \frac{1 + z^{-N}}{N} \sum_{k=0}^{N-1} \frac{\tilde{H}(k)}{1 - e^{j(\pi/N)} W_N^{-k} z^{-1}}$$

where $W_N^{-k} = e^{j(2\pi/N)k}$.

(c) If we express the type 2 frequency samples in polar form as

$$\tilde{H}(k) = |\tilde{H}(k)| e^{j\theta(k)}$$

determine the constraints on $|\tilde{H}(k)|$ and $\theta(k)$ to ensure that $h(n)$ is real.

(d) Use the constraints of part (c) to show that for N even

$$H(z) = \frac{1 + z^{-N}}{N} \left\{ \sum_{k=0}^{(N/2)-1} \frac{2|\tilde{H}(k)|[\cos(\theta(k)) - z^{-1}\cos(\theta(k) - (2\pi/N)(k + \tfrac{1}{2}))]}{1 - 2z^{-1}\cos[(2\pi/N)(k + \tfrac{1}{2})] + z^{-2}} \right\}$$

(e) How is the above equation modified if N is odd?

(f) If the system has linear phase with delay $(N-1)/2$ samples, write an expression for $\theta(k)$ when N is even.

(g) Use the result of part (f) to show that for N even and linear phase,

$$H(z) = \frac{1 + z^{-N}}{N} \left[\sum_{k=0}^{(N/2)-1} \frac{2|\tilde{H}(k)|(-1)^k(1 + z^{-1})\sin(\pi(k + \tfrac{1}{2})/N))}{1 - 2z^{-1}\cos(2\pi(k + \tfrac{1}{2})/N) + z^{-2}} \right]$$

(h) How is the result of part (g) modified if N is odd?

This problem and problem 16 are based on a paper by L. Rabiner and R. Schafer, "Recursive and Nonrecursive Realizations of Digital Filters Designed by Frequency Sampling Techniques, *IEEE Trans. Audio Electroacoust.*, Vol. AU-19, No. 3, Sept. 1971.

18. (a) Verify that Tellegen's theorem in the form of Eq. (4.62) holds for the two networks of Fig. 4.39.

(b) Repeat part (a) for Tellegen's theorem in the form of Eq. (4.65).

19. In Sec. 4.7.3 an expression for the change in the transfer function of a linear signal flow graph due to a large-scale change in a branch transmittance was presented [Eq. (4.87)]. In this problem we wish to derive that result.

To derive this relation we begin by expressing ΔT_{ab} in a Taylor's series expansion as

$$\Delta T_{ab} = \sum_{k=1}^{\infty} \frac{1}{k!} \frac{\partial^k T_{ab}}{\partial F_{nm}^k} [\Delta F_{nm}]^k \tag{P4.19-1}$$

where T_{ab} denotes the transfer function from node a to node b when the branch transmittance from node n to node m is given by F_{nm} and ΔT_{ab} denotes the change in the transfer function when the branch transmittance changes by ΔF_{nm}.
(a) Use Eq. (4.85) to show that $\partial^2 T_{ab}/\partial F_{nm}^2$ can be expressed as

$$\frac{\partial^2 T_{ab}}{\partial F_{nm}^2} = 2T_{mn}T_{an}T_{mb}$$

(b) Show in a similar manner that

$$\frac{\partial^3 T_{ab}}{\partial F_{nm}^3} = 3!T_{mn}^2 T_{an}T_{mb}$$

The results in parts (b) and (c) generalize to

$$\frac{\partial^k T_{ab}}{\partial F_{nm}^k} = k!T_{mn}^{(k-1)} T_{an}T_{mb}, \qquad n \geq 1 \tag{P4.19-2}$$

(c) Combining Eqs. (P4.19-1) and (P4.19-2), show that

$$\Delta T_{ab} = \frac{T_{an}T_{mb}\Delta F_{nm}}{1 - T_{mn}\Delta F_{nm}} \tag{P4.19-3}$$

20. Use Eq. (4.87) to show that for a nonrecursive digital filter, a large-scale change in one branch transmittance generates a proportional change in the overall transfer function; i.e., show that ΔT_{ab} is proportional to ΔF_{nm}.

5

Digital Filter Design Techniques

5.0 Introduction

In the most general sense, a digital filter is a linear shift-invariant discrete-time system that is realized using finite-precision arithmetic. The design of digital filters involves three basic steps: (1) the specification of the desired properties of the system; (2) the approximation of these specifications using a causal discrete-time system; and (3) the realization of the system using finite-precision arithmetic. Although these three steps are certainly not independent, we find it convenient to focus our attention in this chapter primarily on the second step, the first being highly dependent on the application, and the third being discussed in Chapters 4 and 9.

In a practical setting, it is often the case that the desired digital filter is to be used to filter a digital signal that is derived from an analog signal by means of periodic sampling. The specifications for both analog and digital filters are often (but not always) given in the frequency domain, as, for example, in the case of frequency selective filters such as lowpass, bandpass, and highpass filters. Given the sampling rate, it is straightforward to convert from frequency specifications on an analog filter to frequency specifications on the corresponding digital filter, the analog frequencies being in terms of Hertz and the digital frequencies being in terms of radian frequency or angle around the unit circle with the point $z = -1$ corresponding to half the sampling frequency. There are, however, applications in which a digital signal to be filtered is not derived by means of periodic sampling of an analog time function, and, as was noted in Sec. 1.7, there are a variety of means besides periodic sampling for representing analog time functions in terms of sequences. Furthermore, in most of the design techniques that we shall discuss, the sampling period plays no role whatsoever in the approximation procedure. Therefore, the least confusing point of view toward digital filter design is to consider the filter as being specified in terms of angle around the unit circle rather than in terms of analog frequencies.

A separate problem is that of determining an appropriate set of specifications on the digital filter. In the case of a lowpass filter, for example, the specifications often take the form of a tolerance scheme, such as depicted in Fig. 5.1.† The dashed curve represents the frequency response of a system

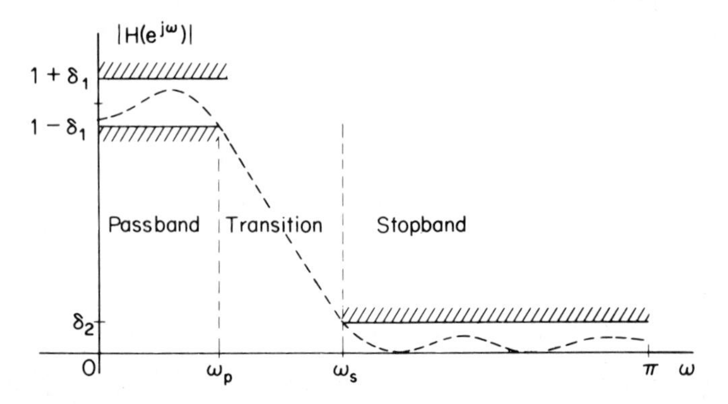

Fig. 5.1 Tolerance limits for approximation of ideal lowpass filter.

that meets the prescribed specification. In this case, there is a passband wherein the magnitude of the response must approximate 1 with an error of $\pm\delta_1$; i.e.,

$$1 - \delta_1 \leq |H(e^{j\omega})| \leq 1 + \delta_1, \qquad |\omega| \leq \omega_p$$

There is a *stopband* in which the magnitude response must approximate zero with an error less than δ_2; i.e.,

$$|H(e^{j\omega})| \leq \delta_2, \qquad \omega_s \leq |\omega| \leq \pi$$

The passband cutoff frequency ω_p and the stopband cutoff frequency ω_s are given in terms of z-plane angle. To make it possible to approximate the ideal lowpass filter in this way we must also provide a *transition band* of nonzero width $(\omega_s - \omega_p)$ in which the magnitude response drops smoothly from the passband to the stopband. Many of the filters used in practice are

† In this figure, the limits of tolerable approximation error are denoted by the shaded horizontal lines. Note also that it is sufficient to plot the specifications only for $0 \leq \omega \leq \pi$, since the remainder can be inferred from symmetry properties.

specified by such a tolerance scheme, with no constraints on the phase response other than those imposed by stability and causality requirements; i.e., the poles of the system function must lie inside the unit circle.

Given a set of specifications in the form of Fig. 5.1, the next step is to find a discrete-time linear system whose frequency response falls within the prescribed tolerances. At this point the filter design problem becomes a problem in approximation. In the case of IIR systems we must approximate the desired frequency response by a rational function, while in the FIR case we are concerned with polynomial approximation. For convenience our discussion is organized so as to distinguish between those design techniques that are appropriate for IIR filters and those that are appropriate for FIR filters. We shall discuss a variety of design techniques for both types of filters. These techniques range from closed-form procedures, which involve only substitution of design specifications into design formulas, to algorithmic techniques, where a solution is obtained by an iterative procedure.

5.1 Design of IIR Digital Filters from Analog Filters

The traditional approach to the design of IIR digital filters involves the transformation of an analog filter into a digital filter meeting prescribed specifications. This is a reasonable approach because:

1. The art of analog filter design is highly advanced and, since useful results can be achieved, it is advantageous to utilize the design procedures already developed for analog filters.
2. Many useful analog design methods have relatively simple closed-form design formulas. Therefore, digital filter design methods based on such analog design formulas are rather simple to implement.
3. In many applications it is of interest to use a digital filter to simulate the performance of an analog linear time-invariant filter.

Consider an analog system function,

$$H_a(s) = \frac{\sum_{k=0}^{M} d_k s^k}{\sum_{k=0}^{N} c_k s^k} = \frac{Y_a(s)}{X_a(s)} \tag{5.1}$$

where $x_a(t)$ is the input and $y_a(t)$ is the output and $X_a(s)$ and $Y_a(s)$ are their respective Laplace transforms. It is assumed that $H_a(s)$ has been obtained through one of the established approximation methods used in analog filter design. (Examples are discussed in Sec. 5.2.) The input and output of such a system are related by the convolution integral,

$$y_a(t) = \int_{-\infty}^{\infty} x_a(\tau) h_a(t - \tau)\, d\tau \tag{5.2}$$

where $h_a(t)$, the impulse response, is the inverse Laplace transform of $H_a(s)$. Alternatively, an analog system having a system function $H_a(s)$ can be described by the differential equation

$$\sum_{k=0}^{N} c_k \frac{d^k y_a(t)}{dt^k} = \sum_{k=0}^{M} d_k \frac{d^k x_a(t)}{dt^k} \tag{5.3}$$

The corresponding rational system function for digital filters has the form

$$H(z) = \frac{\sum_{k=0}^{M} b_k z^{-k}}{\sum_{k=0}^{N} a_k z^{-k}} = \frac{Y(z)}{X(z)} \tag{5.4}$$

The input and output are related by the convolution sum

$$y(n) = \sum_{k=-\infty}^{\infty} x(k)h(n-k) \tag{5.5}$$

or, equivalently, by the difference equation

$$\sum_{k=0}^{N} a_k y(n-k) = \sum_{k=0}^{M} b_k x(n-k) \tag{5.6}$$

In transforming an analog system to a digital system we must therefore obtain either $H(z)$ or $h(n)$ from the analog filter design. In such transformations we generally require that the essential properties of the analog frequency response be preserved in the frequency response of the resulting digital filter. Loosely speaking, this implies that we want the imaginary axis of the s-plane to map into the unit circle of the z-plane. A second condition is that a stable analog filter should be transformed to a stable digital filter. That is, if the analog system has poles only in the left-half s-plane, then the digital filter must have poles only inside the unit circle. These constraints are basic to all the techniques to be discussed in this section.

5.1.1 *Impulse Invariance*

One procedure for transforming an analog filter design to a digital filter design corresponds to choosing the unit-sample response of the digital filter as equally spaced samples of the impulse response of the analog filter [1–4]. That is,

$$h(n) = h_a(nT)$$

where T is the sampling period.

It can be shown as a generalization of Eq. (1.29) that the z-transform of $h(n)$ is related to the Laplace transform of $h_a(t)$ by the equation

$$H(z)\big|_{z=e^{sT}} = \frac{1}{T} \sum_{k=-\infty}^{\infty} H_a\left(s + j\frac{2\pi}{T}k\right) \tag{5.7}$$

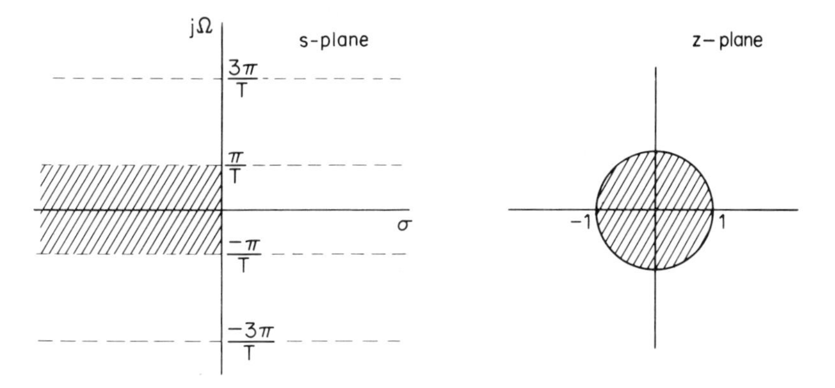

Fig. 5.2 Representation of periodic sampling.

From the relationship $z = e^{sT}$ it is seen that strips of width $2\pi/T$ in the s-plane map into the entire z-plane as depicted in Fig. 5.2. The left half of each s-plane strip maps into the interior of the unit circle, the right half of each s-plane strip maps into the exterior of the unit circle, and the imaginary axis of the s-plane maps onto the unit circle in such a way that each segment of length $2\pi/T$ is mapped once around the unit circle. From Eq. (5.7) it is clear that each horizontal strip of the s-plane is overlayed onto the z-plane to form the digital system function from the analog system function. Thus the impulse invariance method *does not* correspond to a simple algebraic mapping of the s-plane to the z-plane.

The frequency response of the digital filter is related to the frequency response of the analog filter as

$$H(e^{j\omega}) = \frac{1}{T} \sum_{k=-\infty}^{\infty} H_a\left(j\frac{\omega}{T} + j\frac{2\pi}{T} k \right) \tag{5.8}$$

From our previous discussion of the sampling theorem in Sec. 1.7 it is clear that if and only if

$$H_a(j\Omega) = 0, \qquad |\Omega| \geq \pi/T$$

then

$$H(e^{j\omega}) = \frac{1}{T} H_a\left(j\frac{\omega}{T} \right), \qquad |\omega| \leq \pi$$

Unfortunately, any practical analog filter will not be bandlimited, and consequently there is interference between successive terms in Eq. (5.8) as illustrated in Fig. 5.3.

Because of the aliasing that occurs in the sampling process, the frequency response of the resulting digital filter will not be identical to the original

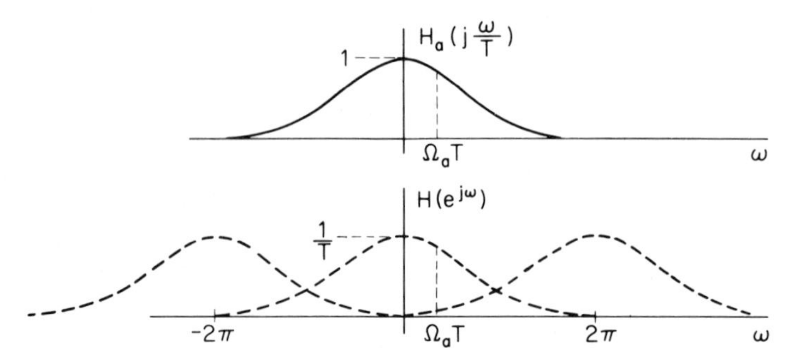

Fig. 5.3 Graphical representation of the effects of aliasing in the impulse invariance design technique.

analog frequency response. It is important to note that if we consider the filter specifications to be in terms of specifications on a digital filter, then a change in the value of T has no effect on the amount of aliasing in the impulse invariant design procedure. For example, referring to Fig. 5.3, we may consider that the cutoff frequency of the digital filter has been specified to be at the frequency labeled $\Omega_a T$. That point in the frequency response is then constrained as the cutoff frequency of a lowpass digital filter, and if T is reduced, then Ω_a in the analog filter must be correspondingly increased in such a way that $\Omega_a T$ remains constant and equal to the cutoff frequency specified for the digital filter. Thus if T is made smaller in an effort to reduce the effect of aliasing, Ω_a must be made correspondingly larger. With the attitude that the digital filter to be designed is specified in terms of frequencies on the unit circle, T is therefore an irrelevant parameter in impulse invariant design and could just as well be considered to be equal to unity. While it is common practice to include the parameter T in discussing impulse invariant design, it is important to keep in mind that the parameter plays a minor role in the design procedure.

To investigate the interpretation of impulse invariant design in terms of a relationship between the s-plane and the z-plane, let us consider the system function of the analog filter expressed in terms of a partial-fraction expansion, so that

$$H_a(s) = \sum_{k=1}^{N} \frac{A_k}{s - s_k} \tag{5.9}$$

The corresponding impulse response is

$$h_a(t) = \sum_{k=1}^{N} A_k e^{s_k t} u(t)$$

where $u(t)$ is a continuous-time unit step function.

and the unit-sample response of the digital filter is then

$$h(n) = h_a(nT) = \sum_{k=1}^{N} A_k e^{s_k nT} u(n)$$

$$= \sum_{k=1}^{N} A_k (e^{s_k T})^n u(n)$$

The system function of the digital filter $H(z)$ is consequently given by

$$H(z) = \sum_{k=1}^{N} \frac{A_k}{1 - e^{s_k T} z^{-1}} \tag{5.10}$$

In comparing Eqs. (5.9) and (5.10) we observe that a pole at $s = s_k$ in the s-plane transforms to a pole at $e^{s_k T}$ in the z-plane and the coefficients in the partial-fraction expansion of $H_a(s)$ and $H(z)$ are equal.† If the analog filter is stable, corresponding to the real part of s_k less than zero, then the magnitude of $e^{s_k T}$ will be less than unity, so that the corresponding pole in the digital filter is inside the unit circle, and consequently the digital filter is also stable. While the poles in the s-plane map to poles in the z-plane according to the relationship $z_k = e^{s_k T}$, it is important to recognize that the impulse invariant design procedure does not correspond to a mapping of the s-plane to the z-plane by that relationship or in fact by any relationship. In particular, the zeros of the digital transfer function are a function of the poles and the coefficients A_k in the partial-fraction expansion and they will not in general be mapped in the same way the poles are mapped.

EXAMPLE. As an example of the determination of a digital filter from an analog filter by means of impulse invariance, let us consider the analog system function $H_a(s)$ given by

$$H_a(s) = \frac{s + a}{(s + a)^2 + b^2}$$

$$= \frac{\frac{1}{2}}{s + a + jb} + \frac{\frac{1}{2}}{s + a - jb}$$

The corresponding transfer function of the impulse invariant digital filter is then

$$H(z) = \frac{\frac{1}{2}}{1 - e^{-aT}e^{-jbT}z^{-1}} + \frac{\frac{1}{2}}{1 - e^{-aT}e^{jbT}z^{-1}}$$

$$= \frac{1 - (e^{-aT} \cos bT)z^{-1}}{(1 - e^{-aT}e^{-jbT}z^{-1})(1 - e^{-aT}e^{jbT}z^{-1})}$$

The digital filter consequently has one zero at the origin and a zero at $z = e^{-aT} \cos bT$.

† See Problem 5 of this chapter for a consideration of the modifications required for multiple-order poles.

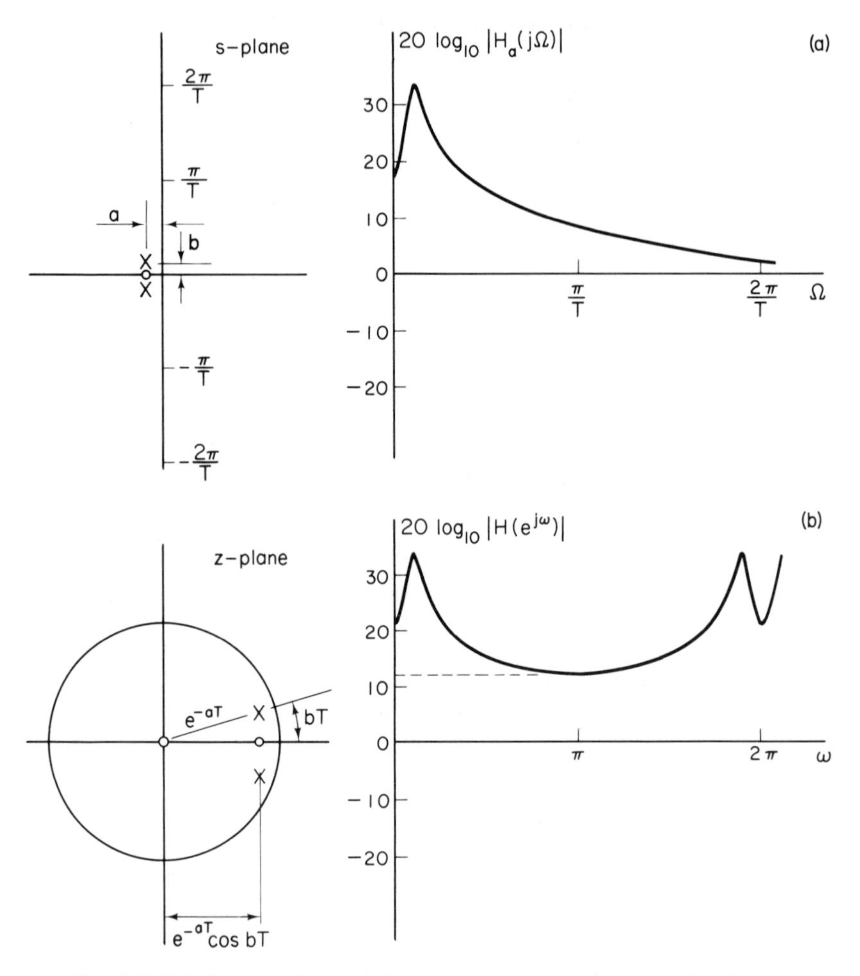

Fig. 5.4 (a) Pole–zero plot and frequency response of a second-order analog system; (b) pole–zero plot and frequency response of the discrete-time system obtained by sampling the impulse response of the above system.

Figure 5.4 shows the s-plane pole-zero plot for $H_a(s)$ and the z-plane pole-zero plot of $H(z)$, along with the corresponding analog and digital frequency response functions. In this case the frequency response of the analog system falls off rather slowly relative to the sampling frequency and thus the effects of aliasing are apparent in the digital frequency response.

It should be noted that when the analog filter is "sufficiently bandlimited," the above procedure produces a digital filter whose frequency response is, from Eq. (5.8),

$$H(e^{j\omega}) \approx \frac{1}{T} H_a\left(j \frac{\omega}{T}\right)$$

Thus, for high sampling rates (T small) the digital filter may have an extremely high gain. For this reason it is generally advisable to use, instead of Eq. (5.10),

$$H(z) = \sum_{k=1}^{N} \frac{TA_k}{1 - e^{s_kT}z^{-1}} \tag{5.11}$$

That is, the unit-sample response is $h(n) = Th_a(nT)$ [3,4].

The basis for impulse invariance as described above is to choose a unit-sample response for the digital filter that is similar in some sense to the impulse response of the analog filter. The use of this procedure is often not motivated so much by a desire to maintain the impulse response shape, but by the knowledge that if the analog filter is bandlimited, then the digital filter frequency response will closely approximate the analog frequency response. However, in some filter design problems, a primary objective may be to control some aspect of the time response such as the impulse response or the step response. In such cases a natural approach would be to design the digital filter by impulse invariance or a *step invariance procedure*. In the latter case, the response of the digital filter to a sampled unit step function is chosen to be samples of the analog step response. In this way, if the analog filter has good step response characteristics, such as small rise time and low peak overshoot, these characteristics would be preserved in the digital filter. Clearly, this concept of waveform invariance can be extended to the preservation of the output waveshape for a variety of inputs. A more detailed consideration of extensions of the impulse invariant procedure is given in Problems 3 and 4 of this chapter.

Although in the impulse invariance design procedure, distortion in the frequency response is introduced due to aliasing, the relationship between analog and digital frequency is linear and consequently, except for aliasing, the shape of the frequency response is preserved. This is in contrast to the procedures to be discussed next, which correspond to the use of algebraic transformations. It should be noted in conclusion that the impulse invariance technique is obviously only appropriate for essentially bandlimited filters. For example, highpass or bandstop filters would require additional band-limiting to avoid severe aliasing distortion.

5.1.2 Designs Based on Numerical Solution of the Differential Equation

A second approach to deriving a digital filter is to approximate the derivatives in Eq. (5.3) by finite differences. This is a standard procedure in numerical analysis [5] and in digital simulations of analog systems. This procedure can be motivated by the intuitive notion that the derivative of an analog time function can be approximated by the difference between consecutive samples of the function to be differentiated. We might expect that as the sampling rate is increased, i.e., the samples are closer together, then the approximation to the derivative would be increasingly accurate. For example, suppose that the first derivative is approximated by the first backward difference [5]

$$\left.\frac{dy_a(t)}{dt}\right|_{t=nT} \rightarrow \nabla^{(1)}[y(n)] = \frac{y(n) - y(n-1)}{T} \tag{5.12}$$

where $y(n) = y_a(nT)$. Approximations to higher-order derivatives are obtained by repeated application of Eq. (5.12); i.e.,

$$\left.\frac{d^k y_a(t)}{dt^k}\right|_{t=nT} = \left.\frac{d}{dt}\left(\frac{d^{k-1}}{dt^{k-1}} y_a(t)\right)\right|_{t=nT} \rightarrow \nabla^{(k)}[y(n)] = \nabla^{(1)}[\nabla^{(k-1)}[y(n)]] \tag{5.13}$$

For convenience we define

$$\nabla^{(0)}[y(n)] = y(n) \tag{5.14}$$

Applying Eqs. (5.12)–(5.14) to (5.3) we obtain

$$\sum_{k=0}^{N} c_k \nabla^{(k)}[y(n)] = \sum_{k=0}^{M} d_k \nabla^{(k)}[x(n)]$$

where $y(n) = y_a(nT)$ and $x(n) = x_a(nT)$. We note that the operation $\nabla^{(1)}[\ ]$ is a linear shift-invariant operator and that $\nabla^{(k)}[\ ]$ can be viewed as a cascade of (k) operators $\nabla^{(1)}[\ ]$. In particular,

$$\mathfrak{Z}[\nabla^{(1)}[x(n)]] = \left[\frac{1 - z^{-1}}{T}\right] X(z)$$

and

$$\mathfrak{Z}[\nabla^{(k)}[x(n)]] = \left[\frac{1 - z^{-1}}{T}\right]^k X(z)$$

Thus taking the z-transform of each side we obtain

$$H(z) = \frac{\displaystyle\sum_{k=0}^{M} d_k \left[\frac{1 - z^{-1}}{T}\right]^k}{\displaystyle\sum_{k=0}^{N} c_k \left[\frac{1 - z^{-1}}{T}\right]^k} \tag{5.15}$$

Comparing Eq. (5.15) to (5.1), we observe that the digital transfer function can be obtained directly from the analog transfer function by means of a substitution of variable

$$s = \frac{1 - z^{-1}}{T} \tag{5.16}$$

so that the process of replacing derivatives by differences does indeed truly correspond to a mapping of the s-plane to the z-plane, according to Eq. (5.16). We previously indicated that the imaginary axis in the s-plane should map into the unit circle in the z-plane and that stable analog filters should map into stable digital filters. To investigate these questions for the transformation of Eq. (5.16) we must express z as a function of s, obtaining

$$z = \frac{1}{1 - sT}$$

Substituting $s = j\Omega$,

$$z = \frac{1}{1 - j\Omega T} \tag{5.17}$$

Clearly, the locus of the $j\Omega$-axis in the s-plane is not the unit circle in the z-plane since $|z| \neq 1$ for all values of Ω in Eq. (5.17). In fact, we can write Eq. (5.17) as

$$\begin{aligned} z &= \tfrac{1}{2}\left[1 + \frac{1 + j\Omega T}{1 - j\Omega T}\right] \\ &= \tfrac{1}{2}[1 + e^{j2\tan^{-1}(\Omega T)}] \end{aligned} \tag{5.18}$$

which corresponds to a circle whose center is at $z = \frac{1}{2}$ and radius is $\frac{1}{2}$, as shown in Fig. 5.5. It is easily verified that the left half of the s-plane maps

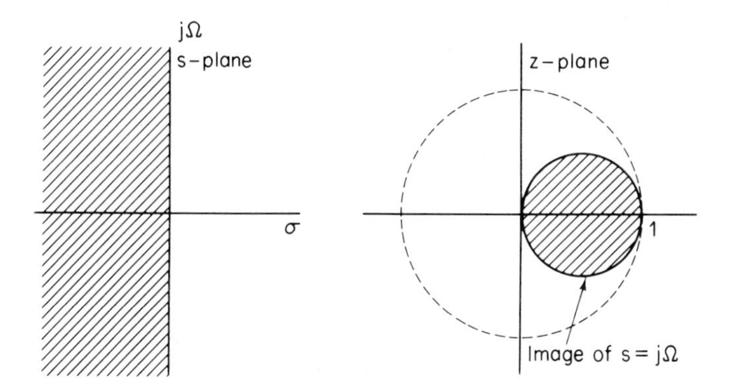

Fig. 5.5 Mapping of the s-plane to the z-plane corresponding to first backward-difference approximation to the derivative.

into the inside of the small circle and the right half of the s-plane maps into the outside of the circle. Therefore, although the requirement of mapping the $j\Omega$-axis to the unit circle is not satisfied, this mapping does satisfy the stability condition since poles in the left half s-plane map to the inside of the small circle, which is inside the unit circle.

It is worth correlating this result with a common intuitive notion. It is generally assumed that the simulation on a digital computer of the processing of a continuous-time signal by a differential equation can be accomplished by replacing derivatives by differences if the continuous signal is sampled at a high-enough rate. For example, if we wish to differentiate a continuous signal we intuitively expect that an approximation to the derivative can be accomplished by sampling the continuous-time function with a small-enough spacing between samples and forming the first difference of the resulting sequence. To show that in fact this intuition is consistent with the results that we just obtained, we remark first of all that if a bandlimited analog signal is sampled at the Nyquist rate, then the spectrum is nonzero over the entire unit circle. As we increase the sampling rate from the Nyquist rate, that is, as we decrease the sampling period, the nonzero part of the spectrum of the digital signal is confined to a smaller and smaller portion of the unit circle, and, in particular, if we choose the sampling period to be sufficiently small, we can concentrate the nonzero part of the spectrum in the vicinity of $z = 1$ in the z-plane. Correspondingly, if T is sufficiently small in Eq. (5.13), then the frequency response of the digital filter will be concentrated on the small circle in Fig. 5.5 in the vicinity of $z = 1$. This is, of course, is the point at which the small circle and the unit circle are tangent, and if both the filter response and the signal spectrum are concentrated in that region, then we can expect that the digital filter will accurately approximate the analog filter.

In the above procedure, derivatives were replaced by backward differences. An alternative approximation to the derivative is a forward difference. The first forward difference is defined as

$$\Delta^{(1)}[y(n)] = y(n+1) - y(n)$$

The mapping corresponding to this approximation is examined in Problem 2 of this chapter, where it is shown that unstable digital filters can result from this approximation.

The major point in the previous example and also in Problem 2 of this chapter is that, in contrast to the impulse invariance technique, decreasing the sampling period theoretically produces a better filter since the spectrum tends to be concentrated in a very small region of the unit circle. In general, however, there is little to recommend the use of forward or backward differences in digital signal processing, since the high sampling rates required result in a very inefficient representation of the filter and the input signal. Furthermore, it is clear that these procedures are highly unsatisfactory for anything but lowpass filters. Thus we are led to consider other mappings that avoid the aliasing problems of the impulse invariance method.

5.1.3 *Bilinear Transformation*

In the previous section a digital filter was derived by approximating derivatives by differences. An alternative procedure is based on integrating the differential equation and then using a numerical approximation to the integral. For example, consider the first-order equation

$$c_1 y_a'(t) + c_0 y_a(t) = d_0 x(t) \tag{5.19}$$

where $y_a'(t)$ is the first derivative of $y_a(t)$. The corresponding analog system function is

$$H_a(s) = \frac{d_0}{c_1 s + c_0}$$

We can write $y_a(t)$ as an integral of $y_a'(t)$, as in

$$y_a(t) = \int_{t_0}^{t} y_a'(t)\,dt + y_a(t_0)$$

In particular, if $t = nT$ and $t_0 = (n-1)T$,

$$y_a(nT) = \int_{(n-1)T}^{nT} y_a'(\tau)\,d\tau + y_a((n-1)T)$$

If the integral is approximated by a trapezoidal rule [5], we can write

$$y_a(nT) = y_a((n-1)T) + \frac{T}{2}\,[y_a'(nT) + y_a'((n-1)T)] \tag{5.20}$$

However, from Eq. (5.19),

$$y_a'(nT) = \frac{-c_0}{c_1}\,y_a(nT) + \frac{d_0}{c_1}\,x_a(nT)$$

Substituting into Eq. (5.20) we obtain

$$[y(n) - y(n-1)] = \frac{T}{2}\left[\frac{-c_0}{c_1}(y(n) + y(n-1)) + \frac{d_0}{c_1}(x(n) + x(n-1))\right]$$

where $y(n) = y_a(nT)$ and $x(n) = x_a(nT)$. Taking the z-transform and solving for $H(z)$ gives

$$H(z) = \frac{Y(z)}{X(z)} = \frac{d_0}{c_1 \dfrac{2}{T}\dfrac{1 - z^{-1}}{1 + z^{-1}} + c_0} \tag{5.21}$$

From Eq. (5.21) it is clear that $H(z)$ is obtained from $H_a(s)$ by the substitution

$$s = \frac{2}{T}\frac{1 - z^{-1}}{1 + z^{-1}} \tag{5.22}$$

That is,

$$H(z) = H_a(s)\big|_{s=(2/T)((1-z^{-1})/(1+z^{-1}))} \tag{5.23}$$

This can be shown to hold in general since an Nth-order differential equation of the form of Eq. (5.3) can be written as a set of N first-order equations of the form of Eq. (5.19). Solving Eq. (5.22) for z gives

$$z = \frac{1 + (T/2)s}{1 - (T/2)s} \tag{5.24}$$

The invertible transformation of Eq. (5.22) is recognized as a bilinear transformation [1–4,6]. To demonstrate that this mapping has the property that the imaginary axis in the s-plane maps onto the unit circle, consider $z = e^{j\omega}$. Then, from Eq. (5.22), s is given by

$$\begin{aligned} s &= \frac{2}{T}\frac{1 - e^{-j\omega}}{1 + e^{-j\omega}} \\ &= \frac{2}{T}\frac{j \sin(\omega/2)}{\cos(\omega/2)} \\ &= \frac{2}{T} j \tan(\omega/2) \\ &= \sigma + j\Omega \end{aligned}$$

Thus for z on the unit circle, $\sigma = 0$ and Ω and ω are related by

$$\frac{T\Omega}{2} = \tan(\omega/2)$$

This relationship is plotted in Fig. 5.6. From the figure it is clear that the

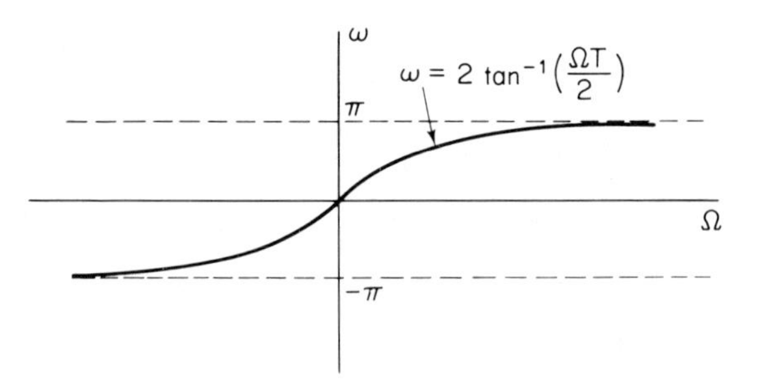

Fig. 5.6 Mapping of the analog frequency axis onto the unit circle using the bilinear transformation.

positive and negative imaginary axes of the s-plane are mapped, respectively, into the upper and lower halves of the unit circle in the z-plane.

In addition to the fact that the imaginary axis in the s-plane maps to the unit circle in the z-plane, the left half of the s-plane maps to the inside of the unit circle and the right half of the s-plane maps to the outside of the unit circle, as depicted in Fig. 5.7.

This can be seen by referring to Eq. (5.24). For the real part of s negative, the magnitude of the factor $(1 + sT/2)/(1 - sT/2)$ is less than unity, corresponding to the inside of the unit circle. Conversely, for the real part of s positive, the magnitude of that ratio is greater than unity, corresponding to the outside of the unit circle. Thus we see that use of the bilinear transformation yields

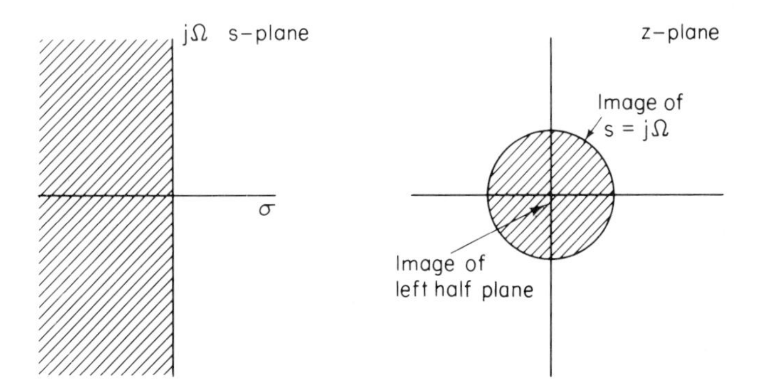

Fig. 5.7 Mapping of the s-plane into the z-plane using the bilinear transformation.

stable digital filters from stable analog filters. Also the bilinear transformation avoids the problem of aliasing encountered with the use of impulse invariance, because it maps the entire imaginary axis in the s-plane onto the unit circle in the z-plane. The price paid for this, however, is the introduction of a distortion in the frequency axis. Consequently, the design of digital filters using the bilinear transformation is only useful when this distortion can be tolerated or compensated. A particular class of filters in which this is true is filters that are chosen to approximate an ideal piecewise constant filter characteristic. For example, if we wish to design a lowpass filter, then we are looking for an approximation to the ideal lowpass characteristic shown in Fig. 5.8. If we are able to design an ideal lowpass filter in the s-plane with a cutoff frequency $\Omega_c = (2/T) \tan (\omega_c/2)$, then when that design is mapped to the z-plane by means of the bilinear transformation, the ideal characteristic of Fig. 5.8 would result. Of course, neither in the analog case nor in the digital case are we able to exactly realize an ideal filter of this type. In general, we would approximate such a filter characteristic, allowing some deviation from one in the passband and some deviation from zero in the stopband with a transition band of nonzero width. Figure 5.9 depicts the

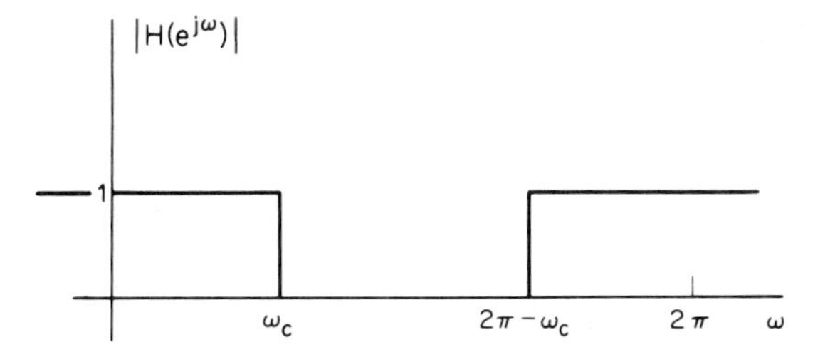

Fig. 5.8 Frequency response of an ideal lowpass filter.

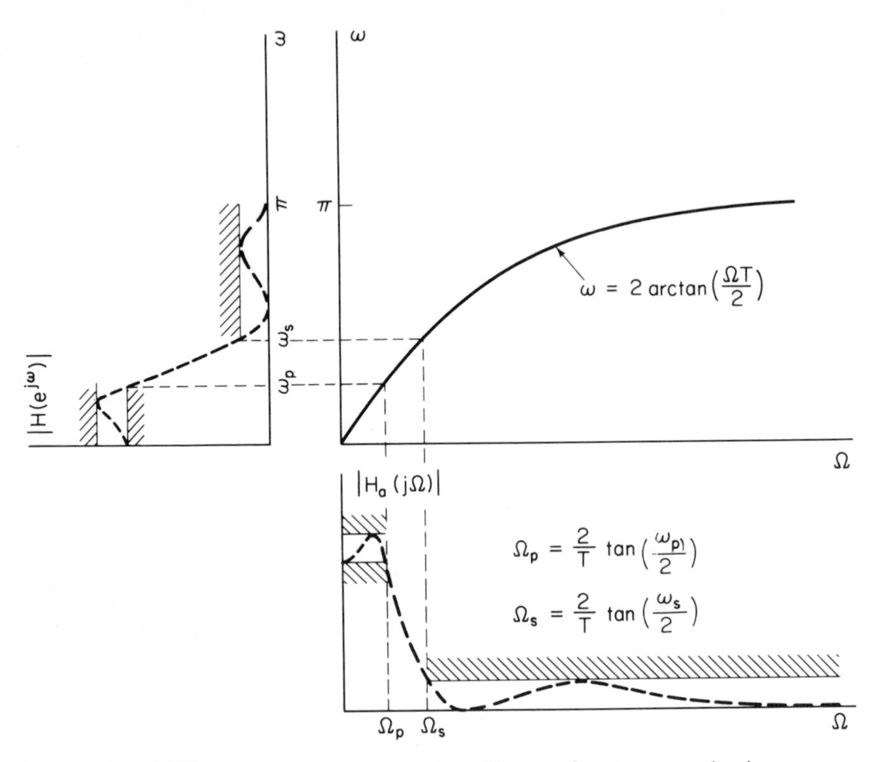

Fig. 5.9 Frequency warping encountered in transforming an analog lowpass filter to a digital lowpass filter. To achieve the desired digital cutoff frequency, the analog cutoff frequencies must be prewarped as indicated.

mapping of an analog frequency response and tolerance scheme to a corresponding digital frequency response and tolerance scheme. If the critical frequencies of the analog filter are prewarped as shown, then, when the analog filter is transformed to the digital filter using Eq. (5.23), the digital filter will meet the desired specifications.

Typical frequency selective analog filter designs are Butterworth, Chebyshev, and elliptic filters [7,8]. As we shall see in the next section, these analog approximation methods have closed-form design formulas which make the design procedure rather straightforward. A Butterworth analog filter is monotonic in the passband and in the stopband. A Chebyshev filter has an equiripple characteristic in the passband and monotonic in the stopband. An elliptic filter is equiripple in both the passband and the stopband. Clearly, these properties will be preserved when the filter is mapped to a digital filter with the bilinear transformation. This is illustrated by the dashed frequency-response curves in Fig. 5.9.

Although the bilinear transformation can be used effectively to map a piecewise constant magnitude characteristic from the *s*-plane to the *z*-plane, the distortion in the frequency axis will manifest itself in terms of distortion in the phase characteristic associated with the filter. If, for example, we were interested in a digital lowpass filter with a linear-phase characteristic, we could not obtain such a filter by applying the bilinear transformation to an analog lowpass filter with a linear-phase characteristic.

5.2 Design Examples: Analog–Digital Transformation

The techniques of the previous section rely upon the availability of appropriate analog filter designs [7,8]. In this section we discuss examples of several analog lowpass approximation techniques, including Butterworth, Chebyshev, and elliptic approximations. The discussion is organized as follows: First, we present the basic design formulas for a particular approximation method. Then, using the same lowpass filter specifications for each approximation method, we carry out the design of a digital filter using both impulse invariance and bilinear transformation [9].

5.2.1 Digital Butterworth Filters

Butterworth filters are defined by the property that the magnitude response is maximally flat in the passband. For an Nth-order lowpass filter, this means that the first $2N - 1$ derivatives of the squared magnitude function are zero at $\Omega = 0$. Another property is that the approximation is monotonic in the passband and the stopband. The squared magnitude function for an analog Butterworth filter is of the form

$$|H_a(j\Omega)|^2 = \frac{1}{1 + (j\Omega/j\Omega_c)^{2N}} \tag{5.25}$$

as sketched in Fig. 5.10.

As the parameter N in Eq. (5.25) increases, the filter characteristics become sharper; that is, they remain closer to unity over more of the passband and become close to zero more rapidly in the stopband, although the

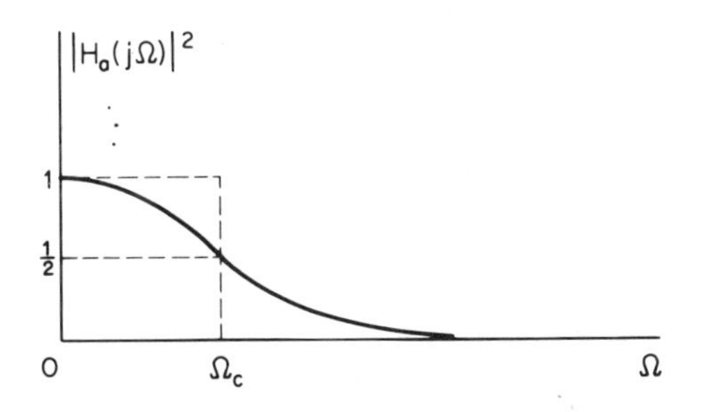

Fig. 5.10 Squared magnitude for analog Butterworth filter.

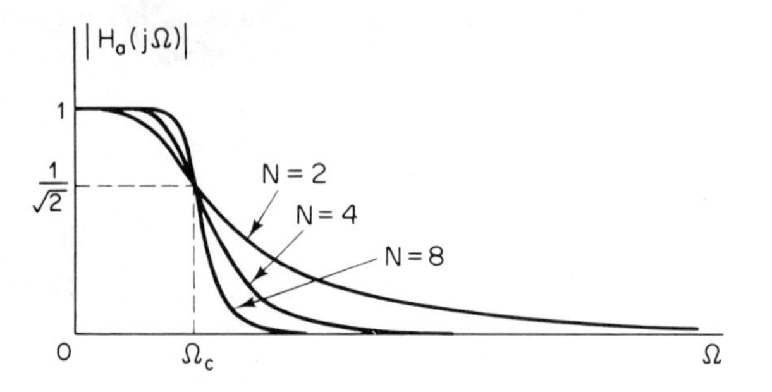

Fig. 5.11 Dependence of Butterworth magnitude characteristic on the order *N*.

magnitude function at the cutoff frequency Ω_c will always be $1/\sqrt{2}$ because of the nature of Eq. (5.25). The dependence of the Butterworth filter characteristic on the parameter N is indicated in Fig. 5.11.

From the squared magnitude function in Eq. (5.25), we observe that $H_a(s)H_a(-s)$ must be of the form

$$H_a(s)H_a(-s) = \frac{1}{1 + (s/j\Omega_c)^{2N}} \tag{5.26}$$

The roots of the denominator polynomial (the poles of the squared magnitude function) are then at

$$s_p = (-1)^{\frac{1}{2N}}(j\Omega_c)$$

Thus there are $2N$ poles equally spaced in angle on a circle of radius Ω_c in the s-plane. The poles are symmetrically located with respect to the imaginary axis. A pole never falls on the imaginary axis and one occurs on the real axis for N odd but not for N even. The angular spacing between the poles on the circle is π/N radians. For example, for $N = 3$, the poles are spaced by $\pi/3$ radians or by 60 degrees, as indicated in Fig. 5.12. To determine the transfer function of the analog filter to associate with the Butterworth squared magnitude function, we wish to perform the factorization $H_a(s)H_a(-s)$. We observe that the poles of the Butterworth squared magnitude function occur in pairs, so that if there is a pole at $s = s_p$, then a pole also occurs at $s = -s_p$. Consequently, to construct $H_a(s)$ from the squared magnitude function, we would choose one pole from each such pair. If we restrict the filter to be stable and causal, which is generally the case, then these poles will correspond to the poles on the left-half-plane part of the Butterworth circle. If we obtain a digital Butterworth filter from an analog Butterworth filter by mapping the pole pattern from the s-plane to the z-plane using the bilinear transformation, then in the z-plane the corresponding squared magnitude function has $2N$ zeros at $z = -1$. The Butterworth circle in the s-plane then maps to a circle

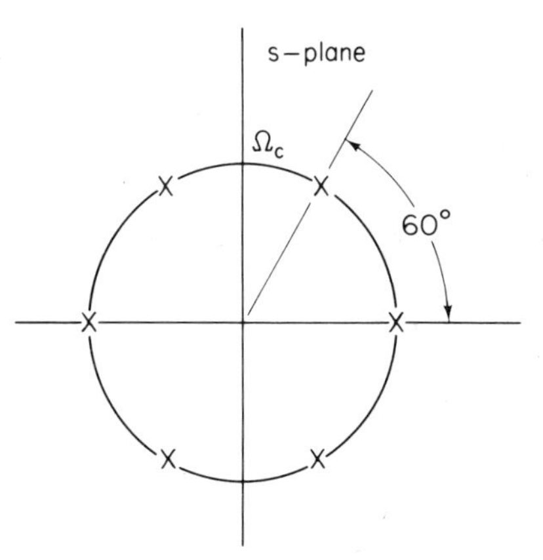

Fig 5.12 s-Plane pole locations for a third-order Butterworth filter.

in the z-plane, since the bilinear transformation is a conformal mapping. However, the Butterworth circle in the z-plane is not centered at the origin. This circle in the z-plane is indicated in Fig. 5.13.

While the poles in the s-plane were equally spaced in angle on the Butterworth circle, that is no longer true in the z-plane. In fact, pole pairs at s_p and $-s_p$ in the s-plane map to pole pairs at z_p and $1/z_p$ in the z-plane. For the previous example with $N = 3$, the poles of the squared magnitude function in the z-plane are as indicated in Fig. 5.14.

Generally, in designing a Butterworth filter using the bilinear transformation, the most straightforward procedure is to first determine the location of the poles in the s-plane and then map the left-hand plane poles to the z-plane with the appropriate transformation, rather than attempt to locate the poles directly in the z-plane.

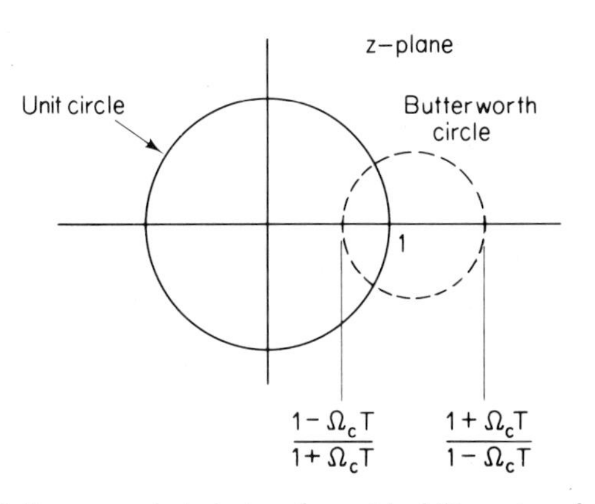

Fig. 5.13 Butterworth circle transformed by bilinear transformation.

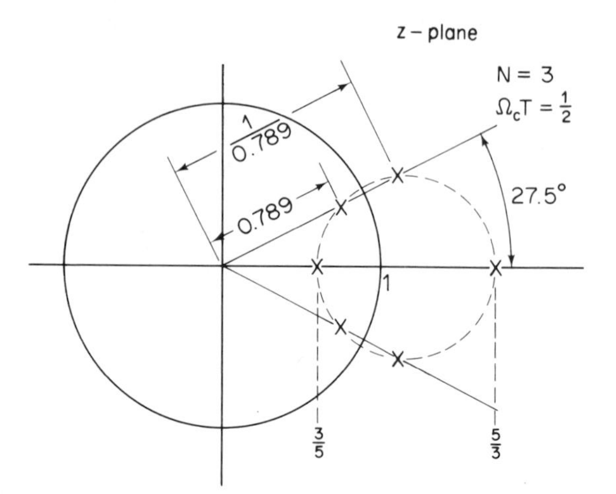

Fig. 5.14 z-Plane pole locations for a third-order Butterworth filter transformed by bilinear transformation.

As an example of the design of a Butterworth lowpass digital filter, assume that we require a filter such that the passband magnitude is constant to within 1 dB for frequencies below 0.2π and the stopband attenuation is greater than 15 dB for frequencies between 0.3π and π. Thus, if the passband magnitude is normalized to unity at $\omega = 0$, then we require that

$$20 \log_{10} |H(e^{j.2\pi})| \geq -1 \qquad \text{and} \qquad 20 \log_{10} |H(e^{j.3\pi})| \leq -15$$

Using these specifications, we will now design a digital filter from an analog Butterworth filter using both impulse invariance and the bilinear transformation. These same specifications will be used for later examples of other approximation methods.

Impulse Invariant Design. In designing the desired filter using impulse invariance on an analog Butterworth filter, we must first transform the specifications to specifications in terms of analog frequency. We recall that impulse invariance corresponds to a linear mapping from analog frequency to digital frequency in the absence of aliasing. A convenient procedure for designing the digital filter on the basis of impulse invariance is to assume that the effect of aliasing is negligible. After the design is carried out, the performance of the resulting filter can then be evaluated.

For the required design, we will assume for convenience that the parameter T is unity. Then we desire an analog Butterworth filter with squared magnitude function $|H_a(j\Omega)|^2$ for which

$$20 \log_{10} |H_a(j.2\pi)| \geq -1$$

and

$$20 \log_{10} |H_a(j.3\pi)| \leq -15$$

Since the form of a Butterworth filter is

$$|H_a(j\Omega)|^2 = \frac{1}{1 + (\Omega/\Omega_c)^{2N}}$$

the filter design consists essentially of determining the parameters N and Ω_c to meet the desired specifications. Let us initially determine these parameters to meet the specifications with equality so that

$$1 + \left(\frac{0.2\pi}{\Omega_c}\right)^{2N} = 10^{0.1} \tag{5.27}$$

and

$$1 + \left(\frac{0.3\pi}{\Omega_c}\right)^{2N} = 10^{1.5} \tag{5.28}$$

The solution of these two equations leads to the values $N = 5.8858$ and $\Omega_c = 0.70474$. The parameter N, however, must be an integer and consequently, in order that the specifications are met or exceeded, we round N up to the nearest integer so that $N = 6$. Now both passband and stopband specifications cannot be met exactly. As the value of Ω_c varies, there is a tradeoff in the amount by which the stopband and passband specifications are exceeded. If we substitute $N = 6$ into Eq. (5.27) we obtain $\Omega_c = 0.7032$. With this value, the passband specifications will be met exactly and the stopband specifications will be exceeded (for the analog filter). This allows some margin for aliasing in the digital filter. With this value of Ω_c and with $N = 6$, there are three pole pairs in the left half of the s-plane with coordinates,

$$\begin{aligned}
&\text{pole pair 1:} \quad -0.1820 \pm j.6792\\
&\text{pole pair 2:} \quad -0.4972 \pm j.4972\\
&\text{pole pair 3:} \quad -0.6792 \pm j.1820
\end{aligned}$$

so that

$$H_a(s) = \frac{0.12093}{(s^2 + 0.3640s + 0.4945)(s^2 + 0.9945s + 0.4945)(s^2 + 1.3585 + 0.4945)}$$

If we express $H_a(s)$ as a partial-fraction expansion and perform the transformation of Eq. (5.11), the resulting system function of the digital filter is

$$H(z) = \frac{0.2871 - 0.4466z^{-1}}{1 - 0.1297z^{-1} + 0.6949z^{-2}} + \frac{-2.1428 + 1.1454z^{-1}}{1 - 1.0691z^{-1} + 0.3699z^{-2}} + \frac{1.8558 - 0.6304z^{-1}}{1 - 0.9972z^{-1} + 0.2570z^{-2}}$$

As is evident from the above equation, the system function resulting from the impulse invariant design procedure can be realized directly in parallel

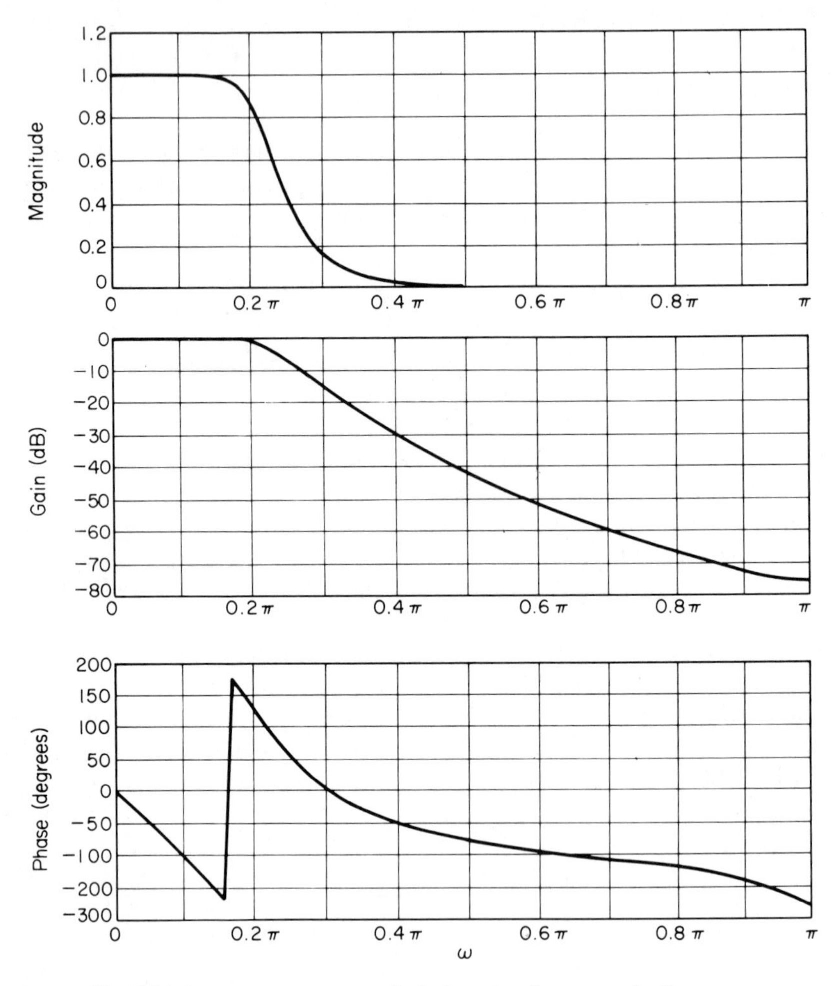

Fig. 5.15 Frequency response of sixth-order Butterworth filter transformed by impulse invariance.

form. If cascade or direct form is desired, the separate second-order terms must be combined in an appropriate way.

The frequency response of the above system is shown in Fig. 5.15. We recall that the filter was designed to exactly meet the specifications at the passband edge and to exceed the specifications at the stopband edge, and, in fact, this is the case. This is an indication that the analog filter was sufficiently bandlimited so that aliasing presented no problem. Sometimes this is not the case. If the resulting digital filter does not meet the specifications, we can try again with a higher-order filter or a different adjustment of the filter parameters, holding the order fixed.

Design Using the Bilinear Transformation. As discussed previously, in carrying out the design using the bilinear transformation, the digital frequency specifications must be prewarped to the corresponding analog frequencies so that with the frequency distortion inherent in the bilinear transformation, the critical analog frequencies will map to the correct critical digital frequencies. For the specific filter that we are considering here, with $|H_a(j\Omega)|^2$ representing the squared magnitude function of the analog filter, we require that

$$20 \log_{10} \left| H_a\left(j2 \tan\left(\frac{0.2\pi}{2}\right)\right) \right| \geq -1$$

and

$$20 \log_{10} \left| H_a\left(j2 \tan\left(\frac{0.3\pi}{2}\right)\right) \right| \leq -15$$

where we have again conveniently assumed that $T = 1$. Solving the equations with equality,

$$1 + \left(\frac{2 \tan(0.1\pi)}{\Omega_c}\right)^{2N} = 10^{0.1} \tag{5.29}$$

and

$$1 + \left(\frac{2 \tan(0.15\pi)}{\Omega_c}\right)^{2N} = 10^{1.5} \tag{5.30}$$

so

$$N = \frac{1}{2} \frac{\log [(10^{1.5} - 1)/(10^{-1} - 1)]}{\log [\tan(0.15\pi)/\tan(0.1\pi)]}$$
$$= 5.30466$$

In order to meet the specifications, N must be chosen as 6. If we determine Ω_c by substituting $N = 6$ into Eq. (5.30) we obtain $\Omega_c = 0.76622$. For this value of Ω_c, the passband specifications are exceeded and the stopband specifications are met exactly. This is reasonable for the bilinear transformation since we do not have to be concerned with aliasing. That is, with proper prewarping, we can be certain that the resulting digital filter will meet the specifications exactly at the desired stopband edge.

In the s-plane the 12 poles of the squared magnitude function are uniformly distributed in angle on a circle of radius 0.76622, as shown in Fig. 5.16. The transfer function in the s-plane corresponding to the left-half-plane poles is

$$H_a(s) = \frac{0.20238}{(s^2 + 0.396s + 0.5871)(s^2 + 1.083s + 0.5871)(s^2 + 1.4802s + 0.5871)}$$

The transfer function $H(z)$ for the digital filter is then obtained by applying

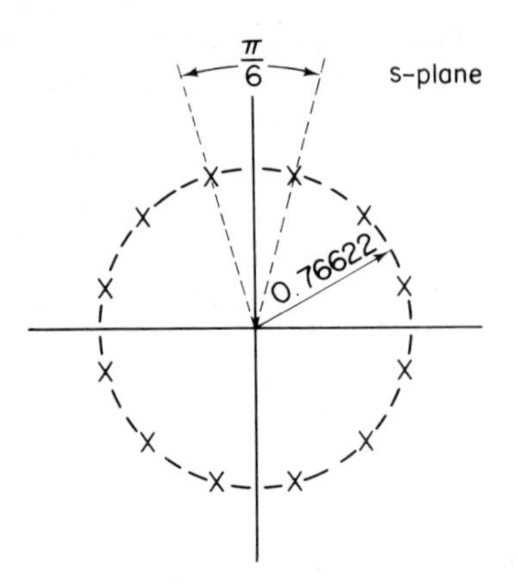

Fig. 5.16 s-Plane pole locations for sixth-order Butterworth filter.

the bilinear transformation to $H_a(s)$ with T chosen as unity, with the result that

$$H(z) = \frac{0.0007378(1 + z^{-1})^6}{(1 - 1.2686z^{-1} + 0.7051z^{-2})(1 - 1.0106z^{-1} + 0.3583z^{-2})} \times \frac{1}{(1 - 0.9044z^{-1} + 0.2155z^{-2})}$$

The magnitude and phase of the digital frequency response is shown in Fig. 5.17. We note that at $\omega = 0.2\pi$ the magnitude is down 0.5632 dB and at $\omega = 0.3\pi$ the magnitude is down exactly 15 dB.

It should also be noted that the magnitude function in Fig. 5.17 falls off much more rapidly than the one in Fig. 5.15. This is because the bilinear transformation maps the entire $j\Omega$ axis of the s-plane onto the unit circle. Since the analog Butterworth filter has a 6th order zero at $s = \infty$, the resulting digital filter has a 6th order zero at $z = -1$.

5.2.2 *Digital Chebyshev Filters*

In a Butterworth filter the frequency characteristic is monotonic in both the passband and the stopband. Consequently, if the filter specifications are in terms of, let us say, maximum passband approximation error, the specifications are exceeded toward the low-frequency end of the passband. A more efficient approach, which usually leads to a lower-order filter, is to distribute the accuracy of the approximation uniformly over the passband or the stopband or both. This is accomplished by choosing an approximation that has an equiripple behavior rather than a monotonic behavior. The class of Chebyshev filters has the property that the magnitude of the frequency response is either equiripple in the passband and monotonic in the stopband or monotonic in the passband and equiripple in the stopband.

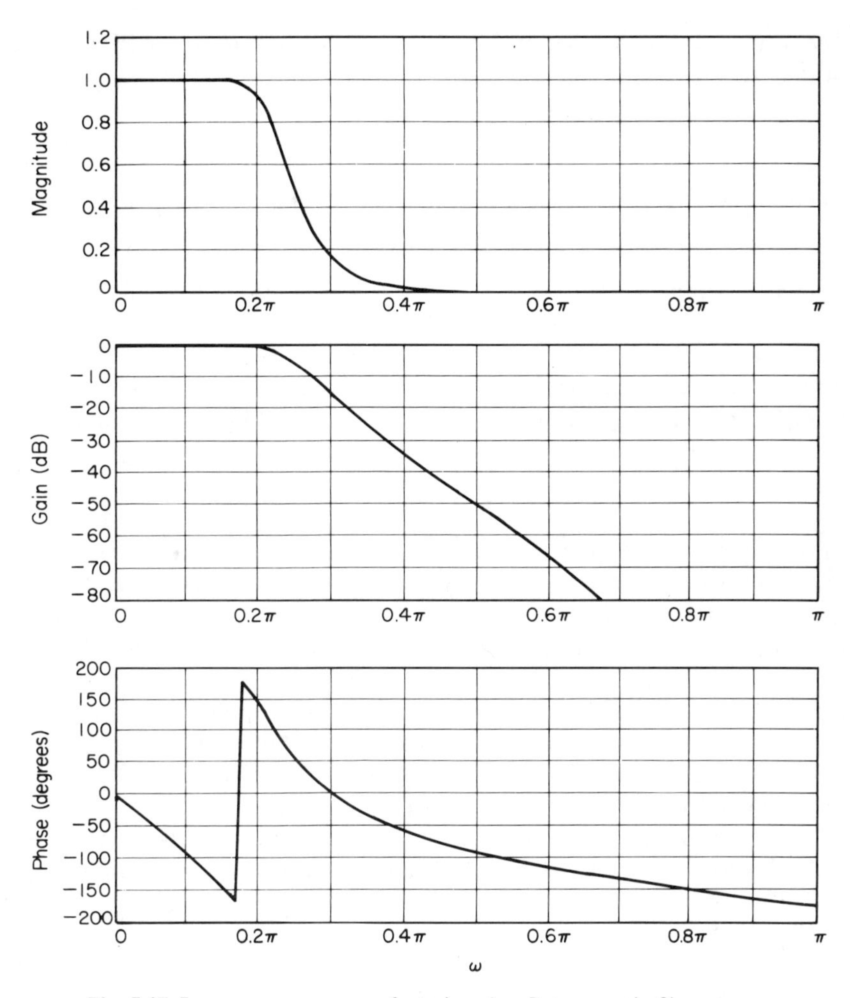

Fig. 5.17 Frequency response of sixth-order Butterworth filter transformed by bilinear transformation.

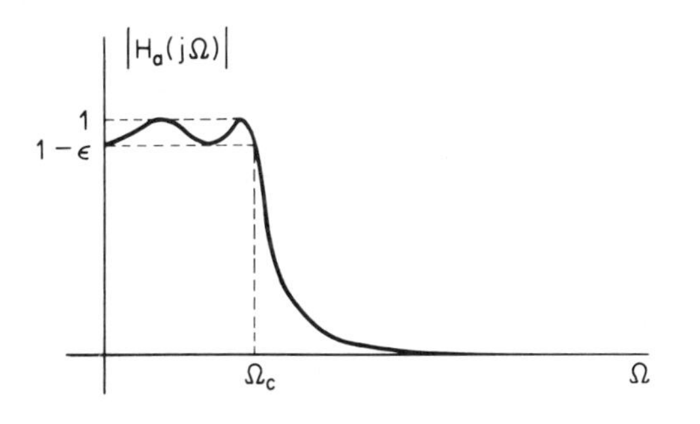

Fig. 5.18 Chebyshev lowpass filter approximation.

The former case is sketched in Fig. 5.18. The analytic form for the squared magnitude function is

$$|H_a(\Omega)|^2 = \frac{1}{1 + \varepsilon^2 V_N^2(\Omega/\Omega_c)} \tag{5.31}$$

where $V_N(x)$ is the Nth-order Chebyshev polynomial defined as

$$V_N(x) = \cos(N \cos^{-1} x) \tag{5.32}$$

For example, for $N = 0$, $V_N(x) = 1$; for $N = 1$, $V_N(x) = \cos(\cos^{-1} x) = x$; for $N = 2$, $V_N(x) = \cos(2\cos^{-1} x) = 2x^2 - 1$; etc.

From Eq. (5.32), which defines the Chebyshev polynomials, it is straightforward to obtain a recurrence formula from which $V_{N+1}(x)$ can be obtained from $V_N(x)$ and $V_{N-1}(x)$, by applying trigonometric identities to Eq. (5.32), with the result that

$$V_{N+1}(x) = 2xV_N(x) - V_{N-1}(x) \tag{5.33}$$

From Eq. (5.32) we note that $V_N^2(x)$ varies between zero and unity for x between zero and unity. For x greater than unity, $\cos^{-1} x$ is imaginary, so that $V_N(x)$ behaves as a hyperbolic cosine and consequently increases monotonically for x greater than unity. Referring to Eq. (5.31), then, $|H_a(\Omega)|^2$ ripples between 1 and $1/(1 + \varepsilon^2)$ for $0 \leq \Omega/\Omega_c \leq 1$ and decreases monotonically for $\Omega/\Omega_c > 1$. Three parameters are required to specify the filter: ε, Ω_c, and N. In a typical design, ε is specified by the allowable passband ripple and Ω_c is specified by the desired cutoff frequency. The order N is then chosen so that the stopband specifications are met.

The poles of the Chebyshev filter lie on an ellipse in the s-plane [2,7,8]. Referring to Fig. 5.19 the ellipse is defined by two circles corresponding to

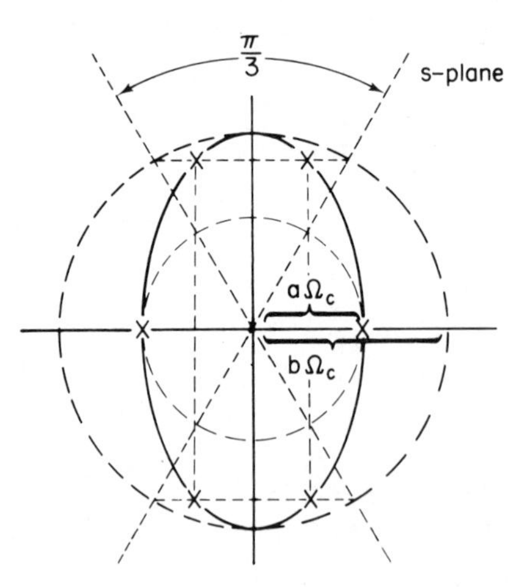

Fig. 5.19 Location of poles for a third-order Chebyshev filter.

the minor and major axes of the ellipse. The radius of the minor axis is $a\Omega_c$, where

$$a = \tfrac{1}{2}(\alpha^{1/N} - \alpha^{-1/N}) \tag{5.34}$$

with

$$\alpha = \varepsilon^{-1} + \sqrt{1 + \varepsilon^{-2}} \tag{5.35}$$

The radius of the major axis is $b\Omega_c$, where

$$b = \tfrac{1}{2}(\alpha^{1/N} + \alpha^{-1/N}) \tag{5.36}$$

To locate the poles of the Chebyshev filter on the ellipse, we first identify the points on the major and minor circles equally spaced in angle with a spacing of π/N in such a way that the points are symmetrically located with respect to the imaginary axis and such that a point never falls on the imaginary axis and a point occurs on the real axis for N odd but not for N even. This division of the major and minor circles corresponds exactly to the manner in which the circle is divided in locating the poles of a Butterworth filter. The poles of a Chebyshev filter fall on the ellipse with the ordinate specified by the points identified on the major circle and the abscissa specified by the points identified on the minor circle. In Fig. 5.19 the poles are shown for $N = 3$.

As an example of the design of a Chebyshev filter, let us consider the same specifications as for the Butterworth filter and again compare the impulse invariant design with the design using the bilinear transformation.

Impulse Invariant Design. We desire an analog Chebyshev filter for which the squared magnitude function $|H_a(j\Omega)|^2$ satisfies the specifications

$$20 \log_{10} |H_a(j.2\pi)| \geq -1$$

and

$$20 \log_{10} |H_a(j.3\pi)| \leq -15$$

We will choose the design to exactly meet the specifications at 0.2π with the frequency response equiripple between $\Omega = 0$ and $\Omega = 0.2\pi$. Consequently, $\Omega_c = 0.2\pi$ and $\delta_1 = 10^{-0.05} = \varepsilon = 0.50885$. For $N = 3$, $20 \log_{10} |H_a(j.3\pi)| = -13.4189$, and for $N = 4$, $20 \log_{10} |H_a(j.3\pi)| = -21.5834$. Thus we choose the larger value of N. For this value of N, the parameters α, a, and b are $\alpha = 4.1702$, $a = 0.3646$, and $b = 1.0644$. Thus

$$H_a(s) = \frac{0.038286}{(s^2 + 0.4233s + 0.1103)(s^2 + 0.1753s + 0.3894)}$$

The transfer function of the digital filter using impulse invariance is

$$H(z) = \frac{0.08327 + 0.0239z^{-1}}{1 - 1.5658z^{-1} + 0.6549z^{-2}} - \frac{0.08327 + 0.0246z^{-1}}{1 - 1.4934z^{-1} + 0.8392z^{-2}}$$

It is worth noting that due to aliasing the attenuation at the stopband edge $\Omega = 0.3\pi$ is slightly worse than for the analog filter. However, since the analog design provided for more attenuation than specified due to the fact that N is required to be an integer, the resulting digital filter meets the specifications. A plot of the resulting digital frequency response magnitude and phase is shown in Fig. 5.20.

Design Using the Bilinear Transformation. In this case the specifications on the analog filter are that

$$20 \log_{10} \left| H_a\left(j2 \tan \left(\frac{0.2\pi}{2}\right)\right) \right| \geq -1$$

and

$$20 \log_{10} \left| H_a\left(j2 \tan \left(\frac{0.3\pi}{2}\right)\right) \right| \leq -15$$

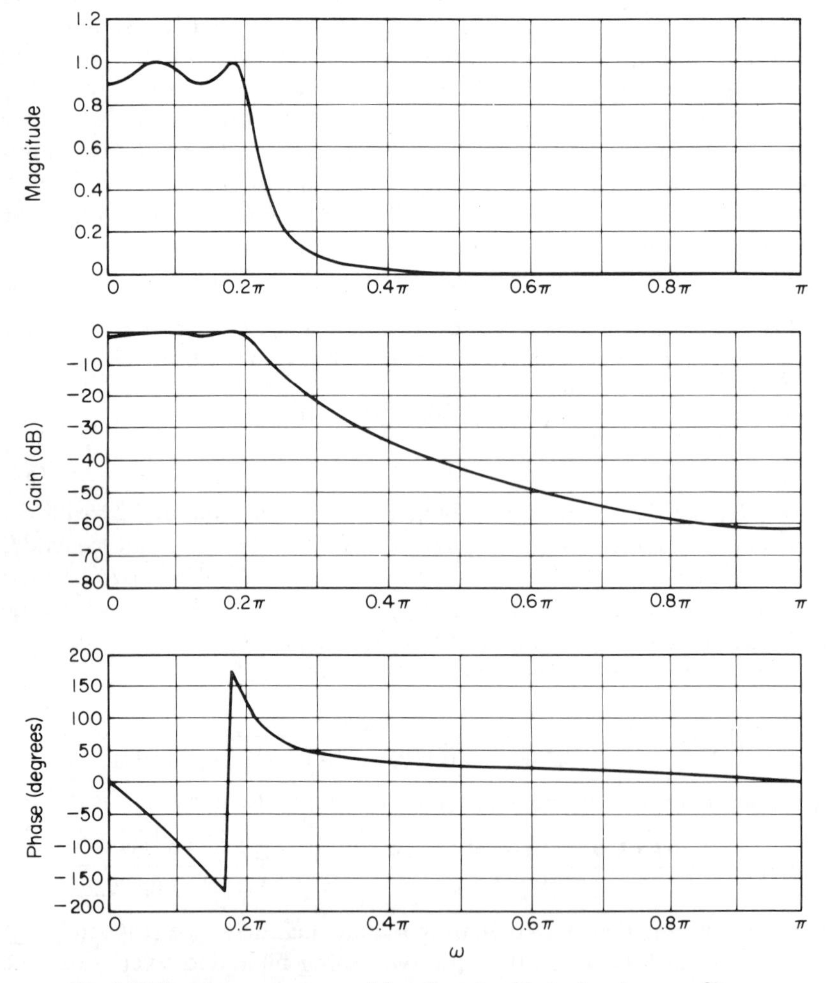

Fig. 5.20 Frequency response of fourth-order Chebyshev lowpass filter transformed by impulse invariance.

Thus the parameter Ω_c is $\Omega_c = 2 \tan (0.2\pi/2)$ and, as before, $\delta_1 = 10^{-0.05} = \varepsilon = 0.50885$. The minimum integer value of N for which the stopband specifications are met is $N = 4$. The transfer function of the resulting analog filter is

$$H_a(s) = \frac{0.04381}{(s^2 + 0.1814s + 0.4166)(s^2 + 0.4378s + 0.1180)}$$

The corresponding digital filter, using the bilinear transformation, is

$$H(z) = \frac{0.001836(1 + z^{-1})^4}{(1 - 1.4996z^{-1} + 0.8482z^{-2})(1 - 1.5548z^{-1} + 0.6493z^{-2})}$$

A plot of the resulting digital frequency response magnitude and phase is shown in Fig. 5.21.

5.2.3 *Elliptic Filters*

We have seen that if we distribute the error uniformly across the entire passband as in the Chebyshev case, we are able to meet the design specifications with a lower-order filter than if we permit a monotonically increasing error in the passband as in the Butterworth case. We note that in both the Chebyshev and Butterworth designs, the stopband error decreases monotonically with frequency, raising the possibility of further improvements if we distribute the stopband error uniformly across the stopband. That is, let us consider, for example, a lowpass filter approximation as in Fig. 5.22. Indeed it can be shown [10] that this type of approximation, i.e., equiripple in the passband and the stopband, is the best that can be achieved for a given filter order N, in the sense that for given values of Ω_p, δ_1, and δ_2, the transition band $(\Omega_s - \Omega_p)$ is as small as possible. That is, this type of approximation yields the sharpest cutoff frequency selective filter.

For analog filters, such an approximation is obtained as

$$|H_a(j\Omega)|^2 = \frac{1}{1 + \varepsilon^2 U_N^2(\Omega)}$$

where $U_N(\Omega)$ is a Jacobian elliptic function. To obtain equiripple error in both the passband and stopband, elliptic filters must have both poles and zeros. As can be seen from Fig. 5.22, the zeros of such a filter lie on the $j\Omega$-axis of the s-plane. A discussion of elliptic filter design, even on a superficial level, is beyond the scope of this section, which is only meant to be an introduction. The reader is referred to the texts by Guillemin [7], Storer [8], and Gold and Rader [1] for a more detailed discussion. Our purpose here is simply to point out that this method of approximation leads to the best amplitude response in the sense previously stated. Since the bilinear transform only distorts the frequency axis, it is clear that digital filters obtained from analog filters by bilinear transformation (with prewarping of Ω_p and Ω_s) are also optimum in the sense of having the smallest transition region for

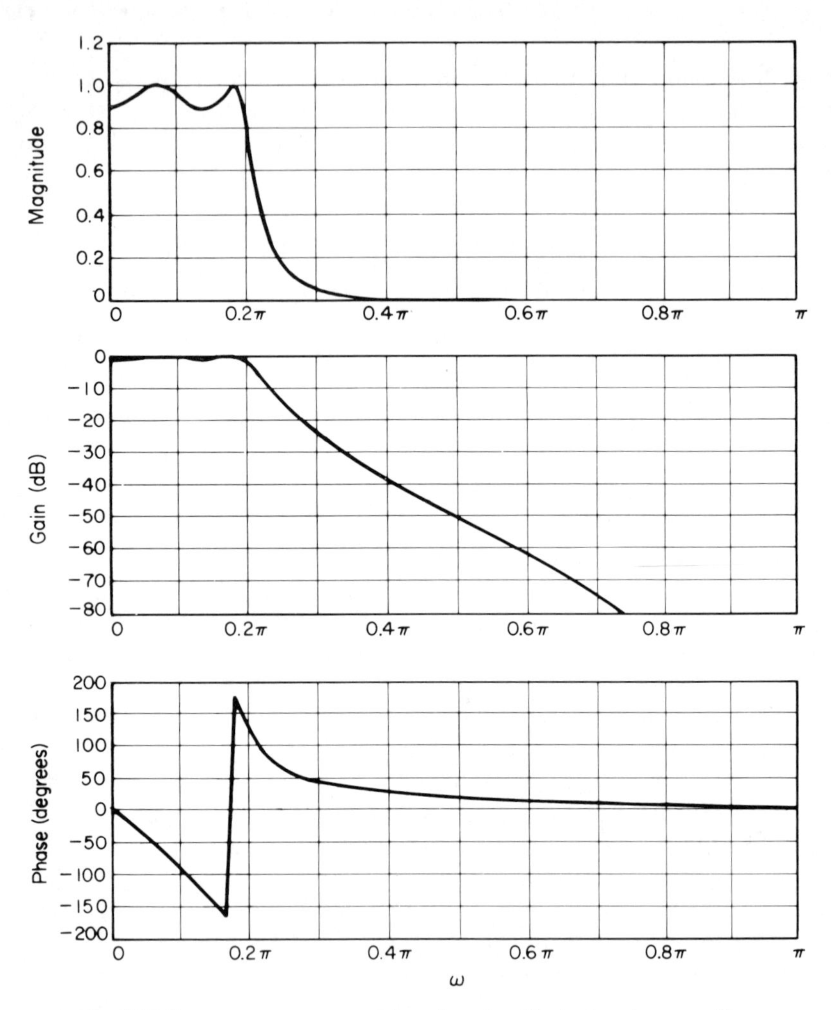

Fig. 5.21 Frequency response of fourth-order Chebyshev lowpass filter transformed by bilinear transformation.

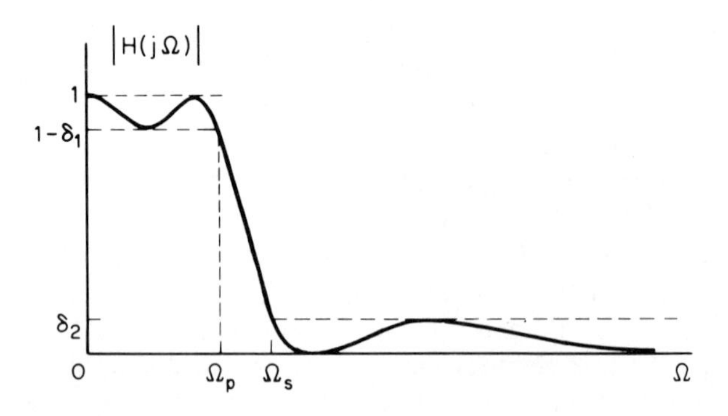

Fig. 5.22 Equiripple approximation in both passband and stopband.

given values of N, ω_p, δ_1, and δ_2. On the other hand, the impulse invariant technique would destroy the optimality of the filter.

An example of an elliptic filter that meets the specifications of the previous examples is shown in Fig. 5.23. In this case the parameters of Fig. 5.22 are $\delta_1 = 10^{-0.05}$, $\omega_p = 0.2\pi$, $\omega_s = 0.3\pi$, and $20 \log_{10} (\delta_2) = -15$ dB. If we fix δ_1, ω_p, and ω_s, it turns out that with $N = 3$, $20 \log_{10} (\delta_2) = -26.71$ dB, thus exceeding the specifications. Prewarping the critical frequencies so that $\Omega_p = 2 \tan (0.2\pi/2)$ and $\Omega_s = 2 \tan (0.3\pi/2)$, we obtain the system function

$$H_a(s) = \frac{0.12460(s^2 + 1.3040)}{(0.6498s + 0.2448)(s^2 + 0.2521s + 0.4313)}$$

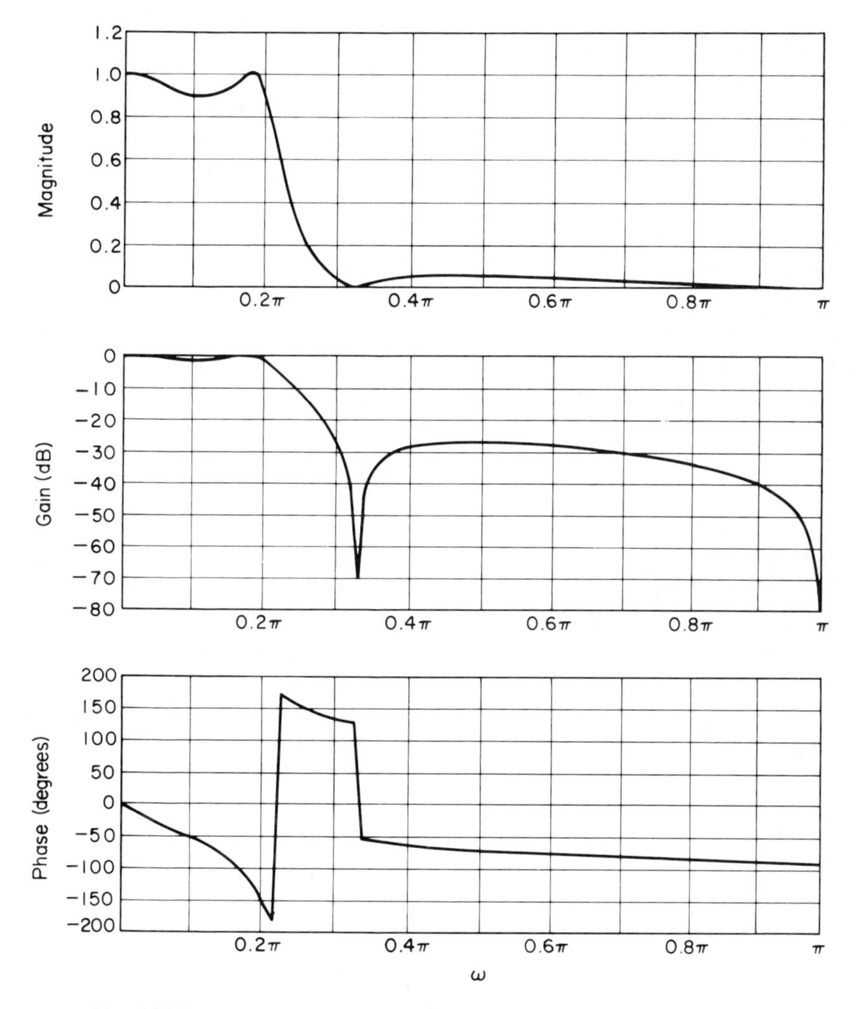

Fig. 5.23 Frequency response of a third-order elliptic filter transformed by bilinear transformation.

and using the bilinear transformation, the digital filter system function is

$$H(z) = \frac{0.05634(1 + z^{-1})(1 - 1.0166z^{-1} + z^{-2})}{(1 - 0.6830z^{-1})(1 - 1.4461z^{-1} + 0.7957z^{-2})}$$

Thus we see that for the example that we have been discussing, the elliptic filter yields the lowest-order filter that meets the specifications.

5.2.4 Frequency Transformations of Lowpass IIR Filters

The previous examples have illustrated the use of the impulse invariance and bilinear transformation methods for the design of IIR digital filters from analog system functions having lowpass frequency selective properties. The ideal frequency responses of the four commonly used types of frequency selective filters are shown in Fig. 5.24. Figures 5.24(a), (b), (c), and (d) depict the ideal frequency response of lowpass, highpass, bandpass, and bandstop filters, respectively. The traditional approach to the design of such frequency selective analog filters is to first design a frequency normalized prototype lowpass filter and then, by an algebraic transformation, derive the desired lowpass, highpass, bandpass, or bandstop filter from the prototype lowpass filter [7]. In the case of digital frequency selective filters, we can design an analog frequency selective filter of the desired type and then transform this filter to a digital filter. A disadvantage of this procedure is that we cannot transform highpass and bandstop filters using the impulse invariance technique, owing to the aliasing distortion that would result. An alternative procedure is to design a digital prototype lowpass filter and then perform an algebraic transformation on the digital lowpass filter to obtain the desired frequency selective digital filter [11–13]. This procedure can be applied regardless of the design procedure used to obtain the digital lowpass filter.

Frequency selective filters of the lowpass, highpass, bandpass, and bandstop types can be obtained from a lowpass digital filter by use of rational transformations very similar to the bilinear transformation that we have used to transform analog system functions into digital system functions. To see how this is done, let us associate the complex variable z with the lowpass system function, $H_l(z)$, and the complex variable Z with the desired system function $H_d(Z)$. Then let us define a mapping from the z-plane to the Z-plane of the form

$$z^{-1} = G(Z^{-1})$$

such that

$$H_d(Z) = H_l(G^{-1}(Z^{-1}))$$

where $G^{-1}(\ )$ denotes the inverse mapping; i.e.,

$$Z^{-1} = G^{-1}(z^{-1})$$

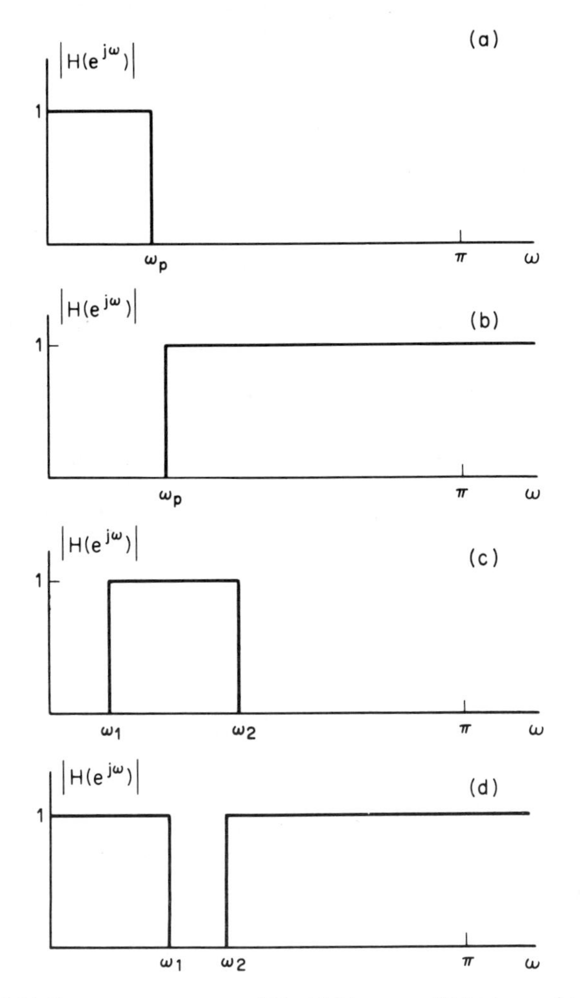

Fig. 5.24 Frequency responses of ideal (a) lowpass, (b) highpass, (c) bandpass, and (d) bandstop filters.

We must insist that the mapping be such that a rational system function $H_l(z)$ corresponding to a stable and causal lowpass digital filter be transformed into a rational system function, $H_d(Z)$ corresponding again to a stable and causal digital filter. Therefore, we require that

1. $G(Z^{-1})$ must be a rational function of Z^{-1} (or Z).
2. The inside of the unit circle of the z-plane must map into the inside of the unit circle of the Z-plane.

Thus, if θ and ω are the frequency variables in the z-plane and Z-plane, respectively, i.e., $z = e^{j\theta}$ and $Z = e^{j\omega}$, then

$$e^{-j\theta} = |G(e^{-j\omega})|\, e^{j \arg [G(e^{-j\omega})]}$$

and we require that

$$|G(e^{-j\omega})| = 1$$

and

$$\theta = -\arg [G(e^{-j\omega})]$$

The above equation specifies the relationship between frequencies in the z-plane and the Z-plane. It has been shown [11–13] that the most general form of the function $G(Z^{-1})$ that satisfies all the above requirements is

$$G(Z^{-1}) = \pm \prod_{k=1}^{N} \frac{Z^{-1} - \alpha_k}{1 - \alpha_k Z^{-1}}$$

where $|\alpha_k| < 1$ for stability. By choosing appropriate values for N and the constants α_k, a variety of mappings can be obtained. The simplest is one that transforms a lowpass filter into another lowpass filter. For this case,

$$z^{-1} = G(Z^{-1}) = \frac{Z^{-1} - \alpha}{1 - \alpha Z^{-1}}$$

If we substitute $z = e^{j\theta}$ and $Z = e^{j\omega}$, we obtain

$$e^{-j\theta} = \frac{e^{-j\omega} - \alpha}{1 - \alpha e^{-j\omega}}$$

from which we can show that

$$\omega = \arctan \left[\frac{(1 - \alpha^2) \sin \theta}{2\alpha + (1 + \alpha^2) \cos \theta} \right]$$

The nature of this relationship is depicted in Fig. 5.25 for different values of

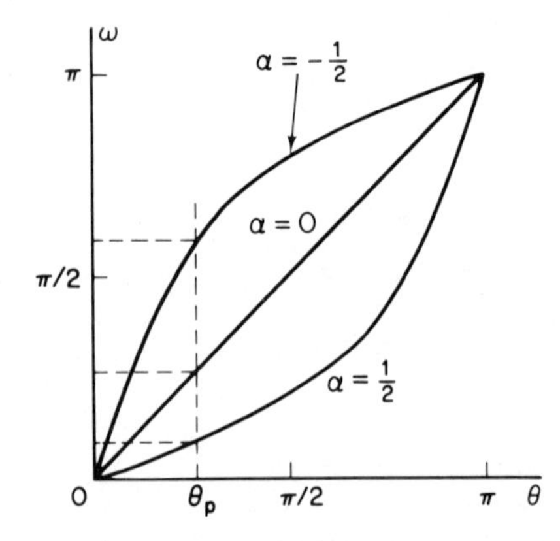

Fig. 5.25 Warping of the frequency scale in lowpass–lowpass transformation.

α. Although a warping of the frequency scale is evident in Fig. 5.25 (except in the case $\alpha = 0$), if the original system has a piecewise constant, lowpass frequency response with cutoff frequency θ_p, then the transformed system will likewise have a similar lowpass response with cutoff frequency ω_p determined by the choice of α. Solving for α in terms of θ_p and ω_p, we obtain

$$\alpha = \frac{\sin((\theta_p - \omega_p)/2)}{\sin((\theta_p + \omega_p)/2)}$$

Thus to use these results to obtain a lowpass filter $H_d(Z)$ with cutoff frequency ω_p from an already available lowpass filter $H_l(z)$, with cutoff frequency θ_p, we would use the above equation to determine α in the expression

$$H_d(Z) = H_l(z)\big|_{z^{-1}=(Z^{-1}-\alpha)/(1-\alpha Z^{-1})}$$

We can derive transformations from a lowpass filter to highpass, bandpass, and bandstop filters in a similar manner. These are summarized in Table 5.1 [11–13].

As an example of the use of these transformations, let us obtain a highpass filter from the Chebyshev lowpass filter of Sec. 5.2.2. We recall that the cutoff frequency of the lowpass filter was $\theta_p = 0.2\pi$. Using the bilinear transformation we obtained

$$H_l(z) = \frac{0.001836(1 + z^{-1})^4}{(1 - 1.5548z^{-1} + 0.6493z^{-2})(1 - 1.4996z^{-1} + 0.8482z^{-2})}$$

The frequency response of this filter is shown in Fig. 5.21. Suppose that we desire a highpass filter with cutoff frequency $\omega_p = 0.6\pi$. From Table 5.1 we obtain

$$\alpha = -\frac{\cos[(0.6\pi + 0.2\pi)/2]}{\cos[(0.6\pi - 0.2\pi)/2]} = -0.38197$$

Thus, using the lowpass–highpass transformation indicated in Table 5.1, we obtain

$$H_d(Z) = H_l(z)\big|_{z^{-1}=-((Z^{-1}-0.38197)/(1-0.38197Z^{-1}))}$$

$$= \frac{0.02426(1 - Z^{-1})^4}{(1 - 1.0416Z^{-1} + 0.4019Z^{-2})(1 - 0.5561Z^{-1} + 0.7647Z^{-2})}$$

The frequency response of this system is shown in Fig. 5.26. Note that except for some distortion of the frequency scale, the highpass frequency response appears very much as if the lowpass response were shifted in frequency by π. Also note that the 4th order zero at $z = -1$ for the lowpass filter now appears at $Z = 1$ for the highpass filter.

Table 5.1

TRANSFORMATIONS FROM A LOWPASS-DIGITAL-FILTER PROTOTYPE OF CUTOFF FREQUENCY θ_p

Filter Type	*Transformation*	*Associated Design Formulas*
Lowpass	$z^{-1} = \dfrac{Z^{-1} - \alpha}{1 - \alpha Z^{-1}}$	$\alpha = \dfrac{\sin\left(\dfrac{\theta_p - \omega_p}{2}\right)}{\sin\left(\dfrac{\theta_p + \omega_p}{2}\right)}$ $\omega_p =$ desired cutoff frequency
Highpass	$-\dfrac{Z^{-1} + \alpha}{1 + \alpha Z^{-1}}$	$\alpha = -\dfrac{\cos\left(\dfrac{\omega_p + \theta_p}{2}\right)}{\cos\left(\dfrac{\omega_p - \theta_p}{2}\right)}$ $\omega_p =$ desired cutoff frequency
Bandpass	$-\dfrac{Z^{-2} - \dfrac{2\alpha k}{k+1}Z^{-1} + \dfrac{k-1}{k+1}}{\dfrac{k-1}{k+1}Z^{-2} - \dfrac{2\alpha k}{k+1}Z^{-1} + 1}$	$\alpha = \dfrac{\cos\left(\dfrac{\omega_2 + \omega_1}{2}\right)}{\cos\left(\dfrac{\omega_2 - \omega_1}{2}\right)}$ $k = \cot\left(\dfrac{\omega_2 - \omega_1}{2}\right)\tan\dfrac{\theta_p}{2}$ $\omega_2, \omega_1 =$ desired upper and lower cutoff frequencies
Bandstop	$\dfrac{Z^{-2} - \dfrac{2\alpha}{1+k}Z^{-1} + \dfrac{1-k}{1+k}}{\dfrac{1-k}{1+k}Z^{-2} - \dfrac{2\alpha}{1+k}Z^{-1} + 1}$	$\alpha = \dfrac{\cos\left(\dfrac{\omega_2 + \omega_1}{2}\right)}{\cos\left(\dfrac{\omega_2 - \omega_1}{2}\right)}$ $k = \tan\left(\dfrac{\omega_2 - \omega_1}{2}\right)\tan\dfrac{\theta_p}{2}$ $\omega_2, \omega_1 =$ desired upper and lower cutoff frequencies

5.3 Computer-Aided Design of IIR Digital Filters

In the previous section we have seen that digital filters can be designed by transforming an appropriate analog filter design into a digital filter. This approach is reasonable when we can take advantage of analog designs that are given in terms of formulas or extensive design tables; e.g., frequency selective filters such as Butterworth, Chebyshev, or elliptic filters. In general, however, analytical procedures do not exist for the design of either analog

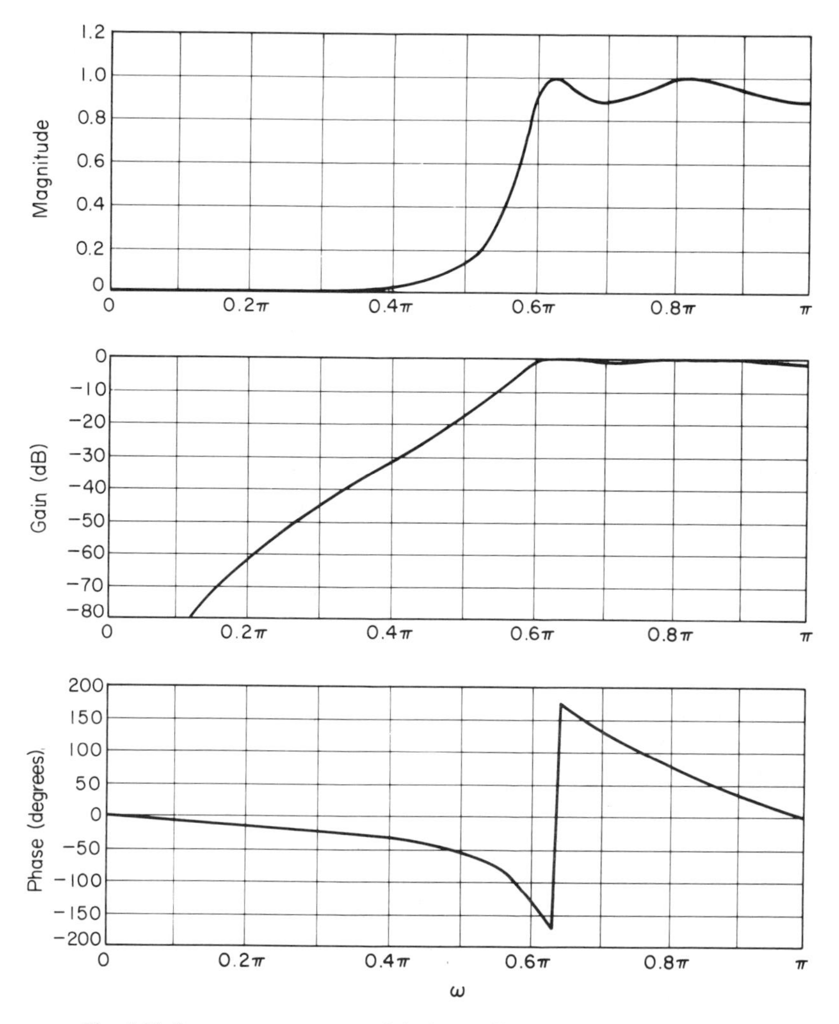

Fig. 5.26 Frequency response of highpass filter obtained by frequency transformation.

or digital filters to match arbitrary frequency response specifications or other types of specifications. In these more general cases, design procedures have been developed that are algorithmic in nature, generally relying on the use of a computer to solve sets of linear or nonlinear equations. In most cases the computer-aided design techniques apply equally well to the design of either analog or digital filters with only minor modification. Therefore, nothing is gained by first obtaining an analog design and then transforming this design to a digital filter.

There are a number of computer-aided design techniques for approximating an arbitrary frequency characteristic. In this section we shall discuss the nature of several of these procedures so as to illustrate the possibilities of computer-aided design of IIR digital filters. In our discussion we shall focus on the formulation of the design equations rather than the details of the numerical procedures involved in carrying out the solution.

5.3.1 *Minimization of Mean-Square Error*

Steiglitz [14,15] has proposed an IIR filter design procedure based on minimization of the mean-square error in the frequency domain. The procedure requires that the desired frequency response $H_d(e^{j\omega})$ be prescribed at a discrete set of frequencies $\{\omega_i\}$, $i = 1, 2, \ldots, M$. The mean-squared error at these frequencies is defined as

$$E = \sum_{i=1}^{M} [|H(e^{j\omega_i})| - |H_d(e^{j\omega_i})|]^2 \tag{5.37}$$

It is assumed that the filter transfer function is of the form

$$H(z) = A \prod_{k=1}^{K} \frac{1 + a_k z^{-1} + b_k z^{-2}}{1 + c_k z^{-1} + d_k z^{-2}} = AG(z) \tag{5.38}$$

The cascade form is chosen because of its relatively low coefficient sensitivity and for convenience in computation of derivatives required in the optimization procedure.

The error, expressed by Eq. (5.37), can be thought of as a function of the parameters $(a_1, b_1, c_1, d_1, a_2, b_2, \ldots, d_K, A)$. Since we wish to find values of these parameters that minimize the error E, we take partial derivatives of E with respect to each parameter and set these derivatives equal to zero thus obtaining $4K + 1$ equations in the $4K + 1$ unknowns.

The equation for A is particularly simple since

$$\frac{\partial E}{\partial |A|} = \sum_{i=1}^{M} \{2[|A| \cdot |G(e^{j\omega_i})| - |H_d(e^{j\omega_i})|]\, |G(e^{j\omega_i})|\} = 0$$

Solving this equation for $|A|$ gives†

$$|A| = \frac{\sum_{i=1}^{M} |G(e^{j\omega_i})| \cdot |H_d(e^{j\omega_i})|}{\sum_{i=1}^{M} |G(e^{j\omega_i})|^2} \tag{5.39}$$

Differentiating with respect to the remaining $4K$ unknown parameters given by the vector

$$\boldsymbol{\phi} = [a_1, b_1, c_1, d_1, a_2, b_2, \ldots, d_K]$$

† The sign of A does not enter into the minimization since only the magnitude error is considered.

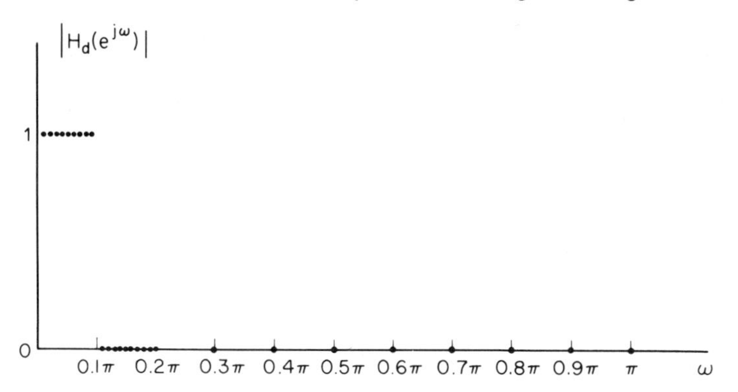

Fig. 5.27 Fixed values of the frequency response for an example of the use of Steiglitz's design procedure.

leads to $4K$ nonlinear equations,

$$\frac{\partial E(\boldsymbol{\phi}, A)}{\partial \phi_n} = 0, \qquad n = 1, 2, \ldots, 4K$$

where ϕ_n represents the nth component of $\boldsymbol{\phi}$. These equations can be solved algorithmically using, for example, the Fletcher–Powell method [16]. We note that this procedure deals only with the magnitude function. As a result the optimization algorithm can yield values for the parameters corresponding to an unstable filter; that is, the poles and zeros of each second-order section are unconstrained. Rather than place constraints on the parameters, Steiglitz tested the roots of each second-order factor after completion of the minimization and, if a pole (or zero) was outside the unit circle, he replaced it by its reciprocal, thus leaving the magnitude function unchanged. He found that continuing the optimization procedure sometimes resulted in further minimization of the error.

The following example [14], illustrates the use of the above procedure. An ideal lowpass filter was specified with cutoff frequency 0.1π as depicted in Fig. 5.27. That is,

$$|H_d(e^{j\omega})| = \begin{cases} 1, & \omega = 0, 0.01\pi, 0.02\pi, \ldots, 0.09\pi \\ 0.5, & \omega = 0.1\pi \\ 0, & \omega = 0.11\pi, 0.12\pi, \ldots, 0.19\pi \\ 0, & \omega = 0.2\pi, 0.3\pi, \ldots, 0.9\pi, \pi \end{cases}$$

Note that there is no requirement that the specification frequencies $\{\omega_i\}$ be equally spaced. Figure 5.28 shows the result of the optimization procedure for $K = 1$ and $K = 2$.

5.3.2 *Minimization of a p-Error Criterion*

Deczky [17] has generalized the procedure of the previous section in a number of ways. Instead of minimizing the average squared error, a weighted

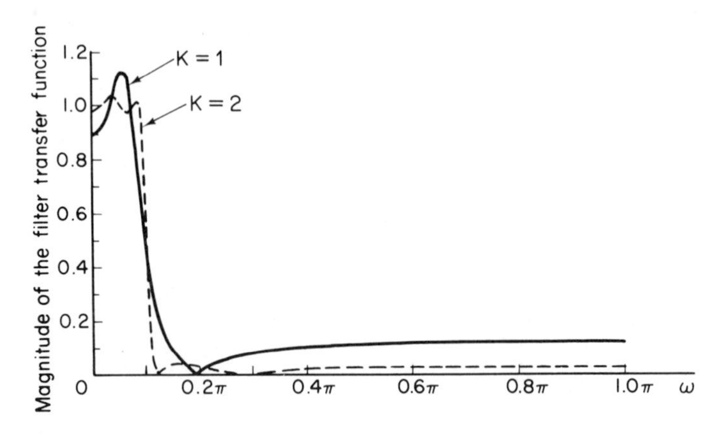

Fig. 5.28 Example of frequency responses obtained by minimization of mean-square error. (After Steiglitz [14].)

average of the error raised to the pth power was minimized. Also, this technique was applied to both the magnitude and the group delay. Therefore, the error expression minimized was a discrete approximation to either

$$E_p = \int_0^{\pi} W(\omega)[|H(e^{j\omega})| - |H_d(e^{j\omega})|]^p \, d\omega \tag{5.40}$$

or

$$E_p = \int_0^{\pi} W(\omega)[\tau(\omega) - \tau_d(\omega)]^p \, d\omega \tag{5.41}$$

where the group delay τ is defined as

$$\tau(\omega) = -\frac{d}{d\omega}\{\arg [H(e^{j\omega})]\} \tag{5.42}$$

The desired digital filter is assumed to be represented as in Eq. (5.38). Proceeding as in Sec. 5.3.1 to minimize either Eq. (5.40) or (5.41), we are faced with the solution of $4K + 1$ nonlinear equations in the $4K + 1$ unknown parameters. It is shown [17] that if one begins with a stable approximating solution, then a local minimum of the error function exists for $p \geq 2$ such that the optimum parameters correspond to a stable transfer function. This result depends on the observation that both the magnitude and group delay of each second-order section in Eq. (5.39) approach infinity as the poles approach the unit circle. Thus the error function becomes unbounded as the poles tend to move across the unit circle into the instability region. This condition serves as a barrier to the movement of the poles in solving the nonlinear equations using the Fletcher–Powell procedure [16].

As an example of the application of this technique, a fourth-order elliptic lowpass digital filter was designed. The attenuation characteristic ($-20 \log_{10} |H(e^{j\omega})|$) of this filter is shown in Fig. 5.29(a) and the group delay is shown in

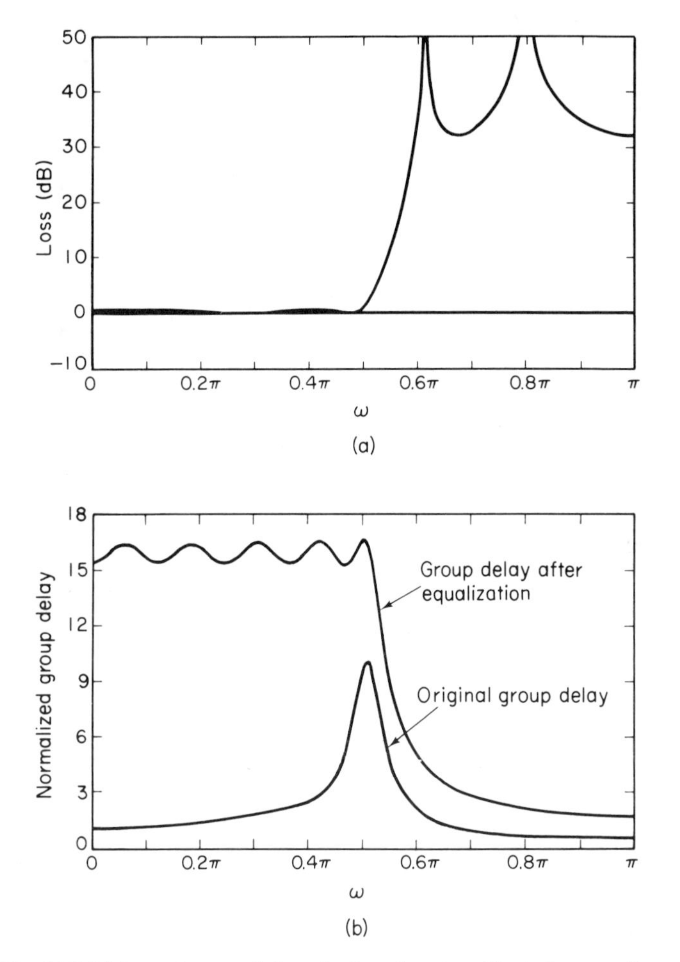

Fig. 5.29 (a) Loss characteristics of a fourth-order elliptic lowpass filter; (b) group delay of the fourth-order elliptic lowpass filter equalized by a four-section allpass filter ($p = 10$). (After Deczky [17].)

Fig. 5.29(b). Since the elliptic filter has optimum magnitude response but rather nonlinear phase, an allpass equalizer was sought such that when cascaded with the elliptic filter, it would improve the phase curve, i.e., flatten the group-delay curve over the passband. Incorporating constraints on the coefficients in Eq. (5.39) for an allpass filter, and using a value $p = 10$ in Eq. (5.41), the group-delay curve of Fig. 5.29(b) resulted. It can be seen that the group delay exhibits an equiripple behavior over the passband. This is due to the fact that a relatively large value of p was used. In fact, it can be shown in general that as $p \to \infty$, the optimum solution approaches a true equiripple approximation.

5.3.3 Least-Squares Inverse Design

In the two procedures discussed above, the filter was specified in the frequency domain and the resulting set of equations was nonlinear in the filter parameters. An alternative procedure, based on a least-squares approximation to the *inverse* of the desired filter, leads to a set of linear equations. In this procedure the filter is specified in terms of the first L samples of the desired impulse response:

$$\{h_d(n)\}, \qquad n = 0, 1, \ldots, L-1$$

In our discussion we shall assume that the filter transfer function is of the form

$$H(z) = \frac{b_0}{1 - \sum_{k=1}^{N} a_k z^{-k}} \tag{5.43}$$

Generalizations of this procedure allowing both zeros and poles are discussed by Shanks [19] and Burrus and Parks [18]. In this simplified case the filter design is based on the criterion that the output of the inverse of $H(z)$ must approximate a unit sample when the input is $h_d(n)$. If $v(n)$ denotes the output of the inverse system with transfer function $1/H(z)$, then

$$V(z) = \frac{H_d(z)}{H(z)}$$

Thus we can write the recursion formula

$$b_0 v(n) = h_d(n) - \sum_{r=1}^{N} a_r h_d(n-r) \tag{5.44}$$

Recall that we require that $v(n)$ be a unit sample. Thus it is reasonable to require that

$$b_0 = h_d(0)$$

and that $v(n)$ be as small as possible for $n > 0$. Therefore, choose the remaining coefficients so as to minimize

$$E = \sum_{n=1}^{\infty} (v(n))^2$$

From Eq. (5.44),

$$E = \frac{1}{b_0^2} \sum_{n=1}^{\infty} (h_d(n))^2 - 2 \sum_{n=1}^{\infty} h_d(n) \sum_{r=1}^{N} a_r h_d(n-r) + \sum_{n=1}^{\infty} \left[\sum_{r=1}^{N} a_r h_d(n-r) \right]^2$$

The coefficients a_i that minimize E satisfy the equations

$$\frac{\partial E}{\partial a_i} = 0, \qquad i = 1, 2, \ldots, N$$

Differentiating with respect to a_i and setting the derivative equal to zero results in

$$\sum_{r=1}^{N} a_r \sum_{n=1}^{\infty} h_d(n-r)h_d(n-i) = \sum_{n=1}^{\infty} h_d(n)h_d(n-r)$$

If we define $\phi(i, r)$ as

$$\phi(i, r) = \sum_{n=1}^{\infty} h_d(n-r)h_d(n-i)$$

Then the coefficients a_i satisfy the set of linear equations

$$\sum_{r=1}^{N} a_r\phi(i, r) = \phi(i, 0), \qquad i = 1, 2, \ldots, N \tag{5.45}$$

These equations can be solved by any conventional technique. It can be shown [18, 19] that the matrix of quantities $\phi(i, r)$ is positive definite. Therefore, a particularly efficient procedure is given by Levinson [20].

5.4 Properties of FIR Digital Filters

The previous sections have been concerned with design techniques for infinite-duration impulse response filters. Although such filters have many attractive features, they also have a number of disadvantages. For example, if one wishes to take advantage of the computational speed of the FFT and implement a filter as discussed in Chapter 3, a finite-duration impulse response is essential. Also, the examples of the previous section make it clear that IIR filters generally achieve excellent amplitude response at the expense of nonlinear phase. In contrast, FIR filters can have exactly linear phase. Thus design techniques for FIR filters are of considerable interest.

The system function of a causal FIR filter is of the form

$$H(z) = \sum_{n=0}^{N-1} h(n)z^{-n}$$

i.e., $H(z)$ is a polynomial in z^{-1} of degree $N-1$. Thus $H(z)$ has $N-1$ zeros that can be located anywhere in the finite z-plane and $N-1$ poles, all of which lie at $z = 0$. The frequency response $H(e^{j\omega})$ is the trigonometric polynomial

$$H(e^{j\omega}) = \sum_{n=0}^{N-1} h(n)e^{-j\omega n} \tag{5.46}$$

We recall that any finite-duration sequence is completely specified by N samples of its Fourier transform, so that the design of an FIR filter may be accomplished by finding either its impulse response coefficients or N samples of its frequency response. In the following sections we discuss examples of both methods.

If the impulse response satisfies the condition

$$h(n) = h(N - 1 - n) \tag{5.47}$$

then the filter has linear phase. As we discussed in Chapter 4, this is easily shown by substituting Eq. (5.47) into (5.46), thereby obtaining

$$H(e^{j\omega}) = \begin{cases} e^{-j\omega((N-1)/2)}\left[h\left(\frac{N-1}{2}\right) + \sum_{n=0}^{(N-3)/2} 2h(n)\cos\left(\omega\left(n - \frac{N-1}{2}\right)\right)\right], & N \text{ odd} \quad (5.48a) \\ e^{-j\omega((N-1)/2)}\left[\sum_{n=0}^{N/2-1} 2h(n)\cos\left(\omega\left(n - \frac{N-1}{2}\right)\right)\right], & N \text{ even} \quad (5.48b) \end{cases}$$

It is seen from these equations that the condition of Eq. (5.47) implies a linear phase shift corresponding to a delay of $(N - 1)/2$ samples. We note that for the case of N odd, the phase shift corresponds to an integer number of samples delay, while for N even, the delay is an integer plus one-half sample.† This distinction between odd and even values of N is often of considerable importance in design and realization of FIR filters. Examples of impulse responses having linear phase are shown in Fig. 5.30.

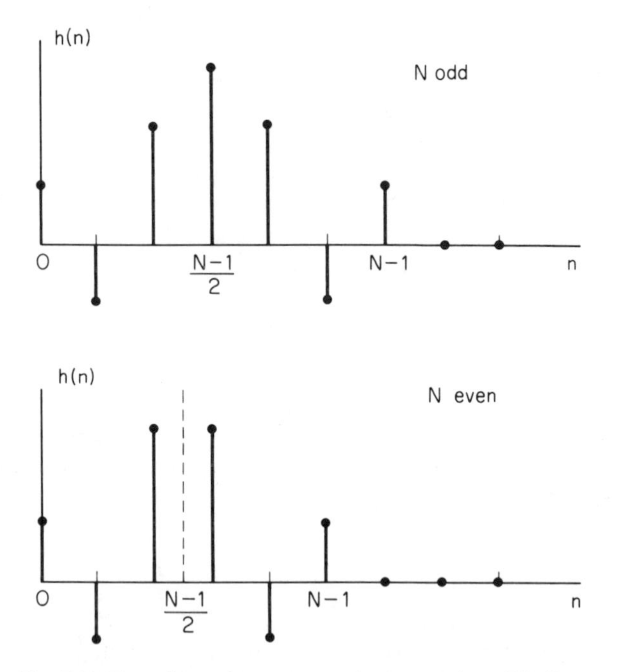

Fig. 5.30 Typical impulse responses for linear-phase FIR filters.

† It can be shown [6] that Eq. (5.47) is both necessary and sufficient for a causal digital filter to have linear phase. Thus only FIR filters can have linear phase.

Since linear phase is generally desirable and often a necessity, and since linear phase often simplifies a design procedure, most of our discussion will center on linear-phase filters.

5.5 Design of FIR Filters Using Windows

The most straightforward approach to FIR filter design is to obtain a finite-length impulse response by truncating an infinite-duration impulse response sequence. If we suppose that $H_d(e^{j\omega})$ is an ideal desired frequency response, then

$$H_d(e^{j\omega}) = \sum_{n=-\infty}^{\infty} h_d(n)e^{-j\omega n} \tag{5.49a}$$

where $h_d(n)$ is the corresponding impulse response sequence, i.e.,

$$h_d(n) = \frac{1}{2\pi}\int_{-\pi}^{\pi} H_d(e^{j\omega})e^{j\omega n}\, d\omega \tag{5.49b}$$

In general, $H_d(e^{j\omega})$, for a frequency selective filter may be piecewise constant with discontinuities at the boundaries between bands. In such cases the sequence $h_d(n)$ is of infinite duration and it must be truncated to obtain a finite-duration impulse response. As we have pointed out before, Eqs. (5.49) can be thought of as a Fourier series representation of the periodic frequency response $H_d(e^{j\omega})$, with the sequence $h_d(n)$ playing the role of the "Fourier coefficients." Thus the approximation of an ideal filter specification by truncation of the ideal impulse response is identical to the study of the convergence of Fourier series, a subject that has received a great deal of study since the middle of the eighteenth century. The most familiar concept from this theory is the *Gibbs phenomenon.* In the following discussion we shall see how this nonuniform convergence phenomenon manifests itself in the design of FIR filters.

If $h_d(n)$ has infinite duration, one way to obtain a finite-duration causal impulse response is to simply truncate $h(n)$, i.e., define

$$h(n) = \begin{cases} h_d(n), & 0 \le n \le N-1 \\ 0, & \text{otherwise} \end{cases} \tag{5.50}$$

In general, we can represent $h(n)$ as the product of the desired impulse response and a finite-duration "window" $w(n)$; i.e.,

$$h(n) = h_d(n)w(n) \tag{5.51}$$

where in the example of Eq. (5.50),

$$w(n) = \begin{cases} 1, & 0 \le n \le N-1 \\ 0, & \text{otherwise} \end{cases} \tag{5.52}$$

Using the complex convolution theorem derived in Chapter 2 we see that

$$H(e^{j\omega}) = \frac{1}{2\pi}\int_{-\pi}^{\pi} H_d(e^{j\theta})W(e^{j(\omega-\theta)})\, d\theta \tag{5.53}$$

That is, $H(e^{j\omega})$ is the periodic continuous convolution of the desired frequency response with the Fourier transform of the window. Thus the frequency response $H(e^{j\omega})$ will be a "smeared" version of the desired response $H_d(e^{j\omega})$. Figure 5.31(a) depicts typical functions $H_d(e^{j\theta})$ and $W(e^{j\omega-\theta)})$ as

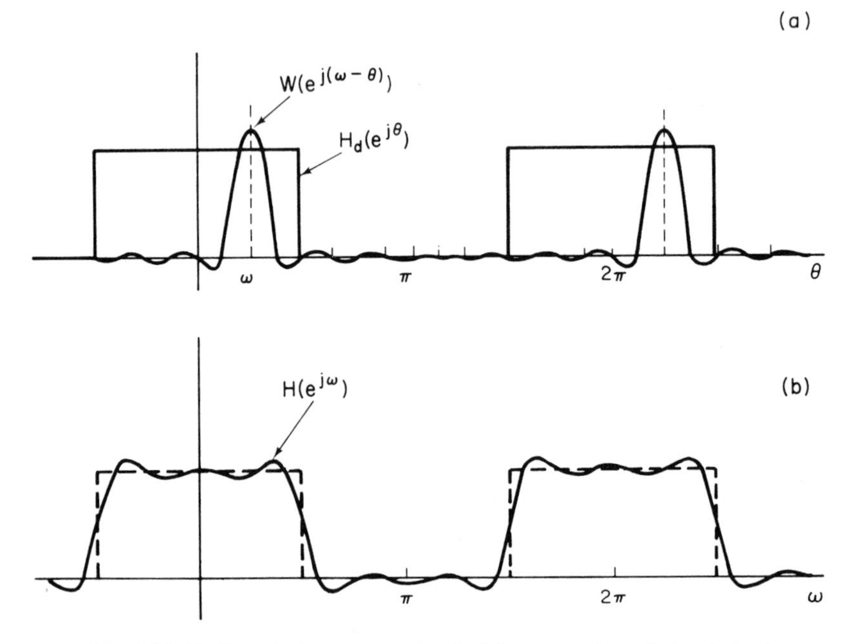

Fig. 5.31 (a) Convolution process implied by truncation of the desired impulse response; (b) typical approximation resulting from windowing the desired impulse response.

required in Eq. (5.53). (Both are shown as real functions only for convenience in depicting the convolution process.)

From Eq. (5.53) we see that if $W(e^{j\omega})$ is narrow compared to variations in $H_d(e^{j\omega})$, then $H(e^{j\omega})$ will "look like" $H_d(e^{j\omega})$. Thus the choice of window is governed by the desire to have $w(n)$ as short as possible in duration so as to minimize computation in the implementation of the filter, while having $W(e^{j\omega})$ as narrow as possible in frequency so as to faithfully reproduce the desired frequency response. These are conflicting requirements, as can be seen in the case of the rectangular window of Eq. (5.52), where

$$\begin{aligned} W(e^{j\omega}) &= \sum_{n=0}^{N-1} e^{-j\omega n} = \frac{1 - e^{-j\omega N}}{1 - e^{-j\omega}} \\ &= e^{-j\omega((N-1)/2)} \frac{\sin(\omega N/2)}{\sin(\omega/2)} \end{aligned} \tag{5.54}$$

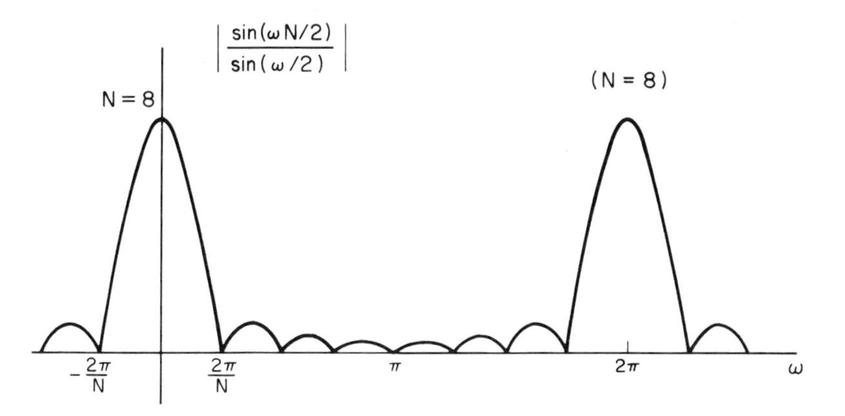

Fig. 5.32 Magnitude of the Fourier transform of a rectangular window ($N = 8$).

The magnitude of $W(e^{j\omega})$ is sketched in Fig. 5.32 for $N = 8$ and the phase is linear, as can be seen from Eq. (5.54). As N increases, the width of the "main lobe" decreases. (The main lobe is arbitrarily defined as the region between $\omega = -2\pi/N$ and $+2\pi/N$.)

However, for the rectangular window, the "side lobes" are not insignificant and, in fact, as N increases, the peak amplitudes of the main lobe and the side lobes grow in a manner such that the area under each lobe is a constant, while the width of each lobe decreases with N. The result of this is that as $W(e^{j(\omega-\theta)})$ slides by a discontinuity of $H_d(e^{j\theta})$ with increasing ω, the integral of $W(e^{j(\omega-\theta)})H_d(e^{j\theta})$ will vary in an oscillatory manner as each lobe of $W(e^{j(\omega-\theta)})$ moves past the discontinuity. This is depicted in Fig. 5.31(b). Since the area under each lobe remains constant with increasing N, the oscillations only occur more rapidly but do not decrease in amplitude as N increases. In the theory of Fourier series, it is well known that this non-uniform convergence, the Gibbs phenomenon, can be moderated through the use of a less abrupt truncation of the Fourier series.

By tapering the window smoothly to zero at each end, the height of the side lobes can be diminished; however, this is achieved at the expense of a wider main lobe and thus a wider transition at the discontinuity. Examples of some commonly used windows are shown in Fig. 5.33. These windows are specified by the following equations [21]:

Rectangular:

$$w(n) = 1, \qquad 0 \le n \le N-1 \tag{5.55a}$$

Bartlett:

$$w(n) = \begin{cases} \dfrac{2n}{N-1}, & 0 \le n \le \dfrac{N-1}{2} \\[2ex] 2 - \dfrac{2n}{N-1}, & \dfrac{N-1}{2} \le n \le N-1 \end{cases} \tag{5.55b}$$

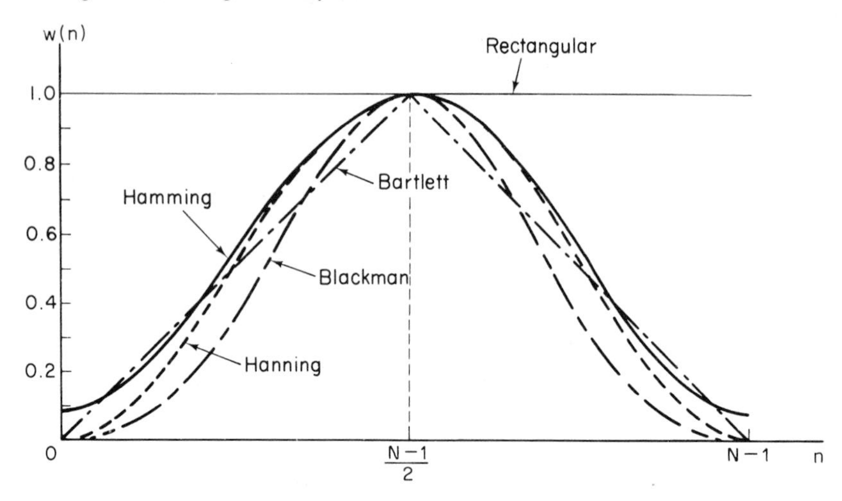

Fig. 5.33 Commonly used windows for FIR filter design.

Hanning:

$$w(n) = \frac{1}{2}\left[1 - \cos\left(\frac{2\pi n}{N-1}\right)\right], \qquad 0 \le n \le N-1 \tag{5.55c}$$

Hamming:

$$w(n) = 0.54 - 0.46\cos\left(\frac{2\pi n}{N-1}\right), \qquad 0 \le n \le N-1 \tag{5.55d}$$

Blackman:

$$w(n) = 0.42 - 0.5\cos\left(\frac{2\pi n}{N-1}\right) + 0.08\cos\left(\frac{4\pi n}{N-1}\right), \qquad 0 \le n \le N-1 \tag{5.55e}$$

The function $20\log_{10}|W(e^{j\omega})|$ is plotted in Fig. 5.34 for each of these windows for $N = 51$. Note that since these windows are all symmetrical, the phase is linear. The rectangular window clearly has the narrowest main lobe and thus, for a given length, N should yield the sharpest transitions of $H(e^{j\omega})$ at a discontinuity of $H_d(e^{j\omega})$. However, the first side lobe is only about 13 dB below the main peak, resulting in oscillations of $H(e^{j\omega})$ of considerable size at a discontinuity of $H_d(e^{j\omega})$. By tapering the window smoothly to zero, the side lobes are greatly reduced; however, it is clear that the price paid is a much wider main lobe and thus wider transitions at discontinuities of $H_d(e^{j\omega})$.

Kaiser [4] has proposed a flexible family of windows defined by

$$w(n) = \frac{I_0\left[\omega_a\sqrt{\left(\frac{N-1}{2}\right)^2 - \left[n - \left(\frac{N-1}{2}\right)\right]^2}\right]}{I_0\left[\omega_a\left(\frac{N-1}{2}\right)\right]}$$

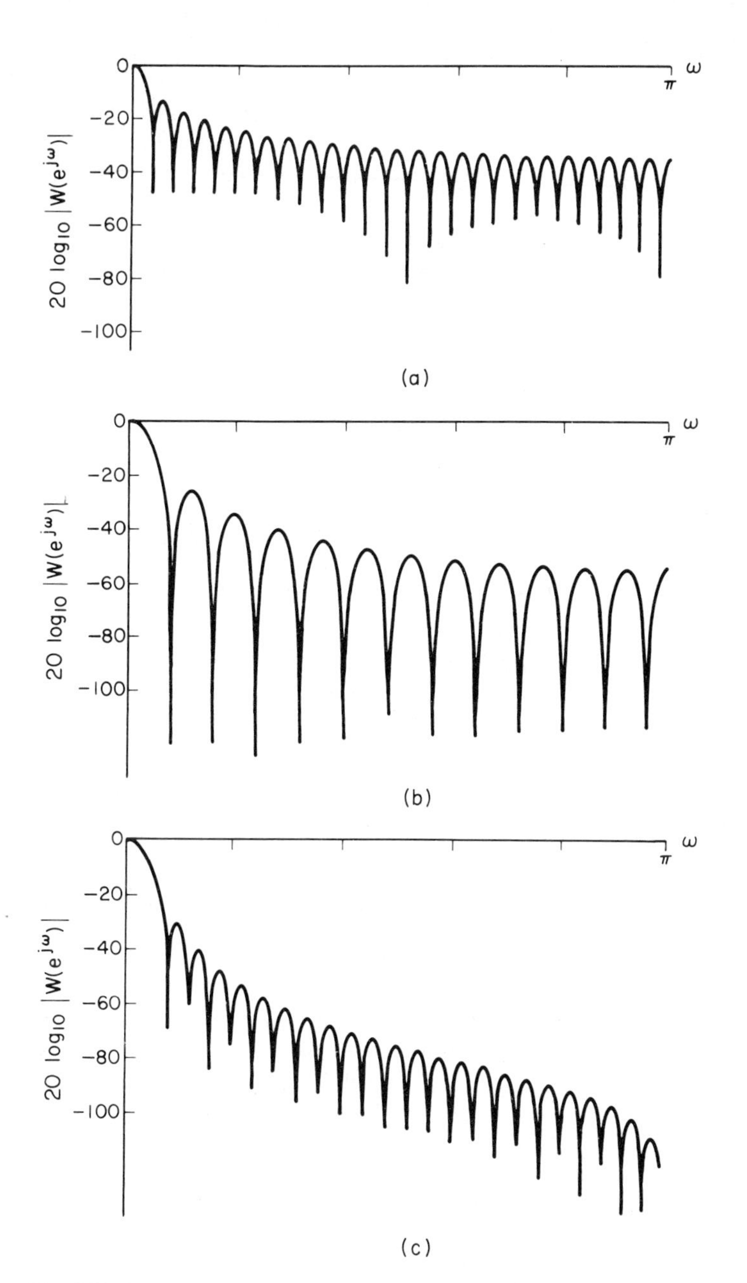

Fig. 5.34 Fourier transforms of windows of Fig. 5.33: (a) rectangular; (b) Bartlett (triangular); (c) Hanning; (d) Hamming; (e) Blackman.

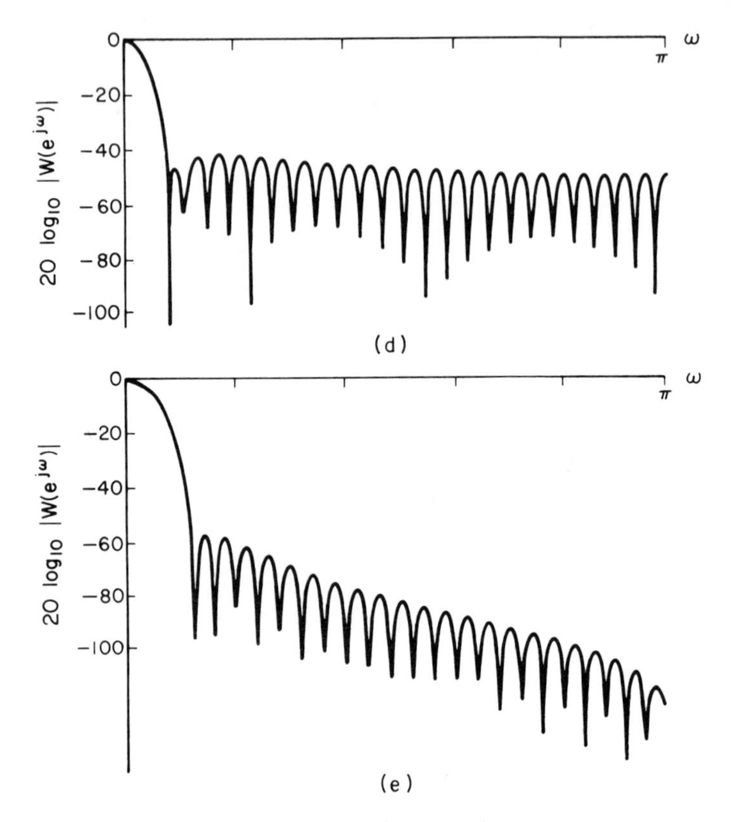

Fig. 5.34 *continued*

where $I_0(\ )$ is the modified zeroth order Bessel function of the first kind. Kaiser has shown that these windows are nearly optimum in the sense of having the largest energy in the main lobe for a given peak side lobe amplitude. The parameter ω_a can be adjusted so as to trade off main-lobe width for side-lobe amplitude. Typical values of $\omega_a((N-1)/2)$ are in the range $4 < \omega_a((N-1)/2) < 9$.

As an illustration of the use of windows in filter design, consider the design of a lowpass filter. Anticipating the need for delay in achieving a causal linear-phase filter, the desired frequency response is defined as

$$H_d(e^{j\omega}) = \begin{cases} e^{-j\omega\alpha}, & |\omega| \leq \omega_c \\ 0, & \text{otherwise} \end{cases}$$

The corresponding impulse response is

$$h_d(n) = \begin{cases} \dfrac{1}{2\pi}\displaystyle\int_{-\omega_c}^{\omega_c} e^{j\omega(n-\alpha)}\,d\omega \\ \dfrac{\sin[\omega_c(n-\alpha)]}{\pi(n-\alpha)}, & n \neq \alpha \end{cases}$$

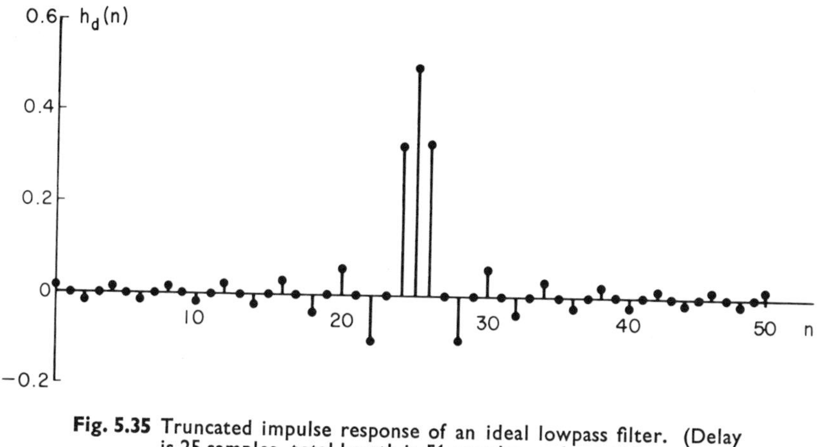

Fig. 5.35 Truncated impulse response of an ideal lowpass filter. (Delay is 25 samples, total length is 51 samples, and cutoff frequency is $\omega_c = \pi/2$.)

Clearly, $h_d(n)$ has infinite duration. To create a finite-duration linear-phase causal filter of length N, we define

$$h(n) = h_d(n)w(n)$$

where

$$\alpha = \frac{N-1}{2}$$

It can easily be verified that if $w(n)$ is symmetrical, this choice of α results in a sequence $h(n)$ satisfying Eq. (5.47). Figure 5.35 shows a plot of $h_d(n)$ for a rectangular window, $N = 51$, and $\omega_c = \pi/2$. Figure 5.36 shows $20 \log_{10} |H(e^{j\omega})|$ for the impulse response of Fig. 5.35 weighted by each of five windows of Fig. 5.34. Note the increasing transition width, corresponding to increasing main-lobe width, and the increasing stopband attenuation, corresponding to decreasing side-lobe amplitude.

From Eq. (5.54) we note that the width of the central lobe is inversely proportional to N. This is generally true and is illustrated for a Hamming window in Fig. 5.37, where it is clearly evident that as N is doubled, the width of the central lobe is halved. Figure 5.38 illustrates the effect of increasing N on the transition region in a lowpass filter design. Clearly, the minimum stopband attenuation remains essentially constant, being dependent on the shape of the window, while the width of the transition region at the discontinuity of $H_d(e^{j\omega})$ depends on the length of the window.

The examples that we have given illustrate the general principles of the windowing method of FIR filter design. Through the choice of the window shape and duration, we can exercise some control over the design process.

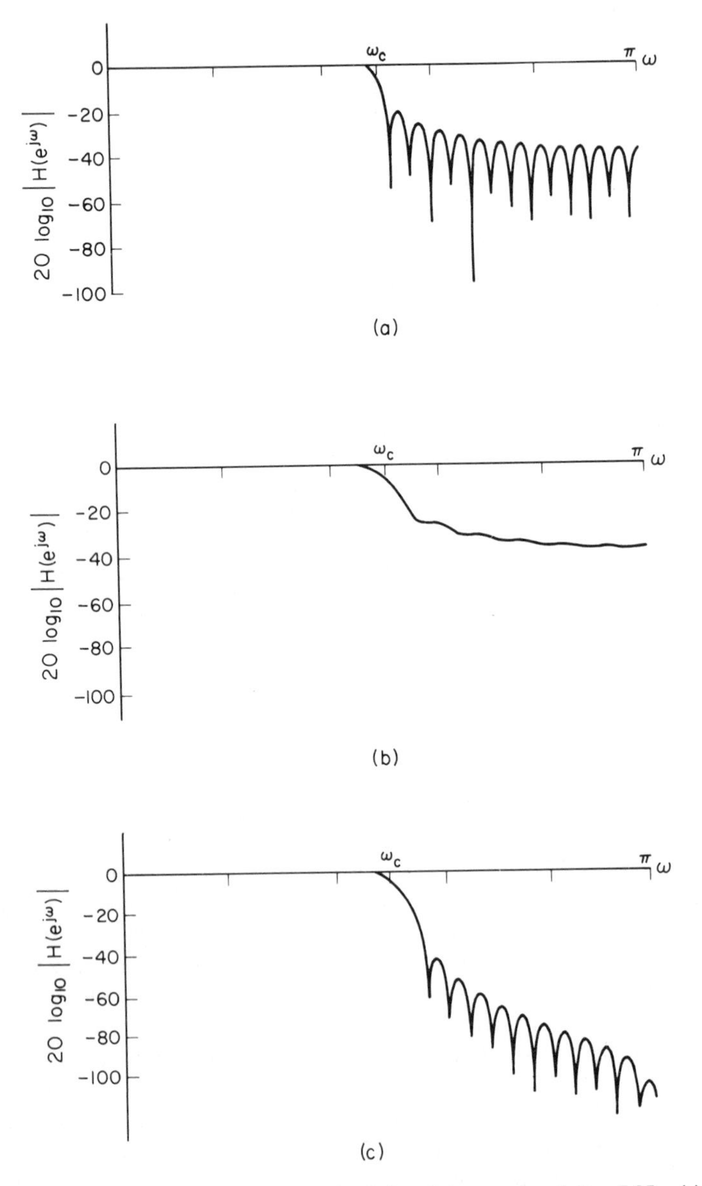

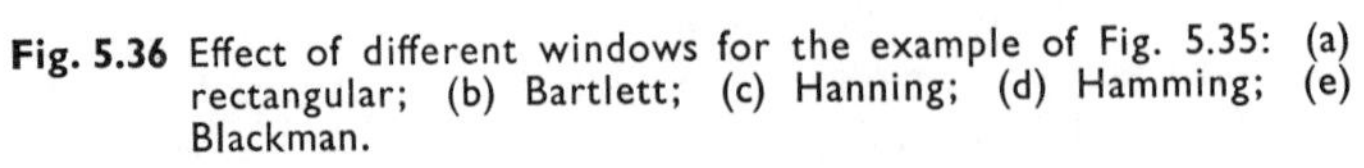

Fig. 5.36 Effect of different windows for the example of Fig. 5.35: (a) rectangular; (b) Bartlett; (c) Hanning; (d) Hamming; (e) Blackman.

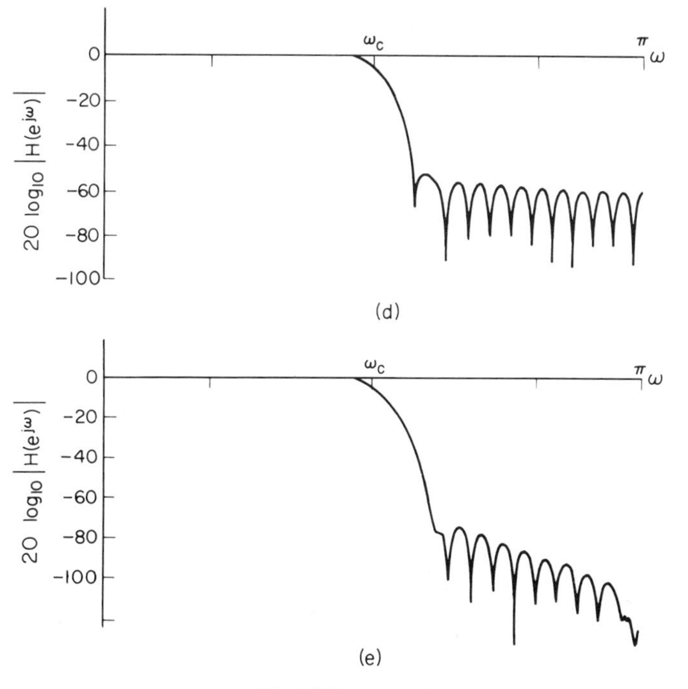

Fig. 5.36 *continued*

For example, for a given stopband attenuation, it is generally true that N satisfies an equation of the form

$$N = \frac{A}{\Delta\omega}$$

where $\Delta\omega$ is the transition width [roughly the width of the main lobe of $W(e^{j\omega})$] and A is a constant that is dependent upon the window shape. As we have seen, the window shape is essential in determining the minimum stopband attenuation. For the windows that we have discussed, the basic parameters for lowpass filter design are summarized in Table 5.2. It should be noted that the values in Table 5.2 are approximate; they depend somewhat upon N and the cutoff frequency of the desired filter. Kaiser's windows, Eq. (5.55e), have a variable parameter, ω_a, whose choice controls the tradeoff between side-lobe amplitude and side-lobe width. Tables and curves governing the use of these windows are given by Kaiser [4,22].

The basic principles illustrated by our examples are true in general and can be applied in the design of any filter for which we can define a desired frequency response. In this sense, the technique has considerable generality. However, a difficulty with the technique is in the evaluation of the integral in Eq. (5.49b). If $H_d(e^{j\omega})$ cannot be expressed in terms of simple functions

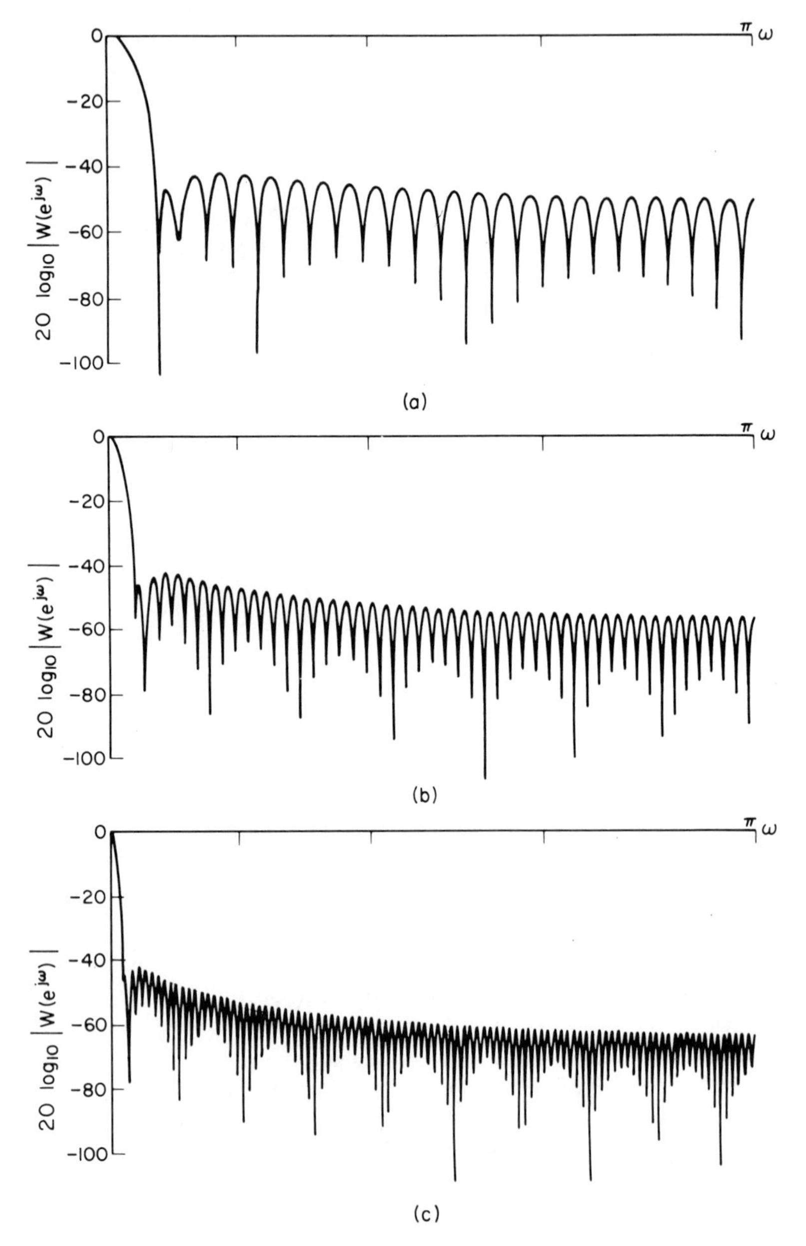

Fig. 5.37 Dependence of the Fourier transform of a Hamming window upon window length: (a) $N = 51$; (b) $N = 101$; (c) $N = 201$.

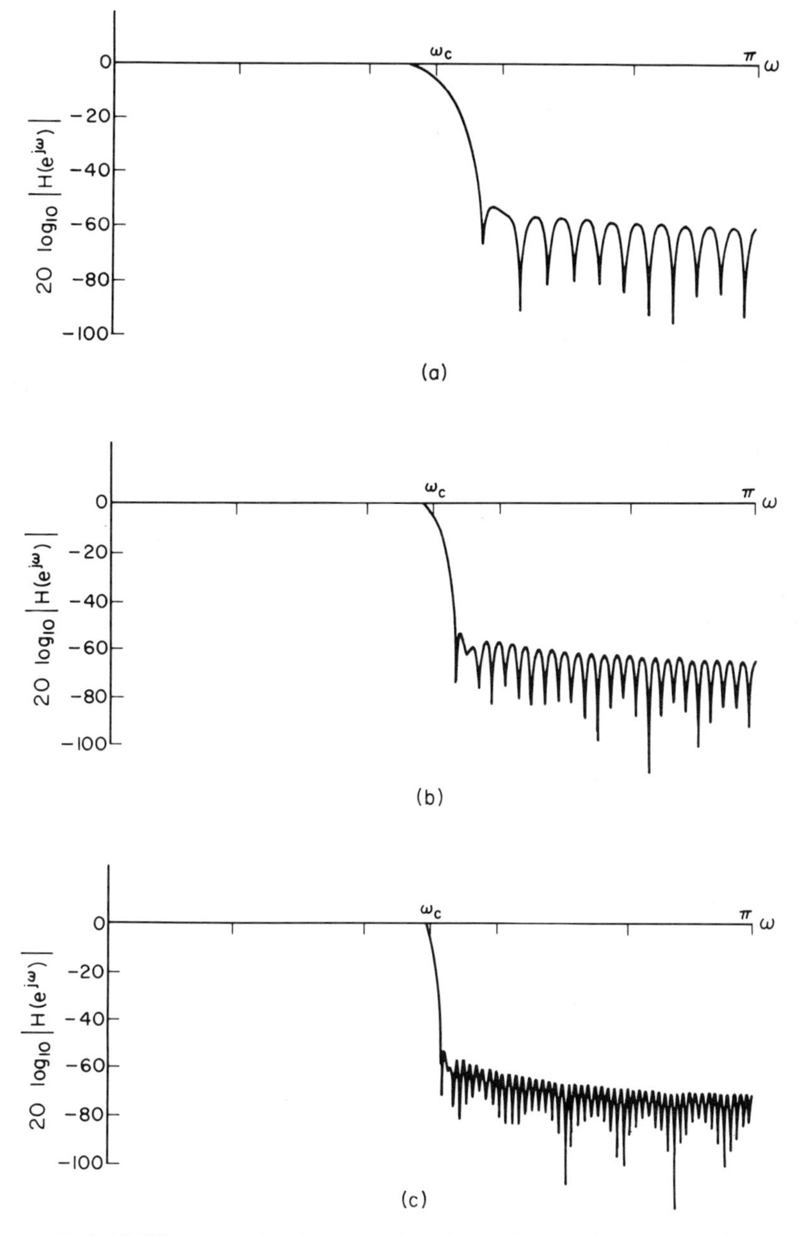

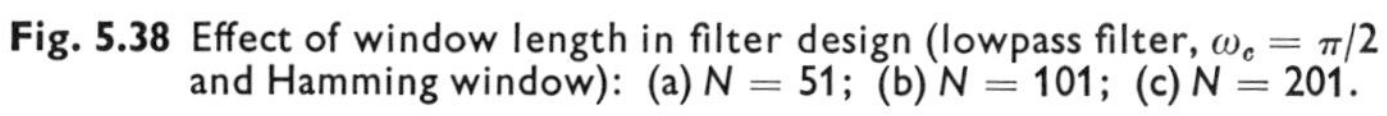

Fig. 5.38 Effect of window length in filter design (lowpass filter, $\omega_c = \pi/2$ and Hamming window): (a) $N = 51$; (b) $N = 101$; (c) $N = 201$.

Table 5.2

Window	*Peak Amplitude of Side Lobe (dB)*	*Transition Width of Main Lobe*	*Minimum Stopband Attenuation (dB)*
Rectangular	−13	$4\pi/N$	−21
Bartlett	−25	$8\pi/N$	−25
Hanning	−31	$8\pi/N$	−44
Hamming	−41	$8\pi/N$	−53
Blackman	−57	$12\pi/N$	−74

for which the integration can be performed, an approximation to $h_d(n)$ must be obtained by sampling $H_d(e^{j\omega})$ and using the inverse discrete Fourier transform to compute

$$\begin{aligned}\tilde{h}_d(n) &= \frac{1}{M}\sum_{k=0}^{M-1} H_d(e^{j(2\pi/M)k})e^{j(2\pi/N)kn} \\ &= \sum_{r=-\infty}^{\infty} h_d(n + rM)\end{aligned}$$

If M is large, $\tilde{h}_d(n)$ can be expected to be a good approximation to $h_d(n)$ in the interval of the window. Another limitation of the procedure is that it is somewhat difficult to determine, in advance, the type of window and the duration N required to meet a given prescribed frequency response specification.† However, a very simple digital computer program can be used to make such a determination by a trial-and-error procedure. Thus, design of digital filters using a window is often a convenient and satisfactory approach.

5.6 Computer-Aided Design of FIR Filters

The window technique is straightforward to apply and is in a sense quite general. However, we often wish to design a filter that is the "best" that can be achieved for a given value of N. It is of course meaningless to discuss this question in the absence of a definition of an approximation criterion. For example, in the case of window designs, it follows from a fundamental result of the theory of Fourier series that the rectangular window provides the best mean-square approximation to a desired frequency response for a given value of N. That is,

$$h(n) = \begin{cases} h_d(n), & 0 \le n \le N-1 \\ 0, & \text{otherwise} \end{cases}$$

minimizes the expression

$$\varepsilon^2 = \frac{1}{2\pi}\int_{-\pi}^{\pi} |H_d(e^{j\omega}) - H(e^{j\omega})|^2 \, d\omega$$

† Kaiser [22] has systematized the process for the window of Eq. (5.55e).

(see Problem 12 of this chapter). However, as we have seen, this approximation criterion leads to adverse behavior at discontinuities of $H_d(e^{j\omega})$. A better criterion for many types of filters is minimization of the maximum absolute error. For example, design procedures have been developed for minimizing the maximum absolute error in one or more frequency bands.

In this section we discuss several iterative design procedures for FIR filters, which in general yield better filters than the window method at the expense of greater complexity in the design procedure.

5.6.1 Frequency-Sampling Design

In Chapter 3 we showed that a finite-duration sequence can be represented by its discrete Fourier transform. Thus an FIR filter has a representation in terms of the "frequency samples"

$$\tilde{H}(k) = H(z)\big|_{z=e^{j(2\pi/N)k}} = \sum_{n=0}^{N-1} h(n)e^{-j(2\pi/N)kn}, \qquad k = 0, 1, \ldots, N-1$$

As we showed in Chapter 3, $H(z)$ can be represented in terms of the samples $\tilde{H}(k)$ by the expression

$$H(z) = \frac{1 - z^{-N}}{N} \sum_{k=0}^{N-1} \frac{\tilde{H}(k)}{1 - e^{j(2\pi/N)k}z^{-1}} \tag{5.56}$$

As was shown in Chapter 4, Eq. (5.56) serves as the basis of the frequency-sampling realization of an FIR filter. If we let $z = e^{j\omega}$, then the frequency response has the representation

$$\begin{aligned} H(e^{j\omega}) &= \frac{1 - e^{-j\omega N}}{N} \sum_{k=0}^{N-1} \frac{\tilde{H}(k)}{1 - e^{j(2\pi/N)k}e^{-j\omega}} \\ &= \frac{e^{-j\omega((N-1)/2)}}{N} \sum_{k=0}^{N-1} \tilde{H}(k) \frac{\sin [N(\omega - (2\pi/N)k)/2]}{\sin [(\omega - (2\pi/N)k)/2]} \end{aligned} \tag{5.57}$$

Equation (5.57) suggests a simple but rather naive approach to filter design, i.e., to specify the filter in terms of samples of one period of the desired frequency response

$$\tilde{H}(k) = H_d(e^{j(2\pi/N)k}), \qquad k = 0, 1, \ldots, N-1$$

relying on the interpolation indicated in Eq. (5.57) to "fill in the gaps" in the frequency response. As an illustration of this approach, consider the approximation of an ideal lowpass filter with cutoff frequency $\omega_c = \pi/2$. Figure 5.39(a) shows the desired frequency response $H_d(e^{j\omega})$ and the samples $\tilde{H}(k)$ for $N = 33$. As can be seen, the magnitude of the frequency response is specified at multiples of $2\pi/33$ radians, with the cutoff frequency $\omega_c = \pi/2$ being between $\omega = 16\pi/33$ and $18\pi/33$. The phase is taken to be linear

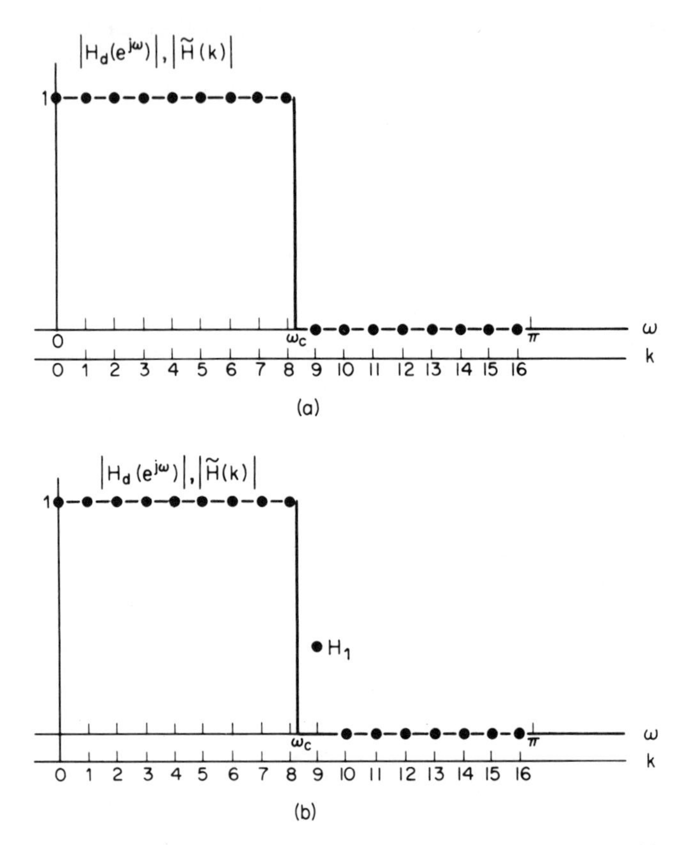

Fig. 5.39 Fixed samples of ideal lowpass filter frequency response: (a) no transition sample; (b) one transition sample H_1.

with delay equal to $(N - 1)/2$ samples. The impulse response can of course be obtained using the inverse discrete Fourier transform, as in

$$
\begin{aligned}
h(n) &= \frac{1}{N} \sum_{k=0}^{N-1} \tilde{H}(k) e^{j(2\pi/N)kn} \qquad n = 0, 1, \ldots, N-1 \qquad (5.58) \\
&= \quad 0 \qquad\qquad\qquad\qquad\qquad \text{otherwise}
\end{aligned}
$$

If we evaluate the frequency response corresponding to such a filter, we obtain the rather disappointing curve shown in Fig. 5.40(a). This figure shows $20 \log_{10} |H(e^{j\omega})|$, with the fixed sample points being indicated by the heavy dots in the passband and the points indicating infinite attenuation at the zero samples in the stopband. We note that there is a smooth transition between $16\pi/33$ and $18\pi/33$; however, the minimum stopband attenuation is somewhat less than 20 dB. This filter would be unsatisfactory for most purposes. As we have repeatedly seen, one way to improve the stopband attenuation is to widen the transition band. This can be easily done in this

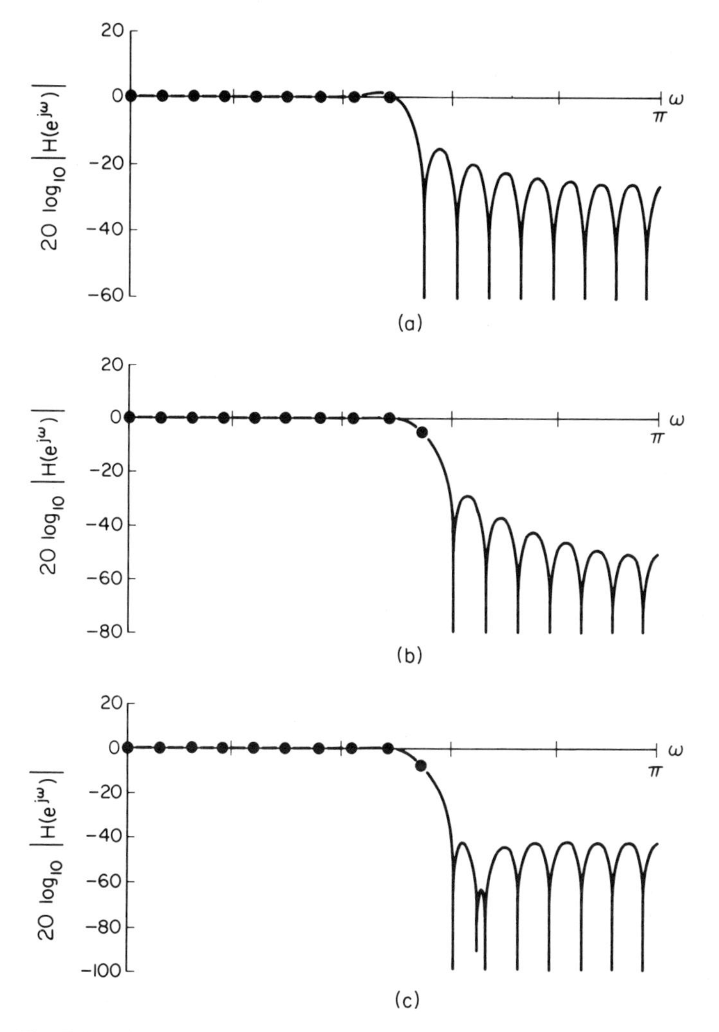

Fig. 5.40 Effect of a single transition sample: (a) $H_1 = 0$ (no transition sample); (b) $H_1 = 0.5$; (c) $H_1 = 0.3904$.

case by allowing a sample at the boundary between passband and stopband to take on a value different from either 1 or 0, as depicted in Fig. 5.39(b). Figure 5.40(b) shows the frequency response for $H_1 = 0.5$. Note that the transition band is now about twice as wide, but the minimum stopband attenuation is considerably greater.

It can be seen from Eq. (5.57) that $H(e^{j\omega})$ is a linear function of the parameters $\tilde{H}(k)$. Thus linear optimization techniques can be used to vary these parameters so as to give the best approximation to the desired filter. This approach, first proposed by Gold and Jordan [23], and developed by Rabiner, et al. [24], has been used to design a variety of filters. For example,

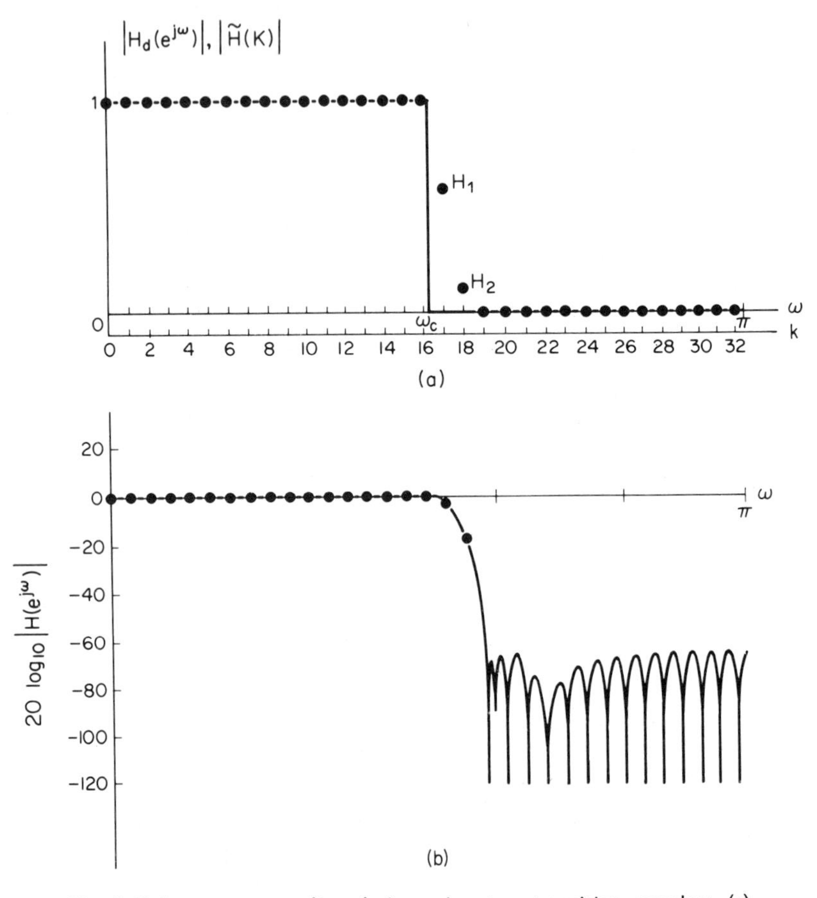

Fig. 5.41 Frequency sampling design using two transition samples: (a) desired frequency response with fixed samples and two transition samples; (b) resulting optimum frequency response for two transition samples.

in the case that we are discussing, a simple gradient search technique can be used to choose the value of H_1 such that the maximum error in either the passband or stopband is minimized. Figure 5.40(c) shows the response for $H_1 = 0.3904$, the value that minimizes the error (maximizes the attenuation) in the stopband

$$\frac{20\pi}{33} \leq |\omega| \leq \pi$$

Thus, it is clear that the stopband attenuation is significantly improved. If further improvement is required, we can broaden the transition region further by allowing a second† sample to differ from 1 or 0. If N is held fixed, this

† Rabiner et al. [24] give results for lowpass filters with up to four variable transition samples.

results in a transition region twice as wide. However, greater attenuation can be achieved. Of course, if we double N, the transition width remains the same, while allowing two transition samples to vary. Figure 5.41(a)shows such a set of samples for the example that we have been discussing for $N = 65$.† Figure 5.41(b) shows $20 \log_{10} |H(e^{j\omega})|$ for $N = 65$ and

$$H_1 = \tilde{H}(17) = H(e^{j(34\pi/65)}) = 0.5886$$

$$H_2 = \tilde{H}(18) = H(e^{j(36\pi/65)}) = 0.1065$$

These are very close to the optimum transition samples that minimize the can maximum absolute error (maximize the attenuation) in the stopband. As be seen, by comparing Fig. 5.41(b) to Fig. 5.40(c), by using two transition samples and increasing N by approximately a factor of 2 (from 33 to 65), the stopband attenuation is increased by about 24 dB, with a transition band that is somewhat narrower ($6\pi/65$ versus $8\pi/66$) than for one transition sample when $N = 33$.

Frequency sampling designs are particularly attractive for narrow-band frequency selective filters where only a few of the samples of the frequency response are nonzero [25,26]. In such cases a frequency sampling realization as discussed in Chapter 4 may be considerably more efficient than either direct convolution or convolution using the DFT. In general, even if more than a few samples are nonzero, the frequency-sampling design method yields excellent results. However, it is clear from the example of lowpass filter design that the method lacks flexibility in specifying the passband and stopband cutoff frequencies since the placement of ones and zeros and transition samples is constrained to integer multiples of $2\pi/N$. By making N large enough, samples can be obtained arbitrarily close to any given frequency; however, this is an inefficient approach. For this reason, particularly if the filter is not to be realized using the frequency sampling structure, other algorithmic design techniques have been developed with more attractive features for general frequency selective filter design.

5.6.2 *Equiripple Approximations for FIR Filters*

The frequency-sampling design technique uses an iterative procedure to arrive at an FIR filter having the smallest maximum stopband approximation error (largest minimum attenuation) for a given duration N, a set of prescribed frequency samples, and a given set of variable frequency samples. In the case of frequency selective filters designed by this technique, there is an undesirable limitation on the choice of cutoff frequencies. Furthermore, the approximation error tends to be highest around the transition region and

† Note that $2 \cdot 33 = 66$ is an even number and would therefore require a noninteger number of samples of delay for linear phase. Although frequency sampling designs can be obtained for N even with no more difficulty than for N odd, there are often practical reasons for choosing N to be an odd integer [25].

smaller in regions that are remote from the region in which the transition samples are located. It seems intuitively reasonable that if the approximation error were spread out uniformly in frequency, a given design specification should be met with a lower-order filter than if the approximation just meets the specification at one frequency and far exceeds it at others. This intuitive notion is confirmed by a theorem to be discussed later in this section.

In all the following discussion we shall be concerned with zero-phase FIR filters with frequency responses of the form

$$H(e^{j\omega}) = \sum_{n=-M}^{M} h(n)e^{-j\omega n}$$

The duration of the impulse response is $N = 2M + 1$, and for zero phase we require

$$h(n) = h(-n)$$

We note that a causal system can be obtained by simply delaying $h(n)$ by M samples. Because of the symmetry of $h(n)$, we can write $H(e^{j\omega})$ as

$$H(e^{j\omega}) = h(0) + \sum_{n=1}^{M} 2h(n) \cos (\omega n) \tag{5.59}$$

From Eq. (5.59) we observe that $H(e^{j\omega})$ is purely real. Suppose that we wish to design a lowpass filter according to the tolerance scheme of Fig. 5.42. That is, we wish to approximate 1 in the band $0 \leq |\omega| \leq \omega_p$ with maximum error δ_1 and we wish to approximate zero in the band $\omega_s \leq |\omega| \leq \pi$ with maximum error δ_2.

It is, of course, not possible to specify independently each of the parameters M, δ_1, δ_2, ω_p, and ω_s. However, design algorithms have been developed in which some of these parameters are fixed and an iterative procedure is

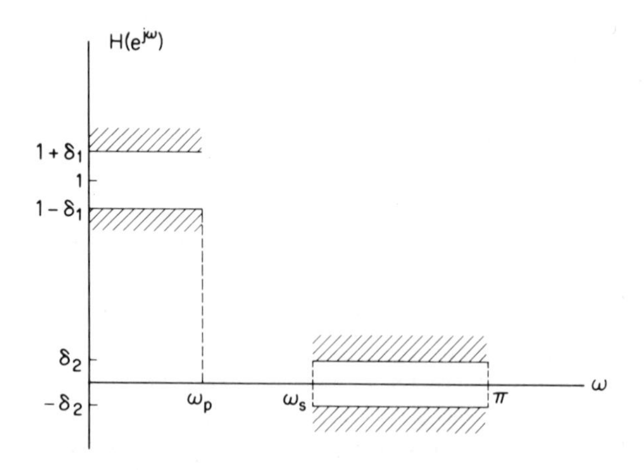

Fig. 5.42 Error tolerance for lowpass filter approximation.

used to obtain optimum adjustments of the remaining parameters. Two rather distinct approaches have been developed. Herrmann and Schuessler [27,28], and later Hofstetter et al. [29–31], developed procedures in which M, δ_1, and δ_2 are fixed and ω_p and ω_s are variable. Parks and McClellan [32,33] and Rabiner [34,35] developed procedures in which M, ω_p, and ω_s are fixed and δ_1 and δ_2 are variable. Since the work of Herrmann and Schuessler was the earliest and stimulated the later approaches to the equiripple design problem, we shall begin by discussing their approach.

Let us assume that we derive an equiripple approximation to a lowpass frequency selective filter as depicted in Fig. 5.43. We shall see later that such

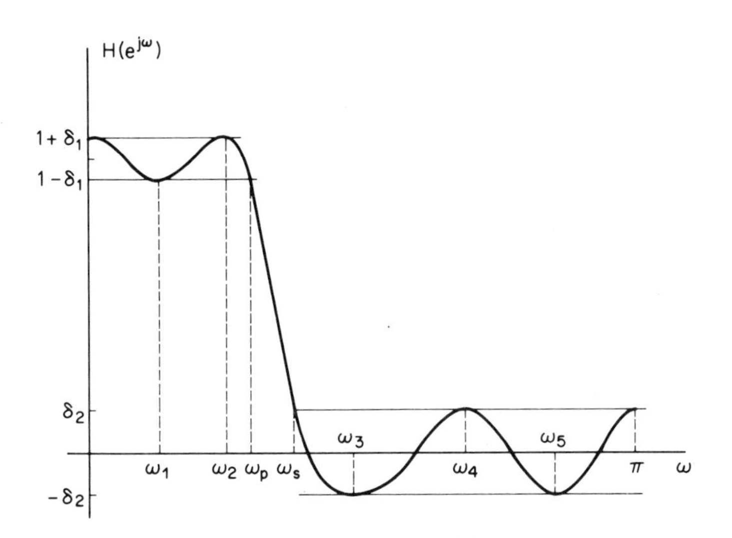

Fig. 5.43 Equiripple approximation of a lowpass filter.

approximations are optimum in the sense of having the smallest values of δ_1 and δ_2 for given values of ω_p and ω_s.

The number of local maxima and minima in the frequency range $0 \leq \omega \leq \pi$ is an important parameter of equiripple approximations. To examine the dependence of this parameter on the length of the impulse response we use the fact that $\cos \omega k$ can be expressed as a sum of powers of $\cos \omega$ to write Eq. (5.59) as

$$H(e^{j\omega}) = \sum_{k=0}^{M} a_k(\cos \omega)^k \tag{5.60}$$

where the a_k's are constants that are related to the values of the unit-sample response. Equation (5.60) places in evidence the fact that $H(e^{j\omega})$ is an Mth-order trigonometric polynomial. As such, there can be at most $M - 1$ local maxima and minima in the interval $0 < \omega < \pi$. Furthermore, if we differentiate Eq. (5.60) with respect to ω we obtain

$$H'(e^{j\omega}) = \frac{dH(e^{j\omega})}{d\omega} = -\sin\omega\left(\sum_{k=1}^{M} k a_k (\cos\omega)^{k-1}\right) \tag{5.61}$$

From Eq. (5.61) we see that $H(e^{j\omega})$ will always have either a maximum or a minimum at $\omega = 0$ and $\omega = \pi$. Thus there will be at most $M + 1$ local extrema in the closed interval $0 \leq \omega \leq \pi$.

Using this fact Herrmann and Schuessler [27,28] showed how to write a set of equations that guarantees the equiripple behavior depicted in Fig. 5.43. The parameters M, δ_1, and δ_2 are fixed and ω_p and ω_s are allowed to be free to vary, being defined by the equations

$$H(e^{j\omega_p}) = 1 - \delta_1$$

$$H(e^{j\omega_s}) = \delta_2$$

For the example of Fig. 5.43 we can write the equations

$$\begin{aligned}
H(e^{j0}) &= 1 + \delta_1, & H(e^{j\pi}) &= \delta_2 \\
H(e^{j\omega_1}) &= 1 - \delta_1, & H'(e^{j\omega_1}) &= 0 \\
H(e^{j\omega_2}) &= 1 + \delta_1, & H'(e^{j\omega_2}) &= 0 \\
H(e^{j\omega_3}) &= -\delta_2, & H'(e^{j\omega_3}) &= 0 \\
H(e^{j\omega_4}) &= \delta_2, & H'(e^{j\omega_4}) &= 0 \\
H(e^{j\omega_5}) &= -\delta_2, & H'(e^{j\omega_5}) &= 0
\end{aligned}$$

In this example, there are three extrema in the passband and four extrema in the stopband. Thus we must have

$$M + 1 = 4 + 3 = 7$$

i.e., $M = 6$ or $N = 13$. There are $M + 1 = 7$ unknown coefficients in either Eq. (5.59) or (5.60), and five unknown frequencies $\omega_1, \omega_2, \ldots, \omega_5$, at which the extrema occur, making a total of 12 unknowns to be determined as a solution to the above set of 12 equations. In general, we can have N_p extrema in the passband and N_s extrema in the stopband, where

$$N_p + N_s = M + 1$$

and we can write $2M$ equations relating the $M + 1$ filter coefficients and the $M - 1$ frequencies at which extrema occur (two extrema occur at $\omega = 0$ and $\omega = \pi$). These equations are unfortunately nonlinear and must be solved by an iterative procedure. Owing to the numerical difficulty of solving nonlinear equations, this approach has only been satisfactory for rather low values of M (on the order of 30). Furthermore, for given values of M, δ_1, and δ_2, there are only M different equiripple filters, corresponding to the fact that only M different values of N_p (or N_s) can be chosen. That is, for fixed values of M, δ_1, and δ_2, only M different choices of ω_p are available. Thus, although the equiripple approximations designed as discussed above will have the narrowest transition region ($\omega_s - \omega_p$) for a given value of M, there is essentially no improvement over the frequency sampling design technique with regard to choice of cutoff frequency.

A technique developed by Hofstetter, Oppenheim, and Siegel [29–31] effectively deals with the computational limitation of the Herrmann–Schuessler approach, but since M, δ_1, and δ_2 are fixed as before, the limitation on choice of ω_p and ω_s remains.

Instead of writing a set of nonlinear equations, Hofstetter et al. used an iterative technique for producing a trigonometric polynomial that has

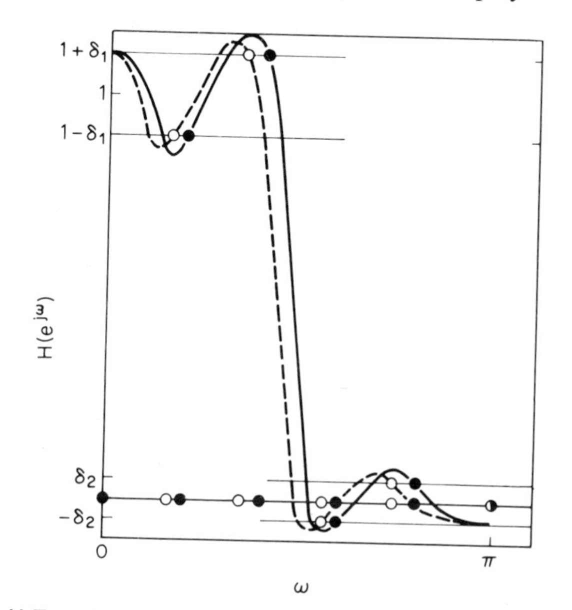

Fig. 5.44 Typical successive approximations in the Hofstetter design procedure. The dashed curve is the approximation obtained by locating the extremum points of the previous approximation (solid curve). (After Hofstetter, et al. [29].)

extrema of the desired value. The procedure begins by choosing N_p and N_s and then estimating the frequencies at which the extrema occur. Then standard Lagrange interpolation methods [5] are used to compute a polynomial that takes on the prescribed extremal values ($1 \pm \delta_1$ in the passband and $\pm\delta_2$ in the stopband) at the estimated frequencies. An example is shown as the solid curve in Fig. 5.44 for $N_p = N_s = 3$. The solid dots are initial estimates of extremum frequencies. The extrema of the resulting polynomial are computed by evaluating the polynomial on a finely spaced set of frequencies and searching for local maxima and minima on the discrete set of frequencies. If the maxima and minima have the prescribed values, the procedure is terminated; otherwise a new polynomial is computed with the new estimate of extremum frequencies being the extrema of the previous polynomial. The dashed curve in Fig. 5.44 shows the resulting polynomial, with the open circles denoting the estimated extremum frequencies.

The fact that this procedure converges to the desired equiripple approximation has been demonstrated computationally, although no proof of convergence was found [29]. An example of a lowpass filter designed by this technique is shown in Fig. 5.45. In this case $M = 125$, $N_p = 32$, and $N_s = 94$, $\delta_1 = 0.01$, and $\delta_2 = 0.00004$. Figure 5.46 shows a bandbass filter in which there are three distinct regions, rather than two as in Fig. 5.45. In this case there is a low-frequency passband containing $N_{pl} = 12$ extrema, in which $H(e^{j\omega}) = 0.25$ is approximated with maximum error of $\delta_{1l} = 0.01$. The upper passband contains $N_{pu} = 31$ extrema and $H(e^{j\omega}) = 1$ is approximated with maximum error of $\delta_{1u} = 0.02$. In the stopband there are $N_s = 18$ extrema and $H(e^{j\omega}) = 0$ is approximated with peak error of $\delta_2 = 0.0001$.

These examples illustrate the flexibility of the equiripple approximation method; however, there remains the problem of lack of precise control over the passband and stopband edges. In order to obtain control over ω_p and ω_s when M is fixed, it is necessary to allow δ_1 and δ_2 to vary. Parks and McClellan [32,33] have shown that with M, ω_p, and ω_s fixed, the frequency selective filter design problem becomes a problem in Chebyshev approximation over disjoint sets, a problem of considerable importance in approximation theory, and one for which a number of very useful theorems and procedures are already available [37]. To formalize the approximation problem in this case, let us define an approximation error function

$$E(\omega) = W(\omega)[H_d(e^{j\omega}) - H(e^{j\omega})] \tag{5.62}$$

where $E(\omega)$ is evaluated over the passbands and stopbands of the desired filter, and $W(\omega)$ is a weighting function.

For example, suppose that we wish to obtain an approximation as in Fig. 5.42, where M, ω_p, and ω_s are fixed. In this case,

$$H_d(e^{j\omega}) = \begin{cases} 1, & 0 \le \omega \le \omega_p \\ 0, & \omega_s \le \omega \le \pi \end{cases}$$

and

$$W(\omega) = \begin{cases} \dfrac{1}{K}, & 0 \le \omega \le \omega_p \\ 1, & \omega_s \le \omega \le \pi \end{cases}$$

This choice of $W(\omega)$ specifies a relationship between the relative sizes of the passband and stopband approximation errors. That is, K should be equal

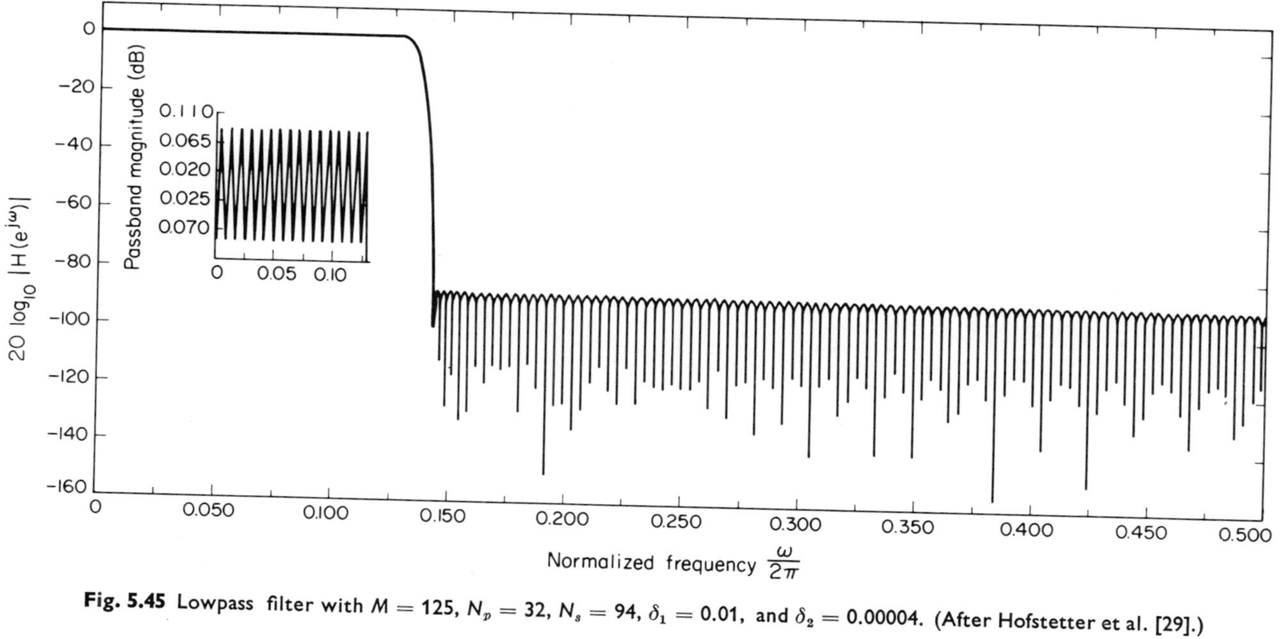

Fig. 5.45 Lowpass filter with $M = 125$, $N_p = 32$, $N_s = 94$, $\delta_1 = 0.01$, and $\delta_2 = 0.00004$. (After Hofstetter et al. [29].)

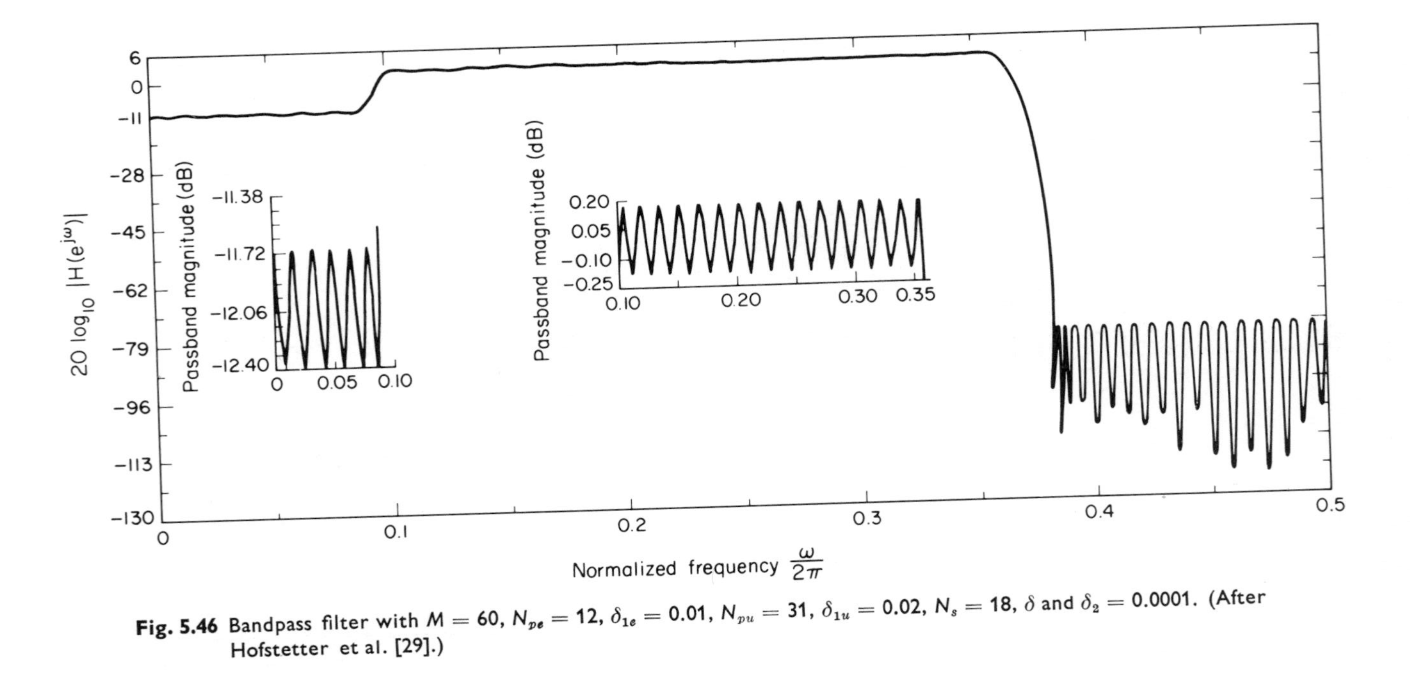

Fig. 5.46 Bandpass filter with $M = 60$, $N_{pe} = 12$, $\delta_{1e} = 0.01$, $N_{pu} = 31$, $\delta_{1u} = 0.02$, $N_s = 18$, δ and $\delta_2 = 0.0001$. (After Hofstetter et al. [29].)

to the desired ratio δ_1/δ_2. In this case, the design procedure requires an algorithm for minimizing

$$\underset{\substack{0 \leq \omega \leq \omega_p \\ \omega_s \leq \omega \leq \pi}}{\text{maximum}} |E(\omega)|$$

which is equivalent to minimizing δ_2.

Parks and McClellan [32] reformulated a theorem of approximation theory [37] in terms of the above filter-design problem, thereby obtaining the following theorem.

ALTERNATION THEOREM Let F be any closed subset of the closed interval $0 \leq \omega \leq \pi$. In order that $H(e^{j\omega})$ of Eq. (5.59) be the unique best approximation on F to $H_d(e^{j\omega})$, it is necessary and sufficient that the error function $E(\omega)$ exhibit on F at least $M + 2$ "alternations," thus: $E(\omega_i) = -E(\omega_{i-1}) = \pm\|E\| = \max |E(\omega)|$ with $\omega_0 \leq \omega_1 \leq \omega_2 \ldots < \omega_{M+1}$ and ω_i contained in F.

It is instructive to interpret this theorem in terms of the lowpass design problem. In this case, the closed subset F is made up of the intervals $0 \leq \omega \leq \omega_p$ and $\omega_s \leq \omega \leq \pi$. Since $H_d(e^{j\omega})$ is piecewise constant, the frequencies ω_i corresponding to peaks in the error function $E(\omega)$ likewise correspond to frequencies at which $H(e^{j\omega})$ just meets the error tolerance. A typical example is sketched in Fig. 5.47.

Recall from our previous discussion that $H(e^{j\omega})$ can have at most $M - 1$ local maxima and minima in the interval $0 < \omega < \pi$, and likewise in the

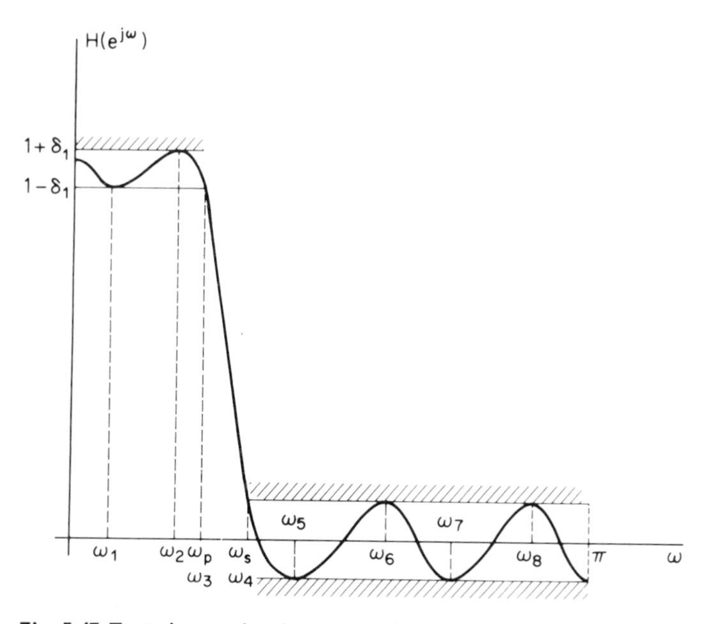

Fig 5.47 Typical example of a lowpass filter approximation that satisfies the alternation theorem ($M = 7$).

combined open intervals $0 < \omega < \omega_p$ and $\omega_s < \omega < \pi$. Furthermore, by definition of the passband and stopband, $H(e^{j\omega})$ is constrained such that

$$H(e^{j\omega_p}) = 1 - \delta_1$$

$$H(e^{j\omega_s}) = +\delta_2$$

Also we recall that $H(e^{j\omega})$ will always have a local maximum or minimum at $\omega = 0$ and $\omega = \pi$. Thus there can be *at most* $M + 3$ frequencies at which the error curve attains its maximum. Therefore, the unique best approximation for the desired lowpass response has either $M + 2$ or $M + 3$ alternations of the error function. Four different possible frequency-response curves for a value of $M = 7$ are shown in Fig. 5.48. Figure 5.48(a) shows the case when the maximum error is attained at both $\omega = 0$ and $\omega = \pi$, and there are $M + 3$ alternations. Figures 5.48(b) and (c) show the cases where the maximum error is attained only at $\omega = \pi$ and $\omega = 0$, respectively. In these two cases there are only $M + 2$ alternations. Figure 5.48(d) shows the case where there are only $M + 2$ alternations and the maximum error is attained at both $\omega = 0$ and $\omega = \pi$. According to the alternation theorem, all of these filters are optimum† approximations for their prescribed passband and stopband cutoff frequencies. Filters of the type depicted in Fig. 5.48(a) were called "extraripple" filters by Parks and McClellan [32]. This terminology was motivated by the fact that such filters have more than the minimum number $(M + 2)$ of alternations of the error required for optimality. If the endpoints ω_p and ω_s are included, the filters designed by the Herrmann–Schuessler and Hofstetter techniques have $M + 3$ points at which the frequency response attains the prescribed tolerance; therefore these filters are identically the extraripple filters.

In addition to providing a clear statement of the conditions for optimum FIR approximation, Parks and McClellan [32] also presented an iterative procedure for obtaining optimum filters. This procedure is similar to the Hofstetter algorithm except in this case, K, M, ω_p, and ω_s are the fixed parameters and δ_2 is allowed to vary. Following the alternation theorem, the procedure begins by estimating $M + 2$ frequencies $\{\omega_i\}$, $i = 0, 1, \ldots, M + 1$, at which the error function $E(\omega)$ attains its maximum value. These frequencies must lie in the regions $0 \leq \omega \leq \omega_p$ and $\omega_s \leq \omega \leq \pi$. Note that since ω_p and ω_s are fixed, it is known that ω_p is equal to one of the ω_k's; i.e., $\omega_p = \omega_l$, where $0 < l < M + 1$ and $\omega_s = \omega_{l+1}$. Assuming that these estimated frequencies are the desired error extremum frequencies we can compute the value of the magnitude of the peak error, which we call ρ, from Eq. (5.62). That is, we can write a set of $M + 2$ equations

$$W(\omega_i)\left[H_d(e^{j\omega_i}) - h(0) - \sum_{n=1}^{M} 2h(n)\cos(\omega_i n)\right] = -(-1)^i\rho, \qquad i = 0, 1, \ldots, M + 1$$

Optimum† means in the sense of minimum maximum error in the passband and stopband.

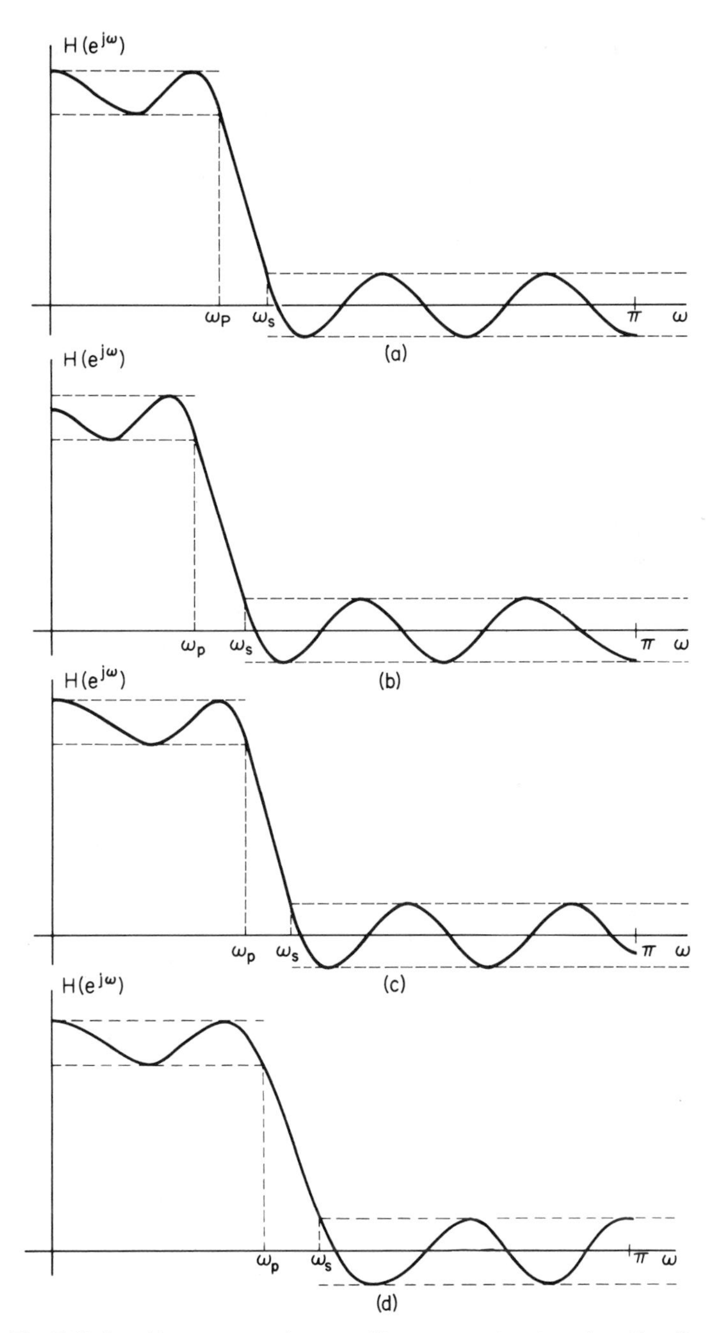

Fig. 5.48 Possible optimum lowpass filter approximations for $M = 7$: (a) $M + 3$ alternations (extraripple case); (b) $M + 2$ alternations (extremum at $\omega = \pi$); (c) $M + 2$ alternations (extremum at $\omega = 0$). (d) $M + 2$ alternations (extremum at both $\omega = 0$ and $\omega = \pi$).

This set of equations can be solved for all the $M + 2$ unknowns $\{h(n)\}$ and ρ, however, it is more efficient to only solve for ρ. Then a trigonometric polynomial is determined which has the correct value at the frequencies $\{\omega_i\}$; i.e., $1 \pm K\rho$ if $0 \le \omega_i \le \omega_p$ and $\pm\rho$ if $\omega_s \le \omega_i \le \pi$. This is depicted in Fig. 5.49, where it is clear that the estimate of extremum points was such

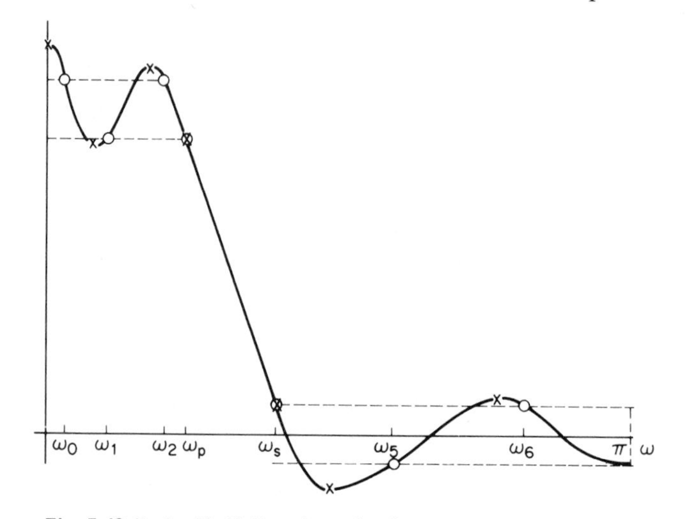

Fig. 5.49 Parks–McClellan algorithm for optimum approximation.

that the computed value of ρ was too small. As in the Hofstetter algorithm, the new estimate of the extremum frequencies is taken to be the peaks of the interpolating polynomial. These peaks are found by searching over a finely divided set of points in the passband and stopband. In this case, ω_p and ω_s are again known to be extremum frequencies of the error function. In addition, there are $M - 1$ local minima and maxima in the open intervals $0 < \omega < \omega_p$ and $\omega_s < \omega < \pi$. The remaining extremum can be either at $\omega = 0$ or $\omega = \pi$. If there is a peak of the error function at both 0 and π, the frequency at which the greatest error occurs is taken as the new estimate of extremum frequency. The cycle—computation of ρ, fitting a polynomial to the assumed error peaks, and then locating the actual error peaks—is repeated until ρ does not change from its previous value. This value of ρ is then the desired minimum of δ_2.

The resulting filter has the minimum $\delta_2(\delta_1 = K\delta_2)$ for a prescribed transition band $(\omega_s - \omega_p)$. If given values of δ_1 and δ_2 are desired, the above algorithm can be employed iteratively to determine a filter with prescribed values of δ_1 and δ_2 by fixing ω_p and varying ω_s until the desired δ_1 and δ_2 are obtained. A summary of results of this type is given in Fig. 5.50, where $\Delta\omega = \omega_s - \omega_p$ is plotted against ω_p for fixed values of M, δ_2, and δ_1 $(K = 1)$. This curve shows that as ω_p increases, the transition width $\Delta\omega$ attains local minima. These points on the curve correspond to the extraripple

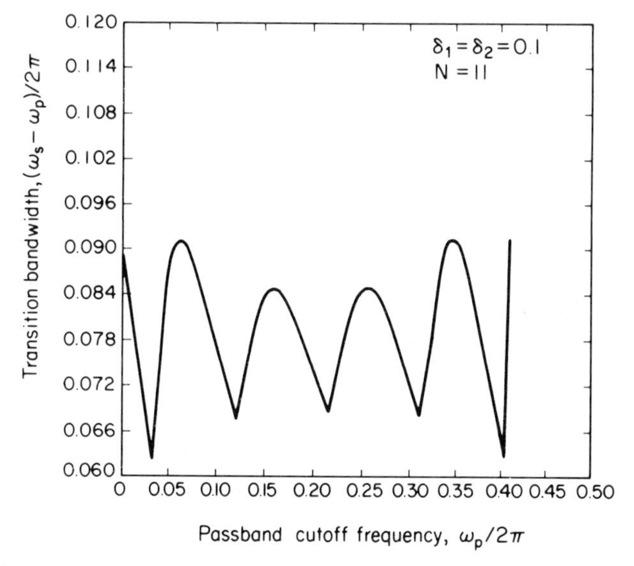

Fig. 5.50 Dependence of transition width upon cutoff frequency for optimum approximation of a lowpass filter. $M = 5, \delta_1 = \delta_2 = 0.1$.

($M + 3$ extrema) filters. All points between the minima correspond to filters that are optimum according to the alternation theorem. Thus the filters obtained by the Herrmann–Schuessler and Hofstetter approaches are special cases of the approximations obtained by the Parks–McClellan algorithm.

In the above algorithm all the unit-sample response values $h(n)$ are implicitly varied on each iteration to obtain the desired optimum approximation. The final step of the algorithm is to compute $h(n)$ by sampling the optimum frequency response at N or more points and computing an inverse discrete Fourier transform.

Rabiner [34,35] has discussed a formulation of the equiripple design problem that is equivalent to the Parks–McClellan formulation but is specially structured to be amenable to solution by linear programming techniques. The essence of this approach is to note that $H(e^{j\omega})$ can be expressed as a linear combination of cosine functions as in Eq. (5.59) or as a linear combination of functions of the form

$$\frac{\sin [N(\omega - (2\pi/N)k)/2]}{\sin [(\omega - (2\pi/N)k)/2]}$$

as in Eq. (5.57). In Eq. (5.59) the coefficients are the unit sample response values $h(n)$ and in Eq. (5.57) the coefficients are frequency samples $\tilde{H}(k)$. In either case, some or all of the coefficients can be varied systematically to meet prescribed design tolerances. If only a few of the parameters are variable, filters of the form of the previous section result. If all the parameters are varied, optimum approximations are obtained. The linear programming solution is slower computationally but is more flexible in that it allows time domain constraints as well as frequency domain constraints [34,35]. Much more detail on FIR design for arbitrary frequency response specifications can be found in Ref. [36].

5.7 A Comparison of IIR and FIR Digital Filters

This chapter has been concerned with design methods for linear shift-invariant digital filters. We have discussed a wide range of methods for the design of both infinite- and finite-duration impulse response filters. Questions that naturally arise are: What type of system is the best, IIR or FIR? Why give so many design methods? Which method yields the best results? The answer to these questions is that we have discussed a wide variety of methods for both IIR and FIR filters because no single type of filter nor single design method is best for all circumstances.

The choice between an FIR filter and an IIR filter depends upon the relative weight that one attaches to the advantages and disadvantages of each type of filter. IIR filters, for example, have the advantage that a variety of frequency selective filters can be designed using closed-form design formulas. That is, once the problem has been specified in terms appropriate for a given kind of filter (e.g., Butterworth, Chebyshev, or elliptic), then the coefficients (or poles and zeros) of the desired digital filter are obtained by straightforward substitution into a set of design equations. This kind of simplicity of the design procedure is attractive if only a few filters are to be designed or if limited computational facilities are available.

In the case of FIR filters, closed-form design equations do not exist. Although the window method can be applied in a rather straightforward manner, some iteration may be necessary to meet a prescribed specification. Most of the other FIR design methods are iterative procedures, requiring rather powerful computational facilities for their implementation. In contrast, it is often possible to design frequency selective IIR digital filters using only a hand calculator and tables of analog filter design parameters. However, the price paid for simplicity of the design process can be measured in terms of loss of flexibility in the attainable filter response. The closed-form IIR designs are primarily limited to lowpass, bandpass, and highpass filters, etc. Furthermore, these designs generally disregard the phase response of the filter. Thus, although we might obtain an elliptic lowpass filter with excellent amplitude response characteristics by a relatively simple computational procedure, its phase response will be very nonlinear (especially at the band edge).

In contrast, FIR filters can have precisely linear phase. Also, the window method and most of the algorithmic methods afford the possibility of approximating rather arbitrary frequency response characteristics with little more difficulty than is encountered in the design of lowpass filters. Also, it appears that the design problem for FIR filters is much more under control than the IIR design problem because there is an optimality theorem for FIR filters that is meaningful in a wide range of practical situations.

Finally, there are questions of economics in implementing a digital filter. Concerns of this type are usually measured in terms of hardware complexity or computational speed. Both of these factors are more or less directly related to the order of the filter required to meet a given specification. If we put aside phase considerations, it is generally true that a given amplitude response specification will be met most efficiently with an IIR filter. However, in many cases, the linear phase available with an FIR filter may be well worth the extra cost, and in some cases [38] one may not need to sacrifice efficiency in choosing an FIR filter. A detailed consideration of these questions is given in Ref. [39].

Thus a multitude of tradeoffs must be considered in designing a digital filter. Clearly, the final choice will most often be made in terms of engineering judgement on such questions as the formulation of specifications, the method of implementation, and computational facilities available for carrying out the design.

Summary

In this chapter we have considered a variety of design techniques for both infinite duration and finite-duration impulse response digital filters. Our emphasis was on frequency-domain specification of the desired filter characteristics since this is most common in practice. Our objective was to give a general picture of the wide range of possibilities available for digital filter design while also giving sufficient detail about some of the techniques so that they may be applied directly without further consultation of the literature on digital filter design. Thus we gave considerable attention to the well-established methods of impulse invariance and bilinear transformation and a much less complete discussion of some of the algorithmic design methods for IIR filters, since these methods are used much less in practice. Similarly, considerable detail is given concerning the windowing and frequency sampling methods of FIR filter design, while not as much discussion is devoted to the algorithmic design methods.

The chapter concludes with some remarks on the choice between the two classes of digital filters. The main point of this discussion is that the choice is not always clear cut and may depend on a multitude of factors that may be difficult to evaluate. However, it should be clear from this chapter and the preceding one that digital filters are characterized by great flexibility in design and implementation; this fact makes it possible to implement rather sophisticated signal processing schemes that in many cases would be difficult, if not impossible, to implement by analog means.

REFERENCES

1. B. Gold and C. M. Rader, *Digital Processing of Signals*, McGraw-Hill Book Company, New York, 1969.
2. C. M. Rader and B. Gold, "Digital Filter Design Techniques in the Frequency Domain," *Proc. IEEE*, Vol. 55, Feb. 1967, pp. 149–171.
3. J. F. Kaiser, "Design Methods for Sampled Data Filters," *Proc. 1st Allerton Conf. Circuit System Theory*, Nov. 1963, pp. 221–236.
4. J. F. Kaiser, "Digital Filters," Chapter 7 in *System Analysis by Digital Computer*, F. F. Kuo and J. F. Kaiser, John Wiley & Sons, Inc., New York, 1966.
5. F. B. Hildebrand, *Introduction to Numerical Analysis*, McGraw-Hill Book Company, New York, 1956.
6. A. J. Gibbs, "On the Frequency-Domain Responses of Causal Digital Filters," Ph.D. Thesis, University of Wisconsin, 1969.
7. E. A. Guillemin, *Synthesis of Passive Networks*, John Wiley & Sons, Inc., New York, 1957.
8. J. E. Storer, *Passive Network Synthesis*, McGraw-Hill Book Company, New York, 1957.
9. R. M. Golden and J. F. Kaiser, "Design of Wideband Sampled-Data Filters," *Bell System Tech. J.*, Vol. 43, No. 4, Pt. 2, July 1964, pp. 1533–1545.
10. A. Papoulis, "On the Approximation Problem in Filter Design," *IRE Conv. Record*, Pt. 2, 1957, pp. 175–185.
11. A. G. Constantinides, "Frequency Transformations for Digital Filters," *Elec. Lett.*, Vol. 3, No. 11, Nov. 1967, pp. 487–489.
12. A. G. Constantinides, "Frequency Transformations for Digital Filters," *Elec. Lett.*, Vol. 4, No. 7, Apr. 1968, pp. 115–116.
13. A. G. Constantinides, "Spectral Transformations for Digital Filters," *Proc. IEE*, Vol. 117, No. 8, Aug. 1970, pp. 1585–1590.
14. K. Steiglitz, "Computer-Aided Design of Recursive Digital Filters," *IEEE Trans. Audio Electroacoust.*, Vol. AU-18, June 1970.
15. L. R. Rabiner and K. Steiglitz, "The Design of Wide-Band Recursive and Nonrecursive Digital Differentiators," *IEEE Trans. Audio Electroacoust.*, Vol. 18, No. 2, June 1970, pp. 204–209.
16. R. Fletcher and M. J. D. Powell, "A Rapidly Convergent Descent Method for Minimization," *Computer J.*, Vol. 6, No. 2, 1963, pp. 163–168.
17. A. G. Deczky, "Synthesis of Recursive Digital Filters Using the Minimum *P* Error Criterion," *IEEE Trans. Audio Electroacoust.*, Vol. AU-20, Oct. 1972, pp. 257–263.
18. C. S. Burrus and T. W. Parks, "Time Domain Design of Recursive Digital Filters," *IEEE Trans. Audio Electroacoust.*, Vol. AU-18, No. 2, June 1970, pp. 137–141.
19. J. L. Shanks, "Recursion Filters for Digital Processing," *Geophys.*, Vol. 32, No. 1, Feb. 1967, pp. 33–51.
20. N. Levinson, "The Wiener rms Error Criterion in Filter Design and Prediction," *J. Math. Phys.*, Vol. 25, No. 4, 1947, pp. 261–278.
21. R. B. Blackman and J. W. Tukey, *The Measurement of Power Spectra*, Dover Publications, Inc., New York, 1958.
22. J. F. Kaiser, "Nonrecursive Digital Filter Design using the I_0–sinh Window Function," *Proc. 1974 IEEE International Symp. on Circuits and Systems*, San Francisco, April, 1974, pp. 20–23.
23. B. Gold and K. L. Jordan, Jr., "A Direct Search Procedure for Designing Finite Duration Impulse Response Filters," *IEEE Trans. Audio Electroacoust.*, Vol. AU-17, No. 1, Mar. 1969, pp. 33–36.

24. L. R. Rabiner, B. Gold, and C. A. McGonegal, "An Approach to the Approximation Problem for Nonrecursive Digital Filters," *IEEE Trans. Audio Electroacoust.*, Vol. AU-18, No. 2, June 1970, pp. 83–106.
25. L. R. Rabiner and R. W. Schafer, "Recursive and Nonrecursive Realizations of Digital Filters Designed by Frequency Sampling Techniques," *IEEE Trans. Audio Electroacoust.*, Vol. 19, No. 3, Sept. 1971, pp. 200–207.
26. L. R. Rabiner and R. W. Schafer, "Correction to Recursive and Nonrecursive Realizations of Digital Filters Designed by Frequency Sampling Techniques," *IEEE Trans. Audio Electroacoust.*, Vol. 20, No. 1, Mar. 1972, pp. 104–105.
27. O. Herrmann, "On the Design of Nonrecursive Digital Filters with Linear Phase," *Elec. Lett.*, Vol. 6, No. 11, 1970, pp. 328–329.
28. O. Herrmann and H. W. Schuessler, "Design of Nonrecursive Digital Filters with Minimum Phase," *Elec. Lett.*, Vol. 6, No. 11, 1970, pp. 329–330.
29. E. Hofstetter, A. V. Oppenheim, and J. Siegel, "A New Technique for the Design of Nonrecursive Digital Filters," *Proc. Fifth Annual Princeton Conf. Inform. Sci. Systems*, 1971, pp. 64–72.
30. E. Hofstetter, A. V. Oppenheim, and J. Siegel, "On Optimum Nonrecursive Digital Filters," *Proc. 9th Allerton Conf. Circuit System Theory*, Oct. 1971.
31. J. Siegel, "Design of Nonrecursive Approximations to Digital Filters with Discontinuous Frequency Responses," Ph.D. Thesis, MIT, June 1972.
32. T. W. Parks and J. H. McClellan, "Chebyshev Approximation for Nonrecursive Digital Filters with Linear Phase," *IEEE Trans. Circuit Theory*, Vol. CT-19, Mar. 1972, pp. 189–194.
33. T. W. Parks and J. H. McClellan, "A Program for the Design of Linear Phase Finite Impulse Response Filters," *IEEE Trans. Audio Electroacoust.*, Vol. AU-20, No. 3, Aug. 1972, pp. 195–199.
34. L. R. Rabiner, "The Design of Finite Impulse Response Digital Filters Using Linear Programming Techniques," *Bell System Tech. J.*, July–Aug. 1972, pp. 1177–1198.
35. L. R. Rabiner, "Linear Program Design of Finite Impulse Response (FIR) Digital Filters," *IEEE Trans. Audio Electroacoust.*, Vol. 20, No. 4, Oct. 1972, pp. 280–288.
36. L. R. Rabiner and B. Gold, *Theory and Application of Digital Signal Processing*, Prentice Hall, Inc., Englewood Cliffs, N.J., 1975.
37. E. W. Cheney, *Introduction to Approximation Theory*, McGraw-Hill Book Company, New York, 1966.
38. R. W. Schafer and L. R. Rabiner, "A Digital Signal Processing Approach to Interpolation," *Proc. IEEE*, Vol. 61, No. 6, June 1973, pp. 692–702.
39. L. R. Rabiner, J. F. Kaiser, O. Herrmann, and M. T. Dolan, "Some Comparisons between FIR and IIR Digital Filters," *Bell Syst. Tech. J.*, vol. 53, No. 2, Febr., 1974, pp. 305–331.

PROBLEMS

1. Digital filters are frequently used to process bandlimited analog signals in the manner depicted in Fig. P5.1. In the ideal case, the analog-to-digital converter samples the analog signal to produce the sequence $x(n) = x_a(nT)$, and the digital-to-analog converter converts the samples $y(n)$ into a bandlimited waveform

$$y_a(t) = \sum_{n=-\infty}^{\infty} y(n) \frac{\sin(\pi/T)(t - nT)}{(\pi/T)(t - nT)}$$

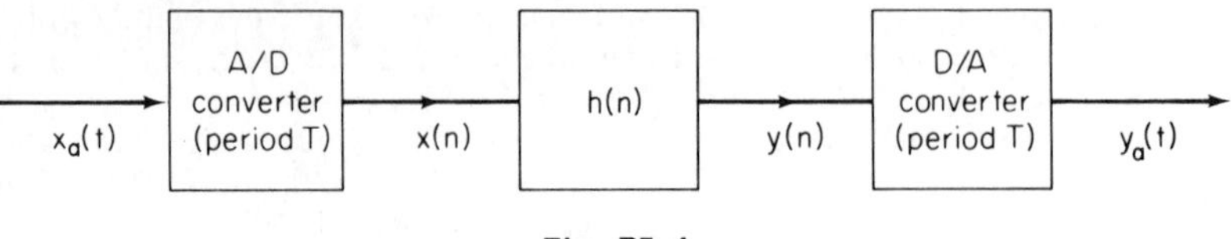

Fig. P5. 1

The overall system is equivalent to a linear time-invariant analog system.

(a) If the system $h(n)$ has a cutoff frequency of $\pi/8$ radians and if $1/T = 10$ kHz, what is the cutoff frequency of the equivalent analog filter?

(b) Repeat part (a) for $1/T = 20$ kHz.

2. The approximation of an analog system using backward difference was discussed in Sec. 5.1.2. An alternative procedure could be based upon forward differences. Assume that $y(n) = y_a(nT)$ and $x(n) = x_a(nT)$ and define the first forward difference as

$$\Delta^{(1)}[y(n)] = \frac{y(n+1) - y(n)}{T}$$

Higher-order differences are defined as

$$\Delta^{(k+1)}[y(n)] = \Delta^{(1)}[\Delta^{(k)}[y(n)]]$$

and

$$\Delta^{(0)}[y(n)] = y(n)$$

Consider the approximation of the differential equation

$$\sum_{k=0}^{N} c_k \frac{d^k y_a(t)}{dt^k} = \sum_{k=0}^{M} d_k \frac{d^k x_a(t)}{dt^k}$$

by the difference equation

$$\sum_{k=0}^{N} c_k \, \Delta^{(k)}[y(n)] = \sum_{k=0}^{M} d_k \, \Delta^{(k)}[x(n)]$$

(a) If $H_a(s) = Y_a(s)/X_a(s)$ and $H(z) = Y(z)/X(z)$, determine the mapping function $s = \psi(z)$ such that

$$H(z) = H_a(\psi(z))$$

(b) What is the contour in the z-plane that is the image of the $j\Omega$ axis in the s-plane for the mapping ψ found in part (a)?

(c) Are stable systems in the s-plane mapped into stable systems in the z-plane?

3. Let $h_a(t)$, $s_a(t)$, and $H_a(s)$ denote the impulse response, step response, and system function of a continuous-time linear time-invariant filter. Let $h(n)$, $s(n)$, and $H(z)$ denote the unit-sample response, step response, and system function of a discrete-time linear shift-invariant digital filter.

(a) If $h(n) = h_a(nT)$, does $s(n) = \sum_{k=-\infty}^{n} h_a(kT)$?

(b) If $s(n) = s_a(nT)$, does $h(n) = h_a(nT)$?

4. Consider a continuous-time system with system function

$$H_a(s) = \frac{s+a}{(s+a)^2 + b^2}$$

Determine the system function $H(z)$ (i.e., the z-transform of the unit-sample response) of a discrete system designed from this system on the basis of:
(a) Impulse invariance, i.e., such that

$$h(n) = h_a(nT)$$

(b) Step invariance, i.e., such that

$$s(n) = s_a(nT)$$

where

$$s(n) = \sum_{k=-\infty}^{n} h(k)$$

and

$$s_a(t) = \int_{-\infty}^{t} h_a(\tau)\, d\tau$$

5. Assume that $H_a(s)$ has an rth-order pole at $s = s_0$, so that $H_a(s)$ can be expressed as

$$H_a(s) = \sum_{k=1}^{r} \frac{A_k}{(s - s_0)^k} + G_a(s)$$

where $G_a(s)$ has only first-order poles.
(a) Give a formula for determining the constants A_k from $H_a(s)$.
(b) Obtain an expression for the impulse response $h_a(t)$ in terms of s_0 and $g_a(t)$, the inverse Laplace transform of $G_a(s)$.
(c) Suppose that we define $h(n) = h_a(nT)$ to be the impulse response of a digital filter. Using the result of part (b) write an expression for the system function $H(z)$.
(d) Discuss a direct procedure for obtaining $H(z)$ from $H_a(s)$.

6. We want to design a digital lowpass filter with a passband magnitude characteristic that is constant to within 0.75 dB for frequencies below $\omega = 0.2613\pi$ and stopband attenuation of at least 20 dB for frequencies between $\omega = 0.4018\pi$ and π.

Determine the transfer function $H(z)$ for the lowest-order Butterworth design which meets these specifications. Draw the cascade-form realization for this filter, including all necessary constants. Use bilinear transformation.

Fundamental Constants

$\log^{-1}(0.75) = 5.6240$	$\log(\tan(0.4018\pi)) = 0.52840$
$\log^{-1}(0.075) = 1.1885$	$\log(525.2) = 2.72032$
$\tan(0.2613\pi) = 1.0736$	$\log(0.1885) = -0.72469$
$\log(\tan(0.2009\pi)) = -0.12683$	$\log(0.23721) = -0.6253$
$\log(\tan(0.1306\pi)) = -0.36131$	$\log(2.00000) = 0.30103$
$\log(\tan(0.2613\pi)) = 0.03081$	$\tan(0.1\pi) = 0.32492$

7. Many IIR digital filters are designed by mapping analog designs using the bilinear transformation. The digital specifications are mapped to analog specifications and the corresponding analog filter is then either determined from a filter table, a computer program, or hand calculation. Before any of these techniques can be used, however, the filter order N must be determined. (N equals the number of poles.) This problem is addressed to that problem. Despite the length of the problem, it is straightforward. Examine carefully the problem to be solved and the information given you. You will find that you have been given far more information than you need. Careful use of the charts minimizes hand calculation. (Figures P5.7-3 – P5.7-7 are reprinted from Ref. [39]).

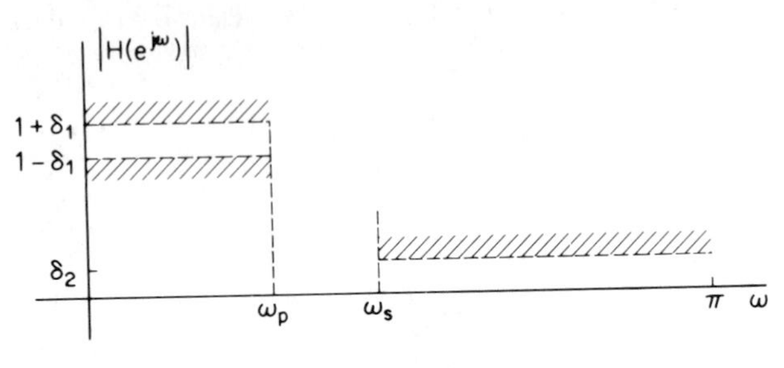

Fig. P5. 7-1

We wish to design a digital filter that meets the following specifications (see Fig. P5.7-1):

$$1 + \delta_1 \geq |H(e^{j\omega})| \geq 1 - \delta_1, \qquad 0 \leq \omega \leq \omega_p$$

$$\delta_2 \geq |H(e^{j\omega})| \geq 0, \qquad \omega_s \leq \omega \leq \pi$$

where $\delta_1 = 0.01$, $\delta_2 = 0.01$, $\omega_p = 0.3\pi$, and $\omega_s = 0.4\pi$. Unfortunately, for historical reasons analog filters are normally specified in terms of a different set of parameters. In addition, they are normally specified with a peak gain of 1 (Fig. P5.7-2). The peak gain constraint can be countered by multiplying the

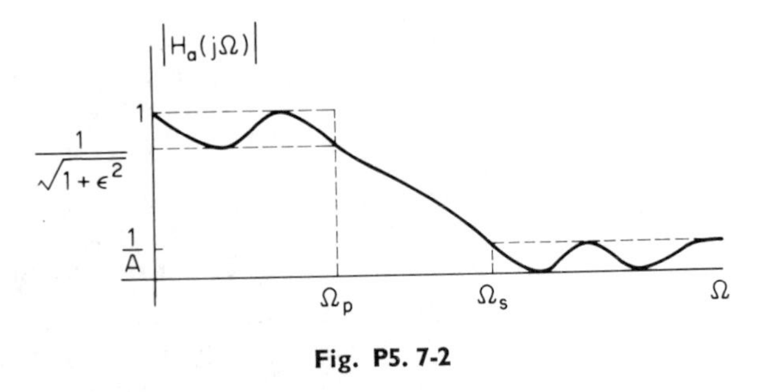

Fig. P5. 7-2

digital magnitude specifications by a factor of $1/(1 + \delta_1)$. Thus the digital and analog specifications are related by

$$\frac{1}{\sqrt{1 + \varepsilon^2}} = \frac{1 - \delta_1}{1 + \delta_1}$$

and

$$\frac{1}{A} = \frac{\delta_2}{1 + \delta_1}$$

Ω_s, Ω_p, ω_s, and ω_p are related by the bilinear transformation warping function. It is also useful to define the additional filter parameter

$$\eta = \frac{\varepsilon}{\sqrt{A^2 + 1}}$$

The parameter η has been shown to be a basic analog filter parameter which is used in the filter design curves. Other definitions that might be useful include

$$k = \text{transition ratio} = \frac{\Omega_p}{\Omega_s}$$

$$F_p = \text{passband cutoff frequency} = \frac{\omega_p}{2\pi}$$

$$F_s = \text{stopband cutoff frequency} = \frac{\omega_s}{2\pi}$$

$$\nu = F_s - F_p = \text{transition bandwidth}$$

$$N = \text{filter order}$$

For elliptic filters,

$$N = \frac{K(k)K(\sqrt{1 - \eta^2})}{K(\eta)K(\sqrt{1 - k^2})},$$

where $K(\ )$ is the complete elliptic integral of the first kind.
For Chebyshev filters,

$$N = \frac{\cosh^{-1}(1/\eta)}{\ln \beta},$$

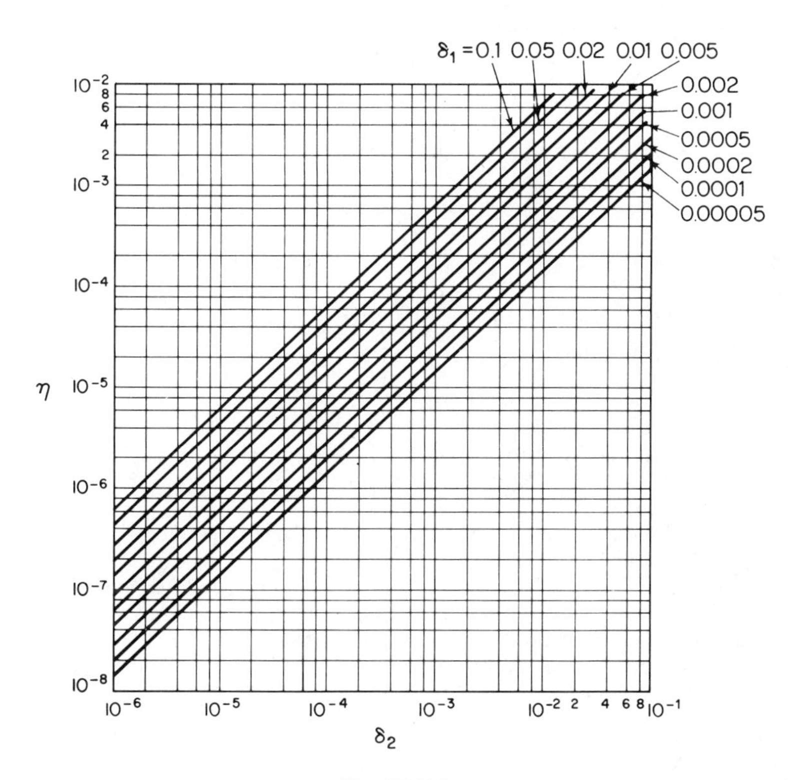

Fig. P5. 7-3

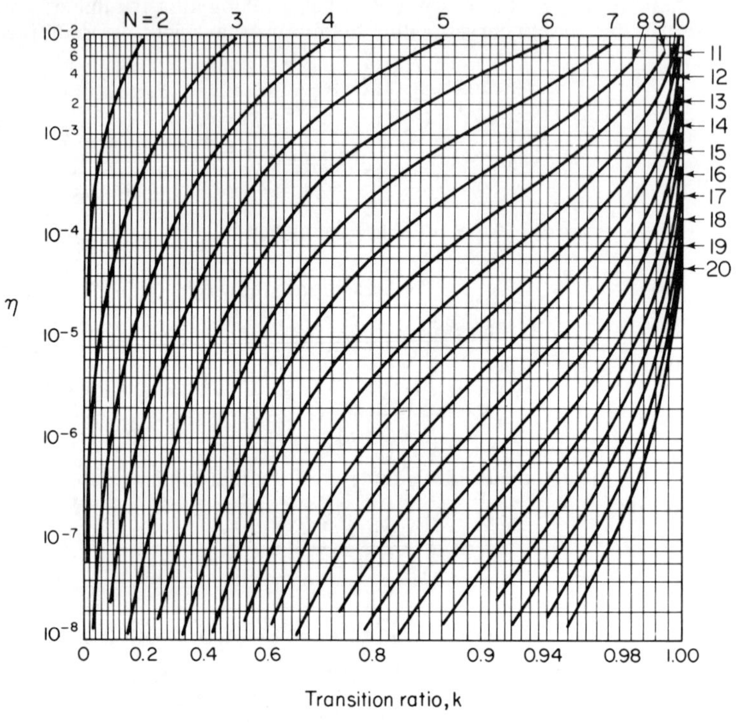

Fig. P5. 7-4

where

$$\beta = \frac{1 + \sqrt{1 - k^2}}{k}$$

For Butterworth filters

$$N = \frac{\ln \eta}{\ln k}$$

Armed with all this information (and see Figs. P5.7-3 through P5.7-7):

(a) What is the order of the minimum-order elliptic filter that meets the specifications?
(b) What is the order of the minimum-order Chebyshev filter that meets the specifications?
(c) What is the order of the minimum-order Butterworth filter that meets the specifications?
(d) Suppose that we are willing to use a filter of twelfth order. For the given values of δ_1, δ_2, and ω_p, how small can ω_s be made with a filter of twelfth order or less for each of the three filter types?

8. Let $H(z)$ denote the system function for a digital lowpass filter. The filter corresponds to a linear shift-invariant causal system and can be implemented by means of a digital network comprised of adders, gains, and unit delays.

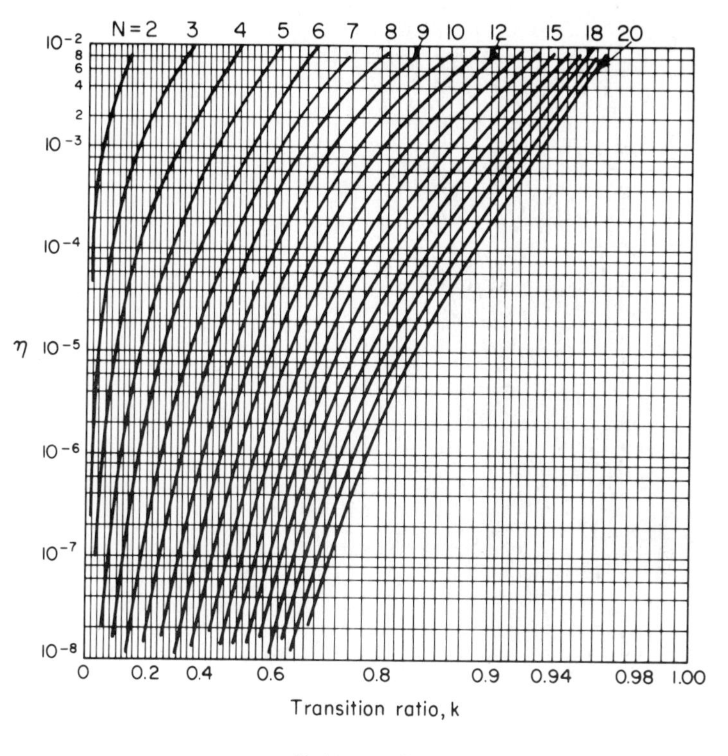

Chebyshev filters

Fig. P5. 7-5

From this lowpass filter we want to implement a lowpass filter for which the cutoff frequency can be varied by changing one parameter. The proposed strategy is to replace each unit delay in the network for $H(z)$ by a network R represented by the difference equation

$$y(n) = x(n-1) - \alpha[x(n) - y(n-1)] \tag{P5.8-1}$$

where α is real, $|\alpha| < 1$, and $x(n)$ and $y(n)$ represent the input and output of R, respectively.

(a) Let $G(z)$ denote the system function for the filter resulting when the network corresponding to Eq. (P5.8-1) is substituted for each unit delay in the original filter. Show that the frequency response associated with $G(z)$ is related to the frequency response associated with $H(z)$ by a mapping of the frequency axis; i.e., if $G(e^{j\omega})$ and $H(e^{j\theta})$ denote the two frequency responses, then ω can be expressed as a real function of θ. Sketch ω as a function of θ.

If the cutoff frequency associated with $H(z)$ is θ_p, determine as a function of the parameter, α, the cutoff frequency ω_p associated with $G(z)$.

(b) Instead of replacing each unit delay by a network described by Eq. (P5.8-1), let us consider replacing each unit delay by a network represented

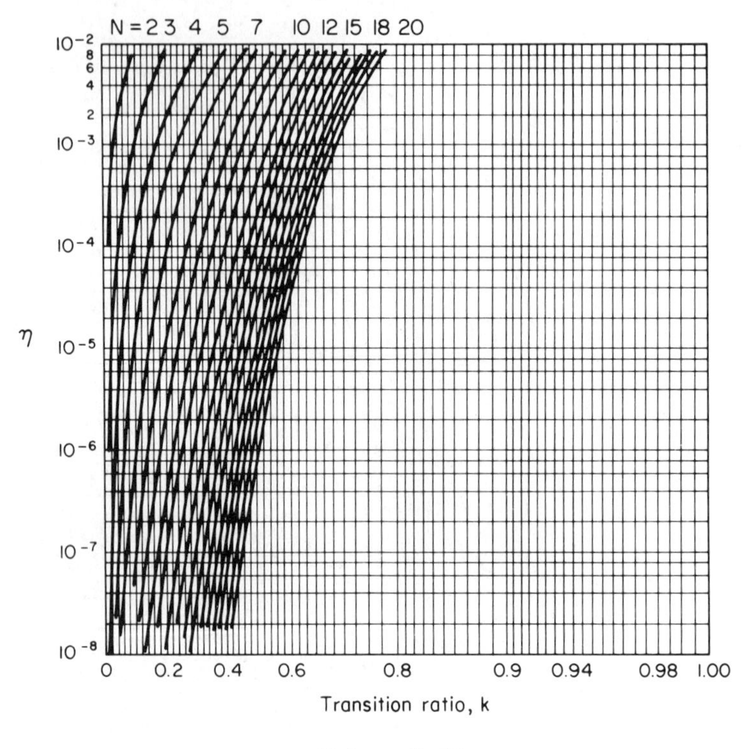

Butterworth filters

Fig. P5. 7-6

by the difference equation

$$y(n) = x(n-2) - a[x(n-1) - y(n-1)] \qquad \text{(P5.8-2)}$$

The network of Eq. (P5.8-2) is the network of Eq. (P5.8-1) in cascade with a unit delay. It will also be true in this case that the response associated with $G(z)$ is related to the frequency response associated with $H(z)$ by a mapping in the frequency axis. Determine this mapping and show that in this case, if $H(z)$ corresponds to a lowpass filter, $G(z)$ will not correspond to a lowpass filter.

9. An analog highpass filter can be obtained from an analog lowpass filter by replacing s by $1/s$ in the transfer function; i.e., if $G_a(s)$ is the transfer function for a lowpass filter, then $H_a(s)$ is the transfer function for a highpass filter if

$$H_a(s) = G_a\left(\frac{1}{s}\right)$$

Also, a digital filter can be obtained by mapping an analog filter by means of the bilinear transformation

$$s = \frac{z-1}{z+1}$$

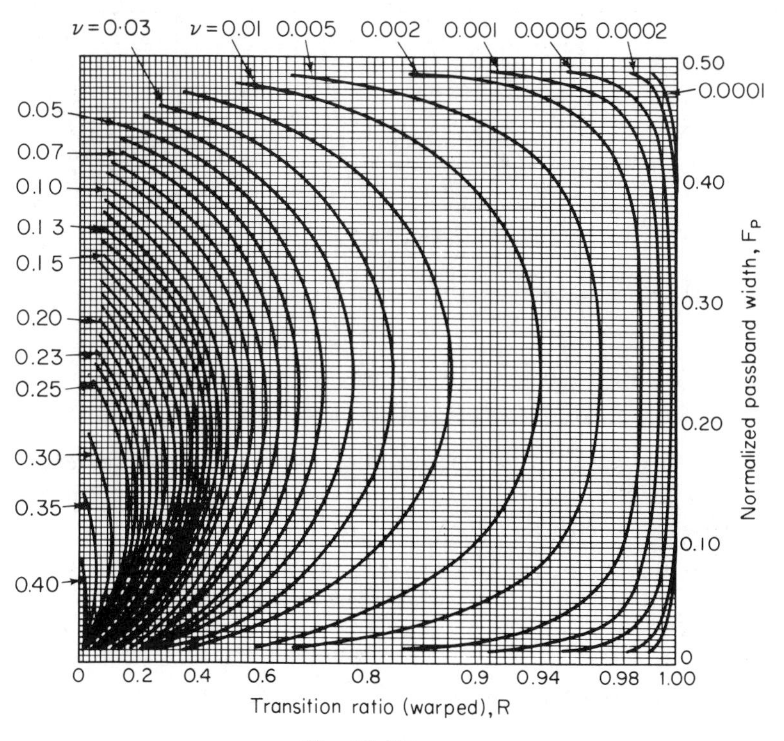

Fig. P5. 7-7

[Assume for convenience that $T = 2$ in Eq. (5.22).] This mapping preserves the character of the magnitude characteristic, although the frequency scale is distorted. The network in Fig. P5.9 depicts a lowpass filter with cutoff frequency $\omega_L = \pi/2$.

The constants A, B, C, and D are real. Determine how to modify the coefficients to obtain a highpass filter with cutoff frequency $\omega_H = \pi/2$.

10. Assume that the continuous-time filter is a lowpass filter and that $H(z) = H_a((z+1)/(z-1))$. Then the passband of the digital filter is centered at
(a) $\omega = 0$ (lowpass).
(b) $\omega = \pi$ (highpass).
(c) A frequency other than 0 or π (bandpass).
(Choose the correct answer.)

11. Let $h(n)$ be the unit-sample response of an FIR filter so that $h(n) = 0$ for $n < 0$, $n \geq N$. Assume that $h(n)$ is real. We can guarantee that the filter will have linear phase by imposing certain symmetry conditions on its unit-sample response $h(n)$.
(a) The frequency response of this filter can be represented in the form

$$H(e^{j\omega}) = \hat{H}(e^{j\omega})\, e^{j\theta(\omega)}$$

where

$$\hat{H}(e^{j\omega}) \text{ is real}$$

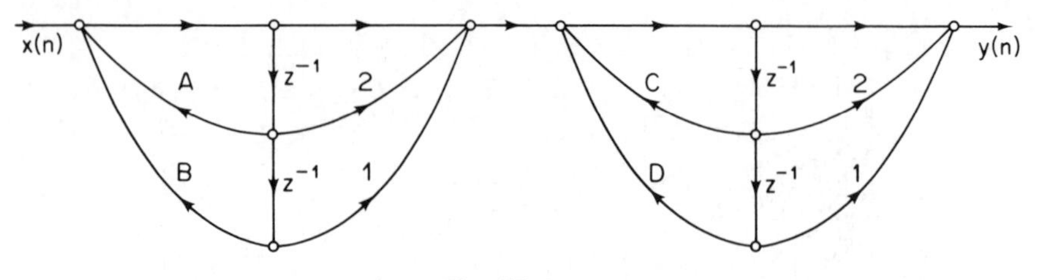

Fig. P5.9

(1) Find $\theta(\omega)$ for $0 \leq \omega \leq \pi$ when $h(n)$ satisfies the condition

$$h(n) = h(N - 1 - n)$$

(2) Find $\theta(\omega)$ for $0 \leq \omega \leq \pi$ when

$$h(n) = -h(N - 1 - n)$$

(Be careful; it may be necessary to treat cases N even and N odd separately.)

(b) Denote by $H(k)$ the N-point DFT of $h(n)$.

(1) If $h(n)$ satisfies

$$h(n) = -h(N - 1 - n)$$

show that

$$H(0) = 0$$

(2) If N is even, show that

$$h(n) = h(N - 1 - n)$$

implies that

$$H\left(\frac{N}{2}\right) = 0$$

12. Let $h_d(n)$ denote the unit-sample response of an ideal desired system with frequency response $H_d(e^{j\omega})$ and let $h(n)$ denote the unit-sample response, of length N samples, for an FIR system with frequency response $H(e^{j\omega})$. In Sec. 5.6 it was asserted that a rectangular window of length N samples applied to $h_d(n)$ will produce a unit-sample response $h(n)$ such that the mean-square error

$$\varepsilon^2 = \frac{1}{2\pi}\int_{-\pi}^{\pi} |H_d(e^{j\omega}) - H(e^{j\omega})|^2 \, d\omega$$

is minimized.

(a) The error function $E(e^{j\omega}) = H_d(e^{j\omega}) - H(e^{j\omega})$ can be expressed as the power series

$$E(e^{j\omega}) = \sum_{n=-\infty}^{\infty} e(n)e^{-j\omega n}$$

Find the coefficients $e(n)$ in terms of $h_d(n)$ and $h(n)$.

(b) Express the mean-square error ε^2 in terms of the coefficients $e(n)$.

(c) Show that for a unit-sample response $h(n)$ of length N samples, ε^2 is minimized when

$$h(n) = \begin{cases} h_d(n), & 0 \leq n \leq N - 1 \\ 0, & \text{otherwise} \end{cases}$$

That is, the rectangular window gives the best mean-square approximation to a desired frequency response for a fixed value of N.

13. An ideal bandlimiting differentiator with delay τ is defined by the frequency response

$$H_a(j\Omega) = \begin{cases} (j\Omega)e^{-j\Omega\tau}, & |\Omega| \leq \Omega_c \\ 0, & \text{otherwise} \end{cases}$$

(a) Find the unit-sample response of an ideal "digital differentiator with delay" by impulse invariance; i.e., find $h_d(n)$ such that $h_d(n) = Th_a(nT)$. (Assume that $\Omega_c = \pi/T$.)

(b) Find and sketch the corresponding frequency response $H_d(e^{j\omega})$. What is the delay of this system in samples?

(c) One way to obtain a causal digital differentiator approximation is to use the window method. Suppose that the unit-sample response $h(n)$ of the approximation is to be nonzero only in the interval $0 \leq n \leq N - 1$. How should we choose τ if (1) N is even and (2) N is odd? What is the delay in samples in each case? Sketch a typical unit-sample response for each case.

(d) Let $N = 2$ and $\tau = T/2$ and choose

$$h(n) = \begin{cases} h_d(n), & n = 0, 1 \\ 0, & \text{otherwise} \end{cases}$$

(1) Express the output of this system in terms of the input.
(2) What is the frequency response $H(e^{j\omega})$ in this case?
(3) Obtain an expression for the relative error

$$E_r(\omega) = \frac{H_d(e^{j\omega}) - H(e^{j\omega})}{H_d(e^{j\omega})}$$

in this case. Plot $E_r(\omega)$ for $0 \leq \omega \leq \pi$.

(e) Repeat part (d) for $N = 3$ and $\tau = T$.

14. A finite impulse response filter has frequency response

$$H(e^{j\omega}) = |H(e^{j\omega})|e^{-j\omega n_0}$$

where n_0 is not necessarily an integer. Let N be the length of the unit-sample response. Recall that the unit-sample response is completely specified by N samples of $H(e^{j\omega})$ taken at $\omega = 2\pi k/N$, $k = 0, 1, \ldots, N - 1$.

(a) Sketch $|H(e^{j\omega})|$ for the case $N = 15$, $n_0 = 0$, and

$$|\tilde{H}(k)| = |H(e^{j(2\pi/15)k})| = \begin{cases} 1, & k = 0 \\ \frac{1}{2}, & k = 1, 14 \\ 0, & \text{elsewhere} \end{cases}$$

(b) Write a general expression for $h(n)$ in terms of the $\tilde{H}(k)$'s. (Do not assume that $n_0 = 0$.)

(c) Sketch $h(n)$ for the cases (1) $n_0 = (N - 1)/2 = 7$, and (2) $n_0 = N/2 = 15/2$, for $|\tilde{H}(k)|$ as in part (a).

(d) Draw a complete block diagram (i.e., digital network) of a realization of this system when $N = 15$, $n_0 = \frac{15}{2}$, and the $|\tilde{H}(k)|$ are as in part (a). This realization should be recursive, i.e., a frequency-sampling filter. Compare the number of additions and multiplications required to the number required by a direct-form realization.

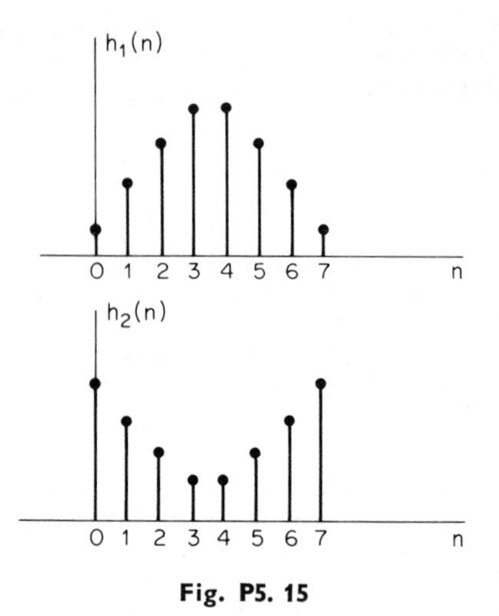

Fig. P5. 15

15. Two finite-duration sequences, $h_1(n)$ and $h_2(n)$, of length 8 are sketched in Fig. P5.15. They are related by a circular shift.

(a) The magnitude of the eight-point DFTs are equal. (Yes or no?)

(b) We wish to implement a lowpass nonrecursive filter and must use either $h_1(n)$ or $h_2(n)$ as the unit-sample response. Which one of the following statements is correct?

(1) $h_1(n)$ is a better lowpass filter than $h_2(n)$.

(2) $h_2(n)$ is a better lowpass filter than $h_1(n)$.

(3) They are both about equally good (or bad) as lowpass filters.

16. The Parks–McClellan algorithm as originally written determined the sequence $a(n)$ such that

$$\min_{\{a(n)\}} \left\{ \max_{\omega} \left[W(\omega) \left(H_d(e^{j\omega}) - \sum_{n=0}^{M} a(n) \cos \omega n \right) \right] \right\} \tag{P5.16-1}$$

where

$$h(n) = h(-n) = a(n)/2 \qquad 1 \le n \le M$$
$$h(0) = a(0)$$

It thus required that (1) N = length of unit-sample response $= 2M + 1$ be odd, and (2) unit-sample response is symmetric. There are three other interesting cases:

A: Positive symmetry, even length: $(N = 2M)$

$$H(e^{j\omega}) = \sum_{n=1}^{M} b(n) \cos [(n - \tfrac{1}{2})\omega]$$

B: Negative symmetry, odd length:

$$H(e^{j\omega}) = \sum_{n=1}^{M} c(n) \sin (\omega n)$$

C: Negative symmetry, even length:

$$H(e^{j\omega}) = \sum_{n=1}^{M} d(n) \sin (\omega(n - \tfrac{1}{2}))$$

Given an algorithm that computes $a(n)$ according to (P5.16-1), show how it can be used to design filters corresponding to the other three cases.

(a) Find a function $W(e^{j\omega})$ that can be used to solve case A and show how to determine $b(n)$ from $a(n)$.

(b) Find a function $W(e^{j\omega})$ that can be used to solve case B and show how to determine $c(n)$ from $a(n)$.

(c) Find a function $W(e^{j\omega})$ that can be used to solve case C and show how to determine $d(n)$ from $a(n)$.

6

Computation of the Discrete Fourier Transform

6.0 Introduction

We have seen in previous chapters that the discrete Fourier transform plays an important role in the analysis, the design, and the implementation of digital signal processing algorithms and systems. Later chapters will add further support to this assertion. One of the reasons that Fourier analysis is of such wide-ranging importance in digital signal processing is because of the existence of efficient algorithms for computing the discrete Fourier transform[1].

From Chapter 3 we recall that the discrete Fourier transform (DFT) is

$$X(k) = \sum_{n=0}^{N-1} x(n) W_N^{kn}, \qquad k = 0, 1, \ldots, N-1 \tag{6.1}$$

where $W_N = e^{-j(2\pi/N)}$. The inverse discrete Fourier transform (IDFT) is

$$x(n) = \frac{1}{N} \sum_{k=0}^{N-1} X(k) W_N^{-kn}, \qquad n = 0, 1, \ldots, N-1 \tag{6.2}$$

In Eqs. (6.1) and (6.2), both $x(n)$ and $X(k)$ may be complex. The expressions of Eqs. (6.1) and (6.2) differ only in the sign of the exponent of W_N and in a scale factor $1/N$. Thus a discussion of computation procedures for Eq. (6.1) applies with straightforward modifications to Eq. (6.2).

To indicate the importance of efficient computation schemes, it is instructive to consider the direct evaluation of the DFT equations. Since $x(n)$

may be complex we can write

$$X(k) = \sum_{n=0}^{N-1} \{(\mathrm{Re}\,[x(n)]\,\mathrm{Re}\,[W_N^{kn}] - \mathrm{Im}\,[x(n)]\,\mathrm{Im}\,[W_N^{kn}]) + j(\mathrm{Re}\,[x(n)]\,\mathrm{Im}\,[W_N^{kn}] + \mathrm{Im}\,[x(n)]\,\mathrm{Re}\,[W_N^{kn}])\},$$

$$k = 0, 1, \ldots, N-1 \quad (6.3)$$

From Eq. (6.3) it is clear that for each value of k, the direct computation of $X(k)$ requires $4N$ real multiplications and $(4N - 2)$ real additions.† Since $X(k)$ must be computed for N different values of k, the direct computation of the discrete Fourier transform of a sequence $x(n)$ requires $4N^2$ real multiplications and $N(4N - 2)$ real additions or, alternatively, N^2 complex multiplications and $N(N - 1)$ complex additions. In addition to the multiplications and additions called for by Eq. (6.3), the implementation of the computation of the DFT on a general-purpose digital computer or with special-purpose hardware of course requires provision for storing and accessing the input sequence values $x(n)$ and values of the coefficients W_N^{kn}. Since the amount of accessing and storing of data in numerical computation algorithms is generally proportional to the number of arithmetic operations, it is generally accepted that a meaningful measure of complexity, or, of the time required to implement a computational algorithm, is the number of multiplications and additions required. Thus, for the direct computation of the discrete Fourier transform, a convenient measure of the efficiency of the computation is the fact that $4N^2$ real multiplications and $N(4N - 2)$ real additions are required. Since the amount of computation, and thus the computation time, is approximately proportional to N^2, it is evident that the number of arithmetic operations required to compute the DFT by the direct method becomes very large for large values of N. For this reason, computational procedures that reduce the number of multiplications and additions are of considerable interest.

Most approaches to improving the efficiency of the computation of the DFT exploit one or both of the following special properties of the quantities W_N^{kn}:

1. $W_N^{k(N-n)} = (W_N^{kn})^*$.
2. $W_N^{kn} = W_N^{k(n+N)} = W_N^{(k+N)n}$

For example, using the first property, i.e., the symmetry of the cosine and sine functions, we can group terms in Eq. (6.3) as

$$\mathrm{Re}\,[x(n)]\,\mathrm{Re}\,[W_N^{kn}] + \mathrm{Re}\,[x(N-n)]\,\mathrm{Re}\,[W_N^{k(N-n)}] = (\mathrm{Re}[x(n)] + \mathrm{Re}\,[x(N-n)])\,\mathrm{Re}\,[W_N^{kn}]$$

† Throughout the discussion the figure for the number of computations is only approximate. Multiplication by W_N^0, for example, does not in fact require a multiplication. Nevertheless, the general dependency of computational complexity on N obtained by including such multiplications is sufficiently accurate to permit comparisons between different classes of algorithms.

and

$$-\mathrm{Im}\,[x(n)]\,\mathrm{Im}\,[W_N^{kn}] - \mathrm{Im}\,[x(N-n)]\,\mathrm{Im}\,[W_N^{k(N-n)}]$$
$$= -(\mathrm{Im}\,[x(n)] - \mathrm{Im}\,[x(N-n)])\,\mathrm{Im}\,[W_N^{kn}]$$

Similar groupings can be found for the other terms in Eq. (6.3). By this method, the number of multiplications can be reduced by approximately a factor of 2. Also we can take advantage of the fact that for certain values of the product kn, the sine and cosine functions take on the values 1 or 0, thereby eliminating the need for multiplications. However, reductions of this type still leave us with an amount of computation that is approximately proportional to N^2. Fortunately, the second property, i.e., the periodicity of the complex sequence W_N^{kn}, can be employed in achieving significantly greater reductions of the computation.

Computational algorithms that exploit both the symmetry and the periodicity of the sequence W_N^{kn} were known long before the era of high-speed digital computation. At that time, any scheme that reduced hand computation by even a factor of 2 was welcomed. Runge [2] and later Danielson and Lanczos [3] described algorithms for which computation was roughly proportional to $N \log N$ rather than N^2. However, the distinction was not of great importance for the small values of N that were feasible for hand computation.† The possibility of greatly reduced computation was generally overlooked until about 1965, when Cooley and Tukey [1] published an algorithm for the computation of the discrete Fourier transform that is applicable when N is a composite number; i.e., N is the product of two or more integers. The publication of this paper touched off a flurry of activity in the application of the discrete Fourier transform to signal processing and resulted in the discovery of a number of computational algorithms which have come to be known as *fast Fourier transform*, or simply FFT, *algorithms*. Collectively, the entire set of such algorithms are often loosely referred to as "the FFT" [5].

The fundamental principle that all these algorithms are based upon is that of decomposing the computation of the discrete Fourier transform of a sequence of length N into successively smaller discrete Fourier transforms. The manner in which this principle is implemented leads to a variety of different algorithms, all with comparable improvements in computational speed. In this chapter we shall be concerned with two basic classes of FFT algorithms. The first, called *decimation-in-time*, derives its name from the fact that in the process of arranging the computation into smaller transformations, the sequence $x(n)$ (the index n is often associated with time) is decomposed into successively smaller subsequences. In the second general class of algorithms, the sequence of discrete Fourier transform coefficients

† An interesting article by Cooley, Lewis, and Welch [4] traces the history of efforts directed toward improving the efficiency of DFT computations.

$X(k)$ is decomposed into smaller subsequences, hence the name *decimation-in-frequency*.

In this chapter we shall consider a number of algorithms for computing the discrete Fourier transform. The algorithms to be discussed vary in efficiency, but all of them are more efficient than direct evaluation of Eq. (6.3). We shall begin with a discussion of the Goertzel method [6,7], which requires computation proportional to N^2 but with a smaller constant of proportionality than the direct method. Most of our effort will be devoted to a discussion of FFT algorithms, i.e., algorithms for which computation is roughly proportional to $N \log N$. In this discussion we shall not attempt to be exhaustive in our coverage of algorithms, but shall illustrate the general principles common to all algorithms of this type by considering in detail only a few of the more commonly used schemes.

6.1 Goertzel Algorithm

A computation procedure that is somewhat more efficient than the direct method is called the *Goertzel algorithm* [6]. This method is an example of how the periodicity of the sequence W_N^{kn} can be used to reduce computation. Specifically, we shall see that the discrete Fourier transform can be viewed as the response of a digital filter which can be structured in a manner that reduces the number of arithmetic operations.

To derive the Goertzel algorithm, we begin by noting that

$$W_N^{-kN} = e^{j(2\pi/N)Nk} = e^{j2\pi k} = 1 \tag{6.4}$$

This is, of course, a direct result of the periodicity of W_N^{-kn}. Because of Eq. (6.4) we may multiply the right side of Eq. (6.1) by W_N^{-kN} without affecting the equation. Thus

$$\begin{aligned} X(k) &= W_N^{-kN} \sum_{r=0}^{N-1} x(r) W_N^{kr} \\ &= \sum_{r=0}^{N-1} x(r) W_N^{-k(N-r)} \end{aligned} \tag{6.5}$$

For convenience, let us define the sequence

$$y_k(n) = \sum_{r=0}^{N-1} x(r) W_N^{-k(n-r)} \tag{6.6}$$

From Eqs. (6.5) and (6.6) it follows that

$$X(k) = y_k(n)|_{n=N}$$

Equation (6.6) is evidently a discrete convolution of the finite-duration sequence $x(n)$, $0 \leq n \leq N - 1$, with the sequence W_N^{-kn}. Consequently,

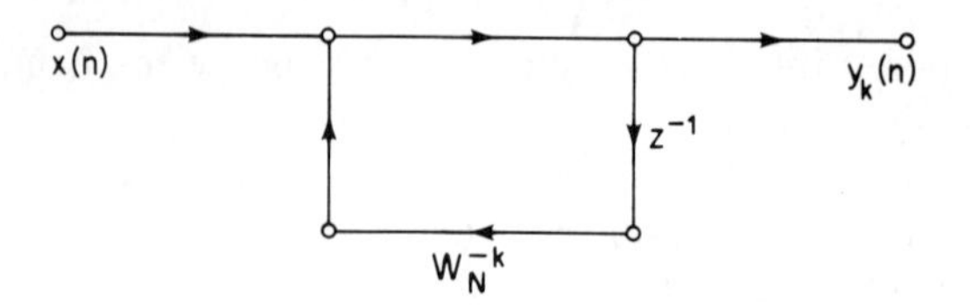

Fig. 6.1 Flow graph of first-order complex recursive computation of $X(k)$.

$y_k(n)$ can be viewed as the response of a system, with unit-sample response W_N^{-kn}, to an input $x(n)$. In particular, $X(k)$ is the value of the output when $n = N$. A system with unit-sample response W_N^{-kn} is depicted in Fig. 6.1.

Since both the input $x(n)$ and W_N^{-k} are complex, the computation of each new value of $y_k(n)$ requires four real multiplications and four real additions. Since all the intervening values $y_k(1), y_k(2), \ldots, y_k(N-1)$ must be computed in order to compute $y_k(N) = X(k)$, the use of the scheme depicted by Fig. 6.1 requires $4N$ real multiplications and $4N$ additions to compute $X(k)$ for a particular value of k. Thus this scheme is slightly less efficient than the direct method. However, we note that the method of Fig. 6.1 does not require the computation of, nor the storage of, the coefficients W_N^{kn} since these quantities are effectively computed by the recursion process implied by Fig. 6.1.

It is possible to retain this simplification while reducing the number of multiplications by a factor of 2. To see how this may be done, note that the system function of the system of Fig. 6.1 is

$$H_k(z) = \frac{1}{1 - W_N^{-k}z^{-1}} \tag{6.7}$$

Multiplying both the numerator and denominator of $H_k(z)$ by the factor $(1 - W_N^k z^{-1})$, we obtain

$$\begin{aligned} H_k(z) &= \frac{1 - W_N^k z^{-1}}{(1 - W_N^{-k}z^{-1})(1 - W_N^k z^{-1})} \\ &= \frac{1 - W_N^k z^{-1}}{1 - 2\cos((2\pi/N)k)z^{-1} + z^{-2}} \end{aligned} \tag{6.8}$$

The signal flow graph of Fig. 6.2 corresponds to the system function of Eq. (6.8).

Only two multiplications are required to implement the poles of this system, since the coefficients are real and (-1) need not be counted as a multiplication. As before, four additions are required to implement the poles. Since we only need to bring the system to a state where $y_k(N)$ can be computed, the complex multiplication by $-W_N^k$ required to implement the zero need not be performed at every iteration of the difference equation, but

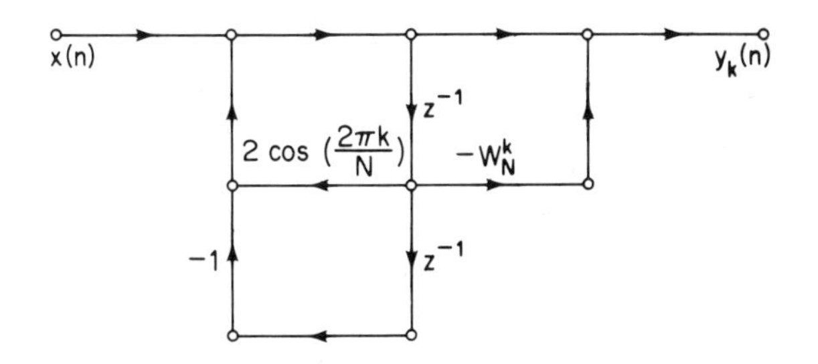

Fig. 6.2 Flow graph of second-order recursive computation of $X(k)$ (Goertzel algorithm).

only after the Nth iteration. Thus the total computation is $2N$ real multiplications and $4N$ real additions for the poles plus four real multiplications and four real additions for the zero. The total computation is therefore $2(N + 2)$ real multiplications and $4(N + 1)$ real additions, about half the number of real multiplications required with the direct method. In this more efficient scheme, we still have the advantage that $\cos((2\pi/N)k)$ and W_N^k are the only coefficients that must be computed and stored, the set of coefficients W_N^{kn} being again computed implicitly in the iteration of the recursion formula implied by Fig. 6.2.

As an additional advantage of the use of this network, let us consider the computation of the z-transform of $x(n)$ at conjugate locations on the unit circle, that is, the computation of $X(k)$ and $X(N - k)$. It is straightforward to verify that the network in the form of Fig. 6.2 required to compute $X(N - k)$ has exactly the same poles as that in Fig. 6.2, but the coefficient for the zero is the complex conjugate of that in Fig. 6.2 (see Problem 6.1). Since the zero is only implemented on the last iteration, the $2N$ multiplications and $4N$ additions required for the poles can be used for the computation of two DFT values. Thus, for the computation of all N points of the discrete Fourier transform using the Goertzel algorithm, the number of multiplications required is approximately N^2 and the number of additions is approximately $2N^2$. However, as with the direct computation of the discrete Fourier transform, the amount of computation is still proportional to N^2.

In the direct method or the Goertzel method we do not need to evaluate all N different values of $X(k)$. Indeed, we can in general evaluate $X(k)$ for *any* M values of k. In this case the total computation is proportional to NM. These schemes are attractive when M is small; however, more sophisticated algorithms are available for which the computation is proportional to $N \log_2 N$ when N is a power of 2. Thus, when M is less than $\log_2 N$, either the Goertzel or the direct method may in fact be the most efficient method, but when all N values of $X(k)$ are required, the algorithms to be considered next are roughly $(N/\log_2 N)$ times more efficient than the direct method or the Goertzel method.

6.2 Decimation-in-Time FFT Algorithms

To achieve the dramatic increase in efficiency to which we have alluded, it is necessary to decompose the DFT computation into successively smaller DFT computations. In this process we exploit both the symmetry and the periodicity of the complex exponential $W_N^{kn} = e^{-j(2\pi/N)kn}$. Algorithms in which the decomposition is based on decomposing the sequence $x(n)$, into successively smaller subsequences, are called *decimation-in-time algorithms.* The principle of decimation-in-time is most conveniently illustrated by considering the special case of N an integer power of 2; i.e.,

$$N = 2^\nu$$

Since N is an even integer, we can consider computing $X(k)$ by separating $x(n)$ into two $N/2$-point† sequences consisting of the even-numbered points in $x(n)$ and the odd-numbered points in $x(n)$. With $X(k)$ given by

$$X(k) = \sum_{n=0}^{N-1} x(n)W_N^{nk}, \qquad k = 0, 1, \ldots, N-1 \tag{6.9}$$

and separating $x(n)$ into its even- and odd-numbered points we obtain

$$X(k) = \sum_{n \text{ even}} x(n)W_N^{nk} + \sum_{n \text{ odd}} x(n)W_N^{nk}$$

or with the substitution of variables $n = 2r$ for n even and $n = 2r + 1$ for n odd,

$$\begin{aligned} X(k) &= \sum_{r=0}^{(N/2)-1} x(2r)W_N^{2rk} + \sum_{r=0}^{(N/2)-1} x(2r+1)W_N^{(2r+1)k} \\ &= \sum_{r=0}^{(N/2)-1} x(2r)(W_N^2)^{rk} + W_N^k \sum_{r=0}^{(N/2)-1} x(2r+1)(W_N^2)^{rk} \end{aligned} \tag{6.10}$$

But $W_N^2 = W_{N/2}$ since

$$W_N^2 = e^{-2j(2\pi/N)} = e^{-j2\pi/(N/2)} = W_{N/2}$$

Consequently Eq. (6.10) can be written as

$$\begin{aligned} X(k) &= \sum_{r=0}^{(N/2)-1} x(2r)W_{N/2}^{rk} + W_N^k \sum_{r=0}^{(N/2)-1} x(2r+1)W_{N/2}^{rk} \\ &= G(k) + W_N^k H(k) \end{aligned} \tag{6.11}$$

Each of the sums in Eq. (6.11) is recognized as an $N/2$-point DFT, the first sum being the $N/2$-point DFT of the even-numbered points of the original

† When discussing FFT algorithms, we shall use the words *sample* and *point* interchangeably to mean sequence value. Also, we shall refer to a sequence of length N as an N-point sequence, and the DFT of a sequence of length N will be called an N-point DFT.

sequence and the second being the $N/2$-point DFT of the odd-numbered points of the original sequence. Although the index k ranges over N values, $k = 0, 1, \ldots, N-1$, each of the sums need only be computed for k between 0 and $N/2 - 1$, since $G(k)$ and $H(k)$ are each periodic in k with period $N/2$. After the two DFTs corresponding to the two sums in Eq. (6.11) are computed, they are then combined to yield the N-point DFT, $X(k)$. Figure 6.3 indicates the computation involved in computing $X(k)$

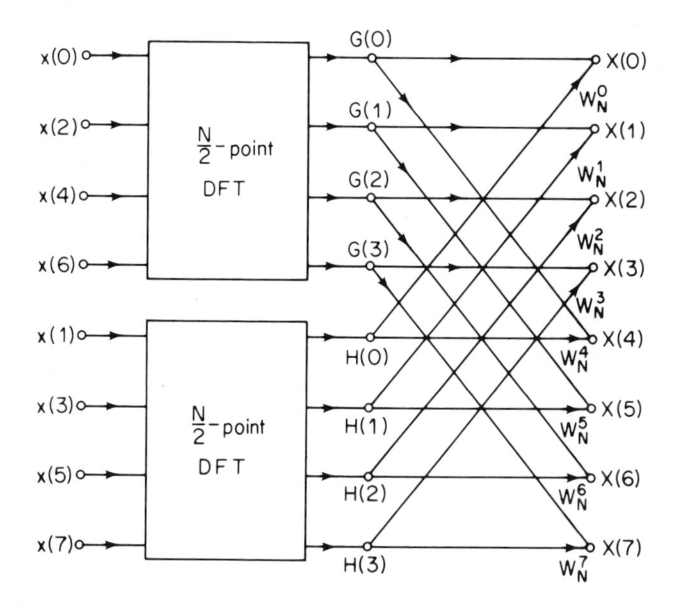

Fig. 6.3 Flow graph of the decimation-in-time decomposition of an N-point DFT computation into two $N/2$-point DFT computations ($N = 8$).

according to Eq. (6.11) for an eight-point sequence, i.e., for $N = 8$. In this figure we have used the signal flow graph conventions that were introduced in Chapter 4 for representing difference equations [5,7]. That is, branches entering a node are summed to produce the node variable. When no coefficient is indicated, the branch transmittance is assumed to be one. For other branches, the transmittance of a branch is an integer power of W_N. Thus we note in Fig. 6.3 that two four-point DFTs are computed, with $G(k)$ designating the four-point DFT of the even-numbered points and $H(k)$ designating the four-point DFT of the odd-numbered points. $X(0)$ is then obtained by multiplying $H(0)$ by W_N^0 and adding the product to $G(0)$. $X(1)$ is obtained by multiplying $H(1)$ by W_N^1 and adding that result to $G(1)$. For $X(4)$ we would want to multiply $H(4)$ by W_N^4 and add the result to $G(4)$. However, since $G(k)$ and $H(k)$ are both periodic in k with period 4, $H(4) = H(0)$ and $G(4) = G(0)$. Thus $X(4)$ is obtained by multiplying $H(0)$ by W_N^4 and adding the result to $G(0)$.

With the computation restructured according to Eq. (6.11), we can compare the number of multiplications and additions required with those required for a direct computation of the DFT. Previously we saw that for direct computation without exploiting symmetry, N^2 complex multiplications and additions were required.† By comparison, Eq. (6.11) requires the computation of two $N/2$-point DFTs, which in turn requires $2(N/2)^2$ complex multiplications and approximately $2(N/2)^2$ complex additions. Then the two $N/2$-point DFTs must be combined, requiring N complex multiplications, corresponding to multiplying the second sum by W_N^k and then N complex additions, corresponding to adding that product to the first sum. Consequently, the computation of Eq. (6.11) for all values of k requires $N + 2(N/2)^2$ or $N + (N^2/2)$ complex multiplications and complex additions. It is easy to verify that for $N > 2$, $N + N^2/2$ will be less than N^2.

Equation (6.11) corresponds to breaking the original N-point computation into two $N/2$-point computations. If $N/2$ is even, as it always is when N is equal to a power of 2, then we can consider computing each of the $N/2$-point DFTs in Eq. (6.11) by breaking each of the sums in Eq. (6.11) into two $N/4$-point DFTs, which would then be combined to yield the $N/2$-point DFTs. Thus $G(k)$ and $H(k)$ in Eq. (6.11) would be computed as indicated below:

$$G(k) = \sum_{r=0}^{(N/2)-1} g(r)W_{N/2}^{rk} = \sum_{l=0}^{(N/4)-1} g(2l)W_{N/2}^{2lk} + \sum_{l=0}^{(N/4)-1} g(2l+1)W_{N/2}^{(2l+1)k}$$

or

$$G(k) = \sum_{l=0}^{(N/4)-1} g(2l)W_{N/4}^{lk} + W_{N/2}^{k} \sum_{l=0}^{(N/4)-1} g(2l+1)W_{N/4}^{lk} \tag{6.12}$$

Similarly,

$$H(k) = \sum_{l=0}^{(N/4)-1} h(2l)W_{N/4}^{lk} + W_{N/2}^{k} \sum_{l=0}^{(N/4-1)} h(2l+1)W_{N/4}^{lk} \tag{6.13}$$

Thus if the four-point DFTs in Fig. 6.3 are computed according to Eqs. (6.12) and (6.13), then that computation would be carried out as indicated in Fig. 6.4. Inserting the computation indicated in Fig. 6.4 into the flow graph of Fig. 6.3, we obtain the complete flow graph of Fig. 6.5. Note that we have used the fact that $W_{N/2} = W_N^2$.

For the eight-point DFT that we have been using as an illustration, the computation has been reduced to a computation of two-point DFTs. The two-point DFT of, for example, $x(0)$ and $x(4)$, is depicted in Fig. 6.6. With the computation of Fig. 6.6 inserted in the flow graph of Fig. 6.5, we obtain the complete flow graph for computation of the eight-point DFT, as shown in Fig. 6.7.

† For simplicity we shall henceforth assume that N is large, so that $N - 1$ can be approximated by N.

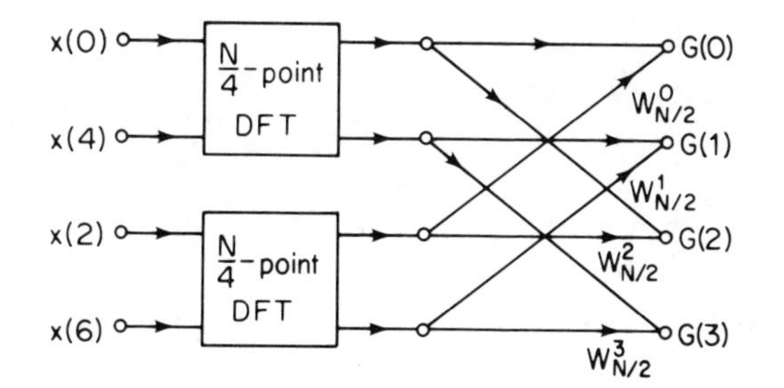

Fig. 6.4 Flow graph of the decimation-in-time decomposition of an $N/2$-point DFT computation into two $N/4$-point DFT computations ($N = 8$).

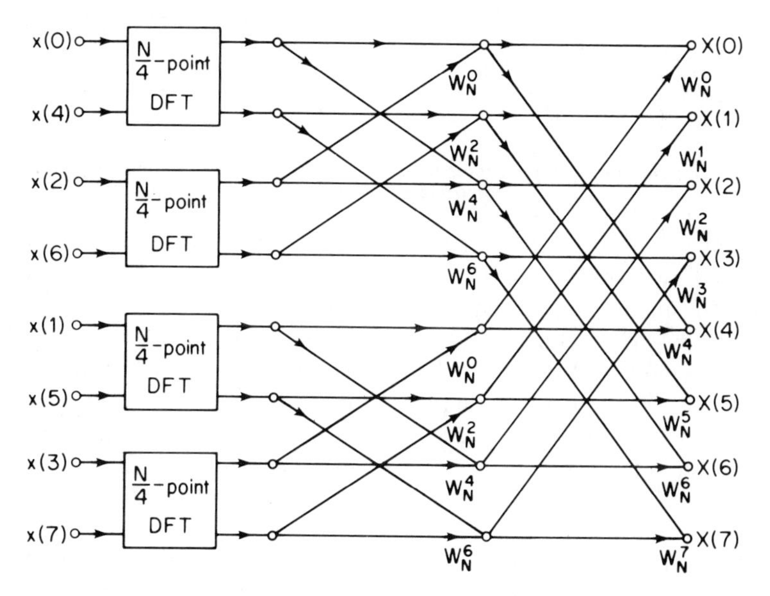

Fig. 6.5 Result of substituting Fig. 6.4 into Fig. 6.3.

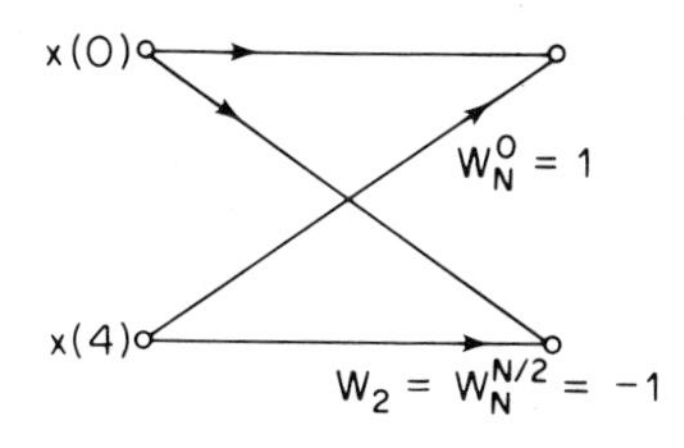

Fig. 6.6 Flow graph of a two-point DFT.

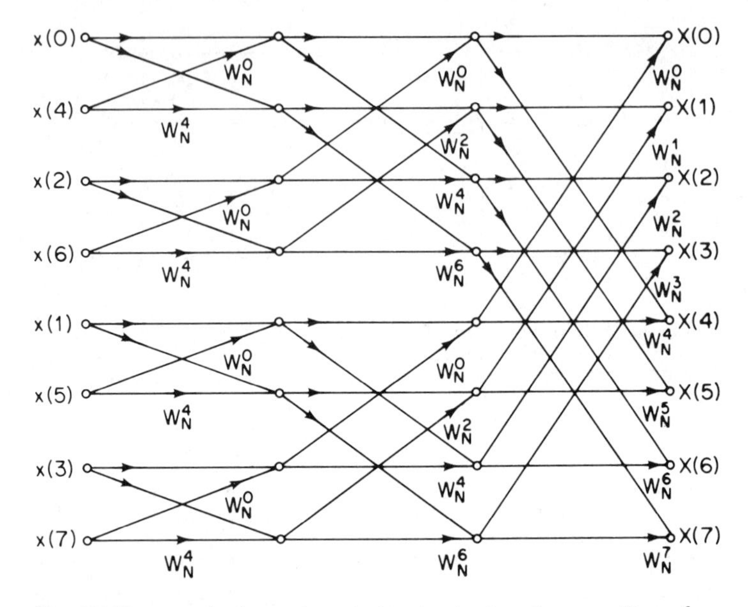

Fig. 6.7 Flow graph of complete decimation-in-time decomposition of an eight-point DFT computation.

For the more general case with N a power of 2 greater than 3, we would proceed by decomposing the $N/4$-point transforms in Eqs. (6.12) and (6.13) into $N/8$-point transforms, and continue until left with only two-point transforms. This requires ν stages of computation, where $\nu = \log_2 N$. Previously we found that in the original decomposition of an N-point transform into two $N/2$-point transforms, the number of complex multiplications and additions required was $N + 2(N/2)^2$. When the $N/2$-point transforms are decomposed into $N/4$-point transforms, then the factor of $(N/2)^2$ is replaced by $N/2 + 2(N/4)^2$, so the overall computation then requires $N + N + 4(N/4)^2$ complex multiplications and additions. If $N = 2^\nu$, this can be done at most $\nu = \log_2 N$ times, so that after carrying out this decomposition as many times as possible the number of complex multiplications and additions is equal to $N \log_2 N$.

The flow graph of Fig. 6.7 displays the operations explicitly. By counting branches with transmittances of the form W_N^r, we note that each stage has N complex multiplications and N complex additions. Since there are $\log_2 N$ stages, we have, as before, a total of $N \log_2 N$ complex multiplications and additions. This is the substantial computational savings that we have previously indicated was possible. We shall see that the symmetry and periodicity of W_N^r can be exploited to obtain further reductions in computation.

6.2.1 *In-Place Computations*

The computational flow graph of Fig. 6.7 describes an algorithm for the computation of the discrete Fourier transform. What is important in the flow graph of Fig. 6.7 are the branches connecting the nodes and the transmittance of each of these branches. No matter how the nodes in this flow-graph are rearranged, it will always represent the same computation provided that the connections between the nodes and the transmittances of the connections are maintained. The particular form for the flow graph in Fig. 6.7 arose out of deriving the algorithm by separating the original sequence into the even-numbered and odd-numbered points, and then continuing to create smaller and smaller subsequences in the same way. An interesting by-product of this derivation is that the flow graph in Fig. 6.7, in addition to describing an efficient procedure for computing the discrete Fourier transform, also suggests a useful way of storing the original data and storing the results of the computation in the intermediate arrays.

To see that this is so, it is useful to note that according to Fig. 6.7, each stage of the computation takes a set of N complex numbers and transforms them into another set of N complex numbers. This process is repeated $\nu = \log_2 N$ times, resulting in the computation of the desired discrete Fourier transform. When implementing the computations depicted in Fig. 6.7 we can imagine the use of two arrays of (complex) storage registers, one for the array being computed and one for the data being used in the computation. For example, in computing the first array in Fig. 6.7, one set of storage registers would contain the input data and the second set of storage registers would contain the computed results for the first stage. While the validity of Fig. 6.7 is not tied to the order in which the input data are stored, let us order the set of complex numbers in the same order that they appear in Fig. 6.7 (from top to bottom). We shall denote the sequence of complex numbers resulting from the mth stage of computation as $X_m(l)$, where $l = 0, 1, \ldots, N - 1$, and $m = 1, 2, \ldots, \nu$. Furthermore, for convenience, let us define the set of input samples as $X_0(l)$. We can think of $X_m(l)$ as the input array and $X_{m+1}(l)$ as the output array for the $(m + 1)$st stage of computations: Thus for the case of $N = 8$ as in Fig. 6.7,

$$
\begin{aligned}
X_0(0) &= x(0) \\
X_0(1) &= x(4) \\
X_0(2) &= x(2) \\
X_0(3) &= x(6) \\
X_0(4) &= x(1) \\
X_0(5) &= x(5) \\
X_0(6) &= x(3) \\
X_0(7) &= x(7)
\end{aligned}
\tag{6.14}
$$

Using this notation and ordering, it can be seen that the basic computation in the flow graph of Fig. 6.7 is as shown in Fig. 6.8. The equations

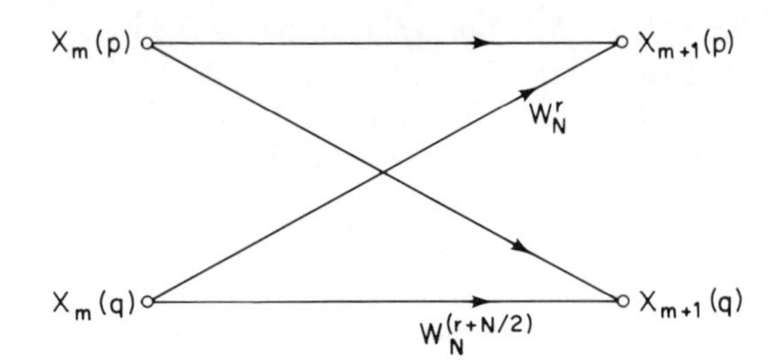

Fig. 6.8 Flow graph of basic butterfly computation in Fig. 6.7.

represented by this flow graph are of the form

$$\begin{aligned} X_{m+1}(p) &= X_m(p) + W_N^r X_m(q) \\ X_{m+1}(q) &= X_m(p) + W_N^{r+N/2} X_m(q) \end{aligned} \tag{6.15}$$

Because of the appearance of the flow graph of Fig. 6.8, this computation is referred to as a *butterfly* computation.

Equations (6.15) suggest a means of reducing the number of complex multiplications by a factor of 2. To see this we note that

$$W_N^{N/2} = e^{-j(2\pi/N)\cdot N/2} = e^{-j\pi} = -1$$

so that equations (6.15) become

$$\begin{aligned} X_{m+1}(p) &= X_m(p) + W_N^r X_m(q) \\ X_{m+1}(q) &= X_m(p) - W_N^r X_m(q) \end{aligned} \tag{6.16}$$

Equations (6.16) are depicted by the flow graph of Fig. 6.9. Therefore, since there are $N/2$ "butterflies" of the form of Fig. 6.9 per stage and $\log_2 N$ stages, the total number of multiplications required is $(N/2) \log_2 N$, instead of $N \log_2 N$ as it seems from Fig. 6.7. Using the basic flow graph of Fig. 6.9 as a replacement for butterflies of the form of Fig. 6.8, we obtain from Fig.

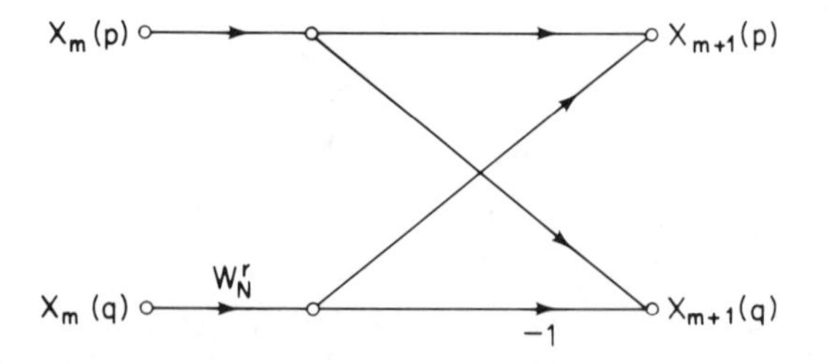

Fig. 6.9 Flow graph of simplified butterfly computation requiring only one complex multiplication.

6.7 the flow graph of Fig. 6.10. The total number of complex multiplications is evident by inspection of Fig. 6.10.

In Eq. (6.16), p, q, and r vary from stage to stage in a manner that is readily inferred from Fig. 6.10 and from Eqs. (6.10), (6.12), (6.13), etc. It is clear from Figs. 6.9 and 6.10 that the complex numbers in locations p and q of the mth array are required to compute the elements p and q of the $(m + 1)$st array. Thus only one complex array of N storage registers is

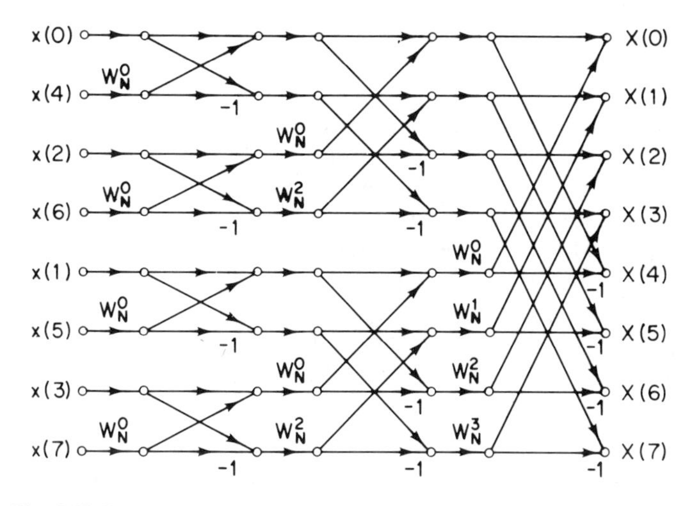

Fig. 6.10 Flow graph of eight-point DFT using the butterfly computation of Fig. 6.9.

physically necessary to implement the complete computation if $X_{m+1}(p)$ and $X_{m+1}(q)$ are stored in the same storage registers as $X_m(p)$ and $X_m(q)$, respectively. This kind of computation is commonly referred to as an "in-place" computation, since it has the advantage that as a new array is computed the results can be stored in the same storage locations as the original array. The fact that the flow graphs of Fig. 6.7 or 6.10 represent an in-place computation is tied to the fact that we have associated nodes in the flow graph that are on the same horizontal line with the same storage location and that the computation between two arrays consists of a butterfly computation in which the input nodes and the output nodes are horizontally adjacent.

In order that the computation may be done in place as discussed above, we note that with the flow graph of Fig. 6.7 (or 6.10) the input data must be stored in a nonsequential order. In fact, the order in which the input data are stored is in *bit-reversed* order. To see what is meant by this terminology, we note that for the eight-point flow graph that we have been discussing, three binary digits are required to index through the data. If we write the indices in Eq. (6.14) in binary form, we obtain the set of equations

$$\begin{aligned}
X_0(000) &= x(000)\\
X_0(001) &= x(100)\\
X_0(010) &= x(010)\\
X_0(011) &= x(110)\\
X_0(100) &= x(001)\\
X_0(101) &= x(101)\\
X_0(110) &= x(011)\\
X_0(111) &= x(111)
\end{aligned} \tag{6.17}$$

If $(n_2n_1n_0)$ is the binary representation of the index of the sequence $x(n)$, then the sequence value $x(n_2n_1n_0)$ is stored in the array position $X_0(n_0n_1n_2)$. That is, in determining the position of $x(n_2n_1n_0)$ in the input array, we must reverse the order of the bits of the index n.

To see why bit-reversed order is necessary for in-place computation, let us recall the process that resulted in Fig. 6.7. The sequence $x(n)$ was first divided into the even-numbered samples and the odd-numbered samples with the even-numbered samples occurring in the top half of Fig. 6.3 and the odd-numbered samples occurring in the bottom half. Formally, such a separation of the data can be carried out by examining the least significant bit (n_0) in the index n. If the least significant bit is zero, the sequence value corresponds to an even-numbered sample and therefore will appear in the top half, and if the least significant bit is 1, the sequence value corresponds to an odd-numbered sample and consequently will appear in the bottom half of the

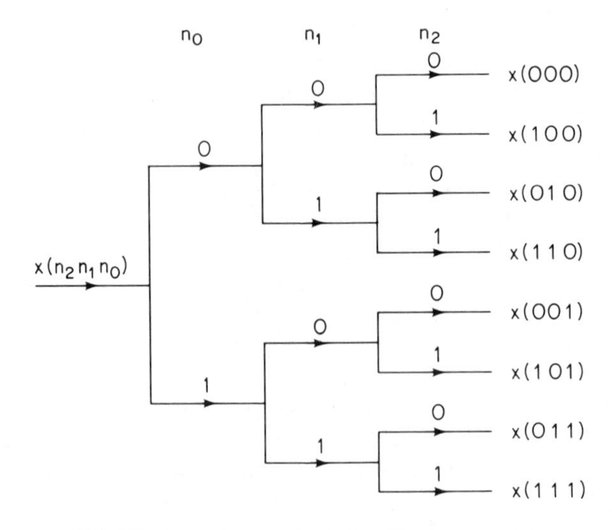

Fig. 6.11 Tree diagram depicting bit-reversed sorting.

array $X_0(l)$. Next the even and odd subsequences are each sorted into their even and odd parts, and this can be done by examining the second least significant bit in the data index. Considering first the even-numbered subsequence, if the second least significant bit is zero, the sequence value is an even-numbered term in this subsequence and if the second least significant bit is 1, then the sequence value has an odd-numbered index in this subsequence. The same process is carried out for the subsequence formed from the original odd indexed sequence values. This process is repeated until N subsequences of length 1 are obtained. This sorting into even and odd indexed subsequences is depicted by the tree diagram of Fig. 6.11.

Thus the necessity for bit-reversed ordering of the sequence $x(n)$ is seen to result from the manner in which the DFT computation is decomposed into successively smaller DFT computations.

6.2.2 *Alternative Forms*

Although it is reasonable to store the results of each stage of the computation in the order in which the nodes appear in Fig. 6.10, it is certainly not necessary to do so. No matter how the nodes of Fig. 6.10 are rearranged, the result will always be a valid computation of the discrete Fourier transform of $x(n)$ as long as the branch transmittances are unchanged. Only the order in which data are accessed and stored will change. If we associate the nodes with complex storage locations and associate the ordering of the nodes with indexing of an array of complex storage locations, it is clear from our previous discussion that a flow graph corresponds to an "in-place" computation only if the rearrangement of nodes is such that the input and output nodes for each butterfly computation are horizontally adjacent. Otherwise two complex storage arrays will be required. Figure 6.10 is, of course, such an arrangement. Another is depicted in Fig. 6.12. In this case the input sequence is in normal order and the transform is in bit-reversed order. Figure 6.12 can be obtained from Fig. 6.10 by interchanging all the nodes that are horizontally adjacent to $x(4)$ in Fig. 6.10 with all the nodes horizontally adjacent to $x(1)$. Similarly, all the nodes that are horizontally adjacent to $x(6)$ in Fig. 6.10 are exchanged with those that are horizontally adjacent to $x(3)$. The nodes horizontally adjacent to $x(2)$, $x(5)$, and $x(7)$ are not disturbed. The resulting flow graph in Fig. 6.12 corresponds to the form of the decimation-in-time algorithm originally given by Cooley and Tukey [1].

It can be seen that the only difference between Figs. 6.10 and 6.12 is in the ordering of the nodes. The branch transmittances (powers of W_N) remain the same. There are, of course, a large variety of possible orderings; however, most do not make much sense from a computational viewpoint. As another useful possibility, suppose that the nodes are ordered such that the input and output both appear in normal order. A flow graph of this type is shown in Fig. 6.13. In this case, however, the computation cannot be

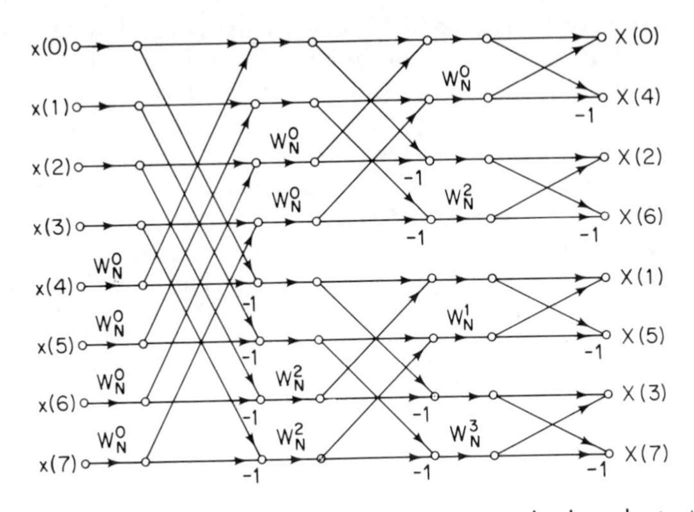

Fig. 6.12 Rearrangement of Fig. 6.10 with input in normal order and output in bit-reversed order

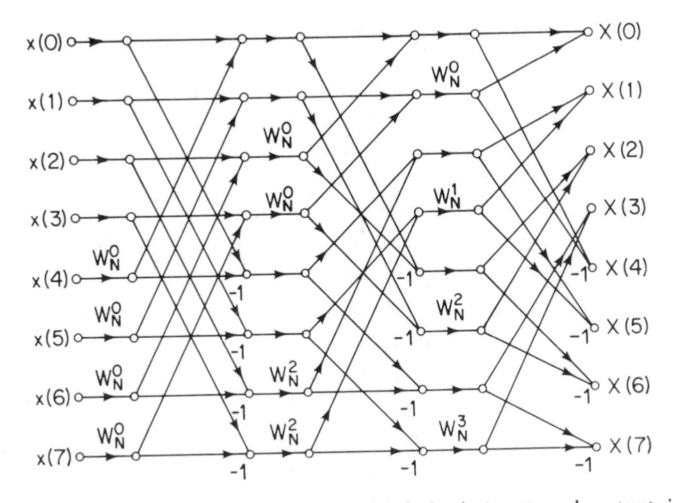

Fig. 6.13 Rearrangement of Fig. 6.10 with both input and output in normal order.

carried out in place. Thus, two complex arrays of length N would be required to perform the computation depicted in Fig. 6.13.

In realizing the computations depicted by Figs. 6.10, 6.12, and 6.13, it is clearly necessary to access elements of intermediate arrays in nonsequential order. Thus, for greater computational speed, the complex numbers must be stored in random access memory. For example, referring to Fig. 6.10, to compute the first array from the input array, the inputs to each butterfly computation are adjacent node variables which are thought of as being stored in adjacent storage locations. In computing the second intermediate array from the first, the inputs to a butterfly are separated by two storage locations, and in computing the third array from the second, the inputs to a butterfly computation are separated by four storage locations. If N is larger than 8, the separation between butterfly inputs is 8 for the fourth stage, 16 for the fifth stage, etc. The separation in the last (νth) stage is $N/2$.

In Fig. 6.12 the situation is similar in that computing the first array from the input data we use data separated by 4, to compute the second array from the first array we use input data separated by 2, and then finally in computing the last array we use adjacent data. Although it is straightforward

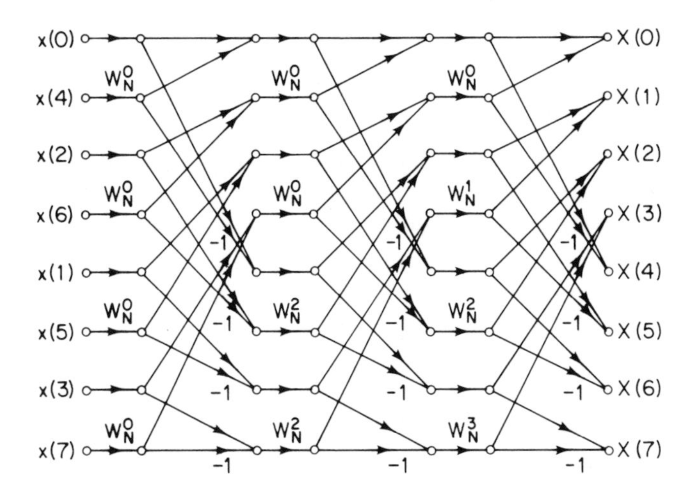

Fig. 6.14 Rearrangement of Fig. 6.10 having the same geometry for each stage, thereby permitting sequential data accessing and storage.

to imagine simple algorithms for modifying index registers to access the data in either the flow graph of Fig. 6.10 or 6.12, the data are not accessed sequentially so that random-access memory would be very desirable. In the flow graph of Fig. 6.13 the data are accessed nonsequentially, the computation is not in place, and a scheme for indexing the data is considerably more complicated than in either of the two previous cases.

A rearrangement of the flow graph in Fig. 6.10 that is particularly useful when random-access memory is not available is shown in Fig. 6.14. This flow graph represents the decimation-in-time algorithm originally given by Singleton [8]. Note first that in this flow graph the input is again in bit-reversed order and the output in normal order. The important feature of this flow graph is that the geometry is identical for each stage; only the branch transmittances change from stage to stage. This makes it possible to access data sequentially. Suppose that we have four magnetic-tape units (or four sequential areas of disc storage) and suppose that the first half of the input data (in bit-reversed order) is stored on one tape and the second half is stored on another tape. Then data can be accessed sequentially on tapes 1 and 2 and the results written sequentially on tapes 3 and 4 with the first half of the new array being written on tape 3 and the second half on tape 4. Then at the next stage of computation tapes 3 and 4 are the input and the output is written on tapes 1 and 2. This is repeated for each of ν stages.

6.3 Decimation-in-Frequency FFT Algorithms

The decimation-in-time FFT algorithms were all based upon the decomposition of the DFT computation by forming smaller and smaller subsequences of the input sequence, $x(n)$. Alternatively we can consider dividing the output sequence, $X(k)$, into smaller and smaller subsequences in the same manner. The class of FFT algorithms based on this procedure is commonly referred to as decimation-in-frequency. To derive the decimation-in-frequency forms of the FFT algorithm for N a power of 2, we can first divide the input sequence into the first half and the last half of the points so that

$$X(k) = \sum_{n=0}^{(N/2)-1} x(n)W_N^{nk} + \sum_{n=N/2}^{N-1} x(n)W_N^{nk}$$

or

$$X(k) = \sum_{n=0}^{(N/2)-1} x(n)W_N^{nk} + W_N^{(N/2)k} \sum_{n=0}^{(N/2)-1} x\left(n + \frac{N}{2}\right)W_N^{nk} \tag{6.18}$$

It is important to observe that while Eq. (6.18) contains two summations over $N/2$ points, each of these summations is not an $N/2$-point DFT since W_N^{nk} rather than $W_{(N/2)}^{nk}$ appears in each of the sums. Combining the two summations in Eq. (6.18) and using the fact that $W_N^{(N/2)k} = (-1)^k$ we obtain

$$X(k) = \sum_{n=0}^{(N/2)-1} \left[x(n) + (-1)^k x\left(n + \frac{N}{2}\right)\right] W_N^{nk} \tag{6.19}$$

Let us now consider k even and k odd separately, with $X(2r)$ and $X(2r + 1)$ representing the even-numbered points and the odd-numbered points, respectively, so that

$$X(2r) = \sum_{n=0}^{(N/2)-1} \left[x(n) + x\left(n + \frac{N}{2}\right)\right] W_N^{2rn} \tag{6.20}$$

$$X(2r + 1) = \sum_{n=0}^{(N/2)-1} \left[x(n) - x\left(n + \frac{N}{2}\right)\right] W_N^{n} W_N^{2rn},$$

$$r = 0, 1, \ldots, (N/2 - 1) \tag{6.21}$$

Equations (6.20) and (6.21) can be recognized as $N/2$-point DFTs; in the case of Eq. (6.20), of the sum of the first half and the last half of the input sequence, and in the case of Eq. (6.21), of the product of W_N^n with the difference of the first half and the last half of the input sequence. As distinguished from Eq. (6.19), the two summations in Eqs. (6.20) and (6.21) correspond to $N/2$-point DFTs because

$$W_N^{2rn} = W_{N/2}^{rn}$$

Thus on the basis of Eqs. (6.20) and (6.21) with $g(n) = x(n) + x(n + N/2)$ and $h(n) = x(n) - x(n + N/2)$, the DFT can be computed by first forming

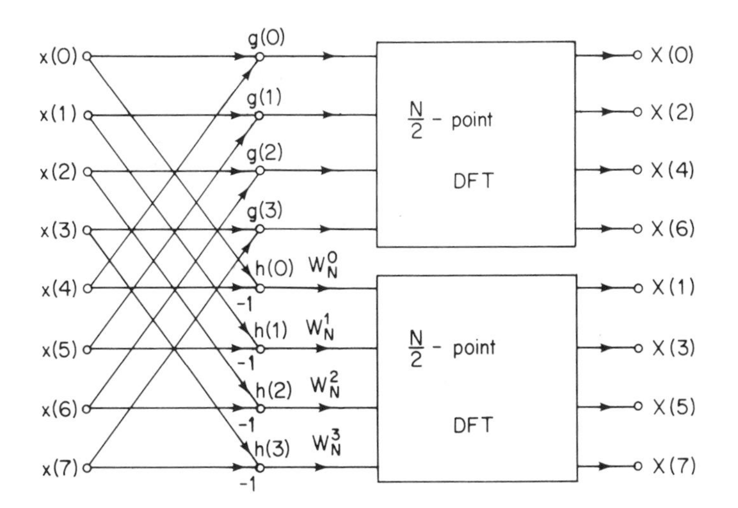

Fig. 6.15 Flow graph of the decimation-in-frequency decomposition of an N-point DFT computation into two $N/2$-point DFT computations $N = 8$).

the sequences $g(n)$ and $h(n)$, then computing $h(n)W_N^n$, and finally computing the $N/2$-point DFTs of these two sequences to obtain the even-numbered output points and the odd-numbered output points, respectively. The procedure suggested by Eqs. (6.20) and (6.21) is illustrated for the case of an eight-point DFT in Fig. 6.15.

Proceeding in a manner similar to that followed in deriving the decimation-in-time algorithm, we note that since N is a power of 2, $N/2$ is even, and consequently, the $N/2$-point DFTs can be computed by computing the even-numbered and odd-numbered output points for those DFTs separately. As in the case of the original decomposition leading to Eqs. (6.20) and (6.21), this is accomplished by combining the first half and the last half of the input points for each of the $N/2$-point DFTs and then computing $N/4$-point DFTs. The flow chart resulting from taking this step for the eight-point example is shown in Fig. 6.16. For the eight-point example, the computation has now been reduced to the computation of two-point DFTs, which, as was discussed previously, are implemented by adding and subtracting the input

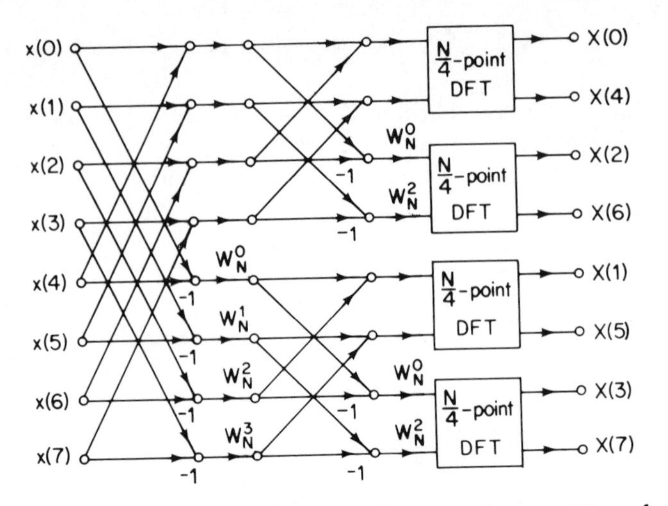

Fig. 6.16 Flow graph of decimation-in-frequency decomposition of an eight-point DFT into four two-point DFT computations.

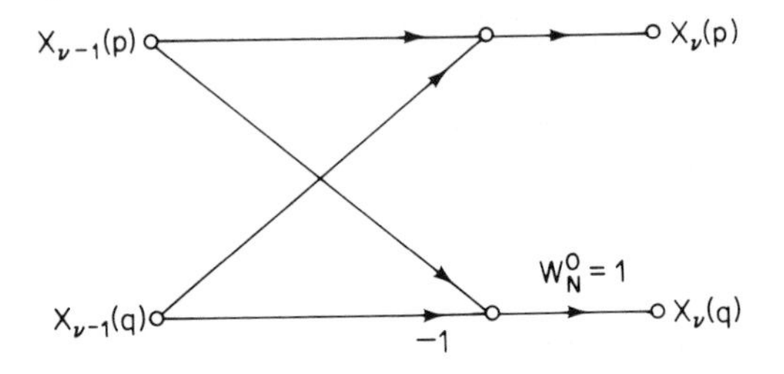

Fig. 6.17 Flow graph of a typical two-point DFT as required in the last stage of decimation-in-frequency decomposition.

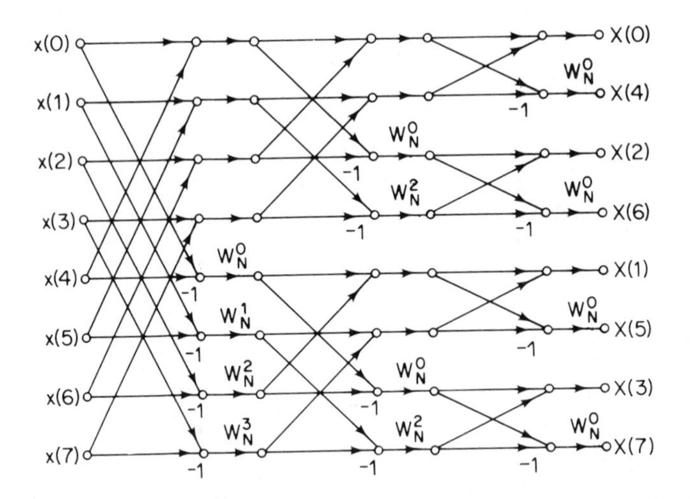

Fig. 6.18 Flow graph of complete decimation-in-frequency decomposition of an eight-point DFT computation.

points. Thus the two-point DFTs in Fig. 6.16 can be replaced by the computation shown in Fig. 6.17, so the computation of the eight-point DFT becomes that shown in Fig. 6.18.

By counting the arithmetic operations in Fig. 6.18, and generalizing to $N = 2^\nu$, we see that the computation of Fig. 6.18 requires $N/2 \log_2 N$ complex multiplications and $N \log_2 N$ complex additions. Thus the total computation is the same for the decimation-in-frequency and the decimation-in-time algorithms.

6.3.1 *In-Place Computation*

The flow graph in Fig. 6.18 depicts one FFT algorithm based on decimation-in-frequency. We can observe a number of similarities and also a number of differences in comparing this graph with the flow graphs derived on the basis of decimation-in-time. As with decimation-in-time, of course, the flow graph of Fig. 6.18 corresponds to a computation of the discrete Fourier transform independently of how the flow graph is drawn as long as the same nodes are connected to each other with the proper branch transmittances. In other words, as with the derivation of the decimation-in-time algorithms, the flow graph of Fig. 6.18 is not based on any a priori assumption about the order in which the input data are stored. However, as was done with the decimation-in-time algorithms, we can interpret successive vertical nodes in the flow graph of Fig. 6.18 as corresponding to successive storage registers in memory, in which case we observe that the flow graph in Fig. 6.18 begins with the input data in normal order and provides the output points in bit-reversed order. We note also that the basic computation is of the form of a butterfly computation, although the butterfly is different than that arising in the decimation-in-time algorithms. However, because of the butterfly nature of the computation we note that the flow graph of Fig. 6.18 can be interpreted as an in-place computation of the discrete Fourier transform. As before, we denote the sequence of complex numbers resulting from the mth stage of the computation as $X_m(l)$, where $l = 0, 1, \ldots, N-1$, and $m = 0, 1, 2, \ldots, \nu$. Then the basic butterfly computation is shown in Fig. 6.19, and the corresponding equations are

$$
\begin{aligned}
X_{m+1}(p) &= X_m(p) + X_m(q) \\
X_{m+1}(q) &= (X_m(p) - X_m(q))W_N^r
\end{aligned}
\qquad (6.22)
$$

By comparing Figs. 6.9 and 6.19 or Eqs. (6.16) and (6.22) we see that the butterfly computations are distinctly different for the two classes of FFT algorithms. However, we also note a striking resemblance between the basic butterfly computations and between Figs. 6.10 and 6.18. Specifically, we note that Fig. 6.18 can be obtained from Fig. 6.10 by reversing the direction of signal flow and interchanging the input and output. That is, in the terminology of Chapter 4, Fig. 6.18 is the transpose of the flow graph in

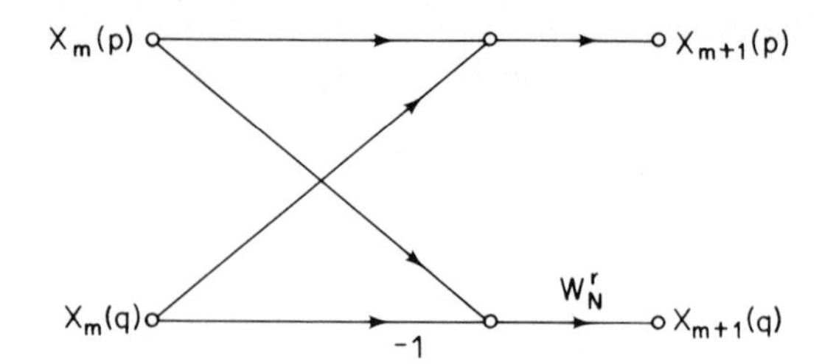

Fig. 6.19 Flow graph of a typical butterfly computation required in Fig. 6.18.

Fig. 6.10 and consequently from the transposition theorem, the input-output characteristics of the two flow graphs must be the same. To see this another way consider a butterfly from a decimation-in-time algorithm as depicted in Fig. 6.9. The corresponding equations are

$$\begin{aligned} X_{m+1}(p) &= X_m(p) + W_N^r X_m(q) \\ X_{m+1}(q) &= X_m(p) - W_N^r X_m(q) \end{aligned} \tag{6.23}$$

If we begin with the output $X(k)$ in normal order as in Fig. 6.10, we can invert the computations of Eq. (6.23) by solving for X_m in terms of X_{m+1}; i.e.,

$$\begin{aligned} X_m(p) &= \tfrac{1}{2}(X_{m+1}(p) + X_{m+1}(q)) \\ X_m(q) &= \tfrac{1}{2}(X_{m+1}(p) - X_{m+1}(q))W_N^{-r} \end{aligned} \tag{6.24}$$

Thus, since

$$X_\nu(k) = X(k)$$

and $X_0(k)$ is equal to $x(n)$ in bit-reversed order, we can compute $x(n)$ in bit-reversed order by repeated application of Eq. (6.24). The flow graph for $N = 8$ is shown in Fig. 6.20.

Figure 6.20 depicts an inverse fast Fourier transform algorithm (IFFT). Now we note that the inverse discrete Fourier transform is

$$x(n) = \frac{1}{N}\sum_{k=0}^{N-1} X(k)W_N^{-kn}$$

so that an FFT algorithm can be used to compute the IDFT if we divide the result by N and use powers of W_N^{-1} instead of powers of W_N. Similarly, an IFFT algorithm can be used to compute the DFT if we multiply the output by N and use powers of W_N^{-1} instead of powers of W_N. Thus the flow graph of Fig. 6.20, representing an IFFT algorithm, can be changed to an FFT algorithm by simply changing $\frac{1}{2}W_N^{-r}$ to W_N^r, since dropping the factor of $\frac{1}{2}$ at each stage is equivalent to multiplying the output by N. With this change and with $x(n)$ in normal order at the input in Fig. 6.20 and $X(k)$ in bit-reversed order at the output, we obtain Fig. 6.18.

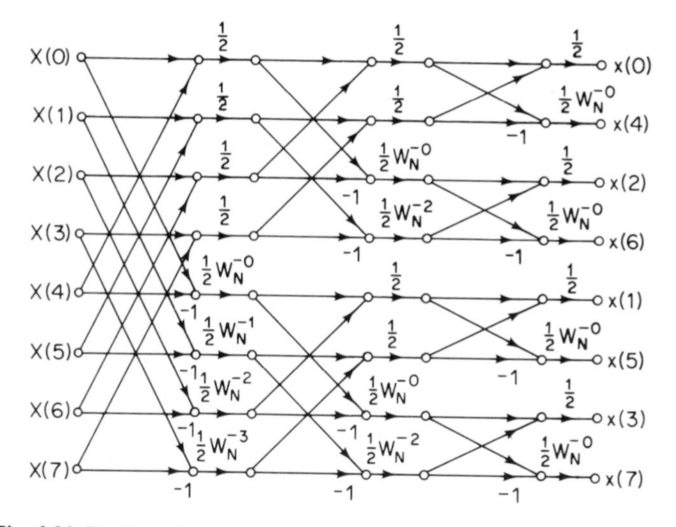

Fig. 6.20 Flow graph of an inverse DFT computation obtained by inverting the butterfly computations of Fig. 6.10.

Thus we see that to any decimation-in-time FFT algorithm there corresponds an inverse transform algorithm that is a decimation-in-frequency algorithm. Or, since inverse transform algorithms are simply related to direct transform algorithms, we can state in general that to each decimation-in-time FFT algorithm there exists a decimation-in-frequency FFT algorithm that corresponds to interchanging the input and output and reversing the direction of all the arrows in the flow graph.

6.3.2 *Alternative Forms*

The previous result implies that all the flow graphs of Sec. 6.2 have counterparts of the decimation-in-frequency type. This, of course, also corresponds to the fact that, as before, it is possible to rearrange the nodes of a flow graph without altering the final result.

If we apply the transposition procedure to Fig. 6.12 we obtain Fig. 6.21. In this flow graph, the output is in normal order and the input is in bit-reversed order. Alternatively, the transpose of the flow graph of Fig. 6.13 results in the flow graph of Fig. 6.22, where both the input and the output are in normal order. As in the case of Fig. 6.13, this flow graph does not correspond to an in-place computation.

The transpose of Fig. 6.14 is shown in Fig. 6.23. Each stage of Fig. 6.23 has the same geometry as is desired for computation involving sequential data storage as discussed before.

6.4 FFT Algorithms for N a Composite Number

Our previous discussion has illustrated the basic principles of decimation-in-time and decimation-in-frequency for the important special case of N a

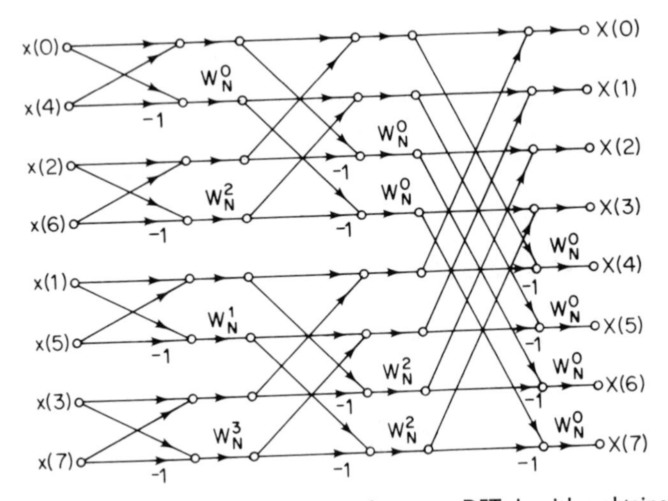

Fig. 6.21 Flow graph of a decimation-in-frequency DFT algorithm obtained from Fig. 6.20. Input in bit-reversed order and output in normal order. (Transpose of Fig. 6.12.)

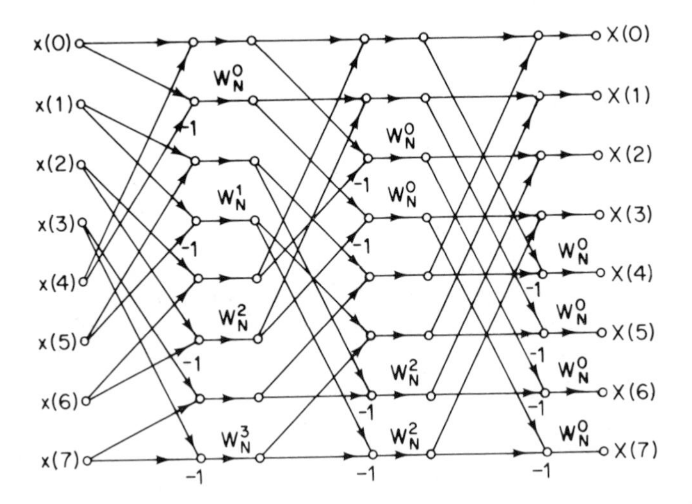

Fig. 6.22 Rearrangement of Fig. 6.18 with both input and output in normal order. (Transpose of Fig. 6.13.)

power of 2; i.e., $N = 2^\nu$. More generally, the efficient computation of the discrete Fourier transform is tied to the representation of N as a product of factors [1,7,9,10]; i.e., suppose that

$$N = p_1 p_2 \dots p_\nu \tag{6.25}$$

As we have seen in the case of N a power of 2 (where all the factors can be taken to be equal to 2), such a decomposition leads to a highly efficient

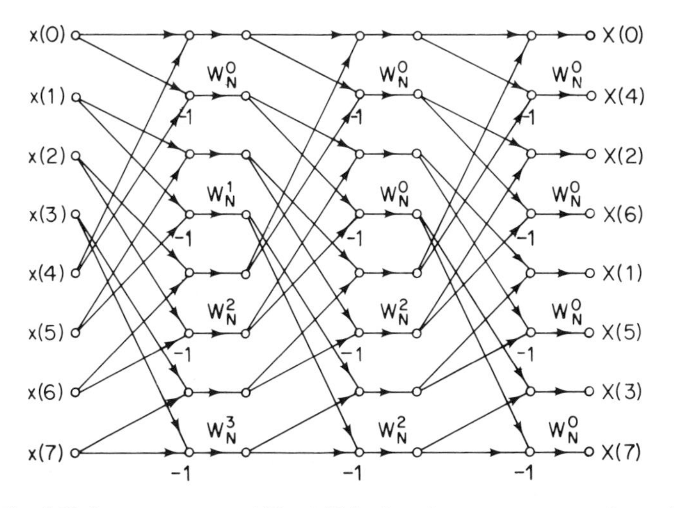

Fig. 6.23 Rearrangement of Fig. 6.18 having the same geometry for each stage, thereby permitting sequential data accessing and storage. (Transpose of Fig. 6.14.)

computational algorithm. Furthermore, all the required computations are butterfly computations that correspond essentially to two-point DFTs. For this reason the power-of-2 algorithms are particularly simple to implement, and often in applications it is advantageous to always deal with sequences whose length is a power of 2. This can be done in many cases by simply augmenting a finite-length sequence with zero samples if necessary. However, in some cases it may not be possible to choose N to be a power of 2, thus making it necessary to consider the more general situation of Eq. (6.25). Let us therefore consider the application of the decimation-in-time principle in the case where N is a product of factors that are not all necessarily equal to 2.

Let us define

$$q_1 = p_2 p_3 \cdots p_v$$

so that

$$N = p_1 \cdot q_1$$

If N is a power of 2, we could choose $p_1 = 2$ and $q_1 = N/2$. Using decimation-in-time we would then decompose the sequence $x(n)$ into two sequences, each $(N/2)$ samples in length, consisting of the even- and odd-numbered samples, respectively. When $N = p_1 \cdot q_1$, we can divide the input sequence into p_1 sequences of q_1 samples each by associating every p_1th sample with a given subsequence. For example if $p_1 = 3$ and $q_1 = 4$, so that $N = 12$, then we can decompose $x(n)$ into three sequences of length 4, with the first sequence consisting of the samples $x(0)$, $x(3)$, $x(6)$, $x(9)$; the second sequence consisting of $x(1)$, $x(4)$, $x(7)$, $x(10)$; and the third sequence consisting of $x(2)$,

$x(5)$, $x(8)$, and $x(11)$. In general we can write $X(k)$ as

$$X(k) = \sum_{n=0}^{N-1} x(n) W_N^{kn}$$

$$= \sum_{r=0}^{q_1-1} x(p_1 r) W_N^{p_1 rk} + \sum_{r=0}^{q_1-1} x(p_1 r + 1) W_N^k W_N^{p_1 rk} + \ldots$$

$$+ \sum_{r=0}^{q_1-1} x(p_1 r + p_1 - 1) W_N^{(p_1-1)k} W_N^{p_1 rk}$$

or

$$X(k) = \sum_{l=0}^{p_1-1} W_N^{lk} \sum_{r=0}^{q_1-1} x(p_1 r + l) W_N^{p_1 rk} \tag{6.26}$$

The inner sums can be expressed as the q_1-point DFTs

$$G_l(k) = \sum_{r=0}^{q_1-1} x(p_1 r + l) W_{q_1}^{rk} \tag{6.27}$$

since, as is easily verified,

$$W_N^{p_1 rk} = W_{q_1}^{rk} \qquad \text{for} \qquad N = p_1 \cdot q_1 \tag{6.28}$$

Thus Eq. (6.26) expresses $X(k)$ in terms of p_1 discrete Fourier transforms of sequences of length q_1 samples. To determine the number of complex multiplications and additions required in implementing the DFT according to Eq. (6.26), let us consider, as we did in the original discussion of decimation-in-time, that the q_1-point DFTs are implemented by means of the direct computation. From Eq. (6.26) we observe that the number of q_1-point DFTs to be evaluated is p_1. Thus a total of $p_1 \cdot q_1^2$ complex multiplications and additions are required. The outer sum in Eq. (6.26) is implemented by multiplying the q_1-point DFTs by the factor W_N^{lk} and adding the results together. Since the double summation in Eq. (6.26) is to be implemented for N values of k, a total of $N(p_1 - 1)$ complex multiplications and additions are required to combine the p_1 q_1-point DFTs.† Therefore, the total number of complex multiplications and additions required to compute the discrete Fourier transform in the form of Eq. (6.26) is $N(p_1 - 1) + p_1 q_1^2$. Now the q_1-point DFTs can be decomposed in a similar manner. In particular, if we now represent q_1 as

$$q_1 = p_2 \cdot q_2$$

† Summing p_1 terms requires p_1-1 additions and we do not need to multiply by W_N^{kl} when $l = 0$. We remind the reader that throughout this chapter we have generally counted multiplication by W_N^{kl} even when W_N^{kl} is unity or j. In interpreting Eq. (6.26), however, it is convenient to recognize W_N^{kl} as unity for $l = 0$ in order that the result which we obtain be consistent with the discussion in Section 6.2.

then the q_1-point sequences in the inner sum of Eq. (6.26) can be broken into p_2 subsequences, each being q_2-points long so that the inner sum in Eq. (6.26) can be replaced by a double summation in the same way that we began. When this is done, the number of operations required in computing the q_1-point DFTs in Eq. (6.26) is, instead of q_1^2,

$$q_1(p_2 - 1) + p_2 q_2^2 \tag{6.29}$$

Consequently, the factor q_1^2 in the expression $N(p_1 - 1) + p_1 q_1^2$ is replaced by Eq. (6.29), and thus the total number of complex multiplications and additions required is

$$N(p_1 - 1) + N(p_2 - 1) + p_1 p_2 q_2^2 \tag{6.30}$$

If we continue this procedure by further decomposing the q_2-point DFTs, then when the original sequence has been decomposed as much as possible the number of complex multiplications and additions will be

$$N(p_1 + p_2 + \ldots + p_\nu - \nu) \tag{6.31}$$

For example, when $p_1 = p_2 = \ldots = p_\nu = p$, the number of complex multiplications and additions is $N(p - 1)\nu$. When $p = 2$, this number is $N \cdot \nu$, as discussed before.† In general, it can be seen from Eq. (6.31) that it is preferable to carry out the decomposition on the basis of as many factors as possible for a given N. Formally, there is no advantage to choosing anything but prime factors since if $p_i = r_i \cdot s_i$, with r_i and $s_i > 1$, then $p_i > r_i + s_i$, except when $r_i = s_i = 2$, in which case $p_i = r_i + s_i$. However, there are examples (notably $p_i = 4$ or 8) where additional economies result which are not accounted for by Eq. (6.31).

To illustrate the decimation-in-time procedure for N not a power of 2, let us consider computation of an 18-point DFT; i.e., $N = 3 \cdot 3 \cdot 2$. Letting $p_1 = 3$ and $q_1 = 6$, and following the preceding discussion, we first divide the original sequence into three sequences, each six points long:

Sequence 1:	$x(0)$	$x(3)$	$x(6)$	$x(9)$	$x(12)$	$x(15)$
Sequence 2:	$x(1)$	$x(4)$	$x(7)$	$x(10)$	$x(13)$	$x(16)$
Sequence 3:	$x(2)$	$x(5)$	$x(8)$	$x(11)$	$x(14)$	$x(17)$

By dividing the original sequence into these three subsequences, we can express $X(k)$ as

$$\begin{aligned} X(k) &= \sum_{l=0}^{2} W_{18}^{lk} \sum_{r=0}^{5} x(3r + l) W_{18}^{3rk} \qquad (6.32) \\ &= G_0(k) + W_{18}^{k} G_1(k) + W_{18}^{2k} G_2(k) \end{aligned}$$

† Recall that the number of multiplications can be further reduced by exploiting symmetry.

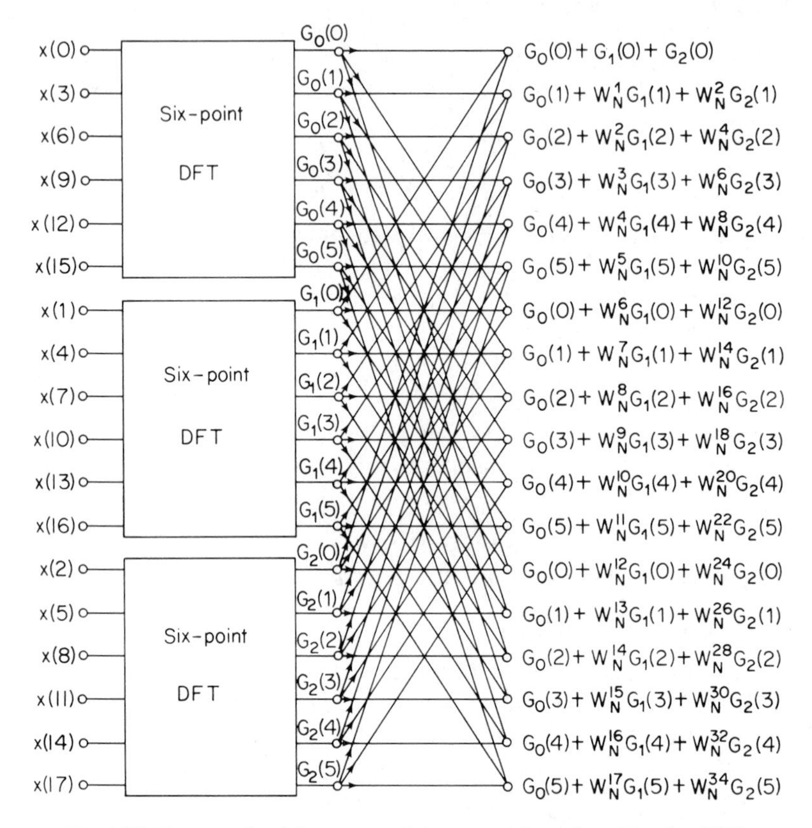

Fig. 6.24 Flow graph of first stage of decomposition of an 18-point DFT.

The inner sum is a six-point DFT with $l = 0$ corresponding to sequence 1, $l = 1$ corresponding to sequence 2, and $l = 2$ corresponding to sequence 3. In this case the six-point DFTs $G_l(k)$ are periodic with period 6. The computation of Eq. (6.32) is illustrated in Fig. 6.24.†

The six-point DFTs corresponding to the inner sum in Eq. (6.32) can be further decomposed by breaking the sequences $x(3r + l)$ into three sequences, each two points long or, alternatively, two sequences, each three points long. Choosing the former, i.e., breaking each of the six-point subsequences into three sequences two points long, we replace the inner sum by

$$G_l(k) = \sum_{r=0}^{5} x(3r + l)W_6^{rk} = \sum_{s=0}^{2} W_6^{sk} \sum_{p=0}^{1} x(9p + 3s + l)W_6^{3pk}$$

† In this figure, the transmittances on each of the branches are to be inferred from the algebraic expression associated with each output node.

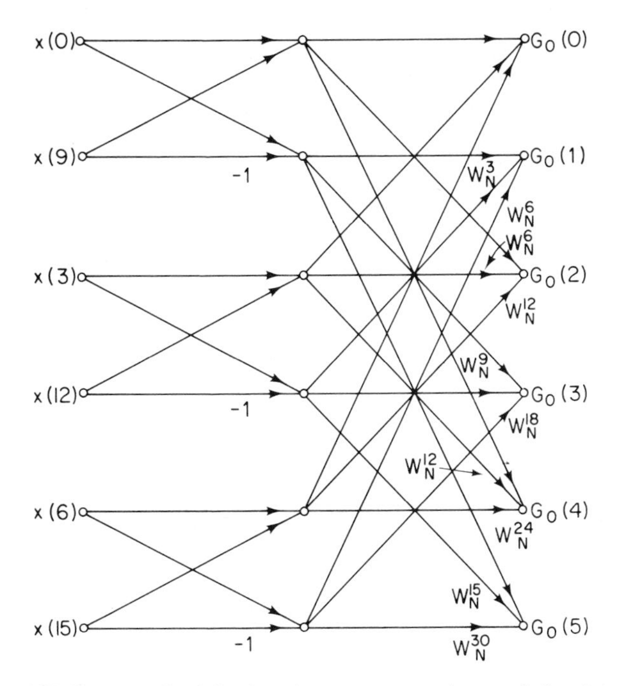

Fig. 6.25 Flow graph of further decomposition of one of the six-point DFTs in Fig. 6.24.

so that the overall computation of $X(k)$ becomes

$$X(k) = \sum_{l=0}^{2} W_{18}^{lk} \sum_{s=0}^{2} W_6^{sk} \sum_{p=0}^{1} x(9p + 3s + l) W_2^{pk} \tag{6.33}$$

One of the six-point DFTs ($G_0(k)$) is shown in detail in Fig. 6.25. The other two have identical form. When Fig. 6.25 and the corresponding flow graphs for $G_1(k)$ and $G_2(k)$ are placed in appropriate positions in Fig. 6.24, the input sequence is in the order: $x(0)$, $x(9)$, $x(3)$, $x(12)$, $x(6)$, $x(15)$, $x(1)$, $x(10)$, $x(4)$, $x(13)$, $x(7)$, $x(16)$, $x(2)$, $x(11)$, $x(5)$, $x(14)$, $x(8)$, and $x(17)$. We observe from Figs. 6.24 and 6.25 that for this ordering of the inputs, the computation can be done in place. The two-point transforms are shown as the familiar butterflies in the first stage of Fig. 6.25; the basic three-point DFT operation is somewhat more complicated but still obviously an in-place computation. Instead of bit reversal, the ordering of the input is somewhat more complicated. Specifically, if we denote $X_0(\,)$ as the input array, then it can be shown that

$$X_0(6l + 2s + p) = x(9p + 3s + l)$$

where $p = 0, 1$; $s = 0, 1, 2$; and $l = 0, 1, 2$. That is, the input must be stored in a generalized "digit-reversed" order in order to carry out the computation in place. As is evident from Fig. 6.24, the resulting output is in normal order. We note from Fig. 6.24 that the basic computation in the last stage (as for factors of 3 in general) is as depicted in Fig. 6.26 for $N = 3 \cdot q_1$.

We recall that in the case of factors of 2, we were able to reduce the number of multiplications by a factor of 2 by exploiting symmetry. In the case of factors

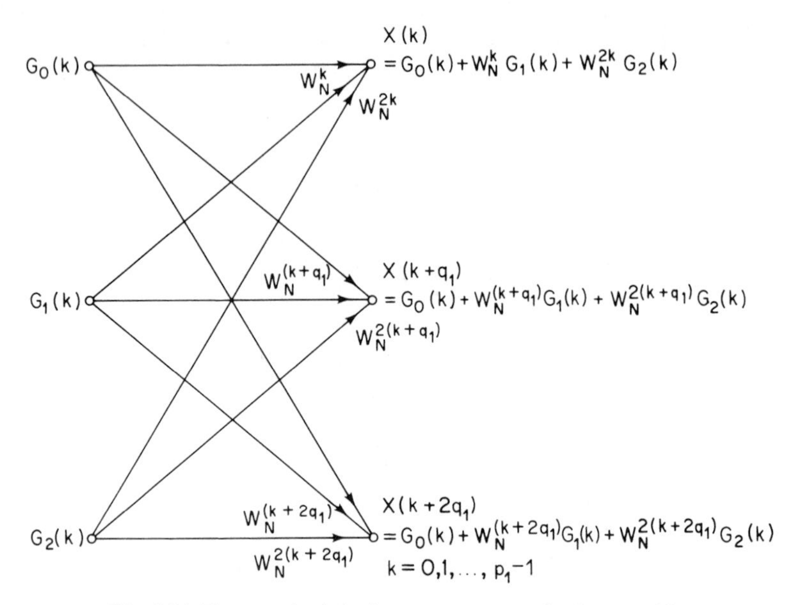

Fig. 6.26 Flow graph of the basic computation for factors of 3.

of 3, in Fig. 6.26, the comparable manipulation of the flow graph yields Fig. 6.27. Since $N = 3q_1$, the basic complex multiplier $W_N^{q_1}$ is

$$W_N^{q_1} = e^{-j(2\pi/3)}$$

Therefore, $W_N^{q_1}$ and all powers thereof are complex coefficients that require multiplications. Thus Fig. 6.27 is no more efficient than Fig. 6.26.

In contrast to factors of 3, it can be shown (see Problem 7 of this chapter) that the basic DFT computation for a factor of 4 (i.e., $N = 4q_1$) is as shown in Fig. 6.28. In this case, the flow graph of Fig. 6.28 can be redrawn as in Fig. 6.29, with a concomitant saving of at least 9 complex multiplications out of the 12 shown in Fig. 6.28. Similar savings result for factors of 8, 16, etc. [11]. Thus, even if $N = 2^\nu$, it is sometimes advantageous to base the computation on factors of 4, using one stage based on a factor of 2 if ν is odd.

Our discussion in this section, although paralleling the earlier discussion, has been far from complete in the sense that we have only attempted to indicate some of the advantages and disadvantages of using values of N with factors other than 2. The basic advantages are increased flexibility and speed in some cases; the basic disadvantage is greatly increased complexity of the computational algorithm. Although we have only discussed decimation-in-time, a similar discussion clearly holds for decimation-in-frequency as well. For a more detailed discussion of FFT algorithms for N a general composite number, see Gentleman and Sande [9] and Singleton [10].

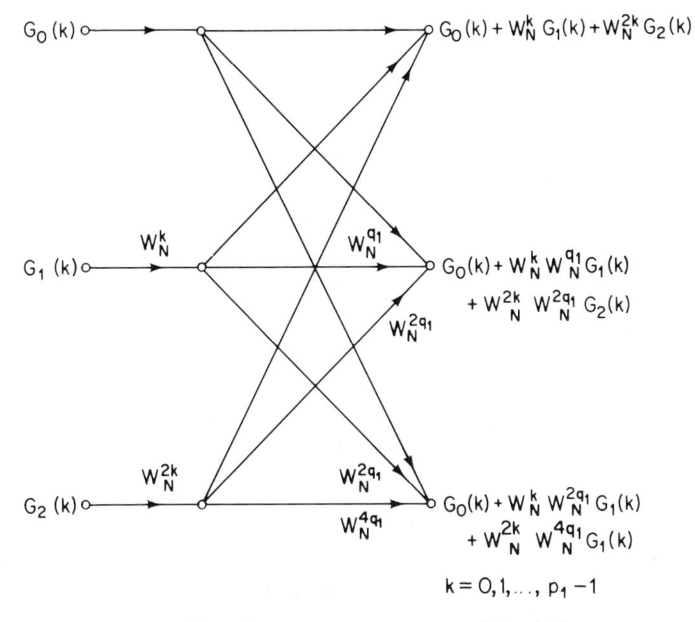

Fig. 6.27 Alternative arrangement of Fig. 6.26.

6.5 General Computational Considerations in FFT Algorithms

We have discussed the basic principles of efficient computation of discrete Fourier transforms. In this discussion we have favored the use of signal flow graph representations rather than explicitly writing out in detail the equations that such flow graphs represent. Of necessity we have shown flow graphs for specific values of N. The justification for this approach lies in the fact that flow graphs of even the simple case of $N = 8$ are easily generalized to larger values of N. That is, by considering a flow graph such as Fig. 6.10, it is possible to see how to structure a general computational algorithm that would apply to any $N = 2^\nu$.

It is abundantly clear from a consideration of the flow graphs of previous sections that there are two major considerations in any FFT algorithm. The first is the accessing and storing of data in the intermediate arrays. The second is the actual implementation of the butterfly computation once the appropriate data have been obtained. Although it is true that the flow graphs of the previous section capture the essence of the FFT algorithms that they depict, there are a multitude of details related to these two basic concerns that must be considered in the implementation of a given algorithm. In this section we discuss some of the details that pertain to implementation of FFT algorithms. As before, our emphasis is on algorithms for $N = 2^\nu$, although much of the discussion applies to the more general case as well.

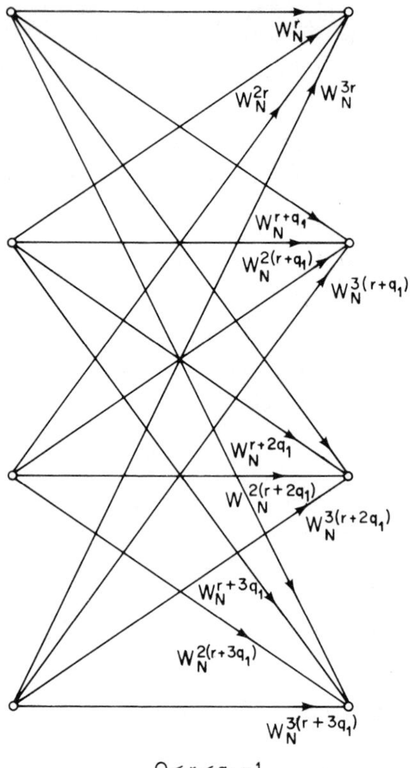

Fig. 6.28 Flow graph of the basic computation for factors of 4.

6.5.1 *Indexing*

Let us consider as an example the algorithm depicted by Fig. 6.10. In this case the input must be in bit-reversed order so that the computation can be performed in place. The result is then in normal order. Generally, sequences are not available in bit-reversed order so the first step in the implementation of Fig. 6.10 would be a process of permutation of the samples of the input sequence. This process is depicted in Fig. 6.30 for $N = 8$.

From the figure it is clear that the permutation of $x(n)$ into bit-reversed order can be done "in place." This is conveniently done by using an index that "counts" in bit-reversed order. (A flow graph of such a bit-reversed counter is given by Gold and Rader [7].) Suppose that n is the normal index and l is the bit-reversed index. Then the initial array is

$$X_0(l) = x(n)$$

We observe from Fig. 6.30 that when $n = l$ no exchange is necessary, but when $n \neq l$ we must exchange $x(n)$ and $x(l)$. However, we must ensure that

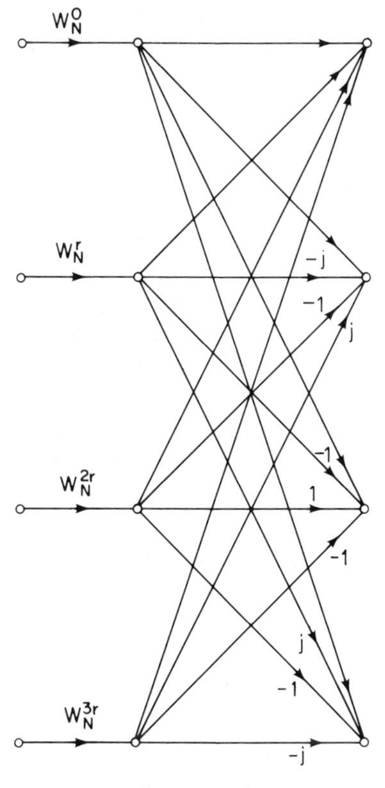

Fig. 6.29 Alternative arrangement of Fig. 6.28 resulting in savings of multiplications.

the exchange is only done once. This can be assured by comparing n and l and only making the exchange when $l > n$. Thus a simple algorithm for bit-reversed sorting is: Index through the sequence $x(n)$, allowing n to go from 0 to $N - 1$ normal order, while l is going from 0 to $N - 1$ in bit-reversed order. Whenever $l > n$, interchange $x(n)$ and $x(l)$.

Once the input is in bit-reversed order, we can proceed with the first stage of computation. In this case the multipliers are all $W_N^0 = 1$ and the

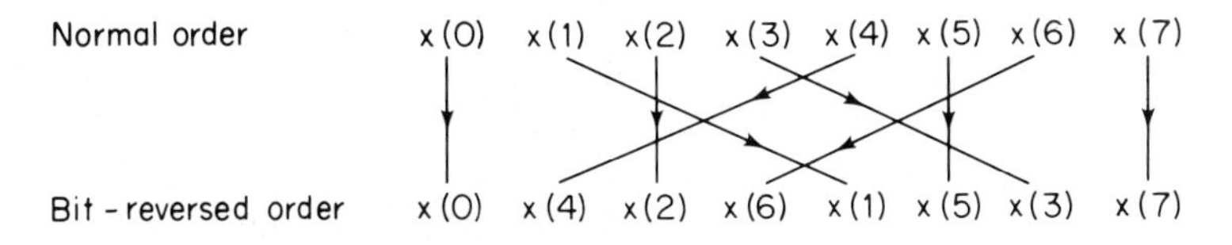

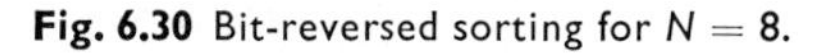

Fig. 6.30 Bit-reversed sorting for $N = 8$.

inputs to the butterflies are adjacent elements of the array $X_0(\cdot)$. In the second stage the multipliers are all either W_N^0 or powers of $W_N^{N/4}$, and the inputs to the butterflies are separated by 2. In the mth stage the multipliers are a power of $W_N^{(N/2^m)}$ and the butterfly inputs are separated by 2^{m-1}. We note that the powers of W_N are required in normal order if computation of butterflies begins at the top of the flow graph of Fig. 6.10. The above statements define the manner in which data must be accessed in a given stage. This of course is dependent upon the flow graph being implemented. For example, in the mth stage of Fig. 6.12 the butterfly spacing is $2^{(\nu-m)}$, the coefficients are again all powers of $W_N^{(N/2^m)}$, and in this case the powers of W_N are required in bit-reversed order. The input is in normal order; however, the output is in bit-reversed order, so it might be necessary to sort the output as discussed previously.

In general, if we consider all the flow graphs in Secs. 6.2 and 6.3, we see that each algorithm has its own characteristic indexing problems. The choice of a particular algorithm depends on a number of factors. The algorithms utilizing an in-place computation have the advantage of making efficient use of memory; on the other hand, the kind of memory required is random access rather than sequential memory. These algorithms have the additional disadvantage that either the input data or the output points are in bit-reversed order. Furthermore, depending on whether a decimation-in-time or decimation-in-frequency algorithm is chosen and whether the inputs or the outputs are in bit-reversed order, the coefficients are required to be accessed in either normal order or in bit-reversed order. If random-access memory is not available but sequential memory is, we have demonstrated fast Fourier transform algorithms that utilize such memory, but again either the inputs or the outputs must be in bit-reversed order. While the flow graph for the algorithm can be arranged so that both the inputs, the outputs, and the coefficients are in normal order, the indexing structure required to implement these algorithms is complicated and twice as much random access memory is required. Consequently, the use of these algorithms does not appear to be advantageous.

The most common FFT algorithms used are those of Figs. 6.10, 6.12, 6.18, and 6.21, for which the computation is in place. If data are to be transformed only once, then clearly bit-reversed sorting must be implemented either at the input or on the output. However, in some situations data are transformed, the result modified in some way, and then inverse-transformed. For example, in implementing FIR digital filters by means of a computation of the discrete Fourier transform, a section of the input data is transformed, multiplied by the DFT of the filter unit-sample response, and the result is inverse-transformed. As another example, in computing an autocorrelation function or cross-correlation function using the discrete Fourier transform, data will be transformed, the DFTs multiplied, and then the resulting product will be inverse-transformed.

When two transforms are cascaded in this way, it is possible by appropriate choice of the FFT algorithms to avoid the need for bit reversal. For example, in implementing an FIR digital filter using the DFT, we can choose an algorithm for the direct transform which utilizes the data in normal order and provides a DFT in bit-reversed order. Either the flow graph corresponding to Fig. 6.12, based on decimation-in-time, or that of Fig. 6.18, based on decimation-in-frequency, could be used in this way. The difference between these two forms is that the decimation-in-time form requires the coefficients in bit-reversed order, whereas the decimation-in-frequency form requires the coefficients in normal order.

In using either of these algorithms, the transform occurs in bit-reversed order and, consequently, we will have stored the DFT corresponding to the frequency response of the filter in bit-reversed order. For the inverse DFT we can then choose a form of the algorithm that has bit-reversed data at the input and provides normally ordered results. Here either the flow graph of Fig. 6.10, based on decimation-in-time, or that of Fig. 6.21, based on decimation-in-frequency, can be used. Figure 6.10, however, utilizes coefficients in normal order, whereas Fig. 6.21 requires the coefficients in bit-reversed order. In order that the coefficients need only be made available either in normal order (which is preferable) or in bit-reversed order, if the decimation-in-time form of the algorithm is chosen for the direct transform, then the decimation-in-frequency form of the algorithm should be chosen for the inverse transform, requiring coefficients in bit-reversed order, or the decimation-in-frequency algorithm for the direct transform should be paired with the decimation-in-time algorithm for the inverse transform, which would then utilize normally ordered coefficients.

6.5.2 *Coefficients*

We have observed that the coefficients W_N^r may be required in either bit-reversed order or in normal order. In either case we must either store a table sufficient to look up all required values or we must compute the values as needed. The first alternative has the advantage of speed but of course requires extra storage. We observe from the flow graphs that we require W_N^r for $r = 0, 1, \ldots, N/2 - 1$. Thus we require $(N/2)$ complex storage registers for a complete table of values of W_N^r.† In the case of algorithms where the coefficients are required in bit-reversed order, it is necessary to store the table in bit-reversed order.

The computation of the coefficients as they are needed saves storage but is less efficient. The greatest efficiency is obtained using a recursion formula. Note that, in general, at a given stage the required coefficients are all powers of some power of W_N; i.e., W_N^q, where q depends on the algorithm

† This number can be reduced using symmetry at the cost of greater effort in looking up desired values.

and the stage. Thus, if the coefficients are required in normal order, we can use the recursion formula

$$W_N^{ql} = W_N^q \cdot W_N^{q(l-1)} \tag{6.34}$$

to obtain the lth coefficient from the $(l-1)$st coefficient. Clearly, algorithms that require coefficients in bit-reversed order are not well suited to this approach. It should be noted that Eq. (6.34) is essentially the coupled form oscillator of Problem 8 of this chapter. When using finite-precision arithmetic, errors can build up in the iteration of this difference equation. Therefore, it is generally necessary to reset the value at prescribed points (e.g., $W_N^{N/4} = -j$) so that errors do not become unacceptable.

6.5.3 Multidimensional Fast Fourier Transforms

The two-dimensional discrete Fourier transform was defined in Chapter 3 as

$$X(k, l) = \sum_{m=0}^{M-1} \sum_{n=0}^{N-1} x(m, n) W_M^{km} W_N^{ln}, \qquad k = 0, 1, \ldots, M-1,\ l = 0, 1, \ldots, N-1 \tag{6.35}$$

We observe that Eq. (6.35) is very similar to Eq. (6.26). In fact, a useful interpretation of an FFT algorithm can be phrased in terms of multidimensional DFTs (see Gold and Rader [7]). If we are concerned with computation of Eq. (6.35), we can observe that the computation involves M DFTs of the form

$$A(m, l) = \sum_{n=0}^{N-1} x(m, n) W_N^{ln}, \qquad \begin{aligned} m &= 0, 1, \ldots, M-1 \\ l &= 0, 1, \ldots, N-1 \end{aligned} \tag{6.36}$$

followed by N DFTs of the form

$$X(k, l) = \sum_{m=0}^{M-1} A(m, l) W_M^{km}, \qquad \begin{aligned} k &= 0, 1, \ldots, M-1 \\ l &= 0, 1, \ldots, N-1 \end{aligned} \tag{6.37}$$

Clearly, we can evaluate Eqs. (6.36) and (6.37) by any one of the previously described algorithms. If M and N are powers of 2, the number of complex multiplications required will be

$$M \cdot \frac{N}{2} \log_2 N + N \cdot \frac{M}{2} \log_2 M = \frac{NM}{2} (\log_2 N \cdot M)$$

in contrast to approximately $N^2 \cdot M^2$ complex multiplications for direct evaluation of Eq. (6.35).

A fundamental difficulty in two-dimensional DFT computations is the large amount of storage required to implement an in-place computation. For complex data, $2(N \cdot M)$ real storage registers are required to store either the input or the resulting transform. For example, if $N = M = 256$, then

131,072 registers are required. If the required amount of random-access memory is unavailable, then some form of sequential-access mass memory, such as disc or tape, can be used, with either rows or columns stored end to end. This increases the complexity of implementing either Eq. (6.36) or (6.37). For example, if the data are stored as rows, the row transforms [Eq. (6.36)] can be easily implemented, but if the results of the row transformations are stored sequentially as rows, then it is difficult to access the data required for the column transforms. One way to deal with this difficulty is to perform several row transforms, save the results, and then store the resulting values of $A(m, l)$ in transposed order. After the entire transposed array $A(l, m)$ has been formed, the desired column transforms are computed by computing row transforms of the transposed array. The resulting two-dimensional transform is stored in transposed order [12].

6.6 Chirp z-Transform Algorithm

We have seen that it is possible to compute the DFT in a very efficient manner. Equivalently, this corresponds to efficient computation of samples of the z-transform of a finite-length sequence taken at equally spaced points around the unit circle. In order to achieve this efficiency in evaluating the z-transform, N is required to be a highly composite number. Also, we may be interested in sampling the z-transform on some other contour, or we may not require samples of the z-transform over the entire unit circle. Thus schemes for increasing the flexibility of DFT computations are of considerable interest.

Suppose that we are interested in obtaining samples of the z-transform of a finite-length sequence on a circle that is concentric with the unit circle, and the samples are to be equally spaced in angle around this circle. Then, with a minor modification, a fast Fourier transform algorithm can be used. Specifically, if we have a finite-duration sequence $x(n)$ of length N, then the discrete Fourier transform of the sequence $x(n)\alpha^{-n}$ will provide N samples equally spaced in angle around a circle of radius α in the z-plane. If we are interested in obtaining frequency samples equally spaced over a small portion of the unit circle, the most efficient approach may often be to use a fast Fourier transform algorithm to compute frequency samples with the desired spacing, but obtaining samples outside the frequency range of interest. For example, if we were considering a 128-point sequence and we were interested in obtaining 128 samples of the z-transform on the unit circle between $\omega = -\pi/8$ and $\omega = +\pi/8$, the most efficient procedure may be to compute a 1024-point DFT by augmenting the original sequence with zeros and retain only the 128 spectral points desired.

An alternative procedure, which in many situations is the most efficient, is the use of the chirp z-transform (CZT) algorithm [13]. This algorithm is directed toward computation of samples of the z-transform on a spiral contour equally spaced in angle over some portion of the spiral. Specifically, let $x(n)$ designate an N-point sequence and $X(z)$ designate its z-transform. Using the chirp z-transform algorithm, $X(z)$ can be computed at the points z_k given by

$$z_k = AW^{-k}, \qquad k = 0, 1, \ldots, M-1 \tag{6.38}$$

where

$$W = W_0 e^{-j\phi_0}$$
$$A = A_0 e^{j\theta_0}$$

with W_0 and A_0 positive real numbers. Consequently, the contour along which the samples are obtained is that indicated in Fig. 6.31.

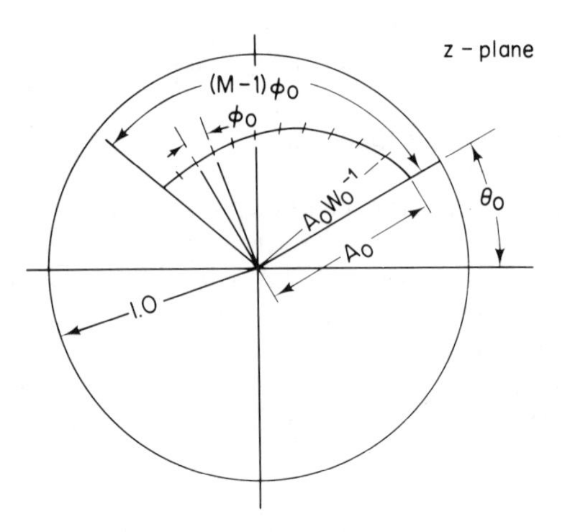

Fig. 6.31 z-plane contour for the chirp z-transform.

This contour is a spiral in the z-plane. The parameter W_0 controls the rate at which the contour spirals; if W_0 is greater than unity, the contour spirals toward the origin as k increases, and if W_0 is less than unity, the contour spirals outward as k increases. The parameters A_0 and θ_0 are the location in radius and angle, respectively, of the first sample, i.e., for $k = 0$. The remaining samples are located along the spiral contour with an angular spacing of ϕ_0. Consequently, if $W_0 = 1$, the spiral is, in fact, a circular arc, and if $A_0 = 1$, this circular arc is part of the unit circle.

With the values of z_k given by Eq. (6.38) we wish to compute

$$X(z_k) = \sum_{n=0}^{N-1} x(n) A^{-n} W^{nk}, \qquad k = 0, 1, \ldots, M-1 \tag{6.39}$$

where N is the length of the sequence $x(n)$. Using the identity†

$$nk = \tfrac{1}{2}[n^2 + k^2 - (k-n)^2] \tag{6.40}$$

† This trick was originated by Bluestein [14].

Eq. (6.39) can be written as

$$X(z_k) = \sum_{n=0}^{N-1} x(n)A^{-n}W^{n^2/2}W^{k^2/2}W^{-(k-n)^2/2}$$

or

$$X(z_k) = W^{k^2/2}\sum_{n=0}^{N-1} x(n)A^{-n}W^{n^2/2}W^{-(k-n)^2/2}$$

Letting

$$g(n) = x(n)A^{-n}W^{n^2/2}$$

we can then write

$$X(z_k) = W^{k^2/2}\sum_{n=0}^{N-1} g(n)W^{-(k-n)^2/2}, \qquad k = 0, 1, \ldots, M-1 \quad (6.41)$$

With $X(z_k)$ expressed in the form of Eq. (6.41), we recognize the summation as corresponding to the convolution of the sequence $g(n)$ with the sequence $W^{-n^2/2}$. Thus the computation of Eq. (6.41) is as depicted in Fig. (6.32), where

$$h(n) = W^{-n^2/2}$$

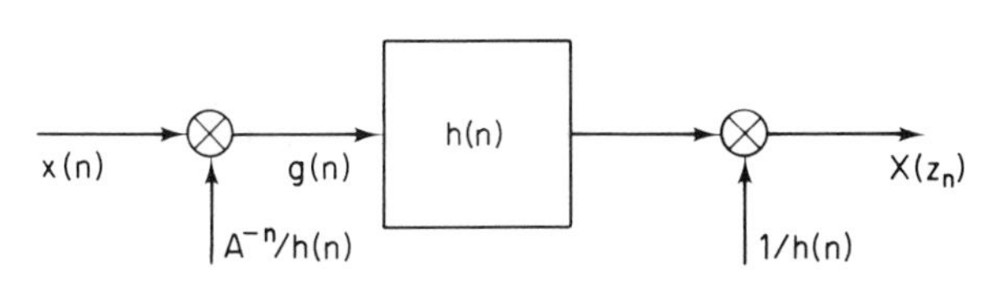

Fig. 6.32 Interpretation of Eq. (6.41) in terms of a linear system.

When A and W_0 are unity, the sequence $h(n)$ can be thought of as a complex exponential sequence with linearly increasing frequency. In radar systems such signals are called *chirp signals;* hence the name *chirp z-transform.* A system similar to Fig. 6.32 is commonly used for spectrum analysis in radar problems.

Since the sequence $g(n)$ is of finite duration, we recall from the discussion in Chapter 3 that the convolution in Eq. (6.41) can be carried out by means of the discrete Fourier transform, computed, of course, using a fast Fourier transform algorithm.†

Whereas the sequence $g(n)$ is of finite duration, the sequence $W^{-n^2/2}$ is of infinite duration; consequently, if the convolution is to be implemented using the discrete Fourier transform, then it is necessary to section the sequence $W^{-n^2/2}$. We note also that, while the result of the convolution is of indefinite length, we are only interested in the result of the convolution for $k = 0, 1, \ldots, M-1$. Consequently, in sectioning the sequence $W^{-n^2/2}$, it would be advantageous to choose the sections in such a way that the result of the

† Bluestein [14] showed that a recursive realization of Fig. 6.32 can be obtained for the case $z_k = e^{j(2\pi/N)k}$ and N a perfect square (see Problem 16 of this chapter).

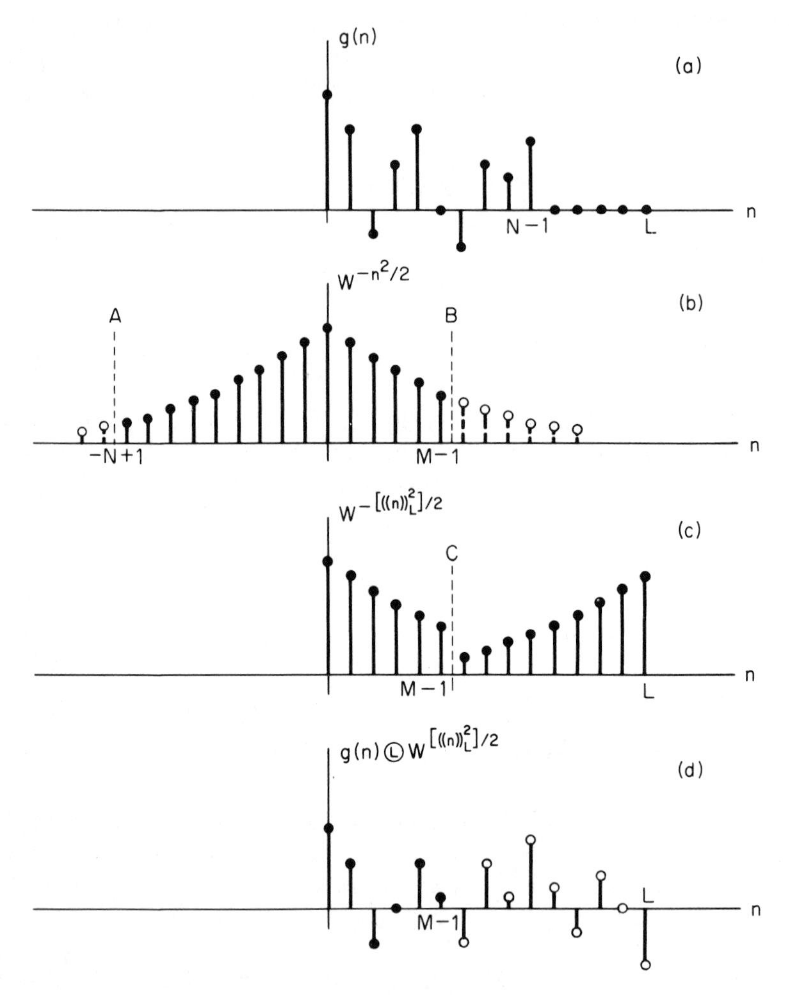

Fig. 6.33 Sequences involved in the CZT algorithm ($L = N + M - 1$).

computation of one section results in the M desired output points. Figure 6.33 depicts the sequences involved in this process for the case $N = 10$ and $M = 6$. The sequences $g(n)$ and $W^{-n^2/2}$ are depicted in Fig. 6.33(a) and (b), respectively.

In implementing the convolution of $g(n)$ with $W^{-n^2/2}$, the only part of $W^{-n^2/2}$ that is required to compute the result of the convolution in the interval 0 to $M - 1$ is that part from $-N + 1$ to $M - 1$, including both of these endpoints. That part of the sequence $W^{-n^2/2}$ is between the dashed lines labeled A and B in Fig. 6.33(b). Consequently, the convolution can be implemented by computing the $(M + N - 1)$-point DFT of $g(n)$ (augmented of course with $M - 1$ zeros) and the $(M + N - 1)$-point DFT of the part of the sequence $W^{-n^2/2}$ in the region from A to B in Fig. 6.33(b). The inverse transform of the product of these two discrete Fourier transforms will be the circular convolution of the sequence $g(n)$ with the section of $W^{-n^2/2}$. As discussed in considering the overlap-save method of implementing a convolution, part of the circular convolution will correspond to a linear convolution and part will not. We can arrange for the "good" or desired points to occur in the region $0 \leq n \leq M - 1$, by interpreting the index n modulo $(N + M - 1)$. This means that we would compute the DFT of the sequence

$$h(n) = \begin{cases} W^{-n^2/2}, & 0 \leq n \leq M - 1 \\ W^{-(N+M-1-n)^2/2}, & M \leq n \leq N + M - 2 \end{cases}$$

as depicted in Fig. 6.33(c). If we multiply the discrete Fourier transforms of $g(n)$ and $h(n)$, the first M values of the corresponding inverse transform are the desired values of the convolution of $g(n)$ with $W^{-n^2/2}$. To obtain the desired M values of $X(k)$ as in Eq. (6.41), we must multiply these values by $W^{k^2/2}$.

In the above discussion the size of the DFTs computed was $(M + N - 1)$. If we wish to compute the discrete Fourier transform using a power-of-2 algorithm, this can easily be accomplished by augmenting the $(M + N - 1)$-point sequences with a sufficient number of zeros so that their total length is a power of 2.† Since the number of complex multiplications required for the computation of each DFT is on the order of

$$(N + M - 1) \cdot \log_2 (N + M - 1)$$

it is clear that the total computation required to implement the evaluation of Eq. (6.39) using the CZT algorithm is proportional to $(N + M - 1) \cdot \log_2 (N + M - 1)$. In contrast, the direct evaluation of Eq. (6.39) requires computation proportional to $N \cdot M$. Clearly, the direct method will be most efficient for small enough values of N or M; however, it is also true that for sufficiently large M and N (on the order of 50), the CZT algorithm will be most efficient.

In addition to increased efficiency, the CZT also offers added flexibility in computation of samples of the z-transform of a finite-length sequence. We do not require $N = M$ as in the FFT algorithms that we discussed, and neither N or M need be highly composite numbers; in fact, they may be primes, if desired. The parameter ϕ_0 is required to be $2\pi/N$ in an FFT algorithm, whereas ϕ_0 is arbitrary in the CZT. Furthermore, the samples of the z-transform are taken on a slightly more general contour that includes the unit circle as a special case.

The CZT algorithm is an example of the way that Fourier analysis can be performed using linear filtering. (The Goertzel algorithm is another.)

† The zeros must be added at point B in Fig. 6.33(c).

Rader [15] has shown that a similar procedure can be used to evaluate the discrete Fourier transform when N is a prime.

EXAMPLE. An example of the use of the CZT algorithm to sharpen resonances by evaluating the z-transform off the unit circle is shown in Fig. 6.34. The signal to be analysed corresponds to a finite-length segment of a synthetic speech signal. The speech signal was generated by exciting a five-pole system with a periodic impulse train. The system was simulated to correspond to a 10 kHz sampling frequency. The poles were located at center frequencies of 270, 2290, 3010, 3500 and 4500 Hz with bandwidths of 30, 50, 60, 87, and 140 Hz respectively.

Figure 6.34(a) depicts the z-plane plot indicating the location of the poles used to generate the signal. The CZT algorithm was applied to one period of the steady-state data for five different choices of $|W|$ with the results shown in Fig. 6.34(b). The first two spectra correspond to spiral contours outside the unit circle with a resulting broadening of the resonance peaks. $|W| = 1$ corresponds to evaluating the z-transform on the unit circle. As $|W|$ increases past unity the contour spirals inside the unit circle and closer to the pole locations resulting in a sharpening of the resonance peaks.

Summary

In this chapter we have considered techniques for computation of the discrete Fourier transform. Our objective has been to show how the periodicity and symmetry of the complex factor $e^{-j(2\pi/N)kn}$ can be exploited to increase the efficiency of DFT computations.

We considered the Goertzel algorithm and the direct evaluation of the DFT expression because of the importance of these techniques when not all N DFT values are required.

Our major concern was with the fast Fourier transform (FFT) algorithms. We described two classes of FFT algorithms, decimation-in-time and decimation-in-frequency, in some detail. Signal flow graphs were used to depict the organization of an FFT algorithm.

Most of the detailed discussion concerned algorithms that require that N be a power of 2; however, some discussion was devoted to applications of the basic decimation principles to cases where N was a product of two or more factors. As a final example of a fast algorithm with wider applicability, the chirp z-transform was discussed. We showed how an available FFT algorithm could be used to evaluate the transform of a finite-length sequence at an arbitrary number of points on a spiral contour in the z-plane.

In all the discussion, the aim was to present the basic principles of efficient computation of the DFT. Our approach was to illustrate the basic principles with examples of actual FFT algorithms, and therefore much of the discussion bears directly on the details of actually implementing FFT algorithms. With the material presented in this chapter, there should be little difficulty in programming a power-of-2 FFT algorithm. In fact, two simple programs (with errors) are given in Problems 4 and 5 of this chapter.

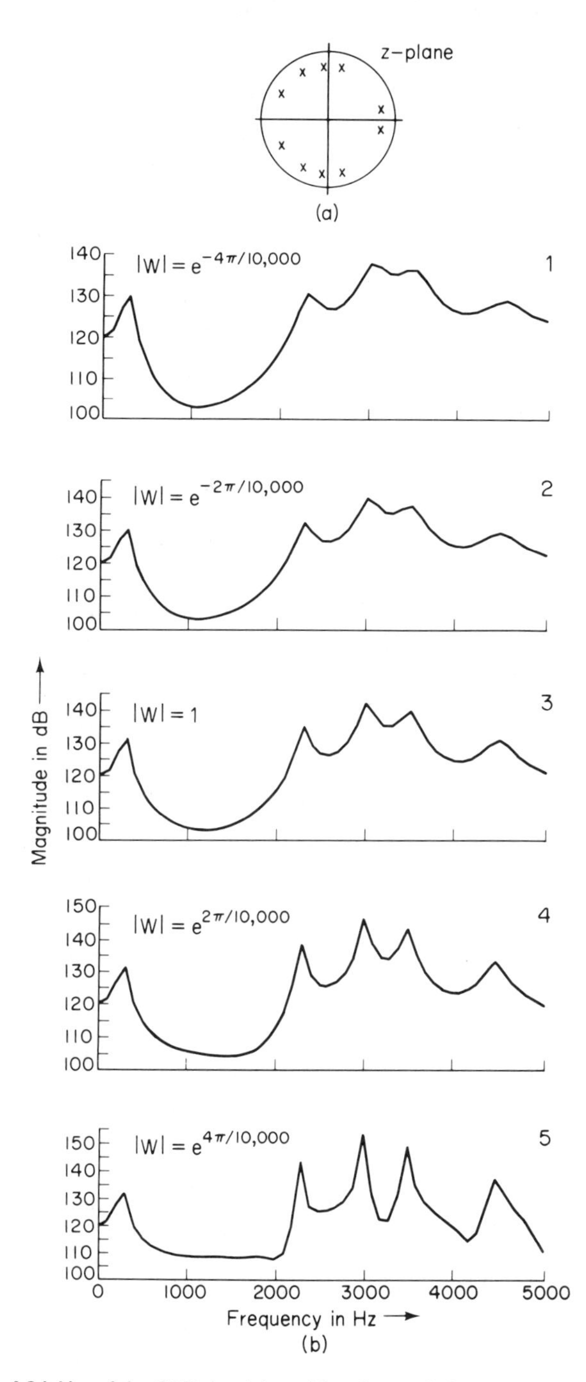

Fig. 6.34 Use of the CZT algorithm. (a) z-plane pole locations for synthetic speech signal. (b) Evaluation of z-transform for several spiral contours. (After Rabiner, Schafer, and Rader [13].)

REFERENCES

1. J. W. Cooley and J. W. Tukey, "An Algorithm for the Machine Calculation of Complex Fourier Series," *Math. Computation*, Vol. 19, 1965, pp. 297–301.
2. C. Runge, *Z. Math. Physik*, Vol. 48, 1903, p. 443; also Vol. 53, 1905, p. 117.
3. G. C. Danielson and C. Lanczos, "Some Improvements in Practical Fourier Analysis and Their Application to X-Ray Scattering from Liquids," *J. Franklin Inst.*, Vol. 233, pp. 365–380, 435–452.
4. J. W. Cooley, P. A. W. Lewis, and P. D. Welch, "Historical Notes on the Fast Fourier Transform," *IEEE Trans. Audio Electroacoust.*, Vol. AU-15, June 1967, pp. 76–79.
5. W. T. Cochran et al., "What is the Fast Fourier Transform?" *IEEE Trans. Audio Electroacoust.*, Vol. AU-15, June 1967, pp. 45–55.
6. G. Goertzel, "An Algorithm for the Evaluation of Finite Trigonometric Series," *Amer. Math. Monthly*, Vol. 65, Jan. 1958, pp. 34–35.
7. B. Gold and C. M. Rader, *Digital Processing of Signals*. McGraw-Hill Book Company, New York, 1969.
8. R. C. Singleton, "A Method for Computing the Fast Fourier Transform with Auxiliary Memory and Limited High-Speed Storage," *IEEE Trans. Audio Electroacoust.*, Vol. AU-15, June 1967, pp. 91–97.
9. W. M. Gentleman and G. Sande, "Fast Fourier Transforms—for Fun and Profit," in *Proc. 1966 Fall Joint Computer Conf.*, *AFIPS Conf. Proc.*, Vol. 29, pp. 563–578, Spartan Books, Washington, D.C., 1966.
10. R. C. Singleton, "An Algorithm for Computing the Mixed Radix Fast Fourier Transform," *IEEE Trans. Audio Electroacoust.*, Vol. AU-17, June 1969, pp. 93–103.
11. G. D. Bergland, "A Fast Fourier Transform Algorithm Using Base 8 Iterations," *Math. Computation*, Vol. 22, Apr. 1968, pp. 275–279.
12. J. O. Eklundh, "A Fast Computer Method for Matrix Transposing," *IEEE Trans. on Computers*, Vol. C-21, No. 7, July, 1972, pp. 801–803.
13. L. R. Rabiner, R. W. Schafer, and C. M. Rader, "The Chirp *z*-Transform Algorithm," *IEEE Trans. Audio Electroacoust.*, Vol. AU-17, June 1969, pp. 86–92.
14. L. I. Bluestein, "A Linear Filtering Approach to the Computation of Discrete Fourier Transform," *IEEE Trans. Audio Electroacoust.*, Vol. AU-18, Dec. 1970, pp. 451–455.
15. C. M. Rader, "Discrete Fourier Transforms When the Number of Data Samples is Prime," *Proc. IEEE*, Vol. 56, June 1968, pp. 1107–1108.

PROBLEMS

1. In Sec. 6.1 we used the fact that $W_N^{-kN} = 1$ to derive a recurrence algorithm for computing the DFT value $X(k)$ for a finite-length sequence $x(n)$.
   (a) Using the fact that $W_N^{kN} = W_N^{Nn} = 1$, show that $X(N-k)$ can be obtained as the output after N iterations of the difference equation depicted in Fig. P6.1-1. That is, show that

$$X(N-k) = y_k(N)$$

   (b) Show that $X(N-k)$ is also equal to the output after N iterations of the difference equation depicted in Fig. P6.1-2. It should be noted that the above system has the same poles as Fig. 6.2, but the coefficient required to

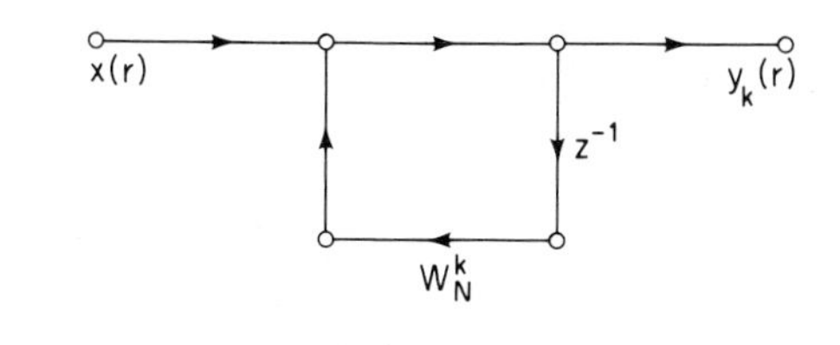

Fig. P6.1-1

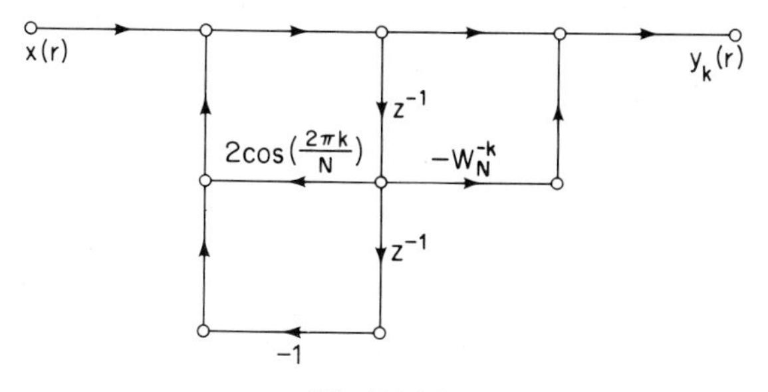

Fig. P6.1-2

implement the zero is the complex conjugate of that in Fig. 6.2. That is, $W_N^{-k} = (W_N^k)^*$.

2. When implementing a decimation-in-time FFT algorithm, the basic butterfly computation is as shown in the flow graph of Fig. P6.2.

$$X_{m+1}(p) = X_m(p) + W_N^r X_m(q)$$

$$X_{m+1}(q) = X_m(p) - W_N^r X_m(q)$$

In using fixed-point arithmetic in implementing the computations it is commonly assumed that all numbers are scaled to be less than unity. Thus we must be concerned with overflow in the butterfly computations.

(a) Show that if we require

$$|X_m(p)| < \tfrac{1}{2} \quad \text{and} \quad |X_m(q)| < \tfrac{1}{2}$$

then overflow cannot occur in the butterfly computation; i.e.,

$$|\text{Re}\,[X_{m+1}(p)]| < 1, \qquad |\text{Im}\,[X_{m+1}(p)]| < 1, \qquad |\text{Re}\,[X_{m+1}(q)]| < 1,$$

and

$$|\text{Im}\,[X_{m+1}(q)]| < 1.$$

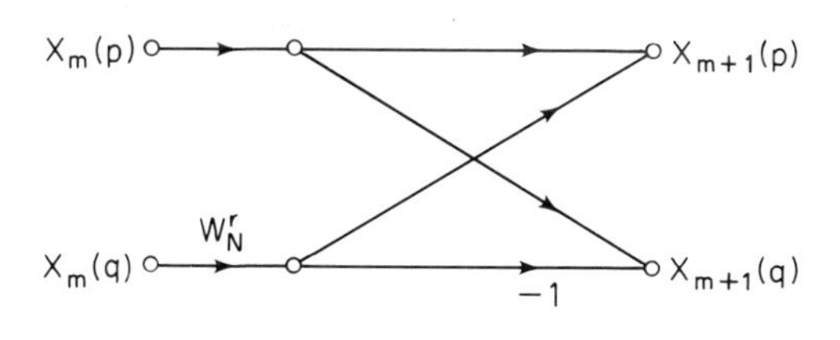

Fig. P6.2

(b) In practice, it is easier and more convenient to require

$$|\text{Re}\,[X_m(p)]| < \tfrac{1}{2}, \qquad |\text{Im}\,[X_m(p)]| < \tfrac{1}{2}$$

$$|\text{Re}\,[X_m(q)]| < \tfrac{1}{2}, \qquad |\text{Im}\,[X_m(q)]| < \tfrac{1}{2}$$

Are these conditions sufficient to ensure that overflow cannot occur in the butterfly computation? Justify your answer.

3. In implementing the FFT algorithm, it is sometimes useful to generate the powers of W_N recursively, using either a canonic-form or coupled-form oscillator. In the following discussion we shall consider the FFT algorithm for N a power of 2, implemented in the form characterized by Fig. 6.10. We are assuming that N is an arbitrary power of 2, *not* that $N = 8$.

To generate the coefficients efficiently, the frequency of the oscillator would change as the array being considered changed. The arrays are numbered 0 through $\log_2 N$, so that, for example, the array holding the initial data is the zeroth array, the next the first array, etc. In computing butterflies within an array, all butterflies requiring the same coefficients are evaluated before obtaining new coefficients. In indexing through the data in an array, we shall assume that points in an array are stored in consecutive complex (double) registers numbered 0 through $N - 1$. All the following questions relate to the computation of the mth array, where $1 \leq m \leq \log_2 N$. The answer should be in terms of m.

(a) How many butterflies are to be computed?
(b) What is the frequency of the oscillator used to generate the coefficients; i.e., how many times can it be iterated before repeating?
(c) Assuming that a coupled-form oscillator is used to generate the powers of W_N, what are the coefficients in the difference equation for the oscillator?
(d) What is the difference between the addresses of the two complex input points to a butterfly?
(e) What is the difference between the addresses of the first points of butterflies utilizing the same coefficients?

4. The FORTRAN program shown in Fig. P6.4 implements one of the FFT algorithms discussed in the text. The program is designed to compute the DFT,

$$X(k) = \sum_{n=0}^{N-1} x(n)e^{-j(2\pi/N)kn}, \qquad k = 0, 1, \ldots, N-1$$

In the subroutine FFT(X, M), X is a complex array of dimension N that contains initially the input sequence $x(n)$ and finally contains the transform $X(k)$. The quantity M is an integer, $M = \log_2 N$.

(a) From a cursory inspection of the program indicate which lines of code are concerned with (1) bit reversal, (2) recursive computation of the complex exponential multipliers, and (3) the basic butterfly computation.
(b) Determine which one of the flow diagrams in the chapter this program is based upon.
(c) Three errors have been inserted into the program as it is given here. Find these errors and make appropriate corrections to the FORTRAN code.

5. The FORTRAN program shown in Fig. P6.5 is an implementation of the decimation-in-frequency algorithm depicted in Fig. 6.18. The program evaluates the DFT:

$$X(k) = \sum_{n=0}^{N-1} x(n)e^{-j(2\pi/N)kn}, \qquad k = 0, 1, \ldots, N-1$$

```
      SUBROUTINE  FFT(X,M)                                       0001
      COMPLEX X(1024),U,W,T                                      0002
      N=2**M                                                     0003
      NV2=N/2                                                    0004
      NM1=N-1                                                    0005
      J=1                                                        0006
      DO 7 I=1,NM1                                               0007
      T=X(J)                                                     0008
      X(J)=X(I)                                                  0009
      X(I)=T                                                     0010
    5 K=NV2                                                      0011
    6 IF(K.GE.J) GO TO 7                                         0012
      J=J-K                                                      0013
      K=K/2                                                      0014
      GO TO 6                                                    0015
    7 J=J+K                                                      0016
      PI=3.14159265358979                                        0017
      DO 20 L=1,M                                                0018
      LE=2**L                                                    0019
      LE1=LE/2                                                   0020
      U=(1.0,0.0)                                                0021
      W=CMPLX(COS(PI/FLOAT(LE1)),SIN(PI/FLOAT(LE1)))             0022
      DO 20 J=1,LE1                                              0023
      DO 10 I=J,N,LE                                             0024
      IP=I+LE                                                    0025
      T=X(IP)*U                                                  0026
      X(IP)=X(I)-T                                               0027
   10 X(I)=X(I)+T                                                0028
   20 U=U*W                                                      0029
      RETURN                                                     0030
      END                                                        0031
```

Fig. P6.4

In the subroutine FFT(X, M), X is a complex array of dimension N that contains initially the input sequence $x(n)$ and finally contains the transform $X(k)$. The quantity M is an integer, $M = \log_2 N$.

This program is a straightforward implementation of the flow graph of Fig. 6.18. The program is very elegant but not as efficient as it could be. Greater efficiency can be obtained at the cost of a more complex program.

A significant increase in efficiency is suggested by noting that in the last stage of the flow graph in Fig. 6.18, the complex multipliers are all unity. Thus if the last stage is implemented separately, we can eliminate $N/2$ complex multiplications.

(a) What is the percentage reduction in multiplications that results?
(b) Modify the program to implement this saving in multiplications.
(c) Many small computers have FORTRAN compilers without the capability of complex arithmetic. Modify the given program so that only real operations are involved. That is, using the present subroutine as a guide, write a subroutine

FFT(XR, XI, M)

where XR and XI are real arrays of dimension N which initially contain the real part and the imaginary part of the input and finally the real and imaginary parts of the transform.

6. Draw the flow diagram for a nine-point (i.e., 3×3) decimation in time FFT algorithm.

```
      SUBROUTINE  FFT(X,M)                                    0001
      COMPLEX X(1024),U,W,T                                   0002
      N=2**M                                                  0003
      PI=3.14159265358979                                     0004
      DO 20 L=1,M                                             0005
      LE=2**(M+1-L)                                           0006
      LE1=LE/2                                                0007
      U=(1.0,0.0)                                             0008
      W=CMPLX(COS(PI/FLOAT(LE1)),-SIN(PI/FLOAT(LE1)))         0009
      DO 20 J=1,LE1                                           0010
      DO 10 I=J,N,LE                                          0011
      IP=I+LE1                                                0012
      T=X(I)+X(IP)                                            0013
      X(IP)=(X(I)-X(IP))*U                                    0014
10    X(I)=T                                                  0015
20    U=U*W                                                   0016
      NV2=N/2                                                 0017
      NM1=N-1                                                 0018
      J=1                                                     0019
      DO 30 I=1,NM1                                           0020
      IF(I.GE.J) GO TO 25                                     0021
      T=X(J)                                                  0022
      X(J)=X(I)                                               0023
      X(I)=T                                                  0024
25    K=NV2                                                   0025
26    IF(K.GE.J) GO TO 30                                     0026
      J=J-K                                                   0027
      K=K/2                                                   0028
      GO TO 26                                                0029
30    J=J+K                                                   0030
      RETURN                                                  0031
      END                                                     0032
```

Fig. P6.5

7. Suppose that N has a factor of 4; i.e., $N = 4 \cdot q_1$.
   (a) Express the DFT $X(k)$ as a combination of four q_1-point DFTs (label these q_1-point transforms $G_l(k)$, $l = 0, 1, 2, 3$) as in Eqs. (6.26) and (6.27).
   (b) Show that the basic computation required to compute $X(k)$ from the q_1-point transforms $G_l(k)$ is as depicted in Fig. 6.28.
   (c) Show that the flow graph of Fig. 6.28 can be simplified to that of Fig. 6.29.
   (d) Compare the number of complex multiplications required to implement a 16-point FFT computation assuming (1) N is decomposed as $N = 2 \cdot 2 \cdot 2 \cdot 2$, and (2) N is decomposed as $N = 4 \cdot 4$. (Assume that coefficients which are integer powers of $W_{16}^4 = j$ do not require any multiplications in either computation.)
8. Suppose that a program is available for evaluating a DFT,

$$X(k) = \sum_{n=0}^{N-1} x(n)e^{-j(2\pi/N)kn}, \qquad k = 0, 1, \ldots, N-1$$

Show how this same program can be used to compute the inverse DFT,

$$x(n) = \frac{1}{N}\sum_{k=0}^{N-1} X(k)e^{j(2\pi/N)kn}, \qquad n = 0, 1, \ldots, N-1$$

9. In this problem we shall consider a procedure for computing the DFT of four real symmetric or antisymmetric N-point sequences from one N-point DFT computation. Let $x_1(n)$, $x_2(n)$, $x_3(n)$, and $x_4(n)$ denote the four real sequences of length N and $X_1(k)$, $X_2(k)$, $X_3(k)$, and $X_4(k)$ denote their DFTs. We shall

first assume that $x_1(n)$ and $x_2(n)$ are symmetric and $x_3(n)$ and $x_4(n)$ are antisymmetric; i.e.,

$$x_1(n) = x_1((N-n))_N \mathcal{R}_N(n)$$

$$x_2(n) = x_2((N-n))_N \mathcal{R}_N(n)$$

$$x_3(n) = -x_3((N-n))_N \mathcal{R}_N(n)$$

$$x_4(n) = -x_4((N-n))_N \mathcal{R}_N(n)$$

(a) Define $y_1(n) = x_1(n) + x_3(n)$ and let $Y_1(k)$ denote the DFT of $y_1(n)$. Determine how $X_1(k)$ and $X_3(k)$ can be recovered from $Y_1(k)$.

(b) $y_1(n)$ as defined in part (a) is real. Similarly, we can define a real sequence $y_2(n) = x_2(n) + x_4(n)$. Let $y_3(n)$ denote the complex sequence

$$y_3(n) = y_1(n) + jy_2(n)$$

First determine how $Y_1(k)$ and $Y_2(k)$ can be determined from $Y_3(k)$ and then, using the results of part (a), show how to obtain $X_1(k)$, $X_2(k)$, $X_3(k)$, and $X_4(k)$ from $Y_3(k)$.

The result of part (b) shows that we can compute the DFT of four real sequences simultaneously if two are symmetric and two are antisymmetric. Now let us consider the case in which all four are symmetric; i.e.,

$$x_i(n) = x_i((N-n))_N \mathcal{R}_N(n), \qquad i = 1, 2, 3, 4$$

(c) Consider a real symmetric sequence $x_3(n)$. Show that the sequence $u_3(n) = x_3((n+1))_N - x_3((n-1))_N$ is an antisymmetric sequence; i.e.,

$$u_3(n) = -u_3((N-n))_N$$

(d) Let $U_3(k)$ denote the DFT of $u_3(n)$. Determine $U_3(k)$ in terms of $X_3(k)$.

(e) By using the procedure of part (c), we are able to represent the symmetric sequence $x_3(n) = x_1(n) + u_3(n)$. Determine how $X_1(k)$ and $X_3(k)$ can be recovered from $Y_1(k)$.

(f) Now let

$$y_3(n) = y_1(n) + jy_2(n)$$

where

$$y_1(n) = x_1(n) + u_3(n)$$

$$y_2(n) = x_2(n) + u_4(n)$$

with

$$u_3(n) = [x_3((n+1))_N - x_3((n-1))_N]\, \mathcal{R}_N(n)$$

$$u_4(n) = [x_4((n+1))_N - x_4((n-1))_N]\, \mathcal{R}_N(n)$$

Determine how to obtain $X_1(k)$, $X_2(k)$, $X_3(k)$, and $X_4(k)$ from $Y_3(k)$. [Note that $X_3(k)$ and $X_4(k)$ cannot be obtained for $k = 0$, and $X_3(N/2)$ and $X_4(N/2)$ cannot be determined if N is even.]

(g) $X_3(k)$ and $X_4(k)$ cannot be determined for $k = 0$ or $k = N/2$ using the preceding method. Show that these points can be evaluated without performing any multiplications.

10. In computing the DFT of real sequences, it is possible to reduce the amount of computation by utilizing the fact that the sequence is real. In this problem we shall discuss several ways of doing this.

(a) Let $x(n)$ be a real-valued sequence with N points and let $X(k)$ represent its DFT, with real and imaginary parts denoted by $X_R(k)$ and $X_I(k)$, respectively, so that

$$X(k) = X_R(k) + jX_I(k)$$

Show that if $x(n)$ is real, then $X_R(k)$ is even and that $X_I(k)$ is odd, i.e., that $X_R(k) = X_R((N - k))_N \mathcal{R}_N(k)$ and $X_I(k) = -X_I((N - k))_N \mathcal{R}_N(k)$.

(b) Consider two real-valued sequences $x_1(n)$ and $x_2(n)$ with DFTs $X_1(k)$ and $X_2(k)$, respectively. Let $g(n)$ be a sequence with complex values, defined as $g(n) = x_1(n) + jx_2(n)$, and let $G(k)$ be its DFT. Let $G_{OR}(k)$, $G_{ER}(k)$, $G_{OI}(k)$, and $G_{EI}(k)$ denote, respectively, the odd part of the real part, the even part of the real part, the odd part of the imaginary part, and the even part of the imaginary part. Determine $X_1(k)$ and $X_2(k)$ in terms of $G_{OR}(k)$, $G_{ER}(k)$, $G_{OI}(k)$, and $G_{EI}(k)$.

The result derived in part (b) can be utilized in several ways: if there are two real sequences for which we want the DFT, we can compute their transforms simultaneously and then separate the transforms using the result in part (b). Another possibility which we will now consider is to initially break a real sequence into two smaller sequences, simultaneously compute the DFT of each of these sequences, and then combine these results to obtain the DFT of the total sequence.

(c) Assume that $x(n)$ is a real-valued sequence with N points and that N is divisible by 2. Let $x_1(n)$ and $x_2(n)$ be the two $N/2$-point sequences defined as

$$x_1(n) = x(2n), \qquad n = 0, 1, 2, \ldots, N/2 - 1$$

$$x_2(n) = x(2n + 1), \qquad n = 0, 1, 2, \ldots, N/2 - 1$$

Determine $X(k)$ in terms of $X_1(k)$ and $X_2(k)$.

11. Consider a finite-length sequence $x(n)$ such that $x(n) = 0$ for $n < n_0$ and $n > N - 1 + n_0$. Suppose that we wish to compute samples of the z-transform of $x(n)$ at the following points in the z-plane:

$$z_k = re^{j(\theta+(2\pi/M)k)}, \qquad k = 0, 1, \ldots, M - 1$$

where $M < N$.

Describe in detail an efficient procedure for computing $X(z)$ at the desired points.

12. Consider a finite-duration sequence $x(n)$ of length M such that $x(n) = 0$ for $n < 0$ and $n > M$. We want to compute samples of the z-transform

$$X(z) = \sum_{n=0}^{M-1} x(n)z^{-n}$$

at n equally spaced points around the unit circle, i.e., at

$$z = e^{j(2\pi/N)k}, \qquad k = 0, 1, \ldots, N - 1$$

Determine and justify procedures for computing the N samples of $X(z)$ using only one N-point DFT for the cases

(a) $N \leq M$.

(b) $N > M$.

13. $X(e^{j\omega})$ denotes the Fourier transform of a finite-duration sequence $x(n)$ of length 10. We wish to compute 10 samples of $X(e^{j\omega})$ at frequencies $\omega_k = (2\pi k^2/100)$, $k = 0, 1, \ldots, 9$, *without* computing more samples of $X(e^{j\omega})$ than required and then discarding some. Discuss the possibility of doing this with each of the following methods:

(a) Directly, using a 10-point FFT algorithm.
(b) Using the Chirp z-transform algorithm.
(c) Using the Goertzel algorithm.

14. One application of the CZT algorithm is to the sharpening of resonance peaks in the spectrum. In general, if we compute the transform of a sequence on a contour in the z-plane near a pole, we expect to observe a resonance. In applying the CZT algorithm, or in computing the DFT, the sequence being analyzed must be of finite-duration. If it is not, the sequence must first be truncated. Whereas the transform of the original sequence had poles, the transform of the truncated sequence can only have zeros (except at $z = 0$ or $z = \infty$). The purpose of this problem is to show that a resonant type of response will still be observed in the transform of the finite-duration sequence.

(a) Let $x(n) = u(n)$. Sketch the pole–zero pattern of its z-transform $X(z)$.
(b) Let

$$\hat{x}(n) = \begin{cases} 1, & 0 \leq n \leq N-1 \\ 0, & \text{otherwise} \end{cases}$$

i.e., $\hat{x}(n)$ is equal to $x(n)$ truncated after N points. Sketch the pole–zero pattern of $\hat{X}(z)$, the z-transform of $\hat{x}(n)$.

(c) Sketch $|\hat{X}(e^{j\omega})|$ as a function of ω. Indicate on your sketch the effect of increasing N.

15. Choose the correct ending to the following sentence. The Chirp z-transform can be used to calculate the z-transform $H(z)$ of a finite-duration sequence $h(n)$ at points $\{z_k\}$ on the *real z-axis of the z-plane* such that

(a) $z_k = a^k$, $k = 0, 1, \ldots, N-1$ for a real, $a \neq \pm 1$.
(b) $z_k = ak$, $k = 0, 1, \ldots, N-1$ for a real, $a \neq 0$.
(c) Both parts (a) and (b).
(d) Neither part (a) nor part (b) i.e., the CZT cannot be used to calculate samples of $H(z)$ for real z.

16. Bluestein [14] has proposed a recursive scheme for evaluating all the DFT values

$$X(k) = \sum_{n=0}^{N-1} x(n)e^{-j(2\pi/N)kn}, \qquad k = 0, 1, \ldots, N-1$$

for N a perfect square; i.e., $N = M^2$.

(a) Using the substitution

$$-nk = \frac{(k-n)^2}{2} - \frac{n^2}{2} - \frac{k^2}{2}$$

show that $X(k)$ can be expressed as the convolution

$$X(k) = h^*(k) \sum_{n=0}^{N-1} (x(n)h^*(n))h(k-n)$$

where

$$h(n) = e^{j(\pi/N)n^2}, \qquad -\infty < n < \infty$$

(b) Show that the desired values of $X(k)$ (i.e., for $k = 0, 1, \ldots, N-1$) can also be obtained by evaluating the convolution of part (a) for $k = N$, $N+1, \ldots, 2N-1$.
(c) Use the result of part (b) to show that $X(k)$ is also equal to the output of the system shown in Fig. P6.16 for $k = N, N+1, \ldots, 2N-1$, where

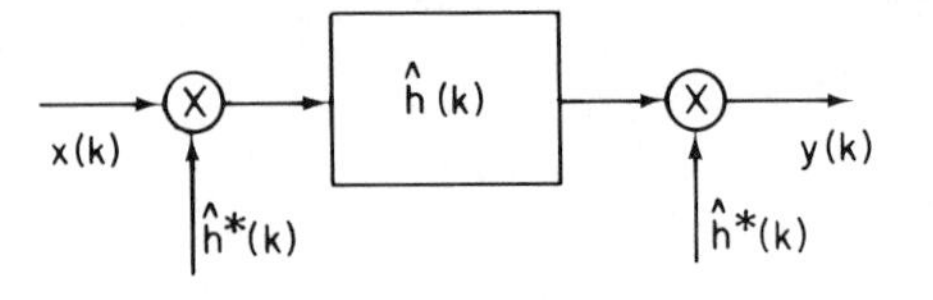

Fig. P6.16

$\hat{h}(k)$ is the finite-duration sequence

$$\hat{h}(k) = e^{j(\pi/N)k^2}, \qquad 0 \le k \le 2N - 1$$

(d) Using the fact that $N = M^2$, show that the system function corresponding to the impulse response $\hat{h}(k)$ is

$$\hat{H}(z) = \sum_{k=0}^{2N-1} e^{j(\pi k^2/N)} z^{-k}$$

$$= \sum_{r=0}^{M-1} z^{-r} e^{j(\pi r^2/N)} \frac{1 - z^{-2M^2}}{1 + e^{j(2\pi/M)r} z^{-M}}$$

(*Hint*: Express k as $k = r + lM$.)

(e) The above expression for $\hat{H}(z)$ suggests a recursive realization of the finite-duration impulse response system. Draw a flow graph of such an implementation.

(f) Use the result of part (e) to determine the total number of complex multiplications and additions required to compute all the desired N values of $X(k)$. Compare with the number required for direct evaluation of $X(k)$.

7

Discrete Hilbert Transforms

7.0 Introduction

In almost every field where Fourier techniques are used to represent and analyze physical processes, one finds that there are situations where there exist relationships between the real and imaginary parts or the magnitude and phase of the Fourier transform. These relationships are known by different names, depending upon the field of interest, but often they are called *Hilbert transform relations*. In this respect the field of digital signal processing is no exception.

We shall see, for example, that if a sequence is causal, then the real and imaginary parts of its Fourier transform are related by a Hilbert transform integral. In this chapter we shall derive a number of such relationships that are important in both the theory and application of digital signal processing. We shall see in Chapter 10, for example, that a thorough understanding of many of the topics of this chapter is essential for gaining insight into homomorphic deconvolution.

The complex functions that arise in the mathematical representation of discrete-time signals and systems are generally very well behaved functions. With few exceptions, the z-transforms that have concerned us have had well-defined regions in which the power series was absolutely convergent. Since a power series represents an analytic function within its region of convergence [1,2], it follows that z-transforms are analytic functions inside their regions of convergence. By the definition of an analytic function, this means that the z-transform has a unique derivative at every point inside the region of convergence. Furthermore, analyticity implies that the z-transform and all its derivatives are continuous functions within the region of convergence.

These properties of analytic functions imply some rather powerful constraints on the behavior of the z-transform within its region of convergence. One such constraint is that the real and imaginary parts satisfy the Cauchy–Riemann conditions. These conditions relate the partial derivatives of the real and imaginary parts of an analytic function. Another is the Cauchy integral theorem, through which the value of a complex function is specified everywhere inside a region of analyticity in terms of the values of the function on the boundary of the region. Based on these relations for analytic functions, it is possible, under certain conditions, to derive explicit integral relationships between the real and imaginary parts of a z-transform on a closed contour within the region of convergence. In the mathematics literature these relations are often referred to as *Poisson's formulas* [2,3]. In the context of system theory, they are known as the *Hilbert transform relations* and they have traditionally played an important role in the theory and practice of signal processing [4,5].

While the Hilbert transform relations can be developed formally from the properties of analytic functions (see Problems 2 and 4 of this chapter), a somewhat more intuitive approach will be taken in this chapter. Specifically, the Hilbert transform relations will be developed from the point of view that the real and imaginary parts (on the unit circle) of the z-transform of a causal sequence are the transforms of the even and odd components of the sequence. As we will show, a causal sequence has the property that it is completely specified by its even part, implying that the z-transform of the original sequence is completely specified by its real part on the unit circle. In addition to applying this argument to specifying the z-transform of a causal sequence in terms of its real part on the unit circle, it can also be applied, under certain conditions, to specify the z-transform of a sequence in terms of its *magnitude* on the unit circle.

The notion of an analytic signal is an important concept in continuous-time signal processing [6]. An analytic signal is a complex time function (which is analytic) having a Fourier transform that vanishes for negative frequencies. A *complex sequence* cannot be considered in a formal sense to be analytic since it is a function of an integer variable. However, it is possible, in a style similar to that described above, to relate the real and imaginary parts of a complex sequence whose spectrum is zero on the unit circle for $-\pi < \omega < 0$. A similar approach can also be taken to relating the real and imaginary parts of the discrete Fourier transform for a periodic or, equivalently, finite-duration sequence. In this case the "causality" condition is that the periodic sequence be zero in the second half of each period.

Thus, in the discussion that follows, a notion of causality will be applied to relate the even and odd components of a function or equivalently the real and imaginary parts of its transforms. This approach will be applied in four situations. The first relates the real and imaginary parts of the Fourier transform $H(e^{j\omega})$ of a sequence $h(n)$ that is zero for $n < 0$. The second relates the real and imaginary parts of the *logarithm* of the Fourier transform under the condition that the inverse transform of the logarithm of the transform is zero for $n < 0$. Relating the real and imaginary parts of the log spectrum corresponds to relating the log magnitude and phase of $H(e^{j\omega})$. In the third situation the real and imaginary parts of the DFT will be related for periodic sequences or, equivalently, for a finite-duration sequence considered to be of length N but with the last $N/2$ points zero. Finally, we relate the real and imaginary parts of a complex sequence whose Fourier transform, considered as a periodic function of ω, is zero in the second half of each period.

7.1 Real- and Imaginary-Part Sufficiency for Causal Sequences

Any sequence can be expressed as the sum of an even sequence and an odd sequence. Specifically, with $h_e(n)$ and $h_o(n)$ denoting the even and odd parts of $h(n)$, then

$$h(n) = h_e(n) + h_o(n) \tag{7.1}$$

where

$$h_e(n) = \tfrac{1}{2}[h(n) + h(-n)] \tag{7.2}$$

and

$$h_o(n) = \tfrac{1}{2}[h(n) - h(-n)] \tag{7.3}$$

Equations (7.1)–(7.3) apply to an arbitrary sequence whether or not it is causal or whether or not it is real. However, if $h(n)$ is causal, then it is possible to recover $h(n)$ from $h_e(n)$ and to recover $h(n)$ for $n \neq 0$ from $h_o(n)$. Consider, for example, the causal sequence $h(n)$ and its even and odd components as shown in Fig. 7.1. Because $h(n)$ is causal, i.e., $h(n)$ is zero for $n < 0$ and $h(-n)$ is zero for $n > 0$, there is no overlap between the nonzero portions of $h(n)$ and $h(-n)$ except at $n = 0$.

It should be clear from Fig. 7.1 and Eqs. (7.2) and (7.3) that for causal sequences

$$h(n) = \begin{cases} 2h_e(n), & n > 0 \\ h_e(n), & n = 0 \\ 0, & n < 0 \end{cases} \tag{7.4}$$

and

$$h(n) = \begin{cases} 2h_o(n), & n > 0 \\ 0, & n \leq 0 \end{cases} \tag{7.5}$$

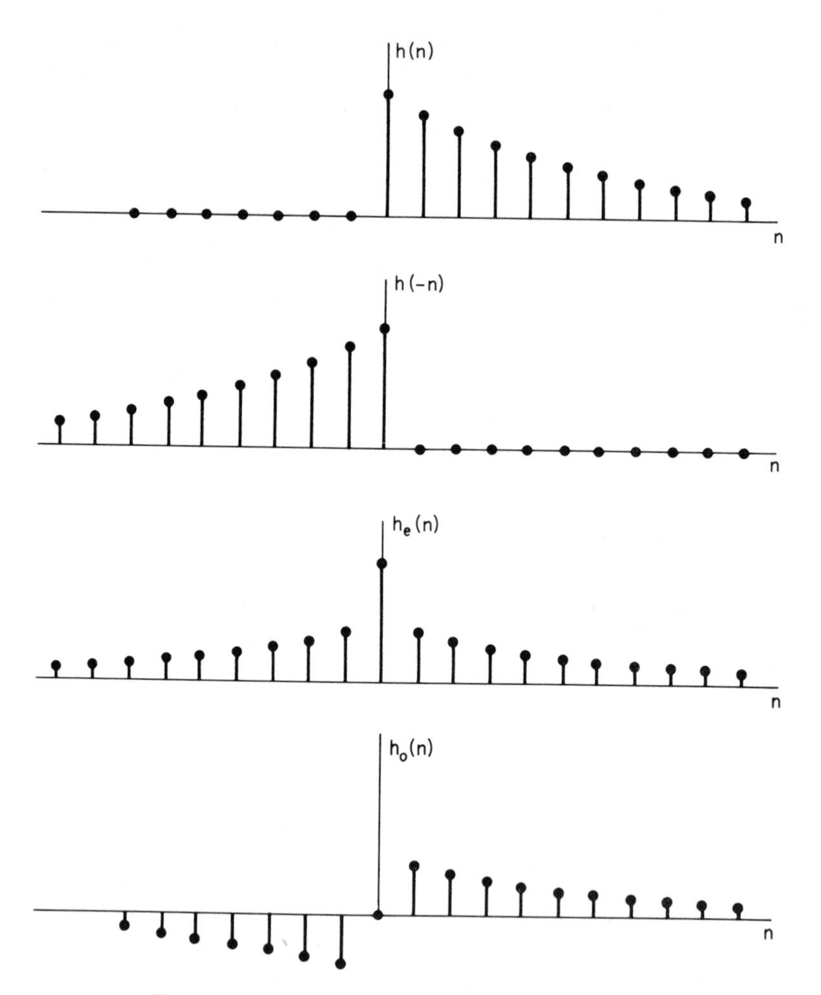

Fig. 7.1 Even and odd parts of a real causal sequence.

Equivalently, if we define

$$u_+(n) = \begin{cases} 2, & n > 0 \\ 1, & n = 0 \\ 0, & n < 0 \end{cases} \tag{7.6}$$

then

$$h(n) = h_e(n)u_+(n) \tag{7.7}$$

and

$$h(n) = h_o(n)u_+(n) + h(0)\,\delta(n) \tag{7.8}$$

We note that $h(n)$ can be completely recovered from $h_e(n)$. On the other hand, $h_o(n)$ will always be zero at $n = 0$, and consequently $h(n)$ can be recovered from $h_o(n)$ only for $n \neq 0$.

An important consequence of Eqs. (7.7) and (7.8) is the implication that the Fourier transform of a real, causal, and stable sequence,

$$H(e^{j\omega}) = H_R(e^{j\omega}) + jH_I(e^{j\omega})$$

is completely known if we know either the real part $H_R(e^{j\omega})$ or the imaginary part $H_I(e^{j\omega})$ and $h(0)$. This is because $H_R(e^{j\omega})$ is the Fourier transform of $h_e(n)$ and $jH_I(e^{j\omega})$ is the Fourier transform of $h_o(n)$. For example, we can compute $h_e(n)$ from $H_R(e^{j\omega})$; then, using Eq. (7.7), we can compute $h(n)$, from which we can compute $H(e^{j\omega})$.

More generally we can show that if $h(n)$ is real, causal, and stable, then $H(z)$ can be determined anywhere in the region outside the unit circle [i.e., the region of convergence of $H(z)$] from a knowledge of either $H_R(e^{j\omega})$ or $H_I(e^{j\omega})$ and $h(0)$ [7–9]. Consider $H(z)$ *outside* the unit circle, i.e., for $z = re^{j\omega}$ with $r > 1$. In this case

$$H(z)\big|_{z=re^{j\omega}} = H(re^{j\omega}) = \sum_{n=0}^{\infty} h(n)r^{-n}e^{-j\omega n}$$

or, using Eq. (7.7),

$$H(re^{j\omega}) = \sum_{n=-\infty}^{\infty} h_e(n)u_+(n)r^{-n}e^{-j\omega n}$$

Alternatively, this expression can be interpreted as the Fourier transform of the product $h_e(n) \cdot [r^{-n}u_+(n)]$. Therefore, $H(re^{j\omega})$ can be obtained as the convolution of the Fourier transform of $h_e(n)$ with the Fourier transform of the sequence $r^{-n}u_+(n)$. The Fourier transform of $h_e(n)$ is $H_R(e^{j\omega})$, and if $r > 1$, the Fourier transform of $r^{-n}u_+(n)$ is $(1 + r^{-1}e^{-j\omega})/(1 - r^{-1}e^{-j\omega})$. [Note that strictly speaking, the Fourier transform of $r^{-n}u_+(n)$ does not exist if $r = 1$.] Now, using the complex convolution theorem of Sec. 2.3.8, we obtain the contour integral relationship

$$H(z)\big|_{z=re^{j\omega}} = \frac{1}{2\pi j}\oint_C \frac{H_R(v)(e^{j\omega} + r^{-1}v)\,dv}{(e^{j\omega} - r^{-1}v)v} \tag{7.9a}$$

or, alternatively,

$$H(z) = \frac{1}{2\pi j}\oint_C \frac{H_R(v)(z + v)\,dv}{(z - v)v}, \qquad |z| > 1 \tag{7.9b}$$

In Eqs. (7.9a) and (7.9b), C must be the unit circle, since it is assumed that only $H_R(e^{j\omega})$ is known. These equations are particularly convenient when $H_R(e^{j\omega})$ can be expressed as a rational function of $e^{j\omega}$ since the integral can then be easily evaluated using the calculus of residues.

EXAMPLE. Suppose that we are given

$$H_R(e^{j\omega}) = \frac{1 - \alpha\cos\omega}{1 - 2\alpha\cos\omega + \alpha^2}, \qquad |\alpha| < 1$$

Let us find $H(z)$ using Eq. (7.9). First we write $H_R(e^{j\omega})$ as a rational function of $e^{j\omega}$,

$$H_R(e^{j\omega}) = \frac{1 - \alpha(e^{j\omega} + e^{-j\omega})/2}{(1 - \alpha e^{-j\omega})(1 - \alpha e^{j\omega})}$$

Then substituting into Eq. (7.9b) we obtain

$$H(z) = \frac{1}{2\pi j} \oint_C \frac{(1 - \alpha(v + v^{-1})/2)}{(1 - \alpha v^{-1})(1 - \alpha v)} \frac{z + v}{z - v} \frac{dv}{v}$$

where C is the unit circle. Writing this equation so as to display the poles of the integrand, we obtain

$$H(z) = \frac{1}{2\pi j} \oint_C \frac{(v - \alpha(v^2 + 1)/2)(z + v)\, dv}{(v - \alpha)(1 - \alpha v)(z - v)v}$$

The poles of the integrand are depicted in Fig. 7.2, where we note that only the

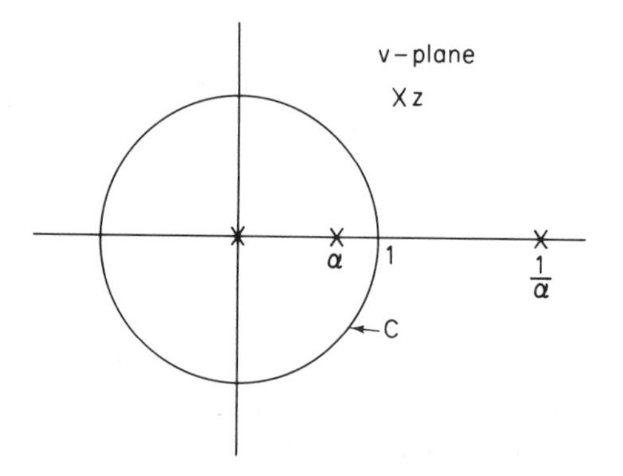

Fig. 7.2 Pole locations in example of computation of the z-transform using contour integration.

poles at $v = 0$ and $v = \alpha$ are inside the unit circle. Therefore, using the residue theorem we obtain

$$H(z) = \frac{-(\alpha/2)z}{-\alpha z} + \frac{(\alpha - \alpha(\alpha^2 + 1)/2)(z + \alpha)}{(1 - \alpha^2)(z - \alpha)\alpha}$$

$$= \frac{1}{2} + \frac{1}{2}\frac{z + \alpha}{z - \alpha} = \frac{z}{z - \alpha}$$

The above expression was derived assuming $|z| > 1$; however, we note that the region of analyticity is $|z| > \alpha$. Thus we have obtained the z-transform directly from its real part on the unit circle.

Equations (7.9a) and (7.9b) express $H(z)$ outside the unit circle in terms of its real part on the unit circle. However, it is also useful to write this expression as a line integral. Therefore, let $v = e^{j\theta}$ in Eq. (7.9a) so that we obtain

$$H(z)\Big|_{z=re^{j\omega}} = \frac{1}{2\pi}\int_{-\pi}^{\pi} H_R(e^{j\theta})P_r(\theta - \omega)\,d\theta + \frac{j}{2\pi}\int_{-\pi}^{\pi} H_R(e^{j\theta})Q_r(\theta - \omega)\,d\theta \tag{7.10}$$

where

$$P_r(\theta) = \text{Re}\left(\frac{1 + r^{-1}e^{j\theta}}{1 - r^{-1}e^{j\theta}}\right) = \frac{1 - r^{-2}}{1 - 2r^{-1}\cos\theta + r^{-2}} \tag{7.11a}$$

and

$$Q_r(\theta) = \text{Im}\left(\frac{1 + r^{-1}e^{j\theta}}{1 - r^{-1}e^{j\theta}}\right) = \frac{2r^{-1}\sin\theta}{1 - 2r^{-1}\cos\theta + r^{-2}} \tag{7.11b}$$

The functions $P_r(\theta)$ and $Q_r(\theta)$ are often referred to as the *Poisson kernel* and the *conjugate Poisson kernel*, respectively [2,3]. Equating the real and imaginary parts in Eq. (7.10), we obtain

$$H_R(re^{j\omega}) = \frac{1}{2\pi}\int_{-\pi}^{\pi} H_R(e^{j\theta})P_r(\theta - \omega)\,d\theta \tag{7.12}$$

and

$$H_I(re^{j\omega}) = \frac{1}{2\pi}\int_{-\pi}^{\pi} H_R(e^{j\theta})Q_r(\theta - \omega)\,d\theta \tag{7.13}$$

Thus we have derived real integral relations for the real and imaginary parts of the z-transform outside the unit circle in terms of the unit circle evaluation of only the real part.

Similar manipulations, beginning with Eq. (7.8), lead to the contour integral representation

$$H(z)\Big|_{z=re^{j\omega}} = \frac{1}{2\pi}\oint_C \frac{H_I(v)(e^{j\omega} + r^{-1}v)\,dv}{(e^{j\omega} - r^{-1}v)v} + h(0) \tag{7.14a}$$

or

$$H(z) = \frac{1}{2\pi}\oint_C \frac{H_I(v)(z + v)\,dv}{(z - v)v} + h(0), \qquad |z| > 1 \tag{7.14b}$$

where the contour C is again the unit circle. Converting Eq. (7.14a) to line integrals and equating real and imaginary parts leads to

$$H_R(re^{j\omega}) = -\frac{1}{2\pi}\int_{-\pi}^{\pi} H_I(e^{j\theta})Q_r(\theta - \omega)\,d\theta + h(0) \tag{7.15}$$

and

$$H_I(re^{j\omega}) = \frac{1}{2\pi}\int_{-\pi}^{\pi} H_I(e^{j\theta})P_r(\theta - \omega)\,d\theta \tag{7.16}$$

where $P_r(\theta)$ and $Q_r(\theta)$ are given by Eq. (7.11a) and (7.11b).

In order to obtain direct relations between the real part on the unit circle and the imaginary part on the unit circle, it is necessary to take the limit as r approaches unity in Eqs. (7.13) and (7.15). This is valid if we perform the integration first. However, if we attempt to obtain a direct relation for $H_I(e^{j\omega})$ in terms of $H_R(e^{j\omega})$, by interchanging the order of integrating and taking the limit, we are confronted with an improper integral since

$$\lim_{r\to 1} Q_r(\theta) = \frac{2 \sin \theta}{2(1 - \cos \theta)} = \cot\left(\frac{\theta}{2}\right)$$

and the function cot $(\theta/2)$ has a singularity at $\theta = 0$. The desired relations can be obtained if we are careful in evaluating the improper integrals in the vicinity of the singular points of the integrand. This can be done formally by interpreting the integrals as *Cauchy principal values* [10]. That is, Eq. (7.13) becomes

$$H_I(e^{j\omega}) = \frac{1}{2\pi} P\int_{-\pi}^{\pi} H_R(e^{j\theta}) \cot\left(\frac{\theta - \omega}{2}\right) d\theta \tag{7.17}$$

and Eq. (7.15) becomes

$$H_R(e^{j\omega}) = h(0) - \frac{1}{2\pi} P\int_{-\pi}^{\pi} H_I(e^{j\theta}) \cot\left(\frac{\theta - \omega}{2}\right) d\theta \tag{7.18}$$

where the symbol P denotes Cauchy principal value. The meaning of Cauchy principal value in Eq. (7.17), for example, is given in Eq. (7.19):

$$H_I(e^{j\omega}) = \frac{1}{2\pi} \lim_{\varepsilon\to 0} \left\{\int_{\omega+\varepsilon}^{\pi} H_R(e^{j\theta}) \cot\left(\frac{\theta - \omega}{2}\right) d\theta + \int_{-\pi}^{\omega-\varepsilon} H_R(e^{j\theta}) \cot\left(\frac{\theta - \omega}{2}\right) d\theta\right\} \tag{7.19}$$

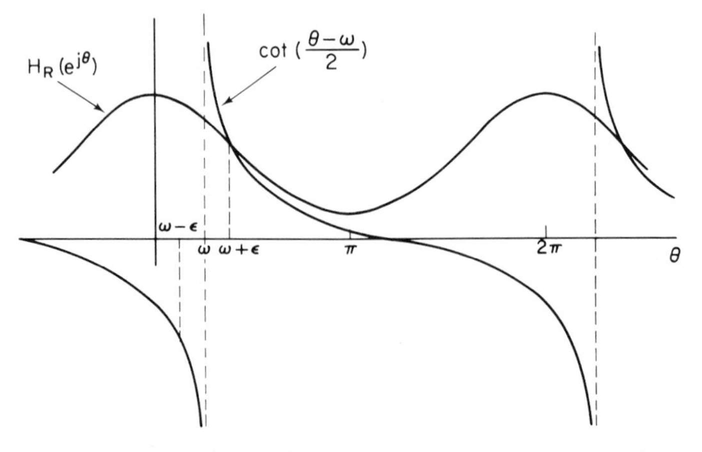

Fig. 7.3 Interpretation of the Hilbert transform as a periodic convolution.

We note that $H_I(e^{j\omega})$ is obtained by the periodic convolution of cot $(-\omega/2)$ with $H_R(e^{j\omega})$, with special care being taken in the vicinity of the singularity at $\theta = \omega$. In a similar manner, Eq. (7.18) involves the periodic convolution of cot $(-\omega/2)$ with $H_I(e^{j\omega})$.

The two functions involved in the convolution integral of Eq. (7.17) [or, equivalently, Eq. (7.19)] are depicted in Fig. 7.3. The fact that the limit in Eq. (7.19) exists depends upon the fact that the function cot $[(\theta - \omega)/2]$ is antisymmetric at the singular point $(\theta = \omega)$ and the gap is symmetrically placed around the singularity.

The evaluation of the integrals in the previous equations is further complicated when $H(z)$ has poles on the unit circle. Our discussion has assumed that the unit circle is entirely within the region of convergence of $H(z)$, and therefore that $H(z)$ has no poles on the unit circle. Poles on the unit circle can be accommodated by allowing impulses in the Fourier transform or by employing an indented contour in the contour integral expressions. However, the mathematical justification of these procedures would take us too far afield and we shall not discuss this question further.

7.2 Minimum-Phase Condition

In the previous section the z-transform of a causal sequence was recovered from its real or imaginary part on the unit circle. In this section conditions will be discussed under which the z-transform can be recovered from its magnitude or phase on the unit circle. These conditions are of considerable importance in many theoretical and practical situations. For example, digital filters are often specified in terms of the magnitude of the frequency response. In such cases the phase response cannot be chosen arbitrarily if a stable and causal system is desired. The results of this section are also very important in the theory of homomorphic systems, which is developed in Chapter 10. Another example arises in the theory and application of inverse filtering, where it is necessary to obtain an appropriate phase curve given only an autocorrelation function (equivalently the squared magnitude of the Fourier transform) [9,11–13].

Suppose that $H(z)$ is specified in polar form (magnitude and phase) as

$$H(z) = |H(z)|\, e^{j\arg[H(z)]}$$

Then, consider the complex logarithm, of $H(z)$, defined as

$$\hat{H}(z) = \log\,[H(z)] = \log|H(z)| + j\arg\,[H(z)] \tag{7.20}$$

If we think of $\hat{H}(z)$ as the z-transform of a sequence $\hat{h}(n)$, then the results of the previous section imply that $\log|H(e^{j\omega})|$ and $\arg\,[H(e^{j\omega})]$ will be Hilbert transforms of each other if and only if $\hat{h}(n)$ is a real, causal, and stable sequence. In considering this question, we must be careful in our interpretation of Eq. (7.20). In particular, the logarithm of zero diverges and the definition of arg $[H(z)]$ is ambiguous since any multiple of 2π can be added to the phase without affecting the value of $H(z)$. Since we wish to interpret $\hat{H}(z)$ as the z-transform of a real, causal, and stable sequence, we shall wish to consider the region of convergence of $\hat{H}(z)$ to include the unit circle, and consequently we require that $\hat{H}(z)$ be analytic in a region that includes the unit circle. Within this region, then, $\hat{H}(z)$ must have a convergent power-series representation

$$\hat{H}(z) = \sum_{n=0}^{\infty} \hat{h}(n) z^{-n}, \qquad R_{\hat{h}-} < |z|$$

where $R_{\hat{h}-} < 1$. Since $\hat{H}(z)$ is infinite at both the poles and the zeros of $H(z)$, we require that within the region of convergence associated with $\hat{H}(z)$, there are no poles or zeros of $H(z)$. Although arg $[H(z)]$ is not generally unique, the ambiguity is resolved by the fact that analyticity of $\hat{H}(z)$ implies that its real and imaginary parts must be continuous functions of z, and consequently, if $\hat{H}(z)$ is to be analytic, we must define arg $[H(z)]$ in Eq. (7.20) to be a continuous function. Furthermore, we will require that for $h(n)$ real, $\hat{H}(z)$ is the z-transform of a real sequence. Consequently, arg $[H(z)]$ will be defined such that for $z = e^{j\omega}$, it is an odd, continuous function of ω.†

Now let us consider a real, stable sequence $\hat{h}(n)$ whose z-transform is $\hat{H}(z)$. From the previous section it should be clear that if $\hat{h}(n)$ is causal, then $\hat{H}(z)$ and consequently $H(z)$ can be recovered from $\hat{H}_R(e^{j\omega}) = \log |H(e^{j\omega})|$ or $\hat{H}_I(e^{j\omega}) = \arg [H(e^{j\omega})]$. Equivalently, if $\hat{h}(n)$ is real, stable, and causal, Eqs. (7.17) and (7.18) can be applied to relate the log magnitude and phase of $H(e^{j\omega})$ so that

$$\log |H(e^{j\omega})| = \hat{h}(0) - \frac{1}{2\pi} P\int_{-\pi}^{\pi} \arg [H(e^{j\theta})] \cot\left(\frac{\theta - \omega}{2}\right) d\theta \tag{7.21}$$

$$\arg [H(e^{j\omega})] = \frac{1}{2\pi} P\int_{-\pi}^{\pi} \log |H(e^{j\theta})| \cot\left(\frac{\theta - \omega}{2}\right) d\theta \tag{7.22}$$

Note that without knowledge of $\hat{h}(0)$, $|H(e^{j\omega})|$ is specified only to within a constant multiplier by arg $[H(e^{j\omega})]$.

The requirement that $\log |H(e^{j\omega})|$ and arg $[H(e^{j\omega})]$ be a Hilbert transform pair is often referred to as the *minimum-phase condition* [4, 5, 14].‡ It corresponds to the requirement that the sequence $\hat{h}(n)$ is causal. As discussed in Chapter 2, then, $\hat{H}(z)$ must be analytic in a region $|z| > R_{\hat{h}-}$ where $R_{\hat{h}-} < 1$; i.e., $\hat{H}(z)$ must be analytic everywhere outside the unit circle. Thus there can be no singularities of $\hat{H}(z)$ outside the unit circle. Since $\hat{H}(z) = \log H(z)$, this requires then that there can be no *poles* or *zeros* of $H(z)$ outside the unit

† Note that we assume that $H(e^{j\omega}) > 0$ at $\omega = 0$.

‡ The motivation of the term "minimum phase" will be made apparent by subsequent discussion.

circle. This requirement on $H(z)$ can be viewed as an alternative expression of the minimum-phase condition. An equivalent condition is that there exist a causal, stable *inverse system* with system function $H^{-1}(z)$ such that

$$H^{-1}(z)H(z) = 1$$

Since $H^{-1}(z) = 1/H(z)$, it is clear that $H(z)$ must have all its poles and zeros inside the unit circle in order for a stable and causal inverse to exist.

Henceforth we shall use the term *minimum-phase system* to denote a system whose frequency response is minimum phase; i.e., the log magnitude and phase are Hilbert transforms of each other. Similarly, a *minimum-phase sequence* is a sequence whose Fourier transform is minimum phase. It should be emphasized at this point that a system (or sequence) can be causal but non-minimum phase. However, all stable, minimum-phase systems (sequences) are causal.

To see the relationship between the causality of $\hat{h}(n)$ and the locations of the poles and zeros of $H(z)$, it is instructive to consider a means for obtaining $\hat{h}(n)$. In particular, we know from Chapter 2 that $-z[d\hat{H}(z)/dz]$ is the z-transform of $n\hat{h}(n)$. But

$$-z\frac{d\hat{H}(z)}{dz} = -z\frac{d}{dz}[\log H(z)] = \frac{-z}{H(z)}\frac{dH(z)}{dz} \tag{7.23}$$

With $H(z)$ a rational function of z, $\hat{H}(z)$ is not rational but its derivative is, and it can consequently be characterized in terms of poles and zeros. With $H(z)$ expressed as a ratio of polynomials

$$H(z) = \frac{P(z)}{Q(z)}$$

we have

$$\frac{-z}{H(z)}\frac{dH(z)}{dz} = \frac{-z\left[Q(z)\dfrac{dP(z)}{dz} - P(z)\dfrac{dQ(z)}{dz}\right]}{P(z)Q(z)}$$

Thus we note that the *poles* of the derivative of $\hat{H}(z)$ are the roots of $P(z)Q(z)$, i.e., the poles *and* zeros of $H(z)$. Since we consider the unit circle to be within the region of convergence, $n\hat{h}(n)$ or, equivalently, $\hat{h}(n)$ will be causal if and only if all poles and zeros of $H(z)$ are inside the unit circle.†

EXAMPLE Consider the sequence $h(n) = \alpha^n u(n)$ for which $H(z) = 1/(1 - \alpha z^{-1})$, $|\alpha| < 1$. $H(z)$ has one zero at $z = 0$ and one pole at $z = \alpha$. Since $|\alpha| < 1$, all poles and zeros are inside the unit circle, and consequently $h(n)$ is minimum phase. To verify that $\hat{h}(n)$ is indeed causal, let us evaluate it according to Eq.

† This includes poles and zeros at infinity; i.e., if $H(z)$ is minimum phase, $\lim_{z\to\infty} H(z)$ must be a nonzero, finite constant.

(7.23). Specifically,

$$\frac{-z}{H(z)}\frac{dH(z)}{dz} = \frac{\alpha}{z - \alpha} = \frac{\alpha z^{-1}}{1 - \alpha z^{-1}}$$

Thus, since we assume that the unit circle is within the region of convergence,

$$n\hat{h}(n) = \alpha^n u(n - 1)$$

and consequently $\hat{h}(n)$ is causal.

The sequence $\hat{h}(n)$ will play a particularly important role in Chapter 10. We shall not consider the properties of the sequence $\hat{h}(n)$ further here but will concentrate on the properties of minimum-phase sequences.

A minimum-phase sequence has the property that all the poles and zeros of its z-transform lie inside the unit circle. In general, a stable causal system has all its poles inside the unit circle, but its zeros need not be inside the unit circle. We shall now show that any system can be represented as the cascade of a minimum-phase system with an allpass system, an allpass system being defined as a system for which the magnitude of the transfer function is unity for all frequencies. Thus if $H_{ap}(z)$ denotes the z-transform of an allpass system, $|H_{ap}(e^{j\omega})| = 1$ for all ω.

The system function of a simple first-order allpass system is

$$H_{ap}(z) = \frac{z^{-1} - a}{1 - az^{-1}} \tag{7.24}$$

The fact that $H_{ap}(e^{j\omega})$ as given by Eq. (7.24) has unity magnitude is considered in Problem 6 of this chapter. With $0 < a < 1$, the corresponding pole–zero plot is shown in Fig. 7.4. More generally, rational allpass system functions consist of a cascade of factors of the form

$$\frac{z^{-1} - a^*}{1 - az^{-1}}$$

and consequently they have the property that their poles and zeros occur at conjugate reciprocal locations.

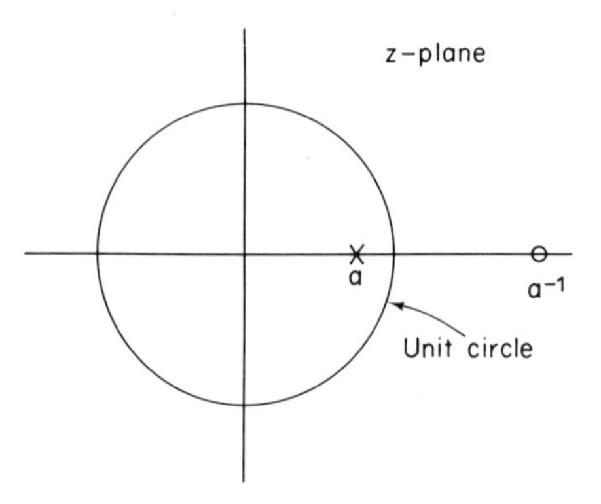

Fig. 7.4 Pole and zero locations for a first-order allpass system.

Consider a non-minimum-phase system $H(z)$, with, for example, one zero outside the unit circle at $z = 1/z_0$, $|z_0| < 1$, and the remainder of its poles and zeros inside the unit circle. Then $H(z)$ can be expressed as

$$H(z) = H_1(z)(z^{-1} - z_0) \tag{7.25}$$

where $H_1(z)$ is minimum phase. Equivalently we can express Eq. (7.25) as

$$\begin{aligned} H(z) &= H_1(z)(z^{-1} - z_0)\frac{1 - z_0^* z^{-1}}{1 - z_0^* z^{-1}} \\ &= H_1(z)(1 - z_0^* z^{-1})\frac{z^{-1} - z_0}{1 - z_0^* z^{-1}} = H_{\min}(z)\frac{z^{-1} - z_0}{1 - z_0^* z^{-1}} \end{aligned}$$

Since $|z_0| < 1$, the factor $H_1(z)(1 - z_0^* z^{-1})$ is minimum phase and the factor $(z^{-1} - z_0)/(1 - z_0^* z^{-1})$ is allpass. The term $H_{\min}(z) = H_1(z)(1 - z_0^* z^{-1})$ differs from $H(z)$ in that the zero of $H(z)$ that was outside the unit circle at $z = 1/z_0$ is reflected inside the unit circle to $z = z_0^*$ in $H_{\min}(z)$. Clearly this example can be generalized to encompass general non-minimum-phase systems with rational system functions. We conclude, therefore, that any rational system function $H(z)$ corresponding to a causal system can be expressed in the form

$$H(z) = H_{\min}(z)H_{ap}(z) \tag{7.26}$$

where $H_{\min}(z)$ is minimum phase and $H_{ap}(z)$ is allpass. Any pole or zero of $H(z)$ that is inside the unit circle also appears in $H_{\min}(z)$. Any pole or zero of $H(z)$ that is outside the unit circle appears in $H_{\min}(z)$ in the conjugate reciprocal location; i.e., it is reflected about the unit circle. Thus we can form a minimum-phase system from a non-minimum-phase system, keeping the *magnitude* of the transfer function the same, by reflecting inside the unit circle those zeros that were outside the unit circle. Alternatively, given a minimum-phase system function we can form a non-minimum-phase system by reflecting zeros outside the unit circle. For example, in the case of finite-length sequences the z-transform is simply a polynomial in z^{-1}, and $H(z)$ has poles only at $z = 0$. For a sequence of length M, $H(z)$ has $M - 1$ zeros. For a given magnitude response, we can have as many as 2^{M-1} different phase curves simply by reflecting zeros about the unit circle.

EXAMPLE. Consider a minimum-phase finite-duration unit-sample response with duration $N = 5$ samples. The unit-sample response of such a system is depicted in Fig. 7.5(a). The system function corresponding to this unit-sample response is

$$H_{\min}(z) = \frac{1}{r^2}(1 - re^{j\theta}z^{-1})^2(1 - re^{-j\theta}z^{-1})^2 \tag{7.27}$$

where $r = 0.55$ and $\theta = 2\pi/3$. The frequency-response functions, log magnitude and phase, are shown in Fig. 7.5(c) and (d), respectively. [Note that $\arg[H_{\min}(e^{j\omega})]$ is plotted modulo 2π and $\log|H_{\min}(e^{j\omega})|$ is normalized to a peak value of 0 dB for convenience in plotting.] According to the previous discussion, we can obtain a new system having the same magnitude response by

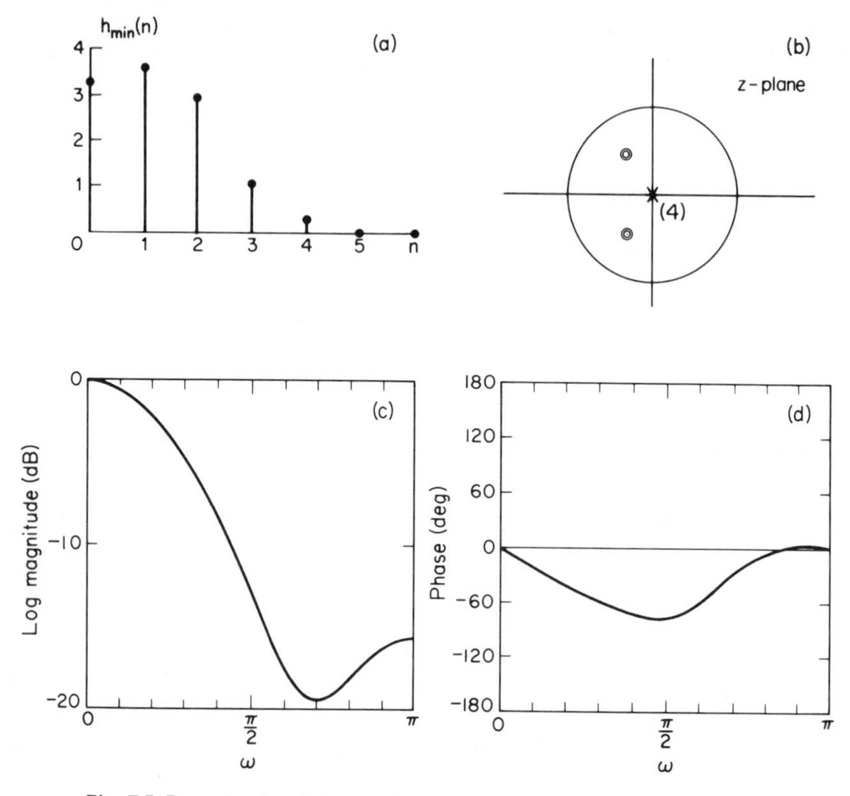

Fig. 7.5 Example of a minimum-phase system: (a) impulse response; (b) z-plane pole–zero plot; (c) $20 \log_{10} |H_{\min}(e^{j\omega})|$; (d) $\arg [H_{\min}(e^{j\omega})]$.

multiplying $H_{\min}(z)$ by an appropriate allpass system function as in Eq. (7.26). In this case, we can reflect one pair of the complex-conjugate zeros outside using the allpass system

$$H_{ap}(z) = \frac{z^{-1} - re^{-j\theta}}{1 - re^{j\theta}z^{-1}} \frac{z^{-1} - re^{j\theta}}{1 - re^{-j\theta}z^{-1}} \tag{7.28}$$

Thus we obtain

$$\begin{aligned} H(z) &= H_{\min}(z)H_{ap}(z) \\ &= (1 - re^{j\theta}z^{-1})(1 - re^{-j\theta}z^{-1})(1 - r^{-1}e^{j\theta}z^{-1})(1 - r^{-1}e^{-j\theta}z^{-1}) \end{aligned} \tag{7.29}$$

We note that the four zeros of $H(z)$ have the conjugate reciprocal symmetry that is a characteristic property of linear-phase systems. Indeed, the unit-sample response $h(n)$ corresponding to $H(z)$ is seen in Fig. 7.6(a) to have symmetry about $n = 2$, implying a linear phase with slope corresponding to a delay of two samples. As seen by comparing Fig. 7.5(c) and 7.6(c), $|H(e^{j\omega})|$ is identical to $|H_{\min}(e^{j\omega})|$; however, the unit-sample response and phase responses corresponding to $H_{\min}(z)$ and $H(z)$ are decidedly different.

Figure 7.7 shows (a) the z-plane plot and (b) $\arg [H_{ap}(e^{j\omega})]$ for the allpass system. The magnitude of $H_{ap}(e^{j\omega})$ is, of course, unity for all values of ω. We

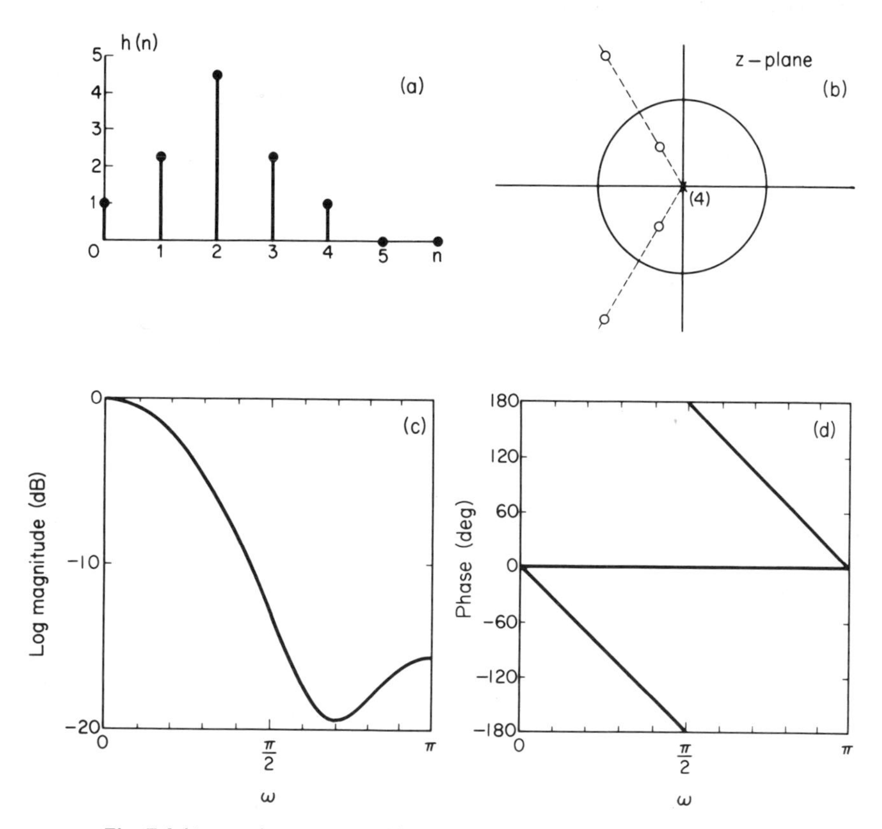

Fig. 7.6 Linear-phase system: (a) impulse response; (b) z-plane pole–zero plot; (c) $20 \log_{10} |H(e^{j\omega})|$; (d) $\arg [H(e^{j\omega})]$ [magnitude response is identical to Fig. 7.5(c)].

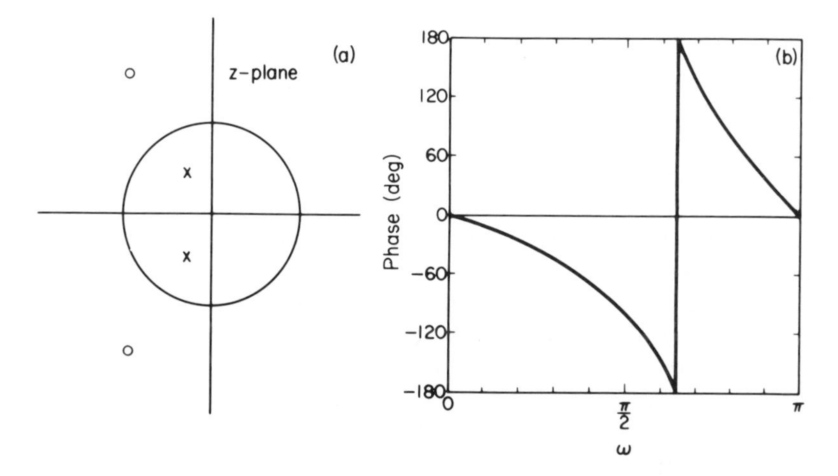

Fig. 7.7 Allpass system for obtaining Fig. 7.6 from Fig. 7.5: (a) z-plane pole–zero plot; (b) $\arg [H_{\rm ap}(e^{j\omega})]$.

again have plotted arg $[H_{ap}(e^{j\omega})]$ modulo 2π simply for convenience. However, it is clear from Fig. 7.7(b) that if the phase were computed as a continuous function of ω, then arg $[H_{ap}(e^{j\omega})]$ is always negative. When this phase curve is added to the phase of the minimum-phase system [Fig. 7.5(d)], the linear phase of Fig. 7.6(d) results.

This simple example illustrates a number of important general properties of minimum-phase systems which are worth emphasizing. First, a comparison of the phase curves in Figs. 7.5(d) and 7.6(d) suggests the reason for the terminology "minimum phase." As discussed above, if we consider the set of causal, real, stable sequences, all having the same magnitude response, then the z-transforms of all these sequences can be expressed as in Eq. (7.26) as the product of a minimum-phase z-transform and an allpass function. As seen in the above example and as discussed in Problem 8 of this chapter, the allpass function has negative phase for $0 < \omega < \pi$ and, consequently, reflection of a zero of the minimum-phase function outside the unit circle algebraically decreases the phase, i.e., increases the negative of phase, which is sometimes called *phase lag*. Thus, in fact, a more precise terminology would be *minimum phase lag;* however, "minimum phase" is the established terminology.

In the case of finite-duration sequences, we have a complementary situation in which all the zeros are *outside* the unit circle. Clearly if all the zeros are reflected outside the unit circle, the system has the maximum phase lag attainable and thus such systems (or sequences) are called *maximum phase.* It can be shown (see Problem 14 of this chapter) that the maximum-phase system has a system function

$$H_{max}(z) = z^{-(N-1)} H_{min}(z^{-1}) \tag{7.30}$$

From Eq. (7.30) it follows that

$$h_{max}(n) = h_{min}(N - 1 - n) \tag{7.31}$$

In the previous example, we note that a maximum-phase system is obtained if we multiply $H(z)$ in Eq. (7.29) by $H_{ap}(z)$ in Eq. (7.28).

A final property of minimum-phase sequences is suggested by a comparison of the minimum-phase unit-sample response $h_{min}(n)$ in Fig. 7.5(a) with the linear-phase unit-sample response $h(n)$ of Fig. 7.6(a). We note that the total energy of the two sequences is the same since the magnitude of their Fourier transforms is the same (by Parseval's Theorem). However, the energy of $h_{min}(n)$ seems to be concentrated around $n = 0$, whereas the energy of $h(n)$ is concentrated around $n = 2$. This property can be formalized by considering the portion of the energy contributed by the first $m + 1$ samples of the sequence; i.e.,

$$E(m) = \sum_{n=0}^{m} |h(n)|^2 \tag{7.32}$$

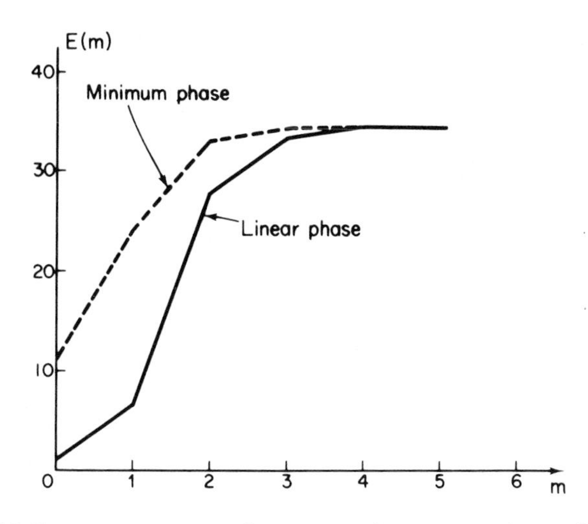

Fig. 7.8 Energy concentration for two impulse responses having Fourier transforms with identical magnitudes.

This quantity is plotted in Fig. 7.8 for $h_{\text{min}}(n)$ and $h(n)$ of the previous example. We note that

$$\sum_{n=0}^{m} |h(n)|^2 \leq \sum_{n=0}^{m} |h_{\text{min}}(n)|^2, \qquad \text{for all } m \tag{7.33}$$

A proof that Eq. (7.33) holds in general for all sequences having the same Fourier transform magnitude is outlined in Problem 11 of this chapter. We can interpret Eq. (7.33) to mean that of all the sequences having the same Fourier transform magnitude, $h_{\text{min}}(n)$ is delayed the least. Thus, minimum-phase sequences are sometimes called *minimum-delay sequences*. Similarly, maximum-phase sequences are called *maximum-delay sequences* [12,13].

7.3 Hilbert Transform Relations for the DFT

We have seen that periodic sequences and finite-length sequences have a representation in terms of the discrete Fourier transform. The results of the preceding sections do not apply directly to the discrete Fourier transform. We can, however, with a suitable definition of causality, relate the real and imaginary parts of the discrete Fourier transform in a manner similar to that developed in Sec. 7.1 [7,8].

To develop these relations, it is most convenient to consider a periodic sequence $\tilde{h}(n)$ of period N. You will recall from Chapter 3 that our discussion, although phrased in terms of periodic sequences, applies just as well to finite-length sequences if we interpret all indices modulo N. Indeed, although our derivation will deal with the properties of discrete Fourier series (DFS) representations, we shall see that the results apply directly to discrete Fourier transform (DFT) representations of finite-length sequences. As in Sec. 7.2, the sequence $\tilde{h}(n)$ can be represented as the sum of an even sequence and an odd sequence so that

$$\tilde{h}(n) = \tilde{h}_e(n) + \tilde{h}_o(n), \qquad n = 0, 1, \ldots, N-1 \tag{7.34}$$

where

$$\tilde{h}_e(n) = \frac{\tilde{h}(n) + \tilde{h}(-n)}{2}, \qquad n = 0, 1, \ldots, N-1 \tag{7.35a}$$

and

$$\tilde{h}_0(n) = \frac{\tilde{h}(n) - \tilde{h}(-n)}{2}, \qquad n = 0, 1, \ldots, N-1 \tag{7.35b}$$

In the remainder of our discussion, we shall assume that N is an even integer. For N odd, similar—but not identical—results can be derived.

A periodic sequence cannot, of course, be causal in the sense used in Sec. 7.1. We will, however, define a "causal" periodic sequence to be one for which $\tilde{h}(n) = 0$ for $N/2 < n < N$. That is, $\tilde{h}(n)$ is identically zero over the last half of the period. We assume that N is even. The case of N odd is considered in Problem 16 of this chapter. Note also that because of the periodicity of $\tilde{h}(n)$, $\tilde{h}(n) = 0$ for $-N/2 < n < 0$. For finite-length sequences we interpret this restriction to mean that the sequence is considered to be of length N, when in fact the last half of the points are zero. In Fig. 7.9 we show an example of a causal periodic sequence and its even and odd parts with $N = 8$. Because $\tilde{h}(n)$ is zero in the second half of each period, $\tilde{h}(-n)$ is zero in the first half of each period, and consequently, except for $n = 0$ and $n = N/2$, there is no overlap between the nonzero portions of $\tilde{h}(n)$ and $\tilde{h}(-n)$. Because of this, it should be clear that for "causal" periodic sequences

$$\tilde{h}(n) = \begin{cases} 2\tilde{h}_e(n), & n = 1, 2, \ldots, (N/2) - 1 \\ \tilde{h}_e(n), & n = 0, N/2 \\ 0, & n = (N/2 + 1), \ldots, N-1 \end{cases}$$

and

$$\tilde{h}(n) = \begin{cases} 2\tilde{h}_o(n), & n = 1, 2, \ldots, (N/2) - 1 \\ 0, & n = (N/2 + 1), \ldots, N-1 \end{cases}$$

Equivalently, if we define $\tilde{u}_N(n)$ as a periodic sequence

$$\tilde{u}_N(n) = \begin{cases} 1, & n = 0, N/2 \\ 2, & n = 1, 2, \ldots, (N/2) - 1 \\ 0, & n = (N/2 + 1), \ldots, N-1 \end{cases}$$

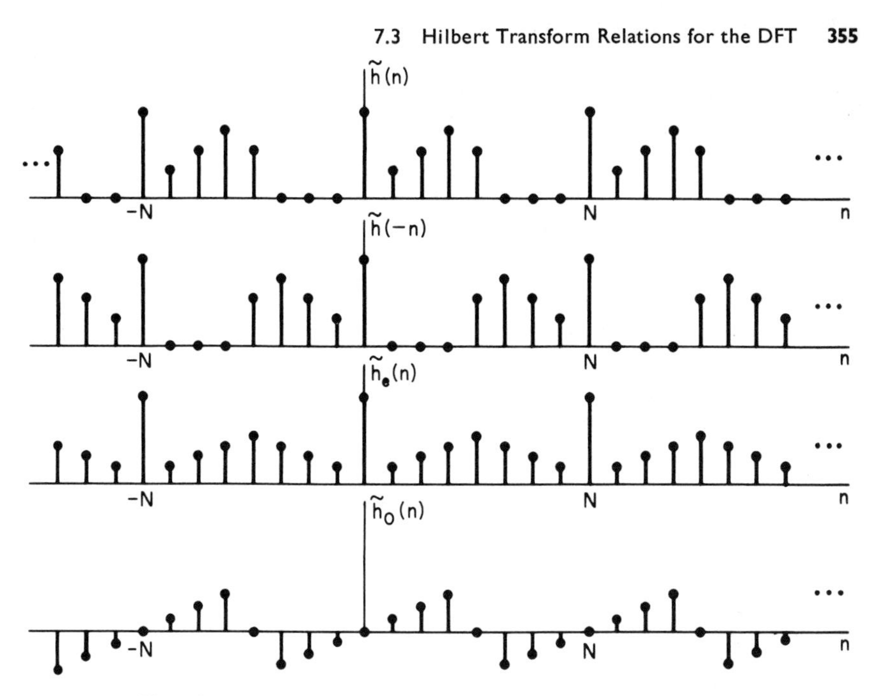

Fig. 7.9 Even and odd parts of a periodic, real, "causal" sequence.

Then it follows that for N even we can express $h(n)$ as

$$\tilde{h}(n) = \tilde{h}_e(n)\tilde{u}_N(n) \tag{7.36}$$

and

$$\tilde{h}(n) = \tilde{h}_o(n)\tilde{u}_N(n) + h(0)\,\delta(n) + h\left(\frac{N}{2}\right)\delta\left(n - \frac{N}{2}\right) \tag{7.37}$$

We note that $\tilde{h}(n)$ can be completely recovered from $\tilde{h}_e(n)$. On the other hand, $\tilde{h}_o(n)$ will always be zero at $n = 0$ and $n = N/2$, and consequently $\tilde{h}(n)$ can be recovered from $\tilde{h}_o(n)$ only for $n \neq 0$ or $n \neq N/2$.

In Chapter 3 we saw that for a real periodic sequence of period N with discrete Fourier series $\tilde{H}(k)$, the real part of $\tilde{H}(k)$, $\tilde{H}_R(k)$ is the DFS of $\tilde{h}_e(n)$ and $j\tilde{H}_I(k)$ is the DFS of $\tilde{h}_o(n)$. Thus an important consequence of Eqs. (7.36) and (7.37) is that they imply that for a periodic (or equivalently finite length) sequence of period N, which is casual in the sense defined above, $\tilde{H}(k)$ can be recovered from its real part or (almost) from its imaginary part. Equivalently, $\tilde{H}_I(k)$ can be constructed from $\tilde{H}_R(k)$ and $\tilde{H}_R(k)$ can be constructed from $\tilde{H}_I(k)$.

Specifically, the DFS of the sequence $\tilde{u}_N(n)$ is

$$\tilde{U}_N(k) = \begin{cases} N, & k = 0 \\ -j2\cot\left(\dfrac{\pi}{N}k\right), & k \text{ odd} \\ 0, & k \text{ even} \end{cases} \tag{7.38}$$

From Eq. (7.36) we note that the DFS of $\tilde{h}(n)$ is the circular convolution of $\tilde{H}_R(k)$ with $\tilde{U}_N(k)$. Thus

$$\begin{aligned}\tilde{H}(k) &= \tilde{H}_R(k) + j\tilde{H}_I(k) \\ &= \frac{1}{N}\sum_{m=0}^{N-1} \tilde{H}_R(m)\tilde{U}_N(k-m) \\ &= \tilde{H}_R(k) + \frac{1}{N}\sum_{m=0}^{N-1} \tilde{H}_R(m)\tilde{V}_N(k-m)\end{aligned}$$

where

$$\tilde{V}_N(k) = \tilde{U}_N(k) - N\delta(k) = \begin{cases} -j2\cot(\pi k/N), & k \text{ odd} \\ 0, & k \text{ even}\end{cases}$$

Equating real and imaginary parts, we obtain

$$j\tilde{H}_I(k) = \frac{1}{N}\sum_{m=0}^{N-1} \tilde{H}_R(m)\tilde{V}_N(k-m) \tag{7.39a}$$

Similarly, beginning with Eq. (7.37), we can show that

$$\tilde{H}_R(k) = \frac{1}{N}\sum_{m=0}^{N-1} j\tilde{H}_I(m)\tilde{V}_N(k-m) + \tilde{h}(0) + \tilde{h}(N/2)(-1)^k \tag{7.39b}$$

Equations (7.39a) and (7.39b) are circular convolutions and can be evaluated using the DFS. For example, Eq. (7.39a) can be evaluated by first computing the inverse DFS of $\tilde{H}_R(k)$, yielding $\tilde{h}_e(n)$. Multiplying $\tilde{h}_e(n)$ by $\tilde{u}_N(n)$ and computing the DFS yields the entire transform $\tilde{H}(k)$.

In Sec. 3.5, we found it useful to introduce special notation to facilitate the interpretation of DFS expressions in the context of finite-length sequences. Thus a finite-length sequence $h(n)$ is thought of as being one period of a periodic sequence $\tilde{h}(n)$; i.e.,

$$h(n) = \tilde{h}(n)\mathscr{R}_N(n)$$

where

$$\begin{aligned}\mathscr{R}_N(n) &= 1 \qquad 0 \le n \le N-1 \\ &= 0 \qquad \text{otherwise}\end{aligned}$$

Alternatively, we obtain the periodic sequence $\tilde{h}(n)$ by interpreting the index n modulo N. For this purpose, we introduced the notation

$$\tilde{h}(n) = h((n))_N$$

These conventions were also applied to the DFS expression $\tilde{H}(k)$ to obtain the DFT expression $H(k)$.

Using this notation, we can write Eq. (7.39a) as

$$\begin{aligned}jH_I(k) &= \frac{1}{N}\sum_{m=0}^{N-1} H_R(m)V_N((k-m))_N \qquad 0 \le k \le N-1 \\ &= 0 \qquad \text{otherwise}\end{aligned} \tag{7.40a}$$

And Eq. (7.39b) can be written as

$$
\begin{aligned}
H_R(k) &= \frac{1}{N}\sum_{m=0}^{N-1} jH_I(m)V_N((k-m))_N \\
&\quad + h(0) + (-1)^k h(N/2) \qquad 0 \le k \le N-1 \\
&= 0 \qquad\qquad\qquad\qquad\qquad\quad \text{otherwise}
\end{aligned} \tag{7.40b}
$$

where

$$
\begin{aligned}
V_N(k) &= -j2\cot(\pi k/N) \qquad 0 < k < N-1,\ k \text{ odd} \\
&= 0 \qquad \text{otherwise}
\end{aligned}
$$

Similar definitions can, of course, be made for u_N and its DFT $U_N(k)$.

When we discussed real-part sufficiency for the z-transform, we were able to apply the results also to relating log magnitude and phase when the sequence was minimum phase. For the discrete Fourier transform it is not possible in general to develop a parallel notion by which the log magnitude and phase of the DFT can be related. The reason for this is that the discussion above applies to sequences which are of finite length so that the z-transform has only zeros. However, the log of a transform $H(z)$ has singularities corresponding to both the poles and the zeros of $H(z)$, and thus its inverse z-transform is of infinite duration. Consequently, the inverse transform of the log of the transform cannot in general be represented by a discrete Fourier transform.

It is possible, of course, to construct a phase function from the log magnitude of a DFT by the process described above. That is, compute the inverse DFT of $\log |H(k)|$; multiply by $u_N(n)$, and compute the DFT of the resulting sequence. The real part of the result is $\log |H(k)|$, and the imaginary part is an approximation to the minimum phase.

To understand this process, let us assume that $H(z)$ is the z-transform of a finite-length sequence $h(n)$. If $H(z)$ has no zeros outside the unit circle, then we can compute $\arg [H(e^{j\omega})]$ knowing only $\log |H(e^{j\omega})|$. Furthermore,

$$
\hat{H}(z) = \log [H(z)]
$$

corresponds to a causal sequence $\hat{h}(n)$, which will in general be infinite in duration. The DFT of $h(n)$ is

$$
H(k) = H(z)\big|_{z=e^{j(2\pi k/N)}}, \qquad k = 0, 1, \ldots, N-1
$$

where N is chosen to be at least the length of the sequence $h(n)$. The DFT

$$
\hat{H}_p(k) = \log [H(k)] = \log |H(k)| + j \arg [H(k)]
$$

corresponds to an aliased sequence

$$
\hat{h}_p(n) = \sum_{r=-\infty}^{\infty} \hat{h}(n + rN)
$$

Clearly, the larger we choose N, the better will be the result of the process of multiplying $\hat{h}_p(n)$ by $u_N(n)$. Computing the DFT to obtain $\log |H(k)|$ for the real part and an approximation to the minimum phase yields results that are very useful in some practical situations (see Chapter 10).

7.4 Hilbert Transform Relations for Complex Sequences

Thus far we have considered Hilbert transform relations for the Fourier transform of causal sequences and the discrete Fourier transform of periodic sequences which are "causal" in the sense that they are zero in the second half of each period. In this section we consider *complex sequences* for which the real and imaginary components can be related through a convolution similar to the Hilbert transform relations derived in the previous sections. These Hilbert transform relations are particularly useful in representing bandpass signals as complex signals in a manner completely analogous to the "analytic signals" of analog signal theory [6].

As in the previous discussions it is possible to base the derivation of the Hilbert transform relations on a notion of causality. Since we are interested in relating the real and imaginary parts of a complex sequence, "causality" will be applied to the Fourier transform of the sequence. We cannot, of course, require that the Fourier transform be zero for $\omega < 0$ since it is periodic. However, we shall define "causality" in this context to mean that the Fourier transform is zero in the second half of each period; i.e., the z-transform is zero on the bottom half ($-\pi \leq \omega < 0$) of the unit circle. Thus, with $s(n)$ denoting the sequence and $S(e^{j\omega})$ its Fourier transform, we require that

$$S(e^{j\omega}) \equiv 0, \qquad -\pi \leq \omega < 0 \tag{7.41}$$

It is clear that the sequence $s(n)$ corresponding to $S(e^{j\omega})$ must be complex since for $s(n)$ to be real requires $S(e^{-j\omega}) = S^*(e^{j\omega})$. Therefore, we express $s(n)$ as

$$s(n) = s_r(n) + js_i(n) \tag{7.42}$$

where $s_r(n)$ and $s_i(n)$ are real sequences.

In analog signal theory, the comparable signal is an analytic function and thus is called an *analytic signal*. Similarly, we shall apply the same terminology to complex sequences like $s(n)$. Although analyticity has no meaning for sequences, we note that to any sequence $s(n)$, there corresponds a band-limited analog signal $s_a(t)$ such that

$$s_a(t)\big|_{t=n} = s(n)$$

Therefore, if

$$S_a(j\omega) = \begin{cases} S(e^{j\omega}), & 0 \leq \omega < \pi \\ 0, & \text{otherwise} \end{cases}$$

then the signal $s_a(t)$ is an analytic function of t. In this sense the sequence $s(n)$ does indeed correspond to an analytic signal.

With $S_r(e^{j\omega})$ and $S_i(e^{j\omega})$ denoting the Fourier transforms of the real sequences $s_r(n)$ and $s_i(n)$, it is easily shown that

$$S_r(e^{j\omega}) = \tfrac{1}{2}[S(e^{j\omega}) + S^*(e^{-j\omega})] \tag{7.43a}$$

and

$$jS_i(e^{j\omega}) = \tfrac{1}{2}[S(e^{j\omega}) - S^*(e^{-j\omega})] \tag{7.43b}$$

The complex transforms $S_r(e^{j\omega})$ and $S_i(e^{j\omega})$ play a role similar to that played in the previous sections by the even and odd parts, respectively, of causal sequences. Note, however, that $S_r(e^{j\omega})$ is not an even function but is conjugate even; i.e., $S_r(e^{j\omega}) = S_r^*(e^{-j\omega})$. Similarly, $jS_i(e^{j\omega})$ is conjugate odd; i.e., $jS_i(e^{j\omega}) = -jS_i^*(e^{-j\omega})$.

If $S(e^{j\omega})$ is zero for $-\pi \le \omega < 0$, then there is no overlap between the nonzero portions of $S(e^{j\omega})$ and $S^*(e^{-j\omega})$. Thus $S(e^{j\omega})$ can be recovered from $S_r(e^{j\omega})$ or $S_i(e^{j\omega})$. Note that since $S(e^{j\omega})$ is assumed to be zero at $\omega = -\pi$, $S(e^{j\omega})$ is totally recoverable from $jS_i(e^{j\omega})$. This is somewhat in contrast with the two previous situations, in which the causal function could be recovered from its odd part except at the endpoints.

In particular,

$$S(e^{j\omega}) = \begin{cases} 2S_r(e^{j\omega}), & 0 \le \omega < \pi \\ 0, & -\pi \le \omega < 0 \end{cases}$$

and

$$S(e^{j\omega}) = \begin{cases} 2jS_i(e^{j\omega}), & 0 \le \omega < \pi \\ 0, & -\pi \le \omega < 0 \end{cases}$$

Alternatively, we can relate $S_r(e^{j\omega})$ and $S_i(e^{j\omega})$ directly; i.e.,

$$S_i(e^{j\omega}) = \begin{cases} -jS_r(e^{j\omega}), & 0 \le \omega < \pi \\ \ \ jS_r(e^{j\omega}), & -\pi \le \omega < 0 \end{cases} \tag{7.44}$$

or

$$S_i(e^{j\omega}) = H(e^{j\omega})S_r(e^{j\omega}) \tag{7.45}$$

where

$$H(e^{j\omega}) = \begin{cases} -j, & 0 \le \omega < \pi \\ \ \ j, & -\pi \le \omega < 0 \end{cases} \tag{7.46}$$

Now $S_i(e^{j\omega})$ is the Fourier transform of $s_i(n)$, the imaginary part of $s(n)$, and $S_r(e^{j\omega})$ is the Fourier transform of $s_r(n)$, the real part of $s(n)$. Thus, according to Eq. (7.45), $s_i(n)$ can be obtained by processing $s_r(n)$ with a discrete system with frequency response $H(e^{j\omega})$ as given by Eq. (7.46). This frequency response has unity magnitude, a phase angle of $-\pi/2$ for ω between 0 and π, and a phase angle of $+\pi/2$ for ω between 0 and $-\pi$. Such a system is often referred to as a *90-degree phase shifter* or a *Hilbert transformer*. From Eq. (7.45) it follows that

$$S_r(e^{j\omega}) = \frac{1}{H(e^{j\omega})} S_i(e^{j\omega}) = -H(e^{j\omega})S_i(e^{j\omega}) \tag{7.47}$$

Thus $-s_r(n)$ can also be obtained from $s_i(n)$ with a 90-degree phase shifter.

The impulse response $h(n)$ of a 90-degree phase shifter, corresponding to the frequency response $H(e^{j\omega})$ given in Eq. (7.46), is

$$\begin{aligned} h(n) &= \frac{1}{2\pi}\int_{-\pi}^{0} je^{j\omega n}\, d\omega - \frac{1}{2\pi}\int_{0}^{\pi} je^{j\omega n}\, d\omega \\ &= \frac{2}{\pi}\frac{\sin^2(\pi n/2)}{n}, \qquad n \neq 0 \\ &= 0, \qquad n = 0 \end{aligned} \tag{7.48}$$

The normalized impulse response is plotted in Fig. 7.10. Using Eqs. (7.45) and (7.47) we obtain the expressions

$$s_i(n) = \sum_{m=-\infty}^{\infty} s_r(n-m)h(m) \tag{7.49}$$

and

$$s_r(n) = -\sum_{m=-\infty}^{\infty} s_i(n-m)h(m) \tag{7.50}$$

Equations (7.49) and (7.50) are the desired Hilbert transform relations between the real and imaginary parts of a discrete-time analytic signal.

An alternative representation of $s(n)$ is in terms of its magnitude and phase; i.e.,

$$s(n) = A(n)e^{j\phi(n)} \tag{7.51}$$

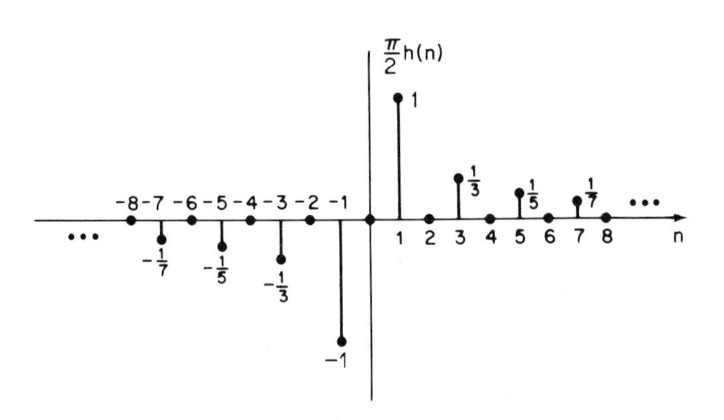

Fig. 7.10 Normalized impulse response of an ideal Hilbert transformer or 90-degree shifter.

where

$$A(n) = (s_r^2(n) + s_i^2(n))^{1/2} \tag{7.52a}$$

and

$$\phi(n) = \arctan \left[\frac{s_i(n)}{s_r(n)}\right] \tag{7.52b}$$

The magnitude sequence, $A(n)$, is often called the *envelope* of the sequence $s(n)$. The concept of minimum phase as discussed in Sec. 7.2 has its counterpart in the theory of analytic signals. The development of this concept leads to rather difficult mathematics, and since we have no need for it in this book, we shall not discuss it further. Details of this nature for bandlimited signals are discussed by Voelcker [15] and also Requicha [16].

7.4.1 Design of Hilbert Transformers

We can see from Eq. (7.48) that the z-transform of $h(n)$ converges *only* on the unit circle. In fact, owing to the discontinuity of the imaginary part, the series

$$H(e^{j\omega}) = \sum_{n=-\infty}^{\infty} h(n)e^{-j\omega n}$$

converges to Eq. (7.46) only in the mean-square sense. Thus the ideal Hilbert transformer or 90-degree phase shifter takes its place alongside the ideal lowpass filter and ideal bandlimited differentiator as valuable theoretical concepts which correspond to noncausal systems and for which the system function exists only in a restricted sense.

Approximations to the ideal Hilbert transformer can, of course, be obtained. In the case of finite-duration approximations, the standard techniques of windowing, frequency sampling, and equiripple approximation can be applied in approximating the ideal characteristics of Eq. (7.46).

Figure 7.11(a) shows an example of a Hilbert transformer designed by windowing Eq. (7.48) with a Blackman window with $N = 27$ (see Chapter 5). The magnitude of the frequency response is shown for $0 \leq \omega \leq \pi$. The phase is $-90°$ for $0 \leq \omega < \pi$ and $+90°$ for $-\pi \leq \omega < 0$. In addition, there is a linear phase shift corresponding to a delay of 13 samples. Figure 7.11(b) shows an equiripple approximation for $N = 27$. This filter was designed to have an equiripple approximation error in a band $0.0874\pi \leq \omega \leq 0.9126\pi$. The phase is as in the previous example. This design is from a table of Hilbert transform designs given by Herrmann [17]. A more detailed discussion and tables are given in Ref. 18.

For systems that admit a recursive realization, it is possible to make use of a considerable amount of work on the design of analog phase splitters. These analog systems take the form of a pair of allpass filters whose phase responses differ by a constant 90 degrees. Using the bilinear transformation method, we can derive a corresponding pair of discrete-time allpass

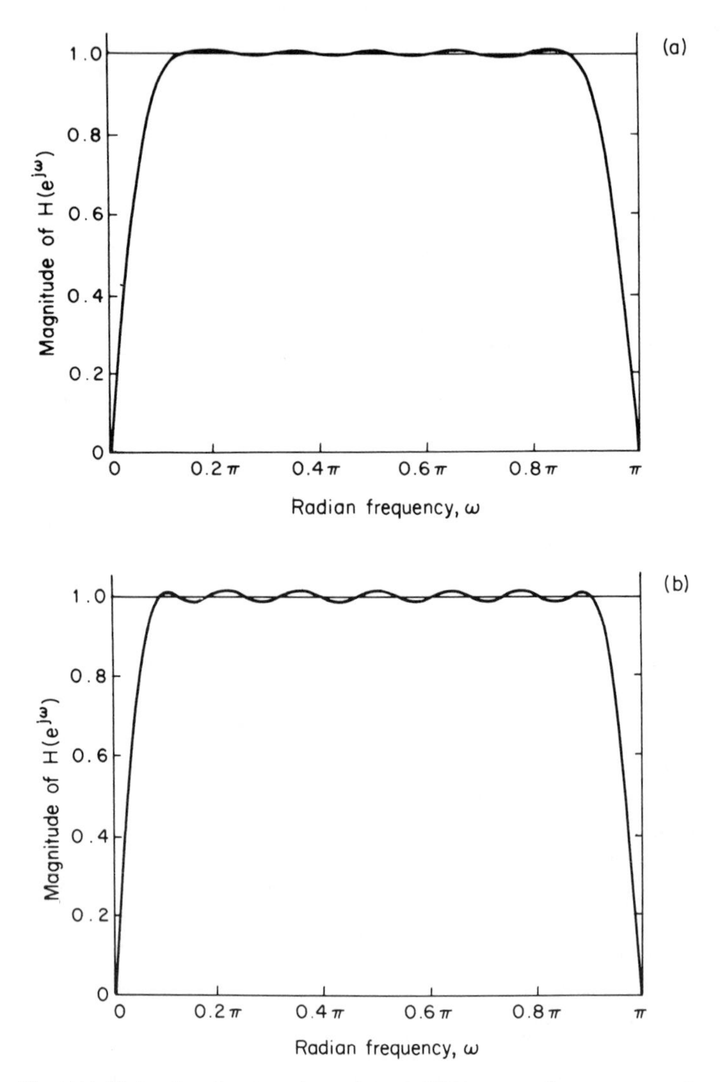

Fig. 7.11 Finite-duration impulse response Hilbert transformer approximations: (a) $N = 27$, design by windowing with Blackman window; (b) $N = 27$, equiripple magnitude approximation error (after Herrmann [17]). No phase error in either case.

systems with the same properties. Such a system, as depicted in Fig. 7.12, does not give an output that is equal to the Hilbert transform of the input, but rather it gives two outputs that are Hilbert transforms of each other. Thus if the input is denoted $x_r(n)$ and $x_i(n)$ denotes its Hilbert transform, then the sequence $y(n) = y_r(n) + jy_i(n)$ has a z-transform that vanishes on the bottom half of the unit circle, and on the upper half $Y(e^{j\omega})$

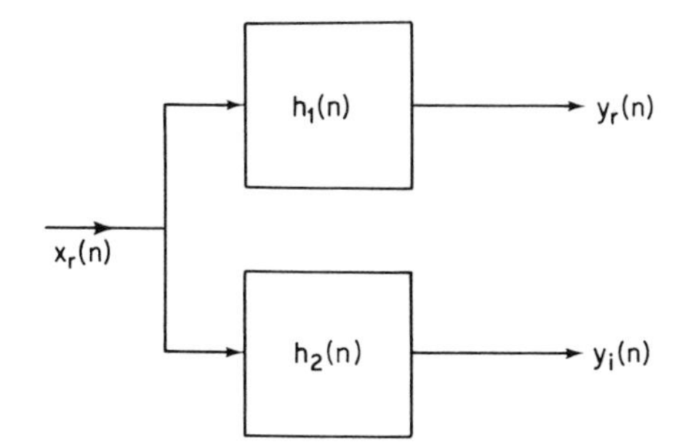

Fig. 7.12 Representation of 90-degree phase splitting system.

differs from the transform of $x(n) = x_r(n) + jx_i(n)$ in phase but not in amplitude. For an example of a recursive realization of such a system, see Gold et al. [8].

7.4.2 Representation of Bandpass Signals

Many of the applications of analytic signals concern narrow-band communications signals. In such applications it is sometimes convenient to represent a bandpass signal in terms of a lowpass signal. To see how this can be done, consider the complex lowpass signal

$$x(n) = x_r(n) + jx_i(n)$$

where $x_i(n)$ is the Hilbert transform of $x_r(n)$, and

$$-X(e^{j\omega}) = 0, \quad -\pi \leq \omega < 0$$

The Fourier transforms $X_r(e^{j\omega})$ and $jX_i(e^{j\omega})$ are depicted in Fig. 7.13(a) and (b), respectively, and the resulting transform $X(e^{j\omega}) = X_r(e^{j\omega}) + jX_i(e^{j\omega})$ is shown in Fig. 7.13(c). (Solid curves are real parts and dashed curves are imaginary parts.) Now consider the sequence

$$s(n) = x(n)e^{j\omega_c n} = s_r(n) + js_i(n) \tag{7.53}$$

where $s_r(n)$ and $s_i(n)$ are real sequences. The corresponding Fourier transform is

$$S(e^{j\omega}) = X(e^{j(\omega-\omega_c)}) \tag{7.54}$$

which is depicted in Fig. 7.13(d). The Fourier transforms $S_r(e^{j\omega})$ and $jS_i(e^{j\omega})$ are shown in Fig. 7.13(e) and (f). It is clear that for bandpass signals, $s_i(n)$ is the Hilbert transform of $s_r(n)$.

Since $x(n)$ can be expressed as

$$x(n) = A(n)e^{j\phi(n)}$$

as in Eq. (7.51), we can write $s(n)$ as

$$s(n) = [x_r(n) + jx_i(n)]e^{j\omega_c n} \tag{7.55a}$$

$$= A(n)e^{j(\omega_c n+\phi(n))} \tag{7.55b}$$

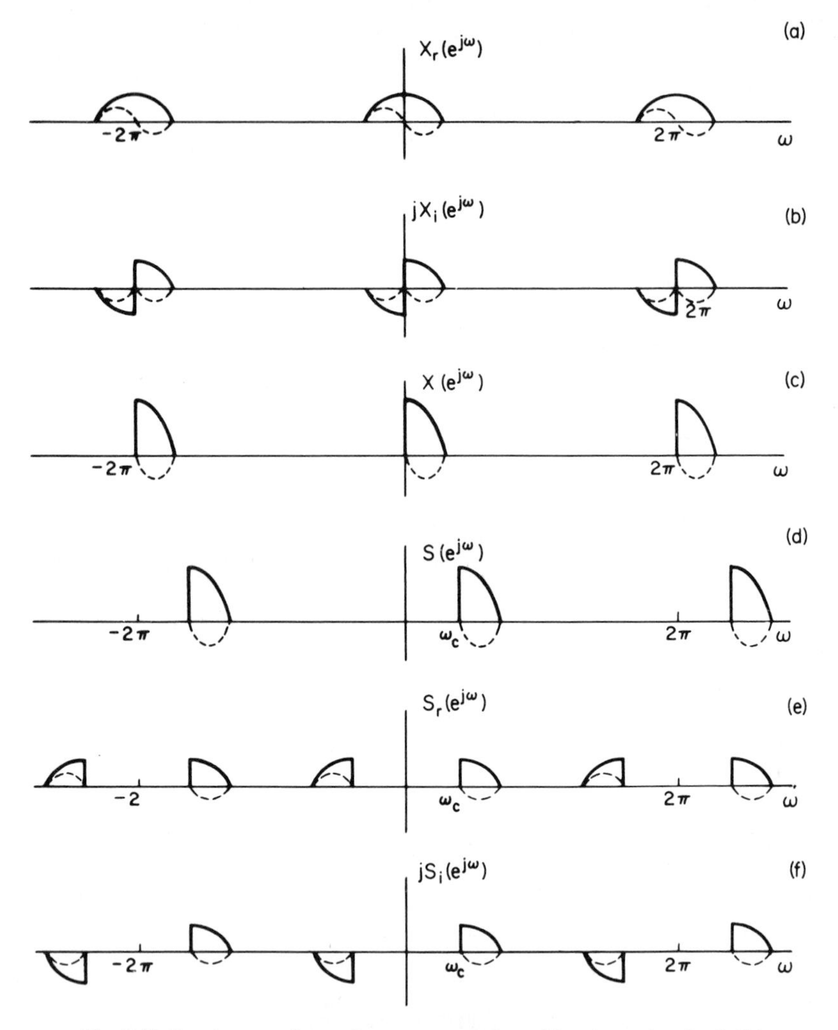

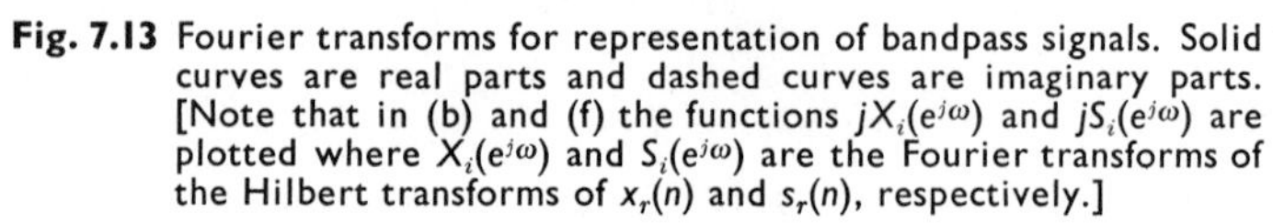

Fig. 7.13 Fourier transforms for representation of bandpass signals. Solid curves are real parts and dashed curves are imaginary parts. [Note that in (b) and (f) the functions $jX_i(e^{j\omega})$ and $jS_i(e^{j\omega})$ are plotted where $X_i(e^{j\omega})$ and $S_i(e^{j\omega})$ are the Fourier transforms of the Hilbert transforms of $x_r(n)$ and $s_r(n)$, respectively.]

Thus we have the expressions

$$s_r(n) = x_r(n) \cos \omega_c n - x_i(n) \sin \omega_c n \tag{7.56a}$$

$$= A(n) \cos [\omega_c n + \phi(n)] \tag{7.56b}$$

and

$$s_i(n) = x_r(n) \sin \omega_c n + x_i(n) \cos \omega_c n \tag{7.57a}$$

$$= A(n) \sin [\omega_c n + \phi(n)] \tag{7.57b}$$

Equations (7.56a) and (7.57a) are the desired representations of bandpass signals in terms of lowpass signals. We note that Eqs. (7.56b) and (7.57b) are in the form of a sinusoid modulated in both amplitude and phase.

Examples of the use of these relations are the representation of bandpass filters and the representation of modulation processes. Another important use of analytic signals is in the theory of bandpass sampling. Specifically, we know that if we have a continuous-time signal with Fourier transform $X_a(j\Omega)$ which is zero for $|\Omega| > (\Omega_0/2)$, then we must sample it at a rate greater than $(\Omega_0/2\pi)$ samples per second in order to recover it from the samples. Equivalently, if we have a real sequence $x_r(n)$ whose Fourier transform is zero for $\omega_0/2 \leq \omega \leq \pi$, it can be resampled; i.e., the sampling rate can be reduced by discarding samples. For example, if $\omega_0 = \pi$, the original sampling rate was too high by a factor of 2, and every other sample can be discarded. More generally, the number of samples per second can be reduced by a factor $2\pi/\omega_0$.

Now let us consider a real bandpass signal $s_r(n)$ as depicted in Fig. 7.13(e). Since the signal is real, the Fourier transform must, of course, be conjugate symmetric. In determining the minimum sampling rate, we must consider the spectrum to have a width $\omega_0 = 2(\omega_c + \Delta\omega)$; i.e., in this case although the actual bandwidth is $\Delta\omega$, the sampling rate of $s_r(n)$ can only be reduced by a factor $2\pi/\omega_0$. However, consider the analytic signal $s(n) = s_r(n) + js_i(n)$, with Fourier transform $S(e^{j\omega})$ as depicted in Fig. 7.13(d). Since $S(e^{j\omega})$ is zero except in the region $\omega_c \leq \omega \leq \omega_c + \Delta\omega$, we can reduce the sampling rate by a factor $2\pi/\Delta\omega$. This can be seen by noting that the complex lowpass analytic signal is

$$x(n) = s(n)e^{-j\omega_c n} \tag{7.58}$$

as depicted in Fig. 7.13(c). This signal can be sampled at a rate that is lower by a factor $2\pi/\Delta\omega$ than the original sampling rate. Further consideration shows that, in fact, it is unnecessary to perform the modulation of Eq. (7.58) before reducing the sampling rate.

Summary

In this chapter we have discussed a variety of relations between the real and imaginary parts of Fourier transforms and complex sequences. These relationships are collectively referred to as *Hilbert transform relationships.*

Our approach to deriving all the Hilbert transform equations was to apply a basic causality principle that allows a sequence or function to be recovered from its even part. An alternative approach, which is developed in several of the problems, is based on the rather special properties of analytic functions. We found that for a causal sequence, the real and imaginary parts of the Fourier transform are related through a convolution-type integral. Also, for the special case when the sequence is causal and both the poles and zeros of its z-transform lie inside the unit circle (the minimum-phase condition), we showed that the logarithm of the magnitude and the phase of the Fourier transform are Hilbert transforms of each other. A number of other important properties of minimum-phase sequences were also discussed.

Hilbert transform relations were derived for periodic sequences that satisfy a modified causality constraint, and for complex sequences whose Fourier transforms vanish on the bottom half of the unit circle.

In discussing Hilbert transform relations, our emphasis was on theoretical concepts rather than applications. Some of the uses of Hilbert transforms are illustrated in the problems and also in Chapter 10, where the results of this chapter play a very important role.

REFERENCES

1. L. V. Ahlfors, *Complex Analysis*, 2nd ed., McGraw-Hill Book Company, New York, 1966.

2. R. V. Churchill, *Complex Variables and Applications*, McGraw-Hill Book Company, New York, 1960.

3. P. M. Morse and H. Feshback, *Methods of Theoretical Physics*, McGraw-Hill Book Company, New York, 1953.

4. H. W. Bode, *Network Analysis and Feedback Amplifier Design*, Van Nostrand Reinhold Company, New York, 1945.

5. E. A. Guillemin, *Theory of Linear Physical Systems*, John Wiley & Sons, Inc., New York, 1963.

6. J. Dugundji, "Envelopes and Pre-Envelopes of Real Waveforms," *Trans. IRE*, Vol. IT-4, Mar. 1958, pp. 53–57.

7. V. Cizek, "Discrete Hilbert Transform," *IEEE Trans. Audio Electroacoust.*, Vol. AU-18, No. 4, Dec. 1970, pp. 340–343.

8. B. Gold, A. V. Oppenheim, and C. M. Rader, "Theory and Implementation of the Discrete Hilbert Transform," *Proc. Symp. Computer Processing in Communications*, Vol. 19, Polytechnic Press, 1970, New York.

9. D. J. Sakrison, W. T. Ford, and J. H. Hearne, "The *z*-Transform of a Realizable Time Function," *IEEE Trans. Geosci. Elect.*, Vol. GE-5, No. 2, Sept. 1967, pp. 33–41.

10. F. B. Hildebrand, *Advanced Calculus with Applications*, Prentice-Hall, Inc., Englewood Cliffs, N.J., 1962.

11. S. Treitel and E. A. Robinson, "The Design of High-Resolution Digital Filters," *IEEE Trans. Geosci. Elect.*, Vol. GE-4, No. 1, June 1966, pp. 25–38.

12. E. A. Robinson, *Random Wavelets and Cybernetic Systems*, Charles Griffin and Co. Ltd., London, 1962.

13. E. A. Robinson, *Statistical Communication and Detection*, Hafner Press, New York, 1967.

14. A. J. Berkhout, "On the Minimum Phase Criterion of Sampled Signals," *IEEE Trans. Geosci. Elect.*, Vol. GE-11, No. 4, Oct. 1973, pp. 186–198.

15. H. B. Voelcker, "Toward a Unified Theory of Modulation," *Proc. IEEE*, Vol. 54, Mar. 1966, pp. 340–353, and May 1966, pp. 735–755.

16. A. A. G. Requicha, "Contributions to a Zero-Based Theory of Bandlimited Signals," Ph.D. Thesis, Department of Electrical Engineering, University of Rochester, 1970.

17. O. Herrmann, "Transversalfilter zur Hilbert-Transformation," *Arch. Electronik Ubertragungstechnik*, Vol. 23, No. 12, 1969, pp. 581–587.

18. L. R. Rabiner and R. W. Schafer, "On the Behavior of Minimax FIR Digital Hilbert Transformers," *Bell Syst. Tech. J.*, Vol. 53, No. 2, Febr., 1974, pp. 361–388.
19. O. Herrmann and H. W. Schuessler, "Design of Nonrecursive Digital Filters with Minimum Phase," *Elect. Letters*, Vol. 6, No. 11, 1970, pp. 329–330.

PROBLEMS

1. In Sec. 7.1 we observed that the z-transform is completely determined outside the unit circle by the value of its imaginary part on the unit circle and the value of $h(0)$.

(a) Beginning with Eq. (7.8), derive Eq. (7.14b).

(b) Use Eq. (7.14b) to find $H(z)$ when

$$H_I(e^{j\omega}) = \frac{-\alpha \sin \omega}{1 + \alpha^2 - 2\alpha \cos \omega}$$

$$h(0) = 1$$

2. In Sec. 7.1 we derived the Hilbert transform of a real causal sequence $h(n)$ by exploiting the property that the real and imaginary parts of $H(z)$ on the unit circle correspond to the even and odd parts of $h(n)$, respectively. Alternatively, since the z-transform is analytic in its region of convergence, we can derive the Hilbert transform directly in the frequency domain, using Cauchy's integral formula. Cauchy's integral formula is the following: If $F(z)$ is analytic everywhere inside and on a simple closed contour C, then

$$\frac{1}{2\pi j} \oint_C \frac{F(\zeta)\, d\zeta}{\zeta - z} = \begin{cases} F(z), & \text{if } z \text{ inside } C \\ 0, & \text{if } z \text{ outside } C \end{cases}$$

Let $H(z)$ be the z-transform of a stable, causal, *complex* sequence $h(n)$.

(a) Let $F(z) = H(1/z)$. $F(z)$ is analytic everywhere in the disc $|z| < R$, where $R > 1$. Why?

(b) Let the contour C be the unit circle $|z| = 1$. If the point ζ lies on C, then show that the point

$$\tilde{z} = \frac{\zeta\zeta^*}{z^*}$$

lies outside C whenever z lies inside C. The points z and $\tilde{z}$ are said to be inverse (or image) points with respect to the circle C.

(c) By considering

$$\frac{1}{2\pi j} \oint_C \frac{F(\zeta)\, d\zeta}{\zeta - z} + \frac{\alpha}{2\pi j} \oint_C \frac{F(\zeta)\, d\zeta}{\zeta - \tilde{z}}$$

for $\alpha = +1$ and $\alpha = -1$, show that we can find $F(z)$ everywhere *inside* the unit circle from $F(e^{j\theta})$ as the convolutions

$$F(re^{j\theta}) = \begin{cases} \dfrac{1}{2\pi} \displaystyle\int_{-\pi}^{\pi} P(r, \theta - \phi) F(e^{j\phi})\, d\phi, & r < 1 \quad \text{(P7.2-1)} \\ F(0) + \dfrac{j}{2\pi} \displaystyle\int_{-\pi}^{\pi} Q(r, \theta - \phi) F(e^{j\theta})\, d\phi, & r < 1 \quad \text{(P7.2-2)} \end{cases}$$

where

$$z = re^{j\theta}, \qquad 0 \le r < 1, \qquad -\pi < \theta \le \pi$$

$$\zeta = e^{j\phi}, \qquad\qquad -\pi < \phi \le \pi$$

and P and Q are real. Find $P(r, \theta)$ and $Q(r, \theta)$.

Hint:

$$\frac{1}{\zeta - z} + \frac{\alpha}{\zeta - \tilde{z}} = \frac{1}{\zeta}\frac{\zeta}{\zeta - z} - \alpha\frac{\zeta^*}{\zeta^* - z^*} + \alpha$$

(d) Let $F(re^{j\theta}) = u(r, \theta) + jv(r, \theta)$. When $r < 1$, use Eqs. (P7.2-1) and (P7.2-2) to find expressions for

(1) $u(r, \theta)$ in terms of $u(1, \theta)$.
(2) $v(r, \theta)$ in terms of $v(1, \theta)$.
(3) $u(r, \theta)$ in terms of $v(1, \theta)$ and $u(0)$.
(4) $v(r, \theta)$ in terms of $u(1, \theta)$ and $v(0)$.
(5) $F(re^{j\theta})$ in terms of $u(1, \theta)$ and $v(0)$.
(6) $F(re^{j\theta})$ in terms of $v(1, \theta)$ and $u(0)$.

You may express your answers to parts (e) and (f) using

$$K(re^{j\theta}) = P(r, \theta) + jQ(r, \theta)$$

(e) Let $H(\rho e^{j\omega}) = H_R(\rho, \omega) + jH_I(\rho, \omega)$. Using $H(\rho e^{j\omega}) = F(\rho^{-1}e^{-j\omega})$, transform your results from part (d) to get the Hilbert transform relations for the unit circle.

[Note that $K(z^{-1}) = U(z)$, where $U(z)$ is the kernel defined in the text.]

(f) You should have found that the constants $H_R(\infty)$ and $H_I(\infty)$ (or their equivalents in the time domain) appear in some of the relations in part (e). Show that if $h(n)$ is real, $H_I(\infty) = 0$ or if $h(n)$ is imaginary $H_R(\infty) = 0$. (*Hint*: Use the initial value theorem, if you have not already done so.)

3. Suppose that $F(z)$ is a rational function; i.e.,

$$F(z) = \frac{N(z)}{D(z)}$$

where $N(z)$ and $D(z)$ are polynomials. Moreover, assume that $F(z)$ has no poles or zeros of multiplicity greater than 1. Let C be a simple closed contour, and let Z and P be, respectively, the number of zeros and poles of $F(z)$ enclosed by the contour. (Assume there are no poles or zeros on C.)

(a) Show that

$$\frac{1}{2\pi j}\oint_C \frac{F'(z)}{F(z)}\,dz = Z - P$$

(b) If $F(z) = |F(z)|e^{j\arg F(z)}$, show that

$$\frac{1}{2\pi}\arg\left[F(z)\right]\Big|_C = Z - P$$

i.e., the change in the phase of F as the contour C is traversed exactly once is $2\pi(Z - P)$.

(It can be shown that these results generalize in the case of multiple-order poles and zeros by counting poles and zeros according to their multiplicity; i.e., a second-order pole is counted twice.)

4. Let $x(n)$ be a real causal sequence for which $|x(n)| < \infty$. The z-transform of $x(n)$ is given by

$$X(z) = \sum_{n=0}^{\infty} x(n)z^{-n}$$

which is a Taylor's series in the variable z^{-1}, and therefore converges to an analytic function everywhere outside some circular disc centered at $z = 0$. [The region of convergence includes the point $z = \infty$ and, in fact, $X(\infty) = x(0)$.] The statement that $X(z)$ is analytic (in its region of convergence) implies strong constraints on X; i.e., its real and imaginary parts each satisfy Laplace's equation and the real and imaginary parts are related by the Cauchy–Riemann equations. We shall now use these properties to determine $X(z)$ from its real part when $x(n)$ is a real, finite-valued causal sequence.

Let $x(n)$ be a real (finite-valued) causal sequence with z-transform

$$X(z) = X_R(z) + jX_I(z)$$

where X_R and X_I are real-valued functions of z.

Suppose that $X_R(z)$ is given by

$$X_R(\rho e^{j\omega}) = \frac{\rho + \alpha \cos \omega}{\rho} \qquad (\alpha \text{ real})$$

for $z = \rho e^{j\omega}$. Then find $X(z)$ (as an explicit function of z) assuming that $X(z)$ is analytic everywhere except at $z = 0$. Do this using both of the methods suggested below.

(a) *Method 1* (frequency domain): Use the fact that the real and imaginary parts of X must satisfy the Cauchy–Riemann equations everywhere that X is analytic. The Cauchy–Riemann equations are the following:

(1) In Cartesian coordinates:

$$\frac{\partial U}{\partial x} = \frac{\partial V}{\partial y}$$

$$\frac{\partial V}{\partial x} = -\frac{\partial U}{\partial y}$$

where $z = x + jy$ and $X(x + jy) = U(x, y) + jV(x, y)$.

(2) In polar coordinates:

$$\frac{\partial U}{\partial \rho} = \frac{1}{\rho}\frac{\partial V}{\partial \omega}$$

$$\frac{\partial V}{\partial \rho} = -\frac{1}{\rho}\frac{U}{\partial \omega}$$

where $z = \rho e^{j\omega}$ and $X(\rho e^{j\omega}) = U(\rho, \omega) + jV(\rho, \omega)$.

Since we know that $U = X_R$, we can integrate these equations to find V and hence X. (Be careful to treat the constant of integration properly.)

(b) *Method 2* (time domain): Use the fact that the sequence $x_e(n)$ whose Fourier transform is $X_R(e^{j\omega})$ must be real and even and the sequence $x_0(n)$ whose Fourier transform is $jX_I(e^{j\omega})$ is real and odd. By linearity,

$$x(n) = x_e(n) + x_0(n)$$

Since we can find $x_e(n)$ directly from $X_R(e^{j\omega})$ and since $x(n)$ is real and causal, we can find $x_0(n)$. Hence we can find $X(z)$.

5. Derive an integral expression for $H(z)$ *inside* the unit circle in terms of Re $[H(e^{j\omega})]$, when $h(n)$ is a real, stable sequence such that $h(n) = 0$ for $n > 0$.

6. Show that the allpass system function

$$H_{ap}(z) = \frac{z^{-1} - a^*}{1 - az^{-1}}, \qquad |a| < 1$$

has unity gain for all frequencies. That is, show that $|H_{ap}(e^{j\omega})| = 1$ for $0 \le \omega \le \pi$.

7. Consider a stable non-minimum-phase causal signal $x(n)$, with z-transform $X(z)$ The zeros of $X(z)$ are z_k, $k = 1, 2, \ldots, M$, where $|z_1| < |z_2| < \ldots < |z_M|$. We propose to obtain a new sequence $y(n)$ that is minimum phase by exponentially weighting the sequence $x(n)$; i.e.,

$$y(n) = \alpha^n x(n)$$

How should α be chosen so that $y(n)$ is minimum phase?

8. In Sec. 7.2 we asserted that minimum-phase sequences are so called because their phase-lag functions (negative of phase) are smallest of all real, causal sequences having the same Fourier transform magnitude. This assertion rests on the statement that the phase of an allpass system which is real, causal, and stable is always less than or equal to zero.

(a) Consider the first-order allpass transfer function

$$H_{ap}(z) = \frac{z^{-1} - a}{1 - az^{-1}}$$

where a is real and $|a| < 1$. Show that the phase of $H_{ap}(e^{j\omega})$ is non-positive for $0 \le \omega < \pi$.

(b) Now consider the second-order allpass transfer function with complex-conjugate poles and zeros; i.e.,

$$H_{ap}(z) = \frac{z^{-1} - a^*}{1 - az^{-1}} \frac{z^{-1} - a}{1 - a^* z^{-1}}$$

where a is complex with $|a| < 1$. Show that the phase of $H_{ap}(e^{j\omega})$ is non-positive for $0 \le \omega < \pi$.

Since the transfer function of any allpass system with real impulse response can always be represented as a product of factors as in parts (a) and (b), the above results show that if

$$H(z) = H_{min}(z) H_{ap}(z)$$

then the phase of $H(e^{j\omega})$ will always be more negative than the phase of $H_{min}(e^{j\omega})$, and thus $H_{min}(e^{j\omega})$ has the minimum phase lag of all systems for which $|H(e^{j\omega})| = |H_{min}(e^{j\omega})|$.

(*Suggestion:* In both parts (a) and (b), the desired result can be shown by algebraic manipulation of arg $[H(e^{j\omega})]$. Alternatively, a simple geometrical argument suffices if the transfer functions are first mapped to the s-plane using the bilinear transformation.)

9. Prove the validity of the following two statements:

(a) The convolution of two minimum-phase sequences is also minimum phase.

(b) The sum of two minimum-phase sequences is not necessarily minimum phase. (Give an example of both a minimum-phase and non-minimum-phase sequence that can be formed as the sum of two minimum-phase sequences.)

10. Let $h_{\min}(n)$ denote a minimum-phase sequence with z-transform $H_{\min}(z)$. If $h(n)$ is a causal, non-minimum-phase sequence whose Fourier transform magnitude is equal to $|H_{\min}(e^{j\omega})|$, show that

$$|h(0)| < |h_{\min}(0)|$$

(*Hint:* Use the initial value theorem.)

11. One of the interesting and important properties of minimum-phase sequences is the minimum-energy delay property; i.e., of all the causal sequences having the same Fourier transform magnitude function $|H(e^{j\omega})|$, the quantity

$$\sum_{n=0}^{m} |h(n)|^2$$

is maximum when $h(n)$ is the minimum-phase sequence. This result is proved as follows: Let $h_{\min}(n)$ be a minimum-phase sequence with z-transform $H_{\min}(z)$. Furthermore, let z_k be a zero of $H_{\min}(z)$ so that we can express $H_{\min}(z)$ as

$$H_{\min}(z) = Q(z)(1 - z_k z^{-1}), \qquad |z_k| < 1$$

where $Q(z)$ is again minimum phase. Now consider another sequence $h(n)$ with z-transform $H(z)$ such that

$$|H(e^{j\omega})| = |H_{\min}(e^{j\omega})|$$

and $H(z)$ has a zero at $z = 1/z_k^*$ instead of at z_k.

(a) Express $H(z)$ in terms of $Q(z)$.

(b) Express $h(n)$ and $h_{\min}(n)$ in terms of the minimum-phase sequence $q(n)$ which has z-transform $Q(z)$.

(c) To compare the distribution of energy of the two sequences, show that

$$\varepsilon = \sum_{n=0}^{m} |h_{\min}(n)|^2 - \sum_{n=0}^{m} |h(n)|^2 = (1 - |z_k|^2)|q(m)|^2$$

(d) Using the result of part (c), argue that

$$\sum_{n=0}^{m} |h(n)|^2 \leq \sum_{n=0}^{m} |h_{\min}(n)|^2, \qquad \text{for all } m$$

12. Shown in Fig. P7.12 are eight different finite-duration sequences. Each sequence is four points long. The magnitude of the Fourier transform is the same for all sequences. Which of the sequences has all the zeros of its z-transform *inside* the unit circle?

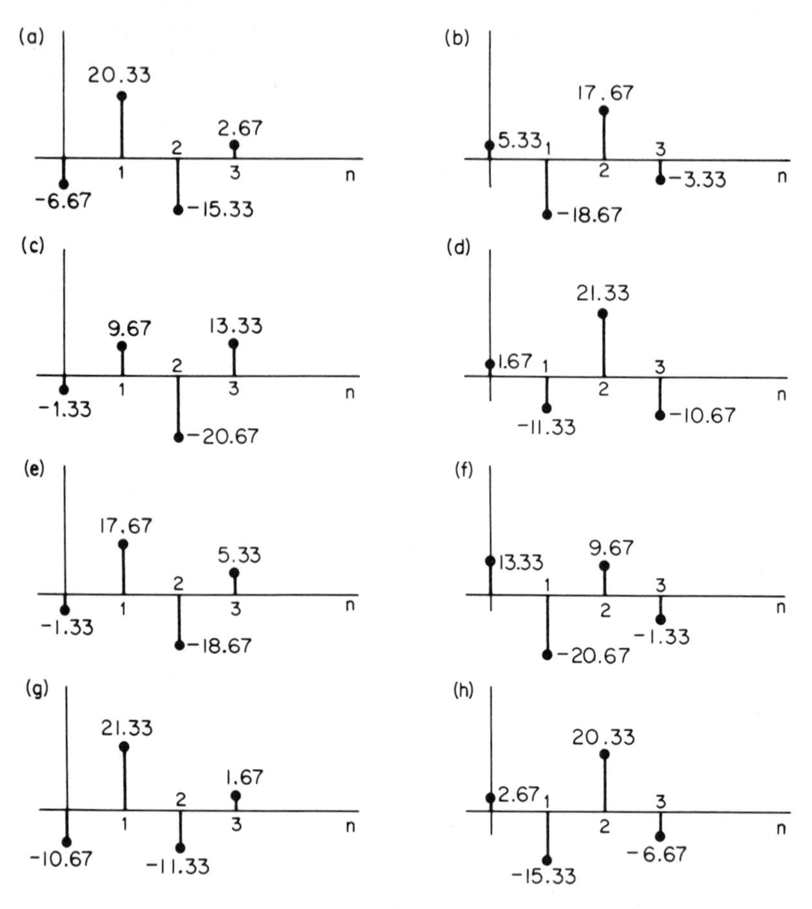

Fig. P7.12

13. Let $h_1(n), h_2(n), \ldots, h_M(n)$ denote M finite-duration sequences, all of duration N, i.e., $h_k(n) = 0$ for $n < 0$ and $n \geq N$. The magnitude of the Fourier transform of each of these sequences is identical. None of the sequences are proportional (with either a real or complex constant of proportionality) to any of the others.

(a) What is the maximum value of M if the sequences are not restricted to be real?

(b) What is the maximum value of M if the sequences are restricted to be real?

14. A maximum-phase sequence is obtained by reflecting all the zeros of the z-transform of a minimum-phase sequence to conjugate reciprocal positions outside the unit circle. That is, we can express the z-transform of a maximum-phase sequence as

$$H_{\max}(z) = H_{\min}(z)H_{\mathrm{ap}}(z)$$

In the case of a finite-duration sequence we can express $H_{\min}(z)$ as

$$H_{\min}(z) = h_{\min}(0) \prod_{k=1}^{N-1} (1 - z_k z^{-1}), \qquad |z_k| < 1$$

(a) Obtain an expression for the allpass function required to obtain $H_{\max}(z)$.
(b) Show that $H_{\max}(z)$ can be expressed as

$$H_{\max}(z) = z^{-(N-1)}H_{\min}(z^{-1})$$

(c) Using the result of part (b), express the maximum-phase sequence $h_{\max}(n)$ in terms of $h_{\min}(n)$.

15. The even part of a sequence $x(n)$ is defined by

$$x_e(n) = \frac{x(n) + x(-n)}{2}$$

Suppose that $x(n)$ is a real finite-duration sequence defined such that $x(n) = 0$ for $n < 0$ and $n \geq N$. Let $X(k)$ denote the N-point DFT of $x(n)$.
(a) Does the DFT of $x_e(n)$ equal Re $[X(k)]$?
(b) What is the inverse DFT of Re $[X(k)]$ in terms of $x(n)$?

16. Consider a real-valued finite-duration sequence $x(n)$ of length N; i.e., $x(n) = 0$, $n < 0$, $n \geq N$, where N is odd. The M-point DFT of $x(n)$ is designated by $X(k)$, so that

$$X(k) = \sum_{n=0}^{N-1} x(n)e^{-j(2\pi/M)nk}$$

Let $X_R(k)$ designate the real part of $X(k)$.
(a) Determine, in terms of N, the smallest value of M (other than the trivial values $M = 1, 2$) that will permit $X(k)$ to be uniquely determined from $X_R(k)$.
(b) With M satisfying the condition determined in part (a), $X(k)$ can be expressed as the circular convolution of $X_R(k)$ with a sequence $U(k)$. Determine $U(k)$.

17. This problem is concerned with exploring one technique for designing digital filters with minimum phase. Such filters have all their poles and zeros inside or on the unit circle. Let us first consider the problem of converting a linear-phase finite-duration unit-sample response (FIR) equiripple lowpass filter to a minimum-phase design. Assume that $H(e^{j\omega})$ is the frequency response of the linear-phase filter, where
(1) $h(n)$ is real. $h(n) = 0$, $n < 0$, $n > N - 1$. Furthermore, assume that N is odd.
(2) The passband ripple is δ_1; i.e., in the passband $|H(e^{j\omega})|$ oscillates between $1 + \delta_1$ and $1 - \delta_1$.
(3) The stopband ripple is δ_2; i.e., in the stopband $|H(e^{j\omega})| \leq \delta_2$ and $|H(e^{j\omega})|$ oscillates between δ_2 and 0 (see Fig. P7.17-1).

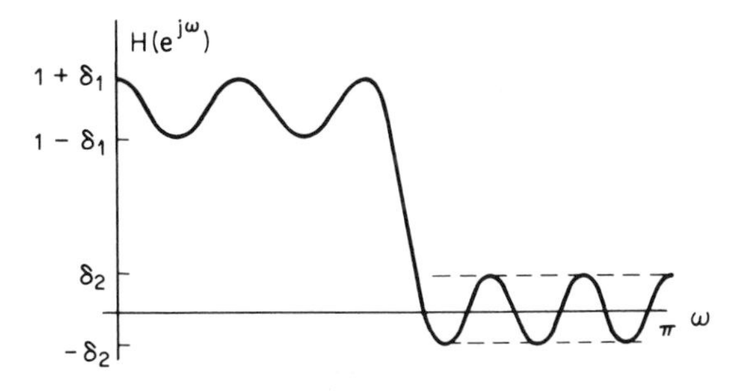

Fig. P7.17-1

(4) $H(e^{j\omega}) = H_0(e^{j\omega})e^{-jn_0\omega}$, where $H_0(e^{j\omega})$ is real and $n_0 = (N-1)/2$.

The following technique has been proposed by Herrmann and Schuessler [19] for converting this linear-phase design into a minimum-phase design that has a system function $H_{\min}(z)$ and unit-sample response $h_{\min}(n)$.

Step A: Create a new sequence

$$h_1(n) = \begin{cases} h(n), & n \neq n_0 \\ h(n_0) + \delta_2, & n = n_0 \end{cases}$$

Step B: Recognize that $H_1(z)$ can be expressed in the form

$$H_1(z) = z^{-n_0}H_2(z)H_2(1/z)$$

for some $H_2(z)$, where $H_2(z)$ has all its poles and zeros inside the unit circle, and $h_2(n)$ is real.

Step C: Set

$$H_{\min}(z) = \frac{H_2(z)}{\sqrt{1+\delta_2}}$$

The denominator is to renormalize the passband so that the resulting $H_{\min}(e^{j\omega})$ will oscillate about a value of unity.

(a) Show that if $h_1(n)$ is chosen as in step A, then $H_1(e^{j\omega})$ can be written as

$$H_1(e^{j\omega}) = e^{-j\omega n_0}H_3(e^{j\omega})$$

where $H_3(e^{j\omega})$ is real and nonnegative for all values of ω.

(b) If $H_3(e^{j\omega}) \geq 0$, as was shown in part (a), show that there exists an $H_2(z)$ such that

$$H_3(z) = H_2(z)H_2(1/z)$$

where $H_2(z)$ is minimum phase and $h_2(n)$ is real (i.e., justify step B).

(c) Demonstrate that the new filter $H_{\min}(e^{j\omega})$ is an equiripple lowpass filter, i.e., that its magnitude characteristic is of the form shown in Fig. P7.17-2, by evaluating δ_1' and δ_2'. What is N', the length of $h_{\min}(n)$?

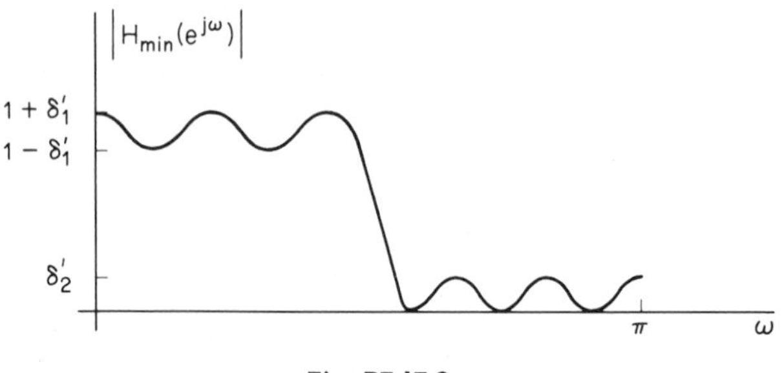

Fig. P7.17-2

(d) In parts (a), (b), and (c) we assumed that we started with a FIR linear-phase filter. Will this technique work if we remove the linear-phase constraints? Explain.

18. Consider a sequence $x(n)$ which has complex values so that $x(n) = x_r(n) + jx_i(n)$, where $x_r(n)$ and $x_i(n)$ are real. The z-transform, $X(z)$, of the sequence $x(n)$ is zero on the bottom half of the unit circle; i.e., $X(e^{j\omega}) = 0$, $\pi \leq \omega \leq 2\pi$.

The real part of $x(n)$ is

$$x_r(n) = \begin{cases} \frac{1}{2}, & n = 0 \\ -\frac{1}{4}, & n = \pm 2 \\ 0, & \text{otherwise} \end{cases}$$

Determine the real and imaginary parts of $X(e^{j\omega})$.

19. Let $H[\ ]$ denote the ideal operation of Hilbert transformation; i.e.,

$$H[x(n)] = \sum_{k=-\infty}^{\infty} h(n-k)x(k)$$

where $h(n)$ is given by Eq. (7.48). Prove the following properties.

(a) $H[H[x(n)]] = -x(n)$.

(b) $\sum_{n=-\infty}^{\infty} x(n)H[x(n)] = 0$.

(*Hint*: Use Parseval's theorem.)

(c) $H[x(n) * y(n)] = H[x(n)] * y(n) = x(n) * H[y(n)]$, where $x(n)$ and $y(n)$ are arbitrary sequences.

8

Discrete Random Signals

8.0 Introduction

The preceding chapters have focused on mathematical representations of discrete-time signals and systems and the insights that derive from such mathematical representations. We have seen that discrete-time signals and systems have both a time-domain and a frequency-domain representation, each having an important place in the theory and design of digital signal processing systems. Until now we have assumed that the signals were deterministic; i.e., each value of a sequence is uniquely determined by a mathematical expression, a table of data, or a rule of some type. In Chapters 1 and 2 we discussed representation of such deterministic signals in terms of their z-transforms or Fourier transforms. Sequences that have a z-transform representation must have finite energy or it must be possible to multiply by an exponential sequence so that the product has finite energy. This finite-energy requirement corresponds essentially to the requirement that the z-transform converges. In Chapter 3 we considered periodic sequences. The z-transform for periodic sequences does not exist since the condition of finite energy cannot be satisfied. However, periodic signals are, by definition, identical from period to period, and consequently they can be uniquely represented in terms of a single period. A single period is a finite-length, and consequently finite-energy, sequence. This property of periodic signals was exploited in Chapter 3 for the representation of periodic signals by means of the Fourier series or discrete Fourier transform.

There are many important examples of signals which either do not have finite energy or are not periodic. Many communication signals, for example, are of indefinite duration and consequently are best modeled in terms of infinite-duration infinite-energy signals. In many situations, signal-generation processes are so complex as to make precise description of a signal extremely difficult, if not impossible. As an example we will see in Chapter 9 that many of the effects encountered in implementing digital signal processing algorithms with finite register length can be represented by additive "noise," which can be conveniently treated as a sequence with infinite energy. Many mechanical systems generate acoustic or vibratory signals which can often be processed to diagnose potential failure; again signals of this type are best modeled in terms of infinite-energy nonperiodic signals. Speech signals to be processed for automatic recognition or bandwidth compression and music to be processed for quality enhancement are two more of many possible examples.

The key to the mathematical representation of such signals lies in their description in terms of averages. As will be developed in this chapter, many (but not all) of the properties of such signals can be summarized in terms of a finite-energy sequence called the *autocorrelation* or *autocovariance sequence*, for which the z-transform or the Fourier transform often exists. As we will see, the Fourier transform of the autocovariance sequence has a useful interpretation in terms of the frequency distribution of the power in the signal. The use of the autocovariance sequence and its transform also has the important advantage that the effect of processing infinite-energy signals with a discrete linear system can be conveniently described in terms of the effect of the system on the autocovariance sequence.

In developing the representation of infinite-energy signals it is convenient to work within the framework of nondeterministic, i.e., random or stochastic, signals. Within this framework the signal is considered to be a member of an ensemble of discrete-time signals which is characterized by a set of probability density functions. The theory of stochastic signals in its most general form is extremely advanced and abstract, and a rigorous treatment requires a degree of mathematical sophistication beyond our scope. Our primary objective in this chapter is to collect and interpret a specific set of results pertaining to representation of random signals (i.e., those having infinite energy) which will be useful in the chapters that follow. Thus we shall avoid a detailed discussion of most of the difficult and subtle mathematical issues of the theory of random processes. Although our approach will not necessarily be rigorous, we shall be careful to summarize the important results and the mathematical assumptions implicit in their derivation.

8.1 A Discrete-Time Random Process

The fundamental concept in the mathematical representation of infinite-energy signals is that of a *random process*. In our discussion of random processes as models for infinite-energy discrete-time signals, we shall assume that the reader is familiar with fundamental concepts of the theory of probability such as random variables, probability distributions, and averages. Those readers who require further background in probability theory are referred to one of the basic texts listed in the references [1–7].

8.1.1 Simple Example: The Bernoulli Process

We shall introduce the concept of a random process through the discussion of a very simple example. Suppose that a sequence of numbers is generated as follows: At a given time n, a coin is tossed, and if the result is heads, the value of the sequence at n is $x(n) = +1$; if the result is tails, the value is $x(n) = -1$. A sequence that might have been generated by this scheme is shown in Fig. 8.1. If we assume that this process has been operating

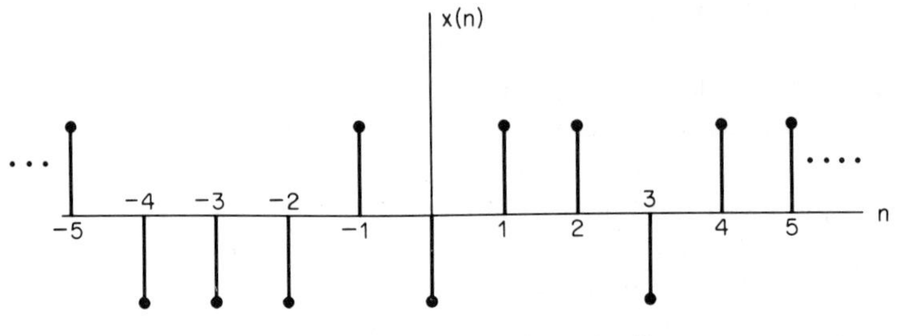

Fig. 8.1 Sequence of $+1$'s and -1's.

for all time, i.e., $-\infty < n < \infty$, then we obtain a sequence of infinite duration. If we attempt to represent this sequence by the methods of previous chapters, we encounter two basic difficulties. First, it is clear that the sequence has infinite energy and that neither the z-transform nor the Fourier transform exists. Second, our personal experience with coin-flipping experiments should suggest that it is impossible to precisely characterize the sequence other than by tabulating the set of sample values, and since it is assumed that the duration is infinite, this is also impossible. Even given a large number of past samples, it would be impossible to determine the next sample in the sequence with any certainty. This uncertainty leads us in general to a characterization of the sequence in terms of probabilities and thus to averages.

In this case let us suppose that, at any time, the probability of heads is p. Then by the fundamental axioms of the theory of probability, the probability of tails must be $1 - p$. The nth-sequence value, $x(n)$, thus has an interpretation as a particular value of a *random variable* x_n, i.e., a function of the outcome of the experiment of tossing a coin.† Specifically, each sequence

† Random variables with this probability law are known as *Bernoulli random variables*.

value can be viewed as the result of assigning a number to the outcome of the coin-tossing experiment. That is, to the event "a head is tossed" the value 1 is assigned. Similarly, to the event "a tail is tossed" the value -1 is assigned. Since the entire set of possible outcomes for the coin-tossing experiment consists of these two mutually exclusive events, the random variable x_n can only take on the two values $x(n) = +1$ and $x(n) = -1$. To each event, we assign a number that specifies the probability of occurrence of that event. In this example the probability of heads is p and thus the probability that $x_n = +1$ is p. Similarly, since the probability of tails is $1 - p$, so is the probability that $x_n = -1$.

The set of random variables $\{x_n\}$ for $-\infty < n < \infty$, together with the probabilistic description of each random variable, constitutes the definition of a random process.† A given sequence of values $\{x(n)\}$, $-\infty < n < \infty$, is a *realization* of the random process and is called a *sample sequence* of the random process. The number of possible sample sequences that could be generated by the scheme of our example is infinite. The collection of all the sequences that could result as realizations of a random process is called an *ensemble* of sample sequences. Several possible sample sequences for the present example are shown in Fig. 8.2. Indeed, unless p equals 0 or 1, any sequence of $+1$'s and -1's is a member of the ensemble of the Bernoulli process.

In applying the random process model in practical signal processing applications, we consider a particular sequence to be one of an ensemble of sample sequences corresponding to a random process. This is the sense in which a random process serves as a representation of an infinite-energy signal. Given a discrete-time signal which we assume to be one of an ensemble of sample sequences, the structure, i.e., the underlying probability law, of the corresponding random process is generally not known and must somehow be inferred. It may be possible to make reasonable assumptions about the structure of the process, or it may be possible to estimate the properties of a random-process representation from a finite segment of a typical sample sequence. For example, it seems plausible that we might infer the underlying probability law of our example by observing a sufficiently long segment of one of the sequences of Fig. 8.2. The conditions under which this can be done are discussed in Sec. 8.2.2. In any case, in order to proceed further with the discussion of random-process models of discrete-time signals, it is necessary to consider a more formal mathematical description of a random process.

8.1.2 Random-Process Descriptions

Formally a random process is an indexed family of random variables $\{x_n\}$ [5–7]. The family of random variables is characterized by a set of probability distribution functions that in general may be a function of the

† Since the random variables are called Bernoulli random variables, it is reasonable in this example to call the corresponding random process a *Bernoulli random process*.

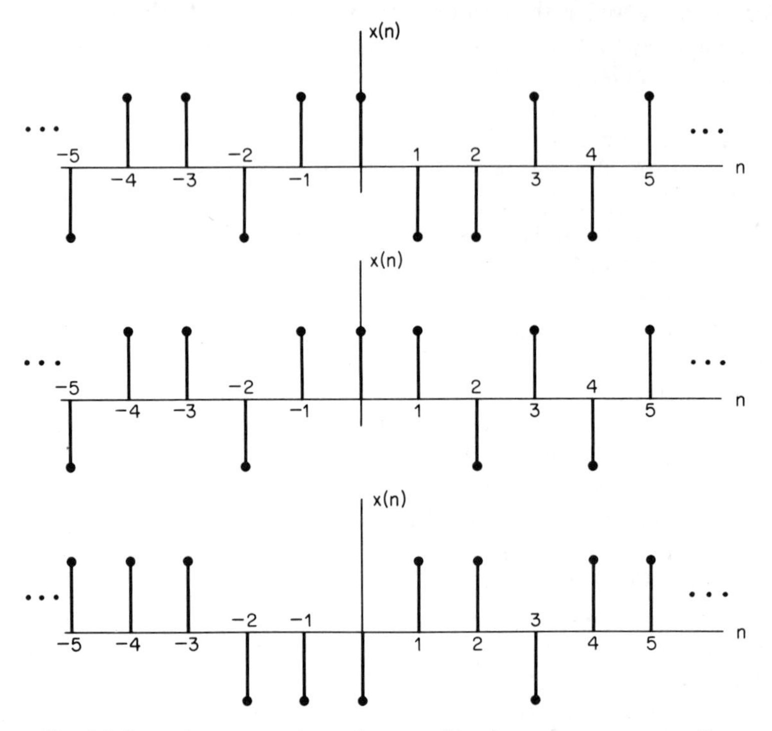

Fig. 8.2 Several sequences from the ensemble of sequences corresponding to the Bernoulli process.

index n. In using the concept of a random process as a model for random signals, the index is associated with time or possibly some other physical dimension. An individual random variable x_n is described by the probability distribution function

$$P_{x_n}(\mathbf{x}_n, n) = \text{Probability } [x_n \leq \mathbf{x}_n] \tag{8.1}$$

where x_n denotes the random variable and $\mathbf{x}_n$ is a particular value of x_n.† If x_n takes on a continuous range of values, it is equivalently specified by the *probability density function*

$$p_{x_n}(\mathbf{x}_n, n) = \frac{\partial P_{x_n}(\mathbf{x}_n, n)}{\partial \mathbf{x}_n} \tag{8.2}$$

or

$$P_{x_n}(\mathbf{x}_n, n) = \int_{-\infty}^{\mathbf{x}_n} p_{x_n}(x, n)\, dx \tag{8.3}$$

† In this chapter, bold face type is used to denote the dummy variables of probability functions, *not* vectors or matrices.

In the previous example the random variables were quantized; i.e., they take on a countable set of values. In this case the distribution is

$$P_{x_n}(\mathbf{x}_n, n) = \begin{cases} 1, & \mathbf{x}_n \geq 1 \\ 1 - p, & -1 \leq \mathbf{x}_n < 1 \\ 0, & \mathbf{x}_n < -1 \end{cases}$$

In such cases the derivative does not exist unless impulse functions are allowed. Rather than adopt this approach we define instead the *probability mass function* of a quantized random variable as

$$p_{x_n}(\mathbf{x}_n, n) = \text{Probability } [x_n = \mathbf{x}_n] \tag{8.4}$$

For quantized random variables the probability distribution is related to the probability mass function by

$$P_{x_n}(\mathbf{x}_n, n) = \text{Probability } [x_n \leq \mathbf{x}_n] = \sum_{x \leq \mathbf{x}_n} p_{x_n}(x, n) \tag{8.5}$$

The probability distribution and the corresponding probability mass function for the Bernoulli process are plotted in Fig. 8.3.

The interdependence of two random variables x_n and x_m of a random process is described by the joint probability distribution function

$$P_{x_n, x_m}(\mathbf{x}_n, n, \mathbf{x}_m, m) = \text{Probability } [x_n \leq \mathbf{x}_n \text{ and } x_m \leq \mathbf{x}_m] \tag{8.6}$$

or, in the case of continuous random variables, by the joint probability density

$$p_{x_n, x_m}(\mathbf{x}_n, n, \mathbf{x}_m, m) = \frac{\partial^2 P_{x_n, x_m}(\mathbf{x}_n, n, \mathbf{x}_m, m)}{\partial \mathbf{x}_n \partial \mathbf{x}_m} \tag{8.7}$$

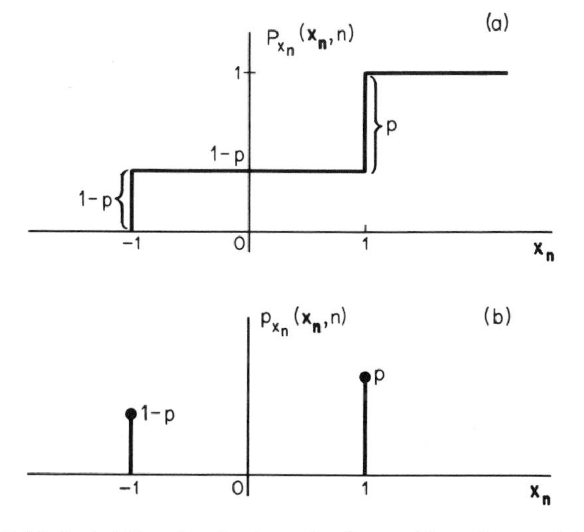

Fig. 8.3 (a) Probability distribution of a Bernoulli random variable; (b) corresponding probability mass function.

In the case of quantized random variables, the joint probability mass function is defined as

$$p_{x_n,x_m}(\mathbf{x}_n, n, \mathbf{x}_m, m) = \text{Probability } [x_n = \mathbf{x}_n \quad \text{and} \quad x_m = \mathbf{x}_m] \tag{8.8}$$

In formulating the Bernoulli process, we assumed that the successive coin flips were independent; i.e., for a given toss of the coin the probability of heads (or tails) did not depend on the outcome of any other toss. In this case the random variables $\{x_n\}$ are *statistically independent*; i.e.

$$P_{x_n,x_m}(\mathbf{x}_n, n, \mathbf{x}_m, m) = P_{x_n}(\mathbf{x}_n, n) \cdot P_{x_m}(\mathbf{x}_m, m)$$

A complete characterization of a random process requires the specification of all possible joint probability distributions. As we have indicated, these probability distribution functions may be a function of the time index n. In the case where all the probability functions are independent of a shift of time origin, the random process is said to be *stationary*. For example, the second-order distribution of a stationary process satisfies

$$P_{x_{n+k},x_{m+k}}(\mathbf{x}_{n+k}, n+k, \mathbf{x}_{m+k}, m+k) = P_{x_n x_m}(\mathbf{x}_n, n, \mathbf{x}_m, m) \tag{8.9}$$

The Bernoulli process is an example of a stationary process, since it was assumed that in flipping the coin, the probability of a head was always equal to p and each random variable was assumed independent of all the others.

In many of the applications of digital signal processing, random processes serve as models for signals in the sense that a particular infinite-energy signal can be considered a sample sequence of a random process. Although the details of such signals are unpredictable—making our previous approach to signal representation inappropriate—certain average properties of the ensemble can be predicted given the probability law of the process. These average properties often serve as a useful, although incomplete, characterization of signals for which the Fourier transform does not exist.

8.2 Averages

It is often useful to characterize a random variable by averages such as mean and variance. Since a random process is an indexed set of random variables, we may likewise characterize the process by statistical averages of the random variables comprising the random process. Such averages are called *ensemble averages*.

8.2.1 *Definitions*

The average or mean of the process is defined as

$$m_{x_n} = E[x_n] = \int_{-\infty}^{\infty} x p_{x_n}(x, n)\, dx \tag{8.10}$$

where E denotes mathematical expectation. We note that in general the mean (expected value) may depend upon n. In general if $g(\ )$ is a single-valued function, then $g(x_n)$ is also a random variable, and the set of random variables $\{g(x_n)\}$ defines a new random process. In order to compute averages of the new random process, we can derive probability distributions of the new random variables. Alternatively, it can be shown that

$$E[g(x_n)] = \int_{-\infty}^{\infty} g(x) p_{x_n}(x, n)\, dx \tag{8.11}$$

If the random variables are quantized, the integrals become summations over all possible values of the random variable.

$$E[g(x_n)] = \sum_{x} g(x) p_{x_n}(x, n) \tag{8.12}$$

In cases where we are interested in the relationship between two (or more) infinite-energy signals, i.e., two random processes, we must be concerned with two sets of random variables $\{x_n\}$ and $\{y_m\}$. For example, the expected value of a function of two random variables is defined as

$$E[g(x_n, y_m)] = \int_{-\infty}^{\infty} \int_{-\infty}^{\infty} g(x, y) p_{x_n, y_m}(x, n, y, m)\, dx\, dy \tag{8.13}$$

where $p_{x_n, y_m}(\mathbf{x}_n, n, \mathbf{y}_m, m)$ is the joint probability density of the random variables x_n and y_m.

There are a number of simple properties of averages that will be useful in the following discussion. In particular, it is easily shown that

1. $E[x_n + y_m] = E[x_n] + E[y_m]$; i.e., the average of a sum is the sum of the averages.
2. $E[ax_n] = aE[x_n]$; i.e., the average of a constant times x_n is equal to the constant times the average of x_n.

In general, the average of a product of two random variables is *not* equal to the product of the averages. If this is the case, however, the two random variables are said to be *linearly independent* or *uncorrelated.* That is, if x_n and y_m are linearly independent,

$$E[x_n y_m] = E[x_m] \cdot E[y_m] \tag{8.14}$$

It can easily be seen from Eq. (8.13) that a sufficient condition for linear independence is

$$P_{x_n y_m}(\mathbf{x}_n, n, \mathbf{y}_m, m) = P_{x_n}(\mathbf{x}_n, n) \cdot P_{y_m}(\mathbf{y}_m, m) \tag{8.15}$$

However, it can be shown that Eq. (8.15) is a stronger statement of independence than Eq. (8.14). As previously stated, random variables satisfying (8.15) are said to be *statistically independent.* If Eq. (8.15) holds for all values of n and m, the random processes $\{x_n\}$ and $\{y_m\}$ are said to be statistically independent. Statistically independent random processes are also linearly independent; however, linear independence does not imply statistical independence.

It can be seen from Eqs. (8.11)–(8.13) that averages are generally functions of time. However, in the case of stationary processes, this is not true. Thus the mean is the same for all the random variables that comprise the process; i.e., the mean of a stationary process is a constant, which we may denote simply m_x.

In addition to the mean of a random process, as defined in Eq. (8.10), there are a number of averages that are particularly important within the context of digital signal processing. These are defined below. [For notational convenience we shall assume that the probability distributions are continuous. Corresponding definitions for quantized random processes can be obtained by applying Eq. (8.12).]

The *mean-square value* of x_n is the average of x_n^2; i.e.,

$$E[x_n^2] = \text{mean square} = \int_{-\infty}^{\infty} x^2 p_{x_n}(x, n)\, dx \tag{8.16}$$

The mean-square value is sometimes referred to as the *average power*.

The *variance* of x_n is the mean-square value of $[x_n - m_{x_n}]$; i.e.,

$$\text{variance} = E[(x_n - m_{x_n})^2] = \sigma_{x_n}^2 \tag{8.17}$$

Since the average of a sum is the sum of the averages it can easily be shown that Eq. (8.17) can be written
as

$$\begin{aligned} \text{variance} &= E[x_n^2] - m_{x_n}^2 \\ &= \text{mean square} - (\text{mean})^2 \end{aligned} \tag{8.18}$$

In general, the mean-square value and the variance are functions of time; however, they are constant for stationary processes.

The mean, mean square, and the variance are simple averages that provide only a small amount of information about the process. A more useful average is the *autocorrelation sequence*, which is defined as

$$\begin{aligned} \phi_{xx}(n, m) &= E[x_n x_m^*] \\ &= \int_{-\infty}^{\infty}\int_{-\infty}^{\infty} \mathbf{x}_n \mathbf{x}_m^* p_{x_n, x_m}(\mathbf{x}_n, n, \mathbf{x}_m, m)\, d\mathbf{x}_n\, d\mathbf{x}_m \end{aligned} \tag{8.19}$$

where * denotes complex conjugation. The autocovariance sequence of a random process is defined as

$$\gamma_{xx}(n, m) = E[(x_n - m_{x_n})(x_m - m_{x_m})^*] \tag{8.20}$$

which can be written as

$$\gamma_{xx}(n, m) = \phi_{xx}(n, m) - m_{x_n} m_{x_m} \tag{8.21}$$

Note that in general, both the autocorrelation and autocovariance are two-dimensional sequences.

The autocorrelation is a measure of the dependence between values of the random process at different times. In this sense it describes the time variation of a random signal. A measure of the dependence between two different random signals is obtained from the cross-correlation sequence. If $\{x_n\}$ and $\{y_m\}$ are two random processes, their cross-correlation is

$$\begin{aligned}\phi_{xy}(n, m) &= E[x_n y_m^*] \\ &= \int_{-\infty}^{\infty}\int_{-\infty}^{\infty} xy^* p_{x_n, y_m}(x, n, y, m)\, dx\, dy\end{aligned} \tag{8.22}$$

where $p_{x_n y_m}(x, n, y, m)$ is the joint probability density of x_n and y_m. The cross-covariance function is defined as

$$\begin{aligned}\gamma_{xy}(n, m) &= E[(x_n - m_{x_n})(y_m - m_{y_m})^*] \\ &= \phi_{xy}(n, m) - m_{x_n} m_{y_m}\end{aligned} \tag{8.23}$$

As we have pointed out, the statistical properties of a random process generally vary with time. However, a stationary process is characterized by an equilibrium condition in which the statistical properties are invariant to a shift of time origin. This means that the first-order probability distribution is independent of time. Similarly, all the joint probability functions are also invariant to a shift in time origin; i.e., the second-order joint probability distribution satisfies Eq. (8.9). From Eq. (8.9) it follows that the second-order joint distribution depends only on the time difference $m - n$. First-order averages such as mean and variance are independent of time; second-order averages, such as the autocorrelation $\phi_{xx}(n, m)$, are dependent on the time difference $m - n$. Thus for a stationary process we can write

$$m_x = E[x_n] \tag{8.24}$$

$$\sigma_x^2 = E[(x_n - m_x)^2] \tag{8.25}$$

independent of n, and if we denote the time difference now by m,

$$\phi_{xx}(n, n + m) = \phi_{xx}(m) = E[x_n x_{n+m}^*] \tag{8.26}$$

That is, the autocorrelation sequence of a stationary random process is a one-dimensional sequence, a function of the time difference m.

In many instances we encounter random processes that are not stationary in the *strict sense;* i.e., their probability distributions are not time invariant, yet the mean is constant and the autocorrelation sequence satisfies Eq. (8.26). Such random processes are said to be *stationary in the wide sense* [5].

EXAMPLE. As an example of the description of a random process by averages, let us consider again the simple Bernoulli process. To begin let us note that the process is stationary since it was assumed that the probabilities of +1 and −1 were independent of time and the random variables $\{x_n\}$ were assumed to be statistically independent. When we use Eq. (8.12) the mean is found to be

$$\begin{aligned} m_x &= (+1) \cdot \text{Probability } [x_n = +1] + (-1) \cdot \text{Probability } [x_n = -1] \\ &= +1 \cdot p + (-1) \cdot (1 - p) \\ &= (2p - 1) \end{aligned}$$

and the mean-square value is

$$\begin{aligned} \text{mean square} &= (+1)^2 \cdot \text{Probability } [x_n = +1] \\ &\quad + (-1)^2 \cdot \text{Probability } [x_n = -1] \\ &= (+1)^2 p + (-1)^2 (1 - p) \\ &= 1 \end{aligned}$$

Thus the variance is

$$\sigma_x^2 = 1 - (2p - 1)^2 = 4p(1 - p)$$

Since we assumed statistical independence, the autocorrelation sequence is

$$\phi_{xx}(m) = \begin{cases} E[x_n^2] = 1, & m = 0 \\ E[x_n] \cdot E[x_{n+m}] = m_x^2, & m \neq 0 \end{cases}$$

In particular, if $p = \frac{1}{2}$, then $m_x = 0$ and

$$\phi_{xx}(m) = \delta(m)$$

In general, such an autocorrelation sequence is obtained whenever all the random variables of a random process are linearly independent. Such processes (called *white noise*) play an important role in many signal-processing problems.

8.2.2 *Time Averages*

As we have mentioned several times, in a signal-processing context, the notion of an ensemble of infinite-energy signals is a convenient mathematical concept that allows us to use the theory of probability in representing infinite-energy signals. However, in a practical sense, we would prefer to deal with a single sequence rather than an infinite ensemble of sequences. For example, we might wish to infer the probability law or certain averages of the random-process representation from measurements on a single member of the ensemble. For the Bernoulli process we recall that the probability distributions are independent of time, and therefore we might intuitively feel that the percentage of +1's and −1's in a long segment of a single sample sequence should be very close to p and $1 - p$, respectively. Similarly, the arithmetic average of a large number of samples of a single sequence should be very close to the mean of the Bernoulli process. To formalize these intuitive notions, we define the time average of a random process as

$$\langle x_n \rangle = \lim_{N \to \infty} \frac{1}{2N + 1} \sum_{n=-N}^{N} x_n \tag{8.27}$$

Similarly, the time autocorrelation sequence is defined as

$$\langle x_n x_{n+m} \rangle = \lim_{N\to\infty} \frac{1}{2N+1} \sum_{n=-N}^{N} x_n x_{n+m}^* \tag{8.28}$$

It can be shown that the above limits exist if $\{x_n\}$ is a stationary process with finite mean. However, the proof of this result is far beyond the scope of our discussion. As defined in Eqs. (8.27) and (8.28), these time averages are functions of an infinite set of random variables and thus are properly viewed as random variables themselves. However, under a condition known as *ergodicity*, the time averages in Eqs. (8.27) and (8.28) are equal to constants in the sense that the time averages of almost all possible sample sequences are equal to the same constant. Furthermore, they are equal to the corresponding ensemble average.† That is, for any single sample sequence $\{x(n)\}$ for $-\infty < n < \infty$,

$$\langle x(n) \rangle = \lim_{N\to\infty} \frac{1}{2N+1} \sum_{n=-N}^{N} x(n) = E[x_n] = m_x \tag{8.29}$$

and

$$\langle x(n)x^*(n+m) \rangle$$
$$= \lim_{N\to\infty} \frac{1}{2N+1} \sum_{n=-N}^{N} x(n)x^*(n+m) = E[x_n x_{n+m}^*] = \phi_{xx}(m) \tag{8.30}$$

The time-average operator $\langle\ \rangle$ has the same properties as the ensemble-average operator $E[\ ]$. Thus we shall often not take the trouble to distinguish between the random variable x_n and its value in a sample sequence, $x(n)$. For example, the expression $E[x(n)]$ should be interpreted as $E[x_n] = \langle x(n) \rangle$. In general, a random process for which time averages equal ensemble averages is called an *ergodic process* [5,6].

In practice, it is common to assume that a given sequence is a sample sequence of an ergodic random process. Thus averages can be computed from a single infinite-energy sequence. Of course, we generally cannot compute the limits in Eqs. (8.29) and (8.30), but the quantities

$$\langle x(n) \rangle_N = \frac{1}{2N+1} \sum_{n=-N}^{N} x(n) \tag{8.31}$$

and

$$\langle x(n)x(n+m) \rangle_N = \frac{1}{2N+1} \sum_{n=-N}^{N} x(n)x^*(n+m) \tag{8.32}$$

or similar quantities are often computed in practice as *estimates* of the mean and autocorrelation [8,9]. The estimation of averages of a random process from a finite segment of data is a problem of statistics, which we shall examine in Chapter 11.

† A more precise statement is that the random variables $\langle x_n \rangle$ and $\langle x_n x_{n+m}^* \rangle$ have means equal to m_x and $\phi_{xx}(m)$, respectively, and their variances are zero [6].

8.3 Spectrum Representations of Infinite-Energy Signals

Although the z-transform of an infinite-energy signal does not exist, the autocovariance and autocorrelation sequences of such a signal are aperiodic sequences for which the z-transform and Fourier transform often do exist. We shall see in the next section that the spectral representation of these averages plays an important role in describing the input–output relations for a linear time-invariant system when the input is an infinite-energy signal. Therefore, it is of interest to consider the properties of correlation and covariance sequences and their corresponding z-transforms.

8.3.1 *Properties of Correlation and Covariance Sequences*

There are a number of useful properties of correlation and covariance functions that follow in a simple way from the definitions. These properties are listed below for future reference. The proof of the validity of some of these properties is considered in Problem 6 of this chapter.

Consider two real, stationary random processes $\{x_n\}$ and $\{y_n\}$ with autocorrelation, autocovariance, cross-correlation, and cross-covariance being given by, respectively,

$$\phi_{xx}(m) = E[x_n x_{n+m}] \tag{8.33}$$

$$\gamma_{xx}(m) = E[(x_n - m_x)(x_{n+m} - m_x)] \tag{8.34}$$

$$\phi_{xy}(m) = E[x_n y_{n+m}] \tag{8.35}$$

$$\gamma_{xy}(m) = E[(x_n - m_x)(y_{n+m} - m_y)] \tag{8.36}$$

where m_x and m_y are the means of the two processes. The following properties are easily derived by simple manipulations of the definitions.

Property 1:

$$\gamma_{xx}(m) = \phi_{xx}(m) - m_x^2 \tag{8.37a}$$

$$\gamma_{xy}(m) = \phi_{xy}(m) - m_x m_y \tag{8.37b}$$

These results follow directly from Eqs. (8.21) and (8.23), and they indicate that the correlation and covariance sequences are identical if $m_x = 0$.

Property 2:

$$\phi_{xx}(0) = E[x_n^2] = \text{mean-square value} \tag{8.38a}$$

$$\gamma_{xx}(0) = \sigma_x^2 = \text{variance} \tag{8.38b}$$

Property 3:

$$\phi_{xx}(m) = \phi_{xx}(-m) \tag{8.39a}$$

$$\gamma_{xx}(m) = \gamma_{xx}(-m) \tag{8.39b}$$

$$\phi_{xy}(m) = \phi_{yx}(-m) \tag{8.39c}$$

$$\gamma_{xy}(m) = \gamma_{yx}(-m) \tag{8.39d}$$

Property 4:

$$|\phi_{xy}(m)| \leq [\phi_{xx}(0)\phi_{yy}(0)]^{1/2} \tag{8.40a}$$

$$|\gamma_{xy}(m)| \leq [\gamma_{xx}(0)\gamma_{yy}(0)]^{1/2} \tag{8.40b}$$

In particular, then,

$$|\phi_{xx}(m)| \leq \phi_{xx}(0) \tag{8.41a}$$

$$|\gamma_{xx}(m)| \leq \gamma_{xx}(0) \tag{8.41b}$$

Property 5: If $y_n = x_{n-n_0}$, then

$$\phi_{yy}(m) = \phi_{xx}(m) \tag{8.42a}$$

$$\gamma_{yy}(m) = \gamma_{xx}(m) \tag{8.42b}$$

Property 6: For many random processes, the random variables become less uncorrelated as they become more separated in time. Thus

$$\lim_{m\to\infty} \phi_{xx}(m) = (E[x_n])^2 = m_x^2 \tag{8.43a}$$

$$\lim_{m\to\infty} \gamma_{xx}(m) = 0 \tag{8.43b}$$

$$\lim_{m\to\infty} \phi_{xy}(m) = m_x m_y \tag{8.43c}$$

$$\lim_{m\to\infty} \gamma_{xy}(m) = 0 \tag{8.43d}$$

The essence of these results is that the correlation and covariance are aperiodic sequences which tend to die out for large values of m. Thus it is often possible to represent these sequences in terms of their z-transforms.

8.3.2 z-Transform Representations

Let us define $\Phi_{xx}(z)$, $\Gamma_{xx}(z)$, $\Phi_{xy}(z)$, and $\Gamma_{xy}(z)$ as the z-transforms of $\phi_{xx}(m)$, $\gamma_{xx}(m)$, $\phi_{xy}(m)$, and $\gamma_{xy}(m)$, respectively. From Eqs. (8.43a) and (8.43c) we note immediately that the z-transforms of $\phi_{xx}(m)$ and $\phi_{xy}(m)$ exist only when $m_x = 0$, in which case $\Phi_{xx}(z) = \Gamma_{xx}(z)$ and $\Phi_{xy}(z) = \Gamma_{xy}(z)$. A number of other properties of these z-transforms follow from the properties of correlation and covariance sequences that were summarized in Sec. 8.3.1. These z-transform properties are summarized below and their proofs considered in Problem 6 of this chapter.

Property 1:

$$\sigma_x^2 = \frac{1}{2\pi j} \oint_C \Gamma_{xx}(z) z^{-1}\, dz \tag{8.44}$$

where C is a closed contour in the region of convergence of $\Gamma_{xx}(z)$.

Property 2:

$$\Gamma_{xx}(z) = \Gamma_{xx}(1/z) \tag{8.45a}$$

$$\Gamma_{xy}(z) = \Gamma_{yx}^*(1/z^*) \tag{8.45b}$$

Equations (8.45) follow directly from property 3 of Sec. 8.3.1. As a consequence it follows that the region of convergence of $\Gamma_{xx}(z)$ must be of the form

$$R_a < |z| < \frac{1}{R_a}$$

Furthermore, since $\gamma_{xx}(m)$ approaches zero at $m = \infty$, the region of convergence must include the unit circle; i.e., $0 < R_a < 1$. In the important case when $\Gamma_{xx}(z)$ is a rational function of z, this implies that its poles and zeros must occur in complex-conjugate reciprocal pairs as depicted in Fig. 8.4.

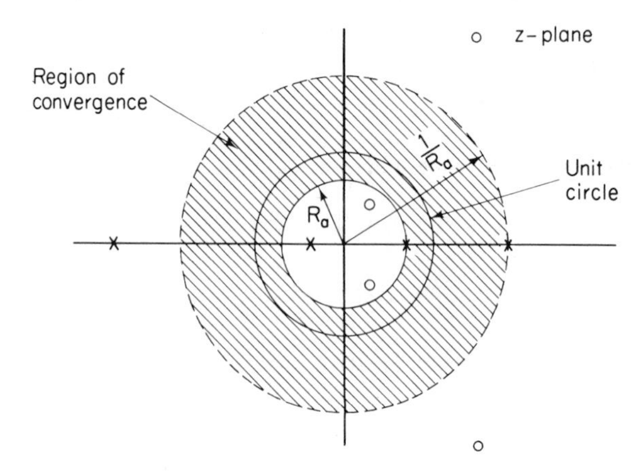

Fig. 8.4 Region of convergence and pole and zero locations of a typical z-transform of a covariance sequence.

8.3.3 Power Spectrum

Since the region of convergence contains the unit circle, we can express Eq. (8.44) as

$$\sigma_x^2 = \frac{1}{2\pi}\int_{-\pi}^{\pi} P_{xx}(\omega)\, d\omega \tag{8.46}$$

where we have defined

$$P_{xx}(\omega) = \Gamma_{xx}(e^{j\omega}) \tag{8.47}$$

When $m_x = 0$, we recall that the variance is equal to the mean-square or average power. Thus the area under $P_{xx}(\omega)$ for $-\pi \leq \omega \leq \pi$ is proportional to the average power in the signal. In fact, as we will see in the next section, the integral of $P_{xx}(\omega)$ over a band of frequencies is proportional to the power in the signal in that band. For these reasons the function $P_{xx}(\omega)$ is called the *power density spectrum*, or simply the *spectrum* [9]. We note that it is also common to define the power spectrum as the Fourier transform of the autocorrelation sequence rather than the autocovariance [5]. This leads to difficulties when $m_x \neq 0$, since $\phi_{xx}(m) \to m_x^2$ when $m \to \infty$. Thus the Fourier transform of the autocorrelation sequence does not exist if $m_x \neq 0$, unless we are willing to extend our definition of the Fourier transform to allow an impulse in the power spectrum at $\omega = 0$. Since we have avoided the use of impulses in this book, we choose to define the power spectrum as in Eq. (8.48). We note that when $m_x = 0$, the autocorrelation and autocovariance sequences are identical, and therefore so are their Fourier transforms.

From property 2 of Sec. 8.3.2, it follows that $P_{xx}(\omega)$ is a symmetric function; i.e., $P_{xx}(\omega) = P_{xx}(-\omega)$. An additional important property is the fact that the power density spectrum is nonnegative. This property will follow in a straightforward way from the results in the next section.

Similarly, the *cross-power density spectrum* is defined as

$$P_{xy}(\omega) = \Gamma_{xy}(e^{j\omega}) \tag{8.48}$$

Again from property 2 of Sec. 8.3.2 it follows that

$$P_{xy}(\omega) = P^*_{yx}(-\omega) \tag{8.49}$$

8.4 Response of Linear Systems to Random Signals

Previous chapters have focused on the development of the theory of linear discrete-time systems for the case when the input is a known function of time. It is clear from these previous discussions that the notion of the frequency response of a linear shift-invariant system and the frequency domain representation of a discrete-time signal are essential concepts in digital signal processing. In this section we shall develop some corresponding results for the case of infinite-energy signals; i.e., signals represented by a random-process model.

Consider a stable linear shift-invariant system with unit-sample response $h(n)$. Let $x(n)$ be a real input sequence that is a sample sequence of a wide-sense stationary discrete-time random process. Then the output of the linear system is a sample function of an output random process related to the input process by the linear transformation

$$y(n) = \sum_{k=-\infty}^{\infty} h(n-k)x(k) = \sum_{k=-\infty}^{\infty} h(k)x(n-k)$$

As we have shown, since the system is stable, $y(n)$ will be bounded if $x(n)$ is bounded. We shall show below that if the input is stationary, then so is the output. The input signal may be characterized by its mean m_x and its autocorrelation function $\phi_{xx}(m)$, or we may also have additional information about first- or even second-order probability distributions. In characterizing the output random process $\{y_n\}$ we desire similar information. For many applications, it is sufficient to characterize both the input and output in terms of simple averages, such as the mean, variance, and autocorrelation. Therefore, we shall derive input–output relationships between these quantities.

The mean of the output process is

$$\begin{aligned} m_y = E[y(n)] &= \sum_{k=-\infty}^{\infty} h(k)E[x(n-k)] \\ &= m_x \sum_{k=-\infty}^{\infty} h(k) \end{aligned} \tag{8.50}$$

where we have used the fact that the expectation of a sum is the sum of the expectations.† In terms of the system function we can write

$$m_y = H(e^{j0})m_x \tag{8.51}$$

Since the input is stationary, we see that the mean of the output is also constant.

Assuming temporarily that the output is nonstationary, the autocorrelation function of the output process is

$$\begin{aligned} \phi_{yy}(n, n+m) &= E[y(n)y(n+m)] \\ &= E\left[\sum_{k=-\infty}^{\infty} \sum_{r=-\infty}^{\infty} h(k)h(r)x(n-k)x(n+m-r)\right] \\ &= \sum_{k=-\infty}^{\infty} h(k) \sum_{r=-\infty}^{\infty} h(r)E[x(n-k)x(n+m-r)] \end{aligned}$$

Since $x(n)$ is assumed to be stationary, $E[x(n-k)x(n+m-r)]$ depends only upon the time difference $m+k-r$. Therefore,

$$\phi_{yy}(n, n+m) = \sum_{k=-\infty}^{\infty} h(k) \sum_{r=-\infty}^{\infty} h(r)\phi_{xx}(m+k-r) = \phi_{yy}(m) \tag{8.52}$$

That is, the output autocorrelation sequence also depends only upon the time difference, m. Thus, for a linear shift-invariant system that is excited by a stationary input, the output is also stationary.

By making the substitution $l = r - k$, Eq. (8.52) can be expressed as

$$\begin{aligned} \phi_{yy}(m) &= \sum_{l=-\infty}^{\infty} \phi_{xx}(m-l) \sum_{k=-\infty}^{\infty} h(k)h(l+k) \\ &= \sum_{k=-\infty}^{\infty} \phi_{xx}(m-l)v(l) \end{aligned} \tag{8.53}$$

where we have defined

$$v(l) = \sum_{k=-\infty}^{\infty} h(k)h(l+k) \tag{8.54}$$

A sequence of the form of $v(l)$ is often called an *aperiodic autocorrelation sequence* or simply the *autocorrelation sequence of* $h(n)$. It should be emphasized that $v(l)$ is the autocorrelation of an aperiodic, i.e., finite-energy,

† Note that we have begun to be less careful about distinguishing between the random variable and its value.

sequence and should not be confused with the autocorrelation of an infinite-energy sequence. Indeed, it can be seen that $v(l)$ is simply the discrete convolution of $h(n)$ with $h(-n)$. Equation (8.53), then, can be interpreted to mean that the autocorrelation of the output of a linear system is the convolution of the input autocorrelation with the autocorrelation of the system impulse response.

Equation (8.53) suggests that z-transforms may be useful in characterizing the response of a linear time-invariant system to an infinite-energy input. Assume for convenience that $m_x = 0$; i.e., the autocorrelation and autocovariance sequences are identical. Then, from Eqs. (8.53) and (8.54),

$$\begin{aligned}\Phi_{yy}(z) &= V(z)\Phi_{xx}(z) \\ &= H(z)H(z^{-1})\Phi_{xx}(z)\end{aligned} \tag{8.55}$$

In terms of the power density spectrum, Eq. (8.55) becomes

$$P_{yy}(\omega) = |H(e^{j\omega})|^2 \, P_{xx}(\omega) \tag{8.56}$$

Equation (8.56) provides the motivation for the term "power density spectrum." To see this, let us assume that $m_x = 0$, so from Eq. (8.51), $m_y = 0$ as well. Therefore,

$$\begin{aligned}\phi_{yy}(0) &= \frac{1}{2\pi}\int_{-\pi}^{\pi} P_{yy}(\omega)\, d\omega \\ &= \text{total average power in output}\end{aligned} \tag{8.57}$$

Substituting Eq. (8.56) into Eq. (8.57),

$$\phi_{yy}(0) = \frac{1}{2\pi}\int_{-\pi}^{\pi} |H(e^{j\omega})|^2 \, P_{xx}(\omega)\, d\omega \tag{8.58}$$

Suppose that $H(e^{j\omega})$ is an ideal bandpass filter, as shown in Fig. 8.5. We recall that $\phi_{xx}(m)$ is an even sequence, so

$$P_{xx}(\omega) = P_{xx}(-\omega)$$

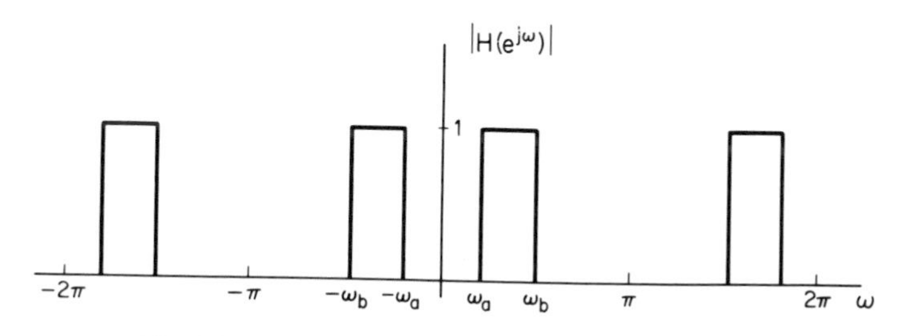

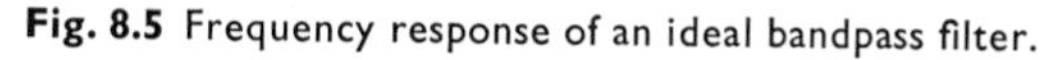
Fig. 8.5 Frequency response of an ideal bandpass filter.

Likewise, $|H(e^{j\omega})|^2$ is an even function of ω. Therefore, we can write

$$\begin{aligned}\phi_{yy}(0) &= \text{average power in output} \\ &= \frac{1}{\pi}\int_{\omega_a}^{\omega_b} P_{xx}(\omega)\,d\omega \end{aligned} \tag{8.59}$$

Thus the area under $P_{xx}(\omega)$ between ω_a and ω_b can be taken to represent the mean-square value of the input in that frequency band. We observe that the output power must remain nonnegative, so

$$\lim_{(\omega_b - \omega_a)\to 0} \phi_{yy}(0) \geq 0$$

This result, together with Eq. (8.59), implies that

$$P_{xx}(\omega) \geq 0 \tag{8.60}$$

Thus we note that the power density function of a real signal is real, even, and positive.

Another interesting result concerns the cross-correlation between the input and output of a linear time-invariant system.

$$\begin{aligned}\phi_{xy}(m) &= E[x(n)y(n+m)] \\ &= E\left[x(n)\sum_{k=-\infty}^{\infty} h(k)x(n+m-k)\right] \\ &= \sum_{k=-\infty}^{\infty} h(k)\phi_{xx}(m-k)\end{aligned} \tag{8.61}$$

In this case we note that the cross-correlation between input and output is the convolution of the unit-sample response with the input autocorrelation sequence.

If we assume $m_x = 0$ so that the z-transforms exist, we can write

$$\Phi_{xy}(z) = H(z)\Phi_{xx}(z) \tag{8.62}$$

or, in terms of power spectra,

$$P_{xy}(\omega) = H(e^{j\omega})P_{xx}(\omega) \tag{8.63}$$

This result has a useful application when the input is white noise; i.e., $\phi_{xx}(m) = \sigma_x^2\delta(m)$. Substituting into Eq. (8.61), we note that

$$\phi_{xy}(m) = \sigma_x^2 h(m) \tag{8.64}$$

i.e., for a white-noise input, the cross-correlation between input and output of a linear system is proportional to the impulse response of the system. Similarly, the power spectrum of a white-noise input is

$$P_{xx}(\omega) = \sigma_x^2, \qquad -\pi \leq \omega \leq \pi$$

Thus from Eq. (8.63),

$$P_{xy}(\omega) = \sigma_x^2 H(e^{j\omega}) \tag{8.65}$$

i.e., the cross power spectrum is in this case proportional to the frequency response of the system. Equations (8.64) and (8.65) may serve as the basis for estimating the impulse response or frequency response of a linear time-invariant system if it is possible to observe the output of the system in response to a white-noise input.

Summary

In this chapter we have attempted to show how the concept of a random process can be used as a representation of discrete-time signals that cannot be directly represented by Fourier methods. We do not intend that this chapter take the place of a formal course in probability and random processes. Thus we have not attempted to give a rigorous discussion but have concentrated our efforts on summarizing and interpreting a variety of basic results that are useful in a digital signal processing context. Specifically, we have focused our attention on the properties of correlation and covariance sequences, the power spectrum, and on input–output relationships for linear shift-invariant discrete systems. In selecting these results for discussion, we have been guided primarily by our needs in Chapters 9 and 11.

REFERENCES

1. W. B. Davenport, *Probability and Random Processes*, McGraw-Hill Book Company, New York, 1970.
2. A. W. Drake, *Fundamentals of Applied Probability Theory*, McGraw-Hill Book Company, New York, 1967.
3. W. Feller, *An Introduction to Probability Theory and Its Applications*, 3rd ed., Vol. 1, 1968, Vol. 2, 1966, John Wiley & Sons, Inc., New York.
4. E. Parzen, *Modern Probability Theory and Its Applications*, John Wiley & Sons, Inc., New York, 1960.
5. W. B. Davenport and W. L. Root, *An Introduction to the Theory of Random Signals and Noise*, McGraw-Hill Book Company, New York, 1958.
6. A. Papoulis, *Probability, Random Variables, and Stochastic Processes*, McGraw-Hill Book Company, New York, 1965.
7. E. Parzen, *Stochastic Processes*, Holden-Day, Inc., San Francisco, 1962.
8. G. E. P. Box and G. M. Jenkins, *Time Series Analysis Forecasting and Control*, Holden-Day, Inc., San Francisco, 1970.
9. G. M. Jenkins and D. G. Watts, *Spectral Analysis and Its Applications*, Holden-Day, Inc., San Francisco, 1968.

PROBLEMS

1. Prove the following properties of averages:

   (a) $E[x_n + y_m] = E[x_n] + E[y_m]$.

   (b) $E[ax_n] = aE[x_n]$.

2. Let $x(n)$ and $y(n)$ be uncorrelated random signals. Show that if

$$w(n) = x(n) + y(n)$$

then

$$m_w = m_x + m_y$$

and

$$\sigma_w^2 = \sigma_x^2 + \sigma_y^2$$

3. In modeling the effects of roundoff and truncation in digital filter implementations, we represent quantized variables as

$$y(n) = Q[x(n)] = x(n) + e(n)$$

where $Q[\ ]$ denotes either rounding or truncation and $e(n)$ is the quantization error. With appropriate assumptions, it is reasonable to suppose that the sequence $e(n)$ is a white-noise sequence; i.e.,

$$E[(e(n) - m_e)(e(n + m) - m_e)] = \sigma_e^2 \delta(m)$$

We will show in Chapter 9 that the first-order probability distribution for rounding is uniform as depicted in Fig. P8.3(a). For truncation the first-order probability distribution is as shown in Fig. P8.3(b).

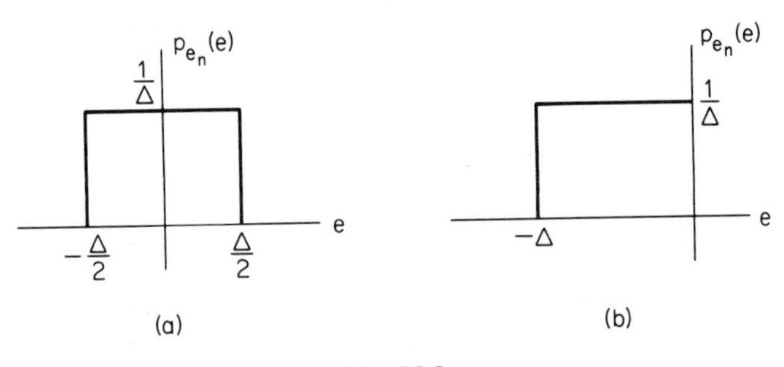

Fig. P8.3

(a) Find the mean and variance for the noise due to rounding.
(b) Find the mean and variance for the noise due to truncation.

4. Let $e(n)$ denote a white-noise sequence and let $s(n)$ denote a sequence that is uncorrelated with $e(n)$. Show that the sequence

$$y(n) = s(n)e(n)$$

is white; i.e.,

$$E[y(n)y(n + m)] = A\delta(m)$$

where A is a constant.

5. Consider a random process in which the sample sequences $x(n)$ are of the form

$$x(n) = \cos(\omega_0 n + \theta)$$

where θ is a uniformly distributed random variable with the probability density function shown in Fig. P8.5. Compute the mean and autocorrelation sequence $\phi_{xx}(m, n)$. Is this a wide-sense stationary process?

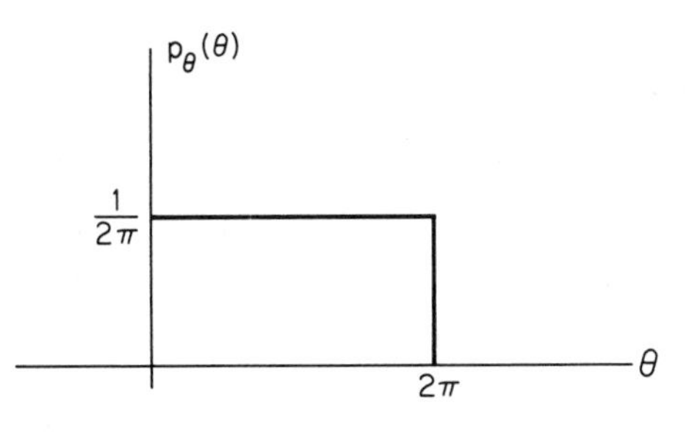

Fig. P8.5

6. Consider two real stationary random processes, $\{x_n\}$ and $\{y_n\}$, with means m_x and m_y, respectively, and variances σ_x^2 and σ_y^2, respectively. Show that

(a) $\gamma_{xx}(m) = \phi_{xx}(m) - m_x^2$

$\gamma_{xy}(m) = \phi_{xy}(m) - m_x m_y$

(b) $\phi_{xx}(0) =$ mean square

$\gamma_{xx}(0) = \sigma_x^2$

(c) $\phi_{xx}(m) = \phi_{xx}(-m)$

$\gamma_{xx}(m) = \gamma_{xx}(-m)$

$\phi_{xy}(m) = \phi_{yx}^*(-m)$

$\gamma_{xy}(m) = \gamma_{yx}^*(-m)$

(d) $|\phi_{xy}(m)| \leq [\phi_{xx}(0)\phi_{yy}(0)]^{1/2}$

$|\gamma_{xy}(m)| \leq [\gamma_{xx}(0)\gamma_{yy}(0)]^{1/2}$

$|\phi_{xx}(m)| \leq \phi_{xx}(0)$

$|\gamma_{xx}(m)| \leq \gamma_{xx}(0)$

Hint: Consider the inequality

$$0 \leq E\left\{\left(\frac{x_n}{(E[x_n^2])^{1/2}} - \frac{y_{n+m}}{(E[y_{n+m}^2])^{1/2}}\right)^2\right\}$$

(e) If $y_n = x_{n-n_0}$, then

$\phi_{yy}(m) = \phi_{xx}(m)$

$\gamma_{yy}(m) = \gamma_{yy}(m)$

(f) Let $\Gamma_{xx}(z)$ and $\Gamma_{xy}(z)$ be the z-transforms of $\gamma_{xx}(m)$ and $\gamma_{xy}(m)$, respectively. Show that

(1) $\sigma_x^2 = \dfrac{1}{2\pi j}\oint_C \Gamma_{xx}(z)z^{-1}\,dz.$

(2) $\Gamma_{xx}(z) = \Gamma_{xx}(1/z)$

$\Gamma_{xy}(z) = \Gamma_{yx}^*(1/z^*)$

7. Consider an *ergodic* random process so that time averages are equal to probability averages. Let $x(n)$ be a particular sample sequence of the random process defined by the set of random variables $\{x_n\}$, $-\infty < n < -\infty$, where $p_x(x)$ is the first-order probability density for all the random variables x_n.

(a) Consider the time average of the function $u(a - x_n)$; i.e., $\langle u(a - x_n)\rangle$. (The function $u(\cdot)$ is the unit step.) Express in words the meaning of this time average.
(b) Consider the probability average of the function $u(a - x_n)$. What is $E[u(a - x_n)]$ in terms of $p_x(x)$?
(c) Is your result in part (b) consistent with your interpretation of part (a); i.e., is it reasonable that

$$E[u(a - x_n)] = \langle u(a - x_n)\rangle?$$

8. We have observed that the autocorrelation sequence serves as an indication of how rapidly a random signal varies. We can formalize this notion by considering the mean-square sample-to-sample variation of a real signal, defined as

$$E[(x(n + 1) - x(n))^2]$$

Consider a zero mean signal $x(n)$ which has a bandlimited power spectrum such that

$$P_{xx}(\omega) = 0, \qquad \omega_c < |\omega| \leq \pi$$

(a) Show that

$$E[(x(n + 1) - x(n))^2] = 2[\phi_{xx}(0) - \phi_{xx}(1)]$$

(b) Show that

$$[\phi_{xx}(0) - \phi_{xx}(1)] \leq \frac{\omega_c^2}{2}\,\phi_{xx}(0)$$

and thus that

$$E[(x(n + 1) - x(n))^2] \leq \omega_c^2 E[x^2(n)]$$

[*Hint*: Use the fact that $\sin^2(\omega/2) \leq \omega^2/4$ for $0 \leq \omega \leq \omega_c$.]

A very important inequality is known as the Chebyshev inequality [1–5]. It states that

$$\text{Probability }[|x(n) - E[x(n)]| \geq \varepsilon] \leq \frac{E[x^2(n)]}{\varepsilon^2}$$

for $\varepsilon > 0$. We note that this inequality gives us a way of expressing how much a signal can be expected to differ from its mean; i.e., the greater the mean-square value the higher the probability that a particular sequence value will differ from the mean by an amount greater than ε.

(c) Use the result of part (b) and the Chebyshev inequality to show that

$$\text{Probability }[|x(n + 1) - x(n)| > \varepsilon] \leq \frac{\omega_c^2 E[x^2(n)]}{\varepsilon^2}$$

for $\varepsilon > 0$. Interpret this result in light of your knowledge of the relationship between the time variation of a signal and its bandwidth.

9. Let $x(n)$ be a real stationary white-noise process, with zero mean and variance σ_x^2. Let $y(n)$ be the corresponding output when $x(n)$ is the input to a linear shift-invariant system with impulse response $h(n)$.
(a) Show that

$$E[x(n)y(n)] = h(0)\sigma_x^2$$

(b) Show that

$$\sigma_y^2 = \sigma_x^2 \sum_{n=-\infty}^{\infty} h^2(n)$$

10. Show that the z-transform of

$$v(n) = \sum_{k=-\infty}^{\infty} h(k)h(k+n)$$

is

$$V(z) = H(z)H(z^{-1})$$

11. An ideal Hilbert transformer with unit sample response

$$h(n) = \frac{2}{\pi}\frac{\sin^2(\pi n/2)}{n} \qquad n \neq 0$$
$$= 0 \qquad n = 0,$$

is excited by a discrete-time random signal $x_r(n)$ as depicted in Fig. P8.11.

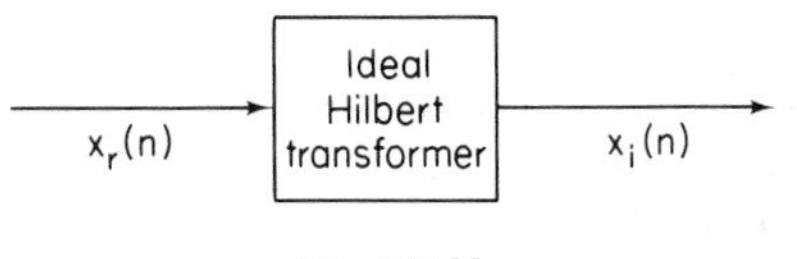

Fig. P8.11

(a) Find an expression for the autocorrelation sequence $\phi_{x_i x_i}(m)$.
(b) Find an expression for the cross-correlation sequence $\phi_{x_r x_i}(m)$. Show that in this case $\phi_{x_r x_i}(m) = -\phi_{x_r x_i}(-m)$.
(c) Find the autocorrelation sequence for the complex analytic signal

$$x(n) = x_r(n) + jx_i(n)$$

(d) Find the power spectrum $P_{xx}(\omega)$ for the above complex analytic signal.

12. Consider a linear shift-invariant digital network with two inputs as depicted in Fig. P8.12. Let $h_1(n)$ and $h_2(n)$ be the unit-sample responses from nodes 1 and 2, respectively, to the output. Show that if $x_1(n)$ and $x_2(n)$ are uncorrelated, then their corresponding outputs are also uncorrelated.

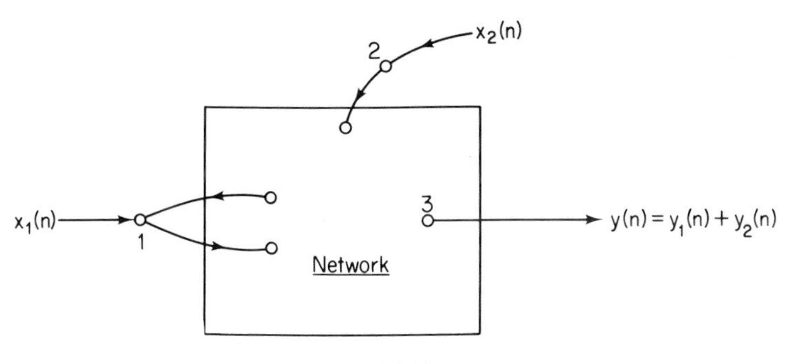

Fig. P8.12

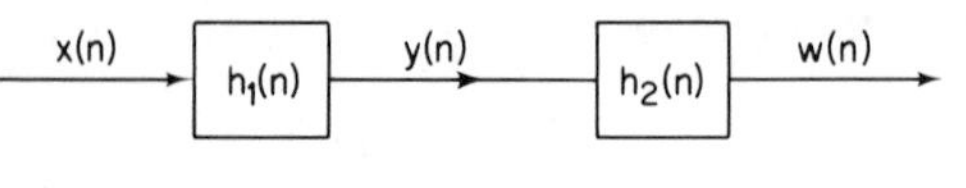

Fig. P8.13

13. Let $x(n)$ be a random white sequence, with zero mean and variance σ_x^2. Let $x(n)$ be the input to the cascade of two linear discrete-time shift-invariant systems as shown in Fig. P8.13.

(a) Is $\sigma_y^2 = \sigma_x^2 \sum_{k=0}^{\infty} h_1^2(k)$?

(b) Is $\sigma_w^2 = \sigma_y^2 \sum_{k=0}^{\infty} h_2^2(k)$?

(c) Let $h_1(n) = a^n u(n)$ and $h_2(n) = b^n u(n)$. Determine the unit-sample response of the overall system in Fig. P8.13, and from this determine σ_w^2. If your answer to part (b) was yes, is that consistent with your answer to part (c)?

14. Since in many applications, discrete-time random signals arise through periodic sampling of continuous-time random signals, we shall be concerned in this problem with a derivation of the sampling theorem for random signals. Consider a continuous-time random process defined by the random variables $\{x_a(t)\}$, where t is a continuous variable. The autocovariance function is defined as (assume zero mean)

$$\gamma_{x_a x_a}(\tau) = E[x_a(t)x_a^*(t + \tau)]$$

and the power spectrum is

$$P_{x_a x_a}(\Omega) = \int_{-\infty}^{\infty} \gamma_{x_a x_a}(\tau) e^{-j\Omega\tau}\, d\tau$$

A discrete-time random process obtained by periodic sampling is defined by the set of random variables $\{x(n)\}$, where $x(n) = x_a(nT)$ and T is the sampling period.

(a) What is the relationship between $\gamma_{xx}(n)$ and $\gamma_{x_a x_a}(\tau)$?

(b) Express the power density spectrum of the discrete-time process in terms of the power density spectrum of the continuous-time process.

(c) Under what condition is the discrete-time power spectrum a faithful representation of the continuous-time power spectrum?

15. Consider a continuous-time random process $\{x_a(t)\}$ with a bandlimited power spectrum as depicted in Fig. P8.15-1. Suppose that we sample $\{x_a(t)\}$ to obtain the discrete-time random process $\{x(n) = x_a(nT)\}$.

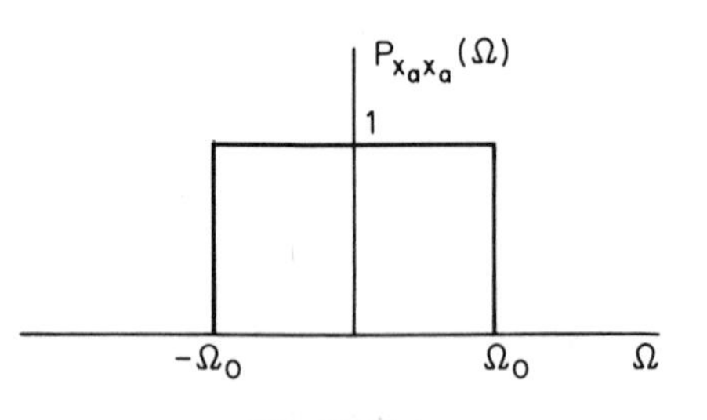

Fig. P8.15-1

(a) What is the autocovariance sequence of the discrete-time random process?
(b) For the analog power spectrum above, how should T be chosen so that the discrete-time process is white?
(c) If the analog power spectrum is as shown in Fig. P8.15-2, how should T be chosen so that the discrete-time process is white?

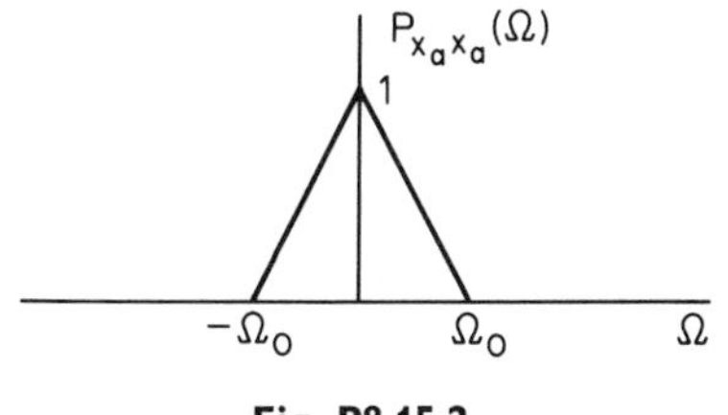

Fig. P8.15-2

(d) What is the general requirement on the analog process and the sampling period such that the discrete-time process be white?

16. It is often appropriate to assume that a wide-sense stationary random process arises as the result of exciting a linear system with a white-noise process. Such processes are called *linear processes*. Consider a stable linear system as depicted in Fig. P8.16, where $x(n)$ is white noise with zero mean and variance σ_x^2.

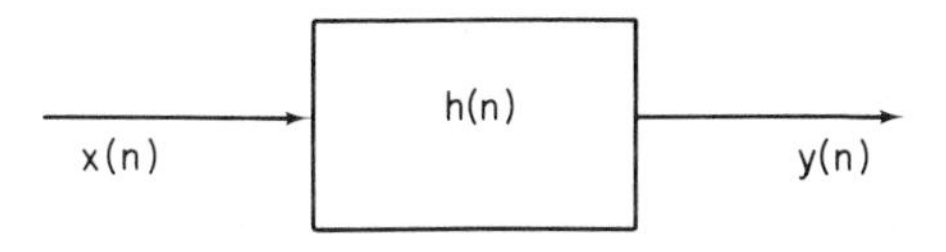

Fig. P8.16

(a) Express the autocovariance of $y(n)$ in terms of the impulse response of the system.
(b) Use the result of part (a) to express the power spectrum of $y(n)$ in terms of the frequency response of the system.

Of particular interest is the case when $H(z)$ is a rational function of the form

$$H(z) = \frac{\sum_{k=0}^{M} b_k z^{-k}}{1 - \sum_{k=1}^{N} a_k z^{-k}}$$

in which case the process $y(n)$ is related to the white-noise process $x(n)$ by the difference equation

$$y(n) = \sum_{k=1}^{N} a_k y(n-k) + \sum_{k=0}^{M} b_k x(n-k)$$

If all that a_k's are zero, y is called a *finite moving-average process*. If all the b_k's, except b_0, are zero and $a_N \neq 0$, then y is called an *autoregressive process*. If both a_N and b_M are nonzero, the process is *mixed*.

(c) Show that the autocovariance sequence $\gamma_{yy}(m)$ of a moving-average process is nonzero only in the interval $|m| \leq M$.
(d) Find a general expression for the autocovariance sequence of an autoregressive process.
(e) Show that if $b_0 = 1$, the autocovariance function of an autoregressive process satisfies the difference equation

$$\gamma_{yy}(0) = \sum_{k=1}^{N} a_k \gamma_{yy}(k) + \sigma_x^2,$$

$$\gamma_{yy}(m) = \sum_{k=1}^{N} a_k \gamma_{yy}(m - k), \qquad m \geq 1$$

(f) Use the result of part (e) and the symmetry of $\gamma_{yy}(m)$ to show that the following set of equations is true:

$$\sum_{k=1}^{N} a_k \gamma_{yy}(|m - k|) = \gamma_{yy}(m), \qquad m = 1, 2, \ldots, N$$

Thus, given the first $N + 1$ covariance values, it can be shown that we can always solve uniquely for the values of a_k and σ_x^2 that characterize the process y.

17. Consider a *mixed* linear process y that satisfies the difference equation

$$y(n) = \sum_{k=1}^{N} a_k y(n - k) + \sum_{k=0}^{M} b_k x(n - k), \qquad b_0 = 1$$

Sometimes it is of interest to "whiten" the spectrum of $y(n)$ by processing $y(n)$ with a linear filter; i.e., we wish to find a system such that its output power spectrum will be flat for input $y(n)$.

Suppose that we know the autocovariance function $\gamma_{yy}(m)$ and its z-transform $\Gamma_{yy}(z)$ but not the a_k's and the b_k's.
(a) Discuss a procedure or find a system function $H_w(z)$ of the whitening filter.
(b) Is the whitening filter unique?
(c) How do the Hilbert transform relations (of Chapter 7) enter into this problem?

18. Sometimes we are interested in the statistical behavior of a linear shift-invariant system when the input is a suddenly applied random signal. Such a situation

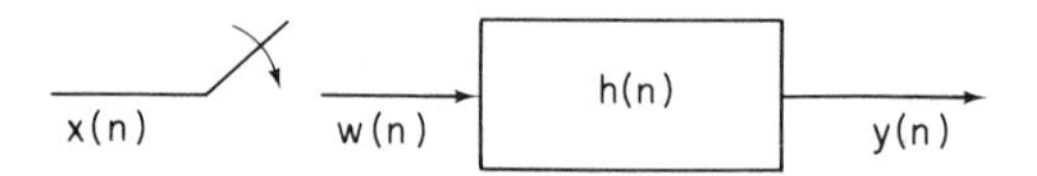

Fig. P8.18

is depicted in Fig. P8.18. Let $x(n)$ be a stationary white-noise process. Then the input to the system

$$w(n) = \begin{cases} x(n), & n \geq 0 \\ 0, & n < 0 \end{cases}$$

is a nonstationary process, as is the output $y(n)$.

(a) Derive an expression for the mean of the output in terms of the mean of the input.
(b) Derive an expression for the autocorrelation sequence $\phi_{yy}(n_1, n_2)$ of the output.
(c) Show that for large n the formulas derived in parts (a) and (b) approach the results for stationary inputs.
(d) Assume that $h(n) = a^n u(n)$. Find the mean and mean-square values of the output in terms of the mean and mean-square values of the input. Sketch the dependence of these parameters as a function of n.

9

Effects of Finite Register Length in Digital Signal Processing

9.0 Introduction

Digital signal processing algorithms such as linear filtering and discrete Fourier transformation are realized either with special-purpose digital hardware or as programs for a general-purpose digital computer. In both cases sequence values and coefficients are stored in a binary format with finite-length registers. This finite-word-length constraint is manifested in a variety of ways.

The parameters of a digital filter designed by one of the techniques of Chapter 5 are generally obtained with high accuracy. When these parameters are quantized, the frequency response of the resulting digital filter may differ appreciably from the original design. In fact, the quantized filter may fail to meet specifications even though the unquantized filter does. In Chapter 4 it was shown that the sensitivity of the filter response to errors in the parameters is dependent on the structure of the filter realization. The related problems of choice of filter structure and the direct design of filters with quantized coefficients remain as important research areas.

When a sequence to be processed is obtained by sampling a band-limited analog signal, the finite-word-length constraint requires that the analog-to-digital conversion process produce only a finite number of possible values for each sample. That is, the samples of the input must be quantized to fit a finite register length. We shall see in this chapter that this effect can often be treated in terms of an additive noise signal.

Even when we start with data representable with a finite word length, the result of processing will naturally lead to numbers requiring additional bits for their representation. For example, a b-bit data sample multiplied by a b-bit coefficient results in a product that is $2b$ bits long. If we do not quantize the result of arithmetic operations in a recursive realization of a digital filter, the number of bits will increase indefinitely, since after the first iteration, $2b$ bits are required; after the second iteration, $3b$ bits are required; etc. Similarly, in an FFT algorithm, if the coefficients are b-bit numbers, then the precise representation of the result of each stage of the FFT computation requires b bits more than is required for the previous stage. The effect of quantization in such examples depends on such factors as whether fixed-point or floating-point arithmetic is used, whether fixed-point numbers represent fractions or integers, and whether quantization is done by rounding or truncation. In this chapter we shall treat the case of fixed-point arithmetic and floating-point arithmetic separately. For fixed-point arithmetic it is natural in a signal processing context to consider a register as representing a fixed-point fraction. In this way the product of two numbers remains a fraction and the limited register length can be maintained by truncating or rounding the least significant bits. With this type of representation the result of addition of fixed-point fractions need not be truncated or rounded. However, the magnitude of the resulting sum can exceed unity. This effect is commonly referred to as *overflow* and can be handled by requiring that the input data be sufficiently small so that the possibility of overflow is avoided. In considering floating-point arithmetic, such dynamic range considerations generally can be neglected, owing to the large range of representable numbers, but quantization is introduced both for multiplication and for addition.

In the following discussion we shall first review the fixed-point and floating-point representations of binary numbers and the one's-complement, two's-complement, and sign-and-magnitude representations of negative numbers. The relationship between the binary representation and truncation or rounding is then discussed. Truncation or rounding of the result of an arithmetic operation has the effect of inserting a nonlinearity in the filter. For simple filters and inputs it is possible to analyze the effect of this nonlinearity in terms of what is referred to as the *limit-cycle behavior* of the filter. For complicated inputs and filters this analysis is very difficult to carry out. In this case it is often useful to perform an approximate analysis by representing the effect of truncation or rounding in terms of an additive error signal, which will be referred to as *roundoff noise*. The filter is then considered to be linear, but the output has a noise component that results from the effect of rounding or truncation. The average properties of this noise can be analyzed using the techniques of Chapter 8. A noise analysis of this type will be discussed for fixed-point and floating-point implementation of digital filters and also the fast Fourier transform.

9.1 Effect of Number Representation on Quantization

9.1.1 *Fixed-Point and Floating-Point Binary Numbers*

Digital computers and special-purpose digital hardware generally use a number representation with a radix of 2, i.e., a binary representation [1]. Therefore, a number is represented by a sequence of binary digits (*bits*) that are either zero or 1. Just as a decimal number is represented as a string of decimal digits with a decimal point dividing the integer part from the fractional part, the sequence of binary digits is divided by a binary point into those representing the integer part of the number and those representing the fractional part. Thus, if Δ denotes the location of the binary point, the binary number $1001_\Delta0110$ has the decimal value of

$$(1 \cdot 2^3 + 0 \cdot 2^2 + 0 \cdot 2^1 + 1 \cdot 2^0) + (0 \cdot 2^{-1} + 1 \cdot 2^{-2} + 1 \cdot 2^{-3} + 0 \cdot 2^{-4}),$$

or 9.375.

The manner in which arithmetic is implemented in a digital computer or in special-purpose hardware depends on the manner of locating the binary point. For fixed-point arithmetic, the implementation is based on the assumption that the location of the binary point is fixed. The manner in which addition is carried out will not depend on the location of the binary point for fixed-point arithmetic as long as the binary point is the same for every register.

For multiplication it is generally most convenient to assume either that all the numbers are integers or that they are all fractions, since the product of integers is an integer and the product of fractions is a fraction. In digital filtering applications, it is usually necessary to approximate the $2b$-bit product of two b-bit numbers by a b-bit result. In integer arithmetic this is difficult. With fractional arithmetic, on the other hand, this can be accomplished by truncating or rounding to the most significant b bits. For multiplication with fractions, overflow can never occur since the product of two fractions is a fraction. For example, if we multiply the two four-bit fractions $_\Delta1001$ and $_\Delta0011$, the eight-bit product $_\Delta00011011$ can be approximated by $_\Delta0001$ (truncation) or $_\Delta0010$ (rounding).

If we add two fixed-point fractions, overflow can occur. For example, the sum of the two four-bit fractions $_\Delta1101$ and $_\Delta1000$ is $1_\Delta0101$. That is, the sum cannot be contained in a four-bit register. This limitation on the range of numbers that can be represented can be essentially removed by using a floating-point representation. In the most common floating-point representation, a positive number F is represented as $F = 2^c \cdot M$, where M, the mantissa, is a fraction, such that

$$\tfrac{1}{2} \leq M < 1$$

and c, the characteristic, can be either positive or negative. When M is in the above range, the floating-point representation is said to be normalized. The product of two floating-point numbers is carried out by multiplying the mantissas as fixed-point fractions and adding the characteristics. Since the product of the mantissas will be between $\frac{1}{4}$ and 1, a normalization of the mantissa and corresponding adjustment of the characteristic may be necessary. That is, if $M < \frac{1}{2}$, M is shifted left one bit and the characteristic is increased by 1.†

The sum of two floating-point numbers is carried out by shifting the bits of the mantissa of the smaller number to the right until the characteristics of the two numbers are equal and then adding the mantissas as illustrated in the following example.

EXAMPLE. Consider the sum of F_1 and F_2 with $F_1 = 4$ and $F_2 = \frac{5}{4}$. Then in floating-point notation, $F_1 = 2^{c_1}M_1$, and $F_2 = 2^{c_2}M_2$, with

$$c_1 = 11_\Delta \qquad (= 3 \text{ in decimal})$$

$$M_1 = {}_\Delta 10000 \qquad (= \tfrac{1}{2} \text{ in decimal})$$

$$c_2 = 01_\Delta \qquad (= 1 \text{ in decimal})$$

$$M_2 = {}_\Delta 10100 \qquad (= \tfrac{5}{8} \text{ in decimal})$$

In order to carry out the addition, c_2 must be changed to equal c_1, and M_2 must be adjusted accordingly. Thus first the representation of F_2 is changed to $F_2 = 2^{\hat{c}_2} \cdot \hat{M}_2$, with

$$\hat{c}_2 = 11_\Delta$$

$$\hat{M}_2 = {}_\Delta 00101$$

in which case the mantissas can now be added. The resulting sum is $F = 2^c \cdot M$, with $c = 11$ and $M = {}_\Delta 10101$. In this case the sum of M_1 and $\hat{M}_2$ is a fraction between $\frac{1}{2}$ and 1, so no further adjustment of c is necessary. In a more general case, the sum may not be in the correct range, and consequently c would be adjusted to bring the mantissa into the proper range.

From this example it should be clear that with floating-point arithmetic the mantissa can exceed the register length for both addition and multiplication and must therefore be truncated or rounded, whereas this is only necessary for multiplication in the fixed-point case. On the other hand, if the result of addition in the fixed-point case exceeds the register length, truncation or rounding will not help; i.e., the dynamic range has been exceeded. Thus, while floating-point introduces error due to arithmetic roundoff of addition as well as multiplication, it provides much greater dynamic range than fixed point. Both of these effects must be considered when comparing fixed-point and floating-point realizations of digital filters.

† Shifting the mantissa one place to the right corresponds to division by 2 and a shift of one place to the left corresponds to multiplication by 2. Thus the characteristic is increased when the mantissa is shifted to the right and decreased when the mantissa is shifted to the left.

9.1.2 Representation of Negative Numbers

In digital signal processing, as in most numerical algorithms, it is necessary to deal with signed numbers. There are three common methods of representing fixed-point negative numbers. The first, and perhaps most familiar, is *sign and magnitude.* In this representation the magnitude (which is of course positive) is represented as a binary number and the sign is represented by the leading binary digit; 0 indicates a positive number and 1 indicates a negative number (or vice versa). For example, in sign and magnitude $0_\Delta 0011$ represents $\frac{3}{16}$ and $1_\Delta 0011$ represents $-\frac{3}{16}$.

Two other common representations of negative numbers are called two's-complement and one's-complement representations. *Two's-complement representation* can be viewed in terms of an interpretation of all the numbers in the register as positive numbers. With a total of $(b + 1)$ bits (one to the left of the binary point and b to the right of the binary point), these numbers, when interpreted as positive numbers, range from zero to $2 - 2^{-b}$. Half these numbers are used to represent positive fractions and half, negative fractions. Specifically, positive fractions are represented as in sign and magnitude. Negative fractions are represented by subtracting the magnitude from 2.0.

EXAMPLE. The sign and magnitude fraction $0_\Delta 0110$ is the two's-complement number $0_\Delta 0110$. However, consider the sign and magnitude number $1_\Delta 0110$. The magnitude is $0_\Delta 0110$ (i.e., $\frac{3}{8}$), which when subtracted from $10_\Delta 0000$ (i.e., 2) results in $1_\Delta 1010$ ($1\frac{5}{8}$). Thus $-\frac{3}{8}$ is represented by $1_\Delta 1010$.

One's-complement representation of negative numbers is similar to two's-complement. Positive fractions are again represented as in sign and magnitude. Negative fractions are represented by subtracting the magnitude from the largest number that fits in the register, i.e., all bits equal to unity. Thus with $(b + 1)$ bits as before (b bits to the right of the binary point and one bit to the left), negative fractions are represented by subtracting the magnitude from the number $2 - 2^{-b}$.

EXAMPLE. Assume that $b = 4$. The positive fraction $+\frac{3}{8}$ is represented in one's-complement as $0_\Delta 0110$. The negative fraction $(-\frac{3}{8})$ has magnitude $(\frac{3}{8})$. Thus the one's-complement representation is generated by subtracting $0_\Delta 0110$ $(\frac{3}{8})$ from $1_\Delta 1111$ $(2 - 2^{-4})$ to yield $1_\Delta 1001$. Consequently, the negative fraction $(-\frac{3}{8})$ is represented by $1_\Delta 1001$ in a one's-complement register.

Table 9.1 shows a comparison of the three number systems for a four-bit word length. It is useful to note that in all three representations the leading bit is zero for a positive fraction and one for a negative fraction (see Problem 1(a) at the end of this chapter). For this reason the leading bit is called the *sign bit.* For sign and magnitude, changing the sign of a number but not the magnitude affects only the leading bit. For the one's-complement and

Table 9.1

	Interpretation		
Binary Number	*Sign and Magnitude*	*Two's-Complement*	*One's-Complement*
$0_\Delta 111$	7/8	7/8	7/8
$0_\Delta 110$	6/8	6/8	6/8
$0_\Delta 101$	5/8	5/8	5/8
$0_\Delta 100$	4/8	4/8	4/8
$0_\Delta 011$	3/8	3/8	3/8
$0_\Delta 010$	2/8	2/8	2/8
$0_\Delta 001$	1/8	1/8	1/8
$0_\Delta 000$	0	0	0
$1_\Delta 000$	−0	−1	−7/8
$1_\Delta 001$	−1/8	−7/8	−6/8
$1_\Delta 010$	−2/8	−6/8	−5/8
$1_\Delta 011$	−3/8	−5/8	−4/8
$1_\Delta 100$	−4/8	−4/8	−3/8
$1_\Delta 101$	−5/8	−3/8	−2/8
$1_\Delta 110$	−6/8	−2/8	−1/8
$1_\Delta 111$	−7/8	−1/8	−0

two's-complement representations, changing the sign of a number affects all the bits. In particular it can be shown that negation of a one's-complement number can be accomplished by conjugating all the bits (see Problem 1(b) of this chapter); negation of a two's-complement number is accomplished by conjugating all the bits, adding 2^{-b}, and ignoring any overflow from the addition (see Problem 1(c) at the end of this chapter). Note also that both $+0$ and -0 are represented in the sign and magnitude and one's-complement systems; in the two's-complement system -1 is represented but $+1$ is not. Each of the representations has its advantages and disadvantages; the choice of representation of negative numbers lies primarily in the hardware or software implementation of arithmetic operations, such as addition, subtraction, and multiplication.

A variety of conventions have been used for the representation of negative floating-point numbers. In this chapter we shall consider the sign of the number to be associated with the mantissa so that the mantissa is a signed fraction. The representation of this signed fraction can of course be in sign and magnitude, one's-complement, or two's-complement notation.

9.1.3 *Effect of Truncation or Rounding*

We shall consider both fixed-point numbers and mantissas of floating-point numbers to be represented as $(b + 1)$-bit binary fractions, with the binary point just to the right of the highest-order bit. This convention represents no loss of generality, and its convenience has been alluded to above. The numerical value (for positive numbers) of a 1 in the least significant bit is 2^{-b}. This quantity will be referred to as the *width of quantization* since the numbers are quantized in steps of 2^{-b}.

As indicated previously, the effect of truncation or rounding depends on whether fixed-point or floating-point arithmetic is used and how negative numbers are represented. Let us consider first the effect of truncation and rounding in the fixed-point case. For sign-and-magnitude, one's-complement, and two's-complement methods, the representation of positive numbers is identical and, consequently, so is the effect of truncation and rounding. Let us denote by b_1 the number of bits to the right of the binary point before truncation and by b the number of bits to the right of the binary point after truncation, with, of course, $b < b_1$. The effect of truncation is to discard the least significant $(b_1 - b)$ bits, and consequently the magnitude of the number after truncation is less than or equal to the magnitude before truncation.

Let us denote the number before truncation and after truncation by x and $Q[x]$, respectively. Then the truncation error is

$$E_T = Q[x] - x$$

This error will be negative or zero for positive numbers. The largest error occurs when all the bits discarded are unity, in which case truncation reduces the value of the register by $(2^{-b} - 2^{-b_1})$. Thus for truncation of positive numbers,

$$-(2^{-b} - 2^{-b_1}) \leq E_T \leq 0 \tag{9.1}$$

For negative numbers, the effect of truncation depends on whether sign-and-magnitude, two's-complement, or one's-complement representation is used. Thus we consider each of the three cases separately.

With sign-and-magnitude representation the effect of truncation, as we discussed above, is to reduce the magnitude of the number. Thus a negative number becomes smaller in magnitude, so that E_T, the value after truncation, minus the value before truncation is positive. That is, for truncation of negative numbers with sign-and-magnitude representation,

$$0 \leq E_T \leq (2^{-b} - 2^{-b_1}) \tag{9.2}$$

For a two's-complement negative number represented by the bit string $1_\Delta a_1 a_2 \ldots a_{b_1}$, the magnitude is given by

$$A_1 = 2.0 - x_1$$

where

$$x_1 = 1 + \sum_{i=1}^{b_1} a_i 2^{-i}$$

Truncation to b bits produces the bit string $1_\Delta a_1 a_2 \ldots a_b$, where now the magnitude is

$$A_2 = 2.0 - x_2$$

with

$$x_2 = 1 + \sum_{i=1}^{b} a_i 2^{-i}$$

The change in magnitude is

$$\Delta A = A_2 - A_1 = \sum_{i=b+1}^{b_1} a_i 2^{-i}$$

and it is easily seen that

$$0 \le \Delta A \le 2^{-b} - 2^{-b_1}$$

Hence the effect of truncation for two's-complement negative numbers is to *increase* the magnitude of the negative number; the truncation error is negative and thus for two's-complement negative numbers

$$-(2^{-b} - 2^{-b_1}) \le E_T \le 0 \tag{9.3}$$

For a one's-complement negative number represented by the bit string $1_\Delta a_1 a_2 \ldots a_{b_1}$, the magnitude is given by $A_1 = 2.0 - 2^{-b_1} - x_1$, and truncation to b bits yields a magnitude $A_2 = 2.0 - 2^{-b} - x_2$, where x_1 and x_2 are as defined above. The change in magnitude is

$$\Delta A = A_2 - A_1 = \sum_{i=b+1}^{b_1} a_i 2^{-i} - (2^{-b} - 2^{-b_1})$$

and now

$$-(2^{-b} - 2^{-b_1}) \le \Delta A \le 0$$

Hence the effect of truncation for one's-complement negative numbers is to *decrease* the magnitude of the negative number; the truncation error is positive and satisfies the inequality

$$0 \le E_T \le (2^{-b} - 2^{-b_1}) \tag{9.4}$$

Note that for two's-complement numbers the range of the error is the same for positive and negative numbers, whereas for one's-complement and sign and magnitude the sign of the error depends on the sign of the number being truncated.

As an alternative to truncation, numbers can be rounded to fit into a finite-length register. Again, let b denote the number of bits to the right of the binary point after rounding. After rounding, the values are quantized in steps of 2^{-b}; i.e., the smallest nonzero difference between two numbers is 2^{-b}. Rounding corresponds to choosing the closest quantization level.†

† It is, of course, possible that the number to be rounded lies exactly halfway between two quantization levels. In this case there are several possible strategies, such as always rounding up, always rounding down, or using random rounding.

Thus the maximum error has a magnitude of $\frac{1}{2} \cdot 2^{-b}$; i.e., the rounding error E_R is in the range†

$$-\tfrac{1}{2}(2^{-b} - 2^{-b_1}) < E_R \leq \tfrac{1}{2}(2^{-b} - 2^{-b_1}) \tag{9.5}$$

Because rounding is based on the magnitude of the number, the error is independent of the way in which negative numbers are represented. Generally, it is reasonable to assume that $2^{-b_1} \ll 2^{-b}$ and consequently that the term 2^{-b_1} can be neglected in the above inequalities. With this approximation, we can summarize the truncation and rounding errors as

Truncation:

$$-2^{-b} < E_T \leq 0 \qquad \text{positive numbers and two's-complement negative numbers} \tag{9.6a}$$

$$0 \leq E_T < 2^{-b} \qquad \text{sign and magnitude and one's-complement negative numbers} \tag{9.6b}$$

Rounding:

$$-\tfrac{1}{2} \cdot 2^{-b} < E_R \leq \tfrac{1}{2} \cdot 2^{-b} \tag{9.6c}$$

These truncation and rounding errors are summarized in Fig. 9.1.

In floating-point arithmetic, the truncation or rounding only affects the mantissa. Thus, for the floating-point representation, relative error is more important than absolute error. This means that floating-point errors are multiplicative rather than additive. In other words, for floating point, if x represents the value before truncation or rounding and $Q[x]$ represents the value after, then

$$Q[x] = x(1 + \varepsilon)$$

where ε is the relative error.

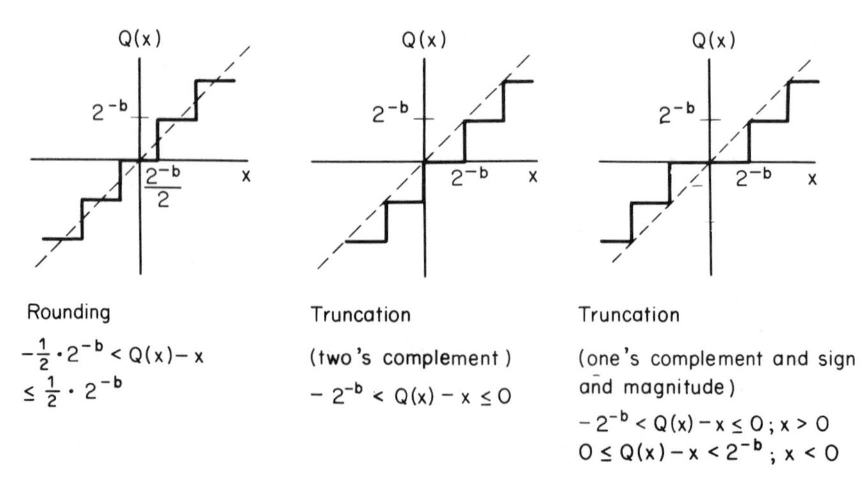

Fig. 9.1 Nonlinear relationships representing rounding and truncation.

† We assume in Eq. (9.5) that a number lying exactly halfway between two quantization steps is rounded up.

For the case of rounding, for example, the error in the mantissa is between $\pm 2^{-b}/2$, and consequently the error in the value of the floating-point word is

$$-2^c \cdot 2^{-b}/2 < Q(x) - x \leq 2^c \cdot 2^{-b}/2$$

or, since $[Q(x) - x] = \varepsilon x$,

$$-2^c \cdot \frac{2^{-b}}{2} < \varepsilon x \leq 2^c \cdot \frac{2^{-b}}{2} \tag{9.7}$$

Then since $2^{c-1} \leq x < 2^c$, we can write that for the case of rounding

$$-2^{-b} < \varepsilon \leq 2^{-b} \tag{9.8a}$$

In a similar manner (see Problem 3, page 466) we can show that for one's-complement and sign-and-magnitude truncation of the mantissa,

$$-2 \cdot 2^{-b} < \varepsilon \leq 0 \tag{9.8b}$$

and for two's-complement truncation,

$$\begin{aligned} -2 \cdot 2^{-b} < \varepsilon \leq 0, \qquad & x > 0 \\ 0 \leq \varepsilon < 2 \cdot 2^{-b}, \qquad & x < 0 \end{aligned} \tag{9.8c}$$

9.2 Quantization in Sampling Analog Signals

The results of Sec. 9.1 can be applied in an analysis of the effects of quantization in sampling an analog signal. When sampling was considered in Sec. 1.7, we assumed that it was possible to obtain a sequence

$$x(n) = x_a(nT), \qquad -\infty < n < \infty$$

where $x_a(t)$ denotes a bandlimited analog signal. That is, it was assumed that the samples of $x_a(t)$ were known with infinite precision. In this case an infinite number of bits is theoretically required to represent each sample. Of course, physical limitations preclude sampling with infinite precision, and therefore each sample must be either truncated or rounded to fit a finite-length register. Thus a somewhat less idealized representation of the sampling process is depicted in Fig. 9.2(a).

The form of the quantizer characteristic depends upon how negative samples are represented and whether rounding or truncation is used. To be specific let us assume that the output samples are represented as two's-complement fixed-point fractions of length $(b + 1)$ bits. Further, assume that the exact input samples $x(n)$ are rounded to the nearest quantization level to obtain the quantized samples $\hat{x}(n)$. To ensure that the unquantized samples are within the range of the $(b + 1)$-bit number, we must also assume that the analog waveform is normalized, so that

$$\left(-1 + \frac{2^{-b}}{2}\right) < x_a(nT) < \left(1 - \frac{2^{-b}}{2}\right) \tag{9.9}$$

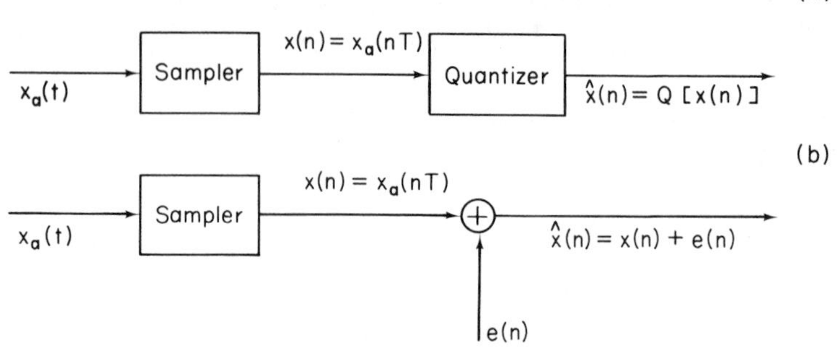

Fig. 9.2 Representation of sampling of an analog signal: (a) nonlinear model; (b) statistical model.

Under these conditions, the quantizer function is as shown in Fig. 9.3, where it is assumed that $b = 2$.

If the exact value of an input sample falls outside the range indicated in Eq. (9.9), then additional distortion results. As indicated in Fig. 9.3, the quantized value $1 - 2^{-b}$ is assigned to all samples exceeding $1 - 2^{-b}/2$, and the quantized value -1 is assigned to all samples less than $-(1 + 2^{-b}/2)$. This clipping of the input is of course generally undesirable, and it must be eliminated by reducing the amplitude of the input until Eq. (9.9) is satisfied.

An equivalent representation of the quantization process is depicted in Fig. 9.2(b). That is, we can express the quantized samples as

$$\hat{x}(n) = Q[x(n)] = x(n) + e(n)$$

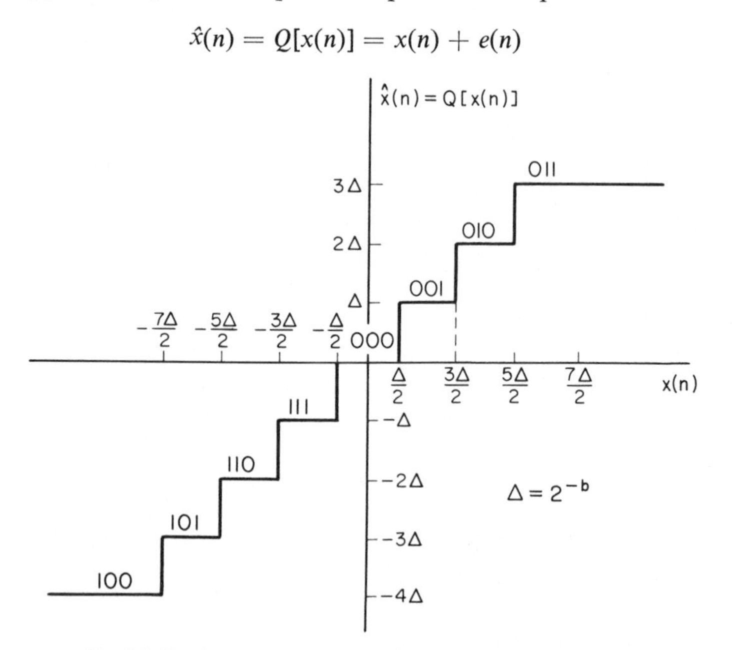

Fig. 9.3 Two's-complement rounding of samples for $b = 2$.

where $x(n)$ is the exact sample and $e(n)$ is called the *quantization error*. Since rounding was assumed,

$$-\frac{\Delta}{2} < e(n) \leq \frac{\Delta}{2}$$

where Δ is the quantization width, i.e., $\Delta = 2^{-b}$.

In order that Fig. 9.2(b) be exactly equivalent to Fig. 9.2(a), we must, of course, know $e(n)$ exactly for all n. In most cases it is reasonable to take the position that $e(n)$ is *not* known, and a statistical model based on Fig. 9.2(b) may then be useful in representing the effects of quantization in sampling. We will also use such a model to describe the effects of quantization in signal processing algorithms. In particular it is common to make the following assumptions:

1. The sequence of error samples $\{e(n)\}$ is a sample sequence of a stationary random process.
2. The error sequence is uncorrelated with the sequence of exact samples $\{x(n)\}$.
3. The random variables of the error process are uncorrelated; i.e., the error is a white-noise process.
4. The probability distribution of the error process is uniform over the range of quantization error.

As we will see, these assumptions, motivated primarily by expediency, lead to a rather simple analysis of quantization effects. It is easy to find examples where these assumptions are clearly not valid. For example, if $x_a(t)$ is a step function, it would be impossible to justify the above assumptions. However, when the sequence $x(n)$ is a complicated signal, such as speech or music, where the signal fluctuates rapidly in a somewhat unpredictable manner, these assumptions are more realistic. Experiments have shown [2–4] that as the signal becomes more complicated, the correlation between the signal and the quantization error decreases, and the error also becomes uncorrelated. In a heuristic sense, the assumptions of the statistical model appear to be valid if the signal is sufficiently complex and the quantization steps sufficiently small so that the amplitude of the signal is likely to traverse many quantization steps in going from sample to sample.

In the case of rounding we assume that the probability distribution of the error is as shown in Fig. 9.4(a). Similarly, for two's-complement truncation the probability distribution is assumed to be uniform over the range of possible quantization errors as shown in Fig. 9.4(b). Furthermore, it is assumed that the error is independent of the signal. This assumption is clearly not valid for one's-complement and sign-and-magnitude truncation, since the sign of the error is always opposite to the sign of the signal. The mean and variance of the quantization noise are easily shown to be (see

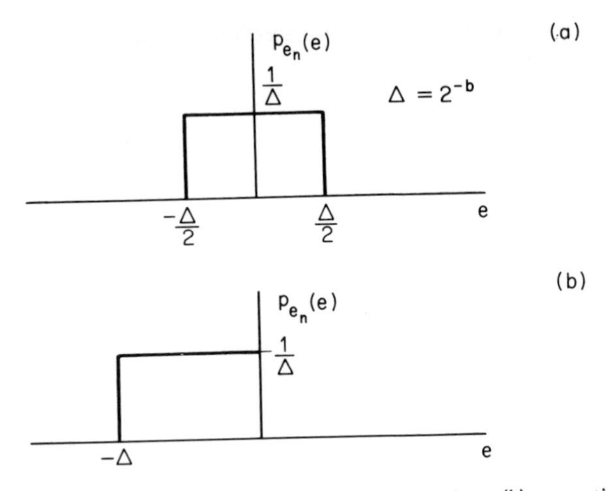

Fig. 9.4 Probability density functions for (a) rounding; (b) truncation.

Problem 3 in Chapter 8)

$$m_e = 0$$

$$\sigma_e^2 = \frac{\Delta^2}{12} = \frac{2^{-2b}}{12}$$

for rounding, and

$$m_e = -\frac{2^{-b}}{2}$$

$$\sigma_e^2 = \frac{2^{-2b}}{12}$$

for two's-complement truncation. The autocovariance sequence of the error is assumed to be

$$\gamma_{ee}(n) = \sigma_e^2 \delta(n)$$

for both rounding and two's-complement truncation.

In digital processing of sampled analog signals, the quantization error is commonly viewed as an additive noise signal. The ratio of the signal power to the noise power is a useful measure of the relative strengths of the signal and the noise. For rounding, the signal-to-noise ratio is

$$\frac{\sigma_x^2}{\sigma_e^2} = \frac{\sigma_x^2}{2^{-2b}/12} = (12 \cdot 2^{2b})\sigma_x^2$$

When expressed on a logarithmic scale, as in

$$\text{SNR} = 10 \log_{10} \left(\frac{\sigma_x^2}{\sigma_e^2}\right) = 6.02b + 10.79 + 10 \log_{10} (\sigma_x^2) \tag{9.10}$$

it is clear that the signal-to-noise ratio increases approximately 6 dB for each bit added to the register length.

Earlier in this section we noted that if the input signal exceeds the dynamic range of the quantization process, we must reduce the input amplitude to eliminate clipping. That is, we quantize samples $Ax(n)$ instead of $x(n)$, where $0 < A < 1$. Since the variance of $Ax(n)$ is $A^2\sigma_x^2$, the signal-to-noise ratio is

$$\text{SNR} = 10 \log_{10} \left(\frac{A^2\sigma_x^2}{\sigma_e^2}\right) = 6b + 10.8 + 10 \log_{10} (\sigma_x^2) + 20 \log_{10} (A) \quad (9.11)$$

A comparison of Eqs. (9.10) and (9.11) shows that reducing the amplitude of the input to reduce clipping distortion reduces the signal-to-noise ratio. Many analog signals, such as speech or music, can be conveniently viewed as a random process. Generally, such signals are characterized by probability distributions that are peaked around zero, and fall off rapidly with increasing amplitude. In such cases the probability that the magnitude of a given sample will exceed three or four times the root-mean-square value of the signal is very low. Thus, if A is set at $\sigma_x/4$, then with high probability, no clipping distortion will occur. In this case the signal-to-noise ratio is

$$\text{SNR} = 6b - 1.24 \text{ dB}$$

Thus to obtain SNR ≥ 80 dB requires $b = 14$ bits. This interrelationship between dynamic range and quantization error is a fundamental characteristic of fixed-point representations of discrete-time signals and signal-processing algorithms. We shall see it recurring throughout the remainder of this chapter.

When quantized signals are processed by some signal-processing algorithm, the input error (or noise) manifests itself as an error (or noise) in the resulting output. For example, if a quantized sequence $\hat{x}(n) = x(n) + e(n)$ is the input to a linear shift-invariant system, then the output can be represented as $\hat{y}(n) = y(n) + f(n)$, where $y(n)$ is the response to $x(n)$ and $f(n)$ is the response to $e(n)$. Since $x(n)$ and $e(n)$ are independent, we can ignore $x(n)$ in computing the output noise power. Using Eqs. (8.50) and (8.53) and the fact that the input noise is white, we can express the mean and variance of the output noise as

$$m_f = m_e \sum_{n=-\infty}^{\infty} h(n) = m_e H(e^{j0}) \quad (9.12)$$

$$\sigma_f^2 = \sigma_e^2 \sum_{n=-\infty}^{\infty} |h(n)|^2 = \frac{\sigma_e^2}{2\pi} \int_{-\pi}^{\pi} |H(e^{j\omega})|^2 \, d\omega \quad (9.13)$$

In deriving these equations, we have implicitly assumed that the system was realized with no error. In fact, this is not true; however, it is reasonable to assume that the errors in the output due to errors introduced in the system realization are independent of errors due to the input quantization noise. Thus errors due to rounding or truncation in realizing a digital filter are considered separately and their effects superimposed on the errors due to input quantization.

9.3 Finite-Register-Length Effects in Realizations of IIR Digital Filters

In this section we discuss the effect of quantizing the results of arithmetic operations in realizations of IIR digital filters. As we noted in Chapter 4, the basic arithmetic operations involved in the implementation of a digital filter are multiplication by a constant (the filter coefficients) and addition. For arithmetic involving fixed-point numbers, the result of a multiplication must be rounded or truncated, but the result of an addition need not be rounded or truncated. However, because the result of an addition can exceed the finite register length, there are important dynamic range considerations in fixed-point implementations of digital filters. This is similar to the problem of quantizing the samples of an analog signal, where we have already seen that wide dynamic range and low quantization error are opposing requirements. In contrast, floating-point implementations have a much less severe dynamic range limitation, but truncation or rounding must be introduced after both multiplication and addition. As we have seen in the previous section, truncation or rounding is a nonlinear process. That is, the effect of quantization in the realization of a shift-invariant digital filter is to introduce nonlinear elements into certain branches of a filter structure.

Strictly speaking, realizations of linear shift-invariant systems are generally nonlinear. It is important to understand the nonlinear effects that arise as a result of quantization, since a common objective is to obtain a realization of a digital filter that is sufficiently accurate and at the same time requires a minimum of hardware complexity. The analysis of nonlinear effects in digital filters is often complex and in many cases is not fully understood. In addition, much of this analysis is very detailed and beyond the scope of this book.

In the following discussion we shall consider first some simple cases with fixed-point arithmetic in which the nonlinearity causes a periodic error at the output when the input is set to zero or when the input is constant or is sinusoidal. In these simple cases it is possible to understand the effects of the nonlinearity in detail. However, when the input is neither constant nor sinusoidal, we find it convenient to use the statistical model of the previous section to convert the nonlinear system into a linear system with internal additive noise sources. Using this model we can compute an average error at the output due to the quantization of the results of arithmetic operations in the realization of a digital filter.

9.3.1 *Zero-Input Limit Cycles in Fixed-Point Realizations of IIR Digital Filters*

If a stable digital filter implemented with infinite-precision arithmetic has an excitation that is zero for n greater than some value n_0, the output for $n > n_0$ will decay asymptotically toward zero. For the same filter, implemented with finite-register-length arithmetic, the output may decay into a nonzero amplitude range, after which it has an oscillatory behavior. This effect is often referred to as *zero-input limit cycle behavior* and is a consequence of the nonlinear quantizers in the feedback loop of the filter. The limit cycle behavior of a digital filter is complex and difficult to analyze, and we shall not attempt to treat the topic in any general sense. For simple first- and second-order filters, it is possible to understand the effect and to

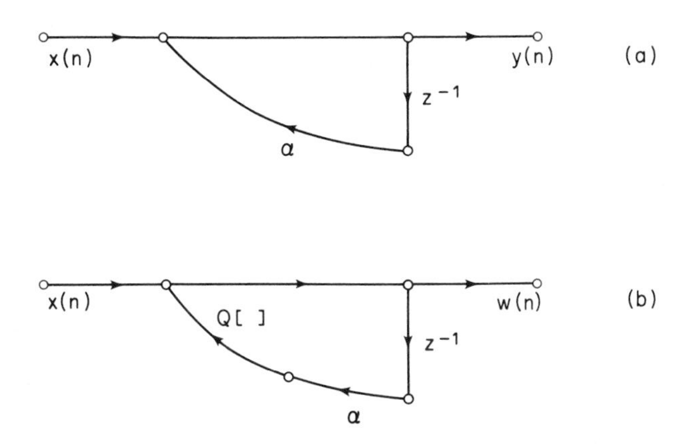

Fig. 9.5 Flow graphs for a first-order IIR system: (a) ideal linear system; (b) nonlinear system due to quantization of product.

develop an interpretation of the oscillations in terms of the effective filter poles moving onto the unit circle [5–10]. The effect is best illustrated by means of an example.

EXAMPLE. As an illustration of limit cycle effects, consider the first-order system characterized by the difference equation

$$y(n) = \alpha y(n-1) + x(n) \tag{9.14}$$

The signal flow graph of this system is shown in Fig. 9.5(a). Let us assume that $\alpha = \frac{1}{2}$ and that the register length for storing the coefficient α, the input $x(n)$, and filter node variable $y(n-1)$ is four bits (i.e., a sign bit to the left of the binary point and three bits to the right of the binary point). Because of the finite-length registers, the product $\alpha y(n-1)$ must be rounded or truncated to four bits before addition to $x(n)$. The flow graph representing the actual realization based on Eq. (9.14) is shown in Fig. 9.5(b). Assuming rounding of the product, the actual output $w(n)$ satisfies the nonlinear difference equation

$$w(n) = Q[\alpha w(n-1)] + x(n) \tag{9.15}$$

where $Q[\ ]$ represents the rounding operation. Let us assume that $\alpha = \frac{1}{2} = 0_\Delta 100$ and that the input is a unit sample of amplitude $\frac{7}{8} = 0_\Delta 111$. Using Eq. (9.15) we see that for $n = 0$, $w(0) = \frac{7}{8} = 0_\Delta 111$. To obtain $w(1)$ we multiply $w(0)$ by α, obtaining the result $\alpha w(0) = 0_\Delta 011100$, a seven-bit number that must be rounded to four bits. This number, $\frac{7}{16}$, is exactly halfway between the two four-bit quantization levels $\frac{4}{8}$ and $\frac{3}{8}$. If we choose always to round upward in such cases, then $0_\Delta 011100$ rounded to four bits is $0_\Delta 100 = \frac{1}{2}$. Since $x(1) = 0$, then $w(1) = 0_\Delta 100 = \frac{1}{2}$. Continuing, $w(2) = Q[\alpha w(1)] = 0_\Delta 010 = \frac{1}{4}$ and $w(3) = 0_\Delta 001 = \frac{1}{8}$. In both these cases no rounding is necessary. However, to obtain $w(4)$, we must round the seven-bit number $\alpha w(3) = 0_\Delta 000100$ to $0_\Delta 001$. The same result is obtained for all values of $n \geq 3$. The output sequence for this example is shown in Fig. 9.6(a). If $\alpha = -\frac{1}{2}$ we can carry out the above

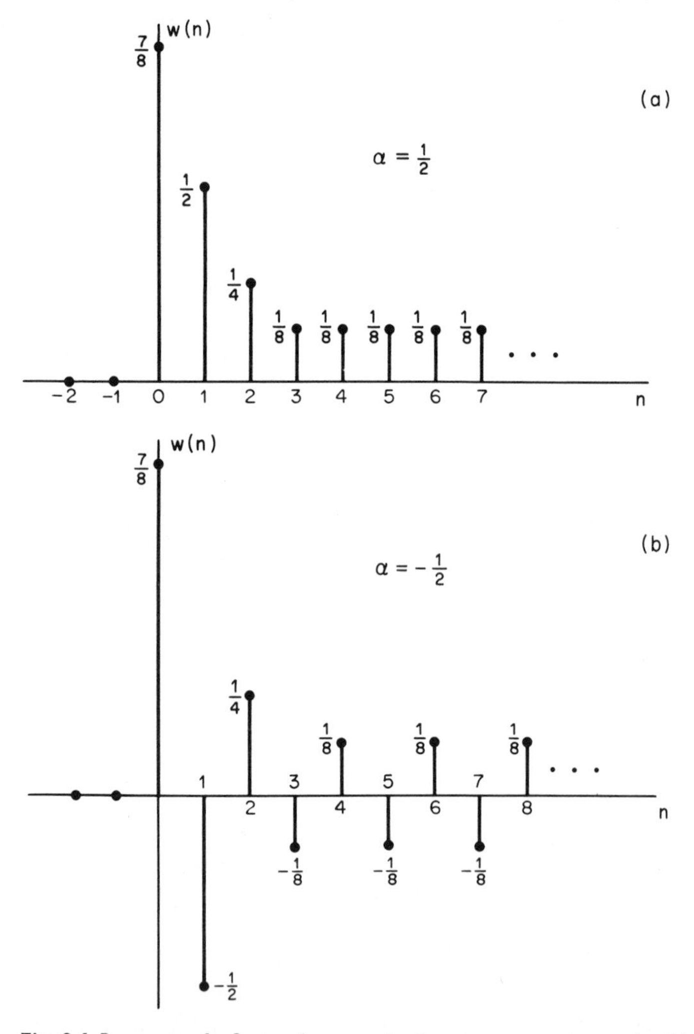

Fig. 9.6 Response of a first-order quantized system to a unit sample: (a) $\alpha = \frac{1}{2}$; (b) $\alpha = -\frac{1}{2}$.

computation again to show that the output is as shown in Fig. 9.6(b). Thus, owing to rounding of the product $\alpha w(n-1)$, the output reaches a constant value of $\frac{1}{8}$ when $\alpha = \frac{1}{2}$ and a periodic steady-state oscillation between $+\frac{1}{8}$ and $-\frac{1}{8}$ when $\alpha = -\frac{1}{2}$. These are periodic outputs similar to those that would be obtained from a first-order filter with a pole at $z = \pm 1$ instead of at $\pm\alpha$. When $\alpha = +\frac{1}{2}$, the period of the oscillation is 1, and when $\alpha = -\frac{1}{2}$, the period of oscillation is 2. Such steady-state periodic outputs are called *limit cycles*, and their existence was first noted by Blackman [11], who referred to the amplitude intervals to which such limit cycles are confined as *dead bands*. In this case the dead band is $-2^{-b} \leq w(n) \leq +2^{-b}$.

The possible existence of a zero-input limit cycle is important in applications where a digital filter is to be in continuous operation, since it is generally desired that the output approach zero when the input is zero. For example, consider a sampled speech signal, being filtered by a digital filter and then converted back to an acoustic signal using a digital-to-analog converter. In such a situation it would be very undesirable for the filter to enter a periodic limit cycle whenever the input is zero.

Jackson [5,6] has considered limit cycle behavior in first- and second-order systems with an analysis based on the above observation; i.e., that in the limit cycle, the system behaves as if the poles of the system were on the unit circle. Specifically, consider the above first-order filter. By the definition of rounding,

$$|Q[\alpha w(n-1)] - \alpha w(n-1)| \leq \tfrac{1}{2} \cdot 2^{-b} \tag{9.16}$$

Furthermore, for values of n in the limit cycle,

$$|Q[\alpha w(n-1)]| = |w(n-1)|$$

i.e., the effective value of α is 1, corresponding to the pole of the filter being on the unit circle. The range of values for which this condition is met is

$$|w(n-1)| - |\alpha w(n-1)| \leq \tfrac{1}{2} \cdot 2^{-b}$$

or, solving for $|w(n-1)|$,

$$|w(n-1)| \leq \frac{\frac{1}{2} \cdot 2^{-b}}{1 - |\alpha|} \tag{9.17}$$

Equation (9.17) defines the dead band for the first-order filter. As a result of rounding, values within the dead band are quantized in steps of 2^{-b}. [Note that Eq. (9.17) gives the correct value for the dead band when $|\alpha| = \frac{1}{2}$.] Whenever the node variable $w(n-1)$ falls within the dead band when the input is zero, the filter enters into a limit cycle and remains in this mode until an input is applied to carry the output out of the dead band.

For a second-order filter, there is a larger variety of limit cycle behavior. Consider the second-order difference equation

$$y(n) = x(n) + \alpha_1 y(n-1) + \alpha_2 y(n-2) \tag{9.18}$$

With $\alpha_1^2 < -4\alpha_2$, the filter poles are complex conjugates, and, with $\alpha_2 = -1$, the poles are on the unit circle. A practical representation of Eq. (9.18) is

$$w(n) = x(n) + Q[\alpha_1 w(n-1)] + Q[\alpha_2 w(n-2)] \tag{9.19}$$

where $Q[\ ]$ again represents rounding of the indicated products. As before, by the definition of rounding,

$$|Q[\alpha_2 w(n-2)] - \alpha_2 w(n-2)| \leq \tfrac{1}{2} \cdot 2^{-b} \tag{9.20}$$

With $x(n) = 0$, the poles of the system will appear to be on the unit circle if

$$Q[\alpha_2 w(n-2)] = w(n-2)$$

Substituting this equation into Eq. (9.20) we obtain

$$|w(n-2)| - |\alpha_2 w(n-2)| \leq \tfrac{1}{2} \cdot 2^{-b}$$

or, solving for $|w(n-2)|$,

$$|w(n-2)| \leq \frac{\frac{1}{2} \cdot 2^{-b}}{1 - |\alpha_2|} \tag{9.21}$$

Thus, if $w(n-2)$ falls in this range when the input is zero, the effective value of α_2 is such that the poles of the system are apparently on the unit circle. Under these conditions, the value of α_1 controls the frequency of oscillation.

In a second mode of limit cycle behavior that can occur in second-order filters, the effect of rounding is to place an effective pole at $z = +1$ and $z = -1$. The dead band corresponding to this mode is bounded by $1/(1 - |\alpha_1| - \alpha_2)$. The amplitude of a limit cycle in this band is of course quantized in steps of 2^{-b} [5,6].

In addition to the above classes of limit cycles, a more severe type of limit cycle can occur due to overflow. The effect of overflow is to insert a gross error in the output, and, in some cases, thereafter the filter output oscillates between the maximum amplitude limits. Such limit cycles have been referred to as *overflow oscillations*. The problem of oscillations caused by overflow is discussed in detail by Ebert et al [12]. A simple example is developed in Problem 9, page 469.

The above discussion was concerned only with zero-input limit cycles, owing to rounding in first- and second-order IIR systems. Although the analysis was somewhat heuristic, the simple formulas that result have been found to be consistent with experimental results and are useful in predicting limit cycle behavior in IIR digital filters. A similar style of analysis can be used for truncation (see, for example, Problem 6 at the end of this chapter). In the case of parallel realizations of higher-order systems, the outputs of the individual second-order systems are independent when the input is zero. Thus the previous analysis can be applied directly. In the case of cascade realizations, only the first section has zero input. Succeeding sections may exhibit their own characteristic limit cycle behavior or they may appear to be simply filtering the limit cycle output of a previous section. For high-order systems realized by other filter structures, the limit cycle behavior becomes more complex, as does the analysis thereof. When the input is nonzero, the quantization effects depend upon the input, and the style of analysis of this section is completely inappropriate except for simple inputs such as a unit sample, a unit step, or a sine wave. In other cases the complexity of the quantization phenomena force us to a statistical model.

In addition to giving an understanding of limit cycle effects in digital filters, the above results are useful when the zero-input limit cycle response of a system is the desired output. This is the case, for example, when one is concerned with digital sine-wave oscillators for signal generation and for generation of coefficients for discrete Fourier transformation.

9.3.2 Statistical Analysis of Quantization in Fixed-Point Realizations of IIR Digital Filters

A precise analysis of truncation or rounding errors is generally not required in practical applications. For example, a common objective of error analysis is to choose the register length necessary to meet some specifications on the relative sizes of signal and errors. The register length can, of course, be changed only in steps of one bit. As we shall see, the addition of one bit to the register length reduces the amplitude of the quantization errors by a factor of approximately one-half. Thus a final decision concerning register length is insensitive to inaccuracies in the error analysis; an analysis correct to within 30–40% is often adequate. Because of this insensitivity, it is possible to use the statistical model developed in Sec. 9.2 in the analysis of quantization errors.

In the following discussion we shall focus primarily on simple first- and second-order systems to illustrate the manner in which the statistical model can be used in estimating the effects of quantization in fixed-point realizations of digital filters. Out of these simple examples will emerge a number of guidelines that can often be generalized in more complex examples. These guidelines are useful in considering the many tradeoffs involved in achieving the most efficient and economical realizations of digital filters.

Consider a fixed-point implementation of a first-order system with rounding being applied to products. Figure 9.7(a) depicts the infinite-precision system and Fig. 9.7(b) depicts the finite-precision realization, with $Q[\ ]$ denoting the rounding operation. In Fig. 9.7(c) the same system is depicted, with the effect of the quantizer being represented by the additive noise source

$$e(n) = Q[\alpha w(n-1)] - \alpha w(n-1)$$

As we discussed in Sec. 9.2, the representations of Fig. 9.7(b) and (c) are identical when $e(n)$ is known. However, let us make the following assumptions concerning the effect of quantization of products.

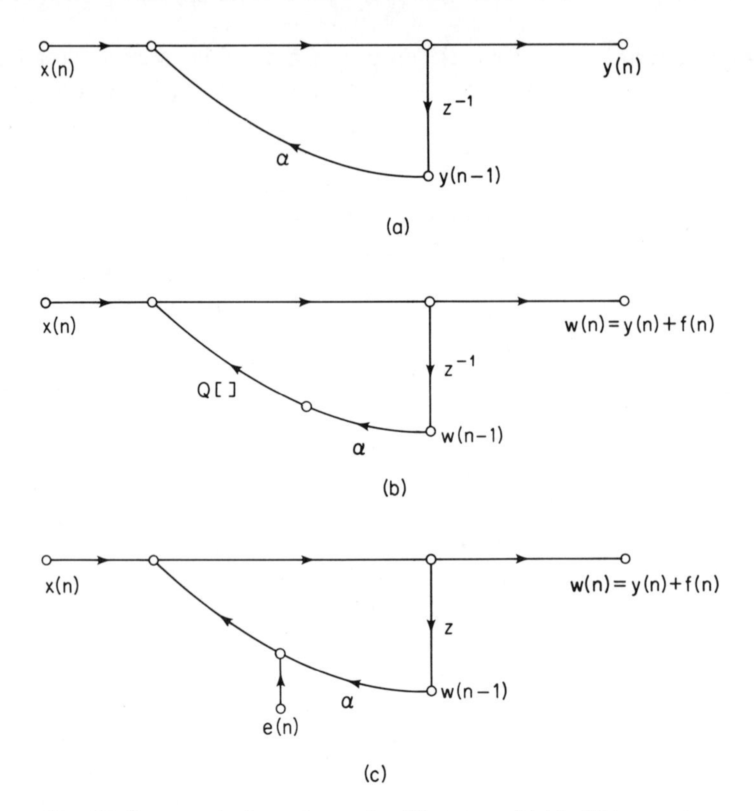

Fig. 9.7 Flow graphs for a first-order IIR system: (a) ideal linear system; (b) nonlinear system; (c) statistical model for fixed-point roundoff noise.

1. The error sequence $e(n)$ is a white-noise sequence.
2. The error sequence has a uniform distribution over one quantization interval.
3. The error sequence $e(n)$ is uncorrelated with the input $x(n)$ and $\alpha w(n - 1)$. [This implies that $e(n)$ is uncorrelated with the output. See Problem 12 in Chapter 8.]

These assumptions are identical to those made for quantization of samples of an analog signal, and the conditions for their validity are much the same. That is, these assumptions hold when the input signal and resulting node variables vary from sample to sample in a sufficiently complex manner. They are clearly not valid for inputs such as a unit sample, a unit step, or a sinusoidal sequence.

If the register length is $(b + 1)$ bits, then, for rounding,

$$-\tfrac{1}{2} \cdot 2^{-b} < e(n) \leq \tfrac{1}{2} \cdot 2^{-b}$$

Assuming a uniform distribution over this range, the mean of $e(n)$ is zero and

$$\sigma_e^2 = \tfrac{1}{12} \cdot 2^{-2b}$$

If $y(n)$ is the output that would be obtained due to $x(n)$ if there were no quantization error, then the actual output can be represented as

$$w(n) = y(n) + f(n)$$

where $f(n)$ represents the output error due to the noise source $e(n)$. If $h_e(n)$ is the unit-sample response of the system from the node at which $e(n)$ enters to the output, then

$$m_f = m_e \sum_{n=-\infty}^{\infty} h_e(n) \tag{9.22a}$$

and since $e(n)$ is assumed to be white noise,†

$$\sigma_f^2 = \sigma_e^2 \sum_{n=-\infty}^{\infty} h_e^2(n) \tag{9.22b}$$

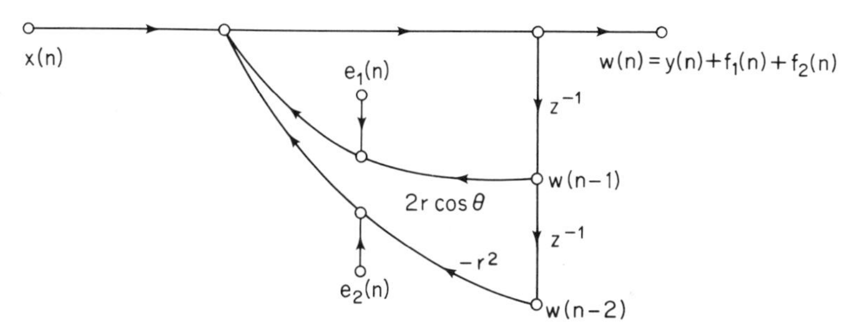

Fig. 9.8 Statistical model for fixed-point roundoff noise in a second-order IIR system.

From Fig. 9.7(c) we see that for this case the unit-sample response from the noise input to the output is the same as for the signal input. This of course is not true in general. For this example, then, $h_e(n) = \alpha^n u(n)$, so

$$\sigma_f^2 = \sigma_e^2 \frac{1}{1-\alpha^2} = \tfrac{1}{12} \cdot 2^{-2b} \frac{1}{1-\alpha^2} \tag{9.23}$$

As a second example consider a second-order filter with one complex pole pair at $z = re^{\pm j\theta}$, as represented by the difference equation

$$y(n) = x(n) + 2r \cos\theta \, y(n-1) - r^2 y(n-2)$$

With rounding of products we obtain the nonlinear difference equation

$$w(n) = x(n) + Q[2r \cos\theta \, w(n-1)] - Q[r^2 w(n-2)]$$

Since there are two multiplications, two noise sources are introduced as depicted in Fig. 9.8. These sources are denoted $e_1(n)$ and $e_2(n)$. Again, the

† We will assume for convenience that $h_e(n)$ is real.

output can be represented as the sum of the ideal output and two error components, $f_1(n)$ and $f_2(n)$, due to $e_1(n)$ and $e_2(n)$, respectively. As before, we assume that $e_1(n)$ and $e_2(n)$ are white-noise sequences, with amplitude densities uniform between $\pm\frac{1}{2}\cdot 2^{-b}$ and that they are both uncorrelated with the input. Furthermore, we shall assume that they are uncorrelated with each other. Since $e_1(n)$ and $e_2(n)$ are uncorrelated, so are $f_1(n)$ and $f_2(n)$ (see Problem 12 of Chapter 8) and consequently (see Problem 2, Chapter 8)

$$\sigma_f^2 = \sigma_{f_1}^2 + \sigma_{f_2}^2$$

or, with $h_1(n)$ and $h_2(n)$ denoting the unit-sample responses from the noise source inputs,

$$\sigma_f^2 = \sigma_{e_1}^2 \sum_{n=-\infty}^{+\infty} h_1^2(n) + \sigma_{e_2}^2 \sum_{n=-\infty}^{+\infty} h_2^2(n)$$

We note that, for this example $h_1(n)$ and $h_2(n)$ are equal and are given by

$$h_1(n) = h_2(n) = \frac{1}{\sin\theta} r^n \sin\,[(n+1)\theta]u(n)$$

It can be verified that

$$\sum_{n=-\infty}^{+\infty} h_1^2(n) = \frac{1+r^2}{1-r^2}\,\frac{1}{r^4+1-2r^2\cos 2\theta}$$

so that with

$$\sigma_{e_1}^2 = \sigma_{e_2}^2 = \tfrac{1}{12}\cdot 2^{-2b}$$

$$\sigma_f^2 = \tfrac{2}{12}\cdot 2^{-2b}\,\frac{1+r^2}{1-r^2}\,\frac{1}{r^4+1-2r^2\cos 2\theta} \tag{9.24}$$

These two examples illustrate a style of analysis that can be applied to obtaining the variance of the output noise due to arithmetic rounding in any fixed-point filter. Some other examples are considered in the problems at the end of the chapter.

The above results are easily modified for two's-complement truncation. We note that the amplitude of the error in two's-complement truncation is in the range

$$-2^{-b} < E_T \leq 0$$

Thus, if we are to represent the effect of truncation of two's-complement numbers in the same style as we did rounding, we would consider an additive noise source $e(n)$ with an amplitude density uniform between -2^{-b} and zero. We again assume that $e(n)$ is linearly independent of itself shifted (white noise) but that it no longer has zero mean. In particular, $E[e(n)] = -\frac{1}{2}\cdot 2^{-b}$. However, the variance of $e(n)$ in this case is identical to that for the case of rounding. Therefore, with two's-complement truncation, the output noise variance is identical to that with rounding. However, the output noise no longer has zero mean value, although the mean value is easily computed using Eq. (9.22a). In many examples the above results can also be considered to apply to one's-complement and sign-and-magnitude truncation, although, as indicated in Sec. 9.2, in those cases the correlation between the signal and the truncation error is, in fact, stronger because the polarity of the error is always opposite to the polarity of the signal to which the truncation is applied.

As indicated previously, another consideration in the implementation of digital filters with fixed-point arithmetic is the possibility of overflow. With the convention that each fixed-point register represents a signed fraction, each node in the filter must be constrained to maintain a magnitude less than unity in order to avoid overflow. Letting $x(n)$ denote the filter input and $y_k(n)$ and $h_k(n)$ denote respectively the output of the kth node and unit-sample response from the input to the kth node in the filter, then

$$y_k(n) = \sum_{r=-\infty}^{\infty} h_k(r)x(n-r)$$

If $x_{\max}$ denotes the maximum of the absolute value of the input, then

$$|y_k(n)| \leq x_{\max} \sum_{r=-\infty}^{\infty} |h_k(r)| \tag{9.25}$$

Thus, since we require that $|y_k(n)| < 1$, Eq. (9.25) requires that

$$x_{\max} < \frac{1}{\sum_{r=-\infty}^{\infty} |h_k(r)|} \tag{9.26}$$

for all nodes in the network. Equation (9.26) thus provides an upper bound on the maximum value of the input to ensure that no overflow occurs at the kth node. In the most general case, scaling of the input according to Eq. (9.26) is required to guarantee that no overflow occurs. This is a consequence of the fact that equality can be achieved in Eq. (9.25) with a sequence $x(n)$ for which at $n = n_0$, $x(n_0 - r) = \text{sgn}\,[h_k(r)]$ [where $\text{sgn}\,(x) = 1$ for $x \geq 0$ and $\text{sgn}\,(x) = -1$ for $x < 0$]. The condition in Eq. (9.26) can be satisfied by applying attenuation to the signal at the filter input.

As an example consider an input $x(n)$ which is a white-noise sequence with a uniform amplitude density. Then we would choose for the case of the first-order filter a maximal input amplitude of $(1 - |\alpha|)$. For this case, if σ_x^2 denotes the variance of the input signal and σ_y^2 denotes the variance of the output signal,

$$\sigma_x^2 = \frac{1}{3}(1 - |\alpha|)^2 \tag{9.27}$$

$$\sigma_y^2 = \frac{1}{3}\frac{(1 - |\alpha|)^2}{1 - |\alpha|^2} \tag{9.28}$$

For this example we can then compute an output *noise-to-signal ratio* as the ratio σ_f^2/σ_y^2, with the result that

$$\frac{\sigma_f^2}{\sigma_y^2} = \tfrac{1}{4}\cdot 2^{-2b}\frac{1}{(1-|\alpha|)^2} \tag{9.29}$$

In a similar manner we can derive a noise-to-signal ratio for the second-order filter considered previously. As in the first-order case, we restrict the input amplitude to guarantee that the dynamic range of the registers is not exceeded. If we consider the input sequence to be uniformly distributed white noise, the resulting output noise-to-signal ratio will be

$$\frac{\sigma_f^2}{\sigma_y^2} = \tfrac{1}{2}\cdot 2^{-2b}\left(\sum_{n=-\infty}^{\infty}|h(n)|\right)^2 = \tfrac{1}{2}\cdot 2^{-2b}\left(\frac{1}{\sin\theta}\sum_{n=0}^{\infty} r^n\,|\sin\,[(n+1)\theta]|\right)^2 \tag{9.30}$$

Although it is difficult to evaluate this expression exactly, it is possible to obtain an upper and lower bound. Since $\sum_{n=-\infty}^{\infty}|h(n)|$ is the largest possible output obtainable with an input that never exceeds unity, it must be larger than the response of the second-order filter to a sinusoid of unity amplitude at the resonant frequency. With this consideration, we obtain

$$\left(\sum_{n=0}^{\infty}|h(n)|\right)^2 \geq \frac{1}{(1-r)^2(1+r^2-2r\cos 2\theta)} \tag{9.31}$$

since the right-hand side of this inequality is the gain at resonance ($\omega = \theta$). Furthermore,

$$\left(\frac{1}{\sin\theta}\sum_{n=0}^{\infty} r^n\,|\sin\,[(n+1)\theta]|\right)^2 \leq \left(\frac{1}{\sin\theta}\sum_{n=0}^{\infty} r^n\right)^2 \tag{9.32}$$

Therefore, for the second-order case,

$$\tfrac{1}{2}\cdot 2^{-2b}\frac{1}{(1-r)^2(1+r^2-2r\cos 2\theta)} \leq \frac{\sigma_f^2}{\sigma_y^2} \leq \tfrac{1}{2}\cdot 2^{-2b}\frac{1}{\sin^2\theta(1-r)^2} \tag{9.33}$$

Sharp cutoff frequency selective filters often require poles that are very close to the unit circle. Since cascade or parallel realizations of such filters require first- and second-order systems, it is important to consider the above expressions for noise-to-signal ratio as the poles approach the unit circle.

For the first-order filter let $\delta = 1 - |\alpha|$ so that as $\delta \to 0$, the pole approaches the unit circle. Then in terms of δ, the noise-to-signal ratio for the first-order filter is

$$\frac{\sigma_f^2}{\sigma_y^2} = \tfrac{1}{4}\cdot 2^{-2b}\frac{1}{\delta^2} \tag{9.34}$$

For the second-order filter, let $\delta = 1 - r$ so that, again, as $\delta \to 0$, the poles approach the unit circle. Then if we assume that $\delta \ll 1$, we can approximate $(1 + r^2 - 2r\cos 2\theta)$ as

$$1 + r^2 - 2r\cos 2\theta \simeq 4\sin^2\theta + \delta^2 \tag{9.35}$$

which for $4 \sin^2 \theta$ large compared with δ^2 we will approximate as $4 \sin^2 \theta$. Consequently, incorporating this approximation,

$$\tfrac{1}{2} \cdot 2^{-2b} \frac{1}{4\delta^2 \sin^2 \theta} \leq \frac{\sigma_f^2}{\sigma_y^2} \leq \tfrac{1}{2} \cdot 2^{-2b} \frac{1}{\delta^2 \sin^2 \theta} \tag{9.36}$$

We observe that the noise-to-signal ratio for the examples above can be considered to be proportional to $2^{-2b}/\delta^2$. We note from this dependence that if δ is halved, then to maintain the same noise-to-signal ratio, b must be increased by 1; i.e., one bit must be added to the register length.

In the above analysis, the filter input was assumed to be uniformly distributed white noise. As δ approaches zero, the frequency response of both the first- and second-order filters becomes more selective, so that more and more of the input energy is out of band. An alternative basis for determining the noise-to-signal ratio is for an input that is sinusoidal. For this choice of inputs, of course, we would not use the general condition of Eq. (9.26) to avoid overflow since we can determine exactly the maximum allowable input amplitude as a function of the filter parameters.

In particular, if the input is of the form $x(n) = A \cos n\omega_0$, then the steady-state output is of the form $y(n) = B \cos (n\omega_0 + \phi)$. To prevent overflow, B must be less than unity, and to maximize the output signal energy, B is chosen to be as large as possible. Thus the maximum noise-to-signal ratio is obtained when A is chosen so that $y(n) = \cos (n\omega_0 + \phi)$. Note that in order to choose A in this way, the frequency of the input signal must be known. For an input sinusoid of unknown frequency, A must be chosen so that overflow will not occur even in the worst case, i.e., for which the frequency of the input coincides with the peak gain in the filter's transfer function [13].

For fixed-point filters, within the validity of the model for roundoff error that we have been using, the output noise is independent of the frequency and amplitude of the input signal. Thus for this choice of inputs, the noise-to-signal ratio obtained for the first-order filter is

$$\frac{\sigma_f^2}{\sigma_y^2} = \frac{1}{24} \cdot 2^{-2b} \frac{1}{1 - |\alpha|^2} \tag{9.37}$$

If, as before, we let $\alpha = 1 - \delta$, then for $\delta \ll 1$,

$$\frac{\sigma_f^2}{\sigma_y^2} = \frac{1}{48} \frac{2^{-2b}}{\delta} \tag{9.38}$$

In this case the noise-to-signal ratio is proportional to $1/\delta$ rather than $1/\delta^2$, so that if δ is multiplied by $\frac{1}{4}$ and the register length is increased by one bit, the noise-to-signal ratio will remain constant. We can consider the second-order case in a similar manner. Again for a sinusoidal input, the output with maximum amplitude has the form $y(n) = \cos(n\omega_0 + \phi)$, so the noise-to-signal ratio in this case is

$$\frac{\sigma_f^2}{\sigma_y^2} = \tfrac{1}{12} \cdot 2^{-2b} \frac{1+r^2}{1-r^2} \frac{1}{1+r^4-2r^2\cos 2\theta} \tag{9.39}$$

Again, choosing $r = 1 - \delta$ with $\delta \ll 1$,

$$\frac{\sigma_f^2}{\sigma_y^2} \cong \frac{2^{-2b}}{4\delta \sin^2 \theta} \tag{9.40}$$

As with the first-order filter, the noise-to-signal ratio is proportional to $1/\delta$ rather than $1/\delta^2$. The comparison of the noise-to-signal ratio for a white-noise input and a sinusoidal input serves to illustrate the dependence of the effect of dynamic range considerations on the particular form of the input. In some sense the two cases considered represent extremes. As the input becomes more confined to a known narrow band of frequencies, the above analysis with a sinusoidal input would be more representative, and as the input becomes more wide band, the analysis with a white-noise input is more representative.

In the above discussion, the noise-to-signal ratio for the case of white-noise input was derived assuming that the input amplitude was small enough to prevent overflow in the most general case. In a practical case a scaling of the input on the basis of Eq. (9.26) can be considered to be pessimistic, since the probability of equality being attained in Eq. (9.25) is extremely small. It is common practice to permit the amplitude to be somewhat larger than dictated by Eq. (9.26) and, if the result of an addition overflows, to clamp the output at the maximum value in the same manner in which we assumed that input samples were clamped in Sec. 9.2. This approach is commonly referred to as *saturation arithmetic*. Saturation of the filter, of course, represents a distortion, and the choice of scaling on the input will depend on how often such distortion is permissible.

The style of analysis illustrated by the previous discussion can be applied to investigate the effects of quantization in systems defined by higher-order linear difference equations. However, it is difficult to obtain results that are widely applicable. This is because the quantization effects are highly dependent on the properties of the desired transfer function and on the specific structure used to realize that transfer function.

As an example consider the two realizations of a second-order system shown in Fig. 9.9. The noise sources in the two systems are manifest in the output in different ways. Thus the two systems will in general have different ooise-to-signal ratios at the output for the same input signal. However, it is not possible to determine which form is to be preferred unless the parameters are known.

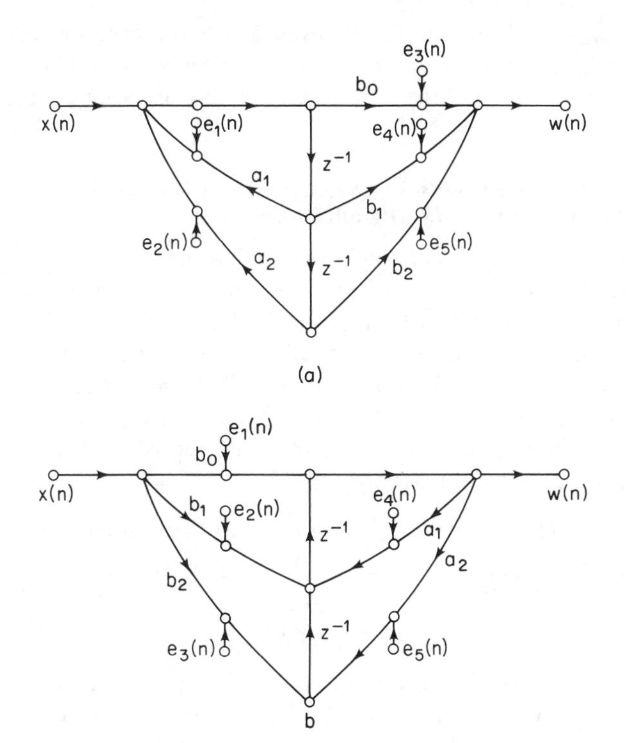

Fig. 9.9 Statistical models for roundoff noise in second-order IIR systems: (a) poles before zeros; (b) zeros before poles.

For higher-order systems, the parallel-form structure is the simplest to analyze. In this case an analysis of first- and second-order systems (including zeros) is adequate to determine the noise-to-signal ratio at the output, since the quantization effects are assumed independent between sections. However, even in this case, the noise-to-signal ratio depends on the structure of the second-order sections. The cascade form presents an even more difficult problem, since the order in which poles and zeros are arranged can have a great effect on the overall noise-to-signal ratio because noise generated in a particular second-order section is filtered by all the succeeding sections. Thus there arises the interesting problem of determining the best pairing of zeros with poles and the best ordering of the resulting second-order sections so that the output noise-to-signal ratio is minimized. This problem is complicated by the fact that signals must be scaled so that overflow does not occur at any point in the chain of second-order systems. A detailed analysis of quantization effects in cascade and parallel realizations is given by Jackson [13,14]. The indication from this work is that for the parallel form, there is little dependence on the form used to realize the second-order sections, i.e., Fig. 9.9(a) or (b). However, for the cascade form, there is some dependence on the form of the second-order sections. Also the parallel structure seems to be slightly better than the best ordering of sections in the cascade form.

9.3.3 *Statistical Analysis of Quantization in Floating-Point Realizations of IIR Digital Filters*

From the preceding discussion it is clear that the limited dynamic range of fixed-point arithmetic results in the need for careful scaling of inputs and intermediate signal levels in fixed-point digital filter realizations. The need for such scaling can be essentially eliminated by using floating-point arithmetic. However, noise is introduced due to both multiplications and additions [15–18]. As discussed in Sec. 9.1, if $Q[x(n)]$ represents the result of applying rounding to the mantissa of a signal represented in floating-point form, then we can express $Q[x(n)]$ as

$$Q[x(n)] = x(n)(1 + \varepsilon(n)) = x(n) + x(n)\varepsilon(n) \tag{9.41}$$

where, with the length of the mantissa as $(b + 1)$ bits, the relative error satisfies

$$-2^{-b} < \varepsilon(n) \leq 2^{-b} \tag{9.42}$$

Thus, in a filter implemented with floating-point arithmetic, we can represent the effect of quantization by an additive error term $e(n) = x(n)\varepsilon(n)$. Again, consider the first-order filter depicted in Fig. 9.10. Figure 9.10(a) shows the first-order system assuming infinite-precision arithmetic. Thus $y(n)$ is the exact output corresponding to input $x(n)$. Figure 9.10(b) depicts the system when floating-point quantizers are inserted after the multiplication and addition to account for the rounding of the mantissa. In this case $w(n)$ represents the noisy output due to input $x(n)$. In Fig. 9.10(c) the quantizers are replaced by additive noise sources appropriate for floating-point rounding. It is important to note that if $e_1(n)$ and $e_2(n)$ are known, Fig. 9.10(b) is equivalent to Fig. 9.10(c). In order to apply statistical analysis, we must make some assumptions about the noise sources $e_1(n)$ and $e_2(n)$. At first consideration, these assumptions may appear difficult to justify. However, as emphasized previously, we are not attempting to develop a precise analysis. Within modest bounds, the results based on the model that we will develop and use have been verified experimentally.

Let us begin by assuming that $x(n)$, the input, is a random-process signal, so that the input and output of the filter can be described in terms of averages. For convenience we assume that $x(n)$ has zero mean. We note that the error sources $e_1(n)$ and $e_2(n)$ are given by

$$e_1(n) = \varepsilon_1(n)\alpha w(n - 1) \tag{9.43}$$

and

$$e_2(n) = \varepsilon_2(n)g(n) \tag{9.44}$$

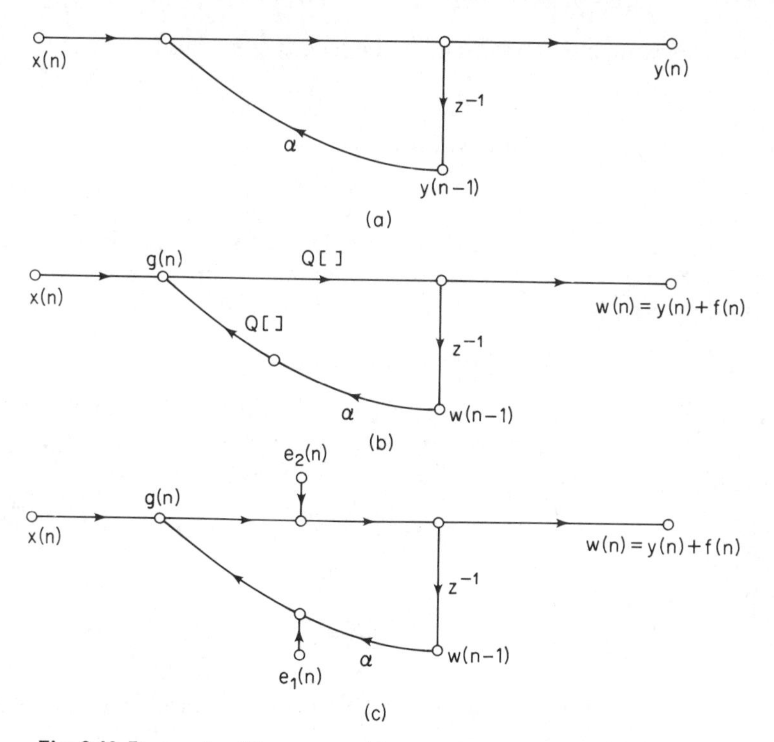

Fig. 9.10 First-order IIR systems: (a) ideal linear system; (b) nonlinear model; (c) statistical model for floating-point roundoff noise.

With no quantization, $w(n-1) = y(n-1)$ and $g(n) = y(n)$, so that if the errors are small, we can approximate $e_1(n)$ and $e_2(n)$ as

$$e_1(n) \approx \alpha\varepsilon_1(n)y(n-1) \tag{9.45}$$

$$e_2(n) \approx \varepsilon_2(n)y(n) \tag{9.46}$$

The consequence of this further approximation is to express the additive error in terms of signals in the ideal unquantized filter rather than in the actual filter. In addition to this approximation, we shall assume, as before, that the relative errors $\varepsilon_1(n)$ and $\varepsilon_2(n)$ are

1. White noise.
2. Uncorrelated with each other.
3. Uncorrelated with the input or any node variable in the system.
4. Uniformly distributed in amplitude in the range -2^{-b} to $+2^{-b}$ [see Eq. (9.8a)].

Since $\varepsilon_1(n)$ is white noise and uncorrelated with $y(n-1)$, it follows (see Problem 4 of Chapter 8) that $e_1(n)$ is white with variance

$$\sigma_{e_1}^2 = \alpha^2\sigma_{\varepsilon_1}^2 \cdot E[y^2(n-1)] \tag{9.47}$$

or, since we assume that $x(n)$, and consequently $y(n)$, has zero mean,

$$\sigma_{e_1}^2 = \alpha^2 \sigma_{\varepsilon_1}^2 \cdot \sigma_y^2 \tag{9.48}$$

Similarly, $e_2(n)$ is white with variance

$$\sigma_{e_2}^2 = \sigma_{\varepsilon_2}^2 \sigma_y^2 \tag{9.49}$$

With $h_1(n)$ and $h_2(n)$ denoting the unit sample responses from the noise source inputs to the output, and $f_1(n)$ and $f_2(n)$ denoting the error components in the output due to $e_1(n)$ and $e_2(n)$, we have

$$f(n) = f_1(n) + f_2(n)$$

Since we assumed that $\varepsilon_1(n)$ and $\varepsilon_2(n)$ are uncorrelated, it follows that $e_1(n)$ and $e_2(n)$ are also uncorrelated, as are the corresponding noise outputs $f_1(n)$ and $f_2(n)$. Thus the variance of the output noise is

$$\sigma_f^2 = \sigma_{f_1}^2 + \sigma_{f_2}^2 \tag{9.50}$$

where

$$\sigma_{f_1}^2 = \sigma_{e_1}^2 \sum_{n=-\infty}^{\infty} h_1^2(n) \tag{9.51a}$$

and

$$\sigma_{f_2}^2 = \sigma_{e_2}^2 \sum_{n=-\infty}^{\infty} h_2^2(n) \tag{9.51b}$$

Since

$$h_1(n) = h_2(n) = \alpha^n u(n) \tag{9.52}$$

it follows from Eqs. (9.48)–(9.52) that

$$\sigma_f^2 = \sigma_y^2 \frac{1}{1-\alpha^2} (\alpha^2 \sigma_{\varepsilon_1}^2 + \sigma_{\varepsilon_2}^2) \tag{9.53}$$

Finally, since $\varepsilon_1(n)$ and $\varepsilon_2(n)$ are both assumed to have a uniform probability density between -2^{-b} and $+2^{-b}$,

$$\sigma_{\varepsilon_1}^2 = \sigma_{\varepsilon_2}^2 = \tfrac{1}{3} \cdot 2^{-2b}$$

Therefore,

$$\sigma_f^2 = \tfrac{1}{3} \cdot 2^{-2b} \sigma_y^2 \frac{1+\alpha^2}{1-\alpha^2} \tag{9.54}$$

or the noise-to-signal ratio at the output is

$$\frac{\sigma_f^2}{\sigma_y^2} = \tfrac{1}{3} \cdot 2^{-2b} \frac{1+\alpha^2}{1-\alpha^2} \tag{9.55}$$

It is interesting to note that the noise-to-signal ratio as given by Eq. (9.55) for the first-order filter was derived without assuming particular spectral properties for the input. Therefore, Eq. (9.55) applies whether the input is wide band, such as white noise, or narrow band, such as a sinusoidal input. For more general filters, however, the noise-to-signal ratio will depend on the form of the input.

In a manner similar to the above, we can analyze a second-order filter. Figure 9.11(a) depicts the ideal second-order filter and Fig. 9.11(b) the filter with the noise sources included. Note that since noise sources must be included due to addition, the order in which the products are added together is important. Figure 9.11(b) depicts the case when the rounded products

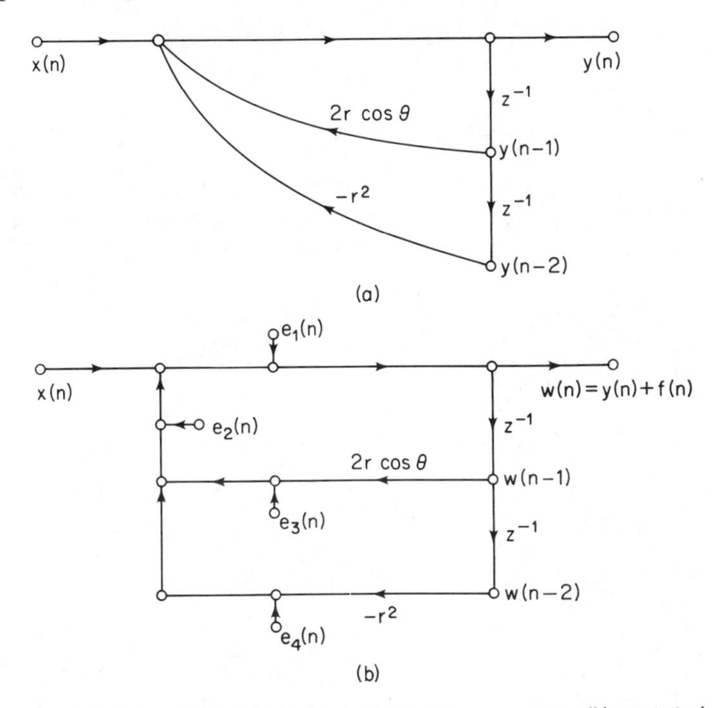

Fig. 9.11 Second-order IIR system: (a) ideal linear system; (b) statistical model for floating-point roundoff noise.

$2r\cos\theta w(n-1)$ and $-r^2w(n-2)$ are added first and then the input $x(n)$ is added to the rounded sum. The noise sources $e_3(n)$ and $e_4(n)$ represent the noise due to the multiplications, and the noise sources $e_1(n)$ and $e_2(n)$ represent the noise due to the additions. With assumptions similar to those above in which we neglected second-order terms, we write that

$$
\begin{aligned}
e_1(n) &= y(n)\varepsilon_1(n) \\
e_2(n) &= (y(n) - x(n))\varepsilon_2(n) \\
e_3(n) &= 2r\cos\theta y(n-1)\varepsilon_3(n) \\
e_4(n) &= -r^2y(n-2)\varepsilon_4(n)
\end{aligned}
\tag{9.56}
$$

Again, we assume that $\varepsilon_1(n)$, $\varepsilon_2(n)$, $\varepsilon_3(n)$, and $\varepsilon_4(n)$ are uncorrelated with each other and with $x(n)$, are white-noise sequences, and have uniform amplitude density functions. Since $\varepsilon_1(n)$, $\varepsilon_2(n)$, $\varepsilon_3(n)$, and $\varepsilon_4(n)$ are white and uncorrelated with $x(n)$ [and consequently also $y(n)$], it follows, as in the first-order example, that $e_1(n)$, $e_2(n)$, $e_3(n)$, and $e_4(n)$ are white. The variances of $e_1(n)$, $e_2(n)$, $e_3(n)$, and $e_4(n)$ are given by

$$\begin{aligned}\sigma_{e_1}^2 &= E[y^2(n)] \cdot \sigma_{\varepsilon_1}^2 \\ \sigma_{e_2}^2 &= E[(y(n) - x(n))^2] \cdot \sigma_{\varepsilon_2}^2 \\ \sigma_{e_3}^2 &= 4r^2 \cos^2 \theta E[y^2(n-1)] \cdot \sigma_{\varepsilon_3}^2 \\ \sigma_{e_4}^2 &= r^4 E[y^2(n-2)] \cdot \sigma_{\varepsilon_4}^2\end{aligned} \tag{9.57}$$

Since each of the four noise sources is assumed uncorrelated, the variance of the output noise sequence, $f(n)$, is the sum of the variances due to each of the roundoff noise sources. Thus the output noise variance is

$$\sigma_f^2 = \sigma_{e_1}^2 \sum_{n=-\infty}^{+\infty} h_1^2(n) + \sigma_{e_2}^2 \sum_{n=-\infty}^{+\infty} h_2^2(n) + \sigma_{e_3}^2 \sum_{n=-\infty}^{+\infty} h_3^2(n) + \sigma_{e_4}^2 \sum_{n=-\infty}^{+\infty} h_4^2(n)$$

From Fig. 9.11 we note that

$$h_1(n) = h_2(n) = h_3(n) = h_4(n) = \frac{1}{\sin \theta} r^2 \sin (n+1)\theta u(n) \tag{9.58}$$

and thus

$$\begin{aligned}\sum_{n=-\infty}^{+\infty} h_1^2(n) = \sum_{n=-\infty}^{+\infty} h_2^2(n) = \sum_{n=-\infty}^{+\infty} h_3^2(n) &= \sum_{n=-\infty}^{+\infty} h_4^2(n) \\ &= \frac{1+r^2}{1-r^2} \frac{1}{r^4 + 1 - 2r^2 \cos 2\theta}\end{aligned} \tag{9.59}$$

Denoting the right-hand side of Eq. (9.59) by G,

$$\sigma_f^2 = G(\sigma_{e_1}^2 + \sigma_{e_2}^2 + \sigma_{e_3}^2 + \sigma_{e_4}^2)$$

Next we take account of the fact that

1. $\sigma_{\varepsilon_1}^2 = \sigma_{\varepsilon_2}^2 = \sigma_{\varepsilon_3}^2 = \sigma_{\varepsilon_4}^2 = \frac{1}{3} \cdot 2^{-2b}$
2. $E[y^2(n)] = E[y^2(n-1)] = E[y^2(n-2)] = \sigma_y^2$
3. $E[(y(n) - x(n))^2] = E[y^2(n)] + E[x^2(n)] - 2E[y(n)x(n)]$
   $= \sigma_y^2 + \sigma_x^2 - 2E[y(n)x(n)]$

to write that

$$\sigma_f^2 = \tfrac{1}{3} \cdot 2^{-2b} G \sigma_y^2 (2 + r^4 + 4r^2 \cos^2 \theta) + \tfrac{1}{3} \cdot 2^{-2b} G(\sigma_x^2 - 2E[y(n)x(n)]) \tag{9.60}$$

Without any further assumptions, we cannot reduce Eq. (9.60) further. However, if we assume that the input is white, then

$$\sigma_y^2 = \sigma_x^2 \sum_{n=-\infty}^{+\infty} h^2(n) = G\sigma_x^2 \tag{9.61}$$

Finally (see Problem 9 of Chapter 8), $E[y(n)x(n)] = h(0)\sigma_x^2$ or, since $h(0) = 1$,

$$E[y(n)x(n)] = \sigma_x^2 \tag{9.62}$$

With these results, Eq. (9.60) can be rewritten as

$$\begin{aligned} \sigma_f^2 &= \tfrac{1}{3} \cdot 2^{-2b} G\sigma_y^2(2 + r^4 + 4r^2 \cos^2 \theta) - \tfrac{1}{3} \cdot 2^{-2b} G\sigma_x^2 \\ &= \tfrac{1}{3} \cdot 2^{-2b} \sigma_y^2 \{(2 + r^4 + 4r^2 \cos^2 \theta)G - 1\} \end{aligned} \tag{9.63}$$

For the high-gain case, it is possible to compare fixed-point and floating-point arithmetic by approximating the expressions for the noise-to-signal ratio. For the first-order case, with $\alpha = 1 - \delta$ and $|\delta| \ll 1$, Eq. (9.55) for the first-order filter with floating-point arithmetic can be approximated as

$$\frac{\sigma_f^2}{\sigma_y^2} \simeq \tfrac{1}{3} \cdot 2^{-2b} \frac{1}{\delta} \tag{9.64}$$

Similarly, for the second-order filter with $r = 1 - \delta$ and $\delta \ll 1$,

$$\frac{\sigma_f^2}{\sigma_y^2} \simeq \tfrac{1}{3} \cdot 2^{-2b} \frac{3 + 4 \cos^2 \theta}{4\delta \sin^2 \theta} \tag{9.65}$$

For fixed-point arithmetic we recall that for a white-noise input, the noise-to-signal ratio behaved as $1/\delta^2$ and, for a sinusoidal input, as $1/\delta$. Comparison of Eqs. (9.64) and (9.65) with Eqs. (9.34) and (9.36) indicates a significantly smaller noise-to-signal ratio for floating-point arithmetic as compared with fixed-point arithmetic with a white-noise input. It is important to keep in mind that the noise-to-signal ratios for the fixed-point filters were computed on the basis that the input signal was as large as possible. If the input signal level decreases, the noise-to-signal ratio will increase since the output noise variance is independent of the input signal level. For floating-point arithmetic, on the other hand, the output noise variance is proportional to the output signal variance, and as the input level is scaled up or down, so is the roundoff noise. It is also important to note that the comparison just discussed assumes that the floating-point mantissa is equal in length to the entire fixed-point word and does not account for the extra bits needed for the characteristic.

As in the fixed-point case, the analysis of higher-order systems becomes very complicated. The quantization noise is dependent on the system parameters and on the structure used to implement the system. In general, the comments at the end of Sec. 9.3.2 apply also to floating-point realizations of digital filters. However, since there is essentially no dynamic range problem with floating-point arithmetic, cascade realizations are probably not too sensitive to the ordering of poles and zeros.

9.4 Finite-Register-Length Effects in Realizations of FIR Digital Filters

The style of analysis developed in the previous section can also be applied to the study of quantization effects in FIR digital filters. In some respects this analysis is simpler than the analysis of IIR filters. For example, there are no limit cycle effects with nonrecursive realizations such as the direct form or cascade form, since these structures have no feedback. If the unit-sample response is of length N samples, the output of a direct or cascade form nonrecursive FIR realization must be zero after the input has been zero for N consecutive samples. However, recursive realizations of FIR systems such as the frequency-sampling structure are subject to the problems discussed in Sec. 9.3, and the analysis of second-order systems applies to the frequency-sampling form [19].

Dynamic range and roundoff noise are important considerations in FIR systems just as they are in IIR systems. In this section we shall examine some of these effects for fixed-point and floating-point realizations in the direct and cascade forms. As before, our intent is to give some insight into some of the quantization effects in FIR digital filters by considering some simple examples and some general results that can be obtained in a simple way.

9.4.1 *Statistical Analysis of Quantization in Fixed-Point Realizations of FIR Digital Filters*

Consider a linear shift-invariant system with unit-sample response $h(n)$, which is nonzero only for $0 \leq n \leq N-1$. The direct-form realization of such a system is a direct realization of the convolution sum relation

$$y(n) = \sum_{k=0}^{N-1} h(k)x(n-k)$$

The flow graph for the direct-form realization is shown in Fig. 9.12(a). Figure 9.12(b) shows the same structure with noise sources added to account for the effects of rounding the products $h(k)x(n-k)$. A constant gain, A, is applied to the input for purposes of preventing overflow. As before we assume that

1. The sources $e_k(n)$ are white-noise sources.
2. The errors are uniformly distributed over one quantization interval.
3. The error sources are uncorrelated with the input and each other.

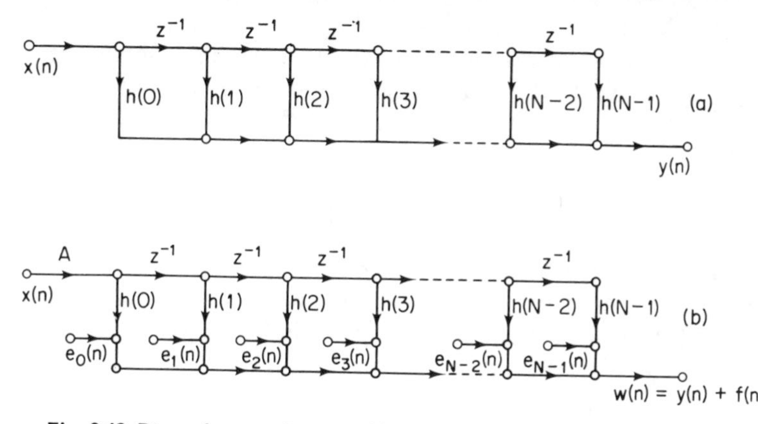

Fig. 9.12 Direct-form realization of an FIR system: (a) ideal linear system; (b) statistical model for fixed-point roundoff noise.

First, assume that $A = 1$. Then it is clear from Fig. 9.12(b) that each noise source adds directly to the output, and therefore the output noise is

$$f(n) = \sum_{k=0}^{N-1} e_k(n)$$

Because the noise sources are assumed independent, the variance of the output noise is (for rounding)

$$\sigma_f^2 = N\frac{2^{-2b}}{12}$$

and the mean is zero. Note that for the direct form, the output noise level is independent of the filter parameters since the noise is not processed by the system at all. Also, note that the noise is proportional to N, the length of the unit-sample response sequence.

The dynamic range limitation of fixed-point arithmetic necessitates scaling of the input so that no overflow occurs. We have seen [Eq. (9.25)] that a least upper bound on the output of a linear shift-invariant system is

$$|y(n)| \le x_{\max} \sum_{n=-\infty}^{\infty} |h(n)|$$

where $x_{\max}$ is the maximum magnitude of the input signal. In order to guarantee no overflow, $|y(n)| < 1$ for all n. This means that the gain at the input should satisfy

$$A < \frac{1}{x_{\max} \sum_{n=0}^{N-1} |h(n)|} \tag{9.66}$$

Such scaling would be appropriate for a wide-band signal, such as white noise, but it would be too conservative for a narrow-band signal, such as a sine wave. In the latter case we have seen that the input should be scaled in terms of the peak of the frequency response of the system. Thus an alternative choice of the input scaling coefficient is

$$A < \frac{1}{x_{\max} \cdot \max_{0 \le \omega \le \pi} [|H(e^{j\omega})|]} \tag{9.67}$$

It should be noted (see Problem 2 of this chapter) that with two's-complement arithmetic, when more than two numbers are to be added together as in Fig. 9.12, overflows can occur in computing partial sums, yet the correct final

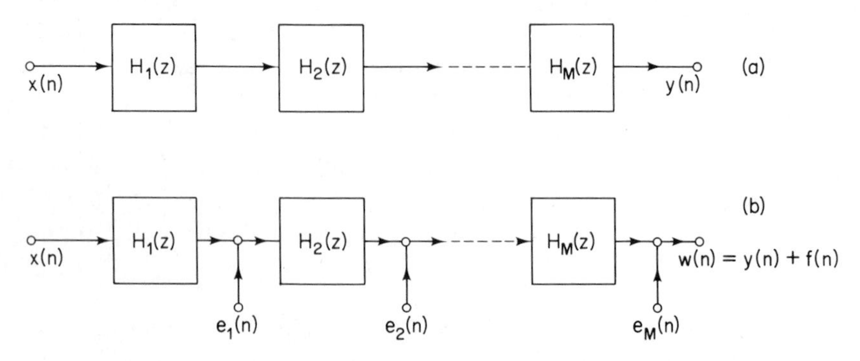

Fig. 9.13 Cascade realization of an FIR system: (a) ideal linear system; (b) statistical model for fixed-point roundoff noise.

result will be obtained if that correct result is less than one. Thus input scaling according to Eq. (9.66) will always give correct outputs for the direct-form realization (except, of course, for roundoff noise).

An FIR digital filter can also be realized as a cascade of second-order sections as in Fig. 9.13(a), where each second-order section $H_k(z)$ is realized in direct form as in Fig. 9.12(a). We assume for convenience that N is odd so that $M = (N - 1)/2$. Since each second-order section has three independent white-noise sources at its output, the effects of quantization can be depicted as in Fig. 9.13(b), where each source $e_i(n)$ has variance $3(2^{-2b}/12) = 2^{-2b}/4$. In this case a given noise source $e_k(n)$ is filtered by succeeding sections, so that the output noise variance will be dependent upon the order of the second-order sections in the chain. If we define $g_i(n)$ to be the unit-sample response from noise source $e_i(n)$ to the output, we can write

$$\sigma_{e_i}^2 = \frac{2^{-2b}}{4}\left(\sum_{n=0}^{N-2i} g_i^2(n)\right) \tag{9.68}$$

and the total output noise variance is

$$\sigma_f^2 = \sum_{i=1}^{M} \sigma_{e_i}^2 = \frac{2^{-2b}}{4}\left(\sum_{i=1}^{M}\sum_{n=0}^{N-2i} g_i^2(n)\right) \tag{9.69}$$

In the cascade structure it is necessary that there be no overflow at the output of any second-order section in order that the final output be correct. Thus it is necessary to introduce scaling at the input to each section. This is done as discussed before for the general direct-form realization.

From Eqs. (9.68) and (9.69) it is clear that the output noise depends on the order of sections. With M sections, there are $M!$ possible orderings, so that for large M, an exhaustive search is impractical. However, based on experimental measurements on low-order filters, Chan and Rabiner [20,21,22] have found that the majority of orderings have reasonably good noise properties, and they have given an algorithm for choosing a good ordering. Their results indicate that a good ordering is one for which the frequency response of the transfer function from each noise source to the output is relatively flat and for which the peak gain is small.

9.4.2 *Statistical Analysis of Quantization Effects in Floating-Point Realizations of FIR Digital Filters*

Floating-point arithmetic can essentially eliminate concern for dynamic range in FIR digital filters. This is often a convenient approach when a

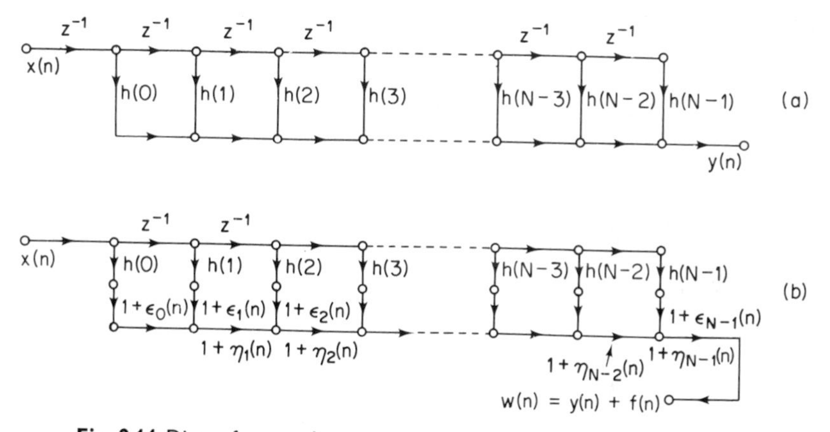

Fig. 9.14 Direct-form realization of an FIR system: (a) ideal linear system; (b) statistical model for floating-point roundoff noise.

filter is to be realized as a general-purpose computer program where ease of programming rather than computational speed is the main consideration. However, the extra complexity of floating-point arithmetic often cannot be justified in special-purpose hardware realizations, where economy is a major concern.

Consider a direct-form realization of an FIR system as depicted in Fig. 9.14(a). Figure 9.14(b) depicts the system that results from using finite-precision floating-point arithmetic. The time-varying coefficients $[1 + \varepsilon_k(n)]$

and $[1 + \eta_k(n)]$ represent the effects of rounding of floating-point products and sums, respectively. As before, the $\varepsilon_k(n)$'s and $\eta_k(n)$'s are assumed to be independent, uniformly distributed, white-noise sequences. In this case the shift-varying model leads to a more compact set of equations than a shift-invariant model with additive noise sources. Our analysis in this section is similar to that of Liu and Kaneko [16], who based a general analysis of direct-form IIR systems on a model of this type.

From Fig. 9.14 we can express the output $w(n)$ as

$$\begin{aligned} w(n) = (1 + \varepsilon_0(n)) \prod_{r=1}^{N-1} (1 + \eta_r(n))h(0)x(n) \\ + (1 + \varepsilon_1(n)) \prod_{r=1}^{N-1} (1 + \eta_r(n))h(1)x(n-1) + \ldots \\ + (1 + \varepsilon_k(n)) \prod_{r=k}^{N-1} (1 + \eta_r(n))h(k)x(n-k) + \ldots \\ + (1 + \varepsilon_{N-1}(n))(1 + \eta_{N-1}(n))h(N-1)x(n-N+1) \\ = \sum_{k=0}^{N-1} A(n, k)h(k)x(n-k) \end{aligned} \tag{9.70a}$$

where

$$A(n, 0) = (1 + \varepsilon_0(n)) \prod_{r=1}^{N-1} (1 + \eta_r(n)) \tag{9.70b}$$

$$A(n, k) = (1 + \varepsilon_k(n)) \prod_{r=k}^{N-1} (1 + \eta_r(n)), \qquad k \neq 0 \tag{9.70c}$$

If we assume that $w(n) = y(n) + f(n)$, then from Eq. (9.70a) it follows that

$$f(n) = \sum_{k=0}^{N-1} [A(n, k) - 1]h(k)x(n-k) \tag{9.71}$$

It can be shown that the quantity $[A(n, k) - 1]$ has zero mean, so that the output noise has zero mean. The general expression for the variance of the output noise is, therefore,

$$\begin{aligned} \sigma_f^2 &= E[f^2(n)] \\ &= E\left[\sum_{k=0}^{N-1} \sum_{l=0}^{N-1} (A(n, k) - 1)(A(n, l) - 1)x(n-k)x(n-l)h(k)h(l)\right] \\ &= \sum_{k=0}^{N-1} \sum_{l=0}^{N-1} E[(A(n, k) - 1)(A(n, l) - 1)]h(k)h(l)\phi_{xx}(l-k) \end{aligned} \tag{9.72}$$

If the input is a random signal with flat power spectrum and variance σ_x^2, then Eq. (9.72) becomes

$$\begin{aligned}\sigma_f^2 &= \sigma_x^2 \sum_{k=0}^{N-1} h^2(k) E[(A(n, k) - 1)^2] \\ &= \sigma_x^2 \sum_{k=0}^{N-1} h^2(k)\{E[A^2(n, k)] - 1\}\end{aligned} \tag{9.73}$$

From Eqs. (9.70b) and (9.70c) and the assumptions on the quantities $\varepsilon_k(n)$ and $\eta_k(n)$,

$$E[A^2(n, 0)] = \left(1 + \frac{2^{-2b}}{3}\right)^N \tag{9.74a}$$

$$E[A^2(n, k)] = \left(1 + \frac{2^{-2b}}{3}\right)^{N+1-k}, \qquad k \neq 0 \tag{9.74b}$$

We note that only a slight error is made by using Eq. (9.74b) for $k = 0$. Also, since $2^{-2b}/3 \ll 1$ in any reasonable situation, we can express Eq. (9.74b) using a binomial approximation as

$$E[A^2(n, k)] = 1 + (N + 1 - k)\frac{2^{-2b}}{3} \tag{9.75}$$

Thus σ_f^2 can be expressed as

$$\sigma_f^2 = (N + 1)\frac{2^{-2b}}{3}\sigma_x^2 \sum_{k=0}^{N-1} h^2(k)\left(1 - \frac{k}{N + 1}\right) \tag{9.76}$$

A number of conclusions can be drawn from Eq. (9.76). First, it is clear that the noise power at the output is (as in the fixed-point case) proportional to N. Second, we note that

$$\sigma_x^2 \sum_{k=0}^{N-1} h^2(k)\left(1 - \frac{k}{N + 1}\right) < \sigma_y^2$$

so that the noise-to-signal ratio at the output for an input with a flat spectrum is bounded as in

$$\frac{\sigma_f^2}{\sigma_y^2} \leq (N + 1)\frac{2^{-2b}}{3} \tag{9.77}$$

Finally, we recall from Fig. 9.14(b) that products were computed and partial sums were accumulated in order of increasing k. Also, it is clear from Eq. (9.76) that those products formed first are the ones that are most heavily weighted by the factor $(1 - k/(N + 1))$. Thus the minimum output noise variance is obtained if the unit-sample response satisfies

$$|h(0)| < |h(1)| < \ldots < |h(N - 1)|$$

Generally this is not the case. However, by adding the products in order of increasing magnitude of the unit-sample response, we can expect to achieve the smallest average error in a floating-point realization. This is in contrast to a fixed-point realization, where the error is independent of the order of the additions. Performing the multiplications and partial sums in nonsequential order generally will result in additional software or hardware complexity; however, it may be worthwhile when the highest possible accuracy is required. Alternative orderings of the additions can also reduce the output noise (see Problem 18 at the end of this chapter).

9.5 Effects of Finite Register Length in Discrete Fourier Transform Computations [19, 24–26]

Since the discrete Fourier transform is widely used in practice for digital filtering and spectrum analysis, it is important to understand the effects of finite register length in DFT calculations. As in the case of digital filters, however, a precise analysis of the effects is difficult, and often a simplified analysis is sufficient for the purpose of choosing the required register length for computing the discrete Fourier transform. The analysis that we shall present is similar in style to that carried out in previous sections. Specifically, we shall analyze arithmetic roundoff by means of an additive noise source at each point in the computation algorithm where roundoff occurs. Furthermore, we shall make a number of assumptions to simplify the analysis. The results that we shall obtain lead to several simplified but useful rules of thumb for the effect of arithmetic roundoff. The analysis that follows is phrased in terms of rounding. As for the analysis of roundoff errors in digital filters, the results can generally be modified for the case of two's-complement truncation with fixed-point arithmetic and for one's-complement or sign-and-magnitude truncation with floating-point arithmetic.

We have seen in Chapter 6 that there are numerous approaches to DFT computation. In this section we shall first consider errors in the direct evaluation of the DFT relation and then illustrate the effects of roundoff in a particular class of FFT algorithms.

9.5.1 *Analysis of Quantization in Computation of the DFT*

The discrete Fourier transform is defined by the equation

$$X(k) = \sum_{n=0}^{N-1} x(n) W_N^{kn}, \qquad k = 0, 1, \ldots, N-1 \tag{9.78}$$

where $W_N = e^{-j(2\pi/N)k}$. Although Eq. (9.78) is generally evaluated by one of the algorithms known collectively as the *fast Fourier transform*, there are instances where straightforward accumulation of the products in Eq. (9.78) is the most reasonable approach (e.g., when only a few values of k are required). The analysis of this computational procedure is quite simple and serves as an introduction to quantization effects in DFT calculations.

We note that for a given value of k, Eq. (9.78) is completely analogous to the convolution sum expression

$$y(n) = \sum_{k=0}^{N-1} h(k)x(n - k)$$

which was the basis for the discussion of the previous section. In this case the quantities W_N^{kn} play the role of the unit-sample response, $X(k)$ plays the role

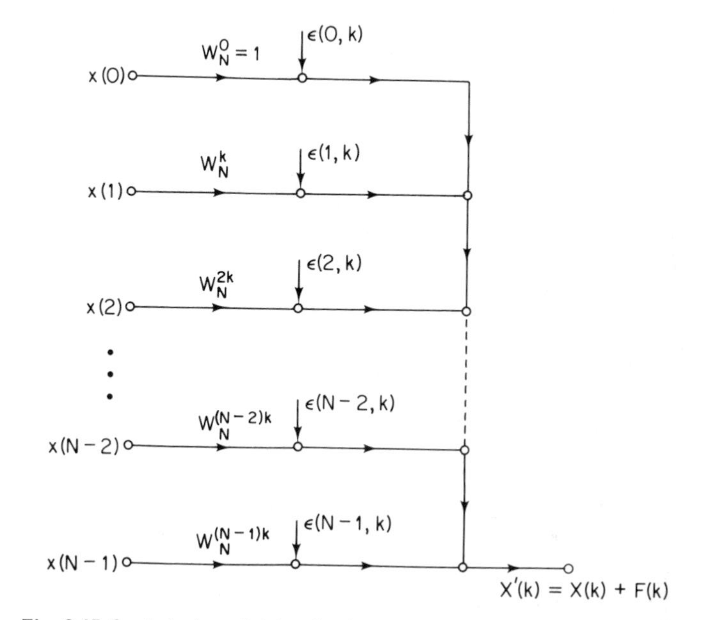

Fig. 9.15 Statistical model for fixed-point roundoff noise in computation of the DFT.

of the output, and $x(n)$ is the input. Note that all these quantities are generally complex. Thus an analysis similar to that of Sec. 9.4 can be used for the direct computation of the DFT, with the additional consideration that the errors, in this case, are complex sequences.

For fixed-point arithmetic, one approach to the direct calculation of $X(k)$ is depicted by the flow graph of Fig. 9.15. In the figure $X'(k)$ represents the result of finite-precision computation of the DFT and $F(k)$ represents the error in computation of the kth value. The complex quantities $\varepsilon(n, k)$ represent the errors due to rounding of the products $x(n)W_N^{kn}$. It can be seen that these complex errors add directly to the output, so that

$$F(k) = \sum_{n=0}^{N-1} \varepsilon(n, k) \tag{9.79}$$

The product $x(n)W_N^{kn}$ is

$$x(n)W_N^{kn} = \text{Re}\,[x(n)]\cos\left(\frac{2\pi}{N}kn\right) + \text{Im}\,[x(n)]\sin\left(\frac{2\pi}{N}kn\right) + j\left\{\text{Im}\,[x(n)]\cos\left(\frac{2\pi}{N}kn\right) - \text{Re}\,[x(n)]\sin\left(\frac{2\pi}{N}kn\right)\right\}$$

With finite-precision fixed-point arithmetic, the rounded complex product can be represented as

$$\begin{aligned} Q[x(n)W_N^{kn}] = {} & \text{Re}\,[x(n)]\cos\left(\frac{2\pi}{N}kn\right) + \varepsilon_1(n, k) \\ & + \text{Im}\,[x(n)]\sin\left(\frac{2\pi}{N}kn\right) + \varepsilon_2(n, k) \\ & + j\,\text{Im}\,[x(n)]\cos\left(\frac{2\pi}{N}kn\right) + \varepsilon_3(n, k) \\ & - j\,\text{Re}\,[x(n)]\sin\left(\frac{2\pi}{N}kn\right) + \varepsilon_4(n, k) \end{aligned}$$

That is, each real multiplication contributes a roundoff error.† To compute the variance of the error in $X'(k)$, we must make a number of assumptions regarding the errors. Specifically, we assume that the errors due to each real multiplication have the following properties:

1. The errors are uniformly distributed random variables over the range $-\frac{1}{2}\cdot 2^{-b}$ to $\frac{1}{2}\cdot 2^{-b}$. Therefore, each error source has variance $2^{-2b}/12$.
2. The errors are uncorrelated with one another.
3. All the errors are uncorrelated with the input and consequently also with the output.

The mean of the error due to rounding a complex multiplication is zero. Since the squared magnitude of the complex error $\varepsilon(n, k)$ is

$$|\varepsilon(n, k)|^2 = [\varepsilon_1(n, k) + \varepsilon_2(n, k)]^2 + [\varepsilon_3(n, k) + \varepsilon_4(n, k)]^2$$

the average value of $|\varepsilon(n, k)|^2$ is

$$E[|\varepsilon(n, k)|^2] = 4\cdot\frac{2^{-2b}}{12} = \frac{1}{3}\cdot 2^{-2b} \tag{9.80}$$

The average magnitude squared of the output error is

$$E[|F(k)|^2] = \sum_{n=0}^{N-1} E[|\varepsilon(n, k)|^2] = \frac{N}{3}2^{-2b} \tag{9.81}$$

† Note that we have assumed that the coefficients W_N^{kn} are represented exactly. The effect of quantizing these coefficients is discussed in Sec. 9.5.3.

As in the case of the direct-form realization of an FIR filter, the output noise is proportional to N.†

As with the fixed-point direct-form realization of an FIR filter, the direct DFT calculation is subject to a dynamic range limitation. From Eq. (9.78) we see that

$$|X(k)| \le \sum_{n=0}^{N-1} |x(n)| < N$$

For no overflow to occur we require that $|X(k)| < 1$. This is assured if

$$\sum_{n=0}^{N-1} |x(n)| < 1 \tag{9.82}$$

Thus, in the worst case, we may need to divide the input by N to prevent overflow. For example, the sequence $x(n) = 1$, $0 \le n \le N - 1$, has the discrete Fourier transform

$$X(k) = \begin{cases} N, & k = 0 \\ 0, & \text{otherwise} \end{cases}$$

On the other hand, $|X(k)|$ may be less than 1, even though Eq. (9.82) is not satisfied. Consider, for example, the sequence $x(n) = A\delta(n)$, which has DFT $X(k) = A$ for $0 \le k \le N - 1$. If $A > 0$ and only slightly less than 1, Eq. (9.82) is not satisfied, but $|X(k)| = A < 1$.

There are a number of solutions to the dynamic range problem. We can divide the input by N, thus increasing the noise-to-signal ratio at the output. We can use a block-floating-point scheme where division by 2 is performed whenever an overflow occurs. We can use floating-point arithmetic, in which case overflow is virtually eliminated. All these will be examined in detail for a class of FFT algorithms. In addition, we shall comment on the error introduced by quantizing the coefficient values.

9.5.2 *Analysis of Quantization Effects in Fixed-Point FFT Algorithms [23]*

There are many different FFT algorithms, and the detailed effects of quantization depend upon the specific algorithm that is used. The most commonly used algorithms are the radix 2 forms, for which the size of the transform that is computed is an integer power of 2. For the most part, the discussion below is phrased in terms of the decimation-in-time form of the radix 2 algorithm. The results, however, are applicable with only minor modification to the decimation-in-frequency form. Furthermore, most of the ideas employed in the error analysis of the radix 2 algorithms can be utilized in other algorithms.

† Note that Eq. (9.81) is a conservative estimate since some of the multiplications (e.g., by W_N^0) can be done without error.

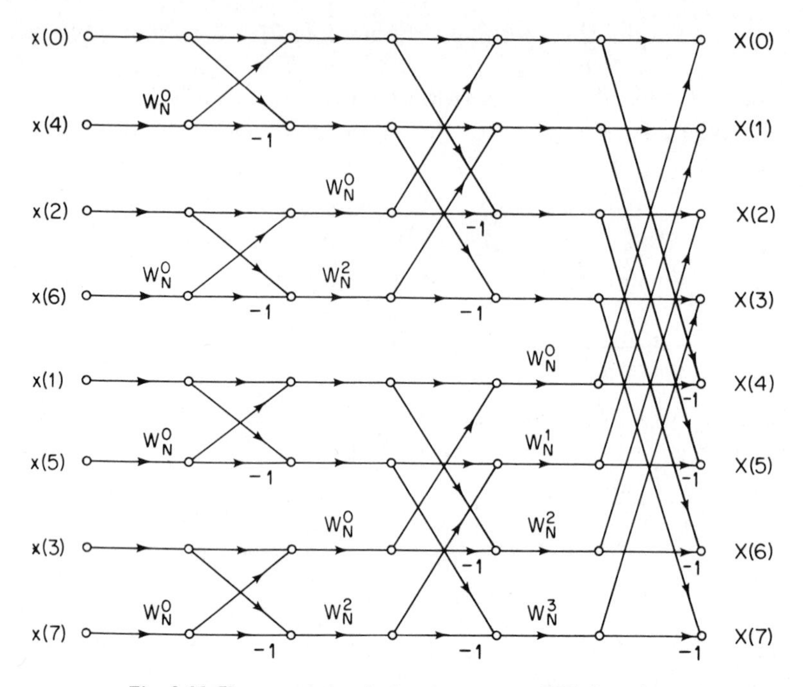

Fig. 9.16 Flow graph for decimation-in-time FFT algorithm.

FFT algorithms are directed toward computing $X(k)$, the DFT of a finite-duration sequence $x(n)$, defined as in Eq. (9.78). A flow chart depicting a decimation-in-time algorithm for $N = 8 = 2^3$ is shown in Fig. 9.16. (An implementation of this particular form of the algorithm was used for the reported experimental work.) There are some key aspects of this diagram, which, as we recall from Chapter 6, are common to all standard radix 2 algorithms. The DFT is computed in $\nu = \log_2 N$ stages. At each stage a new array of N numbers is formed from the previous array by linear combinations of the elements taken two at a time. The νth array contains the desired DFT. The basic numerical computation operates on a pair of numbers in the mth array to produce a pair of numbers in the $(m + 1)$st array. This computation, called a *butterfly*, is

$$\begin{aligned} X_{m+1}(p) &= X_m(p) + W_N^r X_m(q) \\ X_{m+1}(q) &= X_m(p) - W_N^r X_m(q) \end{aligned} \tag{9.83}$$

Here the subscripts m and $m + 1$ refer to the mth array and the $(m + 1)$st array, respectively, and p and q denote the location of the numbers in each array. (Note that $m = 0$ refers to the input array and $m = \nu$ refers to the output array.) A flow graph representing the butterfly computation is shown in Fig. 9.17.

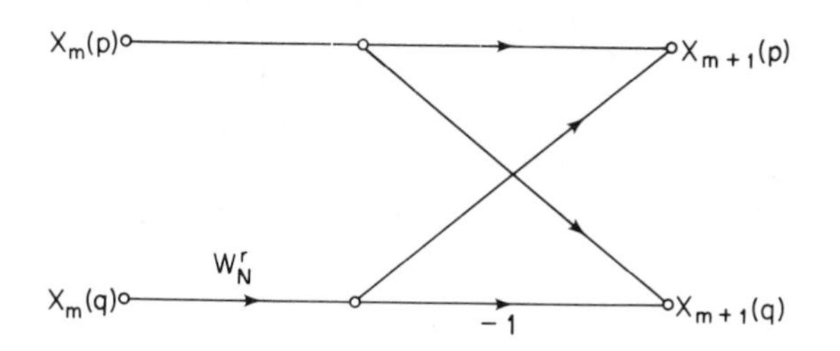

Fig. 9.17 Butterfly computation for decimation-in-time.

The form of the butterfly computation is somewhat different for a radix 2 decimation in frequency algorithm, where the basic computation is

$$\begin{aligned} X_{m+1}(p) &= X_m(p) + X_m(q) \\ X_{m+1}(q) &= [X_m(p) - X_m(q)]W_N^r \end{aligned} \tag{9.84}$$

At each stage, $N/2$ separate butterfly computations are carried out to produce the next array. The integer r varies with p, q, and m in a manner that depends on the specific form of FFT algorithm that is used. Fortunately, our analysis

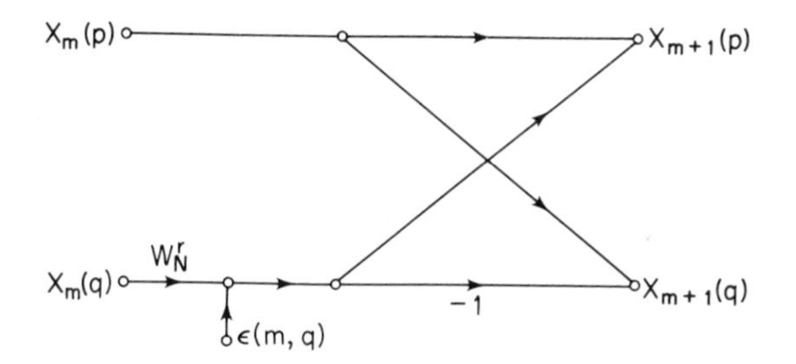

Fig. 9.18 Statistical model for fixed-point roundoff noise in a decimation-in-time butterfly computation.

is not tied to the specific way in which r varies. Also, the specific relationship among p, q, and m, which determines how we index through the mth array, is not important for the analysis. The details of the analysis for decimation-in-time and decimation-in-frequency differ somewhat due to the different butterfly forms, but the basic results do not change significantly. In our analysis we shall assume a butterfly of the form of Eq. (9.83), corresponding to decimation-in-time.

We shall model the roundoff noise by associating an additive noise generator with each fixed-point multiplication. With this model the butterfly of Fig. 9.17 is replaced by that of Fig. 9.18 for analyzing the roundoff noise effects. By the notation $\varepsilon(m, q)$ we have explicitly denoted the fact that this quantity represents a complex error introduced in computing the $(m + 1)$st array from the mth array and specifically in multiplication of the qth element of the mth array by a complex coefficient.

Since we shall assume that in general the input to the FFT is a complex sequence, each of the multiplications is complex and thus, in fact, consists of four real multiplications in exactly the same manner discussed in the previous section. We again make the assumptions:

1. The roundoff noise due to each real multiplication is uniformly distributed in amplitude between $\pm\frac{1}{2}\cdot 2^{-b}$ and thus has a variance of $\sigma_\varepsilon^2 = \frac{1}{12}\cdot 2^{-2b}$.
2. All the noise sources due to each real multiplication are uncorrelated with each other. Thus for any given complex multiplication, the four noise components are uncorrelated with each other. Furthermore, they are uncorrelated with the noise components from the other complex multiplications.
3. All the noise sources are uncorrelated with the input and consequently also with the results of the computation in each array.

Since each of these four sequences is uncorrelated, zero-mean white noise and all have the same variance, we have, as in Eq. (9.80),

$$E[|\varepsilon(m, q)|^2] = \frac{1}{3}\cdot 2^{-2b} \tag{9.85}$$

We shall denote this variance by σ_B^2. To calculate the mean-square value of the output noise at any output node, we must account for the contribution from each of the noise sources that propagate to that node. We can make the following observations from the flow graph of Fig. 9.16:

1. The transmission function from any node in the flow graph to any other node to which it is connected is multiplication by a complex constant of unity magnitude (because each branch transmission is either unity or an integer power of W_N).
2. Each output node connects to $7 = (N - 1)$ butterflies in the flow graph. For example, Fig. 9.19(a) shows the flow graph with all the butterflies removed that do not connect to $X(0)$, and Fig. 9.19(b) shows the flow graph with all the butterflies removed that do not connect to $X(2)$.

The above observations can be generalized to the case of N an arbitrary power of 2.

As a consequence of the first observation, the mean-square value of the magnitude of the component of the output noise due to each elemental noise source is the same and equal to σ_B^2. The total output noise at each output node is equal to the sum of the noise propagated to that node. Since we assume that all the noise sources are uncorrelated, the mean-square value of

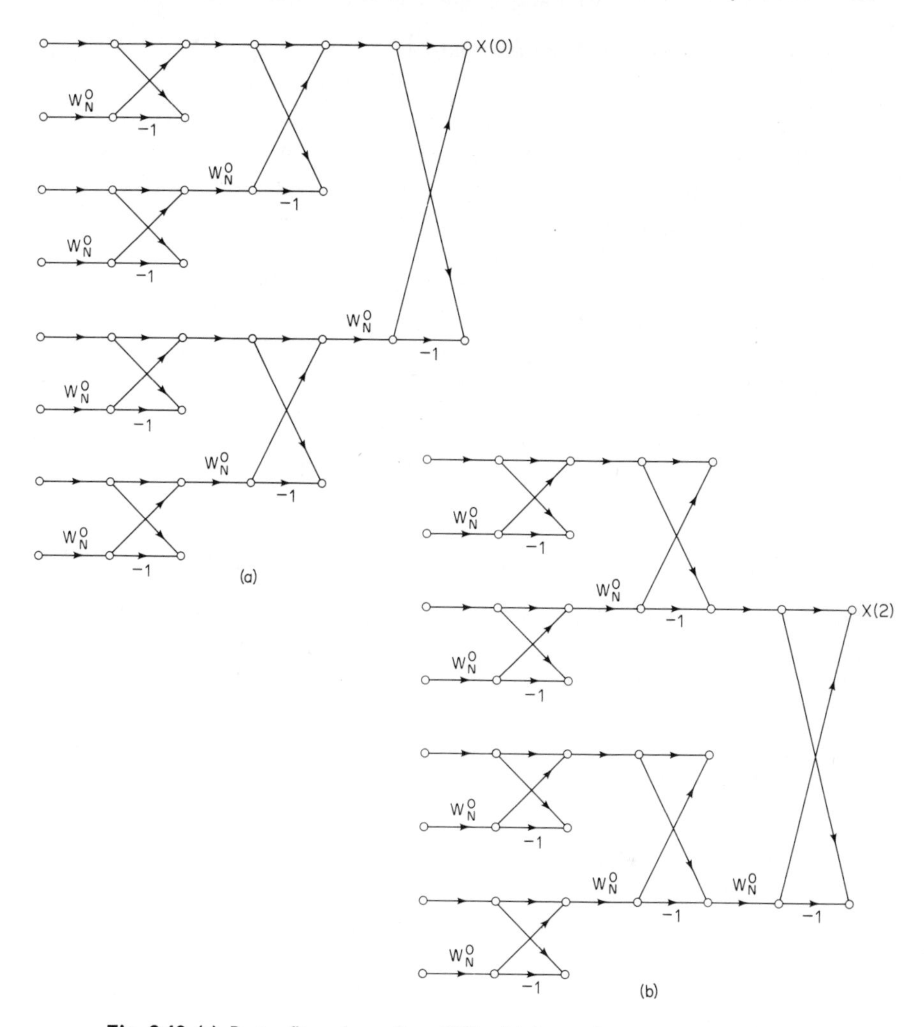

Fig. 9.19 (a) Butterflies that affect $X(0)$; (b) butterflies that affect $X(2)$.

the magnitude of the output noise is equal to σ_B^2 times the number of noise sources that propagate to that node. At most one complex noise source is introduced at each butterfly; consequently, from observation 2 above, at most $(N - 1)$ noise sources propagate to each output node. In fact, all the butterflies do not generate roundoff noise since some (for example, all those in the first and second stages) involve only multiplication by unity. However, if we assume that roundoff occurs for each butterfly, we can consider the result as an upper bound on the output noise. With this assumption, then, the mean-square value of the output noise in the kth DFT value, $F(k)$, is given by

$$E[|F(k)|^2] = (N - 1)\sigma_B^2 \tag{9.86}$$

which, for large N, we shall approximate as

$$E[|F(k)|^2] \cong N\sigma_B^2 \tag{9.87}$$

According to this result, the mean-square value of the output noise is proportional to N, the number of points transformed. The effect of doubling N, or adding another stage in the FFT, is to double the mean-square value of the output noise. In Problem 9.19 we consider the modification of this result when we do not insert noise sources for those butterflies that involve only multiplication by unity or j.

In implementing an FFT algorithm with fixed-point arithmetic we must ensure against overflow. From Eq. (9.83) it follows that (see Problem 2 of Chapter 6)

$$\max(|X_m(p)|, |X_m(q)|) \leq \max(|X_{m+1}(p)|, |X_{m+1}(q)|) \tag{9.88}$$

and also that

$$\max(|X_{m+1}(p)|, |X_{m+1}(q)|) \leq 2\max(|X_m(p)|, |X_m(q)|) \tag{9.89}$$

Equation (9.88) implies that the maximum modulus is nondecreasing from stage to stage so that, if the magnitude of the output of the FFT is less than unity, then the magnitude of the points in each array must be less than unity†; i.e., there will be no overflow in any of the arrays.

To express this constraint as a bound on the input sequence, we recall from the previous section that the condition

$$|x(n)| < \frac{1}{N}, \qquad 0 \leq n \leq N-1 \tag{9.90}$$

is both necessary and sufficient to guarantee that

$$|X(k)| < 1, \qquad 0 \leq k \leq N-1$$

Thus Eq. (9.90) is sufficient to guarantee no overflow for all stages of the algorithm.

To obtain an explicit expression for the noise-to-signal ratio at the output of the FFT algorithm, taking into account the required scaling, consider an input in which successive sequence values are uncorrelated, i.e. a white input signal. Also, assume that the real and imaginary parts of the input sequence are uncorrelated and that each has an amplitude density that is uniform between $-1/(\sqrt{2}N)$ and $+1/(\sqrt{2}N)$. [Note that this signal satisfies Eq. (9.90).]

† Actually one should discuss overflow in terms of the real and imaginary parts of the data rather than the magnitude. However, $|x| < 1$ implies that $|\mathrm{Re}\,(x)| < 1$ and $|\mathrm{Im}\,(x)| < 1$, and only a slight increase in allowable signal level is achieved by scaling on the basis of Re and Im parts.

Then the average squared magnitude of the complex input sequence is

$$E[|x(n)|^2] = \frac{1}{3N^2} = \sigma_x^2 \tag{9.91}$$

The DFT of the input sequence is

$$X(k) = \sum_{n=0}^{N-1} x(n)W^{kn}$$

from which it can be shown that, under the above assumptions on the input,

$$\begin{aligned} E[|X(k)|^2] &= \sum_{k=0}^{N-1} E[|x(n)|^2]\,|W^{kn}|^2 \\ &= N\sigma_x^2 = \frac{1}{3N} \end{aligned} \tag{9.92}$$

Combining Eqs. (9.87) and (9.92) we obtain

$$\frac{E[|F(k)|^2]}{E[|X(k)|^2]} = 3N^2\sigma_B^2 = N^2 2^{-2b} \tag{9.93}$$

Thus, according to the result of Eq. (9.93), the output noise-to-signal ratio is proportional to N^2. Since σ_B^2 is proportional to 2^{-2b}, Eq. (9.93) can be interpreted to imply that the noise-to-signal ratio increases as N^2 or one bit per stage. That is, if N is doubled, corresponding to adding one additional stage to the FFT, then to maintain the same noise-to-signal ratio, one bit must be added to the register length. The assumption of a white signal input is, in fact, not critical here. For a variety of other inputs, the noise-to-signal ratio is still proportional to N^2, with only the constant of proportionality changing.

Equation (9.89) suggests an alternative scaling procedure. Since the maximum modulus increases by no more than a factor of 2 from stage to stage, we can prevent overflow by requiring that $|x(n)| < 1$ and incorporating an attenuation of $\frac{1}{2}$ at the input to each stage. In this case the output will consist not of the DFT as defined by Eq. (9.78) but of $1/N$ times this DFT. Although the mean-square output signal will be $1/N$ times that if no scaling was introduced, the input amplitude can be N times larger without causing overflow. Thus the maximum output magnitude that can be attained (for the white input signal) is the same as before. However, the output noise level will be much less than in Eq. (9.87), since the noise introduced at early stages of the FFT will be attenuated by the scaling that takes place in the later arrays. Specifically with scaling by $\frac{1}{2}$ introduced at the input to each butterfly, we modify the butterfly of Fig. 9.18 to that of Fig. 9.20, where, in particular, there are now two noise sources associated with each butterfly. As before, we assume that the real and imaginary parts of these noise sources are uncorrelated and are also uncorrelated with the other noise sources, and that the real and imaginary parts are uniformly distributed between $\pm\frac{1}{2}\cdot 2^{-b}$.

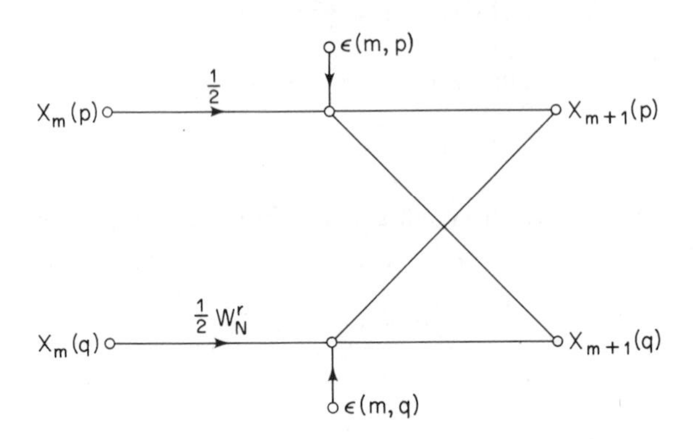

Fig. 9.20 Butterfly showing scaling multipliers and associated fixed-point roundoff noise.

Thus, as before,

$$E[|\varepsilon(m, q)|^2] = \sigma_B^2 = \tfrac{1}{3} \cdot 2^{-2b} = E[|\varepsilon(m, p)|^2]$$

Because the noise sources are all uncorrelated, the mean-square magnitude of the noise at each output node is again the sum of the contributions of each noise source in the flow graph. However, as opposed to the previous case, the attenuation that each noise source experiences through the flow graph depends on the array at which it originates. A noise source originating at the mth array will propagate to the output with multiplication by a complex constant with magnitude $(\frac{1}{2})^{\nu-m-1}$. By examination of Fig. 9.16 we see that for the case $N = 8$, each output node connects to

1 butterfly originating at the $(\nu - 1)$st array
2 butterflies originating at the $(\nu - 2)$nd array
4 at the $(\nu - 3)$rd array, etc.

For the general case with $N = 2^\nu$, each output node connects to $2^{(\nu-m-1)}$ butterflies and therefore to $2^{(\nu-m)}$ noise sources that originate at the mth array. Thus at each output node, the mean-square magnitude of the noise is

$$\begin{aligned} E[|F(k)|^2] &= \sigma_B^2 \sum_{m=0}^{\nu-1} 2^{(\nu-m)} (\tfrac{1}{2})^{(2\nu-2m-2)} \\ &= \sigma_B^2 \sum_{m=0}^{\nu-1} (\tfrac{1}{2})^{(\nu-m-2)} \\ &= \sigma_B^2 \cdot 2 \sum_{k=0}^{\nu-1} (\tfrac{1}{2})^k \\ &= 2\sigma_B^2 \frac{1 - (\frac{1}{2})^\nu}{1 - \frac{1}{2}} = 4\sigma_B^2 (1 - (\tfrac{1}{2})^\nu) \end{aligned} \tag{9.94}$$

For N large we shall assume that $(\frac{1}{2})^{\nu}$ is negligible compared to unity, so that

$$E[|F(k)|^2] \cong 4\sigma_B^2 = \tfrac{4}{3} \cdot 2^{-2b} \tag{9.95}$$

and thus is much less than the noise variance resulting when all the scaling is carried out on the input data.

Now, we can combine Eq. (9.95) with Eq. (9.92) to obtain the output noise-to-signal ratio for the case of step-by-step scaling and white input. We obtain

$$\frac{E[|F(k)|^2]}{E[|X(k)|^2]} = 12N\sigma_B^2 = 4N \cdot 2^{-2b} \tag{9.96}$$

a result proportional to N rather than to N^2. An interpretation of Eq. (9.96) is that the output noise-to-signal ratio increases as N, or by half a bit per stage, a result first obtained by Welch [23]. It is important to note again that the assumption of white signal is not essential in the analysis. The basic result of half-a-bit-per-stage increase holds for a broad class of signals, with only the constant multiplier in Eq. (9.96) being signal dependent.

We should also note that the dominant factor that causes the increase of the noise-to-signal ratio with N is the decrease in signal level (required by the overflow constraint) as we pass from stage to stage. According to Eq. (9.95), very little noise (only a bit or two) is present in the final array. Most of the noise has been shifted off by the scalings.

We have assumed straight fixed-point computation in the above discussion; i.e., only preset attenuations were allowed, and we were not permitted to rescale on the basis of an overflow test. Clearly, if the hardware or programming facility is such that straight fixed-point computation must be used, we should, if possible, incorporate attenuators of $\frac{1}{2}$ at each array rather than use a large attenuation of the input array.

A third approach to avoiding overflow is the use of block floating point. In this procedure the original array is normalized to the far left of the computer word, with the restriction that $|x(n)| < 1$; the computation proceeds in a fixed-point manner, except that after every addition there is an overflow test. If overflow is detected, the entire array is divided by 2 and the computation continues. The number of necessary shifts are counted to determine a scale factor or exponent for the entire final array. The output noise-to-signal ratio depends strongly on how many overflows occur and at what stages of the computation they occur. The positions and timing of overflows are determined by the signal being transformed, and thus in order to analyze the noise-to-signal ratio in a block-floating-point implementation of the FFT, one needs to know the properties of the input signal.

As we have seen, it is possible to find an input that requires no scaling, and it is also possible to find an input that requires division by N to prevent overflow. The case of a white input signal might be expected to provide an example somewhere between these two extremes; i.e., scaling at all stages might not generally be necessary. This problem has been analyzed theoretically [19] but the analysis is quite involved and will not be presented here. Instead, we shall present some experimental results.

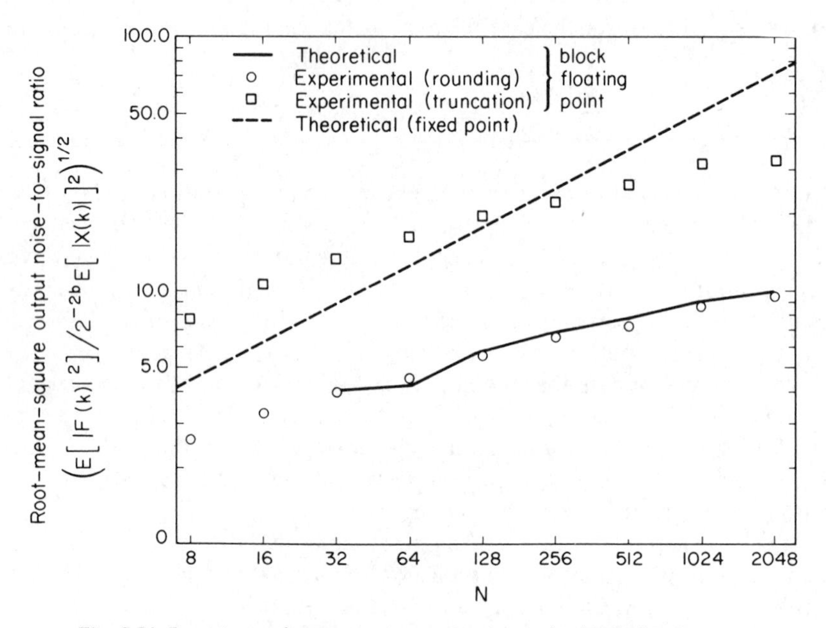

Fig. 9.21 Experimental output noise-to-signal ratio for block-floating-point realization of the FFT.

Figure 9.21 illustrates the dependence of the output noise-to-signal ratio on N. This figure shows experimentally measured values of output noise-to-signal ratio for block-floating-point transforms of white input signals using rounding arithmetic [26]. For comparison, the theoretical curve representing the fixed-point noise-to-signal ratio corresponding to Eq. (9.96) is also shown. We see that for this kind of input, block floating point provides some advantages over fixed point, especially for the larger transforms. For $N = 2048$ the noise-to-signal ratio for block floating point is about $\frac{1}{8}$ that of fixed point, representing a three-bit improvement.

An experimental investigation by Weinstein [26] was used to examine how the results for block floating point change when truncation rather than rounding is used. The results of this experiment are also shown in Fig. 9.21. Noise to-signal ratios are generally slightly worse than for rounding. The rate of increase of noise-to-signal ratio with N seems to be about the same as for rounding.

9.5.3 *Analysis of Quantization Effects in Floating-Point FFT Algorithms*

The effect of arithmetic roundoff in implementing the FFT with floating-point arithmetic has been analyzed theoretically and experimentally by Gentleman and Sande [24], Weinstein [26], and Kaneko and Liu [25]. As with the analysis of roundoff errors for fixed-point arithmetic, noise is introduced due to each butterfly computation. As with floating-point errors in digital filters, we neglect second-order error terms so that noise sources are introduced after each multiplication and addition. These noise sources are assumed to be uncorrelated, and their variance is proportional to the variance of the signal at the node. Kaneko and Liu have obtained detailed formulas for a general statistical model of the input signal. However, unless the input is assumed to be white, the analysis and the results are quite complicated. Therefore, to indicate the nature of roundoff noise for a floating-point implementation of an FFT algorithm, we shall consider only the case of white input signals.

Figure 9.22 depicts the top half of a typical butterfly computation, including real noise sources $e_1(m, q), \ldots, e_8(m, q)$ due to the four real multiplications and four real additions. The complex noise at the output of the

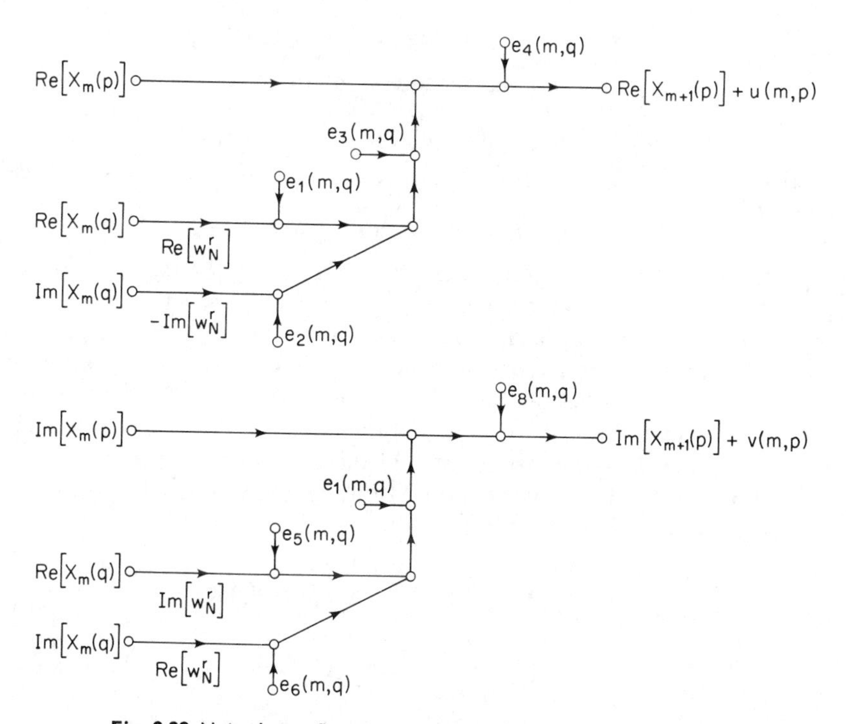

Fig. 9.22 Noisy butterfly computation for floating-point arithmetic.

butterfly is

$$s(m, p) = u(m, p) + jv(m, p)$$

According to our previous discussion of floating-point models, $e_1(m, q), \ldots, e_4(m, q)$ can be expressed as

$$\begin{aligned} e_1(m, q) &= \varepsilon_1(m, q) \operatorname{Re}[W_N^r] \operatorname{Re}[X_m(q)] \\ e_2(m, q) &= -\varepsilon_2(m, q) \operatorname{Im}[W_N^r] \operatorname{Im}[X_m(q)] \\ e_3(m, q) &= \varepsilon_3(m, q) \operatorname{Re}[W_N^r X_m(q)] \\ e_4(m, q) &= \varepsilon_4(m, q)\{\operatorname{Re}[X_m(p)] + \operatorname{Re}[W_N^r X_m(q)]\} \end{aligned} \tag{9.97}$$

Similar expressions hold for e_5, e_6, e_7, and e_8. We assume that the white-noise sources $\varepsilon_i(m, q)$ all have the same variance, which we will denote by σ_ε^2, and that they are uncorrelated with each other. Furthermore, since the input signal is white, with equal variance in the real and imaginary parts, the real and imaginary parts of the signals at each array are white with equal variances. Thus we can write

$$E[(\operatorname{Re}[X_m(q)])^2] = E[\operatorname{Im}[X_m(q)])^2] = \tfrac{1}{2}E[|X_m(q)|^2] \tag{9.98}$$

It follows from Eqs. (9.97) and (9.98) that

$$\begin{aligned} \sigma_{e_1}^2 + \sigma_{e_2}^2 = \sigma_{e_5}^2 + \sigma_{e_6}^2 = \sigma_{e_3}^2 = \sigma_{e_7}^2 &= \tfrac{1}{2}\sigma_e^2 E[|X_m(q)|^2] \\ \sigma_{e_4}^2 = \sigma_{e_8}^2 &= \sigma_e^2 E[|X_m(q)|^2] \end{aligned}$$

The mean-square values of $u(m, p)$ and $v(m, p)$ are then

$$E[(v(m, p))^2] = E[(u(m, p))^2] = 2\sigma_e^2 E[X_m(q)]$$

so that the mean-square magnitude of the complex output noise source $s(m, p)$ is

$$E[|s(m, p)|^2] = 4\sigma_e^2 E[|X_m(q)|^2]$$

Thus the variance of the noise generated in computing the $(m + 1)$st array is $4\sigma_e^2$ times the mean-square magnitude of the signal in the mth array. If the input, i.e., the zeroth array, is white with mean-square magnitude $E[|x(n)|^2]$, then the noise generated in computing the $(m + 1)$st array is $2^m E[|x(n)|^2](4\sigma_\varepsilon^2)$. As before, each output node connects to $2^{(\nu-m-1)}$ butterflies that originate at the mth array and thus to $2^{(\nu-m-1)}$ noise sources that originate at the $(m + 1)$st array. Each of these noise sources propagates to the output through a complex constant of unity magnitude. Thus the mean square magnitude of the output noise is

$$\begin{aligned} E[|F(k)|^2] &= \sum_{m=0}^{\nu-1} 2^{(\nu-m-1)} 2^m E[|x(n)|^2](4\sigma_\varepsilon^2) \\ &= \sum_{m=0}^{\nu-1} \frac{N}{2} E[|x(n)|^2](4\sigma_\varepsilon^2) \\ &= 2\nu N \sigma_\varepsilon^2 E[|x(n)|^2] \end{aligned} \tag{9.99}$$

The mean-square magnitude of the output signal is

$$E[|X(k)|^2] = E[|X(n)|^2]N$$

and thus the output noise-to-signal ratio is

$$\frac{E[|F(k)|^2]}{E[|X(k)|^2]} = 2\nu\sigma_\varepsilon^2 \tag{9.100}$$

We observe from Eq. (9.100) that the noise-to-signal ratio is proportional to ν, while in the fixed-point case it was proportional to $N = 2^\nu$. Since σ_ε^2 is proportional to 2^{-2b}, quadrupling ν (i.e., raising N to the fourth power) results in an increase in noise-to-signal ratio of one bit. Thus, as one would expect, the increase of noise-to-signal ratio as a function of N for the floating-point case is significantly milder than for the fixed-point case.

In the analysis leading to Eq. (9.100), we have not considered the fact that multiplications by unity can be performed noiselessly. For a specified radix 2 algorithm, such as the decimation-in-time algorithm shown in Fig.

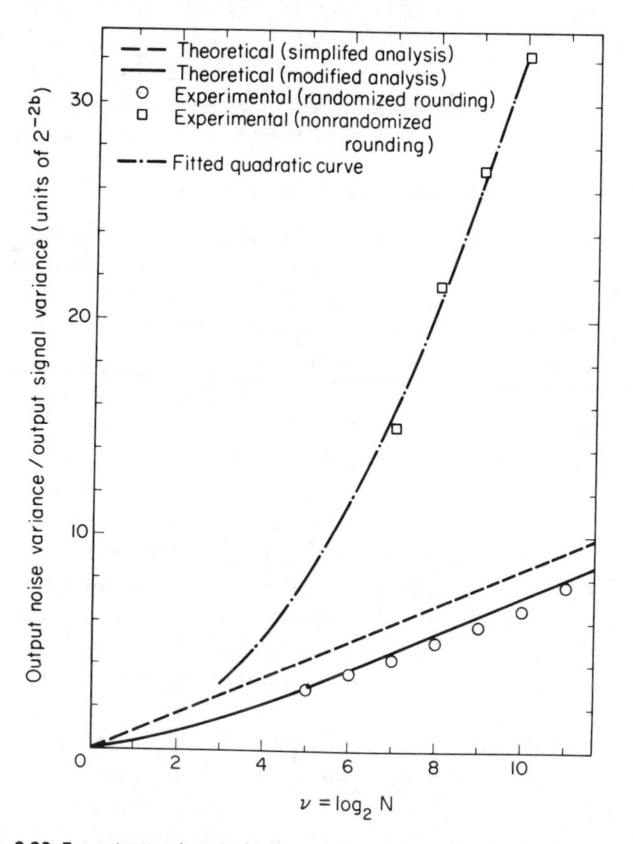

Fig. 9.23 Experimental and theoretical noise-to-signal ratio for floating-point FFT computations.

9.16, these reduced variances for $W_N^r = 1$ and j can be included in the model to obtain a slightly reduced prediction for output noise-to-signal ratio. However, for reasonably large N, this modified noise analysis yields only slightly better predictions of output noise than does the simplified analysis above.

The results discussed above have been verified by Weinstein [26] with excellent agreement, as shown in Fig. 9.23. To obtain this agreement, however, it was necessary to use randomized rounding, i.e., randomly rounding up or down when the value of mantissa was exactly $\frac{1}{2} \cdot 2^{-b}$. The modified theoretical curve shown was obtained by taking into account reduced noise source variances for $W^r = 1$ and $W^r = j$. Also shown are experimental results for nonrandomized rounding. These results were fitted empirically with a curve of the form $a\nu^2$, but this quadratic dependence was not established theoretically.

The present discussion, and all the above experiments, applied to the case of a white signal. Some experimental investigation has been carried out as to whether the predictions are valid beyond this case. Specifically, the noise introduced in computing an FFT was measured for sinusoidal signals of several frequencies, for $\nu = 8, 9, 10$, and 11. The results, averaged over the input frequencies used, were within 15% of those predicted by Eq. (9.100). In these experiments the "randomized" rounding procedure was used.

9.5.4 Effects of Coefficient Quantization in the FFT

As with the implementation of digital filters, the implementation of a fast Fourier transform algorithm requires the use of quantized coefficients. Although the nature of coefficient quantization is inherently nonstatistical, Weinstein [26] has obtained some useful results by means of a rough statistical analysis. For this analysis, jitter is added to each coefficient; i.e., each coefficient is replaced by its true value plus a white-noise sequence. Although the detailed effect of coefficient error due to quantization is different from that due to jitter, it is reasonable to expect that in a gross sense the magnitudes of the errors are comparable. The result obtained is that the ratio of mean-square magnitude output error to mean-square output signal is $(\nu/6) \cdot 2^{-2b}$.

Although this does not predict with great accuracy the error in an FFT algorithm due to coefficient quantization, it is helpful as a rough estimate of the error. The key result, which has been tested experimentally, is that the error-to-signal ratio increases very mildly with N, being proportional to $\nu = \log_2 N$, so that doubling N produces only a slight increase in the error-to-signal ratio.

The experimental results are displayed in Fig. 9.24; the quantity plotted is 2^{2b} times the ratio of the mean-square magnitude output error to mean-square output signal. The theoretical curve is shown, and the circles represent measured output error-to-signal ratio for the fixed-point case. The experimental results generally lie below the theoretical curve. No experimental result differs by as much as a factor of 2 from the theoretical result, and since a factor of 2 in error-to-signal ratio corresponds to only half-a-bit difference in the output error, it seems that the analysis provides a reasonably accurate

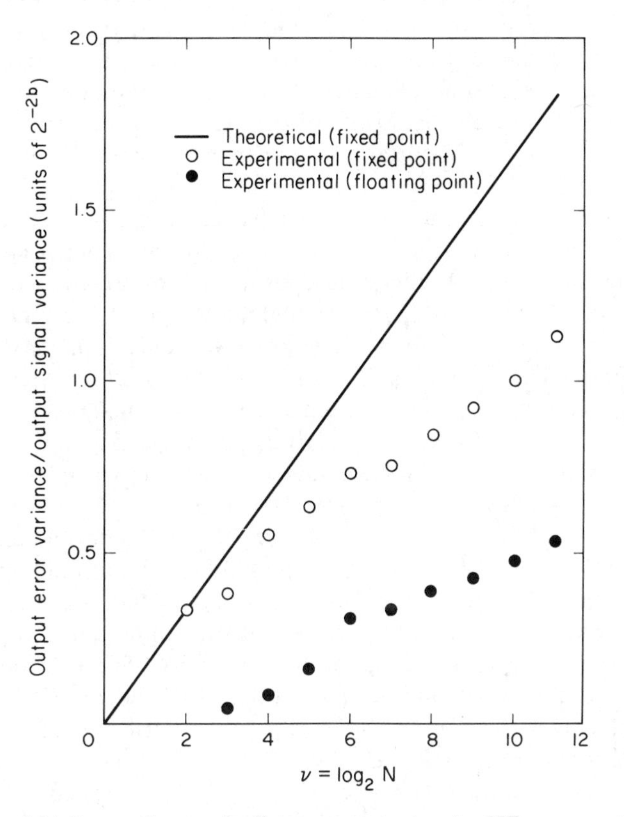

Fig. 9.24 Errors due to coefficient quantization in FFT computations.

estimate of the effect of coefficient errors. The experimental results do seem to increase essentially linearly with ν but with smaller slope than given by the analysis.

In the above experiments fixed-point arithmetic was assumed. However, since a block-floating-point FFT will generally use fixed-point coefficients, the results are also valid for the block-floating-point case. With some slight modifications, it is possible to obtain similar results for the floating-point case. Except for a constant factor, the floating- and fixed-point results are the same. Experimental results for the floating-point case are represented by the solid dots in Fig. 9.24 and are observed to be slightly lower than the results for the fixed-point case.

Summary

In this chapter we have analyzed some of the effects of using finite-precision arithmetic in the implementation of digital filtering and fast Fourier transform algorithms. The underlying theme of our discussion has been the conflict between the desire to obtain fine quantization and wide dynamic range while holding the register length fixed. Register length is an economic factor in hardware implementations, and in software implementations it is generally forced upon us by the characteristics of the available general-purpose computer. Thus it is very important to understand quantization effects in applications of digital signal processing.

Our consideration of finite-register-length effects began with a discussion of the various types of number representations that are commonly used in implementing digital signal-processing algorithms. The subsequent discussion focused on digital filtering algorithms and FFT algorithms. Our principal objective throughout this chapter was to point out some of the problems that result from using finite-precision arithmetic and to illustrate a style of analysis that can be successfully applied in specific problems of quantization in digital signal-processing algorithms. Out of the illustrative examples that constitute this chapter there emerge a number of general principles and guidelines. We demonstrated, for example, the existence of zero-input limit cycles in recursive implementations of IIR digital filters and gave some simple formulas that predict the size of possible zero-input limit cycles. For more complex input signals, we showed how a statistical analysis can yield informative estimates of the effects of quantization in both digital filters and FFT algorithms in terms of noise-to-signal ratios. We considered both fixed-point and floating-point implementations of IIR filters, FIR filters, and FFT algorithms. A basic observation is that for both fixed- and floating-point implementations, quantization effects depend to a great extent upon the form or structure chosen to implement a particular signal-processing algorithm. Also, we found that quantization effects in floating-point implementations are dependent upon the properties of the input data, whereas this is not generally the case in fixed-point implementations.

In addition to the specific results that were obtained, many of which have wide applicability, the entire chapter serves to illustrate a style of analysis that can be applied in a study of quantization effects in a variety of digital signal processing algorithms. The use of statistical methods in problems where processes are unknown or too complex for a deterministic representation is a well-established method in many fields, including digital signal processing. Thus the examples of this chapter indicate the types of assumptions and approximations that are commonly made in studying quantization effects. Noise-analysis methods are also illustrated in review papers by Liu [27] and Oppenheim and Weinstein [28].

REFERENCES

1. I. Flores, *The Logic of Computer Arithmetic*, Prentice-Hall, Inc., Englewood Cliffs, N.J., 1963.
2. W. R. Bennett, "Spectra of Quantized Signals," *Bell System Tech. J.*, Vol. 27, 1948, pp. 446–472.
3. B. Widrow, "A Study of Rough Amplitude Quantization by Means of Nyquist Sampling Theory," *IRE Trans. Circuit Theory*, Vol. CT-3, Dec. 1956, pp. 266–276.
4. B. Widrow, "Statistical Analysis of Amplitude-Quantized Sampled-Data Systems," *AIEE Trans.* (*Appl. Indust.*), Vol. 81, Jan. 1961, pp. 555–568.
5. L. Jackson, "An Analysis of Limit Cycles Due to Multiplication Rounding in Recursive Digital Filters," in *Proc. 7th Allerton Conf. Circuit System Theory*, 1969, pp. 69–78.
6. L. B. Jackson, "An Analysis of Roundoff Noise in Digital Filters," Sc.D. Dissertation, Department of Electrical Engineering, Stevens Instititute of Technology, 1969.
7. A. R. Parker and S. F. Hess, "Limit-Cycle Oscillations in Digital Filters," *IEEE Trans. Circuit Theory*, Vol. CT-8, Nov. 1971, pp. 687–697.
8. I. W. Sandberg, "A Theorem Concerning Limit Cycles in Digital Filters," in *Proc. 7th Annual Allerton Conf. Circuit System Theory*, 1968, pp. 63–68.
9. I. W. Sandberg and J. F. Kaiser, "A Bound on Limit Cycles in Fixed-Point Implementations of Digital Filters," *IEEE Trans. Audio Electroacoust.*, Vol. AU-20, No. 2, June 1972, pp. 110–112.
10. C. Y. Kao," An Analysis of Limit Cycles Due to Sign-Magnitude Truncation in Multiplication in Recursive Digital Filters," *Proc. 5th Ansilomar Conf. Circuits Systems*, 1971.
11. R. B. Blackman, *Linear Data-Smoothing and Prediction in Theory and Practice*, Addison-Wesley Publishing Company, Inc., Reading, Mass., 1965.
12. P. M. Ebert, J. E. Mazo, and M. C. Taylor, "Overflow Oscillations in Digital Filters," *Bell System Tech. J.*, Vol. 48, 1969, pp. 2999–3020.
13. L. B. Jackson, "On the Interaction of Roundoff Noise and Dynamic Range in Digital Filters," *Bell System Tech. J.*, Vol. 49, 1970, pp. 159–184.
14. L. B. Jackson, "Roundoff-Noise Analysis for Fixed-Point Digital Filters Realized in Cascade of Parallel Form," *IEEE Trans. Audio Electroacoust.*, Vol. AU-18, June 1970, pp. 107–122.
15. E. P. F. Kan and J. K. Aggarwal, "Error Analysis of Digital Filter Employing Floating-Point Arithmetic," *IEEE Trans. Circuit Theory*, Vol. CT-18, Nov. 1971, pp. 678–686.
16. B. Liu and T. Kaneko, "Error Analysis of Digital Filters Realized with Floating-Point Arithmetic," *Proc. IEEE*, Vol. 57, Oct. 1969, pp. 1735–1747.
17. I. W. Sandberg, "Floating-Point-Roundoff Accumulation in Digital Filter Realization," *Bell System Tech. J.*, Vol. 46, Oct. 1967, pp. 1775–1791.
18. C. Weinstein and A. V. Oppenheim, "A Comparison of Roundoff Noise in Floating Point and Fixed Point Digital Filter Realizations," *Proc. IEEE* (*Lett.*), Vol. 57, June 1969, pp. 1181–1183.
19. C. J. Weinstein, "Quantization Effects in Digital Filters," *MIT Lincoln Lab. Tech. Rept. 468, ASTIA DOC. DDC AD-706862*, Nov. 21, 1969.
20. D. S. K. Chan and L. R. Rabiner, "Analysis of Quantization Errors in the Direct Form for Finite Impulse Response Digital Filters," *IEEE Trans. Audio Electroacoust.*, Vol. AU-21, No. 4, Aug. 1973, pp. 354–366.

21. D. S. K. Chan and L. R. Rabiner, "Theory of Roundoff Noise in Cascade Realizations of Finite Impulse Response Digital Filters," *Bell System Tech. J.*, Vol. 52, No. 3, Mar. 1973, pp. 329–345.
22. D. S. K. Chan and L. R. Rabiner, "An Algorithm for Minimizing Roundoff Noise in Cascade Realizations of Finite Impulse Response Digital Filters," *Bell System Tech. J.*, Vol. 52, No. 3, Mar. 1973, pp. 347–385.
23. P. D. Welch, "A Fixed-Point Fast Fourier Transform Error Analysis," *IEEE Trans. Audio Electroacoust.*, Vol. AU-17, June 1969, pp. 153–157.
24. W. M. Gentleman and G. Sande, "Fast Fourier Transforms—for Fun and Profit," in *Proc. 1966 Fall Joint Computer Conf., AFIPS Conf. Proc.*, Vol. 29, pp. 563–578, Spartan Books, Washington, D.C., 1966.
25. T. Kaneko and B. Liu, "Accumulation of Roundoff Error in Fast Fourier Transforms," *J. Assoc. Comput. Mach.*, Vol. 17, Oct. 1970, pp. 637–654.
26. C. J. Weinstein, "Roundoff Noise in Floating Point Fast Fourier Transform Computation," *IEEE Trans. Audio Electroacoust.*, Vol. AU-17, Sept. 1969, pp. 209–215.
27. B. Liu, "Effect of Finite Word Length on the Accuracy of Digital Filters—A Review," *IEEE Trans. Circuit Theory*, Vol. CT-18, Nov. 1971, pp. 670–677.
28. A. V. Oppenheim and C. J. Weinstein, "Effects of Finite Register Length in Digital Filtering and the Fast Fourier Transform," *Proc. IEEE*, Aug. 1972, pp. 957–976.
29. A. V. Oppenheim, "Realization of Digital Filters Using Block-Floating-Point Arithmetic," *IEEE Trans. Audio Electroacoust.*, Vol. AU-18, Jan. 1970, pp. 130–136.

PROBLEMS

1. In this problem we wish to consider some of the properties of the various number representations discussed in the text. Thus consider a number x such that $|x| < 1$ and further assume that $|x|$ can be represented as a binary fraction with b bits to the right of the binary point. We shall find it convenient to introduce the notation $\simeq$ to mean "is represented by." Thus, for *sign and magnitude*,

$$x \simeq \begin{cases} |x|, & x \geq 0 \\ 1 + |x|, & x \leq 0 \end{cases}$$

For *one's complement*,

$$x \simeq \begin{cases} |x|, & x \geq 0 \\ 2 - 2^{-b} - |x|, & x \leq 0 \end{cases}$$

For *two's complement*,

$$x \simeq \begin{cases} |x|, & x \geq 0 \\ 2 - |x|, & x < 0 \end{cases}$$

(a) In each of these cases, x is represented by a $(b + 1)$-bit binary number. Show that in all three cases the bit immediately to the left of the binary point (the sign bit) is 0 if $x > 0$ and 1 if $x < 0$.

(b) Show that the following algorithm is sufficient to obtain the one's-complement representation of a negative number x from its magnitude, $|x|$: "Obtain the negative of $|x|$ by changing every 1 to a 0 and every 0 to a 1, including the sign bit."

(c) Show that the following algorithms serve to obtain the two's-complement representation of a negative number x from its magnitude, $|x|$.
(1) "Find the one's complement and add 2^{-b}."
(2) "Starting at the right, examine the bits of $|x|$ in turn. For each 0 in $|x|$, place a 0 in x. When the first 1 is encountered in $|x|$, place a 1 in x. Thereafter, change each 0 to a 1 and each 1 to a zero, for all bits including the sign bit."

2. The two's-complement representation leads to very attractive simplifications of arithmetic processes. To illustrate this we shall consider some details of two's-complement addition. As in Problem 9.1 we use the notation $\simeq$ to mean "is represented by." Thus a number x has the two's-complement representation

$$x \simeq \begin{cases} |x|, & x \geq 0 \\ 2 - |x|, & x < 0 \end{cases}$$

where $|x| < 1$ and $(b + 1)$ bits are used in the representation of x. Two's complement addition is performed as follows:
(1) All numbers are treated as $(b + 1)$-bit *unsigned* binary numbers.
(2) Addition is simple binary addition.
(3) Carries past the sign bit are ignored; i.e., if the sum is greater than 2, the carry is ignored. Thus addition is implemented modulo 2.
(a) Using the above notation and definitions, write complete expressions for the two's-complement addition of two numbers x_1 and x_2, where $|x_1|$ and $|x_2| < 1$. Consider all possible cases; i.e., note that x_1 and x_2 can each be either + or − and also that $|x_1|$ can be either greater than or less than $|x_2|$.
(b) Note that when two numbers of like sign are added, the resultant magnitude may be greater than 1. This condition is called *overflow*. Show that overflow is indicated whenever the result of adding two numbers of like sign has the opposite sign.
(c) Show that two's-complement addition of x_1 and x_2 is equivalent to $f[x_1 + x_2]$, where $f[\ ]$ is depicted in Fig. P9.2.

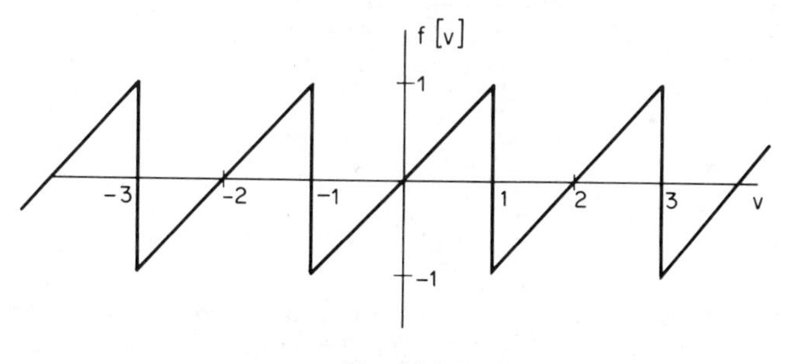

Fig. P9.2

(d) Suppose that $x_1 = \frac{5}{8}$, $x_2 = \frac{3}{4}$, and $x_3 = -\frac{1}{2}$. Find the two's complement representation of each number, and add the two's-complement representations together in the sequence $(x_1 + x_2) + x_3$. Note that overflow occurs in the addition $(x_1 + x_2)$, but the final result is correct. Show that in general any number of overflows can occur in the process of accumulating the sum of three or more two's-complement numbers, and the result will be correct if the correct sum has magnitude less than unity.

3. In a floating-point representation of numbers, the number is represented in the form

$$F = 2^c M, \qquad \tfrac{1}{2} \leq |M| < 1$$

c is commonly called the characteristic and M the mantissa. Since $\frac{1}{2} \leq |M| < 1$, it is a fixed-point fraction. Negative floating-point numbers can be represented by representing the mantissa as a floating-point fraction in either sign-and-magnitude, one's-complement, or two's-complement notation.

Consider a floating-point number F that is to be quantized by quantizing the mantissa to b bits, excluding sign, so that the value of the least significant bit in the mantissa is 2^{-b}.

Let $Q[F]$ represent the quantized value. It is convenient to represent $Q[F]$ as $Q[F] = F(1 + \varepsilon)$, so that the error $E = Q[F] - F$ is

$$E = \varepsilon F$$

(a) Assuming that F is a positive number, show that for rounding $-2^{-b} < \varepsilon \leq 2^{b}$ and for truncation of the mantissa, $-2 \cdot 2^{-b} < \varepsilon \leq 0$.

(b) Assuming that F is a negative number, determine an upper and lower bound on ε for (1) rounding, (2) sign-and-magnitude truncation, (3) one's-complement truncation, and (4) two's-complement truncation.

4. In order to process sequences on a digital computer, we must quantize the amplitude of the sequence to a set of discrete levels. This quantization can be expressed in terms of passing the input sequence $x(n)$ through a quantizer $Q(x)$ that has an input–output relationship as shown in Fig. P9.4-1.

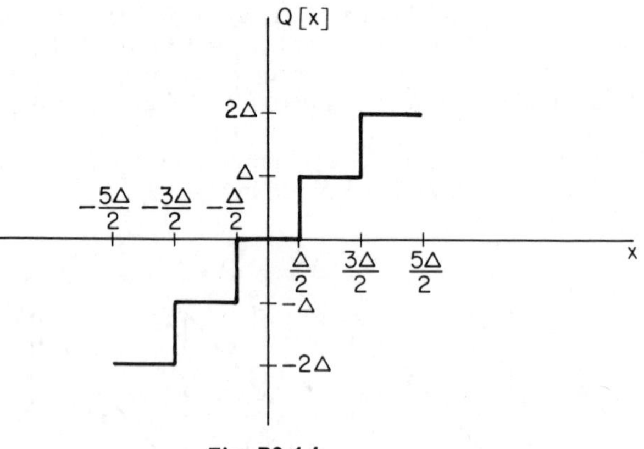

Fig. P9.4-1

If the quantization interval Δ is small compared to changes in the level of the input sequence, we can assume that the output of the quantizer $y(n)$ is of the form

$$y(n) = x(n) + e(n)$$

where $e(n) = Q[x(n)] - x(n)$ and $e(n)$ is a stationary random process with a first-order probability density uniformly distributed between $(-\Delta/2, \Delta/2)$, uncorrelated from sample to sample, and independent of $x(n)$, so that $E[e(n)x(m)] = 0$ for all m and n.

Let $x(n)$ be a stationary white-noise process, with zero mean and variance σ_x^2.

(a) Find the mean, variance, and autocorrelation sequence of $e(n)$.

(b) What is the signal-to-quantizing-noise ratio σ_x^2/σ_e^2?

(c) The quantized signal $y(n)$ is to be filtered by a digital filter with unit-sample response $h(n) = \frac{1}{2}[a^n + (-a)^n]u(n)$. Determine the variance of the noise produced at the output due to the input quantization noise and determine the signal-to-noise ratio at the output.

In some cases we may want to use nonlinear quantization steps, for example, logarithmically spaced quantization steps. This can be accomplished by applying uniform quantization to the logarithm of the input as depicted in Fig. P9.4-2,

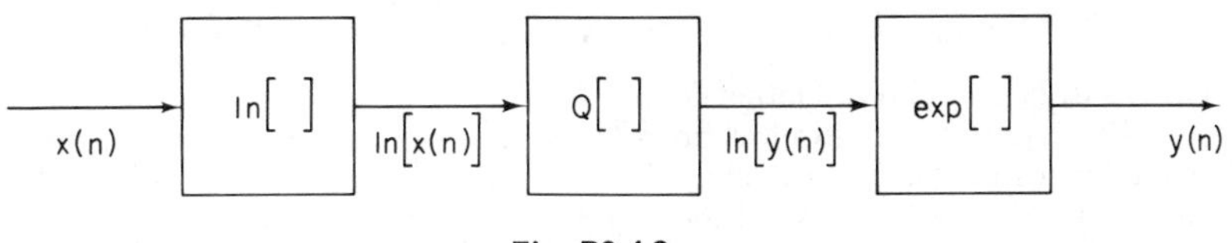

Fig. P9.4-2

where $Q[\ ]$ is a uniform quantizer as defined above. In this case, if we assume that Δ is small compared to changes in the sequence $\ln[x(n)]$, then we can assume that the output of the quantizer is

$$\ln[y(n)] = \ln[x(n)] + e(n)$$

Thus

$$y(n) = x(n) \cdot \exp[e(n)]$$

For small e we can approximate $\exp[e(n)]$ by $(1 + e(n))$, so that

$$y(n) \simeq x(n)[1 + e(n)] = x(n) + f(n)$$

This equation will be used to describe the effect of logarithmic quantization. $e(n)$ is assumed to be a stationary random process, uncorrelated from sample to sample, independent of the signal $x(n)$ and with first-order probability density uniformly distributed between $\pm\Delta/2$.

(d) Determine the mean, variance, and autocorrelation sequence of the *additive* noise $f(n)$ defined above.

(e) What is the signal-to-quantizing-noise ratio, σ_x^2/σ_f^2? Note that in this case σ_x^2/σ_f^2 is independent of σ_x^2, so that within the limits of our assumption, the signal-to-quantizing-noise ratio is independent of the input signal level, whereas for linear quantization the ratio σ_x^2/σ_e^2 depends directly upon σ_x^2.

(f) The quantized signal $y(n)$ is to be filtered by means of a digital filter with unit-sample response $h(n) = \frac{1}{2}[a^n + (-a)^n]u(n)$. Determine the variance of the noise produced at the output due to the input quantization noise and determine the signal-to-noise ratio at the output.

5. The flow graph of a first-order system is shown in Fig. P9.5-1.

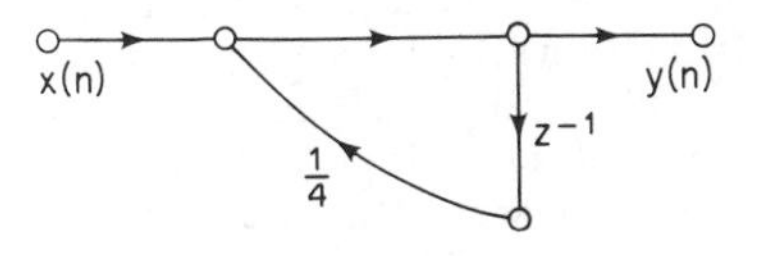

Fig. P9.5-1

(a) Assuming perfect arithmetic, find the response of the system to an input

$$x(n) = \begin{cases} \frac{1}{2}, & n \geq 0 \\ 0, & n < 0 \end{cases}$$

What is the response of the system for large n?

The system is to be implemented with fixed-point arithmetic. The coefficient and all variables in the network are to be represented in sign-and-magnitude notation with five-bit registers. That is, all numbers are to be considered signed fractions of the form

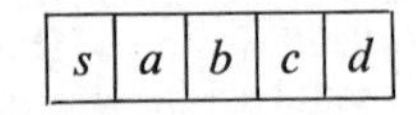

$$s = \text{sign bit}$$

$$\text{register value} = a \times 2^{-1} + b \times 2^{-2} + c \times 2^{-3} + d \times 2^{-4}$$

where a, b, c, and d are either 0 or 1.

The result of a multiplication is truncated; i.e., only the sign and the most significant four bits are retained.

(b) Compute the response of the quantized system to the input of part (a) and plot the responses of both the quantized and unquantized systems for $0 \leq n \leq 5$. How do the responses compare for large n?

(c) Now consider the system depicted in Fig. P9.5-2, where

$$x(n) = \begin{cases} \frac{1}{2}(-1)^n, & n \geq 0 \\ 0, & n < 0 \end{cases}$$

Repeat parts (a) and (b) for this system and input.

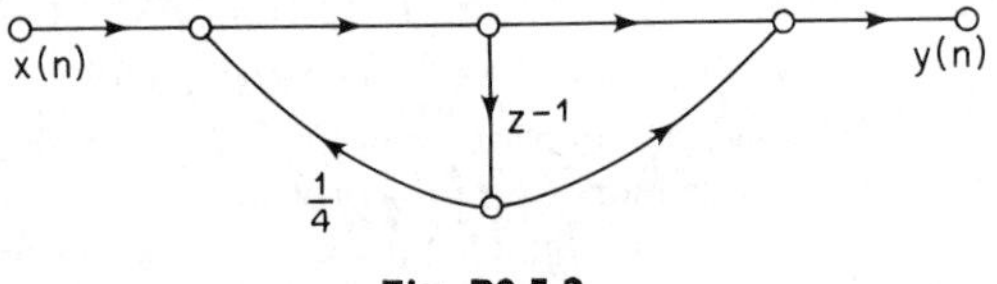

Fig. P9.5-2

6. Consider a first-order system of the form

$$y(n) = \alpha y(n-1) + x(n)$$

Assume that all variables and coefficients are represented in sign-and-magnitude form with the results of multiplications being truncated. Thus the actual difference equation is

$$w(n) = Q[\alpha w(n-1)] + x(n)$$

where $Q[\ ]$ represents sign-and-magnitude truncation.

Consider the possibility of a zero-input limit cycle of the form $|w(n)| = |w(n-1)|$ for all n. Show that if the ideal system is stable, then no zero-input limit cycle can exist. Is the same result true for two's-complement truncation?

7. Consider the first-order filter shown in Fig. P9.7. The quantizer $Q[\ ]$ represents the fact that the result of the multiplication by α is rounded. All numbers are fixed-point fractions with a word length of b bits (excluding the sign bit).

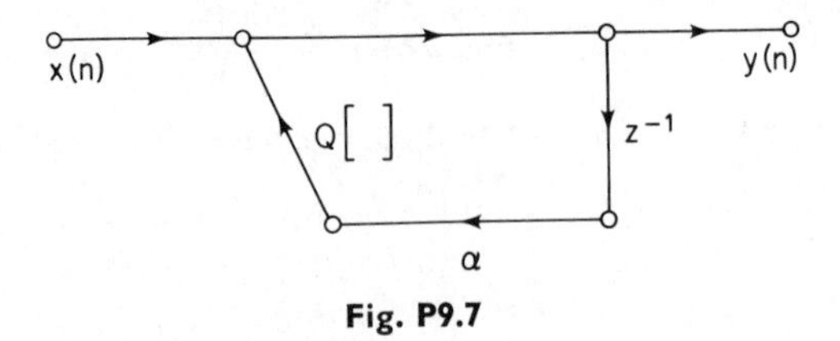

Fig. P9.7

The input is zero, but the filter is started with some initial condition, say $y(-1) = A$. Because of the quantizer there is a range of values of A referred to as the dead band, for which the effective value of the coefficients α is $+$ or $-$ unity; i.e., $|Q[\alpha A]| = A$. Once the output falls into that range, it will either oscillate or remain a constant, depending on whether the effective coefficient is positive or negative.

(a) Determine in terms of α and b the range of values of A corresponding to the dead band.
(b) For $b = 6$ bits and $A = \frac{1}{16}$, sketch $y(n)$ for $\alpha = +\frac{15}{16}$ and $\alpha = -\frac{15}{16}$.
(c) For $b = 6$ and $A = \frac{1}{2}$, sketch $y(n)$ for $\alpha = -\frac{15}{16}$.

8. Consider the second-order system shown in Fig. P9.8. The filter is to be realized with fixed-point arithmetic and the results of all multiplications are rounded. All numbers are fixed-point fractions with a word length of b bits. In a manner similar to Problem 7 of this chapter, there is a dead-band region for $y(n)$ for which the "effective" value of the coefficient $-r^2$ is -1. When this range of $y(n)$ is reached, the effective pole positions can be considered to be on the unit circle and the angular position of the poles change.

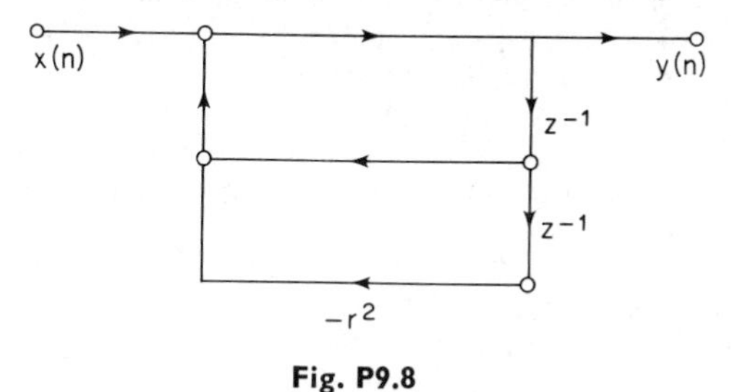

Fig. P9.8

Let $x(n) = 0$, $y(-1) = A$, $y(-2) = 0$, $A \neq 0$.

(a) Find the dead band for A, i.e., the values of A for which $Q[-r^2A] = -A$.
(b) By obtaining a lower bound on A, find the range of values of r for which it is possible to have a dead band.

9. Consider the second-order system depicted in Fig. P9.9-1:

$$y(n) = f[x(n) + ay(n-1) + by(n-2)]$$

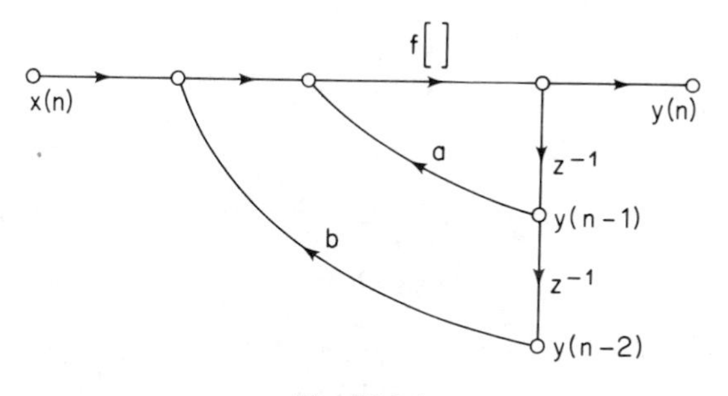

Fig. P9.9-1

The branch transmittance $f[\ ]$ is defined by the function of Fig. P9.9-2, which was shown in Problem 2 above to represent two's-complement addition. In

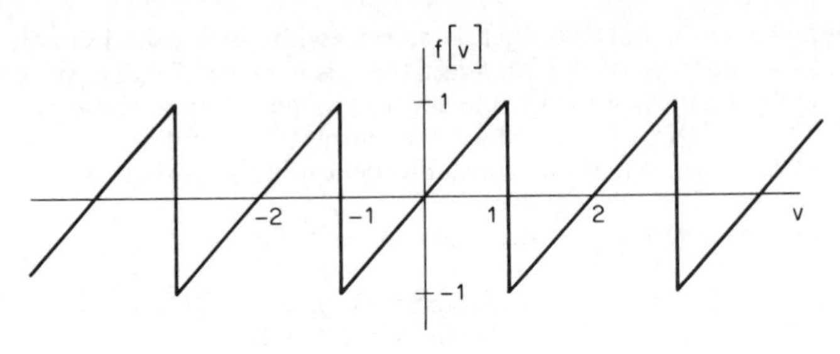

Fig. P9.9-2

other words, the function $f[\ ]$ accounts for possible overflow in forming the sum

$$ay(n-1) + by(n-2) + x(n)$$

A simple example illustrates the possibility of the existence of zero-input limit cycles. We shall neglect rounding of the products $ay(n-1)$ and $by(n-2)$.

(a) First, determine the range of values of a and b for stability of the system when no overflow occurs. Plot the region of the a-b plane corresponding to stability under linear conditions.

(b) Suppose that $x(n) = 0$. What conditions on a and b ensure that no overflow occurs, i.e., $y(n) < 1$? Shade the corresponding region in the a-b plane.

(c) Consider the possibility that with $x(n) = 0$, $y(n) = y_0 > 0$ for all n. What is the value of y_0 when a and b satisfy the requirements for linear stability?

(d) What values of a and b are consistent with the stability constraint and $0 < y_0 < 1$?

(e) Now consider the possibility of a zero-input limit cycle of period 2; i.e., $y(n) = (-1)^n y_0$ for all n. What value of $0 < y_0 < 1$ is possible for a and b satisfying the linear stability constraint?

10. In implementing digital filters with a finite word length for the coefficients, it is not possible to locate the poles and zeros of the filter with arbitrary accuracy. One possible approach to compensating for this is to introduce random jitter in each coefficient in such a way that the *average* value of the coefficient is the desired value.

Consider the implementation of the digital allpass network shown in Fig. P9.10, where $k_2 = -1/k_1$. Throughout this problem it will be assumed that

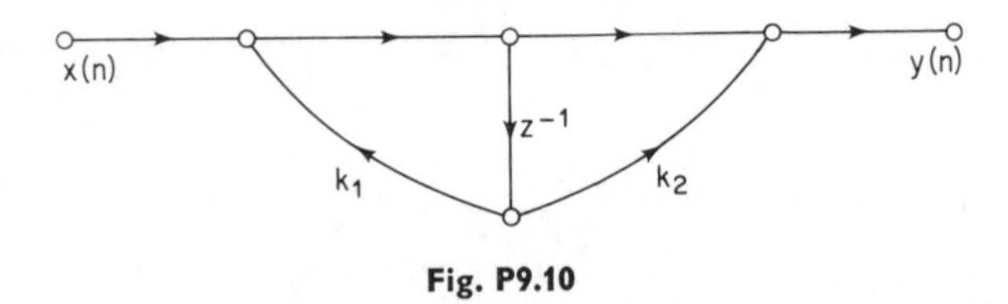

Fig. P9.10

there is no *arithmetic* roundoff noise so that the only noise introduced is due to random jitter introduced into the coefficients. We would like the average value of the coefficient k_1 to be $1/\pi$ and the average value of the coefficient k_2 to be $-\pi$. The coefficients are represented with sign and magnitude for each coefficient and the word length is chosen such that seven bits are reserved for the *fractional* part of the coefficient. Furthermore, the least significant seventh bit is random. For any iteration the probability of the least significant bit in k_1 being unity is p_1, and the probability of it being zero is $(1 - p_1)$. Similarly, the probability of the least significant bit in k_2 being unity is p_2, and the probability of it being zero is $(1 - p_2)$.

(a) Determine p_1 and p_2 so that the expected value of $k_1(n)$ is $1/\pi$ and the expected value of $k_2(n)$ is $-\pi$.

(b) With the above values for p_1 and p_2, we can write that

$$k_1(n) = \frac{1}{\pi} + \varepsilon_1(n)$$

$$k_2(n) = -\pi + \varepsilon_2(n)$$

Let $\varepsilon_1(n)$ and $\varepsilon_2(n)$ be white, statistically independent of each other, and statistically independent of the input $x(n)$. The input is a white random process with zero mean and variance σ_x^2. The output noise, $f(n)$, is defined as the difference between the output of the ideal filter [i.e., with $\varepsilon_1(n) = \varepsilon_2(n) = 0$] and the actual output. Determine the signal-to-noise ratio at the filter output, i.e., the ratio of the variance of the output due to $x(n)$ and the variance of $f(n)$. Ignore second-order error terms.

11. The first-order digital filter shown in Fig. P9.11-1 is to be implemented with floating-point arithmetic, with negative numbers represented in sign-and-magnitude notation. The number of bits in the mantissa for the coefficient, α, is b bits, not including the sign bit.

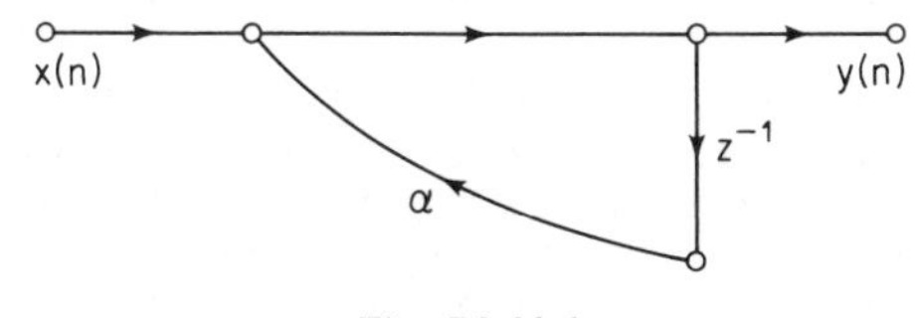

Fig. P9.11-1

It will be assumed that an unlimited number of bits are available for the characteristic. The mantissa resulting from multiplication is *truncated* to b bits. To simplify the analysis, we shall assume that the mantissa resulting from the *addition* is not truncated.

The error introduced due to the truncation of the product will be represented by an additive noise source $e(n)$, as shown in Fig. P9.11-2 with $e(n) =$

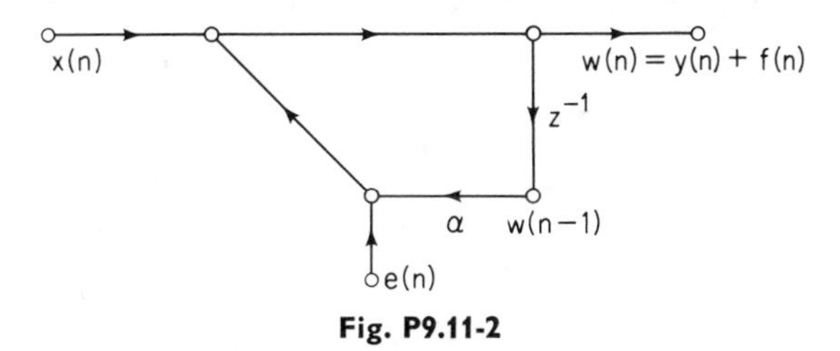

Fig. P9.11-2

$\alpha\varepsilon(n)y(n - 1)$. The output is $y^{(n)} + f(n)$, where $y(n)$ represents the output that would result if no quantization error were introduced, i.e., if $e(n)$ was zero.

The following assumptions are made about $\varepsilon(n)$:

(1) $\varepsilon(n)$ is a stationary random process with a uniform first-order probability density $p(\varepsilon)$ given by

$$p(\varepsilon) = \begin{cases} \frac{1}{2} \cdot 2^b, & 0 < \varepsilon < 2 \cdot 2^{-b} \\ 0, & \text{otherwise} \end{cases}$$

(2) $\varepsilon(n)$ is statistically independent of $x(n)$ and $y(n)$.

(3) $E[\varepsilon(n)\varepsilon(m)] = E[\varepsilon(n)]E[\varepsilon(m)]$ for $n \neq m$.

The input $x(n)$ is a random process with correlation function

$$\phi_{xx}(n) = E[x(r)x(r+n)] = \sigma_x^2 \delta(n)$$

(a) Determine $\phi_{yy}(n)$, the autocorrelation function of $y(n)$.

(b) Let σ_y^2 denote the variance of the output signal $y(n)$ and σ_f^2 denote the variance of the output noise $f(n)$. Determine the noise-to-signal ratio σ_f^2/σ_y^2.

12. We wish to implement a digital filter having a transfer function

$$H(z) = \frac{1 - \frac{1}{4}z^{-1}}{(1 - \frac{1}{2}z^{-1})(1 + \frac{1}{2}z^{-1})}$$

(a) Draw a flow graph of the digital network that implements this filter in canonic form.

(b) Draw flow graphs of the digital networks that implement this filter in each possible cascade realization containing only first-order networks (There are six possibilities, four of which contain the minimum number of delays.)

(c) The filter is to be implemented using fixed-point arithmetic. Data are represented by sign-and-magnitude fractions with a word length of b bits, excluding sign. The multiplier computes a $2b$-bit product and then *rounds* the result to the most significant b bits. Without regard to overflow considerations, determine for each of the seven networks in (a) and (b) the variance of the multiplier roundoff noise at the output. (By inspection you should be able to see that some of them lead to the same answer.) Indicate which of the configurations leads to the lowest output noise.

(d) While fixed-point arithmetic does not introduce roundoff noise into additions, it has limited dynamic range, so the input must be scaled in such a way that no signal value in the filter will exceed the register length. From the convolution sum we can bound the output of a linear system in terms of the maximum input value and the sum of the absolute values of the impulse response as follows:

$$y(n) = \sum_{k=-\infty}^{\infty} x(n-k)h(k)$$

$$|y(n)| \leq \sum_{k=-\infty}^{\infty} |x(n-k)|\,|h(k)|$$

If $x_{\max}$ is the maximum input value and $y_{\max}$ is the maximum output value, then

$$y_{\max} \leq x_{\max} \sum_{k=-\infty}^{\infty} |h(k)|$$

Furthermore, for any filter, we can always find an input that will result in an output that reaches this maximum value. In each of the configurations determined in parts (a) and (b), the signal level reaches a maximum value at some point in the network, which is not necessarily the output. Using the above equations, determine in terms of x_{max} the maximum value that can exist anyplace in the network. Do this for *each* of the seven networks in parts (a) and (b).

(e) Let us assume that the input is a white-noise sequence, uniformly distributed in amplitude between $-x_{max}$ and $+x_{max}$. For each of the filters in parts (a) and (b), we can choose a value for x_{max} on the basis of the answer obtained in part (d), so that the maximum signal level in the filter is unity. (Since we are working with fixed-point fractions, this is effectively the largest value that can fit in a register.) We can then define a noise-to-signal ratio at the output of each filter configuration as the ratio of the variance of the output noise as determined in part (c) to the mean-square value of the output signal. Determine the noise-to-signal ratio for each of the seven configurations. Indicate which of the configurations leads to the lowest output noise-to-signal ratio.

13. The system described by the difference equation

$$y(n) - \alpha y(n-1) = x(n) - \frac{1}{\alpha^*}x(n-1)$$

is an allpass system; i.e., the magnitude of its frequency response is constant, independent of frequency. We want to compare the effect of arithmetic roundoff for a fixed-point and a floating-point realization of the allpass system. Consider all fixed-point numbers as fractions, so that the fixed-point numbers are between ± 1, with a register length of b bits, excluding the sign. For the floating-point case, let t represent the number of bits in the mantissa, excluding sign.

The input to the filter $x(n)$ is considered to be a white random process, with a uniform amplitude distribution between $\pm x_0$. Assume that α is real and $\frac{1}{2} < \alpha < 1$.

(a) The maximum input signal level must be small enough so that there is no overflow in the fixed-point realization of the filter. Taking this into account, determine the output noise-to-signal ratio of the fixed-point realization in both the direct and canonic forms, i.e., the ratio of the variance of the output due to roundoff noise and the variance of the output due to $x(n)$.

(b) Determine the output noise-to-signal ratio for the canonic form of the filter realized with floating-point arithmetic.

(c) Assume that α is close to unity, so that $\alpha = 1 - \delta$, $\delta \ll 1$. Express your results from parts (a) and (b) in terms of δ, making reasonable approximations.

(d) The floating-point realization of a filter requires additional bits for the characteristic. Let B represent the number of bits used for the characteristic in the floating-point word. If $b = B + t$, for each of the two fixed-point realizations (canonic and direct forms), determine B as a function of δ so that the noise-to-signal ratios for fixed point and floating point are equal.

14. We have discussed the noise performance of digital filters realized with fixed-point and with floating-point arithmetic. In this problem an alternative, referred to as *block-floating-point arithmetic*, will be discussed (see [29]). To illustrate the procedure, consider a first-order filter, defined by the difference equation

$$y(n) = x(n) + \alpha y(n-1)$$

corresponding to the digital network shown in Fig. P9.14-1. If the multiplication and addition in the filter are implemented in fixed point, the full register

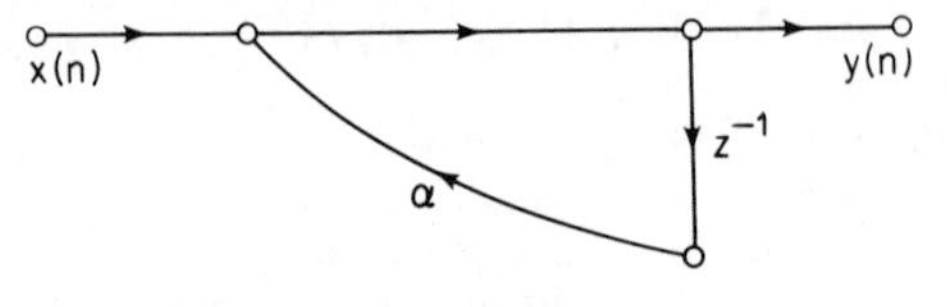

Fig. P9.14-1

length is not utilized on every iteration since the input must be scaled so that the *largest* output value is unity. Block floating point corresponds to multiplying $x(n)$ and $y(n-1)$ by a gain $A(n)$ to jointly normalize them and then dividing $y(n)$ by this gain; i.e.,

$$y(n) = \frac{1}{A(n)}[x(n)A(n) + ay(n-1)A(n)]$$

$A(n)$, of course, changes from iteration to iteration.

Throughout this problem assume that $x(n)$ is a white random process of zero mean and a uniform amplitude distribution between $\pm x_0$.

(a) Show that the network in Fig. P9.14-2 is equivalent to the network in Fig. P9.14-1 if quantization effects are neglected.

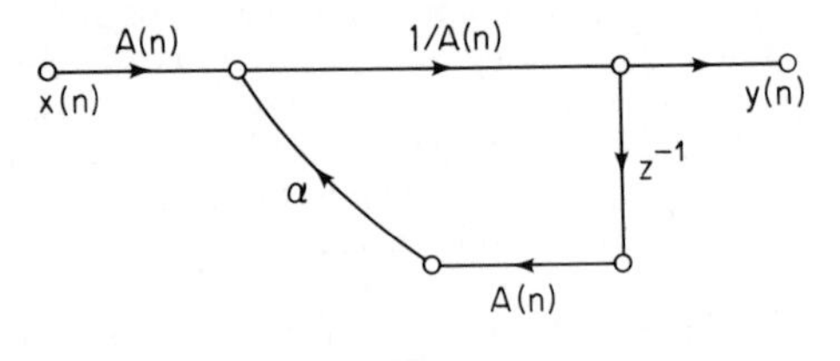

Fig. P9.14-2

(b) The multiplication and addition are carried out with fixed-point arithmetic. The gain $A(n)$ is chosen as

$$A(n) = 2^{c(n)}, \qquad c(n) \geq 0$$

so that it corresponds to a left shift of the registers. Since $A(n)$ is chosen to jointly normalize $x(n)$ and $y(n-1)$, we have

$$\tfrac{1}{2} \leq A(n) \max\{|x(n)|, |y(n-1)|\} < 1$$

where $\max\{|x(n)|, |y(n-1)|\}$ = largest value of $|x(n)|$ and $|y(n-1)|$ for each n.

We must, therefore, require that

$$\frac{\frac{1}{2}}{\max\{|x(n)|, |y(n-1)|\}} \leq A(n) < \frac{1}{\max\{|x(n)|, |y(n-1)|\}} \qquad \text{(P9.14-1)}$$

Since multiplication of a register by a positive power of 2 introduces no roundoff noise, the noise sources introduced due to multiplier roundoff are shown in Fig. P9.14-3.

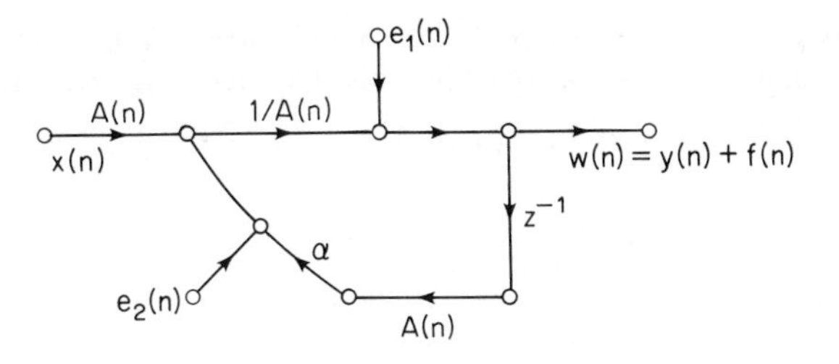

Fig. P9.14-3

Assuming that $e_1(n)$ and $e_2(n)$ are uniformly distributed between $\pm\frac{1}{2}\cdot 2^{-b}$, uncorrelated with each other and with $x(n)$ [and consequently $A(n)$], determine the variance of the output noise $f(n)$ for the network above. Express your answer in terms of $k^2 = E[(1/A(n))^2]$. Show that the output noise variance due to arithmetic roundoff for the network of Fig. P9.14-3 is always *greater* than would be obtained if the first-order filter were implemented directly, as shown in Fig. P9.14-2.

(c) The network in Fig. P9.14-3 can be modified to an equivalent form, shown in Fig. P9.14-4. Let $e_1(n)$, $e_2(n)$, and $e_3(n)$ represent the noise intro-

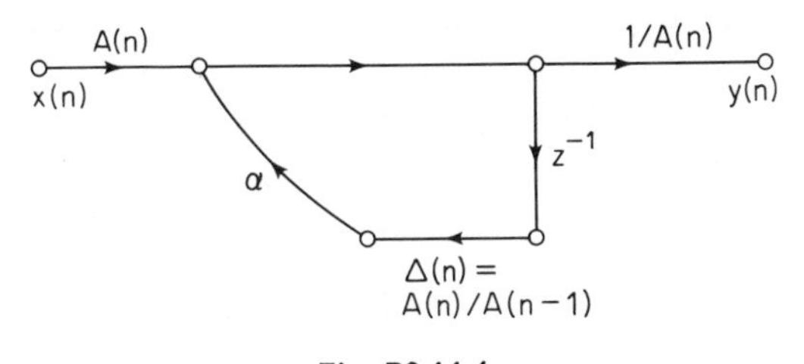

Fig. P9.14-4

duced by rounding the multiplies by $[1/A(n)]$, $\Delta(n)$, and α, respectively. Again assume that $e_1(n)$, $e_2(n)$, and $e_3(n)$ are uniformly distributed between $\pm\frac{1}{2}\cdot 2^{-b}$ and are independent of each other and of $x(n)$ and $y(n)$. Determine the variance of the total output noise due to these three noise sources. Express your answer in terms of $k^2 = E[(1/A(n))^2]$.

(d) Assume that $\alpha = 1 - \delta$, $\delta \ll 1$ (high-gain case). For this value of α, the absolute value of the output $|y(n-1)|$ is assumed to be greater than the absolute value of the input $|x(n)|$. Considering this fact, show that an upper bound upon the quantity $k^2 = E[(1/A(n))^2]$ is $k^2 \leq 4\sigma_y^2$, using Eq. (P9.14-1).

(e) To determine the noise-to-signal ratio, we must require that the maximum value of $x(n)$ be small enough to ensure that $|y(n)| < 1$. Let $x(n)$ be a uniformly distributed white process, and $\alpha = 1 - \delta$, $\delta \ll 1$. With these assumptions and the results of part (d), determine the noise-to-signal ratio for the network of Fig. P9.14-4. Compare this noise-to-signal ratio with the corresponding result for fixed-point arithmetic.

15. Consider the DFT

$$X(k) = \sum_{n=0}^{N-1} x(n) W_N^{kn}, \qquad 0 \leq k \leq N-1$$

where $W_N = e^{-j(2\pi/N)}$. Suppose that the sequence values $x(n)$ are N consecutive values of a stationary white-noise sequence with zero mean; i.e.,

$$E[x(n)x(r)] = \sigma_x^2\delta(n-r)$$

$$E[x(n)] = 0$$

(a) Determine the variance of $|X(k)|^2$.
(b) Determine the cross correlation between values of the DFT; i.e., determine $E[X(k)X^*(r)]$ as a function of k and r.

16. Consider direct computation of the DFT using fixed-point arithmetic with rounding.

Assume that the register length is b bits plus sign and that the roundoff noise introduced by multiplication is independent of that introduced by any other multiplication. Assuming that $x(n)$ is real, determine the variance of the roundoff noise in both the real part and the imaginary part of each spectrum point $X(k)$.

17. In the Goertzel method of computation of the discrete Fourier transform,

$$X(k) = y_k(r)|_{r=N}$$

where $y_k(r)$ is the output of the network shown in Fig. P9.17. Consider the implementation of the Goertzel algorithm using fixed-point arithmetic with

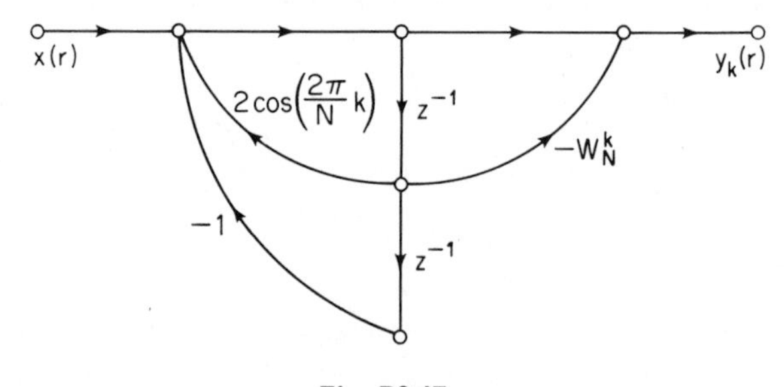

Fig. P9.17

rounding. Assume that the register length is b bits plus sign and that the roundoff noise introduced by a multiplication is independent of that introduced by all other multiplications.

(a) Assuming that $x(r)$ is real, draw a flow-graph representation of the finite-precision computation of the real and imaginary parts of $X(k)$. Assume that multiplication by ± 1 introduces no roundoff noise.
(b) Compute the variance of the roundoff noise in both the real part and the imaginary part of each spectrum point $X(k)$.

18. In Sec. 9.4.2 we showed that in a floating-point direct-form implementation of an FIR filter, the order in which the multiplications and additions are performed is a significant factor in determining the noise-to-signal ratio at the output. In this problem we shall consider an ordering of the additions that results in a significant reduction of the noise-to-signal ratio.

Suppose that $N = 2^\nu$, where ν is an integer. Then we can consider evaluating the convolution sum

$$y(n) = \sum_{k=0}^{N-1} h(k)x(n-k)$$

by adding the indicated products together in pairs, then adding these sums together in pairs, etc. Such an implementation is depicted in Fig. P9.18 for

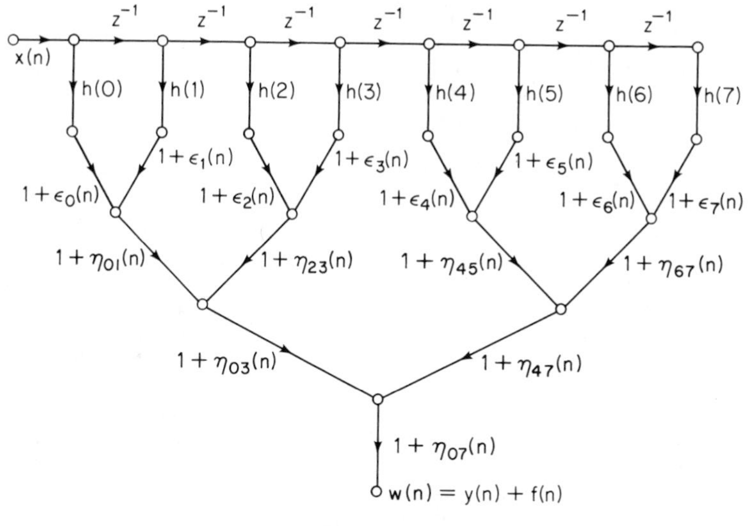

Fig. P9.18

$N = 2^3 = 8$. In this flow graph, the time-varying gains $1 + \varepsilon_k(n)$ represent the errors resulting from quantizing the multiplications $h(k)x(n - k)$, and the time-varying gains $1 + n_{ij}(n)$ represent the errors introduced by quantizing the floating-point additions. The quantities $\varepsilon_k(n)$ and $\eta_{ij}(n)$ are assumed to be uniformly distributed white noise and independent of one another.

(a) Generalize the notation of the above diagram for the general case of $N = 2^\nu$.

(b) Express the output $w(n)$ in the form

$$w(n) = \sum_{k=0}^{N-1} A(n, k)h(k)x(n - k)$$

i.e., express $A(n, k)$ in terms of the quantities $1 + \varepsilon_k(n)$ and $1 + \eta_{ij}(n)$.

(c) Show that

$$E[A(n, k) - 1] = 0$$

$$E[A^2(n, k)] = (1 + \tfrac{2^{-2b}}{3})^{\nu+1}$$

(d) Obtain an expression for the variance of the output noise, σ_f^2, assuming that the input is a white signal with variance σ_x^2 and that $\frac{2^{-2b}}{3} \ll 1$.

(e) Show that the output noise-to-signal ratio is

$$\frac{\sigma_f^2}{\sigma_y^2} = (\nu + 1)\frac{2^{-2b}}{3}$$

(f) Compare the result of part (e) with the bound on σ_f^2/σ_y^2 of Eq. (9.77).

19. In deriving formulas for the noise-to-signal ratio for fixed-point FFT computations, we assumed that each output node was connected to $(N - 1)$ butterfly computations, each of which contributed $\sigma_B^2 = \frac{1}{3} \cdot 2^{-2b}$ to the output noise variance. However, when $W_N^r = \pm 1$ or $\pm j$, the multiplications can, in fact, be done without error. Thus the results derived in Sec. 9.5.2 can be modified so that they are less pessimistic.

(a) For the decimation-in-time algorithm discussed in Sec. 9.5.2, determine for each stage the number of butterflies that involve multiplication by either ± 1 or $\pm j$.

(b) Use the result of part (a) to modify the estimate of the output noise variance, Eq. (9.87), for the case when all scaling is done at the input. Also, obtain a modified expression corresponding to Eq. (9.93) for the noise-to-signal ratio at the output.

(c) Repeat parts (a) and (b) for the case when the output of each stage is attenuated by a factor of $\frac{1}{2}$. That is, derive modified expressions corresponding to Eq. (9.95) for the output noise variance and Eq. (9.96) for the output noise-to-signal ratio, assuming that multiplications by ± 1 and $\pm j$ do not introduce error.

20. In deriving formulas for the noise-to-signal ratio for floating-point FFT computations, we assumed that all butterfly computations introduced the same amount of noise. However, this is not the case, as can be seen by considering multiplications by ± 1 or $\pm j$.

(a) Draw complete flow graphs as in Fig. 9.22 of decimation-in-time floating point butterflies for $W_N^r = 1$ and for $W_N^r = j$, showing all the noise sources that remain.

(b) Assuming that the input signal is white, compute the output noise variance for a butterfly involving $W_N^r = \pm 1$ or $\pm j$.

(c) For the decimation-in-time algorithm discussed in Sec. 9.5.3, determine for each stage the number of butterflies that involve multiplications by either ± 1 or $\pm j$.

(d) Using the results of parts (b) and (c), determine the *average* noise variance for the butterflies involved in computing the $(m + 1)$st array from the mth array.

(e) Using the result of part (d), modify the argument of Sec 9.5.3 to obtain an expression corresponding to Eq. (9.99) for the *average* output noise variance. Use this result to show that the *average* output noise-to-signal ratio is

$$\frac{E[|F(k)|^2]}{E[|X(k)|^2]} = 2\sigma_\varepsilon^2[\nu - \tfrac{3}{2} + (\tfrac{1}{2})^{\nu-1}]$$

21. In Sec. 9.5.2 we considered a noise analysis of the decimation-in-time FFT algorithm of Fig. 6.10. Carry out a similar analysis for the decimation-in-frequency algorithm of Fig. 6.18, obtaining equations for the output noise variance and noise-to-signal ratio for scaling at the input and for scaling of $\frac{1}{2}$ at each stage of computation.

22. Consider the computation of the DFT using quantized coefficients. Let $X(k)$ be the desired DFT and $X'(k)$ be the result using quantized values of W_N^r in a decimation-in-time FFT algorithm.

(a) Show that $X'(k)$ can be expressed as

$$X'(k) = \sum_{n=0}^{N-1} x(n)\Omega_{nk} = X(k) + F(k)$$

where

$$\Omega_{nk} = \prod_{i=1}^{\nu} (W_N^{r_i} + \delta_i)$$

and

$$\prod_{i=1}^{\nu} W_N^{r_i} = W_N^{nk}$$

(b) Assume that the coefficients W_N^r are rounded to b bits, excluding sign, and that the real and imaginary parts of the coefficient errors are uncorrelated and uniformly distributed. Show that the variance of the quantities δ_i is

$$\sigma_\delta^2 = \tfrac{2}{6}^{-2b}$$

(c) The error $F(k)$ can be expressed as

$$F(k) = X'(k) - X(k) = \sum_{n=0}^{N-1} x(n)(\Omega_{nk} - W_N^{nk})$$

Show that the factor $(\Omega_{nk} - W_N^{nk})$ can be expressed as

$$(\Omega_{nk} - W_N^{nk}) = \sum_{i=1}^{\nu} \delta_i \prod_{\substack{j=1 \\ j \neq i}}^{\nu} W_N^{r_i} + \text{higher-order terms}$$

(d) Neglecting the higher-order terms and assuming that the δ_i are mutually uncorrelated, show that the variance of $F(k)$ is

$$\sigma_F^2 = \left(\frac{2\nu}{6}\right) 2^{-2b} \sum_{n=0}^{N-1} |x(n)|^2$$

(e) Use Parseval's theorem to show that

$$\frac{\sigma_F^2}{\frac{1}{N} \sum_{k=0}^{N-1} |F(k)|^2} = \left(\frac{2\nu}{6}\right) 2^{-2b}$$

10

Homomorphic Signal Processing

10.0 Introduction

In previous chapters we have been concerned with the mathematical representation of discrete-time signals and systems. The focus has been primarily on linear shift-invariant systems, and little attention has been given to applications. The purpose of this chapter is twofold. One is to present a class of nonlinear signal-processing techniques. The analysis of this class of techniques will rely on much of the material in the previous chapters. Furthermore, this class of techniques has found application in a number of fields, including image enhancement, speech analysis, and seismic exploration. Consequently, the consideration of this class of techniques provides an opportunity to present a number of applications of some of the results that have been discussed throughout this book.

The class of systems to be discussed is based on a generalization of the class of linear systems. We have seen that linear shift-invariant systems are important because they are relatively easy to analyze and characterize, leading to rather elegant and powerful mathematical representations, and because it is possible to design linear shift-invariant systems to perform a variety of useful signal processing functions. For example, if we are given a signal that is the sum of two component signals whose Fourier transforms occupy different frequency bands, then it is possible to separate the two components with a linear filter. The fact that linear systems are relatively easy to analyze and are useful in separating signals combined by addition is a direct consequence of the property of superposition, which defines the class of linear systems. This observation leads to the consideration of classes of nonlinear systems that obey a generalized principle of superposition. Such systems are represented by algebraically linear transformations between input and output vector spaces and have thus been called *homomorphic systems*.

In this chapter we give a brief introduction to the general theory of homomorphic systems and then proceed to detailed discussions of two classes of homomorphic systems that are especially suited to processing signals combined by multiplication and convolution, these being the two cases where the theory of homomorphic systems has been successfully applied. In these two examples we shall see that the generalized principle of superposition can be exploited in much the same way as it is in characterizing linear systems. In fact, we shall see that the problem of designing homomorphic systems for multiplication and convolution reduces to the problem of designing a linear system.

10.1 Generalized Superposition

The principle of superposition as it is stated for linear systems requires that, if T is the system transformation, then for any two inputs $x_1(n)$ and $x_2(n)$ and any scalar c,

$$T[x_1(n) + x_2(n)] = T[x_1(n)] + T[x_2(n)] \tag{10.1a}$$

and

$$T[cx_1(n)] = cT[x_1(n)] \tag{10.1b}$$

To generalize this principle, let us denote by $\square$ a rule for combining inputs with each other (e.g., addition, multiplication, convolution, etc.) and by $:$ a rule for combining inputs with scalars. Similarly, $\bigcirc$ will denote a rule for combining system outputs and $\urcorner$ a rule for combining outputs with scalars. Then, with H denoting the system transformation, we generalize equations (10.1) by requiring that

$$H[x_1(n) \square x_2(n)] = H[x_1(n)] \bigcirc H[x_2(n)] \tag{10.2a}$$

and

$$H[c : x_1(n)] = c \urcorner H[x_1(n)] \tag{10.2b}$$

Such systems are said to obey a generalized principle of superposition with an input operation $\square$ and an output operation $\bigcirc$. Such systems are depicted as in Fig. 10.1. Clearly, linear systems are a special case for which $\square$ and $\bigcirc$ are addition and $:$ and $\urcorner$ are multiplication.

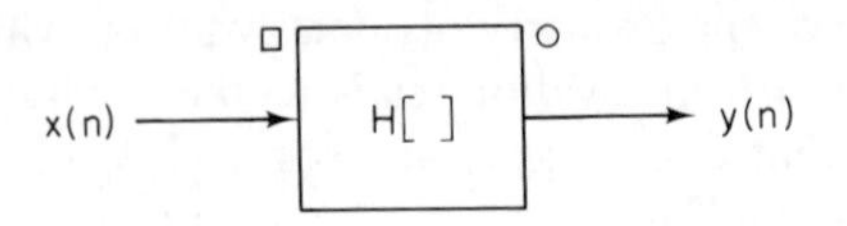

Fig. 10.1 Representation of a homomorphic system with input operation □, output operation ○, and system transformation $H[\,]$.

The theory of linear vector spaces provides the mathematical formalism for representing systems of this class. If we interpret the system inputs and outputs as vectors in vector spaces with the rules □ and ○ corresponding to vector addition and : and ⅂ corresponding to scalar multiplication, then the system transformation H is an algebraically linear transformation from the input vector space to the output vector space.

If we wish to employ the theory of linear vector spaces in this manner, the input and output operations must satisfy the algebraic postulates of vector addition and scalar multiplication, respectively. This means, for example, that the operations □ and ○ must be both commutative and associative; i.e.,

$$\begin{aligned} x_1(n) \,\square\, x_2(n) &= x_2(n) \,\square\, x_1(n) \\ y_1(n) \bigcirc y_2(n) &= y_2(n) \bigcirc y_1(n) \end{aligned} \tag{10.3}$$

and

$$\begin{aligned} x_1(n) \,\square\, [x_2(n) \,\square\, x_3(n)] &= [x_1(n) \,\square\, x_2(n)] \,\square\, x_3(n) \\ y_1(n) \bigcirc [y_2(n) \bigcirc y_3(n)] &= [y_1(n) \bigcirc y_2(n)] \bigcirc y_3(n) \end{aligned} \tag{10.4}$$

There are many such mathematical considerations involved in defining appropriate vector spaces and transformations [1,2]. We shall not proceed further with a discussion of the details of mathematical rigor but will simply describe the central result of applying the theory of vector spaces to systems that obey a generalized principle of superposition.

It can be shown that if the system inputs constitute a vector space with □ and : corresponding to vector addition and scalar multiplication and the system outputs constitute a vector space with ○ and ⅂ corresponding to vector addition and scalar multiplication, then all systems of this class can be represented as a cascade of three systems, as shown in Fig. 10.2. This

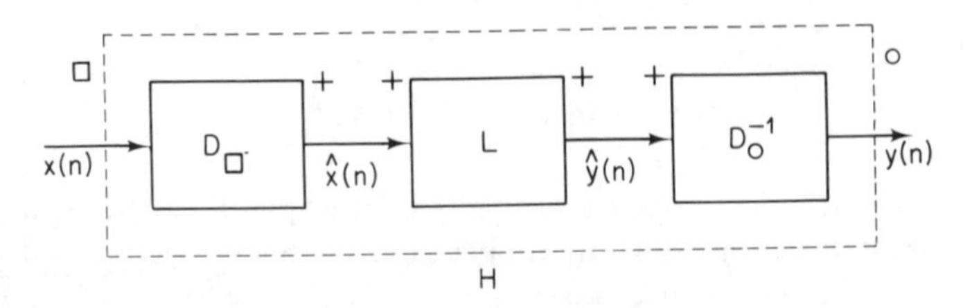

Fig. 10.2 Canonic representation of homomorphic systems.

cascade of Fig. 10.2 is referred to as the *canonic representation of homomorphic systems*. The first system, $D_\square$, has the property that

$$D_\square[x_1(n) \square x_2(n)] = D_\square[x_1(n)] + D_\square[x_2(n)]$$
$$= \hat{x}_1(n) + \hat{x}_2(n) \tag{10.5a}$$
$$D_\square[c : x_1(n)] = cD_\square[x_1(n)] = c\hat{x}_1(n) \tag{10.5b}$$

We observe that $D_\square$ obeys a generalized principle of superposition where the input operation is $\square$ and the output operation is $+$. The effect of the system $D_\square$ is to transform the combination of the signals $x_1(n)$ and $x_2(n)$ according to the rule $\square$ into a conventional linear combination of corresponding signals $D_\square[x_1(n)]$ and $D_\square[x_2(n)]$. The system L is a conventional linear system, so

$$L[\hat{x}_1(n) + \hat{x}_2(n)] = L[\hat{x}_1(n)] + L[\hat{x}_2(n)]$$
$$= \hat{y}_1(n) + \hat{y}_2(n)$$
$$L[c\hat{x}_1(n)] = cL[\hat{x}_1(n)] = c\hat{y}_1(n)$$

Finally, the system $D_\bigcirc^{-1}$ transforms from addition to $\bigcirc$, so

$$D_\bigcirc^{-1}[\hat{y}_1(n) + \hat{y}_2(n)] = D_\bigcirc^{-1}[\hat{y}_1(n)] \circ D_\bigcirc^{-1}[\hat{y}_2(n)]$$
$$= y_1(n) \circ y_2(n)$$
$$D_0^{-1}[c\hat{y}_1(n)] = c \lrcorner\, D_0^{-1}[\hat{y}_1(n)] = c \lrcorner\, y_1(n)$$

Since the system $D_\square$ is fixed by the operations $\square$ and :, it is characteristic of the class and is therefore called the *characteristic system* for the operation $\square$. Similarly, $D_\bigcirc$ is the characteristic system for the operation $\bigcirc$. Furthermore, it is clear that all homomorphic systems with the same input and output operations differ only in the linear part. This result is of fundamental importance, because it implies that once the characteristic systems for the class have been determined, we are left with a linear filtering problem. For example, if we wish to recover $x_1(n)$ from the signal

$$x(n) = x_1(n) \square x_2(n)$$

we must choose the linear system so that its output $\hat{y}(n)$ is

$$\hat{y}(n) = \hat{x}_1(n)$$

Then with $D_\bigcirc = D_\square$,

$$y(n) = D_\square^{-1}[\hat{x}_1(n)] = x_1(n)$$

That is, to perform perfect separation of $x_1(n)$ and $x_2(n)$, we must be able to perfectly separate $\hat{x}_1(n)$ and $\hat{x}_2(n)$ using a linear filter. How well we can approach this ideal situation depends on the operation $\square$ and the properties of the signals $x_1(n)$ and $x_2(n)$. In the remainder of this chapter we shall restrict ourselves to classes of systems for which the input and output operations are identical, and in particular to the two classes of homomorphic systems defined by choosing as this operation either multiplication or convolution. In each case we shall consider representations of the system $D_\square$ and classes of signals for which processing of this type seems to offer advantages over other techniques.

10.2 Multiplicative Homomorphic Systems

There are many signal processing problems where the signal may be represented as a product of two or more component signals. In transmission of a signal over a fading channel, for example, we may model the effect of fading in terms of a slowly varying component multiplying the transmitted signal. As another example, an amplitude-modulated signal is represented by the product of a carrier signal and an envelope function, which we wish to separate at the receiver. Other examples include audio-dynamic range compression and image processing, which we shall discuss in Sec. 10.3. In many problems of this type a linear system may be quite ineffectual in separating or independently modifying the component signals. In contrast, a system that obeys a generalized principle of superposition for multiplication may often be used with impressive results. In this section we discuss the basic theory of homomorphic systems for multiplication, and in the next section we discuss the applications of such systems to digital image processing.

Consider the class of homomorphic systems that obeys a generalized principle of superposition in which the operation $\square$ is multiplication and the operation : is exponentiation. That is, we are concerned with signals of the form

$$x(n) = [x_1(n)]^\alpha \cdot [x_2(n)]^\beta \tag{10.6}$$

It is easily verified that these are appropriate choices for vector addition and scalar multiplication. The characteristic system for multiplication must have the property

$$D_\cdot[[x_1(n)]^\alpha \cdot [x_2(n)]^\beta] = \alpha D_\cdot[x_1(n)] + \beta D_\cdot[x_2(n)] \tag{10.7}$$

A function that formally has this property is the logarithm function. For example, if $x(n) = x_1(n) \cdot x_2(n)$, where $x_1(n) > 0$ and $x_2(n) > 0$ for all values of n, then

$$\log [x_1(n) \cdot x_2(n)] = \log [x_1(n)] + \log [x_2(n)] \tag{10.8}$$

However, the input $x(n)$ may not always be positive; indeed, we may wish to consider signals that are complex. Such cases require the use of the complex logarithm function. Thus the formal representation of canonic systems with multiplication as the input and output operation is as shown in Fig. 10.3, where the sequences $x(n)$, $\hat{x}(n)$, $\hat{y}(n)$, and $y(n)$ are in general complex.

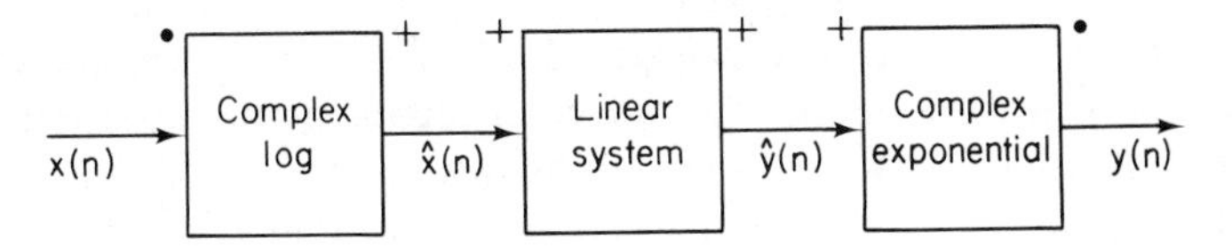

Fig. 10.3 Canonic representation for homomorphic systems with multiplication as the input and output operations.

Let us consider the properties of the complex logarithm function. If we let $x(n) = |x(n)|e^{j\arg[x(n)]}$ denote a complex sequence, then the complex logarithm of $x(n)$ is defined as

$$\log [x(n)] = \log |x(n)| + j \arg [x(n)] \tag{10.9}$$

The inverse of $\log [x(n)]$ is the complex exponential

$$e^{\log[x(n)]} = e^{\log|x(n)|} \cdot e^{j\arg[x(n)]} \tag{10.10}$$

Clearly, any integer multiple of 2π can be added to the imaginary part of the complex logarithm, i.e., to $\arg [x(n)]$, without changing the result of Eq. (10.10). Thus without further restrictions the complex logarithm is not a unique transformation. Since uniqueness is a basic requirement in the definition of a system, we must choose $\arg [x(n)]$ in such a way that this ambiguity is resolved. Furthermore, we have the additional requirement that $\log [x(n)]$ be defined in such a way that generalized superposition holds, i.e., so that with $x(n) = x_1(n) \cdot x_2(n)$, then

$$\log [x(n)] = \log [x_1(n) \cdot x_2(n)] = \log [x_1(n)] + \log [x_2(n)]$$

This implies that

$$\log |x(n)| = \log |x_1(n)| + \log |x_2(n)| \tag{10.11}$$

and

$$\arg [x(n)] = \arg [x_1(n)] + \arg [x_2(n)] \tag{10.12}$$

Thus the ambiguity in $\arg [x(n)]$ must be resolved in such a way that Eq. (10.12) holds.

Although it is common to resolve the ambiguity in the complex logarithm by replacing $\arg [x(n)]$ by its principal value, i.e., its value modulo 2π, this cannot be done here since Eq. (10.12) will then not generally hold. That is, it is not generally true that the principal value of the sum of two angles is equal to the sum of their respective principal values. To resolve the ambiguity and satisfy Eq. (10.12), it is necessary to define $\arg [x]$ so that it is a continuous function of x. This approach is rigorously justified through the theory of Riemann surfaces. We shall not develop the details of this definition of the complex logarithm in this section since in the applications that we shall discuss, the class of signals are nonnegative, in which case the ambiguity does not arise since $\arg [x(n)]$ can always be taken as zero. However, this will not be the case in Sec. 10.4, where we consider homomorphic systems for convolution, and in that section the details of an unambiguous definition of the complex logarithm will be considered. Thus for the present let us assume that an unambiguous definition of the complex logarithm is available for the realization of the characteristic system for multiplicative homomorphic systems. Then a particular homomorphic system of this class differs from all others of this class only in the linear part. If the input is as given by Eq. (10.6), then the output of the complex log is

$$\hat{x}(n) = \alpha\hat{x}_1(n) + \beta\hat{x}_2(n) = \hat{x}_r(n) + j\hat{x}_i(n) \tag{10.13}$$

where

$$\hat{x}_1(n) = \log[x_1(n)] \quad \text{and} \quad \hat{x}_2(n) = \log[x_2(n)] \tag{10.14}$$

The most general form for a linear system for processing complex inputs is shown in Fig. 10.4, where L_{rr}, L_{ri}, L_{ir}, and L_{ii} denote real linear systems.

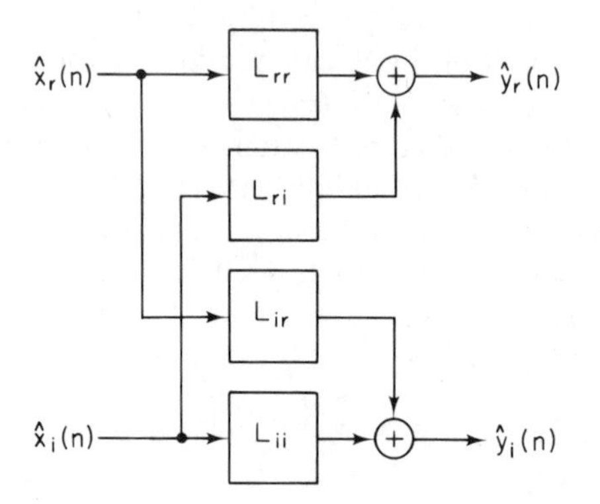

Fig. 10.4 General representation of linear system with complex input and output.

If α and β are complex numbers, the real linear systems must satisfy

$$L_{rr} = L_{ii} \quad \text{and} \quad L_{ri} = -L_{ir} \tag{10.15}$$

If α and β are real, there are no constraints.

In applications of the theory of multiplicative systems, we must make appropriate choices for the linear system. Our choice, of course, depends on the nature of the input signal. Whether or not we can do useful processing of signals of the form

$$x(n) = x_1(n) \cdot x_2(n)$$

depends on the nature of $\hat{x}_1(n)$ and $\hat{x}_2(n)$, the components of the output of the characteristic system. For example, if we wish to separate the two components or perform independent processing of the two components, then their frequency spectra must not overlap significantly. This implies that homomorphic processing of signals combined by multiplication may be useful whenever one component is rapidly varying and the other component is slowly varying. This is indeed the case in the areas of audio companding and image processing, where homomorphic processing has been applied with success [3], [4], [5], [6]. In the next section we shall consider the latter example in some detail.

10.3 Homomorphic Image Processing

One example of the successful application of homomorphic signal processing is in the area of image enhancement. This application is based on a model of images as the product of two basic components. Through homomorphic filtering these components can be modified separately to achieve simultaneous contrast enhancement and dynamic range compression.

10.3.1 Model for Image Formation [6]

Images are formed by reflection of light. That is, an image is formed when light energy from an illumination source is reflected by physical objects. Thus image formation can be modeled as a multiplicative process in which a pattern of illumination is multiplied by a reflectance pattern to produce the brightness image. If the two-dimensional patterns of illumination and reflectance are represented by the functions $f_i(u, v)$ and $f_r(u, v)$, respectively,† where u and v are continuous spatial variables, then the image is expressed as

$$f(u, v) = f_i(u, v) \cdot f_r(u, v) \tag{10.16}$$

Since both the illumination component and the image correspond to patterns of light energy,

$$0 < f(u, v) < f_i(u, v) < \infty \tag{10.17}$$

The reflectance component also is always positive and is further constrained by physical considerations to be less than unity; i.e.,

$$0 < f_r(u, v) < 1 \tag{10.18}$$

In summary, images can be represented as a product of two components. Furthermore, the individual components are always positive. Thus it seems that the structure of images is ideally suited to processing with a homomorphic system for multiplication.

10.3.2 Digital Image Processing

Images are represented discretely by a two-dimensional sequence $x(m, n)$ obtained by periodic sampling of a spatially limited image; i.e.,

$$x(m, n) = f(m\,\Delta u, n\,\Delta v)$$

† In Sec. 10.3 the subscripts r and i refer to reflectance and illumination respectively *not* real and imaginary parts.

where Δu and Δv are chosen so that significant aliasing does not occur. From our previous discussion, $x(m, n)$ has the representation

$$x(m, n) = x_i(m, n) \cdot x_r(m, n) \tag{10.19}$$

where $x_i(m, n)$ and $x_r(m, n)$ are the sampled illumination and reflectance patterns, respectively.

The discrete homomorphic image processor has the canonic form depicted in Fig. 10.5. Since $x(m, n) > 0$, there is no need to be concerned with the

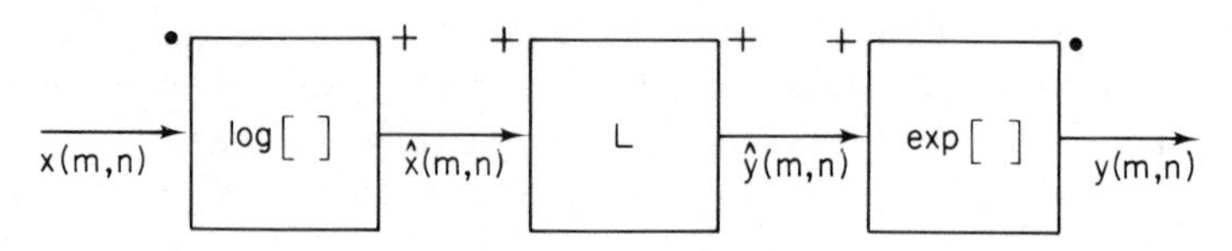

Fig. 10.5 Canonic form for homomorphic image processor.

ambiguity of the complex logarithm. Thus

$$\begin{aligned} \hat{x}(m, n) &= \log [x(m, n)] = \log [x_i(m, n) \cdot x_r(m, n)] \\ &= \log x_i(m, n) + \log x_r(m, n) \\ &= \hat{x}_i(m, n) + \hat{x}_r(m, n) \end{aligned} \tag{10.20}$$

Stockham [6] argues that the output of the characteristic system has a close correspondence to the kind of density pattern that results when an image is represented in a photographic transparency. Thus it is reasonable to call $x(m, n)$ the *intensity representation* and $\hat{x}(m, n)$ the *density representation* of an image. Likewise, $\hat{y}(m, n)$ may be called the *processed density* and $y(m, n)$ the *processed intensity representations.* Clearly, by the properties of linear systems,

$$\hat{y}(m, n) = \hat{y}_i(m, n) + \hat{y}_r(m, n) \tag{10.21}$$

where $\hat{y}_i(m, n)$ and $\hat{y}_r(m, n)$ are the processed illumination and reflectance densities, respectively. Therefore, the processed output intensity is

$$\begin{aligned} y(m, n) &= \exp [\hat{y}_i(m, n) + \hat{y}_r(m, n)] \\ &= \exp [\hat{y}_i(m, n)] \cdot \exp [\hat{y}_r(m, n)] \\ &= y_i(m, n) y_r(m, n) \end{aligned} \tag{10.22}$$

Since $\hat{y}_i(m, n)$ and $\hat{y}_r(m, n)$ are both real (assuming that the impulse response of the linear system is real), $y_i(m, n)$, $y_r(m, n)$, and $y(m, n)$ are all positive. Thus the physical constraint that intensities in a physical image must be strictly positive is satisfied. This behavior is guaranteed for all systems that have the form of Fig. 10.5. This is in contrast to the situation when the intensity representation of an image is processed by a linear system, in which case positive outputs are not generally guaranteed.

Having demonstrated that multiplicative homomorphic systems are well matched to the structure of images, there remains the question of how one chooses the linear system in Fig. 10.5. This choice, of course, depends upon the properties of the density components $\hat{x}_i(m, n)$ and $\hat{x}_r(m, n)$. Fortunately, the illumination and reflectance components have distinctly different characteristics. Illumination generally does not vary rapidly across a scene, although shadows do, of course, correspond to abrupt variations in illumination. On the other hand, if there is significant detail in the scene, the reflectance component will vary rapidly because of sharp edges and changes in texture and size. Thus it is reasonable to assume that the illumination is a low-frequency signal, whereas the reflectance is a high-frequency signal. Similarly, the illumination density is slowly varying, whereas the reflectance density is rapidly varying. Plots of the Fourier transform of an image or the logarithm of an image are typically characterized by a low-frequency peak and a high-frequency plateau, implicitly suggesting that the above assumptions are reasonable, although it is clearly an oversimplification to associate the low-frequency peak entirely with illumination and the high frequencies entirely with reflectance.

Although illumination may vary slowly, it may also vary a great deal across a scene, thus causing a wide dynamic range in the image. This presents significant problems in transmitting images over a communication channel or storing them on a medium such as photographic film. Thus an important problem in image processing—dynamic range reduction—directs our attention to processing the illumination component.

Another important problem in image processing is contrast enhancement, that is, processing the image so as to sharpen the edges of objects in the scene. Since these edges manifest themselves primarily as abrupt variations in reflectance components, we are then concerned with processing the reflectance component in order to alter the contrast of an image.

In order to see how to perform dynamic range reduction and contrast enhancement using the canonic system of Fig. 10.5, let us first assume that the linear system is an ideal frequency-independent gain γ; i.e.,

$$\hat{y}(m, n) = \gamma \hat{x}(m, n)$$

Then the output image will be

$$y(m, n) = [x(m, n)]^\gamma = [x_i(m, n)]^\gamma \cdot [x_r(m, n)]^\gamma \qquad (10.23)$$

The dynamic range of the image can be reduced by reducing the variation of the illumination component. Clearly this implies a choice of $\gamma < 1$. To increase the contrast of the image, we must process the reflectance so that the ratio between two given intensities is increased, implying that $\gamma > 1$.

The process of Eq. (10.23) corresponding to a simple gain for the linear system is available photographically; however, this choice for the linear system is often not a satisfactory solution since for $\gamma < 1$ contrast is reduced along with dynamic range, and for $\gamma > 1$ dynamic range is increased along with contrast. However, if we recall that illumination is a low-frequency signal and reflectance is a high-frequency signal, it is clear that a linear shift-invariant filter that applies a different gain to the high and low frequencies would be a more effective choice. For example, in Fig. 10.6 is indicated a

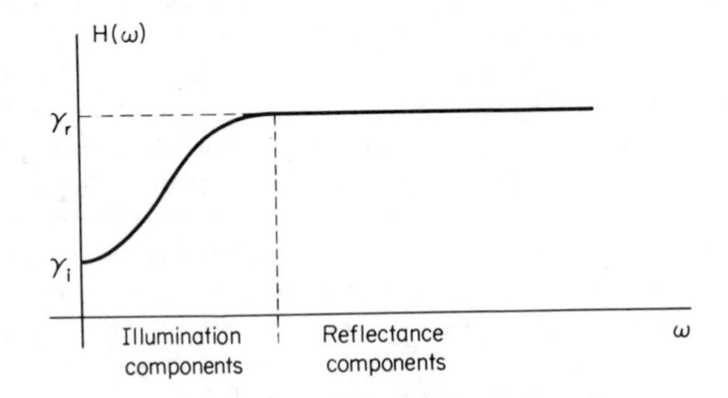

Fig. 10.6 Cross section of a circularly symmetric frequency response to be used for the linear part of a homomorphic image processor to achieve simultaneous contrast enhancement and dynamic range compression.

radial cross section of a circularly symmetric frequency response. At least approximately such a system offers the possibility of simultaneous contrast enhancement and dynamic range reduction; i.e., $y(m, n)$ is approximately of the form

$$y(m, n) = [x_i(m, n)]^{\gamma_i}[x_r(m, n)]^{\gamma_r} \tag{10.24}$$

Thus for simultaneous dynamic range reduction and contrast enhancement, γ_i is chosen less than unity and γ_r is chosen greater than unity.

Two examples of images processed in this way for simultaneous dynamic range compression and contrast enhancement are indicated in Figs. 10.7 and 10.8. The low-frequency gain of the filter, γ_i, was chosen to be 0.5, and the high-frequency gain, γ_r, was chosen to be 2, corresponding to a choice of $\gamma_i = 0.5$ and $\gamma_r = 2$ in Eq. (10.24). Figure 10.7 shows the original scenes and Fig. 10.8 shows the processed scenes. More examples of the enhancement referred to can be found in Ref. [3].

10.4 Homomorphic Systems for Convolution

There are a multitude of signal processing problems where signals are combined by convolution. In communicating or recording in a multipath or reverberant environment, the effect of the distortion introduced can be modeled in terms of noise convolved with the desired signal. In speech processing, it is often of interest to isolate the effects of vocal-tract impulse response and excitation which, at least on a short-time basis, can be considered as having been convolved to form the speech waveform. Other examples are the separation of probability density functons that have been convolved by the addition of independent random processes or processing of seismic signals obtained when an explosion creates a pulse of seismic energy that propagates through the earth.

A common approach to separating the components of such signals, i.e. to deconvolution, is linear inverse filtering. Unfortunately, since linear systems are not matched to the structure of convolutional combinations, inverse filtering requires detailed knowledge of one of the component signals. As an alternative, we are led to consider a class of homomorphic systems which obeys a generalized principle of superposition for convolution [4,7,8].

10.4.1 *Canonic System*

Let us consider sequences combined by discrete convolution; i.e.,

$$x(n) = \sum_{k=-\infty}^{\infty} x_1(k)x_2(n-k) = x_1(n) * x_2(n) \tag{10.25}$$

It is easily shown that discrete convolution satisfies the algebraic postulates for vector addition and consequently is a permissible choice for a class of homomorphic systems. Scalar multiplication by an integer a corresponds to repeated convolution of $x(n)$ with itself a times, and scalar multiplication when a is not an integer is a generalization of this [1,8].

The canonic form for homomorphic filters under convolution is depicted in Fig. 10.9. The characteristic system, D_*, has the property

$$\begin{aligned} D_*[x_1(n) * x_2(n)] &= D_*[x_1(n)] + D_*[x_2(n)] \\ &= \hat{x}_1(n) + \hat{x}_2(n) \end{aligned} \tag{10.26a}$$

$$D_*[c : x_1(n)] = cD_*[x_1(n)] = c\hat{x}_1(n) \tag{10.26b}$$

The system L is a linear system and D_*^{-1} is the inverse of D_*. Thus, if we can determine the system D_*, we have a representation of all the systems that obey a generalized principle of superposition for convolution. For the remainder of this section and in Secs. 10.5 and 10.6 we shall explore in considerable detail the properties of the system D_* and the properties of the signals $\hat{x}(n)$. In Sec. 10.7 we shall then utilize these properties in a number of applications of homomorphic deconvolution.

Fig. 10.7 Two original images.

Fig. 10.8 Images of Fig. 10.7 after processing to achieve simultaneous dynamic range compression and contrast enhancement.

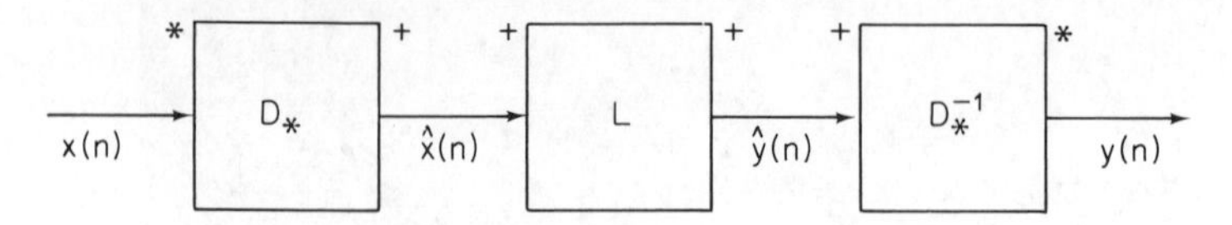

Fig. 10.9 Canonic form for homomorphic filters with convolution as the input and output operations.

10.4.2 Mathematical Representations of the Characteristic System D_*

The key to the mathematical representation of the characteristic system D_* is based on the fact that the z-transform of Eq. (10.25) is

$$X(z) = X_1(z) \cdot X_2(z) \tag{10.27}$$

That is, the z-transform operation $\mathfrak{Z}[x(n)]$ can be viewed as a homomorphic transformation with convolution as the input operation and multiplication as the output operation, as depicted in Fig. 10.10. Using the z-transformation,

Fig. 10.10 z-transform depicted as a homomorphic transformation from convolution to multiplication.

convolutional combinations can be transformed into multiplicative combinations, which can then be processed by a multiplicative homomorphic system. Thus, if we represent signals by their z-transforms, the canonic system of Fig. 10.9 can be replaced by that shown in Fig. 10.11. Since the

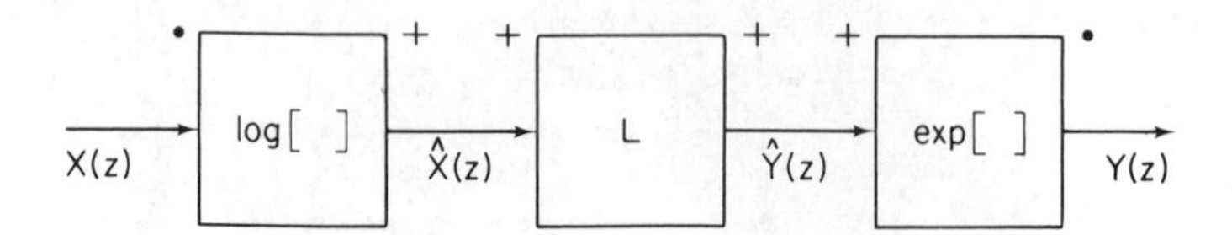

Fig. 10.11 System of Fig. 10.9 with the signals represented in terms of their z-transforms.

function $X(z)$ is normally complex, the complex logarithm must be employed here.

When signals are represented as sequences rather than by their z-transforms, we can formally represent the characteristic system D_* as in Fig. 10.12, where $\mathfrak{Z}^{-1}$ is the inverse z-transform. It is interesting to note that both $\mathfrak{Z}$ and $\mathfrak{Z}^{-1}$ are linear transformations in the conventional sense as well as being

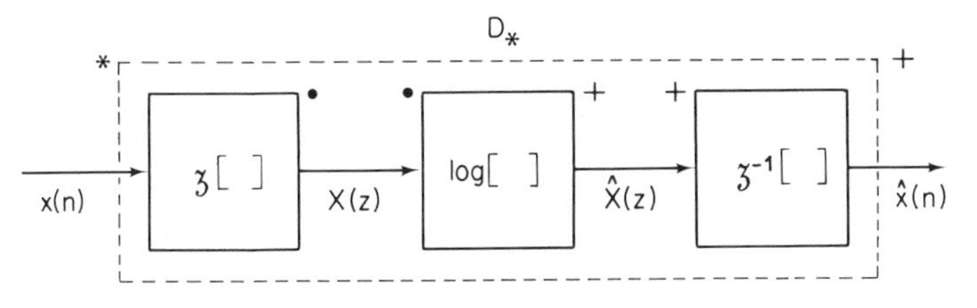

Fig. 10.12 Representation of the characteristic system D_*.

homomorphic transformations between convolutional and multiplicative vector spaces.

There are a number of important assumptions implicit in the representation of the characteristic system D_* as in Fig. 10.12. First, the complex logarithm, log $[X(z)]$, must be uniquely defined so that if

$$X(z) = X_1(z) \cdot X_2(z)$$

then

$$\begin{aligned}\hat{X}(z) &= \log [X_1(z) \cdot X_2(z)] \\ &= \log [X_1(z)] + \log [X_2(z)]\end{aligned}$$

Second, $\hat{X}(z)$ must be a valid z-transform. Third, in order for $\hat{x}(n)$ to be uniquely defined, we must choose a region of convergence for $\hat{X}(z) =$ log $[X(z)]$. Let us consider the choice of region of convergence first. Assume that both $x(n)$ and $\hat{x}(n)$ are real, stable sequences. This assumption, clearly very reasonable in a practical sense, is not really restrictive since only slight modifications are required to allow either $x(n)$ or $\hat{x}(n)$ to be unstable. Therefore, the regions of convergence of both $X(z)$ and $\hat{X}(z)$ must include the unit circle.

If $\hat{X}(z) =$ log $[X(z)]$ is a z-transform, then it must have a Laurent series expansion,

$$\hat{X}(z) = \log [X(z)] = \sum_{n=-\infty}^{\infty} \hat{x}(n) z^{-n}$$

with a region of convergence that includes the unit circle. That is, $\hat{X}(z)$ must be analytic in a region that includes the unit circle. Let us express $\hat{X}(z)$ on the unit circle as

$$\hat{X}(e^{j\omega}) = \hat{X}_R(e^{j\omega}) + j\hat{X}_I(e^{j\omega})$$

Since $\hat{x}(n)$ is real, $\hat{X}_R(e^{j\omega})$ must be an even function of ω, and $\hat{X}_I(e^{j\omega})$ must be an odd function of ω. In addition, $\hat{X}(e^{j\omega})$ must be a periodic function of ω with period 2π. The most important implication of the analyticity of $\hat{X}(z)$ on the unit circle is that $\hat{X}(e^{j\omega})$ must be a continuous function of ω. Since

$$\hat{X}(e^{j\omega}) = \log |X(e^{j\omega})| + j \arg [X(e^{j\omega})]$$

this implies that

$$\hat{X}_R(e^{j\omega}) = \log |X(e^{j\omega})|$$

and

$$\hat{X}_I(e^{j\omega}) = \arg [X(e^{j\omega})]$$

must be continuous functions of ω. Provided that $X(z)$ does not have zeros on the unit circle, the continuity of $\hat{X}_R(e^{j\omega})$ is guaranteed by the fact that $X(e^{j\omega})$ is analytic on the unit circle. However, the continuity of $\hat{X}_I(e^{j\omega})$ is dependent on the definition of the complex logarithm. Thus the requirement that $\hat{X}(z)$ be a valid z-transform and the removal of the ambiguity of the complex logarithm are interrelated.

The problem of uniqueness and analyticity of the complex logarithm is illustrated by Fig. 10.13. In Fig. 10.13(a) we see a typical phase curve for the z-transform $X(z)$ evaluated on the unit circle. If $X(z)$ is the product of two z-transforms, then this curve can be thought of as the sum of the continuous phase curves of the component transforms. Figure 10.13(b) shows the

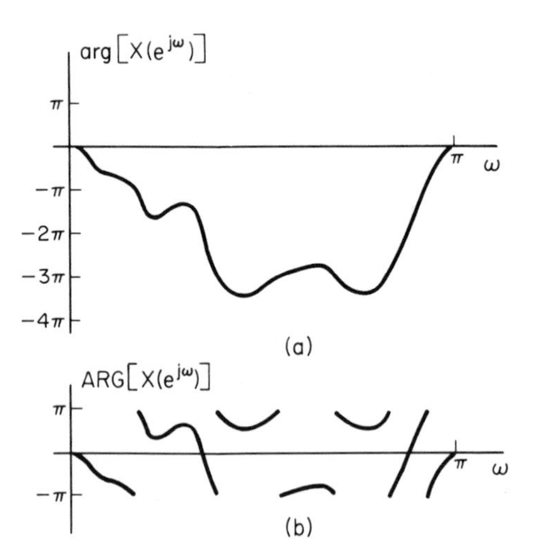

Fig. 10.13 (a) Typical phase curve for a z-transform evaluated on the unit circle (b) principal value of the phase curve in (a).

principal value of the phase of $X(z)$. Both curves are representations of the phase of $X(z)$ since

$$e^{j\arg[X(z)]} = e^{j\text{ARG}X[(z)]}$$

However, it is easily seen that the principal-value curve could not in general correspond to the sum of principal values of the phases of the component transforms, and furthermore ARG $[X(z)]$ is discontinuous and thus cannot satisfy the continuity requirement that results from the fact that $\hat{X}(z)$ must be analytic on the unit circle.

One approach to the problems presented by the complex logarithm is through the concept of the Riemann surface. Another approach to defining the complex logarithm is to assume that the continuous complex logarithm is obtained by integration of its derivative. If we assume a single-valued differentiable complex logarithm, then

$$\frac{d}{dz}\log[X(z)] = \frac{1}{X(z)}\frac{dX(z)}{dz} = \frac{d\hat{X}(z)}{dz} \tag{10.28}$$

If we evaluate this logarithmic derivative on the unit circle, we obtain

$$\hat{X}'(e^{j\omega}) = \frac{X'(e^{j\omega})}{X(e^{j\omega})} = \hat{X}'_R(e^{j\omega}) + j\hat{X}'_I(e^{j\omega})$$

where the prime denotes differentiation with respect to ω. From this expression we obtain

$$\frac{d\hat{X}_I(e^{j\omega})}{d\omega} = \frac{X_R(e^{j\omega})X'_I(e^{j\omega}) - X_I(e^{j\omega})X'_R(e^{j\omega})}{X_R^2(e^{j\omega}) + X_I^2(e^{j\omega})} \tag{10.29}$$

Integrating Eq. (10.29) with respect to ω, we can apply the condition

$$\arg[X(e^{j\omega})]_{\omega=0} = 0$$

to ensure that $\arg[X(e^{j\omega})]$ is an odd, continuous function of ω.

The complex logarithm has been carefully considered for two reasons. First, it is important to fully understand the complex logarithm so that the formal manipulations that we will make in the remainder of this section and in later sections can be done with confidence in their validity. Second, the problems of ambiguity in the definition of the complex logarithm manifest themselves as important computational problems. These computational problems will be considered in a later section on realization of homomorphic systems for convolution.

The mathematical representation of Fig. 10.12 was obtained by recognizing that the multiplicative theory of Sec. 10.2 could be applied to the z-transforms of a convolution. Starting with this representation and the implicit assumption of analyticity of $\log[X(z)]$, we can derive two other representations of the system D_* using the logarithmic derivative.

Assuming that $\log[X(z)]$ is analytic, then

$$\hat{X}'(z) = \frac{X'(z)}{X(z)} \tag{10.30}$$

where $'$ denotes differentiation with respect to z. It is easily shown that

$$z\hat{X}'(z) = \sum_{n=-\infty}^{\infty}[-n\hat{x}(n)]z^{-n} = \frac{zX'(z)}{X(z)} \tag{10.31}$$

so that

$$-n\hat{x}(n) = \frac{1}{2\pi j} \oint_C \frac{zX'(z)}{X(z)} z^{n-1}\, dz$$

where C denotes a closed contour in the region of convergence of $\hat{X}(z)$. Solving for $\hat{x}(n)$ we obtain

$$\hat{x}(n) = \frac{-1}{2\pi jn} \oint_C \frac{zX'(z)}{X(z)} z^{n-1}\, dz, \qquad n \neq 0 \tag{10.32}$$

The value of $\hat{x}(0)$ can be obtained by noting that

$$\begin{aligned}\hat{x}(0) &= \frac{1}{2\pi} \int_{-\pi}^{\pi} \hat{X}(e^{j\omega})\, d\omega \\ &= \frac{1}{2\pi} \int_{-\pi}^{\pi} \hat{X}_R(e^{j\omega})\, d\omega + \frac{1}{2\pi} \int_{-\pi}^{\pi} \hat{X}_I(e^{j\omega})\, d\omega\end{aligned}$$

Since $\hat{X}_I(e^{j\omega})$ is an odd function of ω, this becomes

$$\begin{aligned}\hat{x}(0) &= \frac{1}{2\pi} \int_{-\pi}^{\pi} \hat{X}_R(e^{j\omega})\, d\omega \\ &= \frac{1}{2\pi} \int_{-\pi}^{\pi} \log |X(e^{j\omega})|\, d\omega\end{aligned} \tag{10.33}$$

Beginning with Eq. (10.31) we can also derive a difference equation that represents the system D_*. Rearranging Eq. (10.31) we obtain

$$zX'(z) = z\hat{X}'(z) \cdot X(z)$$

The inverse z-transform of this equation is

$$nx(n) = \sum_{k=-\infty}^{\infty} k\hat{x}(k)x(n-k) \tag{10.34}$$

Dividing by n we obtain

$$x(n) = \sum_{k=-\infty}^{\infty} \left(\frac{k}{n}\right)\hat{x}(k)x(n-k), \qquad n \neq 0 \tag{10.35}$$

Thus we have obtained an implicit relation between $\hat{x}(n)$ and $x(n)$. Under certain conditions, this expression can be rearranged into a recursion formula that can be used in computation. Formulas of this type are discussed in Sec. 10.5.

10.4.3 Inverse Characteristic System D_*^{-1}

The mathematical representation of the inverse characteristic system D_*^{-1} follows simply from the representation of D_* and is depicted in Fig. 10.14. By definition,

$$D_*^{-1}[D_*[x(n)]] = x(n)$$

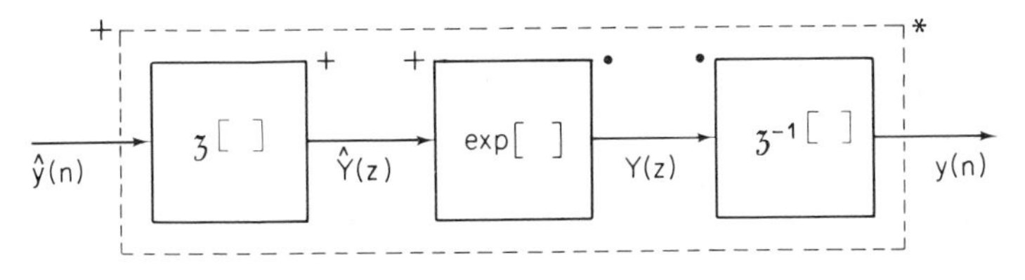

Fig. 10.14 Representation of the system D_*^{-1}.

Therefore, it must be true that both $\hat{y}(n)$ and $y(n)$ are stable sequences, since both $x(n)$ and $\hat{x}(n)$ are assumed to be stable. Thus the regions of convergence of $Y(z)$ and $\hat{Y}(z)$ must include the unit circle. Therefore,

$$y(n) = \frac{1}{2\pi j} \oint_{C'} Y(z) z^{n-1}\, dz$$

where C' is the unit circle and

$$Y(z) = \exp\,[\hat{Y}(z)]$$

Fortunately, the complex exponential function has no uniqueness problems, and if $\hat{Y}(z)$ is analytic on the unit circle, then so is $\exp\,[\hat{Y}(z)]$.

10.4.4 Linear System L

Since we have given a mathematical representation of the characteristic system D_* and its inverse, all that remains to be specified in the canonic system of Fig. 10.9 is the system L. In theory, any system that obeys superposition for addition can be used in the canonic system of Fig. 10.9. However, in practice we have found that a particular class of linear systems is most useful. We recall that in the case of multiplicative systems, a linear time-invariant system was the useful choice for the linear system. Also, we recall from Fig. 10.11 that if signals are represented by their z-transforms, then we can think of a linear system operating on the complex logarithm of the z-transform. In other words, the class of convolutional homomorphic systems is analogous to the class of multiplicative homomorphic systems; however, the roles of the time domain and frequency domain are in a sense interchanged. Thus, although time-invariant linear systems can theoretically be used, it is of particular interest to consider the class of frequency-invariant linear systems for which

$$\hat{Y}(e^{j\omega}) = \frac{1}{2\pi} \int_{-\pi}^{\pi} \hat{X}(e^{j\theta}) L(e^{j(\omega-\theta)})\, d\theta \tag{10.36}$$

For such a system, the output Fourier transform is obtained from the complex logarithm $\hat{X}(e^{j\omega}) = \log\,[X(e^{j\omega})]$, by a periodic continuous-variable convolution. Alternatively, such a system has a time-domain representation as

$$\hat{y}(n) = l(n)\hat{x}(n) \tag{10.37}$$

where $l(n)$ is the inverse transform of $L(e^{j\omega})$. Since $x(n)$, $\hat{x}(n)$, $\hat{y}(n)$, and $y(n)$ are all assumed to be real, stable sequences, it follows that $l(n)$ must be real, and in general it must also be stable. This implies that $L(z)$, the z-transform of $l(n)$, has a region of convergence that includes the unit circle, and that the real and imaginary parts of $L(e^{j\omega})$ are even and odd functions of ω, respectively.

We may naturally inquire further into the question of why this particular class of linear systems is most useful. However, this question and the questions of design criteria for such systems can be answered only in the light of further discussion of the properties of $\hat{x}(n)$ or $\log [X(e^{j\omega})]$. This is the subject of Sec. 10.5.

10.4.5 Note on Terminology

We conclude this section with a brief historical note that will aid in establishing some terminology for the remainder of the chapter. In 1962 Bogert, Healy, and Tukey published a paper with a rather unusual title [9]. In this paper they observed that the logarithm of the power spectrum of a signal containing an echo had an additive periodic component due to the echo, and thus the Fourier transform of the log-power spectrum should exhibit a peak at the echo delay. This function they termed the *cepstrum*, paraphrasing the word spectrum because: "In general, we find ourselves operating on the frequency side in ways customary on the time side and vice versa" [9]. Bogert et al. went on to define a whole vocabulary to complement this new signal processing technique; however, only the term "cepstrum" has been widely used. Since the power spectrum is the Fourier transform of the autocovariance function and is always positive, we can think of the cepstrum as being the output of the characteristic system D_* when the input is an autocorrelation. Since the power spectrum is always positive, only the real logarithm is required. In general, we must use the *complex* logarithm and *complex* Fourier transforms; so to emphasize the relationship while maintaining the distinction, we call the output of the characteristic system the *complex cepstrum*. We hasten to add that the *complex cepstrum* $\hat{x}(n)$ is, of course, real for real inputs $x(n)$. We retain the term "cepstrum" for use when only the real logarithm is used.

10.5 Properties of the Complex Cepstrum

In order to see how we might employ convolutional homomorphic systems, we shall consider the properties of the complex cepstrum—or, equivalently, $\log [X(e^{j\omega})]$—for a useful class of input signals, the class of exponential sequences.

10.5.1 Exponential Sequences

The class of sequences that have rational z-transforms is fortunately both useful and amenable to analysis. Therefore, consider input sequences $x(n)$ whose z-transforms are of the form

$$X(z) = \frac{Az^r \prod_{k=1}^{m_i} (1 - a_k z^{-1}) \prod_{k=1}^{m_o} (1 - b_k z)}{\prod_{k=1}^{p_i} (1 - c_k z^{-1}) \prod_{k=1}^{p_o} (1 - d_k z)} \tag{10.38}$$

where $|a_k|$, $|b_k|$, $|c_k|$, and $|d_k|$ are all less than unity, so that factors of the form $(1 - a_k z^{-1})$ and $(1 - c_k z^{-1})$ correspond to zeros and poles inside the unit circle, and the factors $(1 - b_k z)$ and $(1 - d_k z)$ correspond to zeros and poles outside the unit circle. Such z-transforms are characteristic of sequences composed of a sum of exponential sequences. In the special case when there are no poles, i.e., the denominator of Eq. (10.38) is unity, Eq. (10.38) corresponds to a finite-length sequence.

If $\log [X(z)]$ is computed as assumed in Sec. 10.4, we can write formally that

$$\begin{aligned} \hat{X}(z) = \log [A] + \log [z^r] + \sum_{k=1}^{m_i} \log (1 - a_k z^{-1}) \\ + \sum_{k=1}^{m_o} \log (1 - b_k z) - \sum_{k=1}^{p_i} \log (1 - c_k z^{-1}) - \sum_{k=1}^{p_o} \log (1 - d_k z) \end{aligned} \tag{10.39}$$

If we consider each term in Eq. (10.39) as a z-transform, then the properties of $\hat{x}(n)$ will depend upon the composite properties of the inverse transforms of each term.

For real sequences, A is real and if A is positive, the first term $\log [A]$ simply contributes to $\hat{x}(0)$. Specifically (see Problem 11 of this chapter)

$$\hat{x}(0) = \log |A| \tag{10.40}$$

If A is negative, it is more difficult to determine the contribution to the complex cepstrum due to the term $\log [A]$. Similarly, the term z^r corresponds only to a delay or advance of the sequence $x(n)$. If $r = 0$, this term vanishes from Eq. (10.39). However, if $r \neq 0$, there will be a nonzero contribution to the complex cepstrum. Although the cases of A negative and/or $r \neq 0$ can be formally accommodated [8], doing so seems to offer no real advantage, because if two transforms of the form of Eq. (10.38) are multiplied together, we would not expect to be able to determine how much of either A or r was contributed by each component. This is analogous in linear filtering to the situation in which two signals, each with dc levels, have been added. Thus in practice this question is avoided by measuring the algebraic sign of A and the value of r and then altering the input so that its z-transform is of the form

$$X(z) = \frac{|A| \prod_{k=1}^{m_i} (1 - a_k z^{-1}) \prod_{k=1}^{m_o} (1 - b_k z)}{\prod_{k=1}^{p_i} (1 - c_k z^{-1}) \prod_{k=1}^{p_o} (1 - d_k z)} \tag{10.41}$$

Likewise, Eq. (10.39) becomes

$$\begin{aligned} \hat{X}(z) = \log |A| &+ \sum_{k=1}^{m_i} \log (1 - a_k z^{-1}) + \sum_{k=1}^{m_o} \log (1 - b_k z) \\ &- \sum_{k=1}^{p_i} \log (1 - c_k z^{-1}) - \sum_{k=1}^{p_o} \log (1 - d_k z) \end{aligned} \tag{10.42}$$

With the exception of the term $\log |A|$, which we have already considered, all the terms in Eq. (10.42) are of the form $\log (1 - \alpha z^{-1})$ and $\log (1 - \beta z)$. Bearing in mind that these factors must represent z-transforms with convergence regions that include the unit circle, we can make the power-series expansions

$$\log (1 - \alpha z^{-1}) = -\sum_{n=1}^{\infty} \frac{\alpha^n}{n} z^{-n}, \qquad |z| > |\alpha| \tag{10.43}$$

and

$$\log (1 - \beta z) = -\sum_{n=1}^{\infty} \frac{\beta^n}{n} z^{n}, \qquad |z| < |\beta^{-1}| \tag{10.44}$$

Using these expressions it is clear that for inputs with rational z-transforms as in Eq. (10.41), $\hat{x}(n)$ has the general form

$$\hat{x}(n) = \log |A| \qquad n = 0 \tag{10.45a}$$

$$= -\sum_{k=1}^{m_i} \frac{a_k^n}{n} + \sum_{k=1}^{p_i} \frac{c_k^n}{n}, \qquad n > 0 \tag{10.45b}$$

$$= \sum_{k=1}^{m_o} \frac{b_k^{-n}}{n} - \sum_{k=1}^{p_o} \frac{d_k^{-n}}{n}, \qquad n < 0 \tag{10.45c}$$

Note that for the special case of a finite-length sequence the second term would be missing in each of Eqs. (10.45b) and (10.45c). From Eqs. (10.45) we observe the following properties of the complex cepstrum.

P1. The complex cepstrum decays at least as fast as $1/n$: Specifically

$$|\hat{x}(n)| < C \left| \frac{\alpha^n}{n} \right|, \qquad -\infty < n < \infty$$

where C is a constant and α equals the maximum of $|a_k|$, $|b_k|$, $|c_k|$, and $|d_k|$. This property is also evident from Eq. (10.32).

P2. If $x(n)$ is minimum phase (no poles or zeros outside the unit circle), then

$$\hat{x}(n) = 0, \qquad n < 0$$

P3. If $x(n)$ is maximum phase (no poles or zeros inside the unit circle), then

$$\hat{x}(n) = 0, \qquad n > 0$$

P4. If $x(n)$ is of finite duration, $\hat{x}(n)$ will nevertheless have infinite duration.

10.5.2 Minimum-Phase and Maximum-Phase Sequences

Let us consider some important implications of properties P2–P4. First consider a minimum-phase sequence whose z-transform is of the form

$$X(z) = |A| \frac{\prod_{k=1}^{m_i} (1 - a_k z^{-1})}{\prod_{k=1}^{p_i} (1 - c_k z^{-1})}$$

Clearly

$$x(n) = 0, \qquad n < 0 \tag{10.46a}$$

and from property P2,

$$\hat{x}(n) = 0, \qquad n < 0 \tag{10.46b}$$

Thus, for minimum-phase inputs, the sequence $\hat{x}(n)$ is causal. In this case the mathematical representation of the system D_* can be significantly simplified. We recall from Chapter 7 that the z-transform of a causal sequence is completely determined by the real part of its Fourier transform. Since both $x(n)$ and $\hat{x}(n)$ are causal, we therefore should only have to compute

$$\hat{X}_R(e^{j\omega}) = \log |X(e^{j\omega})|$$

in order to obtain $\hat{x}(n)$. Recall that the inverse Fourier transform of $\hat{X}_R(e^{j\omega})$ is equal to the even part of $\hat{x}(n)$, which we denote by $c(n)$, so

$$c(n) = \frac{\hat{x}(n) + \hat{x}(-n)}{2}$$

Since $\hat{x}(n) = 0$ for $n < 0$ for a minimum-phase sequence, we can write

$$\hat{x}(n) = c(n) \cdot u_+(n) \tag{10.47}$$

where

$$u_+(n) = \begin{cases} 0, & n < 0 \\ 1, & n = 0 \\ 2, & n > 0 \end{cases}$$

These operations are depicted in Fig. 10.15. The sequence $c(n)$ is called the *cepstrum* of the input $x(n)$ because of its close similarity to the original definition of Bogert, et al. Note that only in the case of minimum-phase (or maximum-phase) inputs can we obtain the complex cepstrum from the cepstrum as in Fig. 10.15.

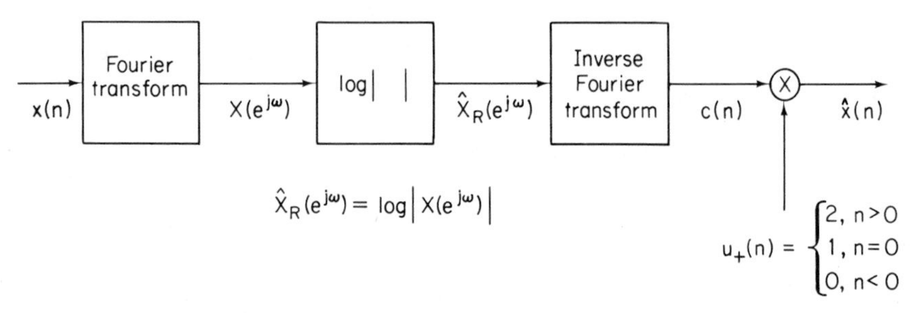

Fig. 10.15 Realization of the system D_* for a minimum-phase input.

An alternative representation is obtained by considering the difference equation in Eq. (10.35). If we apply the conditions of Eqs. (10.46) we obtain

$$\begin{aligned} x(n) &= \sum_{k=0}^{n} \left(\frac{k}{n}\right) \hat{x}(k)x(n-k), \qquad n > 0 \\ &= \hat{x}(n)x(0) + \sum_{k=0}^{n-1} \left(\frac{k}{n}\right) \hat{x}(k)x(n-k) \end{aligned} \tag{10.48}$$

Solving for $\hat{x}(n)$ yields the recursion formula

$$\hat{x}(n) = \begin{cases} 0, & n < 0 \\ \dfrac{x(n)}{x(0)} - \displaystyle\sum_{k=0}^{n-1} \left(\frac{k}{n}\right) \hat{x}(k) \frac{x(n-k)}{x(0)}, & n > 0 \end{cases} \tag{10.49}$$

The value of $\hat{x}(0)$ can easily be shown to be (see Problem 9 of this chapter)

$$\hat{x}(0) = \log [A] = \log [x(0)] \tag{10.50}$$

Therefore, Eqs. (10.49) and (10.50) constitute a representation of the system D_* for minimum-phase systems. It also follows from Eq. (10.49) that the system D_* is a causal system for minimum-phase inputs; i.e., the output for $n < n_0$ is dependent only on the input for $n < n_0$, where n_0 is arbitrary (see Problem 15 at the end of this chapter). Similarly, Eqs. (10.48) and (10.50) represent the inverse characteristic system D_*^{-1}.

In the case of maximum-phase sequences we can parallel the previous discussion. A maximum-phase sequence has no poles or zeros inside the unit circle. Therefore,

$$x(n) = \hat{x}(n) = 0, \qquad n > 0 \tag{10.51}$$

It is likewise true that only $\log|X(e^{j\omega})|$ is required to compute $\hat{x}(n)$, since

$$\hat{x}(n) = u_-(n) \cdot c(n)$$

where

$$u_-(n) = \begin{cases} 2, & n < 0 \\ 1, & n = 0 \\ 0, & n > 0 \end{cases} \tag{10.52}$$

Thus Fig. 10.15 applies to maximum-phase sequences if $u_+(n)$ is replaced by $u_-(n)$.

If we apply Eq. (10.51) to Eq. (10.35), we obtain

$$\begin{aligned} x(n) &= \sum_{k=n}^{0} \left(\frac{k}{n}\right) \hat{x}(k) x(n-k), \qquad n < 0 \\ &= \hat{x}(n)x(0) + \sum_{k=n+1}^{0} \left(\frac{k}{n}\right) \hat{x}(k) x(n-k) \end{aligned} \tag{10.53}$$

Solving for $\hat{x}(n)$,

$$\hat{x}(n) = \begin{cases} \dfrac{x(n)}{x(0)} - \displaystyle\sum_{k=n+1}^{0} \left(\frac{k}{n}\right) \hat{x}(k) \frac{x(n-k)}{x(0)}, & n < 0 \\ \log\,[x(0)], & n = 0 \\ 0, & n > 0 \end{cases} \tag{10.54}$$

These equations then serve as a representation of the characteristic system and its inverse for maximum-phase inputs.

This discussion of minimum- and maximum-phase sequences has an interesting implication for finite-length sequences. Specifically, in spite of property P4, we can show that for an input sequence of length N, we need only N samples of $\hat{x}(n)$ to determine $x(n)$. To see this, consider the z-transform

$$X(z) = X_{\text{min}}(z) \cdot X_{\text{max}}(z)$$

where

$$X_{\text{min}}(z) = A \prod_{k=1}^{m_i} (1 - a_k z^{-1})$$

$$X_{\text{max}}(z) = \prod_{k=1}^{m_0} (1 - b_k z)$$

Correspondingly,

$$x(n) = x_{\text{min}}(n) * x_{\text{max}}(n)$$

where $x_{\text{min}}(n) = 0$ outside the interval $0 \le n \le m_i$ and $x_{\text{max}}(n) = 0$ outside the interval $-m_o \le n \le 0$. Thus the sequence $x(n)$ is nonzero in the interval

$-m_o \leq n \leq m_i$. Using the previous recursion formulas we can write

$$x_{\min}(n) = \begin{cases} 0 & n < 0 \\ e^{\hat{x}(0)} & n = 0 \\ \hat{x}(n)x(0) + \sum_{k=0}^{n-1} \left(\frac{k}{n}\right) \hat{x}(k) x_{\min}(n-k) & n > 0 \end{cases} \tag{10.55}$$

and

$$x_{\max}(n) = \begin{cases} 0 & n > 0 \\ 1 & n = 0 \\ \hat{x}(n) + \sum_{k=n+1}^{0} \left(\frac{k}{n}\right) \hat{x}(k) x_{\max}(n-k) & n < 0 \end{cases} \tag{10.56}$$

Clearly we require $m_i + 1$ values of $\hat{x}(n)$ to compute $x_{\min}(n)$ and m_o values of $\hat{x}(n)$ to compute $x_{\max}(n)$. Thus only $m_i + m_o + 1$ values of the infinite sequence $\hat{x}(n)$ are required to completely recover the finite-length sequence $x(n)$.

10.5.3 Poles and Zeros on the Unit Circle

We have thus far not allowed poles or zeros on the unit circle. From both the theoretical and computational point of view there are good reasons for this. Recall that in the mathematical representation of the characteristic system, we chose the contour of integration to be the unit circle. If $X(z)$ has a pole or a zero on the unit circle, we cannot associate a region of convergence that includes the unit circle with $\log [X(z)]$. It can be shown that a factor $\log (1 - e^{j\theta}e^{-j\omega})$ has a Fourier series

$$\log (1 - e^{j\theta}e^{-j\omega}) = -\sum_{n=1}^{\infty} \frac{e^{j\theta n}}{n} e^{-j\omega n}$$

that converges in some sense. However, the real part of the complex log is infinite and the imaginary part is discontinuous. We prefer to avoid this added difficulty if possible.

Formally, this can be achieved by using a different contour C for the computation of $\hat{x}(n)$ from $\log [X(z)]$. Equivalently, we can multiply the input sequence by an exponential sequence as in

$$w(n) = \alpha^n x(n)$$

where α is real and positive. The resulting sequence has a z-transform

$$W(z) = X(\alpha^{-1}z)$$

Thus the poles and zeros of $X(z)$ are shifted radially by the factor α^{-1}. It is important to note that if $x(n) = x_1(n) * x_2(n)$, then

$$W(z) = X(\alpha^{-1}z) = X_1(\alpha^{-1}z) \cdot X_2(\alpha^{-1}z)$$

so that

$$w(n) = \alpha^n x_1(n) * \alpha^n x_2(n)$$

That is, exponential weighting of a convolution yields a convolution of exponentially weighted sequences.

In addition to providing a means for moving singularities of log $[X(z)]$ off the unit circle, exponential weighting is also a useful technique for converting a mixed-phase signal into either a minimum-phase or a maximum-phase signal.

10.6 Computational Realizations of the Characteristic System D_*

In Sec. 10.4 we gave several mathematical representations of the homomorphic transformation D_*, which we have called the characteristic system for convolution, and whose purpose is to transform a convolutional combination into an additive combination so that linear filtering can be applied. Implicit in all these representations was the assumption of uniqueness and continuity of the complex logarithm, and in two of the representations the Fourier transform was a basic component of the representation. If these mathematical representations are to serve as the basis for computational realizations of the system D_*, then we must deal with the problems of computing the Fourier transform and the complex logarithm.

10.6.1 *Realization Using the Complex Logarithm*

The system D_* is represented by the equations

$$X(e^{j\omega}) = \sum_{n=-\infty}^{\infty} x(n)e^{-j\omega n} \tag{10.57a}$$

$$\hat{X}(e^{j\omega}) = \log [X(e^{j\omega})] \tag{10.57b}$$

$$\hat{x}(n) = \frac{1}{2\pi}\int_{-\pi}^{\pi} \hat{X}(e^{j\omega})e^{j\omega n}\, d\omega \tag{10.57c}$$

Since digital computers perform finite computations, we are limited to finite-length input sequences, and we can compute the Fourier transform at only a finite number of points. That is, instead of using the Fourier transform, we must use the discrete Fourier transform. Thus, instead of Eqs. (10.57), we have the computational realization

$$X(k) = X(e^{j\omega})\big|_{\omega=(2\pi/N)k} = \sum_{n=0}^{N-1} x(n)e^{-j(2\pi/N)kn} \tag{10.58a}$$

$$\hat{X}(k) = \log [X(e^{j\omega})]\big|_{\omega=(2\pi/N)k} = \log [X(k)] \tag{10.58b}$$

$$\hat{x}_p(n) = \frac{1}{N}\sum_{k=0}^{N-1} \hat{X}(k)e^{j(2\pi/N)kn} \tag{10.58c}$$

By the sampling theorem for the z-transform that we derived in Chapter 3 it is clear that $\hat{x}_p(n)$ is related to the desired $\hat{x}(n)$ by

$$\hat{x}_p(n) = \sum_{k=-\infty}^{\infty} \hat{x}(n + kN) \tag{10.59}$$

Since the complex cepstrum in general has infinite duration, $\hat{x}_p(n)$ will be a time-aliased version of $\hat{x}(n)$; however, we noted in property P1 that in general $\hat{x}(n)$ decays faster than an exponential sequence, so that it is to be expected that the approximation would become increasingly better as N gets larger. Thus it may be necessary to append zeros to an input sequence so that the complex logarithm is sampled at a high-enough rate that severe time aliasing does not occur in the computation of the complex cepstrum.

In writing Eqs. (10.58) and (10.59), we have assumed that $\hat{X}(k)$ represents a sampled version of the continuous complex logarithm. Thus we must consider means for computing samples of arg $[X(e^{j\omega})]$ from the DFT, $X(k)$. A simple algorithm is based on the computation of

$$-\pi < \text{ARG}\,[X(k)] \leq \pi$$

which is achieved using standard inverse tangent routines available on most computers. This sampled principal value of the phase is then "unwrapped" to obtain samples of the continuous phase curve. Consider a finite-length input whose Fourier transform is

$$\begin{aligned} X(e^{j\omega}) &= \sum_{n=0}^{M} x(n)e^{-j\omega n} \\ &= Ae^{-j\omega m_o} \prod_{k=1}^{m_i} (1 - a_k e^{-j\omega}) \prod_{k=1}^{m_o} (1 - b_k e^{j\omega}) \end{aligned} \tag{10.60}$$

where $|a_k|$ and $|b_k|$ are less than unity and $M = m_o + m_i$. A continuous phase curve for a sequence of this form is shown in Fig. 10.16(a). The heavy dots indicate samples of $X(e^{j\omega})$ at $\omega = (2\pi/N)k$ as required for computation of $\hat{x}_p(n)$. (N is assumed to be even.) In Fig. 10.16(b) is shown the principal value and its samples as computed from the DFT of the input. It can be seen that in order to obtain the samples of the desired phase, we must add an appropriate integer multiple of 2π to the samples of the principal value. The appropriate multiple of 2π can be determined from ARG $[X(k)]$ if the samples are close enough together so that the discontinuities can be detected. If arg $[X(e^{j\omega})]$ is rapidly varying, we would expect $\hat{x}(n)$ to decay less rapidly than if it were slowly varying. Likewise, if arg $[X(e^{j\omega})]$ is rapidly varying, it requires finer sampling to ensure detection of the discontinuities of ARG $[X(e^{j\omega})]$. Thus the desire to minimize aliasing is consistent with the successful computation of samples of the continuous phase curve. The larger the value of N, the better the computational approximation. This is generally not a severe limitation because of the existence of FFT algorithms. In fact, the

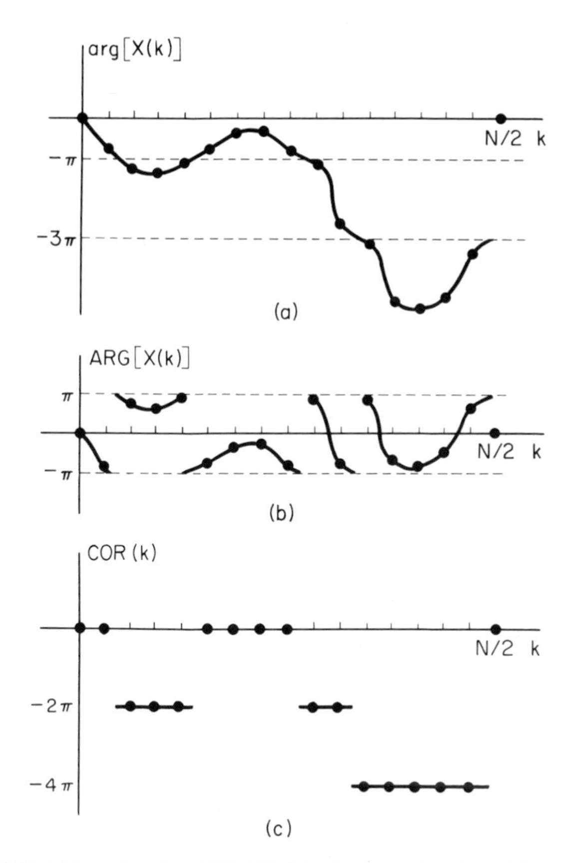

Fig. 10.16 (a) Samples of arg $[X(e^{j\omega})]$; (b) principal value of (a); (c) correction sequence for obtaining arg from ARG.

disclosure of the Cooley–Tukey algorithm was instrumental in stimulating applications of homomorphic systems for convolution.

A final comment about the computation of the continuous complex logarithm concerns the sign of A and the linear-phase component due to the factor $e^{-j\omega m_0}$. The sign of A can be easily determined since it is identical to the sign of $X(k)$ at $k = 0$. The value of m_o can be determined from the result of adding the correction to ARG $[X(k)]$ since it is easily shown from Eq. (10.60) that

$$\arg\,[X(e^{j\pi})] = -m_o\pi$$

This linear-phase component is subtracted from the phase, and the sign of A is effectively made positive before computing the complex cepstrum.

10.6.2 Realization Using the Logarithmic Derivative

As an alternative to the actual computation of the complex logarithm, a mathematical representation based on the logarithmic derivative can be derived. In terms of the Fourier transform this representation is

$$X(e^{j\omega}) = \sum_{n=-\infty}^{\infty} x(n)e^{-j\omega n} \tag{10.61}$$

$$X'(e^{j\omega}) = -j \sum_{n=-\infty}^{\infty} nx(n)e^{-j\omega n} \tag{10.62}$$

$$\hat{x}(n) = \frac{-1}{2\pi nj}\int_{-\pi}^{\pi} \frac{X'(e^{j\omega})}{X(e^{j\omega})} e^{j\omega n}\, d\omega, \qquad n \neq 0 \tag{10.63}$$

$$\hat{x}(0) = \frac{1}{2\pi}\int_{-\pi}^{\pi} \log |X(e^{j\omega})|\, d\omega \tag{10.64}$$

For finite-length sequences and using the DFT instead of the Fourier transform, these equations become

$$X(k) = \sum_{n=0}^{N-1} x(n)e^{-j(2\pi/N)kn} = X(e^{j\omega})\Big|_{\omega=2\pi k/N} \tag{10.65}$$

$$X'(k) = -j\sum_{k=0}^{N-1} nx(n)e^{-j(2\pi/N)kn} = X'(e^{j\omega})\Big|_{\omega=2\pi k/N} \tag{10.66}$$

$$\hat{x}_{dp}(n) = -\frac{1}{2\pi jnN}\sum_{k=0}^{N-1}\frac{X'(k)}{X(k)} e^{j(2\pi/N)kn}, \qquad 1 \leq n \leq N-1 \tag{10.67}$$

$$\hat{x}_{dp}(0) = \frac{1}{N}\sum_{k=0}^{N-1} \log |X(k)| \tag{10.68}$$

where the subscript d refers to the use of the logarithmic derivative and the subscript p anticipates the inherent periodicity of the DFT calculations. In this case we avoid the problems of computing the complex logarithm, at the cost, however, of more severe aliasing since now

$$\hat{x}_{dp}(n) = \frac{1}{n}\sum_{k=-\infty}^{\infty} (n + kN)\hat{x}(n + kN) \tag{10.69}$$

Thus, assuming that the sampled phase curve is accurately computed, we would expect that for a given value of N, $\hat{x}_p(n)$ in Eq. (10.58c) would be a better approximation to $\hat{x}(n)$ than would $\hat{x}_{dp}(n)$ in Eq. (10.67).

For finite-length sequences with Fourier transform as in Eq. (10.60), it can be shown that

$$m_o = \frac{-1}{2\pi j}\int_{-\pi}^{\pi} \frac{X'(e^{j\omega})}{X(e^{j\omega})}\, d\omega$$

Approximating this expression using the inverse DFT, we obtain

$$m_{op} = \frac{-1}{2\pi jN}\sum_{k=0}^{N-1}\frac{X'(k)}{X(k)}$$

The quantity m_{op} will, in general, not be an integer; however, for large N, we would expect m_{op} to approach m_o, the number of zeros of $X(z)$ outside the unit circle.

10.6.3 Minimum-Phase Realizations

In the special case of minimum-phase inputs, the mathematical representation is simplified as in Fig. 10.15. In this case the computational realization is given by the equations

$$(k) = \sum_{n=0}^{N-1} x(n)e^{-j(2\pi/N)kn} \tag{10.70a}$$

$$\hat{X}_R(k) = \log |X(k)| \tag{10.70b}$$

$$c_p(n) = \frac{1}{N}\sum_{k=0}^{N-1} \hat{X}_R(k)e^{j(2\pi/N)kn} \tag{10.70c}$$

In this case it is the cepstrum that is aliased; i.e.,

$$c_p(n) = \sum_{k=-\infty}^{\infty} c(n + kN)$$

To compute the complex cepstrum from $c_p(n)$ in analogy with Fig. 10.15, we write

$$\hat{x}_{cp}(n) = \begin{cases} c_p(n), & n = 0, N/2 \\ 2c_p(n), & 1 \le n < N/2 \\ 0, & N/2 < n \le N - 1 \end{cases}$$

Clearly $\hat{x}_{cp}(n) \neq \hat{x}_p(n)$ since it is the even part of $\hat{x}(n)$ that is aliased rather than $\hat{x}(n)$ itself. Nevertheless, for large N, $\hat{x}_{cp}(n)$ can be expected to differ only slightly from $\hat{x}(n)$. Similarly, if $x(n)$ is maximum phase, an approximation to the complex cepstrum would be obtained from

$$\hat{x}_{cp}(n) = \begin{cases} 0, & 1 \le n < N/2 \\ c_p(n), & n = 0, N/2 \\ 2c_p(n), & N/2 < n \le N - 1 \end{cases}$$

In the case of minimum-phase or maximum-phase sequences, we also have the recursion formulas of Eq. (10.49) through (10.56) as possible realizations of the characteristic system and its inverse. These equations can be quite useful when the input sequence is very short or when only a few samples of the complex cepstrum are desired. With these formulas, there is, of course, no aliasing error.

10.7 Applications of Homomorphic Deconvolution

The theoretical concepts discussed in Secs. 10.4–10.6 have been applied in a number of signal processing problems. These applications are conveniently grouped into two classifications: (1) estimation of speech parameters, and (2) dereverberation, broadly defined as the deconvolution of two or more signals when one is an impulse train.

10.7.1 Estimation of Speech Parameters

Let us begin by considering a model of speech production. Then we shall show how the theoretical results of previous sections can be utilized in estimation of the parameters of this model.

Speech is produced by excitation of an acoustic tube, called the *vocal tract*, which is terminated on one end by the lips and on the other end by the glottis [10]. Insofar as it remains in a fixed configuration, the vocal tract can be modeled as a linear time-invariant system whose output is the convolution of the impulse response of the vocal tract with the excitation waveform. Speech is generated in three basic ways. *Voiced sounds* are produced by exciting the vocal tract with quasi-periodic pulses of air flow caused by vibration of the vocal cords. *Fricative sounds* are produced by forming a constriction in the tract and forcing air through the constriction so that turbulence is created, thereby producing a noise-like excitation. *Plosive sounds* are produced by completely closing off the vocal tract, building up pressure, and abruptly releasing it. All these sources act as a wide-band excitation to the vocal tract, which can be modeled as a slowly time-varying filter that imposes its frequency transmission properties upon the spectrum of the excitation. The vocal tract is characterized by its natural frequencies (called *formants*) which correspond to resonances in the sound-transmission characteristics of the vocal tract.

If we assume that the excitation sources and the vocal tract shape are relatively independent, a reasonable model is as shown in Fig. 10.17. In this discrete-time model, samples of the speech wave are assumed to be the output of a time-varying digital filter that approximates the transmission properties

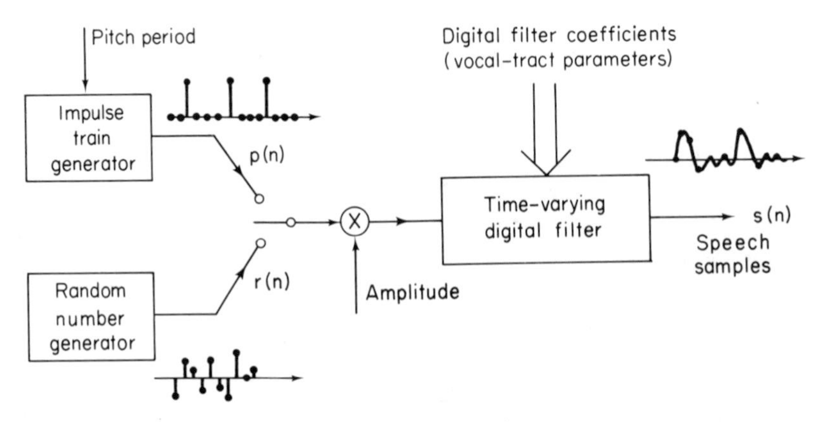

Fig. 10.17 Model of speech production.

of the vocal tract. Since the vocal tract changes shape rather slowly in continuous speech, it is reasonable to assume that this digital filter has fixed characteristics over a time interval on the order of 10 ms. Thus the digital filter may be characterized in each such time interval by an impulse response or a frequency response or a set of coefficients for an infinite duration impulse response filter. Specifically, for voiced sounds (except nasals), the transfer function of the digital filter consists of a vocal tract component

$$V(z) = \frac{A}{\prod_{k=1}^{p} (1 - c_k z^{-1})(1 - c_k^* z^{-1})}, \qquad |c_k| < 1 \tag{10.71}$$

where the c_k's correspond to the natural frequencies of the vocal tract, and an additional component

$$G(z) = B \prod_{k=1}^{m_i} (1 - a_k z^{-1}) \prod_{k=1}^{m_o} (1 - b_k z) \tag{10.72}$$

that accounts for the fact that the finite-duration glottal pulses are not impulses. Thus in Fig. 10.17 the system function of the digital filter is

$$H_v(z) = G(z)V(z)$$

This filter is excited by a train of impulses $p(n)$, in which the spacing between impulses corresponds to the fundamental (or pitch) period of the voice.

For unvoiced speech the theory of acoustic-wave propagation indicates that the vocal-tract transmission has zeros as well as poles. In this instance a reasonable model is

$$H_u(z) = \frac{A \prod_{k=1}^{m} (1 - \alpha_k z^{-1})(1 - \alpha_k^* z^{-1})}{\prod_{k=1}^{p} (1 - c_k z^{-1})(1 - c_k^* z^{-1})} \tag{10.73}$$

where $|c_k| < 1$. In this case the system is excited by a random-noise sequence, $r(n)$. In both voiced and unvoiced cases, an amplitude control regulates the intensity of the input to the digital filter.

Homomorphic deconvolution can be applied to the estimation of the parameters of the speech model if we assume that the model is valid over a short time interval [11]. Thus a short segment of voiced speech can be thought of as a convolution

$$s(n) = p(n) * g(n) * v(n), \qquad 0 \leq n \leq L - 1$$

In order to minimize the effect of "discontinuities" at the beginning and end of the interval, a data window $w(n)$ multiplies $s(n)$ so that the input to the homomorphic system is

$$x(n) = s(n)w(n)$$

If $w(n)$ varies slowly with respect to the term $g(n) * v(n)$, then we can write

$$x(n) \simeq p_w(n) * [g(n) * v(n)] \tag{10.74a}$$

where

$$p_w(n) = w(n) \cdot p(n) \tag{10.74b}$$

Let us examine the contributions to the complex cepstrum of each component in Eq. (10.74a). It is reasonable to assume that over the short time interval of the window, $p(n)$ is a train of equally spaced impulses

$$p(n) = \sum_{k=0}^{M-1} \delta(n - kn_0)$$

so that

$$p_w(n) = \sum_{k=0}^{M-1} w(kn_0)\delta(n - kn_0)$$

where we have assumed that M impulses are spanned by the window. If we define a sequence

$$w_{n_0}(k) = \begin{cases} w(kn_0) & k = 0, 1, \ldots, M-1 \\ 0 & \text{otherwise} \end{cases}$$

then the Fourier transform of $p_w(n)$ is

$$\begin{aligned} P_w(e^{j\omega}) &= \sum_{k=0}^{M-1} w(kn_0)e^{-j\omega kn_0} \\ &= W_{n_0}(e^{j\omega n_0}) \end{aligned} \tag{10.75}$$

Thus $P_w(e^{j\omega})$ and also $\hat{P}_w(e^{j\omega})$ are periodic with period $2\pi/n_0$. The complex cepstrum of $p_w(n)$ is

$$\hat{p}_w(n) = \hat{w}_{n_0}\left(\frac{n}{n_0}\right) \qquad n = 0, \pm n_0, \pm 2n_0, \ldots$$

Thus the periodicity of the complex logarithm manifests itself in the complex cepstrum in terms of impulses spaced at intervals of n_0 samples. If the sequence $w_{n_0}(n)$ is minimum phase, then $\hat{p}_w(n)$ will be zero for $n < 0$. Otherwise, $\hat{p}_w(n)$ will have impulses spaced at intervals of n_0 samples for both positive and negative n. In any case the contribution of $\hat{p}_w(n)$ to $x(n)$ is to be found in the region $|n| \geq n_0$.

The complex cepstrum of $v(n)$ can be obtained from the complex logarithm of $V(z)$:

$$\hat{V}(z) = \log[A] - \sum_{k=1}^{p} \{\log[1 - c_k z^{-1}] + \log[1 - c_k^* z^{-1}]\} \tag{10.76}$$

From this expression it is easily seen that

$$\hat{v}(n) = \begin{cases} 0 & n < 0 \\ \log [A] & n = 0 \\ \frac{1}{n}\sum_{k=1}^{p} [(c_k)^n + (c_k^*)^n] & n > 0 \end{cases} \tag{10.77}$$

or if $c_k = |c_k|\, e^{j\phi_k}$,

$$\hat{v}(n) = \sum_{k=1}^{p} \frac{|c_k|^n}{n} 2 \cos \phi_k n \qquad n > 0 \tag{10.78}$$

The glottal pulse, $g(n)$, is of finite duration and is generally assumed to be nonminimum phase. Therefore, $g(n)$ can be represented as the convolution of a minimum-phase sequence with a maximum-phase sequence as in

$$g(n) = g_{\min}(n) * g_{\max}(n) \tag{10.79}$$

The contribution to the complex cepstrum $\hat{x}(n)$ due to $g(n)$ is

$$\hat{g}(n) = \begin{cases} \hat{g}_{\min}(n) & 0 \le n \\ \hat{g}_{\max}(n) & n < 0 \end{cases} \tag{10.80}$$

where, from our previous discussion, we expect that the primary contribution of $\hat{g}(n)$ to $\hat{x}(n)$ would be in the region around $n = 0$.

In general, the components of the complex cepstrum, $\hat{v}(n)$ and $\hat{g}(n)$, decay rather rapidly, so that for reasonably large values of n_0, vocal tract and glottal pulse contributions do not overlap $\hat{p}_w(n)$.† In other words, in the complex logarithm the vocal tract and glottal components are slowly varying and the pitch components are rapidly varying. This is illustrated by Fig. 10.18. Figure 10.18(a) shows a segment of speech weighted by a Hamming window, and Fig. 10.18(b) shows the complex logarithm of the discrete Fourier transform of Fig. 10.18(a).‡ Note the rapidly varying—almost periodic—component due to $p_w(n)$, and the slowly varying components due to $v(n)$ and $g(n)$. These properties are manifest in the complex cepstrum of Fig. 10.18(c) in the form of impulses at multiples of approximately 8 ms (the period of the input speech) due to $\hat{p}_w(n)$ and in the samples in the region $|nT| < 5$ ms, which we attribute to $\hat{v}(n)$ and $\hat{g}(n)$.

If we wish to separate the components of the speech waveform, the previous discussion suggests that we should lowpass-filter the complex logarithm to obtain $v(n) * g(n)$ and highpass filter to obtain $p_w(n)$. An example is shown in Fig. 10.19. In Fig. 10.19(a) is shown a segment of a vowel sound. After

† For speech sampled at 10 kHz, a typical range for the pitch period is $40 < n_0 < 150$.

‡ In all the figures in this section, the samples of all sequences are connected by straight lines for ease in plotting.

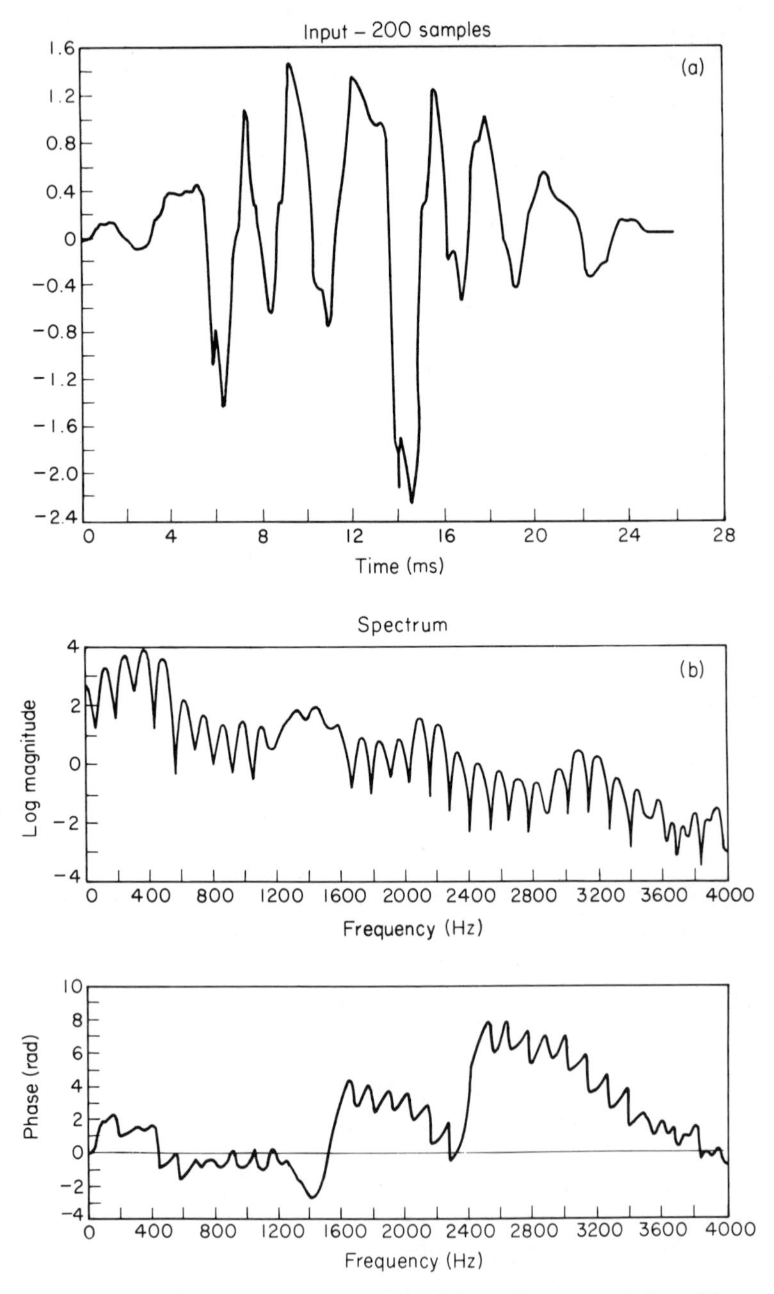

Fig. 10.18(a) Segment of speech weighted by a Hamming window; (b) complex logarithm of the transform of (a); (c) complex cepstrum of (a).

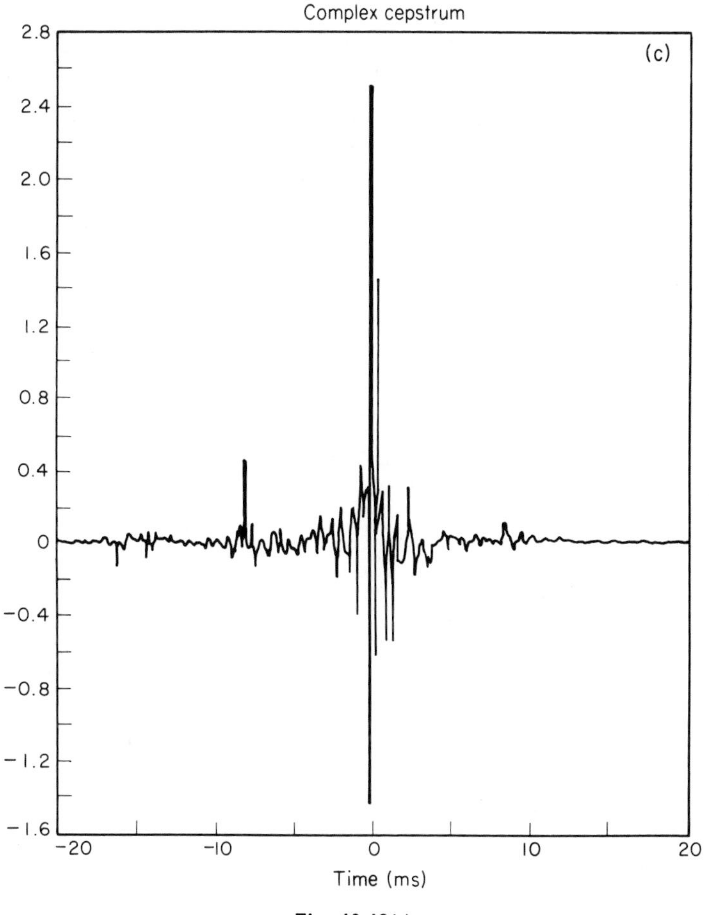

Fig. 10.18(c)

weighting by a Hamming window, the complex cepstrum appears as in Fig. 10.19(b). When the complex cepstrum is multiplied by a sequence

$$l(n) = \begin{cases} 0 & |n| \leq 40 \\ 1 & |n| > 40 \end{cases}$$

and the result is processed by the inverse characteristic system D_*^{-1}, the resulting output is as shown in Fig. 10.19(c).† On the other hand, to recover the pulse $v(n) * g(n)$, we multiply the complex cepstrum by

$$l(n) = \begin{cases} 1 & |n| \leq 40 \\ 0 & |n| > 40 \end{cases}$$

† The sampling rate is 10 kHz so that 40 samples corresponds to 4 ms.

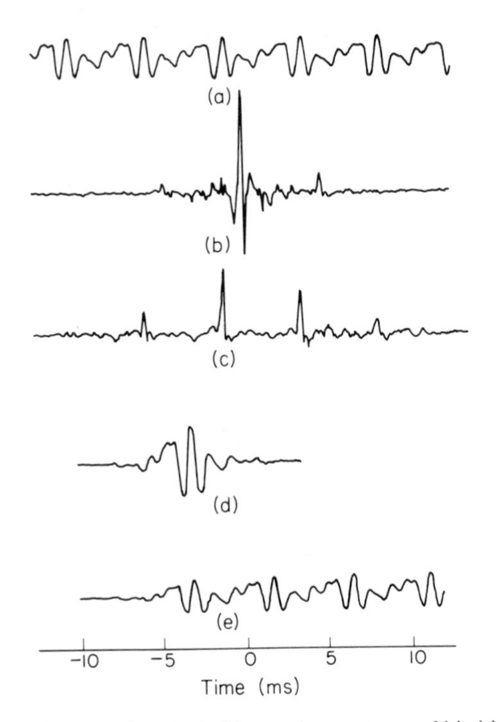

Fig. 10.19 (a) Portion of a vowel; (b) complex cepstrum of (a); (c) recovered train of weighted pitch pulses; (d) recovered vocal-tract impulse response; (e) resynthesized speech, using impulse-response function of (d) and pitch as measured from (e).

In this case the output of D_*^{-1} is shown in Fig. 10.19(d). Figure 10.19(e) shows the result of convolving the waveform of Fig. 10.19(d) with an impulse train of equal amplitude unit samples occurring at the locations of the peak in Fig. 10.19(c).

The previous discussion has shown that homomorphic deconvolution can be successfully applied in *separating* the components of a speech waveform. However, in many applications of speech analysis we are only concerned with *estimating* the parameters of the speech involved rather than recovering the actual component waveforms. For example, it may be sufficient to decide whether a particular speech segment is voiced or unvoiced and if voiced, estimate the pitch period or spectral envelope

$$\log |V(e^{j\omega})G(e^{j\omega})|$$

and, if unvoiced, estimate the spectrum

$$\log |H_u(e^{j\omega})|$$

In such cases we can use the cepstrum rather than the complex cepstrum. Recall that the cepstrum is the inverse Fourier transform of $\log|X(e^{j\omega})|$, and therefore

$$c(n) = \tfrac{1}{2}[\hat{x}(n) + \hat{x}(-n)]$$

Thus we would expect the low-time portion of $c(n)$ to correspond to the slowly varying components in $\log|X(e^{j\omega})|$ that are determined by vocal-tract configuration, and in the case of voiced speech, the even component of $p_w(n)$ should contain impulses at the same places as $\hat{p}_w(n)$. This is depicted in Fig. 10.20. Figure 10.20(a) shows the computations involved in estimating the speech parameters. Figure 10.20(b) shows a typical result for voiced speech. The windowed speech signal is labeled A, $\log|X(k)|$ is labeled C, and the cepstrum $c(n)$ is labeled D. The peak in the cepstrum at about 8 ms indicates the pitch period for this segment of speech. The spectrum envelope, obtained by multiplying $c(n)$ by a "cepstrum window" which only passes the low-time part ($|n| < 40$) and then computing the DFT, is labeled E and is superimposed on $\log|X(k)|$. The situation for unvoiced speech, shown in Fig. 10.20(c), is much the same, with the exception that the random nature of the excitation component of the input speech segment causes a rapidly varying random component in $\log|X(k)|$. Thus in the cepstrum the low-time components correspond to the vocal-tract transfer function; however, since the rapid variations in $\log|X(k)|$ are not periodic, there is no strong peak as for the voiced speech segment. Therefore, the cepstrum serves as an excellent method for determining whether a speech segment is voiced or unvoiced and for estimating the fundamental period of voiced speech [12].

The methods depicted in Fig. 10.20 have been used in speech analysis and synthesis systems. In one approach, an impulse response sequence is computed directly from the low-time part of the cepstrum [13]. Pitch period and a voiced/unvoiced decision is also estimated from the cepstrum. Speech is synthesized from this information by realizing the system of Fig. 10.17 in the form of an explicit convolution of the impulse response with an appropriate excitation sequence. In another approach, the poles and zeros of Eqs. (10.71) and (10.73) are estimated from the cepstrum [14]. In this case, speech is synthesized by realizing the linear system of Fig. 10.17 as a cascade of time-varying second-order digital resonators. In both cases, it is implicitly assumed that the combined vocal tract and glottal pulse response is minimum phase. The fact that this only preserves the short-time magnitude spectrum is not a significant limitation since the ear is known to be relatively insensitive to phase.

10.7.2 Dereverberation

In the previous section we saw that voiced speech can be represented on a short-time basis as the convolution of a pulse with a periodic impulse train.

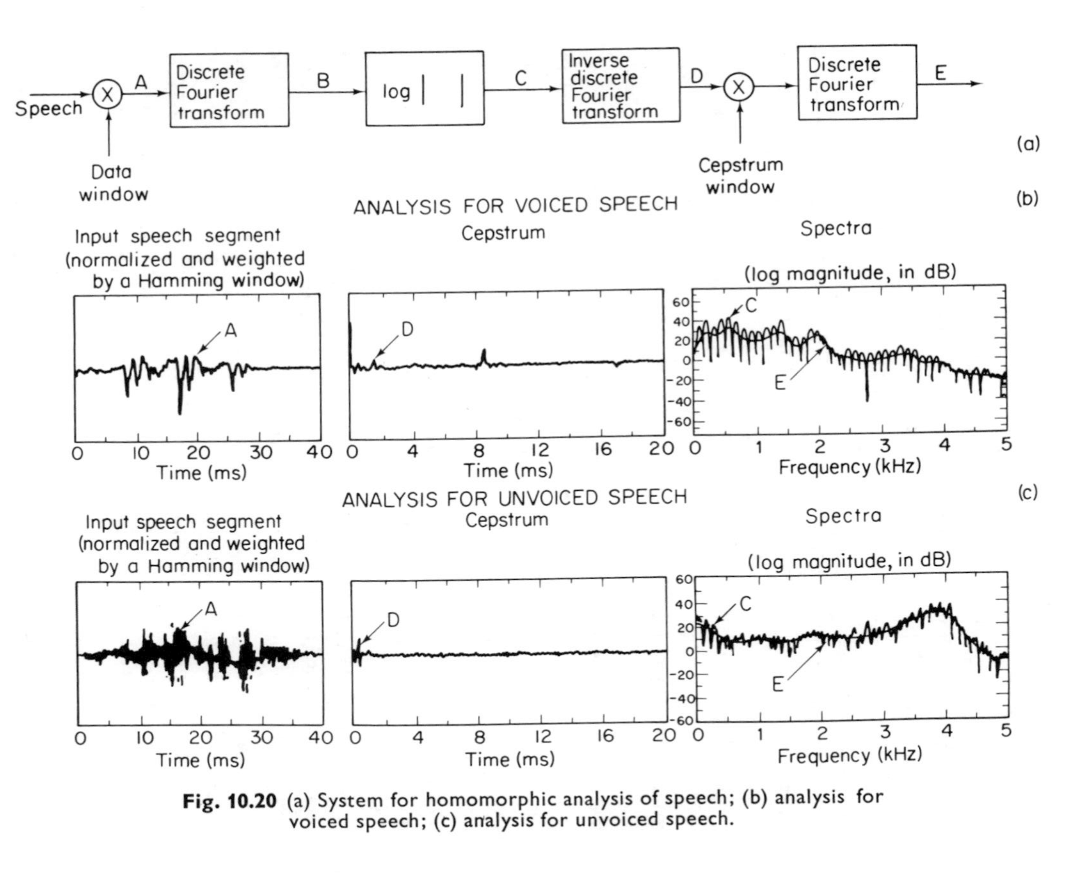

Fig. 10.20 (a) System for homomorphic analysis of speech; (b) analysis for voiced speech; (c) analysis for unvoiced speech.

The physical world is replete with signals that can be represented by very similar models. For example, in many areas of physical measurement and communication, signals are transmitted or recorded in what may broadly be termed a *reverberant environment.*

In such cases a signal can be represented as the sum of a number of overlapping delayed replicas, or echoes, of some basic waveform. Examples include audio recording, conference telephony, radar and sonar detection, seismic measurements, and electrophysiology. In cases where the reverberation is viewed as distortion, we may wish to recover the basic waveform. In other cases, the pattern of echoes may be desired as a characterization of a physical structure or process. In the remainder of this section we shall discuss some examples of applications of homomorphic deconvolution to signals of the above nature.

Let us begin by considering a sequence that is a sum of delayed and scaled replicas of a sequence $s(n)$; i.e.,

$$x(n) = s(n) + \sum_{k=1}^{M} \alpha_k s(n - n_k) \tag{10.81}$$

where $0 < n_1 < n_2 < \ldots < n_M$. Such a signal can be represented as the convolution

$$x(n) = s(n) * p(n) \tag{10.82a}$$

where

$$p(n) = \delta(n) + \sum_{k=1}^{M} \alpha_k \delta(n - n_k) \tag{10.82b}$$

As a simple example to illustrate the use of a homomorphic system for this class of signals, consider the case of a single echo, i.e.,

$$p(n) = \delta(n) + \alpha_1 \delta(n - n_1) \tag{10.83}$$

The Fourier transform of $x(n)$ is

$$X(e^{j\omega}) = S(e^{j\omega})(1 + \alpha_1 e^{-j\omega n_1}) \tag{10.84}$$

Therefore, the contribution to the complex logarithm due to the impulse train is

$$\hat{P}(e^{j\omega}) = \log(1 + \alpha_1 e^{-j\omega n_1}) \tag{10.85}$$

In this simple case, $\hat{P}(e^{j\omega})$ is periodic with period $2\pi/n_1$, and we would therefore expect $\hat{p}(n)$ to be nonzero only at integer multiples of n_1. If $|\alpha_1| < 1$, it is easily shown that

$$\hat{p}(n) = \sum_{k=1}^{\infty} (-1)^{k+1} \frac{\alpha_1^k}{k} \delta(n - kn_1) \tag{10.86}$$

Thus if $\hat{S}(e^{j\omega})$ is slowly varying relative to the variations of $\hat{P}(e^{j\omega})$, it is reasonable to separate these two components with a linear frequency-invariant filter. For example, we could use a filter that passes only the long-time components of the complex cepstrum if we wish to recover $p(n)$.

In the general case,

$$p(n) = \delta(n) + \sum_{k=1}^{M} \alpha_k \delta(n - n_k) \tag{10.87}$$

and

$$P(e^{j\omega}) = 1 + \sum_{k=1}^{M} \alpha_k e^{-j\omega n_k} \tag{10.88}$$

If the echoes are equally spaced, i.e., $n_k = kn_1$, then we saw in Sec. 10.7.1 that the complex cepstrum will have the same form as in the case of a single echo. However, in general, we cannot expect the echoes to be equally spaced. In general, the complex cepstrum of $p(n)$ will consist of impulses at times that are complicated functions of the original delays, the low-time region of $\hat{p}(n)$. However, in the special case when $p(n)$ is minimum phase, we know that $\hat{p}(n) = 0$ for $n < 0$. Furthermore, it can be shown [8] that $\hat{p}(n) = 0$ for $n < n_1$, where n_1 is the shortest delay, and for $n > n_1$ the complex cepstrum will consist of impulses occurring at times

$$n_l = \sum_{k=1}^{M} l n_k, \qquad l = 0, 1, 2, \ldots \tag{10.89}$$

with amplitudes that decrease with increasing n. In view of our discussion in Sec. 10.5.3, it is clear that a nonminimum phase impulse train can be made minimum phase by exponential weighting. That is, if β is small enough we can ensure that the sequence $\beta^n p(n)$ is minimum phase. In many cases it is advantageous to use exponential weighting since it produces a degree of separation between the complex cepstrum components due to $\beta^n s(n)$ and $\beta^n p(n)$.

With these facts concerning the complex cepstrum of an impulse train in mind, let us consider some examples of the use of homomorphic filtering to separate the components of a convolution of the form of Eq. (10.82).

Echoes in Speech Signals. In many communication channels, speech signals may be distorted by echoes or reverberation. Because of the continuing nature of speech signals, the waveform must be processed in pieces of manageable size, with the resulting output segments fitted back together to form the composite output sequence. Figure 10.18(c) shows the complex cepstrum of a segment of a speech waveform. If $s(n)$ in Eq. (10.82) is a speech signal, the complex cepstrum of a segment of $x(n)$ will contain impulses due to $p(n)$ if the window duration is greater than n_M, the longest delay. If the shortest delay, n_1, is greater than the longest pitch period (approximately 15 ms), the contribution due to $p(n)$ will not significantly overlap the complex cepstrum of the speech signal. An example is shown in Fig. 10.21(a). In

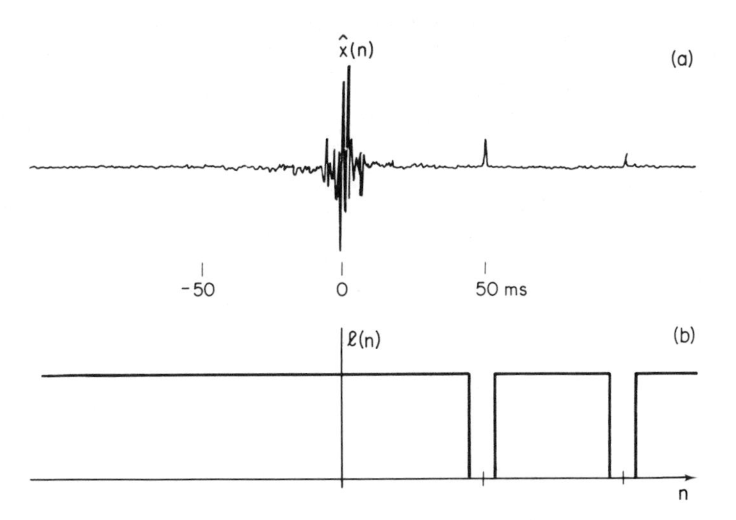

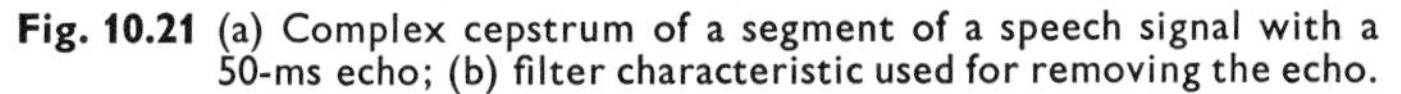

Fig. 10.21 (a) Complex cepstrum of a segment of a speech signal with a 50-ms echo; (b) filter characteristic used for removing the echo.

this case, speech sampled at 10 kHz was delayed and added to itself, producing a signal

$$x(n) = s(n) + \alpha_1 s(n - n_1) \tag{10.90}$$

In order to recover $s(n)$ from $x(n)$, we must remove the contributions to the complex cepstrum due to the echo. This can be done using a frequency invariant filter of the form shown in Fig. 10.21(b). The success of this processing is shown in Fig. 10.22, where Fig. 10.22(a) shows $s(n)$, Fig. 10.22(b)

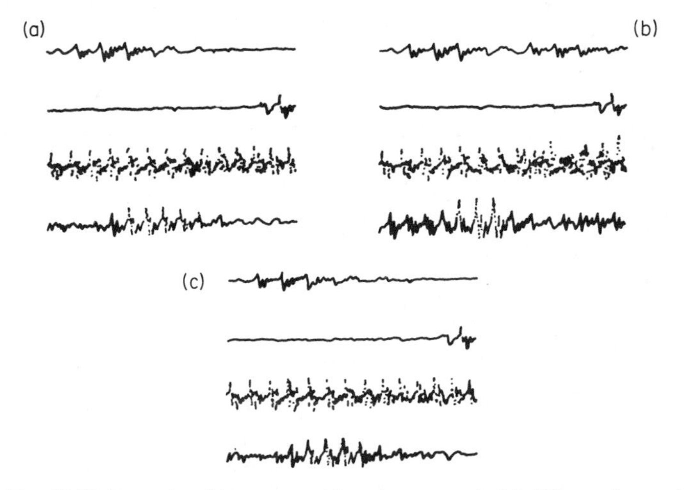

Fig. 10.22 Example of homomorphic echo removal: (a) 410 ms of speech sampled at 10 kHz with the four traces from top to bottom representing contiguous segments of 102.5 ms; (b) speech sample of (a) with a 50-ms echo; (c) speech sample of (b) processed to remove the echo.

shows $x(n)$, and Fig. 10.22(c) shows the output of the system D_*^{-1} for the "comb" filter of Fig. 10.21(b). In this case the speech was processed in segments 2048 samples in length, and the resulting outputs were combined to form the total outputs. The details of the process of combination are described in Ref. [8].

Seismic Signals. Equation (10.82) serves also as a useful model for seismic signals. In this case an explosion creates a pulse of seismic energy that propagates through the earth, being reflected at the boundaries between various layers of the earth's crust. Figure 10.23 shows a model for seismic

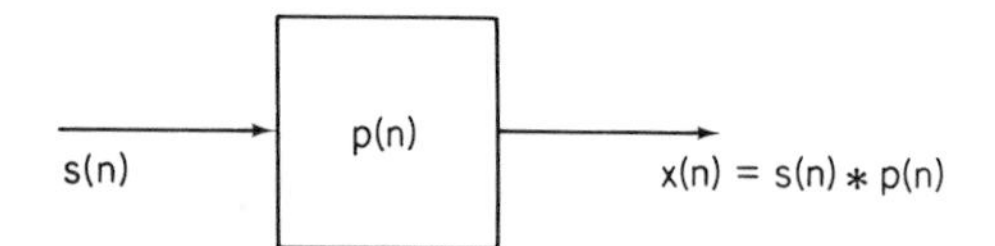

Fig. 10.23 Simple model for seismic waveforms.

waveforms. $p(n)$ is an impulse train that contains information about the structure of the earth's crust; the seismic wavelet $s(n)$ depends on the nature of the excitation disturbance and the dispersion encountered in propagation.

Since, in general, the seismic wavelets overlap in time and disguise the structure of $p(n)$, it is necessary to separate the two components. Ulrych [15] has shown that homomorphic deconvolution can be successfully applied to this problem. Figure 10.24 shows a synthetic example. The signals $p(n)$, $s(n)$, and $x(n)$ are shown in Fig. 10.24(a), (b), and (c), respectively. The complex cepstrum $\hat{x}(n)$ is shown in Fig. 10.24(d); Fig. 10.24(e) and (f) show the result of using frequency-invariant filters that retain the high-time and low-time parts of $\hat{x}(n)$, respectively. Comparison of Fig. 10.24(a) and (e) with Fig. 10.24(c) and (f) indicate considerable success in this example. Figure 10.25(a) shows an actual seismic waveform. The complex cepstrum is shown in Fig. 10.25(b), and the result of low-time filtering is shown in Fig. 10.25(c). This estimate of the seismic wavelet could be useful in estimating attenuation and dispersion properties of the transmission path. Exponential weighting was used in both examples.

10.7.3 Restoration of Acoustic Recordings

A well-known technique for reducing noise involves averaging over a large number of signals wherein the desired waveform is the same but the noise is different. This technique has been applied in estimating the distortion in recordings made by acoustic methods (specifically recordings of Enrico Caruso) [16]. A simplified model for such recordings is shown in Fig. 10.26, where $s(n)$ represents the singer's voice and $h(n)$ represents the impulse

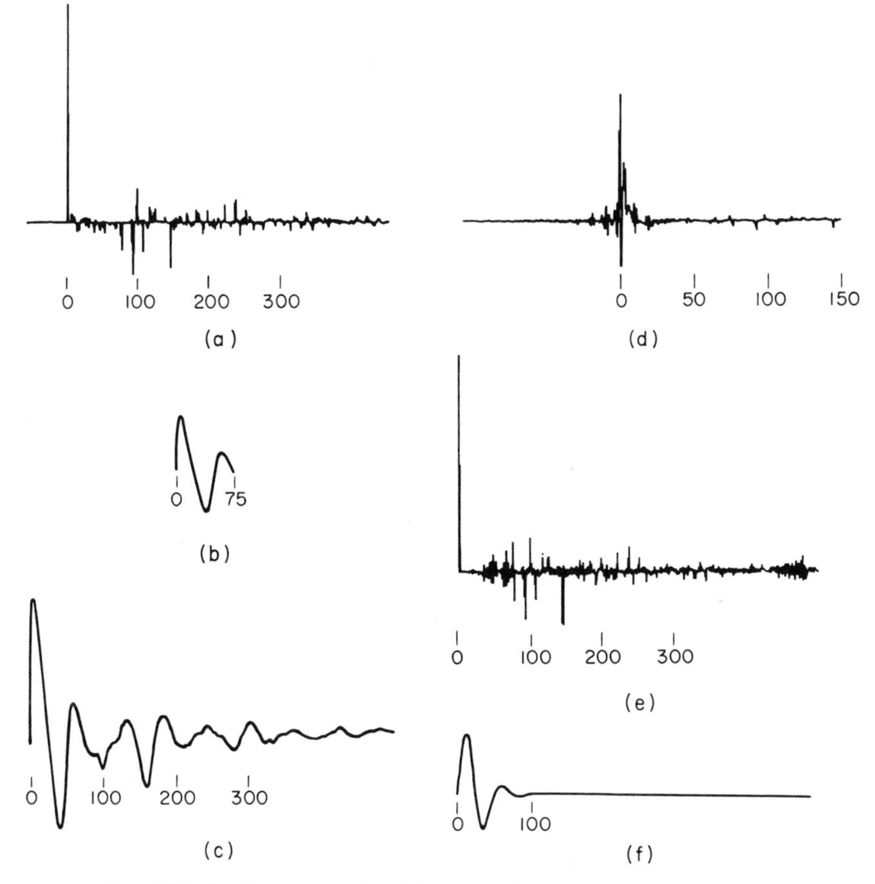

Fig. 10.24 Synthetic example of homomorphic deconvolution of seismic signals: (a) theoretical impulse response of crust near Leduc, Alberta (after O. Jensen); (b) assumed seismic wavelet; (c) synthetic seismogram; (d) complex cepstrum of the trace of (c) exponentially weighted with $\alpha = 0.985$; (e) highpass output; (f) lowpass output. (After Ulrych [15].)

response of the recording system, including the recording horn and mechanical apparatus for cutting the recording. Besides the surface scratch, the most notable distortion is due to the resonances of the recording horn.

In order to estimate the impulse response $h(n)$ so that its effect may be compensated by inverse filtering, the waveform was segmented into sections,

$$x_m(n) = x(n + m), \qquad 0 \leq n \leq N - 1$$

Although it is clearly an approximation, it is assumed that

$$x_m(n) \approx s_m(n) * h(n)$$

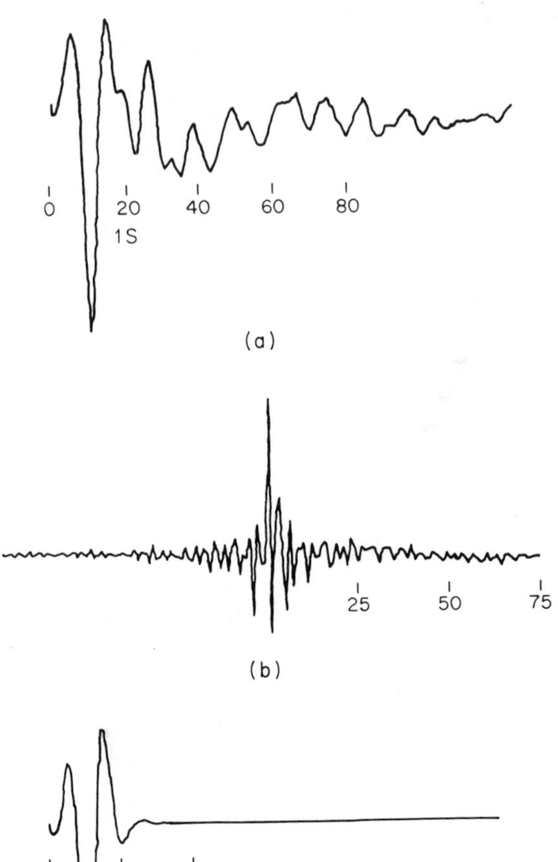

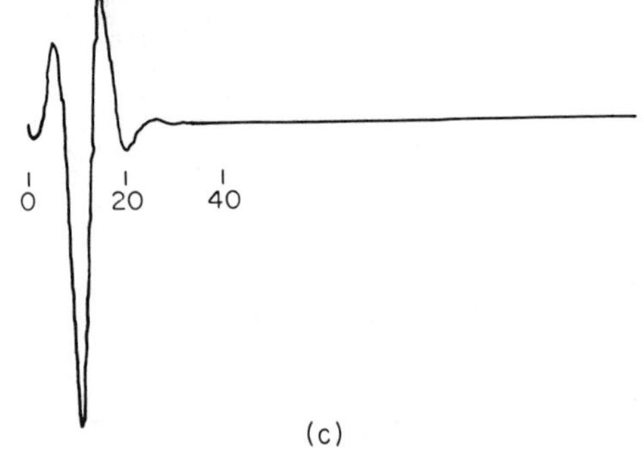

Fig. 10.25 Example of homomorphic deconvolution of an actual teleseismic event: (a) teleseismic event recorded in 1968 at Leduc, Alberta, and originating in Venezuela; (b) complex cepstrum of (a) after exponential weighting with $\alpha = 0.985$, (c) estimate of the seismic wavelet obtained by the low-time filtering (b). (After Ulrych [15].)

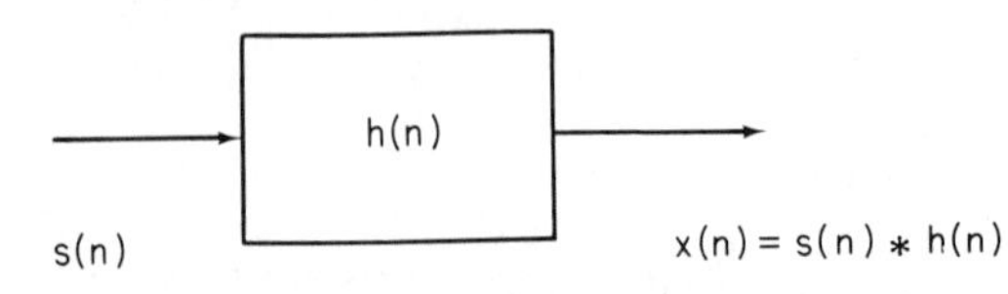

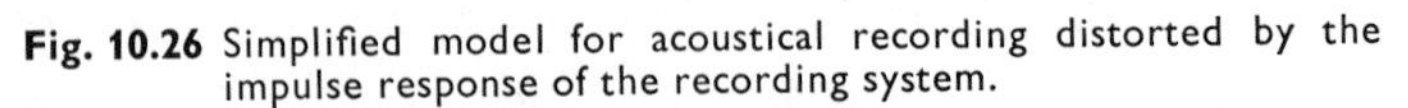

Fig. 10.26 Simplified model for acoustical recording distorted by the impulse response of the recording system.

so that

$$X_m(e^{j\omega}) \approx S_m(e^{j\omega})H(e^{j\omega})$$

and

$$\log |X_m(e^{j\omega})| \approx \log |S_m(e^{j\omega})| + \log |H(e^{j\omega})|$$

Averaging over M sections results in

$$\frac{1}{M}\sum_{m=0}^{M-1} \log |X_m(e^{j\omega})| = \frac{1}{M}\sum_{m=0}^{M-1} \log |S_m(e^{j\omega})| + \log |H(e^{j\omega})|$$

The term

$$\frac{1}{M}\sum_{m=0}^{M-1} \log |S_m(e^{j\omega})| \approx \log |S(e^{j\omega})|$$

is an estimate of the long-time log-power spectrum of speech or singing, and a good approximation to it can be estimated from modern recordings where distortion is minimized. An estimate $He(e^{j\omega})$ of the frequency response of the horn is then obtained from

$$\log |H_e(e^{j\omega})| = \frac{1}{M}\sum_{m=0}^{M-1} \log |X_m(e^{j\omega})| - \overline{\log |S(e^{j\omega})|}$$

An inverse filter for compensating the effect of $h(n)$ is obtained from

$$|H_e^{-1}(e^{j\omega})| = \begin{cases} \dfrac{1}{|H_e(e^{j\omega})|}, & |\omega| \leq \omega_p \\ 0, & \omega_s \leq |\omega| \leq \pi \end{cases} \tag{10.91}$$

where the frequency response drops linearly between ω_p and ω_s. A zero-phase impulse response is obtained by inverse Fourier transformation of Eq. (10.91), and the resulting $h_e^{-1}(n)$ is convolved with $x(n)$ by a fast convolution algorithm [15].

In spite of the great number of approximations involved, the above technique has produced a striking improvement in the subjective quality of the recordings of Caruso. Similar improvements might be expected in processing speech or music recorded in a reverberant environment, as, for example, in conference telephony.

Summary

In this chapter we have discussed a class of nonlinear signal processing techniques and their application in a number of fields, including image enhancement, speech analysis, and seismic exploration. The presentation of this class of techniques provided an opportunity to present a number of applications of results which have been discussed throughout this book.

We first considered the general class of homomorphic systems and then focused on the two subclasses for which a variety of applications have been found. The first was the class of multiplicative homomorphic systems. A number of the details specific to this class were presented, followed by the details of the application to image processing. We then considered the class of homomorphic systems for convolution. A number of important theoretical considerations for this class of systems were based on the fact that the characteristic system for this class required the interpretation of the complex logarithm. Thus we considered in detail the properties of the output of the characteristic system, i.e., the complex cepstrum, and a number of computational realizations of the characteristic system. The application of these ideas to speech processing, removal of echoes in speech signals, analysis of seismic signals, and restoration of acoustic recordings was discussed briefly.

REFERENCES

1. A. V. Oppenheim, "Superposition in a Class of Nonlinear Systems," *Tech. Rept. 432*, Research Laboratory of Electronics, MIT, Cambridge, Mass., Mar. 1965.
2. A. V. Oppenheim, "Generalized Superposition," *Inform. Control*, Vol. 11, Nos. 5–6, Nov.–Dec. 1967, pp. 528–536.
3. A. V. Oppenheim, R. W. Schafer, and T. G. Stockham, Jr., "Nonlinear Filtering of Multiplied and Convolved Signals," *Proc. IEEE*, Vol. 56, No. 8, Aug. 1968, pp. 1264–1291.
4. A. V. Oppenheim, "Generalized Linear Filtering," Chapter 8 in *Digital Processing of Signals*, B. Gold and C. M. Rader, McGraw-Hill Book Company, New York, 1969.
5. T. G. Stockham, Jr., "The Application of Generalized Linearity to Automatic Gain Control," *IEEE Trans. Audio Electroacoust.*, Vol. AU-16, June 1968, pp. 267–270.
6. T. G. Stockham, Jr., "Image Processing in the Context of a Visual Model," *Proc. IEEE*, Vol. 60, No. 7, July 1972, pp. 828–842.
7. A. V. Oppenheim, "Nonlinear Filtering of Convolved Signals," *Quart. Progr. Rept. 54*, Research Laboratory of Electronics, MIT, Cambridge, Mass., Jan. 1965, pp. 168–175.
8. R. W. Schafer, "Echo Removal by Discrete Generalized Linear Filtering," *Tech. Rept. 466*, MIT Research Laboratory of Electronics, MIT, Cambridge, Mass., Feb. 1969. Also Ph.D. Thesis, Department of Elec. Engineering, MIT, Feb. 1968.
9. B. P. Bogert, M. J. R. Healy, and J. W. Tukey, "The Quefrency Alanysis of Time Series for Echoes: Cepstrum, Pseudo-autocovariance, Cross-Cepstrum, and Saphe Cracking," *Proc. Symp. Time Series Analysis*, M. Rosenblatt, Ed., New York, John Wiley & Sons, Inc., New York, 1963, pp. 209–243.
10. J. L. Flanagan, *Speech Analysis, Synthesis and Perception*, 2nd ed., Springer-Verlag, New York, 1972.
11. A. V. Oppenheim and R. W. Schafer, "Homomorphic Analysis of Speech," *IEEE Trans. Audio Electroacoust.*, Vol. AU-16, No. 2, June 1968, pp. 221–226.
12. A. M. Noll, "Cepstrum Pitch Determination," *J. Acoust. Soc. Amer.*, Vol. 41, Feb. 1967, pp. 293–309.
13. A. V. Oppenheim, "A Speech Analysis-Synthesis System Based on Homomorphic Filtering," *J. Acoust. Soc. Amer.*, Vol. 45, Feb. 1969, pp. 458–465.

14. R. W. Schafer and L. R. Rabiner, "System for Automatic Formant Analysis of Voiced Speech," *J. Acoust. Soc. Amer.*, Vol. 47, No. 2, Pt. 2, Feb. 1970, pp. 634–648.
15. T. J. Ulrych, "Application of Homomorphic Deconvolution to Seismology," *Geophys.*, Vol. 36, No. 4, Aug. 1971, pp. 650–660.
16. T. G. Stockham, Jr., "Restoration of Old Acoustic Recordings by Means of Digital Signal Processing," Preprint, 41st Convention, Audio Engineering Society, New York, Oct. 1971.

PROBLEMS

1. Each of the following system transformations is homomorphic, with the input operation indicated. Determine the output operation.

System Transformation $T[x(n)]$	*Input Operation*
$y(n) = T[x(n)] = 2x(n)$	Addition
$y(n) = T[x(n)] = 2x(n)$	Multiplication
$X(z) = T[(x(n)] = \sum_{n=-\infty}^{\infty} x(n)z^{-n}$	Addition
$X(z) = T[x(n)] = \sum_{n=-\infty}^{\infty} x(n)z^{-n}$	Convolution
$X(z) = T[x(n)] = \sum_{n=-\infty}^{\infty} x(n)z^{-n}$	Multiplication
$y(n) = T[x(n)] = x^2(n)$	Multiplication
$y(n) = T[x(n)] = \lvert x(n)\rvert$	Multiplication
$y(n) = e^{x(n)}$	Addition
$y(n) = e^{x(n)}$	Multiplication

2. Two homomorphic systems, H_1 and H_2, are cascaded. H_1 is homomorphic, with multiplication as the input operation and convolution as the output operation. H_2 is homomorphic, with convolution as the input operation and addition as the output operation. Show that the overall system is homomorphic, with multiplication as the input operation and addition as the output operation.
3. Consider the class of homomorphic systems with multiplication as the input and output operations. Show that if the input $x(n)$ is unity for all n, then the output $y(n)$ is unity for all n.
4. Determine which of the following systems cannot be homomorphic, with multiplication as the input and output operations:
(a) $y(n) = 3x(n)$.
(b) $y(n) = x^2(n)$.
(c) $y(n) = [1/x(n)][x(n) - x(n-1)]$.
(d) $y(n) = \lvert x(n)\rvert$.
(e) $y(n) = x(n)/x(n-1)$.
5. Consider the class of homomorphic systems with convolution as the input and output operations. Show that if the input $x(n) = \delta(n)$, then the output $y(n) = \delta(n)$.
6. $\hat{x}(n)$ is the complex cepstrum of $x(n)$. The first column lists properties to be considered for $\hat{x}(n)$. The second column lists properties to be considered for

$x(n)$. For each property in the first column determine the corresponding property in the second column. In all cases assume that $x(n)$ is real. Any property in the second column can only be used once.

(1) $\hat{x}(n)$ real	(a) $x(n) = -x(-n)$
(2) $\hat{x}(n) = -\hat{x}(-n)$	(b) $x(n) = x(-n)$
(3) $\hat{x}(n) = 0,\ n < 0$	(c) $x(n)$ real
	(d) $x(n) = 0,\ n < 0$
	(e) $\sum_{n=-\infty}^{\infty} x^2(n) = 1$
	(f) $\sum_{n=-\infty}^{\infty} \lvert x(n)\rvert = 1/\sqrt{2\pi}$

7. $x_1(n)$ and $x_2(n)$ denote two sequences and $\hat{x}_1(n)$ and $\hat{x}_2(n)$ their complex cepstra. If $x_1(n) * x_2(n) = \delta(n)$, determine the relationship between $\hat{x}_1(n)$ and $\hat{x}_2(n)$.

8. The complex cepstrum $\hat{x}(n)$ of a sequence $x(n)$ has been defined such that if

$$\hat{X}(e^{j\omega}) = \sum_{n=-\infty}^{+\infty} \hat{x}(n)e^{-j\omega n}$$

and

$$X(e^{j\omega}) = \sum_{n=-\infty}^{+\infty} x(n)e^{-j\omega n} = |X(e^{j\omega})|\, e^{j\theta(\omega)}$$

then

$$\hat{X}(e^{j\omega}) = \log X(e^{j\omega}) = \log |X(e^{j\omega})| + j\theta(\omega)$$

where $\theta(\omega)$ is a continuous, odd, periodic function of ω.

A *minimum-phase sequence* is defined as one whose complex cepstrum is zero for $n < 0$, and a *maximum-phase sequence* is defined as one whose complex cepstrum is zero for $n > 0$.

Consider two sequences $x_1(n)$ and $x_2(n)$ with transforms $X_1(e^{j\omega})$ and $X_2(e^{j\omega})$, respectively. $x_1(n)$ is a minimum-phase sequence and $x_2(n)$ is a maximum-phase sequence. If $|X_1(e^{j\omega})| = |X_2(e^{j\omega})|$, determine the relationship between $x_1(n)$ and $x_2(n)$.

9. Let $x(n)$ denote a minimum-phase sequence and $\hat{x}(n)$ its complex cepstrum. Use the initial value theorem (Problem 16 of Chapter 2) to show that $\hat{x}(0) = \log [x(0)]$. Would the same result hold if $x(n)$ were not minimum phase?

10. Let $x(n)$ denote a *maximum-phase sequence* and $\hat{x}(n)$ its complex cepstrum. Show that $\hat{x}(0) = \log [x(0)]$.

11. Consider a sequence $x(n)$ with complex cepstrum $\hat{x}(n)$ and with z-transform $X(z)$ expressed in the form

$$X(z) = \frac{A \prod_{k=1}^{m_i} (1 - a_k z^{-1}) \prod_{k=1}^{m_0} (1 - b_k z)}{\prod_{k=1}^{p_i} (1 - c_k z^{-1}) \prod_{k=1}^{p_0} (1 - d_k z)}$$

where $|a_k|$, $|b_k|$, $|c_k|$, and $|d_k|$ are all less than unity and A is a real, positive number. Show that $\hat{x}(0) = \log A$.

12. Let $\hat{x}(n)$ be the complex cepstrum of $x(n)$. Define a sequence $e(n)$ to be

$$e(n) = \begin{cases} x(n/N), & n = KN,\ K = 0, \pm 1, \pm 2, \ldots \\ 0, & \text{elsewhere} \end{cases}$$

Show that the complex cepstrum of $e(n)$, called $\hat{e}(n)$, is given by

$$\hat{e}(n) = \begin{cases} \hat{x}(n/N), & n = KN, K = 0, \pm 1, \pm 2, \ldots \\ 0, & \text{elsewhere} \end{cases}$$

13. Equation (10.49) represents a recursive relation between $x(n)$ and $\hat{x}(n)$ when $x(n)$ is *minimum phase*. Use Eq. (10.49) to generate recursively the complex cepstrum of the sequence $x(n) = a^n u(n)$.

14. Equation (10.54) represents a recursive relation between $x(n)$ and $\hat{x}(n)$ when $x(n)$ is *maximum phase*. Use Eq. (10.54) to generate recursively the complex cepstrum of the sequence $x(n)$ given by

$$\begin{aligned} x(0) &= 1 \\ x(-1) &= -a \\ x(n) &= 0, \qquad n \neq 0, -1 \end{aligned}$$

15. Equation (10.49) represents a recursive relationship between a sequence $x(n)$ and its complex cepstrum $\hat{x}(n)$. Show from Eq. (10.49) that the characteristic system D_* is a causal system for minimum-phase inputs, i.e., that for minimum-phase inputs, $\hat{x}(n)$ for $n < n_0$ is dependent only on $x(n)$ for $n < n_0$, where n_0 is arbitrary.

11

Power Spectrum Estimation

11.0 Introduction

One of the important areas of application for digital signal processing techniques such as fast Fourier transform algorithms is the estimation of the autocovariance and the power spectrum of a random sequence. The need for power spectrum estimation arises in a variety of contexts, including the measurement of noise spectra for the design of optimal linear filters, the detection of narrow-band signals in wide-band noise, and the estimation of parameters of a linear system by using a noise excitation.

The formal mathematical basis for power spectrum estimation techniques can be found in the more general topic of estimation theory. In practical spectrum estimation, however, it is generally true that optimum estimation techniques, such as maximum likelihood estimation, require more information about the signal than is usually available. For this reason power spectrum estimation as it is currently practiced has a very strong empirical basis, and there are generally tradeoffs involved between different techniques in such a way that there is usually not general agreement on the best method.

The purpose of this chapter is to provide a brief and elementary introduction to power spectrum estimation. The primary intention is to provide some feeling for the style in which spectrum estimation is carried out and the role that can be played by some of the digital signal processing techniques that we have discussed previously. It is a large step from the elementary introduction that follows to a deep understanding of the myriad of tradeoffs, alternatives, and techniques that are involved in the practice of spectrum estimation.

The basic references on spectrum estimation include the books by Bartlett [1], Blackman and Tukey [2], Grenander and Rosenblatt [3], and Hannan [4]. More recent books include works by Jenkins and Watts [5] and Koopmanns [6]. The discussion that follows borrows heavily from these earlier works. We begin with a brief introduction to some of the concepts of the general theory of estimation as it applies to estimating averages of a random process. Then we consider the application of these basic ideas to the problem of estimation of the autocorrelation sequence or the autocovariance sequence of a stationary random process. Next, attention is directed toward the estimation of the power spectrum, with major emphasis being on some of the difficulties that are often encountered in the application of standard techniques. Finally, we discuss the application of some of the digital signal processing techniques that we have discussed in earlier chapters to estimation of correlation sequences and power spectra.

11.1 Basic Principles of Estimation Theory

In Chapter 8 we reviewed the concept of a random process and we discussed characterizations of random processes by averages. In empirically characterizing a signal in terms of a random process model, it is often necessary to estimate averages of the random process model from a single sample sequence of the random process, i.e., a sequence $x(n)$ which is assumed to be a realization of a random process defined by the set of random variables $\{x_n\}$. Furthermore, in order to make the computation of the estimates possible, we must base our estimate on a finite segment of the sample sequence $x(n)$. The fact that we might be able to compute estimates of the various desired averages of the random variables $\{x_n\}$ from a finite segment, $x(n)$ for $0 \le n \le N - 1$, of a single sample sequence $x(n)$ is plausible when we consider ergodic processes, i.e., random processes for which probability averages are equal to time averages. For example, consider a random process for which

$$m_x = E[x_n] = \int_{-\infty}^{\infty} x p_x(x)\, dx, \qquad \text{for all } n \tag{11.1}$$

Furthermore, suppose that

$$m_x = \langle x_n \rangle = \lim_{N\to\infty} \frac{1}{2N+1} \sum_{n=-N}^{N} x_n \tag{11.2}$$

and for each sample sequence generated by the random process,

$$m_x = \langle x(n) \rangle = \lim_{N\to\infty} \frac{1}{2N+1} \sum_{n=-N}^{N} x(n) \tag{11.3}$$

Then it is plausible that the quantity

$$\hat{m}_x = \frac{1}{N}\sum_{n=0}^{N-1} x(n)$$

might be a sufficiently accurate estimate of m_x if N were "large enough." The branch of statistical theory that pertains to such situations is called *estimation theory*. We shall begin our consideration of the empirical determination of averages of a random process model with a brief discussion of the fundamentals of estimation theory. In a more general context than we are considering here, there can be a large number of questions to be answered about a random sequence. For example, if the random sequence has been generated by exciting a discrete linear system with white noise, it may be of interest to estimate the parameters of the linear system. Another possible purpose of analysis may be to simply decide whether the process is white or not white. To characterize the process we may wish to estimate such parameters as the mean, the variance, the autocovariance sequence, or the power density spectrum. It is the estimation of these parameters that we will concentrate on entirely in the next sections.

Let us consider a stationary random process $\{x_n\}$, $-\infty < n < \infty$. Its mean value, m_x, is defined by Eq. (11.1). The time-average mean of the random process, $\{x_n\}$, $-\infty < n < \infty$, is defined in Eq. (11.2). Let us also assume that the time average of each sample sequence, Eq. (11.3), is equal to m_x. The variance of the random process is defined as

$$\sigma_x^2 = E[(x_n - m_x)^2] = \langle (x_n - m_x)^2 \rangle \tag{11.4}$$

The autocovariance sequence is defined as

$$\gamma_{xx}(m) = E[(x_n - m_x)(x_{n+m}^* - m_x^*)] = \langle (x_n - m_x)(x_{n+m}^* - m_x^*) \rangle \tag{11.5}$$

and the power density spectrum is defined as

$$P_{xx}(\omega) = \sum_{m=-\infty}^{\infty} \gamma_{xx}(m) e^{-j\omega m} \tag{11.6}$$

The estimation of a parameter of the random process is based upon a finite segment of a single sample sequence; i.e., we have N values $x(n)$, $0 \le n \le N-1$, from which to estimate some parameter which initially we will denote "α." The estimate $\hat{\alpha}$ of the parameter α is thus a function of the random variables x_n, $0 \le n \le N-1$; i.e.,

$$\hat{\alpha} = F[x_0, x_1, \ldots, x_{N-1}]$$

and therefore $\hat{\alpha}$ is also a random variable. The probability density function of $\hat{\alpha}$ will be denoted $p_{\hat{\alpha}}(\hat{\alpha})$. The functional form and the shape of $p_{\hat{\alpha}}(\hat{\alpha})$ will depend upon the choice of the estimator $F[\ ]$ and the probability densities of the random variables x_n, as indicated in Fig. 11.1. It is reasonable to characterize an estimator as being "good" if there is a high probability that

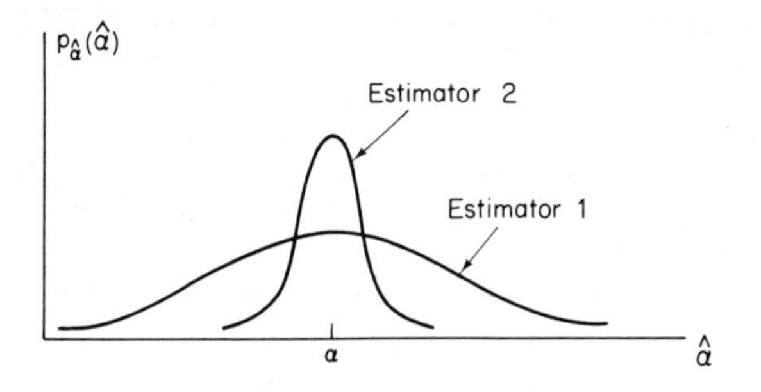

Fig. 11.1 Probability density functions of two estimators.

the estimate will be close to α. On this basis it appears that estimator 2 in Fig. 11.1 is superior to estimator 1 because the probability density of estimator 2 is more concentrated about the true value α.

One way to characterize the concentration of the probability density function of an estimator is in terms of a confidence interval. For example, for the probability density function sketched in Fig. 11.2, the area under the

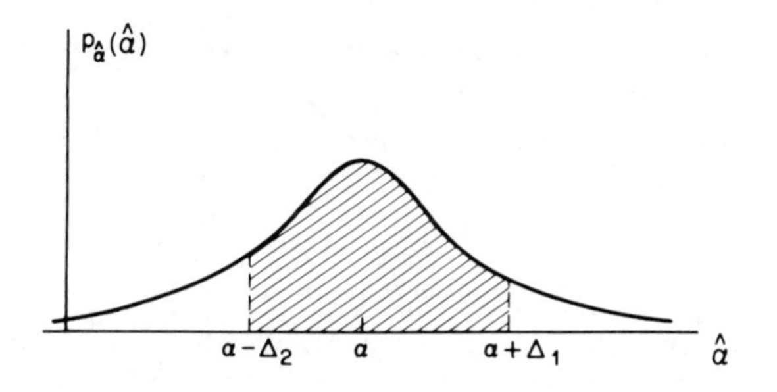

Fig. 11.2. Confidence limits for an estimator.

probability density function for $\alpha - \Delta_2 \leq \hat{\alpha} \leq \alpha + \Delta_1$ represents the probability that the estimate will lie between those two limits. Thus, if we denote that area by $(1 - \beta)$ then

$$\text{Probability } [-\Delta_2 \leq (\hat{\alpha} - \alpha) \leq \Delta_1] = (1 - \beta).$$

Thus, for example, if for a particular estimator we found that for $\Delta_1 = \Delta_2 = 0.1$ the area $(1 - \beta)$ was equal to 0.95, then in that case we might say that with 95% confidence the estimate would be within plus or minus 0.1 of the true value.

Generally speaking, it is plausible that for a good estimator the probability density function $p_{\hat{\alpha}}(\hat{\alpha})$ should be narrow and concentrated around the true value, and we might compare different estimators on that basis. In keeping with this notion two properties of estimators that are commonly used as a basis for comparison are the bias and variance. The bias of an estimator is defined as the true value of the parameter minus the expected value of the estimate, i.e.,

$$\text{bias} = \alpha - E[\hat{\alpha}] \triangleq B \tag{11.7}$$

An unbiased estimator is one for which the bias is 0. This then means that the expected value of the estimate is the true value so that if the probability density $p_{\hat{\alpha}}(\hat{\alpha})$ is symmetrical, then its center would be at the true value α. The variance of the estimator in effect measures the width of the probability density and is defined by

$$\text{var}\,[\hat{\alpha}] = E[(\hat{\alpha} - E[\hat{\alpha}])^2] = \sigma_{\hat{\alpha}}^2 \tag{11.8}$$

A small variance suggests that the probability density $p_{\hat{\alpha}}(\hat{\alpha})$ is concentrated around its mean value, which, if the estimator is also unbiased, will be the true value of the parameter. In many cases a comparison between two estimators will be complicated by the fact that the one with the smaller bias has a larger variance, or vice versa. Consequently, it is sometimes convenient to consider the mean-square error associated with an estimator defined as

$$\text{mean square error} = E[(\hat{\alpha} - \alpha)^2] = \sigma_{\hat{\alpha}}^2 + B^2 \tag{11.9}$$

An estimator is said to be *consistent* if as the number of observations becomes larger, the bias and the variance both tend to zero.

As an illustration of the above ideas let us consider a random process for which the amplitude probability density function is Gaussian; i.e.,

$$p_{x_n}(x) = \frac{1}{\sqrt{2\pi}\,\sigma_x} e^{-(x-m_x)^2/2\sigma_x^2}$$

Also, we shall assume that the random variables $\{x_n\}$ are statistically independent, so that, in particular, $x_0, x_1, \ldots, x_{N-1}$ are real and statistically independent. A class of estimators that is often used are the *maximum likelihood estimates.* A maximum likelihood estimate is based on a consideration of the joint probability of having observed N values as a function of the parameter to be estimated. The maximum likelihood estimate is that value of the parameter for which the probability of having gotten the observed values is a maximum. It is well known [7] that for the problem under consideration here, the maximum likelihood estimate of the mean value m_x of the Gaussian random process is the sample mean defined as

$$\text{sample mean} = \hat{m}_x = \frac{1}{N}\sum_{i=0}^{N-1} x_i \tag{11.10}$$

This is then one choice for the estimator of the parameter m_x. Since $\hat{m}_x$ is a weighted sum of independent Gaussian random variables, the probability density $p_{\hat{m}_x}(\hat{m}_x)$ is also Gaussian.[7] Since $p_{\hat{m}_x}(\hat{m}_x)$ is also Gaussian, it is therefore completely characterized by the bias of the estimator and the variance of the estimator. The expected value of $\hat{m}_x$ is equal to the expected value of x_n and, consequently, the bias is equal to zero. To obtain the variance of the sample mean we need to compute

$$\begin{aligned}
E[\hat{m}_x^2] &= \frac{1}{N^2}\sum_{i=0}^{N-1}\sum_{j=0}^{N-1} E[x_i x_j] \\
&= \frac{1}{N^2}\left[\sum_{i=0}^{N-1} E[x_i^2] + \sum_{i=0}^{N-1}\sum_{\substack{j=0 \\ j\neq i}}^{N-1} E[x_i]\cdot E[x_j]\right] \\
&= \frac{1}{N} E[x_n^2] + m_x^2 \frac{N-1}{N}
\end{aligned}$$

Thus

$$\text{var}\,[\hat{m}_x] = E[\hat{m}_x^2] - \{E[\hat{m}_x]\}^2 = \frac{1}{N}(E[x_n^2] - m_x^2) = \frac{1}{N}\sigma_x^2 \tag{11.11}$$

Equation (11.11) tells us, then, that as the number of observations increases, the variance of the sample mean decreases, and since the bias is zero, the sample mean is a consistent estimator.

If the mean value is known but the variance is to be estimated, then the maximum likelihood estimator is

$$\hat{\sigma}_x^2 = \frac{1}{N}\sum_{i=0}^{N-1}(x_i - m_x)^2 \tag{11.12}$$

It is straightforward to verify that this estimator for the variance is consistent. However, it requires that the parameter m_x be known. If both the mean and the variance are to be estimated, then the maximum likelihood estimate of the mean is, as before, the sample mean, and the maximum likelihood estimate for the variance is the sample variance defined as

$$\hat{\sigma}_x^2 = \frac{1}{N}\sum_{i=0}^{N-1}(x_i - \hat{m}_x)^2 \tag{11.13}$$

where $\hat{m}_x$ is the sample mean. Equation (11.13) differs from Eq. (11.12) in that in the first case the true value of the mean is used, whereas in the second case the estimate of the mean is used. To determine the bias of the sample variance as given by Eq. (11.13), we can first compute the expected value of $\hat{\sigma}_x^2$. Thus

$$\begin{aligned}
E[\hat{\sigma}_x^2] &= \frac{1}{N}\sum_{i=0}^{N-1}(E[x_i^2] + E[\hat{m}_x^2] - 2E[x_i\hat{m}_x]) \\
&= \frac{1}{N}\sum_{i=0}^{N-1} E[x_i^2] + \frac{1}{N^2}\sum_{i=0}^{N-1}\sum_{j=0}^{N-1} E[x_i x_j] - \frac{2}{N^2}\sum_{i=0}^{N-1}\sum_{j=0}^{N-1} E[x_i x_j] \\
&= \frac{N-1}{N} E[x_i^2] - \frac{N-1}{N} m_x^2 \\
&= \frac{N-1}{N}\sigma_x^2
\end{aligned} \tag{11.14}$$

Consequently, the mean value of the sample variance is not equal to the variance, and thus the sample variance is biased. However, as N becomes very large, the mean of the sample variance approaches the variance. Hence this estimate is asymptotically unbiased. To determine the variance of the sample variance let us first, for convenience, assume that the process is zero mean, so that

$$v = \hat{\sigma}_x^2 = \frac{1}{N}\sum_{i=0}^{N-1} x_i^2$$

Then

$$E[v^2] = \frac{1}{N^2}\sum_{i=1}^{N}\sum_{r=1}^{N} E[x_i^2 x_r^2]$$
$$= \frac{1}{N^2}[NE[x_n^4] + N(N-1)\{E[x_n^2]\}^2]$$
$$= \frac{1}{N}[E[x_n^4] + (N-1)\{E[x_n^2]\}^2]$$

We easily see that

$$E[v] = E[x_n^2]$$

so that

$$\begin{aligned}\text{var}\,[\hat{\sigma}_x^2] &= E[v^2] - (E[v])^2 \\ &= \frac{1}{N}\{E[x_n^4] - (E[x_n^2])^2\}\end{aligned} \tag{11.15}$$

Thus from Eqs. (11.14) and (11.15) we note that the sample variance is a consistent estimate.

The previous discussion is intended to illustrate the style of analysis that is employed in describing the properties of estimators. The objective of this analysis is to give some idea of the accuracy of the estimate and how the accuracy depends upon the number of samples employed in the estimate.

To compute confidence limits for estimators we need to know the probability distribution of the random variables x_n. When we do not know these probability distributions, as is typically the case in signal processing applications, a Gaussian probability law is often justifiably assumed. From this assumed probability distribution of the random variables x_n, it is often possible to derive approximate confidence limits on estimates of the mean, variance, and so on. Many times, however, we may be satisfied with expressions for the bias and variance of estimators. Even approximate expressions which display the dependence of the bias and variance upon the length of the sample sequence are useful in guiding application of the digital signal processing techniques that we have discussed in earlier chapters to the problem of estimating averages of random signals. In the next sections we shall therefore be concerned with obtaining expressions for the bias and variance of various estimators of the autocovariance and power spectrum of a stationary random signal. We shall use these expressions to obtain insight into the problems encountered in computing such estimates.

11.2 Estimates of the Autocovariance

The concepts introduced in the previous section can be applied in studying estimators for the autocovariance sequence of a random process. Again we assume a stationary random process $\{x_n\}$, $-\infty < n < \infty$, and for convenience we shall assume that the mean is zero; i.e.,

$$m_x = E[x_n] = 0, \qquad \text{for all } n$$

Then the autocovariance sequence is

$$\gamma_{xx}(m) = E[x_n x^*_{n+m}]$$

which is also equal to the autocorrelation sequence, $\phi_{xx}(m)$. Thus we shall henceforth refer to estimates of the autocorrelation sequence, recognizing that the autocorrelation and autocovariance are identical for a zero mean process. We assume further that

$$\gamma_{xx}(m) = \langle x(n)x^*(n+m)\rangle \tag{11.16}$$

for all sample sequences $x(n)$. We can express Eq. (11.16) as

$$\gamma_{xx}(m) = \langle g_m(n)\rangle$$

where $g_m(n) = x(n)x^*(n+m)$. Consequently, the estimation of the autocovariance of a zero mean process can be thought of as estimating the mean of $g_m(n)$. Given N consecutive values of the sequence $x(n)$, we have $(N - m)$ consecutive samples of $g_m(n)$ from which to estimate the mean of $g_m(n)$. Applying the definition of the sample mean from the previous section, we obtain the estimate of the autocorrelation sequence,

$$c'_{xx}(m) = \frac{1}{N - |m|} \sum_{n=0}^{N-|m|-1} x(n)x^*(n+m) \tag{11.17}$$

where $|m| < N$. If the sequence $g_m(n)$ is Gaussian, then Eq. (11.17) is the maximum likelihood estimate of the autocorrelation sequence. Generally, the formal procedure of determining the maximum likelihood estimate leads to a set of equations that cannot be solved even if the probability law of $g_m(n)$ is known. However, even though Eq. (11.17) may not be formally optimum, it provides a plausible choice for an estimator of the autocorrelation sequence. It is easy to see that $c'_{xx}(m)$ is an unbiased estimate of $\phi_{xx}(m)$ since $E[x(n)x^*(n+m)] = \phi_{xx}(m)$. The variance of $c'_{xx}(m)$ can be found as was done in Sec. 11.1; however, the mathematical manipulations are tedious and will be omitted here. An approximate expression for the variance as given by Jenkins and Watts [5] is

$$\text{var}\,[c'_{xx}(m)] \cong \frac{N}{[N - |m|]^2} \sum_{r=-\infty}^{\infty} [\phi_{xx}^2(r) + \phi_{xx}(r + m)\phi_{xx}(r - m)] \quad (11.18)$$

This expression holds for N much larger than m; however, in general the variance of $c'_{xx}(m)$ is proportional to $1/N$ as in Eq. (11.18). Since the bias is zero and

$$\lim_{N\to\infty} \{\text{var}\,[c'_{xx}(m)]\} \to 0$$

$c'_{xx}(m)$ is a consistent estimate of $\phi_{xx}(m)$.

An alternative estimator for the autocorrelation sequence is

$$c_{xx}(m) = \frac{1}{N} \sum_{n=0}^{N-|m|-1} x(n)x(n + m) \quad (11.19)$$

This estimate differs from the estimate $c'_{xx}(m)$ of Eq. (11.17) only in the multiplying factor in front of the summation. In fact, comparing Eqs. (11.17) and (11.19) we see that

$$c_{xx}(m) = \frac{N - |m|}{N}\, c'_{xx}(m) \quad (11.20)$$

Since the expected value of $c'_{xx}(m)$ is $\phi_{xx}(m)$, the expected value of $c_{xx}(m)$ is

$$E[c_{xx}(m)] = \frac{N - |m|}{N}\, \phi_{xx}(m) \quad (11.21)$$

Consequently, $c_{xx}(m)$ is a biased estimate of the autocorrelation sequence, although it is asymptotically unbiased. In particular, the bias of the estimate $c_{xx}(m)$ is

$$\text{bias} = \phi_{xx}(m)\left[\frac{m}{N}\right] \quad (11.22)$$

From Eq. (11.20) it follows that the variance of $c_{xx}(m)$ is equal to $[(N - |m|)/N]^2$ times the variance of $c'_{xx}(m)$, so that for N large compared with m,

$$\text{var}\,[c_{xx}(m)] \cong \frac{1}{N} \sum_{r=-\infty}^{\infty} [\phi_{xx}^2(r) + \phi_{xx}(r + m)\phi_{xx}(r - m)] \quad (11.23)$$

As the value of m approaches N, the variance of the estimate $c'_{xx}(m)$ becomes exceptionally large. This results from the fact that the estimate $c'_{xx}(m)$ is based on computing the sample mean of the sequence $g_m(n)$. If m is on the order of N, then there are only a few points available to use in computing the sample mean of $g_m(n)$. For this reason, as m approaches the record length, the variance of the estimate $c'_{xx}(m)$ becomes very large. Consequently, we do not obtain a useful estimate. On the other hand, the variance of the biased estimator $c_{xx}(m)$ does not become as large for m on the order of the record length. However, as m approaches N, the bias approaches $\phi_{xx}(m)$. That is, the mean value of the estimate approaches 0. Since the bias is as large as the function that we are estimating, this again could not be considered to be a reasonable estimate when m is on the order of N.

The above conclusions were based upon a consideration of the bias and the variance as the value of the lag m increases with the record length fixed. From another point of view, the value of the lag m can be fixed and the behavior of the bias and the variance can be considered as the record length N increases. With m fixed we see from Eq. (11.18) that the variance of the unbiased estimate $c'_{xx}(m)$ decreases with increasing N. For the biased estimate, we see from Eqs. (11.22) and (11.23) that both the bias and the variance decrease with increasing N. Jenkins and Watts [5] conjecture that, in many cases, the mean-square error for the biased estimator is less than for the unbiased estimator. This conjecture, if valid, provides a rationale for using the biased estimator $c_{xx}(m)$. Both estimators, however, are asymptotically unbiased, so we can generally expect to improve the estimate of the autocorrelation sequence by basing the estimate on an increased number of samples.

11.3 The Periodogram as an Estimate of the Power Spectrum

In the previous section we considered two plausible estimators of the autocovariance sequence. We saw that these estimators provide a consistent asymptotically unbiased estimate of the autocovariance. It is tempting to conclude that the Fourier transform of such an estimate of the autocovariance sequence would provide a good estimate of the power density spectrum. Unfortunately, this is not the case. Specifically, we will show that the Fourier transforms of the consistent estimates of the covariance are not consistent estimates of the power spectrum since the variance does not approach zero as the record length, N, increases. We will see, however, that smoothing the Fourier transform of the estimate of the covariance does produce a good estimate of the power spectrum.

In general, exact expressions for the variance of spectrum estimates become very unwieldy. In order to obtain results that provide insight into spectrum analysis, it is useful to direct the derivation of such results toward approximate expressions which can be easily interpreted. Thus many of the expressions that will be developed in the following discussion are only approximate. In all cases, however, the basis for the approximation will be pointed out.

11.3.1 Definition of the Periodogram

As an estimate of the power density spectrum let us consider the Fourier transform of the biased autocorrelation estimate $c_{xx}(m)$. That is,

$$I_N(\omega) = \sum_{m=-(N-1)}^{N-1} c_{xx}(m)e^{-j\omega m} \tag{11.24}$$

Since the Fourier transform of the real finite-length sequence $x(n)$, $0 \leq n \leq N-1$, is

$$X(e^{j\omega}) = \sum_{n=0}^{N-1} x(n)e^{-j\omega n}$$

it can be shown that (see Problem 1 of this chapter).

$$I_N(\omega) = \frac{1}{N}|X(e^{j\omega})|^2 \tag{11.25}$$

The spectrum estimate $I_N(\omega)$ is often called the *periodogram*.

As before, it is of interest to determine the bias and variance of the periodogram as an estimate of the power spectrum. The expected value of $I_N(\omega)$ is

$$E[I_N(\omega)] = \sum_{m=-(N-1)}^{N-1} E[c_{xx}(m)]e^{-j\omega m} \tag{11.26}$$

Since we have shown that for a zero mean process

$$E[c_{xx}(m)] = \frac{N-|m|}{N}\phi_{xx}(m), \qquad |m| < N$$

then

$$E[I_N(\omega)] = \sum_{m=-(N-1)}^{N-1} \left(\frac{N-|m|}{N}\right)\phi_{xx}(m)e^{-j\omega m} \tag{11.27}$$

Thus because of the finite limits of summation and the factor $(N-|m|)/N$, $E[I_N(\omega)]$ is not equal to the Fourier transform of $\phi_{xx}(m)$, and therefore the periodogram is a biased estimate of the power spectrum, $P_{xx}(\omega)$.

Alternatively, consider the Fourier transform of the estimate $c'_{xx}(m)$; i.e.,

$$P_N(\omega) = \sum_{m=-(N-1)}^{N-1} c'_{xx}(m)e^{-j\omega m} \tag{11.28}$$

The expected value of $P_N(\omega)$ is

$$\begin{aligned} E[P_N(\omega)] &= \sum_{m=-(N-1)}^{N-1} E[c'_{xx}(m)]e^{-j\omega m} \\ &= \sum_{m=-(N-1)}^{N-1} \phi_{xx}(m)e^{-j\omega m} \end{aligned} \tag{11.29}$$

Again, because of the finite limits of summation, this is a biased estimate of $P_{xx}(\omega)$, even though $c_{xx}(m)$ is an unbiased estimate of $\phi_{xx}(m)$.

We can interpret Eqs. (11.27) and (11.29) as Fourier transforms of windowed autocorrelation sequences. In the case of Eq. (11.27) the window is the triangular window

$$w_B(m) = \begin{cases} \dfrac{N - |m|}{N}, & |m| < N \\ 0, & \text{otherwise} \end{cases} \tag{11.30}$$

In Chapter 5 we called this the Bartlett window. For Eq. (11.29) the window is rectangular; i.e.,

$$w_R(n) = \begin{cases} 1, & |m| < N \\ 0, & \text{otherwise} \end{cases} \tag{11.31}$$

Using the concepts introduced in Chapter 5 we can see that Eqs. (11.27) and (11.29) can be interpreted in the frequency domain as the convolutions

$$E[I_N(\omega)] = \frac{1}{2\pi}\int_{-\pi}^{\pi} P_{xx}(\theta) W_B(e^{j(\omega-\theta)})\, d\theta \tag{11.32}$$

and

$$E[P_N(\omega)] = \frac{1}{2\pi}\int_{-\pi}^{\pi} P_{xx}(\theta) W_R(e^{j(\omega-\theta)})\, d\theta \tag{11.33}$$

where

$$W_B(e^{j\omega}) = \frac{1}{N}\left(\frac{\sin[\omega N/2]}{\sin[\omega/2]}\right)^2 \tag{11.34}$$

and

$$W_R(e^{j\omega}) = \frac{\sin[\omega(2N-1)/2]}{\sin[\omega/2]} \tag{11.35}$$

are the Fourier transforms of the Bartlett and rectangular windows, respectively.

11.3.2 *Variance of the Periodogram*

To obtain an expression for the variance of the periodogram, it is convenient to first assume that the sequence $x(n)$, $0 \leq n \leq N - 1$, is a sample of a real, white, zero-mean process with Gaussian probability density functions. The periodogram $I_N(\omega)$ can be expressed as

$$I_N(\omega) = \frac{1}{N}|X(e^{j\omega})|^2$$

$$= \frac{1}{N}\sum_{l=0}^{N-1}\sum_{m=0}^{N-1} x(l)x(m)e^{j\omega m}e^{-j\omega l}$$

To evaluate the covariance of $I_N(\omega)$ at two frequencies ω_1 and ω_2 we first consider

$$E[I_N(\omega_1)I_N(\omega_2)] = \frac{1}{N^2}\sum_{k=0}^{N-1}\sum_{l=0}^{N-1}\sum_{m=0}^{N-1}\sum_{n=0}^{N-1} E[x(k)x(l)x(m)x(n)]e^{j[\omega_1(k-l)+\omega_2(m-n)]} \tag{11.36}$$

To obtain a useful result, we must simplify Eq. (11.36). In general, it is not possible to obtain a very simple result even when $x(n)$ is white, because $E[x(n)x(n+m)] = \sigma_x^2\delta(m)$ does not guarantee a simple expression for $E[x(k)x(l)x(m)x(n)]$ for all combinations of k, l, m, and n. However, in the case of a white Gaussian process, it can be shown [7] that

$$\begin{aligned} E[x(k)x(l)x(m)x(n)] = {} & E[x(k)x(l)]E[x(m)x(n)] \\ & + E[x(k)x(m)]E[x(l)x(n)] \\ & + E[x(k)x(n)]E[x(l)x(m)] \end{aligned}$$

Therefore,

$$E[x(k)x(l)x(m)x(n)] = \begin{cases} \sigma_x^4, & k = l \text{ and } m = n \\ & \text{or } k = m \text{ and } l = n \\ & \text{or } k = n \text{ and } l = m \\ 0, & \text{otherwise} \end{cases} \tag{11.37}$$

For other than Gaussian joint density functions, the result will not necessarily be so simple. However, our objective is to give a result that will lend insight into the problems of spectrum estimation rather than to give a general formula with wide validity which would be difficult to interpret. Thus, if we substitute Eq. (11.37) into Eq. (11.36), we obtain

$$E[I_N(\omega_1)I_N(\omega_2)] = \frac{\sigma_x^4}{N^2}\left\{N^2 + \sum_{m=0}^{N-1}\sum_{n=0}^{N-1} e^{j(m-n)(\omega_1+\omega_2)} + \sum_{m=0}^{N-1}\sum_{n=0}^{N-1} e^{j(n-m)(\omega_1-\omega_2)}\right\}$$

or

$$\begin{aligned} E[I_N(\omega_1)I_N(\omega_2)] = \sigma_x^4\Bigg\{1 & + \left(\frac{\sin[(\omega_1+\omega_2)N/2]}{N\sin[(\omega_1+\omega_2)/2]}\right)^2 \\ & + \left(\frac{\sin[(\omega_1-\omega_2)N/2]}{N\sin[(\omega_1-\omega_2)/2]}\right)^2\Bigg\} \end{aligned} \tag{11.38}$$

(If the signal is non-Gaussian, Eq. (11.38) contains additional terms which are proportional to $1/N$ [4, 8].) The covariance of the periodogram is

$$\text{cov}\,[I_N(\omega_1), I_N(\omega_2)] = E[I_N(\omega_1)I_N(\omega_2)] - E[I_N(\omega_1)]E[I_N(\omega_2)] \tag{11.39}$$

Since $E[I_N(\omega_1)] = E[I_N(\omega_2)] = \sigma_x^2$, from Eqs. (11.38) and (11.39) we obtain

$$\text{cov}\,[I_N(\omega_1), I_N(\omega_2)] = \sigma_x^4\left\{\left(\frac{\sin\,[(\omega_1 + \omega_2)N/2]}{N\sin\,[(\omega_1 + \omega_2)/2]}\right)^2 + \left(\frac{\sin\,[(\omega_1 - \omega_2)N/2]}{N\sin\,[(\omega_1 - \omega_2)/2]}\right)^2\right\} \tag{11.40}$$

From Eq. (11.40) we can draw a number of interesting conclusions about the periodogram. The variance of the estimate of the spectrum at a particular frequency $\omega = \omega_1 = \omega_2$ is

$$\text{var}\,[I_N(\omega)] = \text{cov}\,[I_N(\omega), I_N(\omega)] = \sigma_x^4\left\{1 + \left(\frac{\sin\,[\omega N]}{N\sin\,\omega}\right)^2\right\} \tag{11.41}$$

Clearly, the variance of $I_N(\omega)$ *does not* approach zero as N approaches infinity. Thus the periodogram is not a consistent estimate. In fact, var $[I_N(\omega)]$ is of the order of σ_x^4 no matter how N is chosen.

We also see from Eq. (11.40) that for frequencies $\omega_1 = 2\pi k/N$ and $\omega_2 = 2\pi l/N$, where k and l are integers,

$$\text{cov}\,[I_N(\omega_1), I_N(\omega_2)] = \sigma_x^4\left\{\left(\frac{\sin\,[\pi(k + l)]}{N\sin\,[\pi(k + l)/N]}\right)^2 + \left(\frac{\sin\,[\pi(k - l)]}{N\sin\,[\pi(k - l)/N]}\right)^2\right\} \tag{11.42}$$

which is equal to zero for $k \neq l$. Thus values of the periodogram spaced in frequency by integer multiples of $2\pi/N$ are uncorrelated. As N increases, these uncorrelated frequency samples with zero covariance come closer together. It is reasonable to expect that a good estimate of the power spectrum should approach a constant as N increases since we have assumed that the signal was white. A consequence of the fact that the variance of the periodogram approaches a non-zero constant and that the spacing between spectral samples with zero covariance decreases as N increases is that as the record length becomes longer, the rapidity of the fluctuations in the periodogram increases. This behavior is illustrated in Fig. 11.3, where the periodogram is plotted for record lengths of $N = 14$, 51, 135, and 452 samples.

11.3.3 General Variance Expressions

All the previous discussion was for the case of estimating the spectrum of white noise. If we consider data that are nonwhite but Gaussian, the analysis is considerably more difficult. In evaluating the covariance between spectrum samples in this more general case, it is useful to take a heuristic approach and develop an approximate expression. The approach that we shall take is heuristic; a more rigorous derivation is given by Jenkins and Watts [5]. With the application of some approximations to their result, the approximate results derived here can be obtained. The basis for the

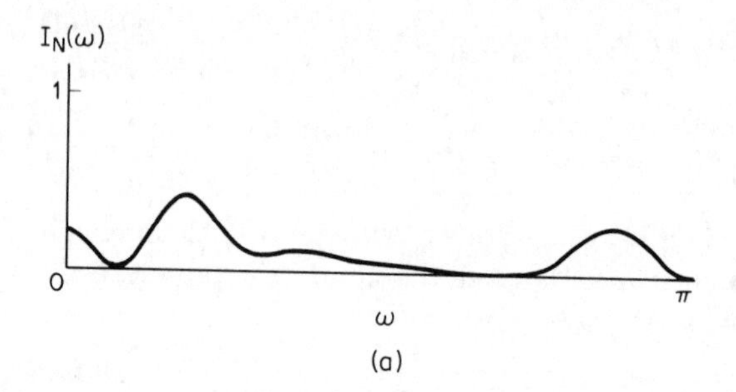

(a)

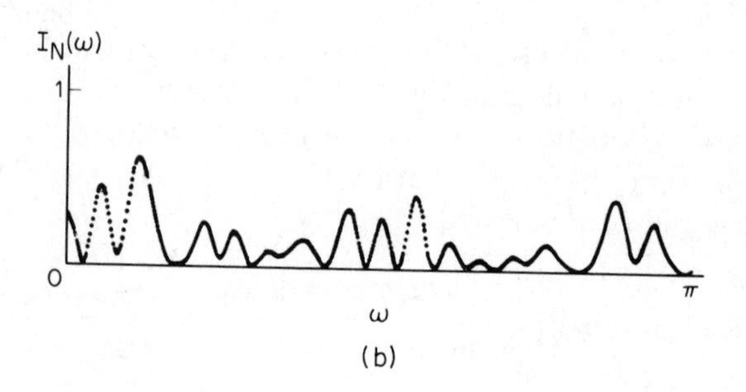

(b)

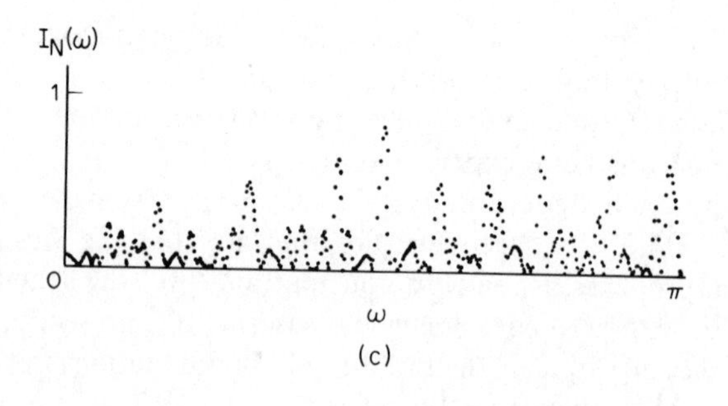

(c)

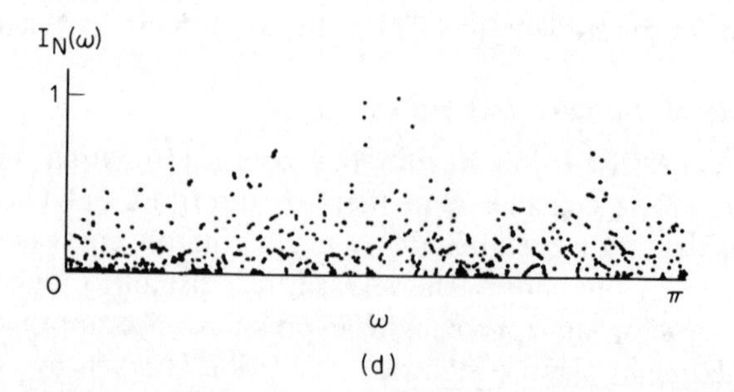

(d)

Fig. 11.3 Periodograms for record lengths N = (a) 14, (b) 51, (c) 135, and (d) 452, showing increasing fluctuations with increasing N.

heuristic approach is that a nonwhite (or colored) random sequence can be generated by processing white noise with a linear system. The power density spectrum of the output noise is the product of the power density spectrum of the input and the magnitude squared of the frequency response of the system. Now let us consider a sample of nonwhite noise. The sample length is denoted as N. It is, of course, not exactly true that a segment of nonwhite noise can be obtained by filtering a segment of white noise with the linear system because of the transient effects at the beginning and the end of the segment. However, if the record length is long compared with the duration of the filter impulse response, it seems at least plausible that a sample of nonwhite noise can be approximated in this way. Now let us consider a nonwhite Gaussian process with a power density spectrum $P_{xx}(\omega)$. Let $x_N(n)$ denote an N-point sample of the nonwhite noise and let $w_N(n)$ denote an N-point sample of white noise with unit variance. Then our approximation is that $x_N(n)$ is the result of processing $w_N(n)$ with a linear system for which the squared magnitude of the frequency response is $P_{xx}(\omega)$. With $I_N(\omega)$ denoting the periodogram of the colored noise and $I_N^w(\omega)$ denoting the periodogram of the white noise, it follows that

$$I_N(\omega) = \frac{1}{N} |X_N(e^{j\omega})|^2$$

$$I_N^w(\omega) = \frac{1}{N} |W_N(e^{j\omega})|^2$$

and since

$$|X_N(e^{j\omega})|^2 \cong P_{xx}(\omega)\, |W_N(e^{j\omega})|^2$$

it follows that

$$I_N(\omega) \cong P_{xx}(\omega) I_N^w(\omega)$$

Consequently, using Eq. 11.40 the covariance of the periodogram at different frequencies can be expressed approximately as

$$\begin{aligned} \text{cov}\,[I_N(\omega_1,)\, I_N(\omega_2)] \cong P_{xx}(\omega_1) P_{xx}(\omega_2) \Bigg\{ &\left(\frac{\sin\,[(\omega_1 + \omega_2)N/2]}{N \sin\,[(\omega_1 + \omega_2)/2]} \right)^2 \\ &+ \left(\frac{\sin\,[(\omega_1 - \omega_2)N/2]}{N \sin\,[(\omega_1 - \omega_2)/2]} \right)^2 \Bigg\} \end{aligned} \tag{11.43}$$

If Eq. (11.43) is evaluated at frequency samples equally spaced by $2\pi/N$, we see again that the covariance between frequency samples is zero. Furthermore, the variance of the periodogram is

$$\text{var}\,[I_N(\omega)] = P_{xx}^2(\omega) \left\{ 1 + \left(\frac{\sin\,[\omega N]}{N \sin \omega} \right)^2 \right\} \tag{11.44}$$

so that as N increases, the variance becomes proportional to the square of the spectrum. Thus, in general, the periodogram is not a consistent estimate and it can be expected to fluctuate rather wildly about the true spectrum value. Although the results that we have derived in this section were based on the assumption of Gaussian probability densities, the qualitative results hold for a rather broad range of distributions.

11.4 Smoothed Spectrum Estimators

Since the periodogram is not a consistent estimate of the spectrum, and since its behavior with increasing N is highly undesirable, it is necessary to study modifications that give better results. In this section we will show how the periodogram can be used to obtain a consistent spectrum estimate. Our continued interest in the periodogram is due to the ease with which it can be computed by use of FFT methods, as will be evident later.

11.4.1 Bartlett's Procedure—Averaging Periodograms

A standard approach to reducing the variance of estimates is to average over a number of independent estimates. The application of this approach to spectrum estimation is often attributed to Bartlett. In this approach a data sequence $x(n)$, $0 \leq n \leq N-1$, is divided into K segments of M samples each so that $N = KM$; i.e., we form the segments

$$x^{(i)}(n) = x(n + iM - M), \qquad 0 \leq n \leq M-1, \quad 1 \leq i \leq K \tag{11.45}$$

and compute the K periodograms

$$I_M^{(i)}(\omega) = \frac{1}{M}\left|\sum_{n=0}^{M-1} x^{(i)}(n)e^{-j\omega n}\right|^2, \qquad 1 \leq i \leq K \tag{11.46}$$

If $\phi_{xx}(m)$ is small for $m > M$, then it is reasonable to assume that periodograms $I_N^{(i)}(\omega)$ are independent of one another. The spectrum estimate is defined as

$$B_{xx}(\omega) = \frac{1}{K}\sum_{i=1}^{K} I_M^{(i)}(\omega) \tag{11.47}$$

The expected value of this spectrum estimate is

$$\begin{aligned} E[B_{xx}(\omega)\,]\,] &= \frac{1}{K}\sum_{i=1}^{K} E[I_M^{(i)}(\omega)] \\ &= E[I_M^{(i)}(\omega)] \end{aligned}$$

From Eqs. (11.32) and (11.34) we note that

$$E[B_{xx}(\omega)] = E[I_M^{(i)}(\omega)] = \frac{1}{2\pi M}\int_{-\pi}^{\pi} P_{xx}(\theta)\left(\frac{\sin\,[(\omega-\theta)M/2]}{\sin\,[(\omega-\theta)/2]}\right)^2 d\theta \tag{11.48}$$

That is, the expected value of the Bartlett estimate is the convolution of the true spectrum $P_{xx}(\omega)$ with the Fourier transform of the triangular window function corresponding to an M-sample periodogram where $M = N/K$. Thus the Bartlett estimate is also a biased estimator. If we assume that the K periodograms averaged in Eq. (11.47) are statistically independent, then $B_{xx}(\omega)$ is the sample mean of a set of K independent observations of the periodogram $I_M(\omega)$. Therefore, from Eqs. (11.11) and (11.44),

$$\begin{aligned} \text{var}\,[B_{xx}(\omega)] &= \frac{1}{K}\,\text{var}\,[I_M(\omega)] \\ &\cong \frac{1}{K}\,P_{xx}^2(\omega)\left\{1 + \left(\frac{\sin\,[\omega M]}{M\,\sin\,[\omega]}\right)^2\right\} \end{aligned} \tag{11.49}$$

From Eq. (11.49) it is clear that the variance of $B_{xx}(\omega)$ is inversely proportional to the number of periodograms averaged, and as K gets large, the variance approaches zero, so the Bartlett estimate is a consistent estimate.

A comparison of Eq. (11.48) for $E[B_{xx}(\omega)]$ with Eq. (11.32) for $E[I_N(\omega)]$ shows that in both cases the expected value of the estimate is in the form of a convolution of the true spectrum with a "spectrum window" of the form

$$W_B(e^{j\omega}) = \frac{1}{N'}\left(\frac{\sin\,[\omega N']}{\sin\,[\omega]}\right)^2$$

where $N' = N$ for the periodogram and $N' = M = N/K$ for the Bartlett estimate. The bias of $B_{xx}(\omega)$ is greater than the bias of $I_N(\omega)$, owing to the increased width of the main lobe of the spectrum window. The bias can therefore be interpreted in terms of the effect on spectrum resolution. For a fixed record length, as the number of periodograms increases, the variance decreases but M decreases, and therefore the spectrum resolution decreases. Thus there is a tradeoff between bias or spectrum resolution and the variance of the estimate in the Bartlett procedure. The actual choice of M and N in a given measurement of the spectrum will generally be guided by prior knowledge of the signal under consideration. For example, if we know that the spectrum has a very narrow peak, and if it is important to resolve that peak, we must choose M large enough to maintain the desired frequency resolution. From the expression for the variance we can determine the record length $N = KM$ for an acceptable variance of the spectrum estimate.

11.4.2 Windowing

We have seen that the variance of the Bartlett spectrum estimate is reduced at the expense of increased bias and decreased spectrum resolution. In the Bartlett procedure the spectrum resolution was decreased by using shorter record lengths. Another approach is to smooth the periodogram by convolution with an appropriate spectrum window [2]. That is, if $S_{xx}(\omega)$ denotes the smoothed periodogram, then

$$S_{xx}(\omega) = \frac{1}{2\pi}\int_{-\pi}^{\pi} I_N(\theta)W(e^{j(\omega-\theta)})\,d\theta \tag{11.50}$$

where $W(e^{j\omega})$ is the spectrum window. Since the periodogram is the Fourier transform of $c_{xx}(m)$, then $S_{xx}(\omega)$ is the Fourier transform of the product of $c_{xx}(m)$ and the inverse Fourier transform of $W(e^{j\omega})$. Thus if

$$w(m) = \frac{1}{2\pi}\int_{-\pi}^{\pi} W(e^{j\omega})e^{j\omega m}\, d\omega$$

is a finite-duration window sequence of length $2M - 1$, then

$$S_{xx}(\omega) = \sum_{m=-(M-1)}^{M-1} c_{xx}(m)w(m)e^{-j\omega m} \tag{11.51}$$

The window $w(m)$ should be an even sequence so that $S_{xx}(\omega)$ will be a real and an even function when the data sequence $x(n)$ is real. Furthermore, we recall that the power spectrum is a nonnegative function of frequency, and therefore it is reasonable to require that $S_{xx}(\omega)$ be nonnegative as well. Note that both the periodogram and the Bartlett estimate are nonnegative functions of frequency. From Eq. (11.50), however, it is clear that a sufficient, although certainly not necessary, condition for $S_{xx}(\omega)$ to be nonnegative is

$$W(e^{j\omega}) \geq 0, \qquad -\pi \leq \omega \leq \pi$$

This is true of the Bartlett or triangular window but not true, for example, of the Hamming or Hanning windows. Thus these latter window sequences, although they provide better frequency resolution and lower side lobes, may yield negative estimates of the power spectrum.

The expected value of Eq. (11.51) is easily seen to be

$$E[S_{xx}(\omega)] = \frac{1}{2\pi}\int_{-\pi}^{\pi} E[I_N(\theta)]W(e^{j(\omega-\theta)})\, d\theta \tag{11.52}$$

Since from Eq. (11.32)

$$E[I_N(\theta)] = \frac{1}{2\pi}\int_{-\pi}^{\pi} P_{xx}(\phi)W_B(e^{j(\theta-\phi)})\, d\phi$$

we see that $E[S_{xx}(\omega)]$ is the frequency-domain convolution of $W_B(e^{j\omega})$ and $W(e^{j\omega})$ with $P_{xx}(\omega)$. Thus, $E[S_{xx}(\omega)]$ is the Fourier transform of $\phi_{xx}(m)$ times the product of the triangular window $w_B(m)$ and $w(m)$; that is,

$$E[S_{xx}(\omega)] = \sum_{m=-(M-1)}^{M-1} c_{xx}(m)w_B(m)w(m)e^{-j\omega m} \tag{11.53}$$

where

$$w_B(m) = 1 - \frac{|m|}{N}, \qquad |m| < N$$

If M is small compared to N, then $W(e^{j\omega})$ will be wide compared to $W_B(e^{j\omega})$ and thus Eq. (11.52) is approximately

$$E[S_{xx}(\omega)] \approx \frac{1}{2\pi}\int_{-\pi}^{\pi} P_{xx}(\theta)W(e^{j(\omega-\theta)})\, d\theta \tag{11.54}$$

From either Eq. (11.52) or (11.54) we see that an increase in the spectrum window width causes additional smoothing of the spectrum and reduces the frequency resolution of the spectrum estimate.

To study the effect of the window on the variance of the spectrum estimate, the covariance of the smoothed periodogram estimate can be computed as before. The covariance between two frequencies ω_1 and ω_2 is

$$\text{cov}\,[S_{xx}(\omega_1), S_{xx}(\omega_2)] = E[(S_{xx}(\omega_1) - E[S_{xx}(\omega_1)])(S_{xx}(\omega_2) - E[S_{xx}(\omega_2)])]$$

From Eqs. (11.50) and (11.52),

$$S_{xx}(\omega) - E[S_{xx}(\omega)] \cong \frac{1}{2\pi}\int_{-\pi}^{\pi}(I_N(\theta) - E[I_N(\theta)])W(e^{j(\omega-\theta)})\,d\theta$$

so that

$$\text{cov}\,[S_{xx}(\omega_1), S_{xx}(\omega_2)] \cong \frac{1}{4\pi^2}\int_{-\pi}^{\pi}\int_{-\pi}^{\pi} W(e^{j(\omega_1-\theta)})W(e^{j(\omega_2-\phi)}) \times \text{cov}\,[I_N(\theta), I_N(\phi)]\,d\theta\,d\phi \quad (11.55)$$

However, from Eq. (11.43),

$$\text{cov}\,[I_N(\theta), I_N(\phi)] \cong P_{xx}(\theta)P_{xx}(\phi)\left[\left(\frac{\sin[(\theta+\phi)N/2]}{N\sin[(\theta+\phi)/2]}\right)^2 + \left(\frac{\sin[(\theta-\phi)N/2]}{N\sin[(\theta-\phi)/2]}\right)^2\right]$$

If we assume that the terms

$$\left(\frac{\sin[(\theta+\phi)N/2]}{N\sin[(\theta+\phi)/2]}\right)^2$$

and

$$\left(\frac{\sin[(\theta-\phi)N/2]}{N\sin[(\theta-\phi)/2]}\right)^2$$

are narrow compared with variations of $P_{xx}(\theta)$ and $W(e^{j\theta})$ and that they are highly concentrated about $\theta = -\phi$ and $\theta = \phi$, respectively (i.e., N is large), then first approximating the integration on θ we obtain†

$$\text{cov}\,[S_{xx}(\omega_1), S_{xx}(\omega_2)] \cong \frac{1}{2\pi N}\int_{-\pi}^{\pi} P_{xx}^2(\phi)W(e^{j(\omega_2-\phi)})[W(e^{j(\omega_1+\phi)}) + W(e^{j(\omega_1-\phi)})]\,d\phi \quad (11.56)$$

If we further assume that the spectrum window is sufficiently narrow that the term $W(e^{j(\omega_1+\phi)})W(e^{j(\omega_2-\phi)})$ is negligible, then Eq. (11.56) becomes

$$\text{cov}\,[S_{xx}(\omega_1), S_{xx}(\omega_2)] \cong \frac{1}{2\pi N}\int_{-\pi}^{\pi} P_{xx}^2(\phi)W(e^{j(\omega_1-\phi)})W(e^{j(\omega_2-\phi)})\,d\phi \quad (11.57)$$

† We use the fact that

$$\frac{1}{2\pi}\int_{-\pi}^{\pi}\left(\frac{\sin[\theta N/2]}{N\sin[\theta/2]}\right)^2 d\theta = \frac{1}{N}$$

(See Problem 3 of this chapter.)

From Eq. (11.57) it is clear that as the width of the spectrum window $W(e^{j\omega})$ increases so that there is greater overlap between $W(e^{j(\omega_1-\phi)})$ and $W(e^{j(\omega_2-\phi)})$, the covariance between estimates at different frequencies increases.

To obtain the variance of the spectrum estimate $S_{xx}(\omega)$, we evaluate Eq. (11.57) for $\omega = \omega_1 = \omega_2$, obtaining

$$\text{var}\,[S_{xx}(\omega)] \cong \frac{1}{2\pi N}\int_{-\pi}^{\pi} P_{xx}^2(\phi)W^2(e^{j(\omega-\phi)})\,d\phi \tag{11.58}$$

Now let us assume that $W(e^{j\omega})$ is narrow with respect to variations of $P_{xx}(\omega)$; i.e., that we have been able to choose the length of the window $w(m)$ great enough to maintain adequate spectrum resolution. Then Eq. (11.58) can be still further approximated by

$$\text{var}\,[S_{xx}(\omega)] \cong \frac{1}{N}P_{xx}^2(\omega)\frac{1}{2\pi}\int_{-\pi}^{\pi} W^2(e^{j\phi})\,d\phi \tag{11.59a}$$

From Parseval's theorem we note that

$$\frac{1}{2\pi}\int_{-\pi}^{\pi} W^2(e^{j\phi})\,d\phi = \sum_{m=-(M-1)}^{M-1} w^2(m)$$

if $w(m) = w(-m)$. Therefore, we obtain the convenient expression

$$\text{var}\,[S_{xx}(\omega)] \cong \left(\frac{1}{N}\sum_{m=-(M-1)}^{M-1} w^2(m)\right)P_{xx}^2(\omega) \tag{11.59b}$$

Equations (11.54) and (11.59) are approximate expressions for the mean and variance of the spectrum estimate $S_{xx}(\omega)$. These expressions are valid under the assumptions that the length $(2M - 1)$ of the window $w(m)$ that is applied to the estimate $c_{xx}(m)$ is such that $W(e^{j\omega})$ is narrow with respect to variations of the spectrum $P_{xx}(\omega)$ while at the same time $W(e^{j\omega})$ is wide compared to $(\sin[\omega N/2]/\sin[\omega/2])^2$. These expressions can be compared to the corresponding expressions for the periodogram to see the improvement due to the window. We see from Eq. (11.27) that the periodogram is asymptotically unbiased; i.e.,

$$\lim_{N\to\infty} E[I_N(\omega)] = P_{xx}(\omega)$$

From Eq. (11.54) we see that as the record length N becomes large, we can also make the window width large so that $W(e^{j\omega})$ will be narrow with respect to variations in $P_{xx}(\omega)$, therefore implying that

$$\lim_{M\to\infty} E[S_{xx}(\omega)] = P_{xx}(\omega)\frac{1}{2\pi}\int_{-\pi}^{\pi} W(e^{j\omega})\,d\omega$$

Thus for the smoothed spectrum estimate to be asymptotically unbiased we require that

$$w(0) = \frac{1}{2\pi}\int_{-\pi}^{\pi} W(e^{j\omega})\, d\omega = 1$$

The variance of the periodogram is seen from Eq. (11.44) to be approximately

$$\text{var}\,[I_N(\omega)] \cong \begin{cases} 2P_{xx}^2(0), & \omega = 0 \\ 2P_{xx}^2(\pi), & \omega = \pi \\ P_{xx}^2(\omega), & \text{otherwise} \end{cases}$$

Therefore, for $0 < \omega < \pi$ the variance of the smoothed periodogram $S_{xx}(\omega)$ differs from the variance of $I_N(\omega)$ by the factor

$$\frac{1}{N}\sum_{m=-(M-1)}^{M-1} w^2(m) = \frac{1}{2\pi N}\int_{-\pi}^{\pi} W^2(e^{j\omega})\, d\omega \tag{11.60}$$

Clearly, one objective should be to choose M and the window shape so that the variance of $S_{xx}(\omega)$ is smaller than the variance of $I_N(\omega)$. That is, the factor of Eq. (11.60) should be less than unity. In Problem 4 of this chapter, this factor is computed for various commonly used windows.

11.4.3 The Welch Method—Averaging of Modified Periodograms

Welch [9] has introduced a modification of the Bartlett procedure that is particularly well suited to direct computation of a power spectrum estimate using the FFT. The data record is again sectioned into $K = N/M$ segments of M samples each as defined in Eq. (11.45). In this case, however, the window $w(n)$ is applied directly to the data segments before computation of the periodogram. Thus we define the K modified periodograms

$$J_M^{(i)}(\omega) = \frac{1}{MU}\left|\sum_{n=0}^{M-1} x^{(i)}(n)w(n)e^{-j\omega n}\right|^2, \qquad i = 1, 2, \ldots, K \tag{11.61}$$

where

$$U = \frac{1}{M}\sum_{n=0}^{M-1} w^2(n) \tag{11.62}$$

and the spectrum estimate is defined as

$$B_{xx}^w(\omega) = \frac{1}{K}\sum_{i=1}^{K} J_M^{(i)}(\omega) \tag{11.63}$$

It can be shown (see Problem 5 of this chapter) that the expected value of $B_{xx}^w(\omega)$ is

$$E[B_{xx}^w(\omega)] = \frac{1}{2\pi}\int_{-\pi}^{\pi} P_{xx}(\theta)W(e^{j(\omega-\theta)})\, d\theta \tag{11.64}$$

where

$$W(e^{j\omega}) = \frac{1}{MU}\left|\sum_{n=0}^{M-1} w(n)e^{-j\omega n}\right|^2 \tag{11.65}$$

The normalizing factor U is required for the estimate $B_{xx}^{w}(\omega)$ to be asymptotically unbiased. Welch [9] shows that if the segments of $x(n)$ are nonoverlapping, then

$$\text{var}\,[B_{xx}^{w}(\omega)] \approx \frac{1}{K}\,P_{xx}^{2}(\omega)$$

as we have already seen for the Bartlett procedure. Welch [9] also considers the case where the segments $x^{(i)}(n)$ overlap each other, so that the modified periodograms are not independent. Thus by windowing the data segments before computing the periodogram, we achieve the variance reduction of the original Bartlett procedure and at the same time achieve smoothing of the spectrum with the concomitant reduction of resolution. In this case the spectrum window is proportional to the squared magnitude of the Fourier transform of the window rather than simply the Fourier transform itself. This means that no matter what data window is used, the spectrum window will always be nonnegative and we can see that the spectrum estimate $B_{xx}^{w}(\omega)$ will always be nonnegative as well.

11.5 Estimates of the Cross Covariance and Cross Spectrum

The methods of the previous sections apply with slight modifications to the estimation of the cross-covariance and cross-spectrum of two different random signals. For example assume that $x(n)$ and $y(n)$ refer to zero-mean random signals so that $\gamma_{xy}(m) = \phi_{xy}(m)$. Then corresponding to the autocovariance estimate $c_{xx}(m)$ of Eq. (11.19) we have the cross-covariance (or cross-correlation) estimate

$$c_{xy}(m) = \frac{1}{N}\sum_{n=0}^{N-m-1} x(n)y(n+m), \qquad 0 \le m < N \qquad (11.66a)$$

$$c_{xy}(-m) = \frac{1}{N}\sum_{n=0}^{N-m-1} x(n+m)y(n), \qquad 0 \le m < N \qquad (11.66b)$$

Note that when $y(n) = x(n)$, Eq. (11.66) reduces to Eq. (11.19).

The expected value of Eq. (11.66) is

$$E[c_{xy}(m)] = \left(1 - \frac{m}{N}\right)\phi_{xy}(m), \qquad 0 \le m < N$$

where $\phi_{xy}(m)$ is the true cross-correlation sequence. Similarly, from Eq. (11.66b),

$$\begin{aligned} E[c_{xy}(-m)] &= \left(1 - \frac{m}{N}\right)\phi_{yx}(m) \\ &= \left(1 - \frac{m}{N}\right)\phi_{xy}(-m) \qquad 0 \le m < N \end{aligned}$$

Combining the above equations we obtain

$$E[c_{xy}(m)] = \left(1 - \frac{|m|}{N}\right)\phi_{xy}(m), \qquad -N < m < N \tag{11.67}$$

The estimate $c_{xy}(m)$ is seen to be an asymptotically unbiased estimate of the cross-covariance $\phi_{xy}(m)$. As in the estimate $c_{xx}(m)$, the variance of the estimate is inversely proportional to N.

To estimate the cross-power spectrum we can take the Fourier transform of $c_{xy}(m)$ and obtain the spectrum estimate

$$C_{xy}(\omega) = \sum_{m=-(N-1)}^{N-1} c_{xy}(m)e^{-j\omega m} \tag{11.68}$$

If $x(n) = y(n)$, it is clear from Eq. (11.24) that Eq. (11.68) reduces to the periodogram. Note that in general $c_{xy}(m)$ has no special symmetry properties, so that $C_{xy}(\omega)$ is generally a complex function. It is easily shown that

$$E[C_{xy}(\omega)] = \sum_{m=-(N-1)}^{N-1} \left(1 - \frac{|m|}{N}\right)\phi_{xy}(m)e^{-j\omega m} \tag{11.69}$$

Although $C_{xy}(\omega)$ is seen from Eq. (11.69) to be an asymptotically unbiased estimate of the cross power spectrum, just as in the case of the periodogram, the variance of $C_{xy}(\omega)$ does not go to zero as N increases. Thus windowing or averaging of estimates based on short segments of the records must be used to reduce the variance and smooth the estimate. For example, we consider the smoothed spectrum estimate

$$S_{xy}(\omega) = \sum_{m=-(M-1)}^{M-1} c_{xy}(m)w(m)e^{-j\omega m} \tag{11.70}$$

Under assumptions similar to those employed in Sec. 11.4, it can be shown that

$$E[S_{xy}(\omega)] \cong \frac{1}{2\pi}\int_{-\pi}^{\pi} P_{xy}(\theta)W(e^{j(\omega-\theta)})\,d\theta \tag{11.71}$$

Similarly, it can be shown that var $[S_{xy}(\omega)]$ decreases with increasing record length and it also decreases with decreasing window width. Thus the tradeoff between spectrum resolution and variance reduction is much the same as we have already discussed for the power spectrum of a single random signal.

We shall not go into further detail regarding cross-covariance and cross-power-spectrum estimates. Several chapters are devoted to such estimates and their applications in reference [5].

11.6 Application of the FFT in Spectrum Estimation

The FFT provides an efficient means of computing estimates of the power spectrum at equally spaced frequencies $\omega_k = (2\pi/M)k$. Also, by using techniques developed in Chapter 3 for computing convolutions, we can use the

FFT to compute covariance estimates. We shall examine the details of such computational procedures in this section.

11.6.1 Implementation of the Bartlett or Welch Methods

Suppose that we wish to compute a spectrum estimate at equally spaced frequencies by averaging periodograms as discussed in Sec. 11.4.3. That is, we wish to compute

$$B_{xx}^{w}\left(\frac{2\pi}{M}k\right) = \frac{1}{K}\sum_{i=1}^{K} J_M^{(i)}\left(\frac{2\pi}{M}k\right), \qquad k = 0, 1, \ldots, M-1$$

where

$$J_M^{(i)}\left(\frac{2\pi}{M}k\right) = \frac{1}{MU}\left|\sum_{n=0}^{M-1} x^{(i)}(n)w(n)e^{-j(2\pi/M)kn}\right|^2$$

for $i = 1, 2, \ldots, K$ and $k = 0, 1, \ldots, M-1$. This can be accomplished as follows: Compute

$$X_M^{(i)}(k) = \sum_{n=0}^{M-1} x^{(i)}(n)w(n)e^{-j(2\pi/M)kn}, \qquad k = 0, 1, \ldots, M-1$$

for each section using an appropriate FFT algorithm.† Next compute $|X_M^{(i)}(k)|^2$ for each section. These results are then added together and when all K of the estimates have been accumulated, we can divide the result by the quantity KMU. For real data $x(n)$, we can compute two transforms at once using symmetry properties as developed in Problem 10 of Chapter 6, thereby significantly reducing the computation.

This is a very simple procedure, resulting in a direct estimate of the power spectrum which will always be nonnegative and which we can interpret as in Secs. 11.4.1 and 11.4.3. If, however, we wish to estimate the correlation function as well as the power spectrum, it is preferable to first compute the correlation estimate $c_{xx}(m)$ and then compute the power spectrum estimate, since the rather simple expedient of computing the inverse DFT of $B_{xx}^{w}((2\pi/M)k)$ produces something that can at best be termed a time-aliased estimate of the covariance sequence. (See Problem 6 of this chapter.) Thus we shall next consider how the FFT can be used to compute covariance estimates.

11.6.2 Computation of Correlation Estimates

The FFT can be used to efficiently compute the autocorrelation estimate

$$c_{xx}(m) = \frac{1}{N}\sum_{n=0}^{N-|m|-1} x(n)x(n+m), \qquad 0 \le m \le M-1 \qquad (11.72)$$

where $M \le N$. [If the sample mean is first subtracted from $x(n)$, $c_{xx}(m)$ is an estimate of the autocovariance.] Recalling that $c_{xx}(-m) = c_{xx}(m)$, it is

† Note that $w(n) = 1$ for $0 \le n \le M-1$ corresponds to the Bartlett method of Sec. 11.4.1, whereas any other choice of $w(n)$ corresponds to the Welch method of Sec. 11.4.3.

clear that Eq. (11.72) need only be evaluated for positive values of m. The key to understanding how fast Fourier transforms can be used to compute $c_{xx}(m)$ is to observe that $c_{xx}(m)$ is a discrete convolution of $x(n)$ with $x(-n)$. Suppose that we compute $X(k)$, the DFT of $x(n)$, and multiply by $X^*(k)$. The inverse DFT of $X(k)X^*(k) = |X(k)|^2$ corresponds to the circular convolution of $x(n)$ with $x(-n)$, i.e., a circular correlation. By augmenting the sequence $x(n)$ with $(L - N)$ zero samples and computing an L-point DFT, we can force the values of the circular correlation to be correct in the interval $0 \le m \le M - 1$.

To see how to choose L, consider Fig. 11.4. Figure 11.4(a) shows the two sequences $x(n)$ and $x(n + m)$ for a particular positive value of m. Figure

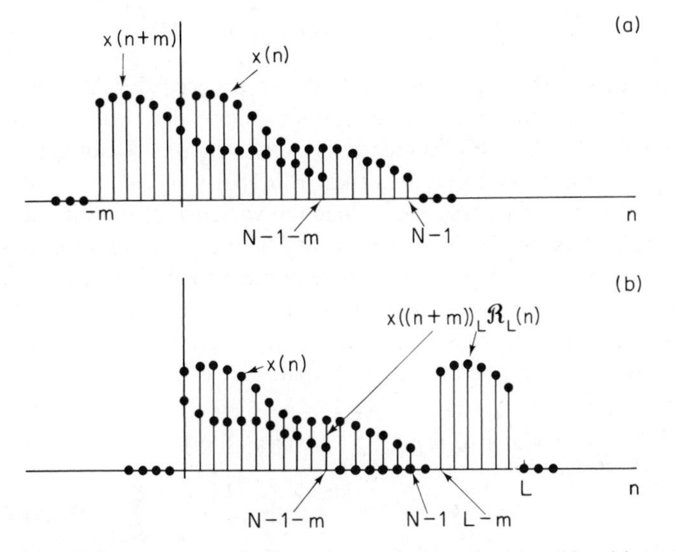

Fig. 11.4 Computation of the autocovariance estimate: (a) $x(n)$ and $x(n + m)$ for a sequence of length N; (b) the periodic sequences $x(n)$ and $x(n + m)$ involved in circular correlation.

11.4(b) shows the sequences $x(n)$ and $x((n + m))_L$ that are involved in the circular correlation corresponding to $|X(k)|^2$. Clearly, the circular correlation will be equal to $Nc_{xx}(m)$ for $0 \le m \le M - 1$ if $x((n + m))_L$ does not wrap around and overlap $x(n)$ for $0 \le m \le M - 1$. From Fig. 11.4(b) we see that this will be the case for $L \ge N + M - 1$.

Thus we can compute $c_{xx}(m)$ for $0 \le m \le M - 1$ by the following process:

1. Form an L-point sequence by augmenting $x(n)$ with $(M - 1)$ zeros
2. Compute the L-point DFT

$$X(k) = \sum_{n=0}^{L-1} x(n)e^{-j(2\pi/L)kn}, \qquad k = 0, 1, \ldots, L - 1$$

3. Compute the L-point inverse DFT

$$v(m) = \frac{1}{L}\sum_{n=0}^{L-1} |X(k)|^2\, e^{j(2\pi/N)km}, \qquad m = 0, 1, \ldots, L-1$$

4. Then

$$c_{xx}(m) = \frac{1}{N}\, v(m), \qquad m = 0, 1, \ldots, M-1$$

If M is small, it may be most efficient to simply evaluate Eq. (11.72). directly. This computation is proportional to $N \cdot M$. In contrast, the above procedure requires an amount of computation proportional to

$$L \log L = (N + M) \log (N + M)$$

so that for sufficiently large values of M the above procedure is more efficient. The precise breakeven value of M is of course dependent upon the particular implementation of the DFT calculations; however, it is reasonable to expect that this value would be much less than 100 [10].

We have seen that in order to reduce the variance of the estimate $c_{xx}(m)$, we must make N large. In such cases it may be inconvenient or impossible to efficiently compute the L-point DFT's required above. However, since M is usually much less than N, we can section the input in a manner similar to the procedures that were discussed in Sec. 3.9 for convolution.

To see how this can be done, let us write Eq. (11.72) as

$$c_{xx}(m) = \frac{1}{N}\left[\sum_{n=0}^{M-1} x(n)x(n+m) + \sum_{n=M}^{2M-1} x(n)x(n+m) + \cdots + \sum_{n=(K-1)M}^{KM-1} x(n)x(n+m)\right]$$

where $N = KM$. In writing Eq. 11.72 in this form, we utilize the fact that with $x(n)$ considered to be zero outside the interval $0 \le n \le N-1$ the upper limit in Eq. 11.72 can be replaced by $N - 1$. By appropriate substitutions we can write

$$c_{xx}(m) = \frac{1}{N}\sum_{i=1}^{K}\sum_{n=0}^{M-1} x(n + (i-1)M)x(n + (i-1)M + m)$$

If we define

$$v_i(n) = \sum_{n=0}^{M-1} x(n + (i-1)M)x(n + (i-1)M + m) \tag{11.73}$$

then

$$c_{xx}(m) = \frac{1}{N}\sum_{i=1}^{K} v_i(m), \qquad 0 \le m \le M-1 \tag{11.74}$$

To evaluate Eq. (11.73) it is convenient to define the L-point sequences

$$x_i(n) = \begin{cases} x(n + (i-1)M), & 0 \le n \le M-1 \\ 0, & M \le n \le L-1 \end{cases} \tag{11.75a}$$

and

$$y_i(n) = x(n + (i-1)M), \qquad 0 \le n \le L-1 \tag{11.75b}$$

Then the circular correlation

$$v_i(m) = \sum_{n=0}^{M-1} x_i(n) y_i((n+m))_L$$

is equal to $v(m)$ for $0 \le m \le M-1$ if $L \ge 2M-1$. Typical sequences are depicted in Fig. 11.5. If $X_i(k)$ and $Y_i(k)$ are the L-point DFT's of $x_i(n)$ and

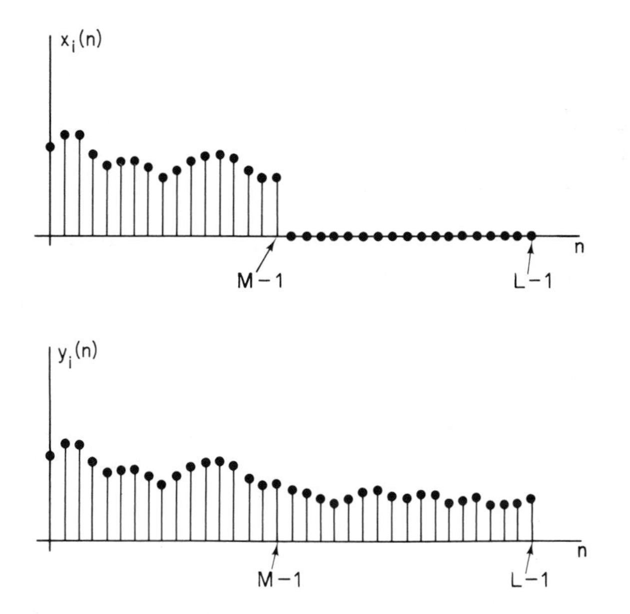

Fig. 11.5 The two L-point sequences required to compute the contribution of the ith segment to the autocovariance estimate.

$y_i(n)$, then if $L \ge 2M-1$,

$$v_i(m) = \frac{1}{L} \sum_{k=0}^{L-1} V_i(k) W_L^{km} \qquad 0 \le m \le M-1$$

where

$$V_i(k) = X_i(k) Y_i^*(k) \tag{11.76}$$

If instead of Eq. (11.74) we compute

$$V(k) = \sum_{i=1}^{K} V_i(k), \qquad k = 0, 1, \ldots, L-1 \tag{11.77}$$

then

$$c_{xx}(m) = \frac{1}{N}\, v(m) = \frac{1}{L}\sum_{k=0}^{L-1} V(k)e^{j(2\pi/L)km}, \qquad 0 \le m \le M-1 \quad (11.78)$$

Thus we can compute M values of $c_{xx}(m)$ by computing $2K$ L-point DFT's and one L-point inverse DFT.

Note that the above procedure applies equally well to computing the cross-correlation estimate $c_{xy}(m)$. Suppose that we have records of length N of two sequences $x(n)$ and $y(n)$. Then we would simply define, as before,

$$x_i(n) = \begin{cases} x(n + (i-1)M), & 0 \le n \le M-1 \\ 0, & M \le n \le L-1 \end{cases} \quad (11.79a)$$

and now

$$y_i(n) = y(n + (i-1)M), \qquad 0 \le n \le L-1 \quad (11.79b)$$

for $i = 1, 2, \ldots, K$ instead of Eq. (11.75). Then if we use Eqs. (11.76) and (11.77) as before we obtain

$$c_{xy}(m) = v(m), \qquad 0 \le m \le M-1 \quad (11.80)$$

For $m < 0$, $c_{xy}(m)$ is computed in the same manner by interchanging the x and y sequences since $c_{yx}(m) = c_{xy}(-m)$.

Rader [10] has shown that for the particular choice of $L = 2M$ and for computing autocorrelation estimates, a significant reduction in computation is possible. Figure 11.6 shows two sets of sequences $x_i(n)$, $y_i(n)$ and $x_{i+1}(n)$, $y_{i+1}(n)$ for $L = 2M$ as required for an autocorrelation computation. It is clear from the figure that

$$y_i(n) = x_i(n) + x_{i+1}(n - M) \quad (11.81)$$

From Eq. (11.81) it follows that

$$Y_i(k) = X_i(k) + (-1)^k X_{i+1}(k), \qquad k = 0, 1, \ldots, 2M-1 \quad (11.82)$$

Therefore, $Y_i(k)$ can be computed using Eq. (11.82) rather than using a separate FFT computation. Furthermore, two transforms, e.g., $X_i(k)$ and $X_{i+1}(k)$, can be computed with one FFT computation using the techniques developed in Problem 10 in Chapter 6. Thus the procedure for computing $c_{xx}(m)$ can be summarized as follows:

1. Form the sequence

$$x_1(n) = \begin{cases} x(n), & 0 \le n \le M-1 \\ 0, & M \le n \le 2M-1 \end{cases}$$

   and compute the $2M$-point transform $X_1(k)$. Define $A_0(k) = 0$.

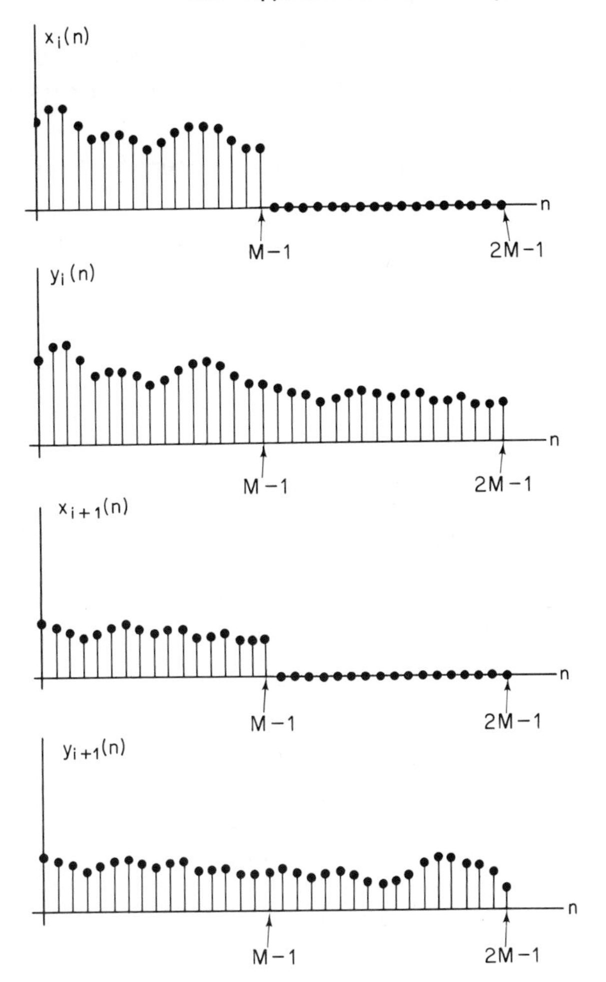

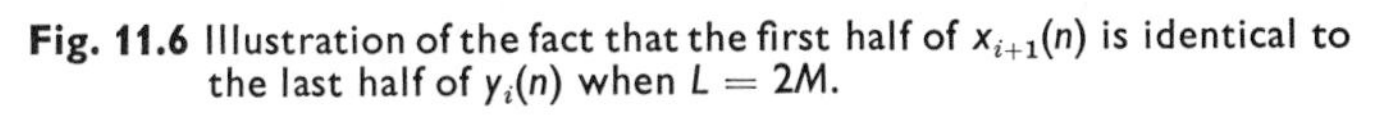

Fig. 11.6 Illustration of the fact that the first half of $x_{i+1}(n)$ is identical to the last half of $y_i(n)$ when $L = 2M$.

2. For $i = 1, 2, \ldots, K$, form

$$x_{i+1}(n) = \begin{cases} x(n + iM), & 0 \le n \le M - 1 \\ 0, & M \le n \le 2M - 1 \end{cases}$$

and compute the $2M$-point transform $X_{i+1}(k)$. [Define $X_{K+1}(k) = 0$.]
Then for $0 \le k \le 2M - 1$ and $i = 1, 2, \ldots, K$, compute

$$A_i(k) = A_{i-1}(k) + X_i(k)[X_i^*(k) + (-1)^k X_{i+1}^*(k)]$$

3. Finally, define $V(k) = A_K(k)$ and let $v(m)$ be the $2M$-point inverse DFT of $V(k)$. Then

$$c_{xx}(m) = \frac{1}{N} v(m), \qquad 0 \le m \le M$$

Thus, for the particular case of $L = 2M$ and for the computation of $c_{xx}(m)$ only, we must compute K transforms $X_i(k)$, and one inverse transform.

11.6.3 Computation of Smoothed Spectrum Estimates from $c_{xx}(m)$

Once $c_{xx}(m)$ has been computed using the previous technique, we can compute samples of the smoothed spectrum estimate $S_{xx}(\omega)$ by forming the sequence

$$s_{xx}(m) = \begin{cases} c_{xx}(m)w(m), & 0 \leq m \leq M - 1 \\ 0, & M \leq m \leq L - M + 1 \\ c_{xx}(L - m)w(L - m), & L - M - 1 \leq m \leq L - 1 \end{cases} \tag{11.83}$$

where $w(m)$ is an appropriate data window. Then the DFT of $s_{xx}(m)$ is

$$S_{xx}(k) = S_{xx}(\omega)\big|_{\omega=(2\pi/L)k}, \qquad k = 0, 1, \ldots, L - 1$$

Note that although L can be chosen as large as is convenient and practical, thereby giving values of $S_{xx}(\omega)$ at closely spaced frequencies, the frequency resolution is still determined by the shape and length of the window $w(m)$.

11.7 Example of Spectrum Estimation

In Chapter 9 we assumed that the error introduced by quantization is a white-noise process. Furthermore, we assumed the quantization noise to be uncorrelated with the original signal. We can test these assumptions by estimating covariance sequences and power spectra using the methods described in this chapter.

As an example consider the experiment depicted in Fig. 11.7. A lowpass filtered speech signal $x_a(t)$ was sampled at a 10-KHz rate, producing the

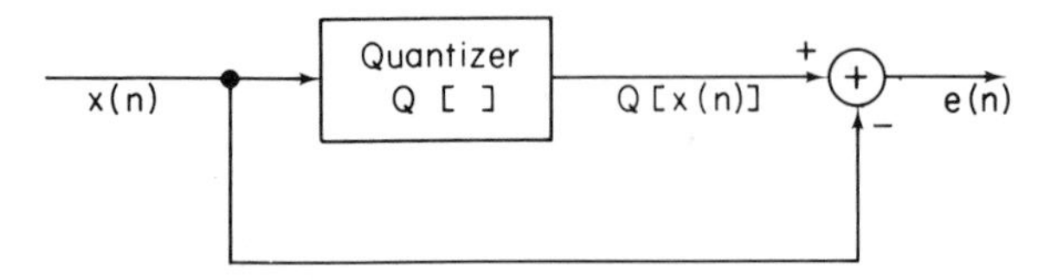

Fig. 11.7 Experiment to determine the properties of quantization noise.

sequence of samples $x(n)$. (We shall assume that the resulting samples have infinite precision.) The amplitude range of the samples was

$$-16{,}000 \leq x(n) \leq 16{,}000$$

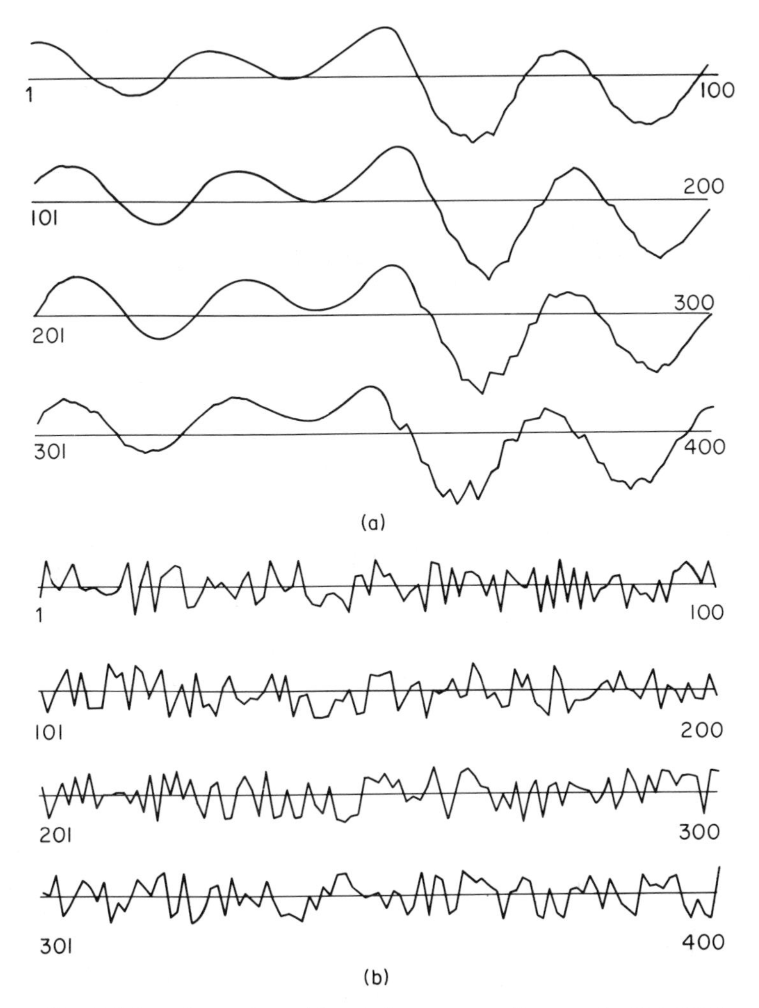

Fig. 11.8 Speech waveform (a) and the corresponding quantization error (b) for eight-bit quantization [magnified 66 times with respect to (a)]. Each line corresponds to 400 consecutive samples connected by straight lines for convenience in plotting.

These samples were first quantized with an eight-bit linear quantizer and the resulting error sequence

$$e(n) = Q[x(n)] - x(n)$$

was computed. Figure 11.8(a) shows 400 consecutive samples of the speech signal and Fig. 11.8(b) shows the corresponding error sequence. (The samples are connected with straight lines for convenience in plotting.) Visual inspection and comparison of these two plots tends to strengthen our belief in the previously stated assumptions, although scrutiny might suggest some degree of correlation.

Figure 11.9 shows an estimate of the normalized autocovariance and power spectrum of the error sequence for a record length of $N = 2000$ samples.

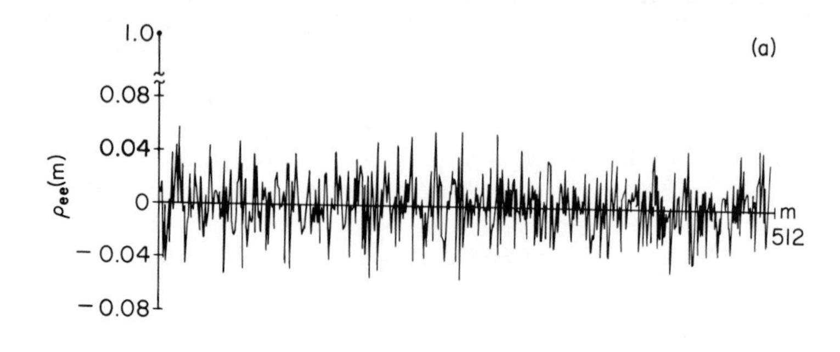

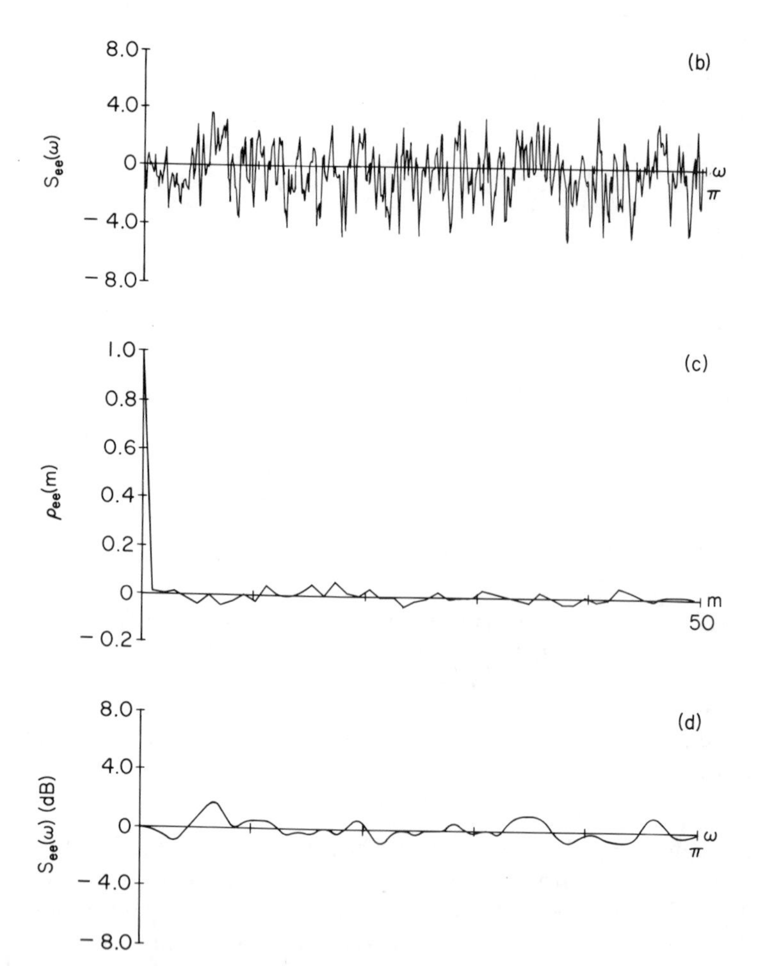

Fig. 11.9 (a) Normalized autocovariance estimate for eight-bit quantization noise; record length, 2000. (b) Power spectrum estimate using Bartlett window $M = 512$. (c) Normalized autocovariance estimate, $0 \leq m \leq 50$. (d) Power spectrum estimate using Bartlett window $M = 50$.

The mean and variance of $e(n)$ were estimated as discussed in Sec. 11.1. These estimates were $\hat{m}_e = -1$ and $\hat{\sigma}_e^2 = 1300$. Then $\hat{m}_e$ was subtracted from $e(n)$ and the covariance estimate calculated for $M = 512$ using the method of Sec. 11.6.2. Finally, the resulting covariance estimate was divided by $\hat{\sigma}_e^2$, the estimate of the variance of $e(n)$. The resulting normalized covariance estimate, denoted $\rho_{ee}(m)$, is shown in Figs. 11.9(a) and 11.9(c). Note that the autocovariance is 1.0 at $m = 0$ and much smaller elsewhere. Indeed, $-0.0548 \leq \rho_{ee}(m) \leq 0.0579$ for $1 \leq m \leq 512$. This seems to support our assumption that the error sequence is uncorrelated from sample to sample.

The power spectrum was estimated by windowing the normalized autocovariance with a Bartlett window ($M = 512$) as discussed in Sec. 11.6.3. The result, shown in Fig. 11.9(b) plotted in dB, shows rather erratic fluctuations about 0 dB (the value of the normalized power spectrum of white noise). A smoother estimate is shown in Fig. 11.9(d). In this case a Bartlett window with $M = 50$ was used. The resulting smoothing, corresponding to a loss of resolution, is very evident in comparing Fig. 11.9(b) with Fig. 11.9(d). We see from Fig. 11.9(d) that the spectrum estimate is between -1.097 dB and $+1.631$ dB for all frequencies. Thus we are again encouraged to believe that the white-noise model is appropriate for this case of quantization.

Although we have computed quantitative estimates of the autocovariance and the power spectrum, our interpretation of these measurements has been only qualitative. The question that arises is: How small would the autocovariance be if $e(n)$ were really a white noise process? In order to give quantitative answers to such questions we can compute confidence intervals for our estimates and apply statistical decision theory, something we are not prepared to do in this introductory summary of power spectrum estimation. (See Jenkins and Watts [5] for some tests for white noise.) In many cases, however, this additional statistical treatment is not necessary. We are often comfortable and content with the observation that the normalized auto covariance for $1 \leq m \leq 512$ is much less than the value at $m = 0$.

One of the most important insights of this chapter is that the estimate of the autocovariance and power spectrum of a stationary random process should improve if the record length is increased. This is illustrated by Fig. 11.10 which corresponds to Fig. 11.9 except that N is increased to 14,000 samples. Recall from Eq. (11.23) that the variance of the autocovariance estimate is proportional to $1/N$. Thus, increasing N from 2000 to 14,000 should bring about a sevenfold reduction in the variance of the estimate. A comparison of Figs. 11.9(a) and 11.10(a) seems to verify this result. For $N = 2000$, the estimate falls between the limits -0.0548 and $+0.0579$, while for $N = 14{,}000$ the limits are -0.0254 and $+0.0239$. Intuitively, this is consistent with the seven-fold variance reduction that we expected. We note from Eq. (11.58) that a similar reduction in variance of the spectrum estimate is

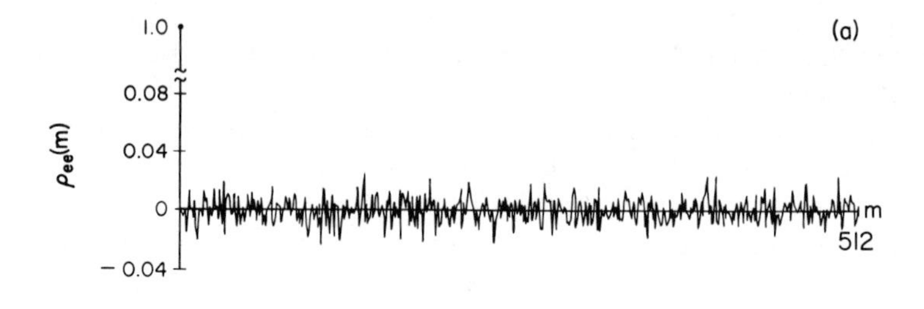

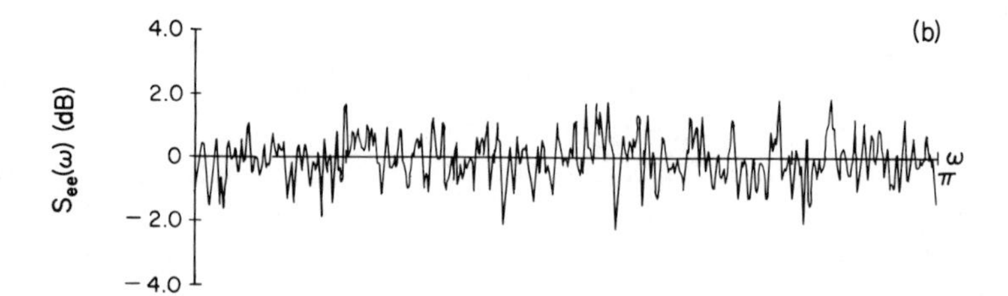

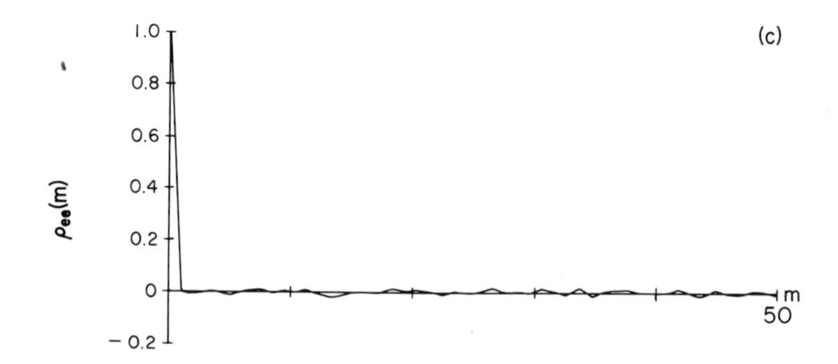

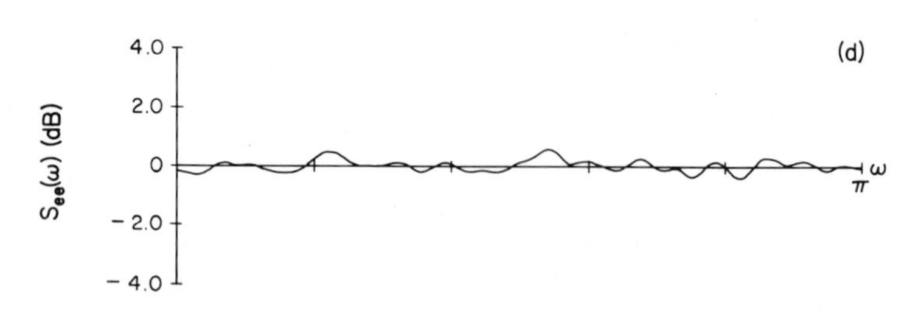

Fig. 11.10 (a) Normalized autocovariance estimate for eight-bit quantization noise; record length, 14,000. (b) Power spectrum estimate using Bartlett window $M = 512$. (c) Normalized autocovariance estimate $0 \leq m \leq 50$. (d) Power spectrum estimate using Bartlett window $M = 50$.

also expected. This is again evident in comparing Figs. 11.9(b) and (d) to Figs. 11.10(b) and (d), respectively.

To test our second assumption about the quantization process, we computed the normalized cross-covariance between $x(n)$ and $e(n)$. In this case the means $\hat{m}_x = 1$ and $\hat{m}_e = -1$ were subtracted before computing the cross-covariance estimates discussed in Secs. 11.5 and 11.6.2. The result was

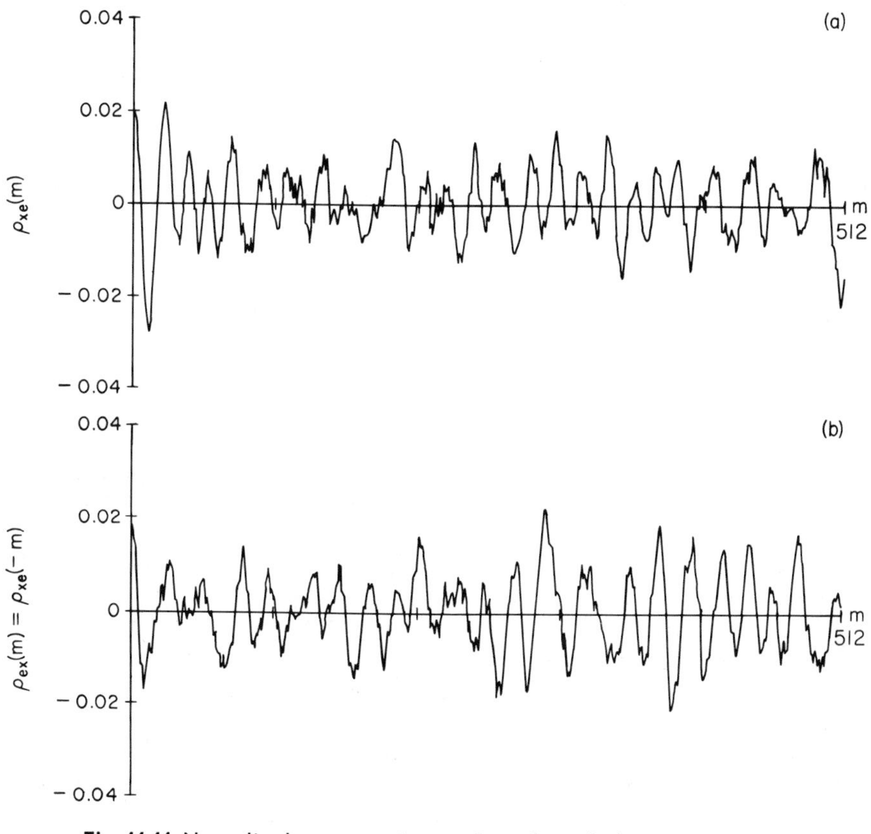

Fig. 11.11 Normalized cross-covariance estimate for eight-bit quantization; record length, 14,000: (a) $\rho_{xy}(m)$ for $0 \leq m \leq 511$; (b) $\rho_{yx}(m) = \rho_{xy}(-m)$ for $0 \leq m \leq 511$.

divided by $\hat{\sigma}_x \cdot \hat{\sigma}_e$, producing a normalized cross-covariance estimate that would be 1.0 for perfect correlation and 0 if $x(n)$ and $e(n)$ are uncorrelated. The normalized estimates $\rho_{xe}(m)$ and $\rho_{ex}(m) = \rho_{xe}(-m)$ are shown in Fig. 11.11(a) and (b), respectively, for record lengths of $N = 14{,}000$. We note that the value of the normalized cross-covariance sequence is between -0.0279 and $+0.0222$ for $-511 \leq m \leq 511$. When we recall that the normalized cross-covariance is equal to unity for perfect correlation, we are again encouraged to accept the assumption that the quantization noise is uncorrelated with the quantizer input.

In Chapter 9 we argued that the white-noise model was reasonable as long as the quantization step size was small. When the number of bits is small, this condition does not hold. To see the effect upon the quantization noise spectrum, the previous experiment was repeated using only eight quantization levels, or three-bits. Figure 11.12 shows the quantization error

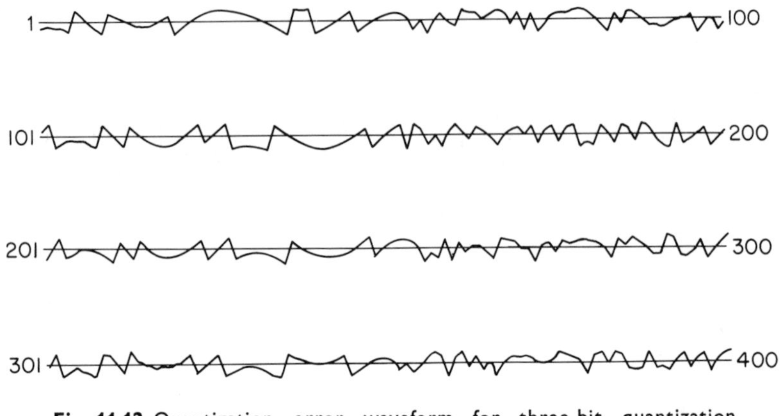

Fig. 11.12 Quantization error waveform for three-bit quantization. [Same scale as original signal, which is shown in Fig. 11.8(a).]

for three-bit quantization of the speech waveform segment shown in Fig. 11.8(a). Note that portions of the error waveform tend to look very much like the original speech waveform. We would expect this to be reflected in the estimate of the power spectrum.

Figure 11.13 shows the autocovariance and power spectrum estimates of the error sequence for three-bit quantization for a record length of 14,000 samples. In this case the autocovariance shown in Fig. 11.1(a) and (c) is much less like the ideal autocovariance for white noise. Table 11.1 gives the first 10 values of $\rho_{xx}(m)$ for this case.

Figure 11.13(b) and (d) shows the power spectrum estimates for Bartlett windows with $M = 512$ and $M = 50$, respectively. In this case, the spectrum is clearly not flat. (In fact, it tends to have the general shape of the speech spectrum.) Thus the white-noise model for quantization noise can only be viewed as a rather crude approximation in this case.

Figure 11.14 shows the normalized cross-covariance between the signal $x(n)$ and the three-bit quantization noise $e(n)$. In this case the correlation is somewhat greater; however, we might still be willing to accept the assumption that the signal and quantization error are uncorrelated.

This example illustrates how covariance and power spectrum estimates

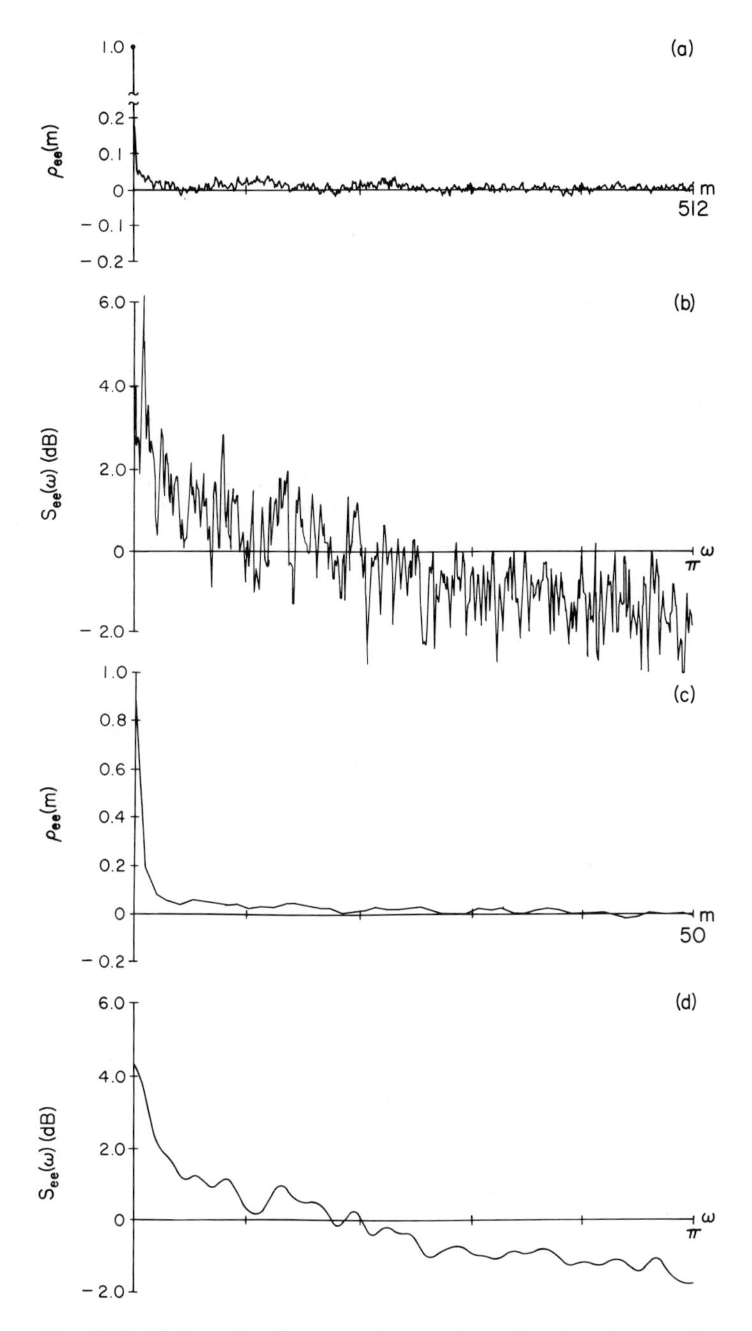

Fig. 11.13 (a) Normalized autocovariance estimate for three-bit quantization; record length, 14,000. (b) Power spectrum estimate using Bartlett window $m = 511$. (c) Normalized autocovariance estimate $0 \leq m \leq 50$. (d) Power spectrum estimate using Bartlett window $m = 50$.

Table 11.1

m	$\rho_{xx}(m)$
0	1.0
1	0.192
2	0.084
3	0.055
4	0.040
5	0.056
6	0.050
7	0.045
8	0.037
9	0.038

are often used to bolster our confidence in theoretical models. Specifically we have demonstrated the validity of some of our basic assumptions in Chapter 9, and we have given an indication of how these assumptions break down for very crude quantization. This is only a rather simple, but useful, example of how the techniques of this chapter are often applied in practice.

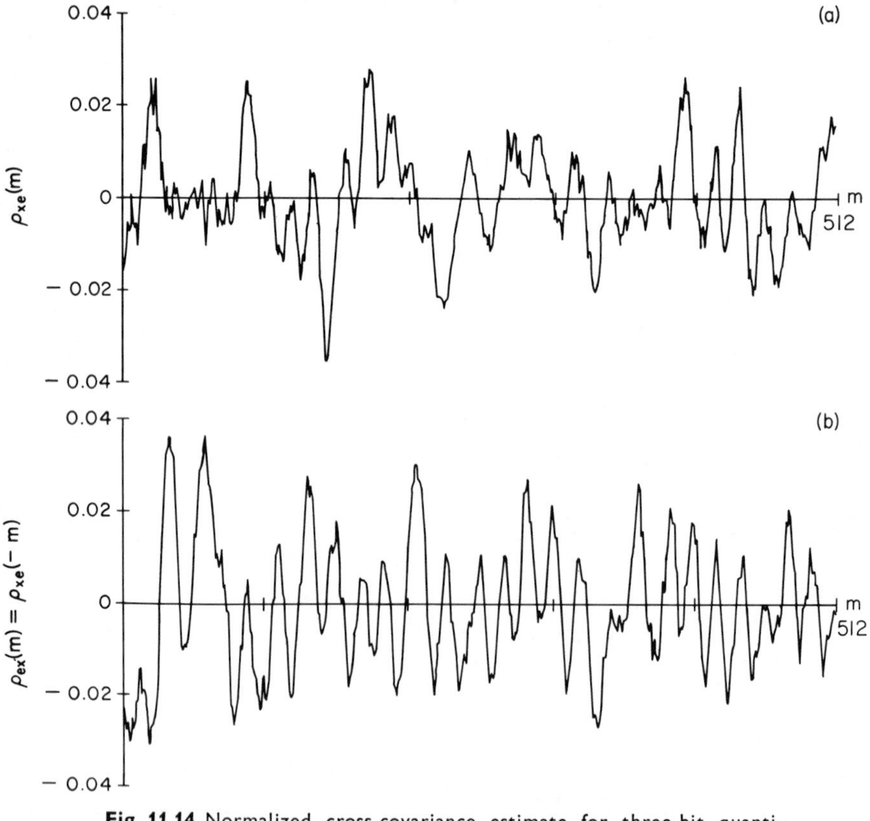

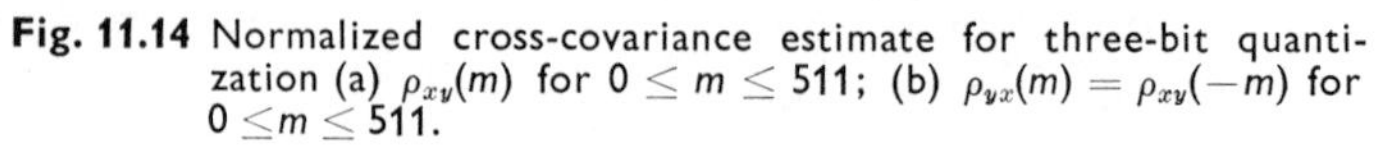
Fig. 11.14 Normalized cross-covariance estimate for three-bit quantization (a) $\rho_{xy}(m)$ for $0 \leq m \leq 511$; (b) $\rho_{yx}(m) = \rho_{xy}(-m)$ for $0 \leq m \leq 511$.

Summary

In this chapter we have discussed techniques for estimating averages of discrete time random processes. Our intent has been to make plausible some results from statistical estimation theory and to illustrate the implications of these results. We have been particularly concerned with approximate results for the mean and variance of estimates of the autocovariance and power spectrum of a stationary random sequence. A primary concern was procedures for computing these estimates and the implications of the FFT for efficient computation. A final section illustrates the application of some of the techniques described in the chapter to a study of quantization noise properties.

REFERENCES

1. M. S. Bartlett, *An Introduction to Stochastic Processes with Special Reference to Methods and Applications*, Cambridge University Press, New York, 1953.
2. R. B. Blackman and J. W. Tukey, *The Measurement of Power Spectra*, Dover Publications, Inc., New York, 1958.
3. U. Grenander and M. Rosenblatt, *Statistical Analysis of Stationary Time Series*, John Wiley & Sons, Inc., New York, 1957.
4. E. J. Hannan, *Time Series Analysis*, Methuen & Company Ltd., London, 1960.
5. G. M. Jenkins and D. G. Watts, *Spectral Analysis and Its Applications*, Holden-Day, Inc., San Francisco, 1968.
6. L. H. Koopmanns, *Spectral Analysis of Time Series*, Academic Press, New York, 1974.
7. W. B. Davenport, *Probability and Random Processes*, McGraw-Hill Book Company, New York, 1970.
8. D. R. Brillinger and M. Rosenblatt, "Asymtotic Theory of Estimates of *k*th Order Spectra," in *Spectral Analysis of Time Series*, B. Harris, ed., John Wiley & Sons, Inc., New York, 1967.
9. P. D. Welch, "The Use of Fast Fourier Transform for the Estimation of Power Spectra," *IEEE Trans. Audio Electroacoust.*, Vol. AU-15, June 1970, pp. 70–73.
10. T. G. Stockham, Jr., "High-Speed Convolution and Correlation," *Spring Joint Computer Conf., AFIPS Proc.*, Vol. 28, Spartan Books, Washington, D.C., 1966, pp. 229–233.
11. C. M. Rader, "An Improved Algorithm for High-Speed Autocorrelation with Applications to Spectral Estimation," *IEEE Trans. Audio Electroacoust.*, Vol. AU-18, Dec. 1970, pp. 439–441.

PROBLEMS

1. Let $X(e^{j\omega})$ be the Fourier transform of a real finite-length sequence $x(n)$ that is zero outside the interval $0 \leq n \leq N-1$. The periodogram $I_N(\omega)$ is defined in Eq. (11.24) as the Fourier transform of the $2N-1$ point autocorrelation estimate

$$c_{xx}(m) = \frac{1}{N} \sum_{n=0}^{N-|m|-1} x(n)x(n+m) \qquad |m| \leq N-1.$$

Show that the periodogram is related to the Fourier transform of the finite length sequence as follows:

$$I_N(\omega) = \frac{1}{N} |X(e^{j\omega})|^2.$$

2. The smoothed spectrum estimate $S_{xx}(\omega)$ is defined as

$$S_{xx}(\omega) = \sum_{m=-(M-1)}^{M-1} c_{xx}(m)w(m)e^{-j\omega m},$$

where $w(m)$ is a window sequence of length $2M - 1$. Show that

$$E[S_{xx}(\omega)] = \frac{1}{2\pi}\int_{-\pi}^{\pi} E[I_N(\theta)]W(e^{j(\omega-\theta)})\,d\theta,$$

where $W(e^{j\omega})$ is the Fourier transform of $w(n)$.

3. In deriving Eq. (11.56) we used the fact that

$$\frac{1}{2\pi}\int_{-\pi}^{\pi}\left(\frac{\sin[\theta N/2]}{N\sin[\theta/2]}\right)^2 d\theta = \frac{1}{N}.$$

(a) First find the Fourier transform of the sequence

$$s(n) = \frac{1}{N} \qquad 0 \le n \le N-1$$
$$= 0 \qquad \text{otherwise.}$$

(b) Use Parseval's theorem to prove the desired result.

4. In Sec. 11.4.2 we studied the method of windowing as it applies to smoothing spectrum estimates. It was shown that the ratio of the variance of a smoothed spectrum estimate to the variance of the periodogram is

$$R = \frac{\text{var}\,[S_{xx}(\omega)]}{\text{var}\,[I_N(\omega)]} = \frac{1}{N}\sum_{m=-(M-1)}^{M-1} w^2(m) = \frac{1}{2\pi N}\int_{-\pi}^{\pi} W^2(e^{j\omega})\,d\omega.$$

where N is the record length and $2M - 1$ is the total window length. Thus, by adjusting the shape and length of the window, the variance of $S_{xx}(\omega)$ can be reduced over that of the periodogram.

Another measure of the amount of smoothing (reduction in resolution) caused by windowing is the width of the main lobe. In this case we define this bandwidth as the symmetric interval between the first negative and positive frequencies at which $W(e^{j\omega}) = 0$.

In this problem we shall study these properties for the following three windows:

Rectangular

$$w_R(m) = 1 \qquad |m| \le M-1$$
$$= 0 \qquad \text{otherwise}$$

Bartlett

$$w_B(m) = 1 - |m|/M \qquad |m| \le M-1$$
$$= 0 \qquad \text{otherwise}$$

Raised Cosine

$$w_H(m) = \alpha + \beta\cos(\pi m/(M-1)) \qquad |m| \le M-1$$
$$= 0 \qquad \text{otherwise.}$$

(If $\alpha = \beta = 0.5$ this is the Hanning window and if $\alpha = 0.54$ and $\beta = 0.46$ this is the Hamming window.)

(a) Find the Fourier transform of each of the above windows; i.e., compute $W_R(e^{j\omega})$, $W_B(e^{j\omega})$, and $W_H(e^{j\omega})$. Sketch each of these functions of ω.
(b) For each of these windows, show that the entries in the following table are correct. (Assume that $M \gg 1$.)

Window Name	*Approximate Width of Main Lobe*	*Approximate Variance Ratio* (*R*)
Rectangular	$2\pi/M$	$2M/N$
Bartlett	$4\pi/M$	$2M/(3N)$
Raised Cosine	$3\pi/M$	$2M(\alpha^2 + \beta^2/2)/N$

5. In the Welch method, a record of length N is sectioned into K finite-length sequences.

$$x^{(i)}(n) = x(n + iM - M) \qquad 0 \leq n \leq M - 1, \qquad 1 \leq i \leq K,$$

and these sections are windowed before computing the modified periodograms

$$J_M^{(i)}(\omega) = \frac{1}{MU}\left|\sum_{n=0}^{M-1} x^{(i)}(n)w(n)e^{-j\omega n}\right|^2 \qquad 1 \leq i \leq K,$$

where

$$U = \frac{1}{M}\sum_{n=0}^{M-1} w^2(n).$$

The spectrum estimate is defined as

$$B_{xx}^w(\omega) = \frac{1}{K}\sum_{i=1}^{K} J_m^{(i)}(\omega).$$

Show that

$$E[B_{xx}^w(\omega)] = \frac{1}{2\pi}\int_{-\pi}^{\pi} P_{xx}(\theta)W(e^{j(\omega-\theta)})\,d\theta$$

where

$$W(e^{j\omega}) = \frac{1}{MU}\left|\sum_{n=0}^{M-1} w(n)e^{-j\omega n}\right|^2$$

Hint: Make use of the fact that if $x(n)$ has zero mean, then

$$\phi_{xx}(m) = \frac{1}{2\pi}\int_{-\pi}^{\pi} P_{xx}(\omega)e^{j\omega m}\,d\omega.$$

6. Let $x(n)$ be a real stationary random signal. The Bartlett spectrum estimate as defined by Eq. (11.47) is

$$B_{xx}(\omega) = \frac{1}{K}\sum_{i=1}^{K} I_M^{(i)}(\omega).$$

Let us consider the inverse Fourier transform of $B_{xx}(\omega)$ which we denote $b_{xx}(m)$.
(a) Show that

$$E[b_{xx}(m)] = \phi_{xx}(m)w_B(m)$$

where

$$\begin{aligned} w_B(m) &= 1 - |m|/M \qquad |m| \leq M - 1 \\ &= 0 \qquad\qquad\qquad \text{otherwise} \end{aligned}$$

(b) Now suppose that we compute $B_{xx}(2\pi k/M)$ using the FFT method of Sec. 11.6.1. Then let $b_{xxp}(m)$ be defined as

$$b_{xxp}(m) = \frac{1}{M} \sum_{k=0}^{M-1} B_{xx}(2\pi k/M) e^{j(2\pi km/M)}$$

Obtain an expression for $E[b_{xxp}(m)]$.

(c) How are the results of (a) and (b) modified for the Welch method? That is, repeat (a) and (b) beginning with $B_{xx}^{w}(\omega)$ as in Eq. (11.63).

Index

Index

A

B

C

D

E

F

M

N

O

P

T

U

V

W

Z